Animals' Influence on the Landscape and Ecological Importance

Friedrich-Karl Holtmeier

Animals' Influence on the Landscape and Ecological Importance

Natives, Newcomers, Homecomers

Springer

Friedrich-Karl Holtmeier
Institute of Landscape Ecology
University of Münster
Havixbeck, Germany

Kind permission was given for the translation of German-language title "Tiere in der Landschaft: Einfluss und ökologische Bedeutung" published in 2002 by Verlag Eugen Ulmer GmbH & Co., Stuttgart (Hohenheim), Germany into English.

ISBN 978-94-017-9293-6 ISBN 978-94-017-9294-3 (eBook)
DOI 10.1007/978-94-017-9294-3
Springer Dordrecht Heidelberg New York London

Library of Congress Control Number: 2014952723

Cover illustration: Mountain goat (*Oreamnos americanus*) above treeline on Mount Evans (4,350 m). F.-K. Holtmeier, 30 June 1984.

Printed on acid-free paper

Springer is part of Springer Science+Business Media (www.springer.com)

It is a sophism that in ecology simple answers can be given to seemingly simple questions.

(H. Remmert 1980)

Preface

My interest in animals has existed through most of my life, originally stimulated at a young age by my grandfather. As a professor, I have regularly given lectures on 'Animals as ecological agents' to students in physical geography and landscape ecology. In 1999, I published a first book on that issue. A second, extended edition followed 3 years later (Holtmeier 2002). Both editions, however, were published in the German language, thus automatically restricting its readership. Both books are now out of print. In view of the numerous new publications relevant to the book's topic, and recalling the very positive response to the books by many reviewers and readers, I have been thinking for some time of preparing a third edition. It was clear however that, given the global importance of the subject, it should be written in the English language, thus guaranteeing a wider circulation. Thus, I was delighted when Springer (Dordrecht) agreed to publish an English edition.

The book is addressed to a broader readership rather than simply to experts in the field. It is based on an extensive review of the literature, on my own field experiences and on contributions of my former cooperators. The influences of wild and feral mammals, particularly of herbivores, as well as of birds, insects and soil invertebrates, are demonstrated through examples from all over the world. Their selection out of a wealth of very heterogeneous and often widely scattered literature has been an exacting task. This presentation by example, however, allows us to go deeper into details than is possible in textbooks that aim to present a complete overview (e.g., Kratochwil and Schwabe 2001) and seemed to be an acceptable way to explain the heterogeneous and complex interactions of animal populations with their habitats (quality, spatial and temporal structures, landscape use). On the other hand, the great variety that appears from such various settings may heighten the readers' awareness of problems that often arise when applying results of local studies to other landscapes. Thus, hasty generalization that disguises rather than clarifies might be prevented.

The conception of the book has not much changed from its earlier editions. The text has however been revised, partly re-written, completed and re-arranged where necessary. Ample illustration by photos, drawings and diagrams has been retained as well as an extensive updated reference list. However, in view of the immense wealth of literature on autecology, synecology, population ecology, nature conservation, wildlife and landscape management, this list cannot be complete. Surely, the reader will miss inclusion of one or another familiar publication. Russian publications have only been referred to when translations in the English or German language were available. One may safely assume that a broad readership unable to read Russian would appreciate more high-quality publications such as the monograph on marmots by Bibikov (1996), for example, to be translated into English. The references include many publications that were written in the German language. Anglophone readers seldom or never refer to this relatively extensive literature although it provides much useful interesting information.

I hope this book will contribute to better understanding of the interplay of animals and their habitats and the often important role of animal populations in shaping landscape structures and dynamics.

University of Münster, Germany Friedrich-Karl Holtmeier

References

Bibikov, D. I. (1996). *Die Murmeltiere der Welt* (Die Neue Brehmbücherei 358, 2nd ed.). Magdeburg/Heidelberg.

Holtmeier, F.-K. (2000). *Die Höhengrenze der Gebirgswälder* (Arbeiten aus dem Institut für Landschaftsökologie 8). Münster : Westfälische Wilhelms-Universität.

Kratochwil, A., & Schwabe, A. (2001). *Ökologie der Lebensgemeinschaften*. Stuttgart.

Acknowledgements

My thanks go to those friends and colleagues who essentially contributed to the conception of this book by many good discussions. First of all, I mention Prof. Dr. Gabriele Broll (Institute for Geography, University of Osnabrück). Professor Dr. Hermann Mattes and Dr. Andreas Vogel (Institute for Landscape Ecology, University of Münster) also gave much good advice. I also thank Prof. Dr. Diana F. Tomback (Department of Ecology, University of Colorado, Denver). With her, I share my interest in nutcrackers (*Nucifraga* spp.). Moreover, I am obliged to Dr. Sabine Begall (Department of Biology, University of Essen) for information on coruros (*Spalocopus cyanus*) in Chile. Many colleagues provided me with information, data and photos: W. Albrecht (National Bison Association, Denver), Dr. Torsten Becker (Institut für Allgemeine Botanik, University of Hamburg), Dr. James B. Benedict (Center for Mountain Archeology, Ward, Colorado) and Audrey DeLella-Benedict (Cloudy Naturalists, Ward, Colorado), Dr. Manfred Bölter (Institute for Polar Ecology, Kiel), Dr. Larry E. Burrows (Landcare Research, Christchurch, New Zealand), Dr. David H. Cummings (WWF Programme Office, Zimbabwe, Fil. Maisteri Saino Heino (Kevo Subarctic Research Institute, Finland), Dr. Jim Hone (Faculty of Applied Science, University of Canberra, Australia), Dr. Quentin Hart (Bureau of Rural Sciences, Canberra), Prof. Dr. Norbert Hölzel (Institute of Landscape Ecology, University of Münster, Germany), Prof. Dr. Ragnar Kinzelbach (Allgemeine und Spezielle Zoologie, Zoologische Sammlung Rostock, Germany), Dr. Till Kleinebecker (Institute of Landscape Ecology, University of Münster, Germany), Prof. Dr. Frank Klötzli (Institute of Geobotany, University of Zürich), Dr. Walter Leuthold-Glinz (Zürich), Prof. Dr. Frank Lehmkuhl (Geographical Institute, Technical University of Aachen), Prof. Dr. Ernst Löffler (Chair of Physical Geography Geographie, University of Saarbrücken), Dr. McIllroy (Commonwealth Scientific and Industrial Research Organization CSIRO, Australia), Dipl. Biol. Fank-Uwe Michler (University of Technology, Dresden), Prof. Dr. Seppo Neuvonen (Finnish Forest Research Institute METLA, Joensuu), Prof. Dr. Rolf O. Peterson (Technical University of Michigan, Houghton), Dr. Ralf Peveling (Institute for Nature-, Landscape- and Environmental Conservation, Biogeography, University of Bâle, Switzerland), Prof. Dr. G. Schaller (Wildlife Conservation Society, New York),

Dr. Olle Tenow (Department of Entomology, Swedish University of Agricultural Sciences, Uppsala), Prof. Dr. Michael Tyler (Belair, South Australia), Prof. Dr. Clemens Tysdell (Department of Economics, University of Ola, St. Lucia, Australia) and Dr. Peter Wardle (Landcare Research, Lincoln, New Zealand).

My thanks also go to Prof. Dr. Robert M. M. Crawford (University of St. Andrews, Scotland) for occasional proofreading. Not least I want to thank Dr. Kerstin Anschlag (Institute of Geography, University of Bonn), who was a great help to me in search for literature and preparation of the many figures and graphs.

Contents

List of Figures

List of Photos

List of Tables

Chapter 1
Introduction

This book on animals as ecological agents in landscape ecosystems draws attention to a range of problems that have been only marginally considered in landscape sciences (geography, landscape ecology). These are addressed preferably to vegetation, geomorphology and other physical factors (e.g. Klink 1987). There are exceptions however, such as the studies of Troll (1936) on termite savannas, the monograph on plant geography and ecology of New Zealand by Schweinfurth (1966), and the many studies on beavers, mainly in North America (e.g., Naiman et al. 1986; 1988; Johnston and Naiman 1987, 1990a, b; Barnes and Dibble 1988; Woo and Waddington 1990; Butler and Malanson 1994; Snodgrass 1997; for further references see Sects. 4.2.2, 5.3.1 and 6.1). Schott (1934) considered the influence of beavers on meadow formation and its role in European colonization of Canadian woodlands. In his book on zoogeomorphology, Butler (1995/2008) gives a review of animals' geomorphic influences and also goes back to his own studies on the effects of beavers, grizzly bears and burrowing animals. Price (1971) presented a short but very instructive study on the interrelationships between periglacial landforms and the effects of Arctic ground squirrels on microtopography and vegetation in the alpine zone of an Alaskan mountain range. Löffler (1996) addressed the influence of termites on morphogenesis in the tropics characterized by seasonally alternating wet and dry conditions. A review on the influence of small mammals in context with the mosaic-cycle concept (Remmert 1992) has been presented by Korn (1991). In a study on impact and response of high-latitude and high-altitude ecosystems, Komarkova and Wielgolaski (1999) briefly touched upon disturbances by animals. The comprehensive handbook on global eco-zones by Schultz (2000) gives an idea of the influence of animals on energy transfer in the zonal ecosystems. The 'Handbook of New Zealand mammals'(Kitching 1995) not only covers the history of introduction and aut- and synecological aspects but also highlights the significance of the introduced mammals to the New Zealand environment and native fauna.

Abundant special studies in wildlife-biology, hunting sciences, game management and nature conservation often provide detailed information, which help us to understand the role of animals as landscape-ecological agents better. In these publications

F.-K. Holtmeier, *Animals' Influence on the Landscape and Ecological Importance*,
DOI 10.1007/978-94-017-9294-3_1

however, the influence of habitat structures and quality (e.g., forage supply, shelter from climate and disturbances) on animals is usually in the foreground (e.g., Rammer 1936; Remmert 1980; Kloft and Gruschwitz 1988), whereas possible impacts of animals on their habitats are more marginally considered. This also applies to more recent zoological studies on the spatial behavior of animals at the landscape level (e.g., Barrett and Peles 1999). There are, however, exceptions (e.g., Forman 1995; Johnston 1995; Remmert 1998).

Interest in the functional role of animals in landscape ecosystems and the responses of those systems to the animals' impact has been continuously growing, as reflected in many case studies published in the Journals of BioScience, Animal Ecology, Mammalogy, Journal of Applied Ecology, Wildlife Management, and Oikos, for example. Not least, research on animals' impact on their habitats has a comparatively long tradition in forestry (e.g., USDA Forest Service, New Zealand Forest Service). Much interesting information can be found in the specific research reports. Likewise, the research projects that have been carried out in the Serengeti and other African National Parks must be mentioned in this context (e.g., Sinclair and Norton-Griffith 1979; Huntley and Walker 1987; Owen-Smith 1983; Sinclair and Arcese 1995) as well as 'classic' compilations on ungulate impacts on grasslands, for example (e.g., Hodgson and Illius 1996; Du Toit and Cumming 1999), or reviews on the effects of herbivores on plant communities and ecosystem processes in general (e.g., Gill 1992; Hobbs 1996: Augustine and McNaughton 1998). Valuable reviews on the functional role of large predators in ecosystems have just been published by Estes et al. (2011) and Ripple et al. (2014). A review on the management of wild large herbivores with economic, conservation and environmental objectives was presented by Gordon et al. (2004). Not least, the increasing discussion on causes and function of biodiversity (e.g., Huston 1994) as well as the growing number of studies on the effects of disturbances, for example by introduction of exotic parasites and animals, invasion of alien species, mass-outbreaks of insects and small mammals, and of excessive grazing pressure (e.g., Turner and Stratton 1987; Huston 1994; Hugget 1995; Bayerische Akademie für Naturschutz und Landschaftspflege 2001), have stimulated broader interest in impacts of animals on the environment. The same holds true for the increased numbers of publications on the effects of large herbivores on the Pleistocene and Holocene vegetation as well as for studies on large herbivores as an instrument of landscape management (e.g., Beutler 1992, 1996, 1997; Geissen 1996; Hamilton 1996; Hofmann et al. 1998; Gerken and Görner 1999; Section 6). Danell et al. (2006) have presented a comprehensive compilation with 16 chapters written by experts on large herbivore ecology, ecosystem dynamics and conservation. A comparatively rich literature is available on the influence of insects, in particular of termites and noxious forest insects and fossorial mammals on their habitats. An extensive textbook on the ecology of biocoenoses has been given by Kratochwil and Schwabe (2001).

Now and then I refer in this book also to daily newspapers, television reports and internet news. They confront us almost every day with the latest 'ecological catastrophes'. 'Alarming news' spreads like wildfire. Mass-outbreaks of defoliating insects and bark beetles are often portrayed as being a fatal threat to our forests. Rapidly spreading animal diseases (e.g., foot and mouth disease, avian influenza, etc.) or the irruption of introduced species (Chap. 5) often provoke general helplessness and a dooms day atmosphere. Although the sources of information may sometimes be obscure, such news serve its purpose as they make abroad audience aware that the future of our ecosystems and biocoenoses is not reliably predictable, as the magnitude and quality of future general conditions (natural, economic, political) that will be controlling environmental change (landscape, biocoenoses, and climate) are unknown. The many reports on attacks of 'bloodthirsty' bears and wolves on sheep, cattle, joggers, hikers or mushroom gatherers, and the pros and cons of animal advocates and opponents of reintroduction of large predators, show that such events with wild beasts involved can hardly be discussed unemotionally and that place seems to be restricted for certain wild animals (Sect. 5.2.3).

Nevertheless, studies focusing on the functional role of animals in landscape ecosystems are still rare as compared to the many autecological investigations. In addition, many explanations and hypotheses – for example on the influence of grazing on plant species diversity or on factors controlling population cycles – are plausible but often not convincingly substantiated. It is the great complexity of the interactions of animals with their habitats, which makes a complete experimental analysis so difficult, if not impossible, in particular as the interdependencies assume a 'newquality', dependent on the level of consideration (autecology, synecology, population ecology; Fig. 1.1). Methods of analysis and evaluation have to be adjusted

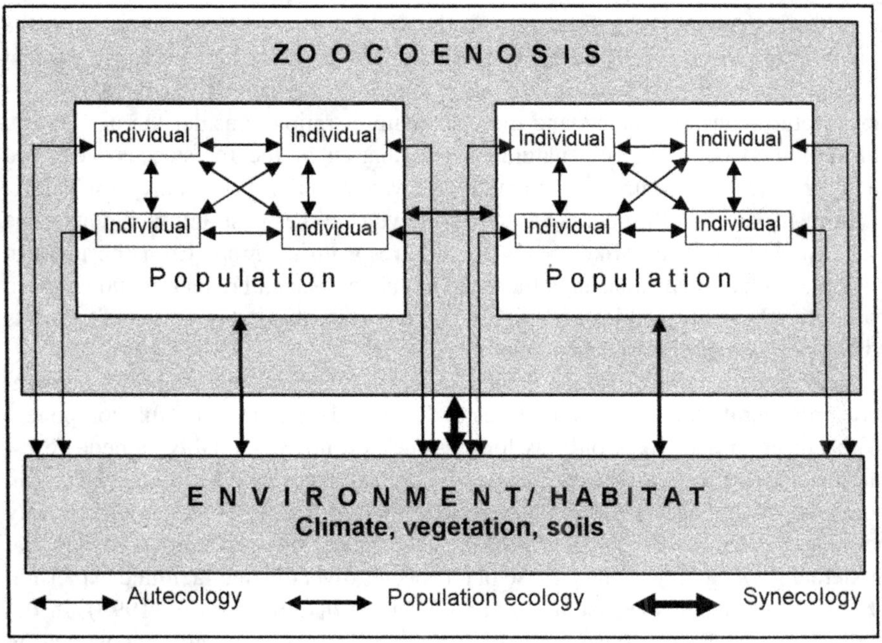

Fig. 1.1 Autecological, population-ecological and synecological interrelationships in an ecosystem (Scheme by F.-K. Holtmeier)

accordingly. Thus, even if the specific response of all components to changing conditions were known at the autecological level, for example, the response at higher levels of the ecosystem would not be 'predictable'. Yet, experimental studies, for example, comparing sampling areas grazed at different grazing intensities with 'exclosures' are essential to understand the interdependencies and to test hypotheses. Generalization, however, of the results of case studies is problematical, as the impact of animals on their habitats depends on the particular local conditions and peculiarities. Thus, the possibilities of transferring results of local studies and models to other areas are limited.

Moreover, the specific attitude to the animals' influences plays an important role (Sect. 6.4). The alleged 'catastrophes' caused by bark beetles (*Ips typographicus*) in the National Park 'Bavarian Forest' during the late 1990s is a good example. Bark beetles belong to the forest biocoenosis, together with woodpeckers and other woodland species, and yet they are usually considered to be noxious insects. Bark beetles usually attack weakened trees already damaged otherwise. Under certain conditions that favor a mass-outbreak (e.g., much dead wood due to wind throw, immissions, warm and dry weather in spring time; Weissbacher 1999) extended forest stands may collapse due to bark beetle attacks, as was the case in the National Park. On the other hand, bark beetle attacks may initiate natural rejuvenation of over-aged forests (Nüsslein et al. 2000). Forest owners and tourist managers who were interested in forests mainly for economic reasons considered the mass-outbreak to be absolutely an 'economic catastrophe' and postulated an effective 'pest control'. They were supported by forest owners from outside the park who were afraid the bark beetles would expand into the economic forest surrounding the forest nature reserve (Weissbacher 1999). In their view, the prospect of a gradual natural rejuvenation of the forests is not a tolerable alternative these days. By contrast, nature conservationists and the park administration consider the bark beetle infestation to be supporting natural self-regulation of the forest ecosystem and therefore reject any control measure that, in a way, would protect nature from nature.

In many areas, wild boars (*Sus scrofa*) have become 'pest animals' during the last few decades. Very likely, wild boars are profiting more from the modern agricultural landscape than any other wild ungulates. Radical reduction, however, in particular shooting females and piglets, seems to be inevitable but usually meets stiff opposition of animals' advocates (Sect. 6.4).

Red deer (*Cervus elaphus*) in great numbers may impede or even prevent natural forest regeneration and profitable forestry. Thus, red deer are a 'useful' component of the forest biocoenosis only as long as they do not sustainably impede forest maintenance or restoration (e.g., Mayer 1975; Neumann 1979; Moser 1988), or in other words, in case of intolerable damage the red deer population will usually be considered to be 'over-sized'. On the other hand, there is some evidence that moderate grazing may even increase plant species diversity and facilitate expansion of forest tree species (Risenhoover and Maass 1987; Krüsi et al. 1995). In the Swiss National Park (Lower Engadine, Switzerland), for example, conservation of botanical diversity will only be possible if browsing pressure is sufficient (Sect. 2.3.3). One should be aware, however, that current great botanical diversity is the result of forest removal and pastoralism for hundreds of years and thus is not natural. Intensive browsing pressure may impede natural regeneration and reduce tree species richness as well as diversity. If conservation of the present plant diversity is the objective of habitat management, red deer population will have to be adjusted accordingly by harvesting. It would be easy to present more examples that highlight problems resulting from different attitudes of those affected by animals' impact on landscape ecosystems in one or the other way. However, we will leave it at that for the time being.

Animals as ecological agents in landscape ecosystems can be considered from different points of view. In landscape planning management, for example, animal populations have to be considered with special regard to the particular spatial and functional landscape units (landscape mosaics) and landscape connectivity (corridors; e.g., Forman 1995). Restoration of wetland for conservation of amphibians, for example, may bring about alterations that were not intentional and unforeseeable right from the beginning of restoration. They may however adversely affect other animals.

Conversely, suppression or eradication of a 'problem species' also may have far-reaching consequences for other members of the biocoenosis and for the entire ecosystem. European moles (*Talpa europaea*), for example, are a 'nightmare' for gardeners. They are usually not tolerated on lawns, sports grounds, golf courses and hay meadows. While being completely protected by law in Germany, for example, moles are poisoned or killed as ever by trapping in some other countries (e.g., England). Both measures are considered to be 'inhuman'. In addition, poisoning may be harmful to soil organisms. Therefore one is looking for alternative non-lethal means and applies repellents for example to keep moles off. Reduction of insects and earthworms, that are the main food of moles, has also been considered to bean appropriate method for mole control (Edwards et al. 1999). Earthworm density could be easily reduced by systematic removal of the litter, by use of herbicides that prevent grasses and thus support of herb-rich vegetation, or by application of acidic nitrogen fertilizer (Schaefer 1981). These measures would however negatively affect not only earthworms and moles but also soil physical and chemical properties, decomposition and nutrient turnover. Control measures may easily run out of control, similar to the spirits called by the desperate sorcerer's apprentice in the well-known poem by Johann Wolfgang von Goethe '*The spirits that I've called I can't get rid of now*'.

In the landscape ecological view the functional role of animals in landscape ecosystems and animals' interactions with physical and biological factors is of primary interest with animal populations being in the focus rather than individuals. On the other hand, the effects of animals on landscape cannot be explained without considering the influence of landscape (habitats) on the animals. Thus, aut- and synecological as well as landscape ecological aspects must be closely linked to each other. The configuration of the ecosystems (vegetation, soils, climate, water balance, biocoenoses), as well as their spatial and temporal structures, are of particular importance in this context.

In the past, the importance of animals was assessed by their influence on landscape physiognomy (e.g., Mertens 1961; De Lattin 1967). On that score, however, animals are of minor importance compared to vegetation or human impact.

Nevertheless, there are exceptions such as beavers or termites. Beavers, for example, are able to completely alter floodplains, riparian vegetation and biocoenosis by dam-building. Therefore they are considered a keystone species (Paine 1969). Not without reason have beavers become a classic example of animal-habitat interactions, and most textbooks on physical geography, biology and landscape ecology refer to the beavers' activities. This also applies to the present book (Sect. 4.2.2).

Just as important are mound-building termites shaping landscape physiognomy and ecology in East African savannas (so-called 'termite savannas': e.g., Troll 1936; Hesse 1955; Lee and Wood 1971; Darlington 1985; Schmitt 1991; Sect. 4.2.1). Moreover, fossorial mammals such as marmots, ground squirrels, pocket gophers and voles may influence landscape physiognomy. Landscapes characterized by 'Mima mounds', for example, can be found in western North America from Canada to Mexico, in Africa (Kenya, South Africa), on the Peruvian Altiplano and in Argentina. 'Mima mounds' have been built by burrowing rodents in the course of thousands of years (Dalquest and Scheffer 1942; Price 1949; Scheffer 1958; Ross et al. 1968; Mielke 1977; Cox 1984, 1990; Gakahu and Cox 1984; Cox and Gakahu 1986; Cox and Roig 1986; Cox and Allen 1987; Cox and Hunt 1990; Schmitt 1991; Bussmann 1994). The name derives from the Mima Prairie in Thurston County, western Washington. In South Africa, so-called 'heuweltjies' ('small mounds' in Afrikaans) can be found looking similar to 'Mima mounds'(Lovegrove 1991). The mounds consist of excavated mineral soil. They are up to 2 m high and 25–50 m in diameter. About 50–100 mounds per hectare can be found. Mounds built by the Bobak (*Marmota bobak*) locally occur as prominent landmarks in grass steppes from Eastern Europe to Kazakhstan (Bibikov 1996).

The long-term disturbances by prairie dogs (*Cynomis ludovicianus*), in concert with grazing, trampling and wallowing of buffalos (*Bison bison*) and other ungulates have created highly varying vegetation mosaics in the North America prairie (Sect. 3.2.2).

Birds may also play a major role. Large numbers of lesser snow geese (*Chen c. caerulescens*), for example, have lasting effects on composition and structure of plant communities and nutrient cycles in their breeding areas on the west coast of Hudson Bay (Bazely and Jefferies 1988; Kerbes et al. 1990) where they stay during summer. They have been considered a keystone factor in these places. Lesser snow geese traditionally grazed during summer in the intertidal salt marsh on the shore of La Pérouse Bay. Numbers of lesser snow geese have dramatically increased in the last decades, however, mainly due to changes in agricultural practices and easily available energy-rich food in their wintering habitats and along their migration routes (Ngai and Jefferies 2004). Geese now exceed the carrying capacity of the salt marsh vegetation in their summer habitat and are forced to forage in the inland fresh-water marshes. Fresh-water marsh vegetation is of less nutritional quality than salt-water marsh vegetation, and does not much profit from fertilization by geese droppings of which the phosphorus content is very low and dropping densities are much less than on the salt water marsh. Ngai and Jefferies (2004) assume that the considerably increased removal of phytomass by foraging geese will reduce primary productivity of the fresh-water marshes in the long-term. Declines in goslings' survival and size seem to be already reflecting impoverishment of the food source.

Several seabird species nesting in large colonies on subantarctic islands are effective geomorphic agents. Their subterranean nesting holes undermine the surface over large areas thus enforcing erosion and vegetation change (Sect. 2.6.1.2.2).

In north-eastern Siberia, the black-billed Capercaillie (*Tetrao urogalloides*) shape the physiognomy of the Dahurian larch forest (*Larix dahurica*). Capercaillies live in large groups (several dozen individuals) during winter and stay in great numbers at leks for many weeks in summer. By recurrent clipping the short shoots of young larches for decades these birdsforce the trees into bonsai-like growth forms (so-called 'gardens') (e.g. Andreev 1991; Klaus and Andreev 2001).

Fossil coral reefs that shape, so to say 'posthumously', the landscape of emerged sea floors as well as guano deposits may be considered as an example of animals' influence on landscape physiognomy at geological dimensions.

In general, however, the influences of animals on their environment are less spectacular. Sometimes they will be noticed only after having caused unexpected and intolerable damage to forests or cultural land. The impacts of mass-outbreaks of defoliating insects, bark beetles, voles, African red-billed weaver birds or locusts, for example, may even assume disastrous proportions. On the other hand, the decline of animal populations may also result in considerable ecosystem changes. The decline of a beaver population resulting in collapsing dams, increased runoff and erosion of the riparian landscape is a good example (Sect. 4.2.2).

By contrast, the influences of many animal species are often hardly recognizable even though they may have far-reaching and lasting impacts on landscape ecosystems. One could also put it like this: many processes and changes in landscape ecosystem would never have happened without animals involved (Vitousek 1986; Naiman 1988). Bioturbation by the soil fauna (e.g., earthworms) and ground-dwelling mammals play an important role in this respect. Under certain conditions, microgeomorphic structures (mounds, holes, dust wallows) are created by animals' activities. In addition, seed dispersal and pollination of many plant species rely on animals. Forage plants respond to browsing and grazing pressure by developing defensive strategies (chemical compounds, mechanical defence by spines or thorns). Not least, selection of plants exposed to regular trampling by hooved herbivores must be mentioned in this context.

In geography and landscape ecology, animals have usually been considered as 'fixtures' of landscapes and landscape zones (e.g. Müller-Hohenstein 1979; Forman 1995; Hugget 1995), as a topic of nature conservation or as indicators of environmental conditions and change (e.g., Jungbluth 1978; Hammelbacher and Mühlenberg 1986; Kronauer 1988; Riecken 1990, 1991, 1992; Meinig 1992; Hanley 1993, 1996; Bastian and Schreiber 1994; Handke 1996; Jedicke 1996; Petrak 1996, 2001; Schultz and Finch 1997; Brinkmann 1988; Schönborn 1999; Schlumprecht et al. 2001). The emphasis, however, is usually on the physical properties of the ecosystems (e.g. size, structure, climate, geomorphology) and on vegetation in particular, because vegetation as primary producer and landscape structuring factor is of pivotal importance. Another and probably the main reason, however, is the great number of animal species exceeding the number of plant species by far. In Germany, for example, 40,000–50,000 animal species contrast with about 1,900 plant species only (Nowak 1982; Röser 1988).

In addition, animals' mobility usually makes observation of animals, collecting data, and assessment of the animal effects more difficult and time-consuming than analyses of plant communities, soils or geomorphic landscape structures. This becomes also apparent when developing landscape-ecological models. The animal effects depend completely on the particular local habitat conditions and temporally varying size of animal populations. The latter again are controlled by habitat quality and 'population internal' conditions such as reproductive behavior and so-called 'self-regulation' (Sect. 2.2.1; see also Schwerdtfeger 1968; Tischler 1990). Models may help to better understand ecosystem functional mechanisms. However, models must simplify and therefore cannot reflect the complex reality. As models are based on local data, generalization and transferability of results to other locations must be considered with due care. Modelling may also involve a certain risk of disguising problems rather than solving them.

On the other hand, practical landscape planning and management often are confronted with questions as to how ecosystems would respond to management, which management practices would be appropriate and how they could be optimized. Expectations and demands of decision-makers for scientific support however may already be far ahead of the possibility to provide well-substantiated reasonable answers. This becomes particularly apparent when animal species have been introduced into areas where they never lived before, or where animals have been reintroduced to their original habitats that have mostly undergone substantial and lasting changes under the growing impact by land use (Sections 5 and 6). But even seemingly simple questions, as for example on the effects of seed caching animals on ecosystem structures and dynamics, often turn out to be highly complex and specific to the local conditions, in a way that universally valid rules can hardly be deduced.

An omnipresent problem is how to manage 'over-sized' game populations. They usually reflect disturbances and malfunction of natural regulation mechanisms in cultural landscape ecosystems. The problem may only be solved by the politically effective implementation of ideas that are well-founded on scientific research and experience. These should overcome the often dubious interests of the relatively small but politically very powerful hunters' lobby. The activities of this pressure group are often driven by emotion and tradition rather than being substantiated by objective facts. Not least, a clear difference should be made between 'natural carrying capacity' and 'tolerable carrying capacity' (Sect. 6.2), highlighting the fact that both are closely related to the local situation. Thus, it is impossible to define a generally tolerable game density.

Over-grazing by domestic mammals is not less problematic. Studies on grazing impact on vegetation and landscape change have a long tradition in geography (e.g., Clark 1956; Simoons 1974; Meurer 1986a, b, 1988). In Middle Europe, over-grazing is rather exceptional nowadays. In the Mediterranean, however, and in many developing countries it is still very common. Over-grazing has often resulted in serious environmental damage (fragmentation or destruction of the plant cover, soil degradation and enforced erosion, reduced carrying capacity) that may jeopardize people's existence in those regions. The problem of over-grazing by cattle, sheep and goats in the developing countries could possibly become alleviated in the long-run by creating awareness of the given ecological constraints. In view of the generally bad economy and the socio-cultural background, however, it is unlikely that people could be convinced to reduce pastoral use drastically, at least not in the near future.

From these examples it should become clear already that general statements on the role of animals as ecological agents cannot be expected, although there is an urgent need for action. To meet it, the animals' role has always to be assessed with special regard to the specific local socio-cultural and economic conditions, thus going beyond a merely landscape-ecological approach. The message of such studies could be that nature conservationists, for example, should desist from re-introducing an animal species into its former habitat that might have too much changed and shrunk due to intense landscape use (agriculture, forestry, road construction, etc.), although one can generally understand the efforts to right the wrongs of the past by restoration of the native fauna. Such activities, however, often have caused lasting conflicts and might do so again in the future. Resultant problems are usually 'solved' at the animals' cost. In particular those who are engaged in nature conservation and being anxious to do something good often appear to be suppressing this aspect. In the present book, selected examples of interactions of native ('natives'), introduced ('newcomers') and re-introduced ('homecomers') animal species with their habitats are considered (Section 5).

This book is neither a textbook nor a hand book reflecting the state of the art. The idea behind it is to introduce students and other interested readers into an extremely complex and interesting set of problems. While system analysts preferably consider the role of animals in energy fluxes, this book focuses mainly on other aspects, as for example on animals' influences on distribution patterns and composition of vegetation, on soils and geomorphic processes, as well as on the particular function of keystone species. The book shall help the reader to become aware of and to better understand the active and passive role of animals in landscape ecosystems and the usually complex problems of wildlife management and nature conservation. Afterwards, the well-meaning reintroduction of animal species, for example, into their former habitats possibly reveals itself in a different light. Thus, a good case will be made for constructive criticism of the often emotional actionism of supposed nature conservationists and advocates of animals' rights, driven by emotion rather than by rational reasons.

References

Augustine, D. J., & McNaughton, S. J. (1998). Ungulate effects on the functional species composition of plant communities: Herbivore seletivits and plant tolerance. *Journal of Wildlife Management, 62*(4), 1165–1183.

Andreev, A. V. (1991). Winter habitat segregation in sexually dimorphic Black-billed Capercaillie Tetrao urogallus. *Ornis Scandinavica, 22*, 287–291.

Barnes, W. J., & Dibble, E. (1988). The effect of beaver in riverbank forest succession. *Canadian Journal of Botany, 66*, 40–44.

Barrett, G. W., & Peles, J. D. (Eds.). (1999). *Landscape ecology of small mammals*. New York: Springer.

Bastian, O., & Schreiber, K.-F. (Eds.). (1994). *Analyse und ökologische Bewertung der Landschaft*. Jena/Stuttgart: G. Fischer.

Bayerische Akademie für Naturschutz und Landschaftspflege. (Ed.). (2001). *Störungsökologie*. Laufener Seminarbeiträge 1/01.

Bazely, D. R., & Jefferies, R. L. (1988). Lesser snow geese and the nitrogen economy of a grazed salt marsh. *Journal of Ecology, 77*, 24–34.

Beutler, A. (1992). Die Großtierfauna Mitteleuropas und ihr Einfluß auf die Landschaft. *Landschaftsökologie, Weihenstephan, 6*, 49–69.

Beutler, A. (1996). Die Großtierfauna Mitteleuropas und ihr Einfluß auf die Landschaft. In B. Gerken & C. Meyer (Eds.), *Wo lebten Pflanzen und Tiere in der Naturlandschaft und der frühen Kulturlandschaft Europas?* (Natur- und Kulturlandschaft 1, pp. 107–112). Höxter: Huxaria Druck und Verlag.

Beutler, A. (1997). Das Weidelandschaftsmodell: Versuch einer Rekonstruktion der natürlichen Landschaft. In: B. Gerken & C. Meyer (Eds.), *Vom Waldesinnsaum zur Hecke Geschichte, Situation und Perspektiven eines Natur-Lebensraum-Gefüges* (Natur und Kulturlandschaft 2, pp. 194–196). Höxter: Huxaria Druck und Verlag.

Bibikov, D. I. (1996). *Die Murmeltiere der Welt* (Die Neue Brehmbücherei 358, 2nd ed.). Magdeburg/Heidelberg.

Brinkmann, R. (1988). Berücksichtigung faunistisch-tierökologischer Belange in der Landschaftsplanung. *Informationsdienst Naturschutz Niedersachsen, 18*(4), 57–128.

Bussmann, R. W. (1994). *The forest of Mount Kenya (Kenya), vegetation, ecology, destruction and management of a tropical mountain forest ecosystem*. Dissertation, University of Bayreuth, Faculty of Biology, Chemistry, and Geosciences, Bayreuth.

Butler, D. R. (1995). *Zoogeomorphology: Animals as geomorphic agents* (2nd ed., 2008). Cambridge: Cambridge University Press.

Butler, D. R., & Malanson, G. P. (1994). Canadian landform examples. Beaver landforms. *Canadian Geographer, 38*, 76–79.

Clark, A. H. (1956). The impact of exotic invasion on the remaining New World mid-latitude grassland. In W. L. Thomas (Ed.), *Man's role in changing the face of the earth* (pp. 737–762). Chicago.

Cox, G. W. (1984). Mounts of mystery. *Natural History, 5*, 36–45.

Cox, G. W. (1990). Soil mining by pocket gophers along topographical gradients in a Mima mound field. *Ecology, 71*(3), 817–843.

Cox, G. W., & Allen, D. W. (1987). Soil translocation by pocket gophers in a Mima mound field. *Oecologia, 72*, 207–210.

Cox, G. W., & Gakahu, C. G. (1986). A latitudinal test of the fossorial rodent hypothesis of Mima mound origin. *Zeitschrift für Geomorphologie, N. F., 30*(4), 485–501.

Cox, G. W., & Hunt, J. (1990). Form of Mima mounds in relation to occupancy by pocket gophers. *Journal of Mammalogy, 71*(1), 90–94.

Cox, G. W., & Roig, V. G. (1986). Argentinian Mima mounds occupied by ctenomiyid rodents. *Journal of Mammalogy, 67*(2), 428–432.

Dalquest, W. W., & Scheffer, V. B. (1942). The origin of the Mima mounds of western Washington. *Journal of Geology, 50*, 68–84.

Danell, K., Bergström, R., Duncan, P., & Pastor, J. (Eds.). (2006). *Large herbivore ecology, ecosystem dynamics and conservation* (Conservation Biology 11). Cambridge: Cambridge University Press.

Darlington, J. P. E. C. (1985). Lenticular soil mounds in the Kenya highlands. *Oecologia, 66*, 116–121.

De Lattin, G. (1967). *Grundriß der Zoogeographie* (Hochschulbücher für Biologie 12). Stuttgart: Gustav Fischer Verlag.

Du Toit, J. T., & Cumming, D. H. M. (1999). Functional significance of ungulate diversity in African savannas and the ecological implications of the spread of pastoralism. *Biodiversity and Conservation, 8*, 1643–1661.

Edwards, G. R., Crawley, M. J., & Heard, M. S. (1999). Factors influencing molehill distribution in grassland: Implications for controlling the damage caused by molehills. *Journal of Applied Ecology, 36*, 434–442.

Estes, J. A., Terborgh, J., Brashares, J. S., Power, M. E., Berger, J., Bond, W. J., Carpenter, S. R., Essington, T. E., Holt, R. D., Jackson, J. B., Marquis, R. J., Oksanen, L., Oksanen, T., Paine, R. T., Pikitch, E. K., Ripple, W. J., Sandin, S. A., Scheffer, M., Schoener, T. W., Shurin, J. B., Sinclair, A. R. E., Soulé, M. E., Virtanen, R., & Wardle, D. A. (2011). Trophic downgrading of Planet Earth. *Science, 333*, 301–306.

Forman, R. T. T. (1995). *Land mosaics. The ecology of landscapes and regions*. Cambridge: Cambridge University Press.

Gakahu, C. G., & Cox, G. W. (1984). The occurrence and origin of Mima mound terrain in Kenya. *African Journal of Ecology, 22*, 31–42.

Geissen, H.-P. (1996). *Welche Natur wollen wir? Naturschutz auf naturgeschichtlicher Grundlage* (Bibliothek Natur und Wissenschaft 9). Solingen: Verlag Natur und Wissenschaft.

Gerken, B., & Görner, B. (Eds.). (1999). *Landschaftsentwicklung mit großen Weidetieren. Geschichte, Modelle und Perspektiven* (Natur- und Kulturlandschaft 3). Höxter/Jena: Huxaria Druck und Verlag.

Gill, R. M. A. (1992). A review of damage by mammals in north temperate forests: 3. Impact on trees and forests. *Forestry, 65*(4), 363–388.

Gordon, I. J., Hester, A. J., & Festa-Bianchet, M. (2004). The management of wild large herbivores to meet economic, conservation and environmental objectives. *Journal of Applied Ecology, 41*, 1021–1031.

Hamilton, R. G. (1996). Using fire and bison to restore a functional tallgrass prairie landscape. In *Transactions 61st North American wildlife and natural resources conference* (pp. 208–214). Oklahoma City.

Hammelbacher, K., & Mühlenberg, M. (1986). Laufkäfer (Carabidae) und Weberknechte (Opiliones) als Bioindikatoren der Skibelastung auf Almflächen. *Natur und Landschaft, 61*(12), 463–466.

Handke, K. (1996). Die Bedeutung tierökologischer Untersuchungen für die Angewandte Landschaftsökologie. *Arbeitsberichte Landschaftsökologie Münster, 2*, 93–112.

Hanley, T. A. (1993). Balancing economic development, biological conservation, and human culture: The sitka black-tailed deer Odocoileus hemionus sitkensis as an ecological indicator. *Biological Conservation, 66*, 61–67.

Hesse, P. R. (1955). A chemical and physical study on the soils of termite mounds in East Africa. *Journal of Ecology, 43*, 449–461.

Hobbs, N. T. (1996). Modification of ecosystems by ungulates. *Journal of Wildlife Management, 60*(4), 695–713.

Hodgson, J., & Illius, A. W. (Eds.). (1996). *The ecology and management of grazing systems*. Wallingford: CAB International.

Hofmann, R., Cornelius, R., Linder, R., & Scheibe, K. (1998). Wiedererstehen des postglazialen Großsäugerspektrums als landschaftsgestaltendes Multi-Spezies-System: ein Pilotversuch zur "Redynamisierung" halboffener Landschaften. *Schriftenreihe für Landschaftspflege und Naturschutz, 56*, 301–315.

Hugget, R. J. (1995). *Geoecology, an evolutionary approach*. London/New York: Routledge.

Huntley, B. J., & Walker, B. H. (Eds.). (1987). *Ecology of tropical savannas* (Ecological studies 42). Berlin/Heidelberg/New York: Springer.

Huston, M. A. (1994). *Biological diversity. The coexistence of species in changing landscapes* (2nd ed.). Cambridge: Cambridge University Press.

Jedicke, E. (1996). Tierökologische Daten in raumbedeutsamen Planungen. *Geographische Rundschau, 48*(11), 623–639.

Johnston, C. A. (1995). Effects of animals in landscape pattern. In L. Hansson, L. Fahrig, & G. Merriam (Eds.), *Mosaic landscapes and ecological processes* (pp. 57–80). London: Chapman & Hall.

Johnston, C. A., & Naiman, R. J. (1987). Boundary dynamics at the aquatic-terrestrial interface: The influence of beaver and geomorphology. *Landscape Ecology, 4*(1), 5–19.

Johnston, C. A., & Naiman, R. J. (1990a). Browse selection by beaver: Effects in riparian forest composition. *Canadian Journal of Forest Research, 20*, 1036–1043.

Johnston, C. A., & Naiman, R. J. (1990b). Aquatic patch creation in relation to beaver population trend. *Ecology, 71*(4), 1617–1621.

Jungbluth, J. H. (1978). *Der tiergeographische Beitrag zur ökologischen Landschaftsforschung*. (Biogeographica, 13), Passages: 352.

Kerbes, R. H., Kotanen, P. M., & Jefferies, R. L. (1990). Destruction of wetland habitats by lesser snow geese: A keystone species on the west coast of Hudson Bay. *Journal of Applied Ecology, 27*, 241–258.

Kitching, R. M. (Ed.). (1995). *The handbook of New Zealand animals*. Oxford University Press New Zealand: Auckland/Oxford/Melbourne/New York.

Klaus, S., & Andreev, A. V. (2001). Raufußhuhn als "Landschaftsgärtner": Steinauerhühner in Nordostsibirien. *Der Falke, 4*, 132–137.

Klink, H.-J. (1987). Ökologische Raumgliederung. In G. Olschowy (Ed.), *Natur- und Umweltschutz in der Bundesrepublik Deutchland* (pp. 55–68). Hamburg/Berlin: Paul Parey Verlag.

Kloft, W. J., & Gruschwitz, M. (1988). *Ökologie der Tiere*. Stuttgart: Verlag Eugen Ulmer.

Komarkova, V., & Wielgolaski, F. E. (1999). Stress and disturbances in cold regions. In L. R. Walker (Ed.), *Ecosystems of disturbed ground* (Ecosystems of the World 16, pp. 39–121). Amsterdam: Elsevier.

Korn, H. (1991). Small mammals and the mosaic-cycle concept of ecosystems. In H. Remmert (Ed.), *The mosaic cycle concept of ecosystems* (Ecological Studies 85, pp. 106–131). Heidelberg: Springer.

Kratochwil, A., & Schwabe, A. (2001). *Ökologie der Lebensgemeinschaften*. Stuttgart: Eugen Ulmer Verlag.

Kronauer, H. (1988). Bodentiere als Indikatoren für Waldschäden. *Allgemeine Forstzeitung, 43*, 1170–1171.

Krüsi, B. O., Schütz, M., Wildi, O., & Grämiger, H. (1995). Huftiere, Vegetationsdynamik und botanische Vielfalt im Schweizerischen Nationalpark. *Cratschla/Mitteilungen aus dem Schweizerischen Nationalpark, 3*(2), 12–25. Zernez.

Lee, K. E., & Wood, T. G. (1971). *Termites and soil*. London/New York: Academic.

Löffler, E. (1996). *Geomorphology of Papua New Guinea*. Canberra: CSIRO/ANU.

Lovegrove, B. G. (1991). Mima-like mound (heuweltjes) of South Africa: The topographical, ecological and economic impact of burrowing animals. *Symposia of the Zoological Society London, 63*, 183–198.

Mayer, H. (1975). Der Einfluß des Schalenwildes auf die Verjüngung und Erhaltung von Naturwaldreservaten. *Forstwissenschaftliches Centralblatt, 94*, 209–222.

Meinig, H. (1992). Möglichkeiten und Grenzen der ökologischen Habitatsbewertung mittels Säugetieren. In R. Eikhorst (Ed.), *Beiträge zur Biotop- und Landschaftsbewertung* (pp. 39–94). Duisburg: Verlag für Ökologie und Faunistik.

Mertens, R. (1961). Tier und Landschaft. Zoologische Unterlagen zur Landschaftskunde. *Frankfurter Geographische Hefte, 37*, 31–38.

Meurer, M. (1986a). Nutzungsbedingte Eingriffe in die Pflanzendecke des tunesichen Mogod-Berglandes und standortangepaßte Meliorationsmaßnahmen. *Giessener Beiträge zur Entwicklungsforschung, Reihe 1, Bd., 13*, 89–103.

Meurer, M. (1986b). Die Auswirkungen eines abgestuften Ziegenbestandes auf die nordtunesiche Macchie und Garigue. *Verhandlungen der Gesellschaft für Ökologie, XIV*, 229–236.

Meurer, M. (1988). Maßstabsbezogene weideökologische Untersuchungen an Fallbeispielen aus Nordtunesien und Nordbenin. *Giessener Beiträge zur Entwicklungsforschung, Reihe 1, Bd., 16*, 171–196.

Mielke, H. (1977). Mound building by pocket gophers (Geomyidae): Their impact on soils and vegetation in North America. *Journal of Biogeography, 4*, 171–180.

Moser, O. (1988). Verbiß durch Schalenwild - tragbar oder waldzerstörend? *Österreichische Forstzeitung, 11*, 51–52.

Müller-Hohenstein, K. (1979). *Die Landschaftsgürtel der Erde*. Stuttgart: Teubner.

Naiman, R. J. (1988). Animal influences on ecosystem dynamics. *BioScience, 38*, 750–752.

Naiman, R. J., Melillo, J. M., & Hobbie, J. E. (1986). Ecostem alteration of a boreal forest stream by beaver (Castor canadensis). *Ecology, 67*, 1254–1269.

Naiman, R. J., Johnston, C. A., & Kelley, J. C. (1988). Alteration of North American streams by beaver. *BioScience, 38*(11), 753–762.

Neumann, M. (1979). *Bestandesstruktur und Entwicklungsdynamik im Urwald Rothwald/NÖ und im Urwald Corkova Uvala/Kroatien*. Dissertation Universität für Bodenkultur Wien 10.

Ngai, J. T., & Jefferies, R. L. (2004). Nutrient limitation of plant growth and forage quality in Arctic coastal marshes. *Journal of Ecology, 92*, 1001–1010.

Nowak, E. (1982). Wie viele Tierarten leben auf der Welt, wie viele davon in der Bundesrepublik Deutschland? *Natur und Landschaft, 57*(11), 383–389.

Nüsslein, S., Faisst, G., Weisbacher, A., Moritz, K., Zimmermann, L., Bittersohl, J., Kennel, M., Troycke, A., & Adler, H. (2000). *Zur Waldentwicklung im Nationalpark Bayerischer Wald 1999*. Buchdrucker-Massenvermehrung und Totholzflächen im Rachel-Lusen Gebiet. Berichte der Bayerischen Landesanstalt für Wald und Forstwirtschaft, LWF 25.

Owen-Smith, N. (Ed.). (1983). *Management of large mammals in African conservation areas*. Pretoria: HAUM.

Paine, R. T. (1969). A note on trophic complexity and community stability. *American Naturalist, 103*, 91–93.

Petrak, M. (1996). Erfassung von Schalenwildarten und deren Bewertung für die Landschaftsplanung. *Schriftenreihe für die Landschaftspflege und Naturschutz, 46*, 69–75.

Petrak, M. (2001). Integration freilebender Huftiere und Maßnahmen zur Landschaftsentwicklung: Anforderungen an Naturschutz, Nutzung und Entwicklungskonzepte. *Natur- und Kulturlandschaft, 4*, 415–424.

Price, W. A. (1949). Pocket gophers as architects of Mima (pimple) mounds of the western United States. *Texas Journal of Science, 1*, 1–17.

Price, L. W. (1971). Geomorphic effect of the Arctic ground squirrel in an alpine environment. *Geografiska Annaler, 53A*, 100–106.

Rammer, W. (1936). *Das Tier in der Landschaft. Die deutsche Tierwelt in ihren Lebensräumen*. Leipzig: Bibliographisches Institut.

Remmert, H. (1980). *Ökologie*. Berlin/Heidelberg/New York: Springer.

Remmert, H. (1992). Das Mosaikzyklus-Konzept und seine Bedeutung für den Naturschutz -eine Übersicht. *Laufener Seminarbeiträge, 2/92*, 45–57.

Remmert, H. (1998). *Spezielle Ökologie. Terrestrische Ökosysteme*. Berlin/Heidelberg/New York: Springer.

Riecken, U. (1990). *Möglichkeiten und Grenzen der Bioindikation durch Tierarten und Tiergruppen im Rahmen umweltrelevanter Planungen* (Schriftenreihe für Landschaftspflege und Naturschutz 32). Bonn: Bundesamt für Naturschutz.

Riecken, U. (1991). Probleme der Raumgliederung aus tierökologischer Sicht. *LÖLF-Mitteilungen, 4*, 37–43.

Riecken, U. (1992). *Planungsbezogene Bioindikation durch Tierarten und Tiergruppen* (Schriftenreihe für Landschaftspflege und Naturschutz 36). Bonn: Bundesamt für Naturschutz.

Ripple, W. J., Estes, J. A., Beschat, R. L., Wilmers, C. C., Ritchie, E. G., Hebblewhite, M., Berger, J., Elmhagen, B., Letnic, M., Nelson, M. P., Schmitz, O. J., Smith, D. W., Wallach, A. D., & Wirsing, A. J. (2014). Status and ecological effects of the world's largest carnivores. *Science, 343*, 1241484. doi:10.1126/science,1241484.

Risenhoover, K. L., & Maass, S. A. (1987). The influence of moose on the composition and structure of Isle Royale forests. *Canadian Journal of Forest Research, 17*, 357–364.

Röser, B. (1988). *Saum- und Kleinbiotope. Ökologische Funktion, wirtschaftliche Bedeutung und Schutzwürdigkeit in Agrarlandschaften*. Landsberg/Lech: Ecomed.

Ross, B. A., Tester, J. R., & Breckenridge, W. J. (1968). Ecology of mima-type mounds in northwestern Minnesota. *Ecology, 49*, 172–177.

Schaefer, V. H. (1981). A test of the possible reduction of the digging activity of moles in pastures by increasing soil nitrogen. *Acta Theriologica, 26*, 118–123.

Scheffer, V. B. (1958). Do fossorial rodents originate Mima-type microrelief? *American Midland Naturalist, 59*, 505–510.

Schlumprecht, H., Schupp, D., & Südbeck, P. (2001). Methoden zur Entwicklung eines Indikators "Bestandesentwicklung ausgewählter Vogelarten". *Naturschutz und Landschaftplanung, 33*(11), 333–343.

Schmitt, K. (1991). *The vegetation of the Aberdare National Park, Kenya* (Hochgebirgsforschung 7). Innsbruck: Universitätsverlag Wagner.

Schönborn, C. (1999). Die Makrolepidopteren der Ilmaue (Thüringen): Analyse differenzierender Umweltfaktoren und Schlußfolgerungen für eine ökologische begründete Sanierungskonzeption. *Zeitschrift für Ökologie und Naturschutz, 8*, 53–69.

Schott, C. (1934). Kanadische Biberwiesen - ein Beitrag zur Frage der Wiesenbildung. *Zeitschrift der Gesellschaft für Erdkunde, Berlin, 9/10*, 370–374.

Schultz, J. (2000). *Handbuch der Ökozonen*. Stuttgart: Ulmer.

Schultz, W., & Finch, O.-D. (1997). Eine Tierartenklassifizierung als Basis für biotopbezogene ökofaunistische Zustandsanalysen und Bewertungen. *Zeitschrift für Ökologie und Naturschutz, 6*, 151–168.

Schweinfurth, U. (1966). *Neuseeland, Beobachtungen und Studien zur Pflanzengeographie und Ökologie der antipodischen Inselgruppe* (Bonner Geographische Abhandlungen 36). Bonn: Ferdinand Dümmler's Verlag.

Schwerdtfeger, F. (1968). *Demökologie: Ökologie der Tiere* (Vol. II). Hamburg/Berlin: Parey.

Simoons, F. J. (1974). Contemporary research themes in the cultural geography of domesticated animals. *The Geographical Review, 64*, 557–576.

Sinclair, A. R. E., & Arcese, P. (1995). *Serengeti II: Dynamics, management and conservation of an ecosystem*. Chicago: University of Chicago Press.

Sinclair, A. R. E., & Norton-Griffith, M. (Eds.). (1979). *Serengeti. Dynamics of an ecosystem*. Chicago/London: University of Chicago Press.

Snodgrass, J. W. (1997). Temporal and spatial dynamics of beaver-created patches as influenced by management practices in a south-eastern North American landscape. *Journal of Applied Ecology, 34*, 1043–1056.

Tischler, W. (1990). *Ökologie der Lebensräume*. Stuttgart: Gustav Fischer Verlag.

Troll, C. (1936). Termitensavannen. In Landeskundliche Forschung, Festschrift Norbert Krebs (pp. 275–312). Stuttgart: J. Engelhorns Nf.

Turner, M. G., & Stratton, S. P. (1987). Fire, grazing, and the landscape heterogeneity of a Georgia barrier island. In M. G. Turner (Ed.), *Landscape heterogeneity and disturbance* (Ecological Studies 64, pp. 85–101). New York: Springer.

Vitousek, P. M. (1986). Biological invasions and ecosystem properties: Can species make a difference? In H. A. Mooney & H. A. Drake (Eds.), *Ecology of biological invasions of North America and Hawaii* (pp. 163–176). New York: Springer.

Weissbacher, A. (1999). Situation und Entwicklung der Borkenkäferpopulation im Nationalpark Bayerischer Wald. *Mitteilungen des Biologischen Bundesamtes, Land- und Forstwirtschaft, Berlin-Dahlem, 362*, 53–62.

Woo, M.-K., & Waddington, J. M. (1990). Effects of beaver dams on subarctic wetland hydrology. *Arctic, 43*(3), 223–230.

Chapter 2
Animal's Functional Role in the Landscape

When assessing the functional role of animal populations in the landscape and their effects on the interrelationships of physical and biotic factors, aut- and synecological as well as landscape ecological aspects must be closely linked to each other. In this context, landscape spatial structures, landscape history, biocoenoses, vegetation, soils, hydrology, local and microclimates as well as natural and human disturbances are of major importance.

2.1 Landscape: Habitat and System of Interacting Abiotic and Biotic Factors

The term 'landscape' has often been discussed, yet without coming to a generally accepted definition (see discussion by Leser 1997, for example). Right from the beginning this could not be expected, however, because 'landscape' may be considered from historical, economical, scientific, artistic and many additional points of view. Thus, in the context of this book landscape is considered to be the space where animals live: rural environments, forests and natural woodlands, lakes and rivers, moorlands and also built-up areas and their spatial and functional interrelationships. Some animal species are restricted to certain habitats, such as lakes and rivers for example (e.g., fishes, crustaceans), others use several habitats (biotopes) (e.g., amphibians, dragonflies, grey herons, sea birds, red deer, hippos) or are migratory. The habitats of different species may overlap, provided that no interspecific competition for essential resources (e.g., food, nesting places) prevents this. In the Brazilian Pantanal Wetland, for example, white-lipped peccary (*Tyassu pecari*), collared peccary (*Pecari tajacu*) and feral pigs (*Sus scrofa*) have coexisted for more than 200 years although their niches partly overlap. Seasonal flooding and periods of drought, however, may reduce food supply. In this case, feral pigs, which are typical generalists and effective foragers, may be at an advantage. As pigs are capable of using also belowground parts of plants such as roots, tubers and knolls, they become

F.-K. Holtmeier, *Animals' Influence on the Landscape and Ecological Importance*,
DOI 10.1007/978-94-017-9294-3_2

a competitor of native peccaries (Sicuro and Olivlira 2002). There is no biotope in landscape occupied by a single species only. We always have to do with zoocoenoses of different taxa. A species may prevent coexistence with another one not only by competition but also by alteration of vegetation structures, for example.

On the other hand, some species may profit from others. In African rain forests, for example, small grazers benefit from African forest elephants (*Loxodonta cyclotis*). The elephants create sunlit gaps by throwing trees down or killing them by bark stripping. Thus, they promote growth of nutrient-rich grasses and herbs that will be an important food resource for other grazers. In some places, elephants also expose mineral-rich soil (salt licks) by churning the soil or destroying termite mounds. Thereby the elephants and also other grazers partly meet their mineral requirements.

Various mammals such as badgers, foxes, prairie dogs, aardvarks, gerbils, kangaroo rats, ground squirrels, moles and rabbits build subterranean burrows and runway systems. After the fossorial animals leave their burrows, other animals such as reptiles, amphibians, birds or bumble bees may move in (Blumenberg 1986; Korn 1991; Hawkins and Nicoletto 1992; Gruttke 1994; Boye 1996). The burrowing owl (*Athene cunicularia*), prairie rattle snakes (*Crotalus viridis*) and various small mammals regularly use prairie dog burrows as nesting places or to hide from predators. In collapsed mole hills, nests of wild bees can often be found (Westrich 1989).

In oligotrophic savannas such as the Kalahari, solitary tall acacia trees (*Acacia erioloba*, *Acacia hämatoxylon*) are attractive to numerous animal species. The trees shade the ground and provide nesting places and also limited food supply (Belsky and Canham 1994). Excreta, nesting material and mortal remains of the animals accumulated under the trees considerably enrich the amount of nutrients. Concentrations of nitrogen and potassium were found to be double as high under the trees as compared to the surrounding oligotrophic grassland. Phosphorus content was even two-and-a-half-fold higher than in the grasslands. Enrichment of nutrients as well as the shade given by tree canopies facilitate the establishment of plants producing fleshy fruits, of which the seeds have reached such places by frugivorous animals. The vegetation under the *Acacia* canopies as well as the savanna vegetation mosaic as a whole – i.e., oligotrophic grassland interspersed with *Acacia* tree islands – appear to be due to these interrelationships between the *Acacia* trees and fruit-consuming animals attracted by the tree's shade and nesting possibilities. If over-grazing had resulted in low *Acacia* shrub, the present biodiversity would have been lost (Dean et al. 1999).

Lakes, ponds, rivers, peatland, forests, and heathland with their characteristic biocoenoses, climates, surface structures, soils, and biogeochemical properties (nutrient turnover) may be considered to be ecosystems. These all have in common fluxes of energy and material cycles. Moreover, ecosystems may regulate themselves to a certain degree (e.g., Ellenberg 1973), at least as long as they are resilient to disturbances. The effects of inner (e.g., a species getting out of control, failure of reproduction) or external disturbances (e.g., wild fires, mass-outbreaks of defoliating insects, introduction of animal and plant diseases) can hardly be reliably predicted because ecosystems are dynamic systems subjected to more or less rapid change. Remmert (1980a) compared ecosystems with a mobile consisting of a number of parts that are linked to each other mechanically and that all will acquire motion when one is touched. In a didactic view, this metaphor may play its part well as it shows the mutual dependence of various mobile components. It will not work however if the response of an ecosystem to a disturbance is considered as, in contrast to a mobile, the components of an ecosystem will usually not respond to impacts simultaneously. Moreover, they may change their quality (e.g., vegetation structure and composition, animal populations) and importance under the influence of disturbances.

Landscape is a mosaic of ecosystems of different size and quality that are at least partly linked by mutual interactions (Forman and Godron 1981; Forman 1995; Sect. 2.1.2). The spatial pattern of the landscape mosaic is usually related to topography. The interrelationships between the physical and biotic components are very complex and often almost inscrutable. Many finely differentiated and seemingly perfect diagrams have been published, showing for example interactions of fluxes of material and energy and their effects within ecosystems (e.g., Ellenberg 1973; Schreiber 1980b, 1989) or within subsystems only (e.g., Leser 1976, 1986, 1997; Mosimann 1984). At close sight, however, these diagrams reveal themselves to be at least partly based on plausibility rather than on well-substantiated results. They may even hide the fact that it is still very difficult or impossible to assess many of the processes and interactions even in simply organized ecosystems. Thus, we are relatively well informed about water balance for example, while our knowledge of the role of animals in landscape ecosystems still is insufficient.

The influence of animals on vegetation (e.g., browsing, grazing, bark-stripping, girdling, trampling), biocoenoses, soils (bioturbation, burrowing, erosion, nutrient enrichment), microtopography (e.g., mole hills, termite mounds) and flows of materials and energy (food uptake, transport, defecation), turnover-rates in decomposition) are the most relevant effects on landscape ecosystems (Fig. 2.1). Herbivores may influence surface-near microclimates by alteration of vegetation (height, density, coverage, albedo). Contrasts in microclimatic conditions near the ground surface may become aggravated. Large numbers of domestic and wild ruminants typical of many tropical grasslands and steppes outside the tropics as well as termites produce large quantities of methane, which enforces the greenhouse effect (Zimmermann et al. 1992; Collins and Wood 1984; Rasmussen and Khalil 1986; Schütz et al. 1990; Fabian 1991; Schönwiese 1992; Rudolph 1994). Moreover, many animal species are important dispersers of diaspores. Zoochoric seed dispersal may not only extend over relatively long distances but often is even more effective than is dispersal by wind or running water. In contrast to herbivores, carnivores have only little influence on spatial and temporal ecosystem structures, with the exception of some burrowing carnivores (e.g., moles). However, predators may influence foraging behavior of grazers and browsers, for example, and thus modify the herbivores' impact on vegetation and soil.

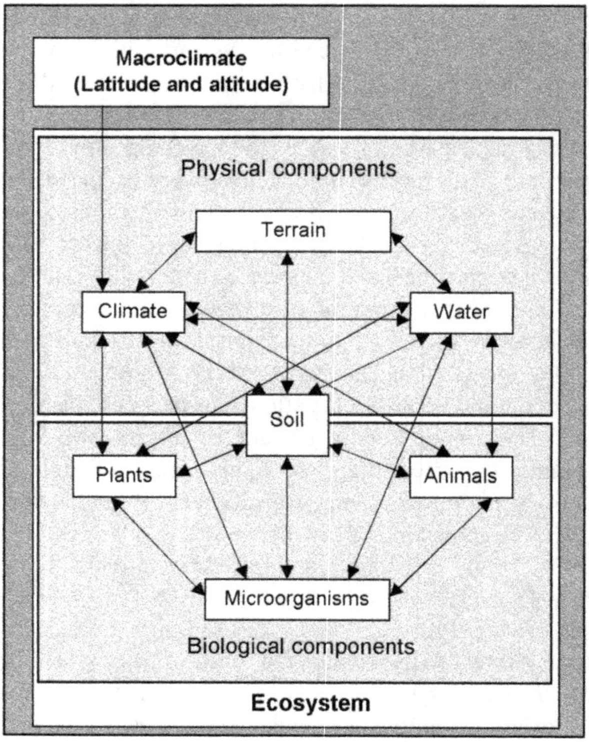

Fig. 2.1 Interactions of physical and biological factors in an ecosystem (strongly simplified). As soil plays an important role in fluxes of matter and energy between the lithosphere, biosphere, atmosphere and hydrosphere it is given a central position in this ecosystem model. Prepared by F.-K. Holtmeier

2.1.1 The Role of Animals in Matter and Energy Fluxes Within Ecosystems

As to the part of animals in flows of matter and energy, animals living in the soil and/or in the litter layer come first. By far, maximum turnover of dead organic matter takes place in the litter and humus layer. About 80–90 % of annual plant production reach the ground without having passed the digestive tract of herbivores. While chemical decomposition of dead organic matter is done by microorganisms (partly during digestation already) soil animals act as so-called primary decomposers (micro-, meso-, macrofauna) breaking down organic matter mechanically (e.g., earthworms, lumbricides). Dead organic matter would also decompose without their assistance, but at a comparatively low speed, however. Experiments with litterbags that excluded the invertebrates showed a considerable delay of decomposition (Madge 1965; Zlotin 1971; Beck 1989; Chew 1994; Fig. 2.2). If soil mesofauna such as springtails (collembola), for example, is excluded decomposition will become even more protracted (Beck 1989). Even in a tropical rain forest, where microbial activity is usually very high, 90 % of the leaf-litter contained in fine-meshed litterbags was not decomposed (Beck 1989).

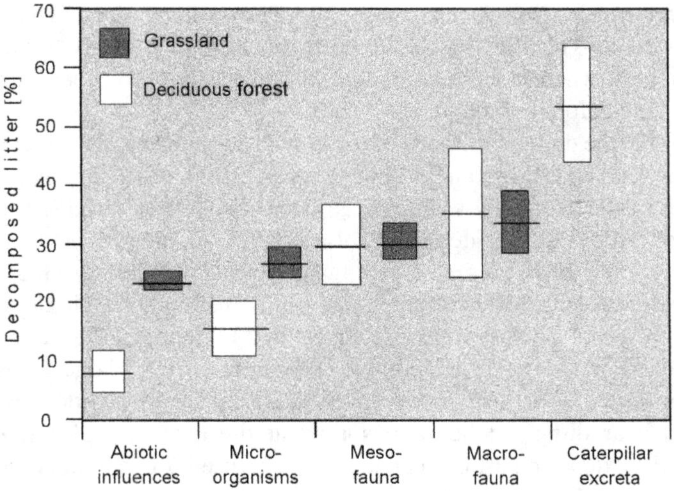

Fig. 2.2 Decomposition of leaf and grass litter (May–October) under the influence (cumulating from the *left* to the *right*) of physical factors, microorganisms, primary decomposers and excreta of caterpillars. The *horizontal bars* show standard deviation. Results of litterbag experiments. After Zlotin 1971, modified from Chew (1994)

Carrion beetles and dung beetles bury dung and carrion as food for their brood and thus accelerate turnover. In East African savannas, for example, more than 2,000 dung beetle species exist. The beetles are of particular importance during the rainy season when they remove the dung dropped by the huge herds of mammalian herbivores. Under favorable conditions, dung is removed very quickly. After 2 weeks already, a good portion of the dung has usually disappeared (Anderson and Coe 1974).

Some beetle species have specialized on the dung of certain mammal species, as for example the giant dung beetle (*Heliocopris dilloni*) which uses elephant dung (Kingston and Coe 1977). Others have developed specific modes of behavior that prevent competition between the different beetle species in the use of dung (Cambefort and Hanski 1991). Some beetles get into the dung heaps and will stay there as long as they will have depleted the food source or the heaps decay. Others dig galleries under the dung heaps where they store the dung. Some species produce dung balls rolling them straight-line on the surface for often more than 10 m before they bury the balls either for food storage or for making a breeding ball. Thus they fertilize and improve soil aeration. Without the dung beetle activities the grassland would probably have become completely blanketed with the dung of the huge herds of grazers (Heinrich and Bartholomew 1980). In the steppes of Central Asia, situation is similar although large grazers are by far not as numerous as in the African savannas (Botkin et al. 1981).

European dung beetles behave in a similar way. Geotrupid beetles (*Geotrupes stercarosus*), for example, dig up 40 cm deep galleries under all kinds of dung to drop their dung pills. The large geotrupid beetle (*Geotrupes steroccararius*) as well as the scarab beetle (*Copris lunaris*) prefer cattle or horse dung. The dor beetle (*Geotrupes vernalis*) carries sheep dung and game droppings into its subterranean brood chambers, whereas the scarab (*Sisyphus schaefferi*) forms beads from cattle and sheep dung before burying it.

In Australia, the dung of millions of cattle and sheep had become a serious threat to the range land (meadows and pastures) because the native dung beetles (about 250 species of the subfamily scarabaeinae) were not able to disintegrate the huge quantities of cattle and sheep dung, both of which contain more water than kangaroo droppings. At best, the beetles used the dung partly and only during a limited period in the course of the year. Most cowpats get dry and hard remaining almost unchanged for months or years if not becoming disintegrated by termites (Ferrar and Watson 1970: Lee and Wood 1971; Waterhouse 1974). Thus, over large areas nutrient cycling is impeded or prevented. Moreover, plants rejected by cattle may establish themselves near dung patches and spread at the cost of real forage plants. Consequently, extensive pastures can no longer be used. In addition, dung patches provide favorable living conditions to parasites such as the Australian bush fly (*Musca vetustissima*) or the blood-feeding buffalo fly (*Haematoba irritans exigua*), which was introduced from Timor. The latter is attracted to large mammals for fluid nourishment and oviposition. The flies cause irritation which can result in reduced beef and milk production. These parasites, partly acting as pathogens, are a serious threat to livestock farming.

To tackle the problem, African dung beetles were introduced in the late 1960s (Waterhouse 1974; Bornemissza 1976; Doube et al. 1991). Scarab beetles (*Ontophagus gazella*, *Euonticellus intermedius*, *Euonticellus africanus*) have spread most successfully. The dung beetles process about 90 % of the dung produced per year. The buried dung and the excrement of the beetles improve soil physical properties (aeration, water-holding capacity, stabilization of soil structure) and counteract compaction. Not least, dung beetles destroy the breeding places of livestock parasites by removing dung heaps. Thus, the dung problem has been solved partly. To achieve complete decomposition of all dung on the continent, approximately 100 additional dung beetle species with different ecological properties and demands must be introduced and integrated into the different habitats (Topp 1981).

According to the ecosystem character, the velocity of energy flow and turnover rates as well as the 'weight' of the components involved in the nutrient cycles differ considerably (e.g., Odum 1971, 1983; Walter 1977; Remmert 1980a, b; Schultz 1988, 2000). In this respect beech (*Fagus sylvatica*) forests, for example, differ much from pine or a spruce forests, streams from lakes, lakes from oxbows, and alpine heaths from subarctic dwarf shrub-lichen heaths or arctic tundra. While there is no question about the very important role of detrivores in ecosystems the role of leaf-eating (phyllophagous) animals is less clear. In a Central European beech forest, for example, phyllophagous animals rarely use more than 1 % of the leaf mass (Remmert 1973). In other words: the system would work without them as well, even though structural changes, for example, caused by leaf-consumers cannot not be excluded. Moreover, what applies to a Central European beech forest must not automatically go for other ecosystems, particularly if these are located in other climatic regions.

In East African savannas, for example, large mammalian herbivores are by far more important when compared to ungulates in Central European forests. The African ungulates consume almost half of the annual grass production, in the Serengeti even up to 90 % (Howe and Westley 1993). They persistently influence composition and structure of the plant cover as well as succession and turnover rates (e.g., Buechner and Dawkins 1961; Laws 1970: Laws et al. 1970; Wing and Buss 1970; Braun 1973; Botkin et al. 1981; Cumming 1982; Owen-Smith 1982; Parker 1983; Hatton and Smart 1984; Belsky 1992).

However, also in temperate forests and grasslands, ungulates may influence turnover rates and soil nutrients more than usually expected. In Yellowstone National Park, for example, net nitrogen mineralization on grassland grazed by large numbers of elk (*Cervus elaphus*) and bison (*Bison bison*) as well as by smaller herds of pronghorn (*Antilocapra americana*), bighorn sheep (*Ovis canadensis*) and muledeer (*Odocoileus hemionus*) was found to be double that of ungrazed fenced plots (Frank and Groffman 1998).

In the tundra, where the poikilothermic primary decomposers are inactive for most part of the year, warm-blooded (homeothermic) animals play a major role in phytomass turnover compared to deciduous forests in the temperate zone. In winter, the stomachs of reindeer, muskox, snow hares, ground squirrels, lemmings, voles and wild geese act as a 'fermentation chamber', where the plant residues become prepared for the further mineralization process (Remmert 1980b; Batzli et al. 1980; Batzli 1981). In addition, nitrogen-rich excreta of the warm-blooded herbivores enhance mineralization and thus nutrient supply to plants, in particular as the bulk of nitrogen in the excreta is already mineralized (ammonium) or easily soluble (urea) (e.g., McKendrick et al. 1980; Cargill and Jefferies 1984; Bazeley and Jefferies 1985). The influences of moose (*Alces alces*) herbivory on aboveground and belowground processes in early successional stands (*Salix* spp., *Alnus tenuifolia*) in interior Alaska were found to accelerate carbon turnover (Kielland et al. 1997). Moreover, addition of phosphorus has a positive effect on decomposition and availability of nutrients (Remmert 1980b). In the alpine tundra, where thermal conditions are to some extent similar to the Arctic, herbivorous mammals possibly are also more important in phytomass turnover than herbivorous invertebrates (e.g., Dearing 2001).

2.1.2 The Role of Animals in Matte and Energy Fluxes Between Ecosystems

Animals also influence flows of matter and energy between ecosystems (e.g., Woodmansee 1978: McNaughton 1985). Atypical high nitrate concentration, for example, that were measured in streams of various watersheds covered with undisturbed deciduous forests in northern Carolina obviously resulted from defoliation during a mass-outbreak of the fall cankerworm (*Alsophila pometaria*, Geometridae). The trees responded to defoliation by reduced production of wooden (structural) tissue while leaf production and thereby the amount of comparatively easily decomposable litter (narrow C/N-ratio) increased. In addition, large quantities of mineral-rich excreta of caterpillars increased nitrification to an extent that nitrate was only partly taken up by the plants and accumulated in the soil. Consequently, a good portion got into lakes and streams (Swank et al. 1981).

Ten thousands of migratory or overwintering gray-lag geese (*Anser anser*) and other Nordic goose species (e.g., barnacle goose, *Branta leucopis*; white-fronted goose (*Anser albifrons*) graze on agricultural land around the large nature conservation area 'Oostvaardersplassen' in Holland. Geese spend the nights in the center of the nature conservation area on shallow ponds surrounded by extend reeds where they are safe from predators. There they deposit vast amounts of nutrient-rich droppings and thus considerably enhance eutrophication (Krüger 1999).

The direction of nutrient flow is opposite, where brown bears (*Ursus arctos*), for example, defecate in riverine forests thus importing nutrients that they ingested when feeding on salmon. Nitrogen input may reach 10–20 % of the normal nitrogen concentration in the forest stands (Hilderbrand et al. 1999). Likewise large quantities of nitrogen accumulated in river otter (*Lutra lutra*) latrines along the shoreline (latrines 25–300 m distant from each other) significantly increase nitrogen concentration in the terrestrial vegetation, as has also been evidenced by tracer experiments (∂^{15}) in Alaska (Ben-David et al. 1998).

European wild boars (*Sus scrofa*) that spread in great numbers over the Great-Smoky-Mountain-National Park during the 1940s have caused considerable change in soil conditions by rooting (bioturbation). Boars also modified the structure of the deciduous forests. High concentrations of nitrogen and potassium in the soil solution down to 2 m depth as well as increased nitrate export into the water courses must be attributed to wild boars' activities (Singer et al. 1984; Table 2.1). Red deer (*Cervus elaphus*) and lactiferous hinds, in particular, prefer nitrogen-rich forage as is common to irrigated and fertilized meadows and pastures, for example. For resting and rumination red deer go for shelter-giving forest where their droppings enrich nutrients by nutrients ingested on the meadows (Schütz et al. 2000).

Table 2.1 Effects of rooting by wild boars (*Sus scrofa*) on soil properties (A-horizon) in high-elevation deciduous forests in Great Smoky Mountains National Park

Soil properties	Without boars	With boars
Without vegetation [%]	0	80
Leaf litter [kg ha^{-1}]	3,095	1,830
Nitrate [ɲg g^{-1}]	19	29
Calcium [ɲg g^{-1}]	90	56
Phosporus [ɲg g^{-1}]	58	32
Leaching of nitrate [mg l^{-1}] into streams	3.8	6.6
Nitrate [mg l^{-1}] in streams	0.7	1.5

Data from Sinclair et al. (1984)

In savannas and steppes, rapidly moving herds of ungulates can carry large amounts of nutrients even between grazing areas far distant from each other (McNaughton 1985). Hippos (*Hippopotamus amphibius*), for example, graze on savanna and defecate in lakes, ponds and rivers (Laws 1968).

On Aldabra Island, a large corral atoll located in the western Indian Ocean (NNW of Madagascar), feral goats (*Capra hircus*) improve the food source of the many native giant tortoise (*Geochelone gigantea*). During the day, the goats are roaming about over large areas. In the evening, they return to their resting places at the sea shore. Their droppings add nutrients to the soil from areas that are beyond the reach of the tortoise (Gould and Swingland 1980).

Moreover, water birds and sea birds nesting on land are involved in flows of materials and energy from lakes, ponds and streams or from the sea to land. Common herons (*Ardea cinera*), for example, feeding on fish and amphibians in shallow waters and on small mammals (e.g., voles) on agricultural land, carry organic matter from their feeding habitats to their breeding colonies where considerable amounts of nutrients accumulate with the food remains and excreta falling from nesting trees. As the excreta are extremely rich in nitrogen (urea and guanine) and very acid (pH about 3) they cause serious change to the tree and herb layer (e.g., Gerken et al. 1992). Cormorants (*Phalacrocorax carbo*) have a similar effect. In a large (12 ha) heron rookery, occupied by about a total of 8,000 individuals of three heron species (night heron, *Nycticorax nycticorax*; great blue heron, *Ardea herodias*; great white egret, *Egretta alba*), vegetation cover considerably declined parallel to the increase of nesting density and input of guano (Weseloh and Brown 1971).

The import of nutrients from the sea to land by petrels, shearwaters, cormorants and penguins, however, is of a much greater order of magnitude. Along the coast of Peru and Chile, example, highly productive up-dwellings provide an inexhaustible food source for seabirds. On the arid coastal mainland, where cormorants (*Phalocrocorax bougainville*), gannets (*Sula variegata* and *Sula nebouxii*), brown pelicans (*Pelecanus occidentalis*), Peruvian penguin (*Sphensicus humbolti*) and other seabirds excrete their droppings, huge guano deposits have accumulated. They were a rich source of fertilizer. Conditions are similar along the coast of Southwest Africa where also gannets, cormorants (*Phalocrocorax capensis*) and black-footed penguins (*Spheniscus demersus*) produce most of the guano (Crocker 1966). Practically, in all seabird rookeries phosphate, potassium and nitrogen are enriched at large quantities. The C/N-ratio is very narrow (e.g., Aston 1909; Leamy and Blakemore 1960); Blakemore and Gibbs 1968; Smith 1976; Wormell 1976; Burger et al. 1978). As the nesting material is strongly enriched with ammonium, Ca- and Mg-ions become leached. Consequently, nitrification of ammonium causes acidification. During their breeding season on Rottnest island (18 km off the Australian west coast), wedge-tailed shearwaters (*Puffinus pacificus*) deposit large amounts of nitrogen-rich guano (>230 kg ha^{-1} $year^{-1}$, dry mass; Bancroft et al. 2004). This is by far more than the amount of nitrogen fertilizer applied to the agricultural areas in western Australia (8–20 kg ha^{-1} $year^{-1}$ (Raupach et al. 2001).

Huge colonies of sea birds can be found in particular on subantarctic and arctic islands and along the coast of the antarctic continent. The ice-free areas are bare

Photo 2.1 Chinstrap penguins (*Pygoscelis antarctica*) in Admirality Bay, King George Island, Southern Shetlands. M. Bölter, 1984

rock, locally sparsely covered with plants. Lichens, mosses, fungi and algae prevail (Walter and Breckle 1991). With the exception of two native and a few adventive phanerogams no other higher organized plants occur. Moreover, no permanent terrestrial animal species are living there, about 50 insect species and various mites and collembolan species disregarded. As soils on ice-free ground generally are almost devoid of organic matter, they were even called 'ahumic soils' (Tedrov and Ugolini 1966). Penguins as well as some flying birds are the only living beings introducing organic substances to the continent (Boyd et al. 1966; Allen et al. 1967; Joly et al. 1987). About 100 millions of penguins are living in the Antarctic (Laws 1985). They represent 90 % of the total bird biomass of the Antarctic (Odening 1984). Adelie penguins (*Pygoscelis adeliae*) and chinstrap-penguins (*Pygoscelis antarctica*) are the most common species (Photo 2.1). They nest everywhere on the Antarctic coasts provided that suitable breeding places are available (Müller-Schwarze 1984).

The arctic oceans belong to the most productive marine ecosystems. These are characterized by a great food web based on an immense production of phytoplankton and microplankton that supports bacteria, protozoa, zooplankton (krill) and higher organisms such as squid and fish. In the past, whales were the main krill consumers. They used about 190 million tons of krill per year. Since whales have declined due to over-harvesting, seabirds and seals now are the most important consumers of the available krill (Everson 1977; Laws 1985). Penguins together with the other seabirds (albatrosses, petrels, shearwaters) consume between 115 and 130 million tons of animal plankton (krill), fish and squid per year

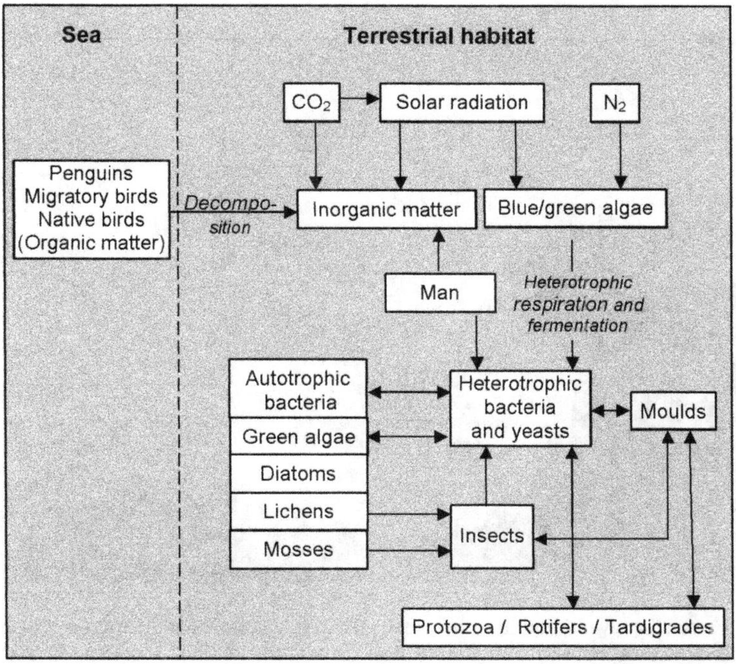

Fig. 2.3 Influence of seabirds on turnover rate, nutrient fluxes and organisms on the Antarctic mainland. Modified from Boyd et al. (1966)

(Croxall 1984; Odening 1984; Laws 1985). During the breeding season these birds deposit huge amounts of waste in their rookeries along the coast. The rookeries are occupied by several ten thousand or even hundred thousand individuals. Occasionally there may be even a few millions in one breeding colony. It has been estimated that about 10 kg m^{-2} $years^{-1}$ (dry weight) of protein-rich excreta accumulate in the rookeries (Tatur and Myrcha 1984; Myrcha and Tatur 1991). This nutrient-rich 'manure' is an important basis for the flows of materials and energy within the comparatively simply-structured terrestrial food web (Fig. 2.3). While at the coast of the Antarctic mainland low temperatures and limited moisture supply prevent mineralization of the guano, it becomes rapidly decomposed on subantarctic islands where thermal and hygric conditions are a little more favorable.

Studies on King George Island showed that penguin guano and other bird excrement together with plant material (some liver mosses, lichens, algae and some higher plants) increase the amount of enzymes in the soil and considerably enhance enzyme activities (urease, phosphatase, xylanase) compared to mineral soils. Microbial biomass to enzyme activity ratios were particularly high in places where penguin guano had accumulated (Tscherko et al. 2003).

Chemically aggressive guano solutions washed down from the rookeries may cause phosphatization of the weathered bedrock (Myrcha et al. 1985; Tatur 1989; see Fig. 2.4). In abandoned rookeries, guano is a rich nutrient source of

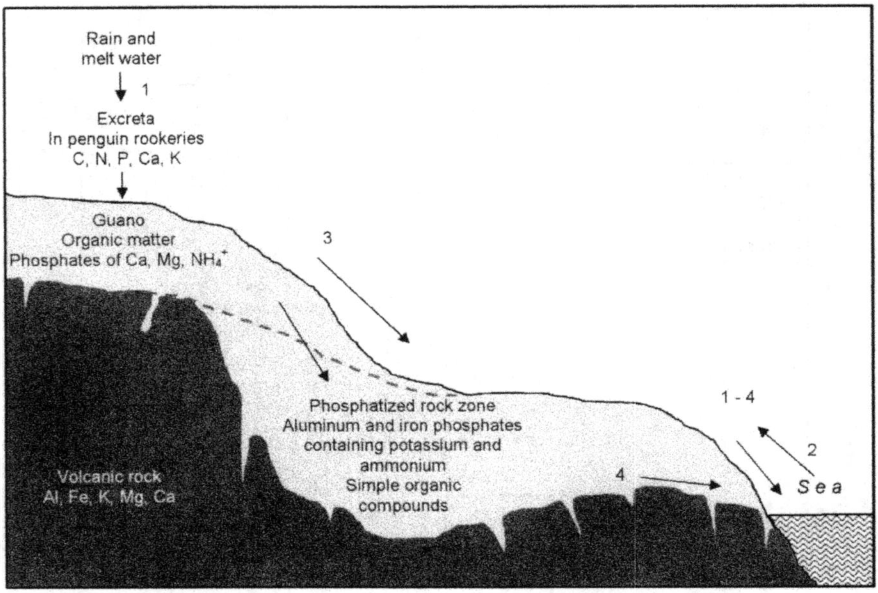

Fig. 2.4 Factors involved in ornithogenic sedimentation. *1* – Rain and meltwater, *2* – Saltwater introduced to the ecosystem with penguin food and as aerosol, *3* – Erosion of the guano surface by runoff. Nutrient-rich, alkaline reaction ($NH_4^+ >> NO_3^-$). Leaching of the guano layer by seepage, nutrient-rich, acid reaction (NO_3^-, NH_4^+). Modified from Myrcha et al. (1985)

plants as long as mineralization is low and leaching is minimal (Myrcha and Tatur 1991). However, studies on arctic Inexpressible Island (Terra Nova Bay, west coast of the Ross Sea) provided evidence that modification of the soils in penguin rookeries may gradually disappear after abandonment, with the exception of small stones that the penguins had collected in the surroundings for nest building (Campbell and Clairidge 1966).

However, breeding colonies of seabirds may strongly influence vegetation and soil also in other regions. On the Isle of Rhum (west coast of Scotland), for example, lush grass vegetation, the so-called 'shearwater greens', has developed at an elevation above 600 m due to regular manuring by great numbers of wedge-tailed shearwaters (*Puffinus puffinus*) from the end of March until mid-October. Grassland comparable to the 'shearwater greens' can otherwise be found in close proximity of the coast only. Repeated fertilization of the nearly un-vegetated ultra-basic soil with NPK fertilizer for consecutive 5 years had the same effect. Vegetation cover increased from 5 % to almost 60 %, and a herb-rich *Agrostis*/*Festuca* grassland has developed (Wormell 1976). The 'shearwater greens' not only provide optimal forage to red deer (*Cervus elaphus*) but have become the basis of a complex biocoenosis that would not have developed without Manx shearwaters (Wormell 1976).

Repeatedly, the intensive green color of resting places of seabirds and of the talus debris at the food zone of bird cliffs has been described (e.g., Summerhayes

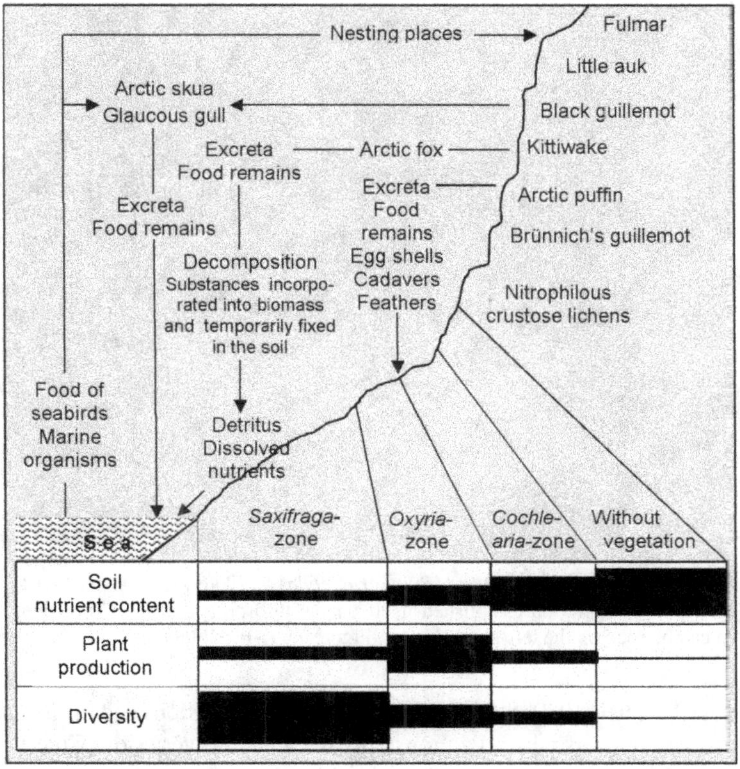

Fig. 2.5 Vegetation zones on a bird cliff (Northwest Svalbard) and their causes. Modified from Thannheiser and Wüthrich (1999)

and Elton 1928; Hinz 1976; Eurola and Hakala 1977; Hartmann 1980; Wüthrich 1992, 1994; Möller 2000). Below the cliffs a catena-like sequence of different nutrient concentrations and corresponding vegetation can be found ('cliff series', Summerhayes and Elton 1928). Immediately below the cliffs, concentrations of phosphorus potassium and magnesia may be a thousandfold higher compared to normal tundra (Wüthrich 1992, 1994). Extremely high nutrient concentrations have a toxic effect and prevent higher plants from colonizing such places. At greater distance from the cliffs nutrient concentrations decrease and vegetation diversity increases (Fig. 2.5; Wüthrich and Thannheiser 1997). As long as seabirds import nutrients to the cliffs the gradient-like distribution of nutrients will continue to exist. After the breeding season, however, when the birds leave their colonies, runoff, percolation and leaching translocate the nutrients downslope where they are stored in the humus-rich ornithogenic soils. The nutrients are only partly cycled on land. As there are no herbivores that would consume living plants or parts of them, the plants die off at the end of the growing season. This is the reason why no already mineralized nutrients (e.g., ammonium) or dissolved nutrients are available as is the case in terrestrial tundra where warm-blooded herbivores play an important role in

Photo 2.2 Breeding colony of kittiwake (*Ryssa tridactyla*) on Ekkerøy, Varanger Fiord, northern Norway). Luxuriant vegetation established itself on the bird cliff and in its footzone. A small potato field can be seen on the left-hand side (cf. Figure 7). F.-K. Holtmeier, 29 August 1001

this respect (Remmert 1980b). Due to adverse climatic conditions and the low number of primary decomposers mineralization of organic matter is very slow. Consequently, a mainly mechanically disintegrated detritus develops, that is transferred together with the dissolved substances to the sea (Klekowsky and Opalinski 1986).

Considerable amounts of nutrients accumulate also on the cliffs, where a luxuriant vegetation may develop. On Ekkerøy for example, a small peninsula located about 12 km east of the town of Vardø (Varanger fjord; northern Norway), horizontal layers of Cambrian shists form a steep cliff (Photo 2.2) occupied by about 22,000 breeding pairs of kittiwake (*Ryssa tridactyla*). The birds have built their nests on different 'floors' under narrow 'roofs' formed by little projecting shist layers, that provide shelter from the rain (Fig. 2.6). Excreta, nesting material, remains of fish and other food as well as carcasses accumulate on the ledges below the nesting places. Dripping water and runoff translocate dissolved nutrients (from above). As a result of rich nutrient supply and soaked material accumulated on the projecting horizontal shist layers, a luxuriant vegetation has developed. It consists mainly of sea mayweed (*Tripleurospermum maritimum*), orphan John (*Sedum telephium*), common yarrow (*Achillea milleflorum*) and grasses. Lush vegetation is also common to the many almost vertical clefts, where moisture is high due to permanent dripping water and runoff. Luxuriant plant bunches hang down from the precipices like a potted geranium from a balcony (cf. Photo 2.2). The foot zone of the bird cliff is densely covered with stinging nettle (*Urtica*) indicating intensive eutrophication from above. As most nutrients however remain on the cliff where they are taken up by plants, toxic concentrations are unlikely in the foot zone. By contrast, local people have cultivated potatoes on a marine terrace immediately below the cliff.

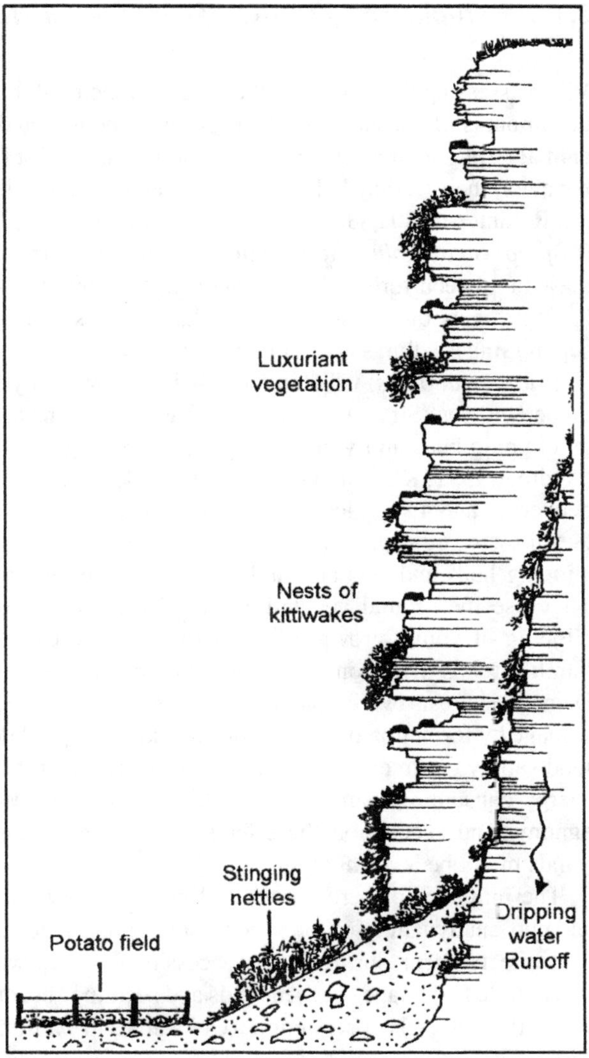

Fig. 2.6 Vertical section (*schematic*) through a breeding colony of kittiwakes (*Ryssa tridactyla*) on Ekkerøy, northern Norway (cf. Photo 2.2). Drawing after a field sketch by F.-K. Holtmeier from 29 August 2001

The plots are small (30–40 m^2) but highly productive due to natural fertilization by seepage and runoff from the cliff. If not sufficiently tended, however, the small 'potato fields' will be rapidly occupied by stinging nettle and common sorrel (*Rumex acetosa*), typical indicators of strong eutrophication. In some years, kittiwakes did not return to this breeding place. In 2011, however, the cliff has been completely occupied again. A similar luxuriant vegetation with Norwegian Archangelica (*Angelica archangelica*) and sea mayweed (*Tripleurospermum maritimum*) supported by ornithogenic 'fertilization' has been described by Vogel (1988) from Iceland. Although sea birds mainly influence the immediate coastal areas they also penetrate far into the tundra where they import large amounts of nutrients that influence nutrient cycles and vegetation (Remmert 1968, 1980b).

2.1.3 Animals as a Cause of Changes in Habitat Quality

When assessing the influence of animals on their habitats not only animal numbers, the amounts of consumed phytomass, excrements and dead organic matter, i.e., the animals' role in nutrient cycles and energy flow, must be considered, but also their behavior, their feeding habits, and seasonal migrations have to be taken into account (cf. Remmert 1980a; Gessaman and MacMahon 1984; Petrak 1993). Thus, roe deer (*Capreolus capreolus*), for example, do not much influence turnover of the phytomass in a beech forest (*Fagus sylvatica*) by consumption, whereas they may have lasting effects on forest structure, plant species and succession by browsing and suppressing seedlings and saplings. The same holds true for black-billed capercaille (*Tetrao urogalloides*) which impede height growth of young larches (*Larix gmelini*) in the northeast-Siberian larch-forest region (Kolyma mountain area near Magadan) by clipping buds and young shoots and breaking branches when landing on them (Andreev 1991; Klaus and Andreev 2001). The weight of a fully grown cock ranges between 3 and 4 kg at least. Therefore, cocks stay mainly on the forest floor, where they browse young growth of larch while lighter hens prefer the crowns of large trees clipping buds and shoots. Larch shoots are the only winter food for capercaillie. They use the ground vegetation from May through October only. By intensive clipping of young growth of larch for decades, capercaillie create and maintain 'larch gardens' with bonsai-like growth forms that lend the larch taiga its character. Damage is most severe in lekking sites. As larch rapidly recovers from mechanical damage by formation of new shoots, this kind of 'gardening' considerably increases food supply per tree and area. Thus, the birds may feed more effectively and avoid energy-consuming searches for food. Compared to these influences on larch physiognomy and succession the effects of capercaillie on turnover within the larch stands might be less important.

The influence of herbivorous and granivorous small mammals on energy flow and nutrient dynamics of forest ecosystems appears to be negligible. At best, significant effects occur during early successional stages when young trees are highly susceptible to damage or when seeds and seedlings become destroyed (Potter 1978). Thus, red squirrels for example (Chickarees, *Tamasciurus sabrinus*) damage the bark of young maples (*Acer saccharum*, *Acer rubrum*) and drink the out-coming sap. Mould and bacteria enter the wounds and cause canker (Shigo 1967). On the other hand, mycophagous small mammals, for example, can influence forest ecosystems persistently even though they generally are unimportant with regard to turnover. Flying squirrels (*Glaucomys sabrinus*), grey squirrels (*Sciurus griseus*), Abert's squirrels (*Sciurus aberti*), Chipmunks (*Tamias* spp.), chickarees and some vole species are living on the fruiting bodies of hypogeous mycorrhizal fungi, in particular of basidiomycetes and ascomycetes (Fogel and Trappe 1978). In places where these small mammals defecate they inoculate the soil with spores of mycorrhizal fungi that are essential for nutrient uptake of the trees. Thus, the small mammals essentially contribute to maintenance of the forests (Maser et al. 1978a, b; Kotter and Farentinos 1984a, b; Maser et al. 1985; Hayes et al. 1986; Maser et al. 1986;

Maser and Maser 1987; Carey et al. 1999). One gram of dried feces has been estimated to contain between 200 million and one billion spores (Kotter 1981a, b). In Australia, similar interrelationships exist between some mycophagous marsupials and ectotrophic eucalypts (Malajczuk et al. 1987). The rat kangaroo (*Bettongia penicillata*) for example is a key factor in succession on burned areas as it disperses large amounts of fruiting bodies of ectotrophic mycorrhizal fungi that are essential for nutrient supply to the trees and shrubs re-colonizing the burns (Lamont et al. 1985). Moreover, small mammals can accelerate decomposition of fallen trunks by inoculation with fungal spores (Maser et al. 1978b; MacMahon 1981). This might be particularly important in tropical rain forests as the endobiontic fungi there are not specialized and are able to grow in all kinds of wood (Malloch et al. 1980).

After the eruption of Mount St. Helens (Washington) on May 18, 1980, northern pocket gophers (*Thomomys talpoides*) and ants that had survived under the volcanic ashes brought buried original soil on top of the sterile volcanic ash blanket (Sect. 2.6.1.1). As the old soil contained mycorrhizal fungi, mycorrhizal associations could gradually recover and improve growing conditions for the plants re-colonizing the area (Andersen and MacMahon 1985; Allen and MacMahon 1988; Allen 1991; Allen and Grisafulli 1994).

The function of animals is usually very complex. For example, consumption of seeds by chipmunks may impede natural regeneration in forest stands. On the other hand, chipmunks ensure the nutrient supply of the trees by dispersal of mycorrhizal fungi. Leaf-eating beetles as well as other phyllophagous insects may reduce leaf mass in deciduous forests for a short while without lowering production automatically. Moderate loss of leaves (up to 30 %, Vorontzov, cited in Perry 1994) may even stimulate photosynthesis in the leaves remaining on the trees. On the forest floor, growth will increase due to better light conditions. Under certain conditions this may compensate for reduced production caused by leaf consumption in the canopy layer. In addition, the defoliators attack mainly trees and shrubs weakened already (Perry 1994), which then die off more rapidly. As a result, more favorable conditions for regeneration are likely because of reduced competition for light and nutrients. The mineral-rich feces of the insects as well as their cadavers, that are rich in nitrogen and phosphorous, stimulate mineralization of the litter and thus improve nutrient availability (Fig. 2.7).

The same holds true for damage caused by the caterpillars of the green oak tortrix (*Tortrix viridiana*), the gipsy moth (*Lymantria dispar*) and other noxious forest insects (e.g., Novak et al. 1989) during their occasional mass outbreaks. Such unexpected mass outbreaks are usually misrepresented by the media as 'ecological catastrophes'. Basically these alleged 'catastrophes' are natural phenomena immanent to the forest ecosystem that, however, does not properly correspond to the popular idea of an 'intact nature'. A few more figures may complement the preceding remarks. In a mixed deciduous forest near Kursk (Russia), the caterpillars of the green oak tortrix reduced the oak leaf area by 70 %. Afterwards the other tree and the herbaceous layer not only received more light but also more water due to the reduced canopy interception of rain. Partly nibbled oak leaves, large amounts of exuvia and cadavers, nitrogen and phosphorous, increased supply to the soil by 17 %. Potassium content rose by 13 %. Consequently, increased mineralization (up to two-and-a-half fold

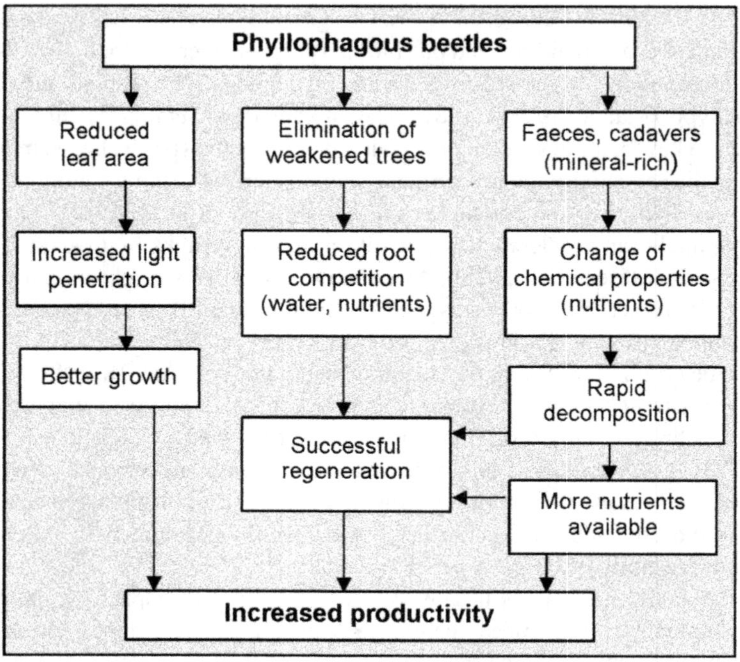

Fig. 2.7 Influence of leaf-eating beetles on plant production in a deciduous forest. Scheme by F.-K. Holtmeier

increase of soil respiration) improved nutrient supply of the trees that were not affected by the caterpillars. As a result, the composition of the herbaceous layer changed (Rabotnov 1995).

In a 110–115 years old oak stand on podsolized Gley with fine-humus-rich raw humus near Münster (Westfalia, Germany), pH values in precipitation were recorded from May through July 1995. Beneath the stand canopy pH increased to 7.1 (=2.6 times higher than in open land precipitation), and annual potassium input (37.91 kg ha^{-1} $year^{-1}$) was considerably higher than in a beech stand nearby (28.95 kg ha^{-1} $year^{-1}$; Höper 1997). Increased values resulted from input of matter from the tree canopies and to a great extend from the excretory products of the green oak tortrix during a mass outbreak.

Studies on the influence of grasshoppers (*Melanoplus frigidus*, northern migratory grasshopper and *Aeropus sibiricus*, Siberian grasshopper) on alpine sedge communities (*Carex curvula*) showed that grasshoppers consumed between 19 and 30 % of the above-ground mass of phanerogams (seed plants), whereas they used only 3 % to produce body substance. High carbon losses probably resulted from the grasshoppers' respiration. As grasshoppers do not leave their habitat, their excreta and carcasses supply the system with other nutrients, particularly with nitrogen. Consequently, the C/N ratio narrows and mineralization of dead organic matter accelerates (Blumer and Diemer 1996).

Aphids (Aphidoidae) may also considerably increase turnover rate. The plant saps they feed on contain great quantities of sugar but very little protein and essential trace elements. Thus, the aphids must take up comparatively more plant sap to meet their requirements of protein and trace elements than they need for metabolism. Aphids excrete the surplus sugar (ca. 78 % glucose, sacrose, fructose and melizitose) with their liquid waste (so-called 'honeydew'). According to conservative estimates (Llewellyn 1972), one kilogram 'honeydew' was excreted by the aphids under lime trees per square meter and year (*Tilia* x *vulgaris* Hayne). The sugar is not only an essential food source of ants, flies and wasps but also an important carbon source of non-symbiontic nitrogen fixing bacteria. Supply of 'honey dew' increases soil nitrogen (Delwiche and Wijler 1956; Graf 1971; Owen and Wiegert 1976; Dighton 1978; Petelle 1980, 1984). Especially in nitrogen-poor ecosystems such as the tundra for example, where nitrogen mineralization is very slow, the influence of aphids may be very important for turnover (Remmert 1980a, 1980b; Sect. 2.1.1). On the other hand, aphids take away great amounts of high-quality assimilates from deciduous trees and may thus reduce growth rates. In the end, the ants living in symbiosis with the aphids are also involved as they 'nurse' the aphids (e.g., protecting them from predators, overwintering the eggs of several aphid species in the nest chambers of the ants) thus supporting population growth (Klimetzek and Wellestein 1978; Stary 1990).

Some cuckoo-spit insects (Cecropidae) tap the sap flow of trees. The sap contains strongly diluted amino acids. The common frog hopper (*Philaenus pumarius*), for example, uses sap to build mucous cocoons (=so-called 'cuckoo-spit') where the nymph lives. Other species excrete the water directly on the ground, as for example *Pleus grossus*, a large gregarious cicada in West Africa. During the growing season cicadas feed on the fruits of arborescent legumes. The root system of the trees extends close to the surface far away from the trunk. A few roots, however, reach the ground water table. Thus, the trees are able to take up water even during the dry season. Nutrients, however, are concentrated in the uppermost top soil as is typical of tropical forests. Thus, nutrients accumulated in the top soil are not available to trees for most of the dry season. Under trees heavily affect by cicadas, these insects supply so much water to the soil that moisture becomes available to plants even during the dry season. Under such conditions, seeds that otherwise would stay in dormancy may germinate, mineralization goes on, and the trees can take up nutrients from the topsoil. However, this process will keep going only as long as water loss by evapotranspiration and cuckoo-spit cicadas does not exceed water supply of the trees from the ground water table (Owen and Wiegert 1976).

Many plants rely on pollination by animals (birds, insects, bats, flying foxes) for regeneration. During evolution, mutualistic relationships developed between many plant species and pollinating animals. Animals were essential for survival of these plants (Faegri and van der Pijl 1971; Feinsinger 1983; Whittaker and Jones 1994). The saguaro or giant cactus (*Carnegia gigantea*), for example, which is a dominant plant in the Sonoran desert, as well as the organ pipe cactus (*Lemaieocerus thurberi*) depends on cross-pollination. The diurnal honey bee (*Apis mellifera*), the nocturnal nectar-feeding Mexican long-nose bat (*Leptonycteris nivalis*) and the white-winged dove (*Zenaida asiatica mearnsi*) are the most important pollinators. *Zenaida*, however, pollinates the giant cactus only (Alcorn et al. 1959, 1961, 1962; McGregor et al. 1962). On the other hand, in the Magaliesberg mountain range (South Africa) opportunistic avian nectarivores, such as the Cape rock thrush (*Monticola rupestris*), have been found to enhance reproduction in an endemic montane aloe (*Aloe peglerae*) more than do specialized bird pollinators (Area et al. 2013).

In the end, it would be no problem to reduce also these complex interactions to the productivity aspect. Animals' influence on productivity, however, is only one aspect when considering the role of animals as ecological agents in landscape. In this broader context, the focus is on the interactions of animal populations, habitat quality and changes in composition, succession and structure of vegetation, soil conditions (e.g., moisture, nutrients, compaction, erosion), microclimates, etc.

Mutualisms between herbivores and their food sources, for example, play an important role in that respect (e.g., Mattes 1978, 1982; McNaughton 1979a; Owen and Wiegert 1981; Tomback 1982; Tomback and Linhart 1990). Under the selection pressure of grazers, grass formations have developed that are well adapted to grazing. They consist of many extended and long-lived clones with many clone series (Haberd 1962, 1967; Owen and Wiegert 1981). Apparently, many grass species are not able to exist without being grazed (McNaughton 1979a). Prevailing vegetative reproduction and proliferation of grasses are considered to be a response to lasting grazing pressure. As grazing reduces seed production (e.g., McNaughton 1979b; Owen and Wiegert 1981), vegetative reproduction ensures the continued existence of clonal grasses and thus the food resource of the grazers. On the other hand, Belsky (1986), who reviewed the specific literature on this issue, found no clear evidence of a positive influence of herbivory on productivity and vitality of forage plants. Moreover, the assumption that saliva of herbivorous mammals and grasshoppers would stimulate the growth of grasses has not been proven (Reardon et al. 1972, 1974; Detling et al. 1980). Insects are supposed to inject grasses with growth-stimulating substances (Dyer and Bokhari 1976). All these processes resist measurement and quantification – if these processes are ever tried. Therefore, they are almost disregarded in conventional ecosystem models.

2.2 The Role of Animal Population Density

The extent of animals' influences on the landscape are closely related to their population density. Population density fluctuates for various reasons. For example, population-internal factors such as territorial behavior and reproduction, or external factors such as lasting periods of drought, extremely snow-rich winters, fire, floods and other natural 'catastrophes', and not least human impact (hunting, alteration of habitats, etc.) may be driving factors. Internal and external factors usually act simultaneously and their specific effects are often hard to distinguish.

2.2.1 *Habitat Conditions and 'Self-Regulation' as Density Influencing Factors*

Fluctuations in population density may extend over short (as much as a year) or long intervals. Rapid fluctuations are typical of so-called r-strategists that promptly respond to unstable or unpredictable environments. Their fecundity is high, maturation sets on early, and regeneration time is short. R-strategists are usually small (e.g., insects, small mammals such as voles or weasels), whereas so-called K-strategists are of large body size (e.g., large grazers and browsers such as elephants or hippos, large top carnivores such as lions or eagles). They are able to compete successfully for limited resources and their life span usually is comparatively long.

Short-term fluctuations may be caused by bad weather conditions, for example, varying forage supply, physical condition of the animals, epidemics, parasites and natality/mortality ratio. After a good seed production of beech (*Fagus sylvatica*) in mixed Bavarian mountain forests, the bank vole (*Clethrionomis glareolus*) and the yellow-necked field mouse (*Apodemus flavicollis*), for example, were able to reproduce very successfully under the snowpack during the consecutive winter (Fig. 2.8), being protected from cold temperatures and their many predators. No reproduction was observed, however, in the following year, when no beech nuts were available (Bäumler and Hohenadl 1980).

Fluctuations may also be caused by long reproductive cycles or continuous population growth over at least two generations (Caughley 1970), often culminating in mass occurrences followed by sudden decline. Thus, in some years cockchafers (May beetles, *Melolontha melolontha*, *Melolontha hippocostani*), voles (*Microtus* spp.) or rabbits (*Oryctolagus cuniculus*) are teeming while they are relatively rare in other years.

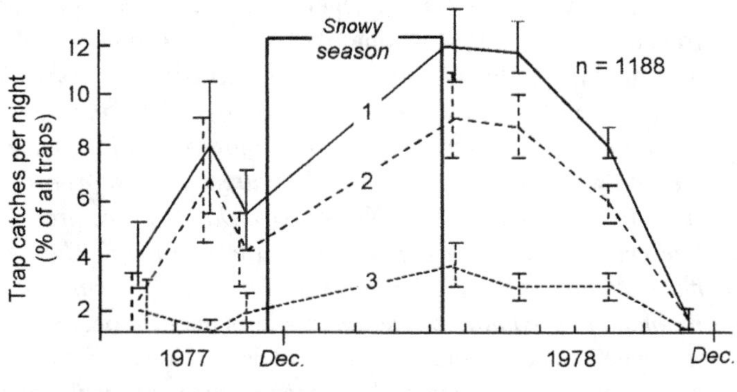

Fig. 2.8 Development of the population densities of bank vole (*Clethrionomys glareolus*) and yellow-necked field mouse (*Apodemus flavicollis*) after a beech mast (*Fagus sylvatica*) in autumn 1977. *1* – Total population, *2* – Bank vole, *3* – Yellow-necked field mouse. Modified from Bäumler and Hohenadl (1980)

In Sub-Saharan Africa, recurrent mass-outbreaks of multimammate rat (mice; *Mastomyx natalensis* and related species) are a serious problem in public health and agriculture (Fiedler 1988). *Mastomyx natalensis*, which is ubiquitious in equatorial Africa, transmits Lassa virus to man. Moreover, extreme damage caused to agricultural crop can result in shortage of human food. The outbreaks usually followed high rainfall at the end of the previous year, although they cannot be explained by the effects of rainfall alone (Leirs et al. 1996). Effective control of *Matomys* population has been impossible so far.

Mass outbreaks of common vole (*Microtus arvalis*) that regularly occur at intervals in grassland and agricultural fields of the northwestern lowlands of Germany have driven people to despair and into the churches where they implored relief from the plague. In northern Spain, for example, common voles in mass outbreaks during the early 2000s (Jacob and Tkadlec 2010) devastated extensive barley and maize fields. In addition, a number of people fell sick with tularemia (*Francisella = Pasteurella tularensis*) probably transmitted by the voles. Common voles are characterized by an almost unbelievable reproductive capacity. They are able to reproduce already when 12 days old. After maturity they produce a litter of five or more young voles every 3 weeks. In Central Europe, the reproduction period usually lasts from May through October. Nevertheless, winter reproduction is not unusual. As population density increases, territory size decreases and a couple of females together rear the brood in a nesting community. At higher population density territorial fights between competing males increase and their numbers decline considerably. Due to the resultant change of the sex ratio a mass outbreak is very likely (e.g., Frank 1986). During the 1949s and 1950s the field vole (*Microtus agrestis*) reproduced to an extent never observed before. The voles profited from the large clear-cuts created during the 2nd World War and during the post-war era. After grasses and herbs had completely overgrown the clear cuts, field voles benefited from optimal conditions and reproduced vigorously. They caused severe damage (girdling, bark-stripping) to afforestation and confronted foresters with great problems.

Such mass occurrences are immediately presented by the media as 'a catastrophe'. In 1990, the German news journal 'Der Spiegel' (no. 32), for example, reported on a mass occurrence of voles in Mongolia: *Voles are overrunning Mongolia – Mongolia is being threatened by a famine caused by disastrous murine plague. Up to 40 million of the Brandt's vole (Microtus brandti) graze down the steppes east of the capital Ulan Bator and threaten the basic food source of cattle, goats and sheep in an area that extends over 400.000 km². Milk has already run short everywhere in the country. The English geographer Nicholas Middleton, lecturer at Oxford University: 'Wherever you go in the steppe – voles are swarming'. Their explosion-like eruption has been favored by the fur hunters who strongly decimated foxes and wolves that are the main predators of the rodents.* It is unquestionable that hunters have almost eradicated many natural predators of vole. However, overgrazing by cattle, sheep and goats and inadequate land management probably had a much stronger influence than lacking predators. Even if there were more foxes and wolves, all predators together would not be able to effectively regulate excessive vole numbers.

In another German news journal (Focus 18, 27 April 1998), one could read: '*Cockchafers – abnormal calamity. The forests in Hesse (Hessian Riet, Germany) suffer from a scourge never observed before*'. Actually, it is the formulation that is 'abnormal' rather than mass occurrence of cockchafers. The information ranges at the same level with an official report that goshawk (*Accipiter gentilis*) had reproduced 'unnaturally'. Another statement made up out of thin air was that hooded crow (*Corvus corone cornix*) and raven (*Corvus corax*) had produced a 'surplus density' (Goretzky 1999). Such boulevard-ecological explanations lacking any justification appear all the time.

Cyclic mass-outbreaks are very common in some herbivorous insects, particularly at high latitudes. Normally, these insects represent only a small proportion of the total number of forest insects (e.g., only 1–2 % of lepidopters). At certain intervals, an outbreak of insects reproduces like an explosion. The mass outbreaks of the autumnal moth (*Epirrita autumnata*) and the winter moth (*Operophtera brumata*) in the subarctic birch forests of Northern Europe (Sect. 2.4.1) as well as of the larch bud moth (*Zeiraphera diniana*) in the Central Alps (Sect. 2.4.2) are typical examples. The population cycles usually extend over 2–3 years. Afterwards, the populations decline usually due to shortage of food of sufficient quality. The mountain pine beetle (*Dendroctonus ponderosae*) and the spruce budworm (*Choristoneura fumifera*) in Western North America (Sects. 2.4.3.1 and 2.4.3.2) are additional examples to mention in this context.

Populations are however regulated not only by food shortage and predators and/or territorial behavior. It may also happen that vegetation recovering from insect impact no longer meets the habitat requirements of animals, as is typical for example of the interrelationships between wainscots (*Nonagria*, *Archanara*) and reed (*Phragmites australis*). Reed is the food source and pupation habitat of wainscots. Wainscots pupate during July/August within the internodes of reed near the culm base. The diameter of the internodes must be at least seven centimeters to allow pupation. Reed responds to wainscot feeding by thriving on quite densely grouped thin culms instead of thick culms at lower density. Reed stands severely attacked by wainscots are characterized by an irregular mosaic of relatively open patches with thick culms and patches with dense stands of thin culms (Vogel 1985). On the one hand, the thin reed culms provide abundant high-quality food to the hatching caterpillars, on the other hand however, pupation fails because of too narrow internodes. Consequently, wainscot population collapses. Thus reed itself regulates its noxious wainscots (Vogel 1985).

Such population cycles represent maximum amplitudes of population dynamics. During the progressive phase, population density increases exponentially as long as no opposite factors such as shortage of food, for example, availability of nesting places or competition, have a limiting effect. Increasing density may result in social stress and population crash, as has happened at high vole density (Frank 1953, 1954). In other cases, surplus individuals emigrate, as is typical of lemmings (*Lemmus lemmus*), for example (e.g. Batzli 1981). After the decline of a population, recovery normally starts at a very low level (see also Fig. 2.10b). Such gradations usually extend over several years. In most cases, however, the amplitude of fluctuations in population density is much narrower.

Fluctuations are the result of a bundle of density-dependent and density-independent factors, often related to each other by mutual interactions (Fig. 2.9). When animal population density is approaching the natural habitat carrying capacity, processes and factors (e.g., delayed maturity, change of sex ratio, increasing competition, social behavior) often subsumed under 'self-regulation', may limit further population growth. The term 'self-regulation' however has to be considered with some reservation as it is not really correct. 'Self-regulation' lags behind population peaks because animals cannot actively control their numbers. In the end external factors (forage supply, habitat structures, climate, etc.) are the main controllers.

On the other hand, favorable conditions usually promote reproduction. During the progressive phase of a Canadian population of white-tailed deer (*Odocoileus virgianus*) in a habitat providing ample forage, 6 months old hinds were already

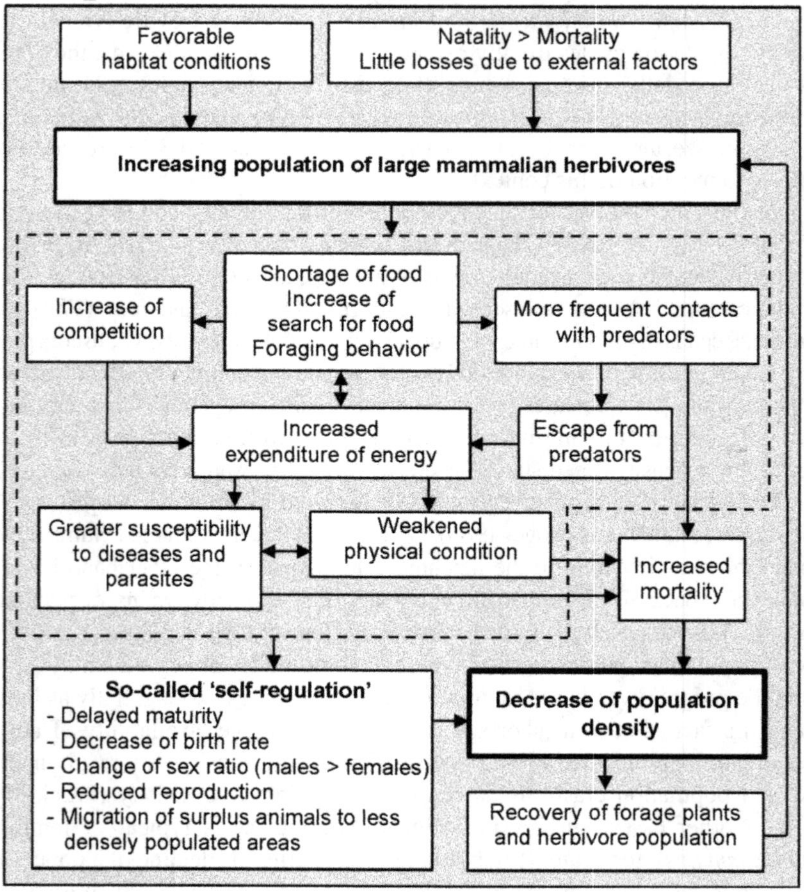

Fig. 2.9 Density-dependent and density-independent factors and their influences on population density. Scheme by F.-K. Holtmeier

pregnant and twin births prevailed (>50 %). Under comparable conditions, does give birth to more fawns than roebucks and may thus produce a surplus within a comparatively short period of time (Ellenberg 1978; Müri 1999a, b). Artificial reduction of mountain goats (*Oreamnos americanus*) in the Rocky Mountains had the same effect (Houston and Stevens 1988). At low population density of American elk (*Cervus canadensis*) about 25 % of the calves often are twins compared to only 1 % when conditions are unfavorable.

Due to 'self-regulation' setting in at a critical animal density, population growth is sigmoid and approaches a habitat's carrying capacity (K-value) asymptotically (Fig. 2.10a). As 'self-regulation' mechanisms become effective with a certain delay,

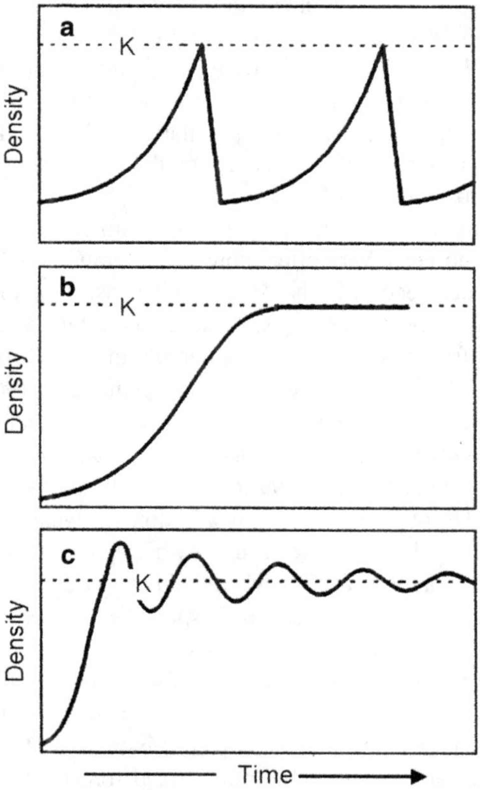

Fig. 2.10 Main types of population dynamics. (**a**) in response to the negative feedbacks of a growing animal population, population growth rate decreases, approaching the natural habitat carrying capacity (K) asymptotically, (**b**) Cyclic mass-outbreaks. Exponential population growth beyond natural habitat carrying capacity. Consequently, population drastically declines to a very low level due to the combined effects of depleted resources, predators, diseases, social stress and intraspecific competition. Exponential growth resumes after recovery of the habitat. (**c**) Exponential population growth exceeds habitat carrying capacity for a while as is typical of initial colonization, for example. After a decline due to overutilization and delayed 'self-regulation', population density gradually levels out fluctuating around the K-value (dynamic equilibrium). Modified from Odum (1991)

the carrying capacity may be exceeded for a while, before animal density will rapidly decrease. In response to the reduced animal impact, habitat conditions may improve again (e.g., more forage, reduced competition) and population will increase. Mass-outbreaks at certain intervals are likely (Fig. 2.10b). In other cases, however, population density may be fluctuating around the K-value for a while until the amplitude levels out towards a dynamic equilibrium (steady state) (Fig. 2.10c). This kind of development can be found in old-established populations responding to disturbances as well as after animals were introduced into an environment where they never existed before (Riney 1964; Caughley 1970; Challies 1975). The curve shape reflects habitat resilience.

Different from the three basic types of continuous (steady) population development presented in Fig. 2.10, development in reality does not take place continuously but at different temporal steps (time-discrete; e.g. succession of generations). In addition, they are controlled by a multitude of partly mutually interacting and mostly non-linear variables. As demonstrated by a simulation (Fig. 2.11) even a slight variation in the density of the initial population and/or of the reproductive rate will result after a certain time in unpredictable population developments (Christiansen and Fenchel 1977; Costantino et al. 1991).

Cyclic fluctuations partly culminating in mass outbreaks (e.g., *Epirrita autumnata*) are natural phenomena. Very often, however, alterations of habitats by humans have favored mass occurrences. In the African savannas, for example, expansion of cereal growing (corn, wheat, rice, millet) and concomitant construction of water reservoirs and tanks for irrigation have considerably improved living conditions for the red-billed weaver bird (*Quelea quelea*) (Schmutterer 1965; Dreiser 1993; Sect. 2.4.4). While in the past the numbers of red-billed weavers were controlled by seed production of native sweet grasses and lack of water during the dry seasons, cereals now provide abundant additional food, and water is available to the weaver birds everywhere and throughout the year. In addition, the weaver birds profited from planting of *Acacia* trees along the river courses which provide safe nesting sites.

In Yellowstone National Park (USA), where wild fires occurring at intervals control long-term forest succession (Romme and Despain 1990), fire fighting has altered the ecological conditions in a way that favors mass occurrences of the mountain pine beetle (*Dendroctonus ponderosae*; see also Amann 1977). Likewise, the magnitude of damage caused by mass outbreaks of the larch-bud moth (*Zeiraphera griseana*) to larch forests in the Central Alps has increased after man transformed many of the original larch-Swiss stone pine mixed forests (*Larix decidua*, *Pinus cembra*) into pure larch stands (Auer 1947; Holtmeier 1967b, 1974, 1995a, b; Sect. 2.4.2).

In most cases, availability of food (quantity, quality) seems to be an integral part of the complex of factors controlling population dynamics (Remmert 1980a). It often has a key function (Verme 1969; Thorne et al. 1976; Remmert 1980a; Caughley 1983; Sinclair 1985; Clutton-Brock et al. 1987; Houston and Stevens 1988; Choquenot 1991). As long as sufficient high-quality food is available and neither territorial behavior nor predators, diseases, lack of nesting places and other habitat properties have a limiting effect, population growth up to the maximum habitat

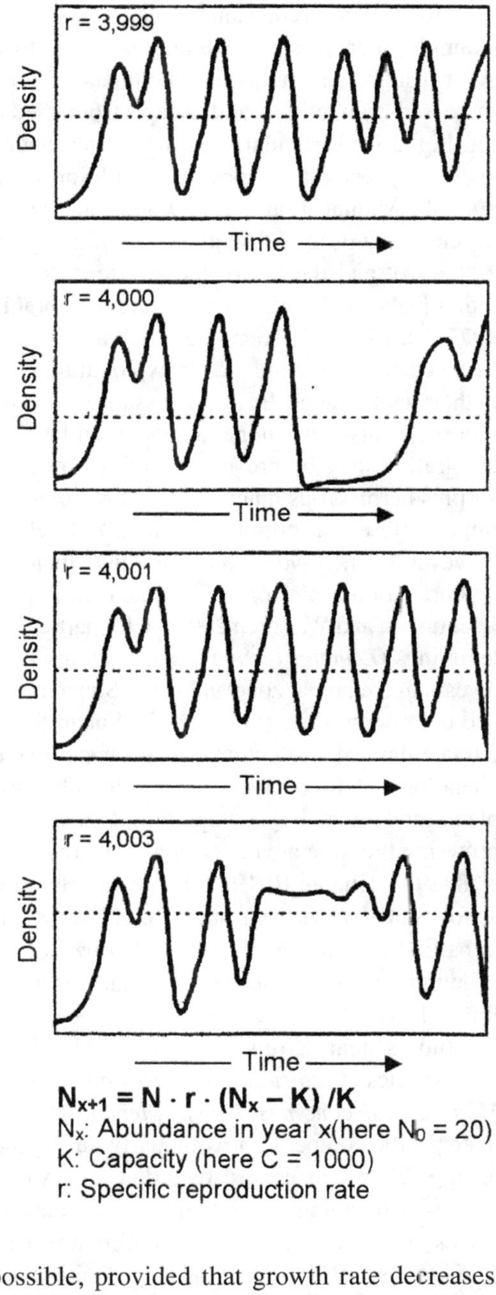

Fig. 2.11 Simulation of population dynamics at slightly different reproduction rates. Prepared by F.-K. Holtmeier and M. Reichstein

carrying capacity is theoretically possible, provided that growth rate decreases parallel with growing population density. The situation becomes critical for animals and their habitats when food is depleted and animal numbers are temporarily exceeding habitat carrying capacity (see Fig. 2.10b, c). This may be due to external, density-independent influences, such as periods of drought or extremely snow-rich winters, or because 'self-regulation' is lagging behind population growth.

Under such circumstances, competition for food between wild ungulates, for example, increases and the animals must travel longer distances when foraging. As a result, more frequent encounters with predators are likely. Search for food, increased competition and escape from predators increase expenditure of energy (cf. Fig. 2.9). Particularly during winter when forage is scarce anyway such influences may result in bad physical condition and increasing mortality of the weakened animals. Malnutrition during winter, for example, may cause too low body weight of females at birth. Subsequent loss of the young animals is likely (e.g., Albon et al. 1983a, b), and the reproduction rate decreases. Surplus and weakly competitive individuals are forced to move to less favorable habitats (see also Klein 1970; Carl 1971; Batzli 1981). There are observations, however, that do not fit into this theory. In Sweden, for example, over-exploitation of food sources by moose (*Alces alces*) in their winter range did not necessarily result in emigration to other areas with lower moose density (Sweanor and Sandegren 1989). Moreover, thinning of population by emigration may be prevented by insurmountable obstacles. Thus, emigration of surplus herbivorous mammals from remote oceanic islands, for example, is usually impossible and a population crash is likely. After recovery of the food source, however, the herbivore population may increase again.

Population cycles of small rodents have probably been studied most intensively (Henttonen and Wallgren 2001), with particular interest on the cyclic fluctuations of lemmings (*Lemmus lemmus*) culminating in mass occurrences at intervals of about 4 years in the boreal zone and in the Subarctic/Arctic. Population cycles of lemmings and other microtines (voles) in the Subarctic/Arctic also are primarily attributed to climate-dependent fluctuations in forage quantity and quality. Thermal deficiency, phenology of food plants, slow mineralization, resultant poor nutrient supply to plants and secondary compounds (chemical defenses) enriched in plants under browsing pressure are considered to be the driving factors (Kalela 1962; Chapin III et al. 1978; Chapin III 1980: Laine and Henttonen 1983). Generalist predators such as foxes (*Vulpes vulpes*), snowy owl (*Nyctea scandiaca*) and specialized predators, in particular least weasels (*Mustela nivalis*) and stoats (*Mustela erminea*), respond to lemming population peaks by increasing reproduction (MacLean et al. 1974; Batzli 1981).

Studies near Kilpisjärvi (NW-Finland) showed that the cycles of various vole species (*Clethrionomys rufocanus, Clethrionomys rutilus, Microtus agrestis, Micrototus oeconomus*, *Lemmus lemmus*) run parallel, a few differences caused by interspecific competition particularly during the progressive phase and culmination of population densities disregarded (cf., Andersson and Jonasson 1986). All vole species populations reach their low simultaneously. Likewise, synchronous rodent cycles have been proved for entire northern Fennoscandia (Myrberget 1973; Christiansen 1983) including Kola Peninsula (Koshinka 1970). Thus, a common overriding factor or complex of factors, possibly climate, may be assumed.

However, the relationships between population dynamics and their controlling factors are not yet completely understood, and it is occasionally difficult to unambiguously distinguish what is cause and effect. Thus, in the view of Laine and Henttonen (1983), for example, flowering cycles dependent of thermal conditions and availability of nutrients are a cause of rodent population increase, whereas Järvinen (1987) considers bud-clipping by lemmings to be the cause of fluctuations in flowering frequency. Moreover an often mentioned direct dependence of population dynamics on the phytomass of food plants could not be proved. Thus, many studies show that regeneration decreases before the food source will be depleted (Kalela 1957; Schaffer and Tamarin 1973; Krebs and Myers 1974; Vitala 1977; Oksanen and Oksanen 1981). Although sounding a little deterministic this seems to make sense as young animals born in late summer have only a little chance of survival. However, it may be questioned whether this has to be considered an adaptation. According to observations by Myrberget (1986) near Kilpisjärvi an influence of weather conditions on population cycles cannot be statistically substantiated.

Andersson and Jonasson (1986) emphasized that the existing hypotheses on causal relationships between fluctuations in food resources (quantity, quality, phenology) and rodent cycles (e.g., Kalela 1962; Tast and Kalela 1971; Laine and Henttonen 1983) do not satisfactorily explain the complex phenomenon (see also Schultz 1972; Rhoades 1983), even if disregarding regional peculiarities (Andersson and Jonasson 1986). Kausrud et al. (2008), however, found that changes in temperature and humidity, and thus conditions in the snowpack, markedly affect the dynamics in populations of lemmings and other small rodents. In a changing climate, which may bring about more precipitation and higher temperatures, lemming cycles will probably cease.

2.2.2 *Influence of Predator–Prey Relationship on Population Density*

A few remarks on the influence of the predator–prey relationships, however, seem to be appropriate; in particular, as ecologically-oriented popular literature as well as nature conservationists and also the hunters' lobby consider population control by predators to be of paramount importance in population dynamics and for balance of nature in general. Nature conservationists emphasize the functional role of predators as regulators in ecosystems to justify reintroduction of large predators into their original habitats. Hunters, on the other hand, do not get tired of claiming that they have to take the role of the large predators that became nearly extinct in our landscapes.

Though there is a close relationship between prey populations and numbers of predators, causes and effects are often confused. The often quoted fluctuations in snowshoe hare (*Lepus americanus*) and lynx (*Lynx canadensis*) with cycles of 8–11 years may serve as an example (Fig. 2.12). Three quarters of lynx food consist of snowshoe hares. As lynx, however, usually kill only a tenth of the hare population, snowshoe hare is more important for lynx than is lynx for snowshoe hare. In other words, prey apparently controls predator and not the other way round (Remmert 1980a).

Nevertheless, the problem is multi-faceted. In general, snowshoe hares killed by lynx are in a bad physical condition (Pease et al. 1979; Keith 1983; Keith et al. 1984;

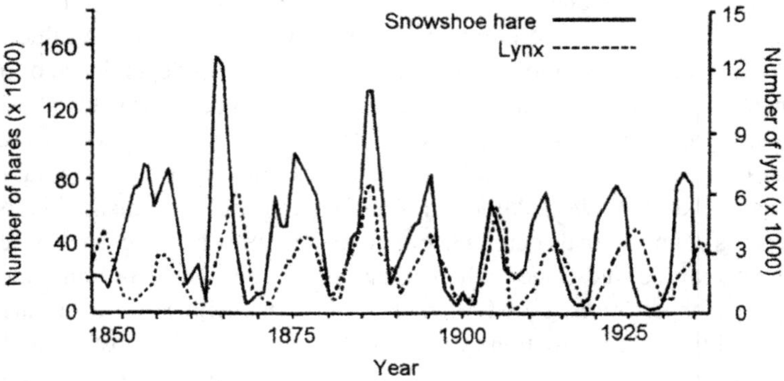

Fig. 2.12 Cyclic population dynamics of snowshoe hare (*Lepus americanus*) and lynx (*Lynx lynx*) inferred from the number of pelts delivered to the Hudson Bay Company (McLulick 1937)

Smith et al. 1988). Thus, snowshoe hare decline is very likely due to shortage of food, in particular of high-quality food. Plants under severe browsing pressure produce defensive chemical compounds (Bryant 1981; Bryant et al. 1983, 1985; Bryant and Chapin III 1986; Bryant et al. 1989) that make plants inedible for snowshoe hare. Smith et al. (1988), on the other hand, could not find evidence of snowshoe hare control by lack of forage. In addition, they denied that chemical defenses play a major role in snowshoe hare decline. Anyway, the authors conceded negative effects of insufficient food quality. Even though there is no absolute lack of food, competition gradually increases due to growing hare numbers. Therefore, hares in search of food are forced to travel relatively long distances. Shelter from predators is often missing. Both search for forage and escape from predators increase expenditure of energy. Studies on demography and ecology of a declining snowshoe hare population in Alberta (Canada; Keith et al. 1984) came to the result that shortage of forage is the primary cause of population decline, but that it is the combined effect of insufficient nutrition during subsequent one to two winters together with predator impact that are responsible for high hare mortality. Later on, during the population low, predation possibly is the only controlling factor (Hodges et al. 1999). In a broad study on mountain hare cycles in northern Europe, Newey et al. (2007) found that different factors may limit or control hare population dynamics with considerable regional differences in periodicity, amplitude, and predator control. In Fennoscandia, generalist predators (e.g., red fox) may have a limiting effect, whereas in Scotland predator influence is unimportant.

The possible role of predators in ecosystems has often been discussed in the context of lemming cycles. Predators, in particular r-strategists such as weasels (*Mustela nivalis*) and stoats (*Mustela erminea*), benefit from lemming cycles and rapidly respond to cyclic mass occurrences by intense reproduction. As a result, rodents will increasingly suffer from strong predator pressure in late winter and during the following summer (Tast and Kalela 1971; Laine and Henttonen 1983; Henttonen and Wallgren 2001), in particular as these predators prey on lemmings and their litter also under the winter snowpack (MacLean et al. 1974; Batzli 1981). As is the case in snowshoe hare population dynamics, the influence of predators increases after the prey population peak.

Comparable observations were made during studies on the influence of weasels and other predators on lemmings at Point Barrow (Alaska) and on Banks Island (Northwest Territories, Canada) (Thompson 1955; Maher 1967; MacLean et al. 1974; Fitzgerald 1981). Predator pressure on lemming density was particularly strong during the regressive phase of the lemming cycles. Weasels, in particular, had a strong impact. Presumably they killed mainly females and young in their winter nests, whereas snowy owls preyed more on males than on females (Pitelka 1957). Pomarine skuas (*Stercorarius pomarius*) killed both females and males equally (Maher 1970).

The situation is similar in mass outbreaks of the field vole. Its most successful predator is the buzzard (*Buteo buteo*). When voles occur at high density they make up 75 % of its prey. Consequently, reproduction in buzzards increases. At low vole numbers the proportion of voles in the prey of buzzards ranges between 25 and 30 % only (Leicht 1979). In the latter case, buzzards, now at great numbers, must take more other prey, such as moles, shrews, reptiles, young rabbits and hares, and birds as well. Buzzard numbers may remain on a relatively high level, at least for a while, although reproduction decreases (Hohmann 1994; Steiner 1999). Neither buzzards nor the many other predators (e.g., other predatory birds, storks, grey herons, carrion crows, weasels, stoats, polecats, foxes, wild boars, feral pigs, hedge hogs, cats, etc.) preying on voles will prevent cyclic mass occurrences.

The project on moose (*Alces alces*) and wolf (*Canis lupus*) relationship and dynamics on Isle Royale is the longest (55 years) continuous study of any predator–prey system in the world (Vucetich and Peterson 2013) and provides insight into the high complexity of predator–prey relationship and interactions with the environment. Isle Royale (>500 km^2) is located in Lake Superior and extends for about 20–30 km almost parallel to the Canadian northern lake shore. Moose reached the island first at the beginning of the twentieth century. They came over the ice in winter or swam across the 24 km wide channel between the mainland and Isle Royale in summer (Fig. 2.13; Allen 1974). The latest arrived during the winter 1912/1913. In 1915,

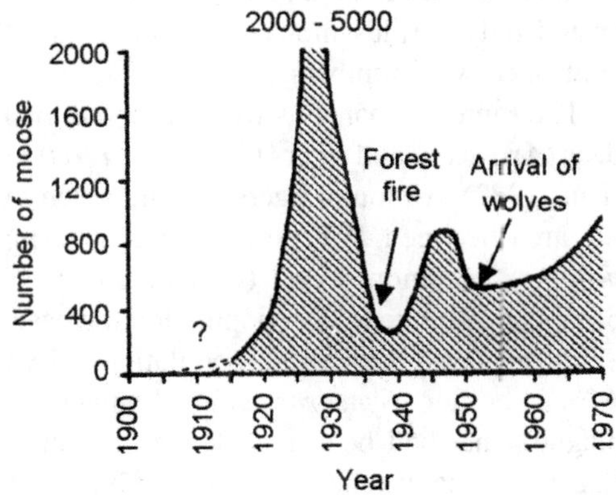

Fig. 2.13 Development of the moose population on Isle Royale (USA) until 1970. Modified from Krefting (1974)

about 200 moose were already living on Isle Royale. The vegetation had not been affected by other ungulates prior to the arrival of moose, with the exception of a few woodland caribou (*Rangifer tarandus*) in the relatively humid coastal strip, where they fed on tree lichens (Allen 1974; Pastor et al. 1988).

Moose, benefitting from the abundant forage supply, rapidly increased. In general, rapid population growth is typical of immigration of animal species into a new environment (Riney 1964; Caughley 1970). Around 1932, moose numbers (2,000–5,000 individuals; four to ten moose per square kilometer) already exceeded the island's habitat carrying capacity. Almost everywhere signs of over-browsing were apparent in aspen, birch, mountain ash, and balsam fir as well as in the aquatic vegetation (Murie 1934; Hickie 1936). Canadian yew (*Taxus canadensis*) prevailing in the lower canopy layer almost became extinct (Slavik and Janke 1987). In 1933/1934, moose population declined to approximately 200 individuals due to malnutrition (Aldous and Krefting 1946). An extreme wildfire that affected about a quarter of the island considerably reduced the food resource for the 3 to 4 coming years. Thereafter moose population gradually recovered, and in 1948 about 800 moose were living again on the island. The recovery was mainly attributed to the luxuriant secondary vegetation (*Populus balsamea*, *Betula papyrifera*) that had colonized the burn areas. Thus, the food source improved fundamentally. Increasing population density in moose and also in other cervids after forest fires or extended clear cutting typically occur also in other woodlands (e.g., Flook 1962; Strandgaard 1982). Due to too severe winters (1948/1949 and 1959/1960) moose numbers declined again to about 500 individuals. This decline has been ascribed to high calf mortality caused by the harsh winter climate (e.g., Mech et al. 1987). At the same time – between 1948 and 1959 – wolves (*Canis lupus*) had arrived at Isle Royale. Before wolves turned up, the island had no predators that could have threatened moose. In spite of pessimistic predictions that a wolf population would eradicate moose, and although food plants had decreased between 1950 and 1970 (with the exception of balsam fir), moose numbers continued to grow up to more than 1,000 individuals (approximately 3 moose km^{-2}) (Brandner et al. 1990). This increase was ascribed to an ample supply of high-quality browse during summer and autumn keeping moose in good physical condition. Consequently, survival rate in winter was relatively high, though wolf population had doubled simultaneously. This, however, means that wolves had harvested only surplus moose (Hansen et al. 1973) and that moose population had controlled wolf numbers.

The same mechanisms were observed, by the way, between populations of Axis deer (*Axis axis*) and tigers (*Panthera tigris*) in Kanha National Park (Central India) (Kurt 1982). Although tigers had increased by 30 % (from 30 to 48 individuals) in the area between 1972 and 1976 and despite the fact that also red dogs (*Cuon alpinus*), common jackal (*Canis aureus*) and leopards (*Panthera pardus*) were preying on axis deer, deer population increased from 5,000 to 13,000 animals. In the Serengeti (East Africa) the populations of wildebeest (*Conochaetes taurinus*) and African buffalo (*Syncerus caffer*) dramatically declined due to the rinderpest (cattle plague), that had been introduced by domestic cattle from Russia and Southern Europe during the end of the twentieth century. In the African grasslands, rinderpest became the main cause of wild ruminants' mortality. After the rinderpest was under control the large ungulate populations rapidly recovered in spite of the many large predators (e.g., lions, leopards, spotted hyenas) (Sinclair 1979a, b).

Coming back to the predator–prey relationships on Isle Royale, one must not ignore the fact that wolves preyed not only on moose but also on beavers (*Castor canadensis*). About 2,000 beavers were estimated to be living on the island. Wolf predation, however, did not prevent beaver numbers from increasing (Allen 1974).

An unexpected development began after extremely snow-rich winters in 1969, 1971 and 1972. In contrast to less snow-rich winters, when moose occurred at relatively low density almost everywhere on the island and wolves came across moose only occasionally, moose kills drastically increased. Almost twice as many calves fell victim to wolves. Because of deep snow on the island's interior, cows and their calves now stayed in the wind-swept, less snow-rich coastal areas where travelling is easier. Thus, the animals could save energy when searching for food. The calves, however, had almost no access to forage as the food trees had been intensively browsed by moose during previous years. Thus, forage above the former browsing line was out of the calves' reach. The calves became completely exhausted, as was reflected in the complete lack of fat in their bone marrow. As the weakened moose calves concentrated within a relatively small area, wolves could easily kill many of them, in particular as also the wolves preferred the costal terrain because of the relatively thin snowpack. Losses of calves to wolves remained at a relatively high level also during the consecutive years. This is possibly due to the fact that the physical condition of the calves that were born after one of these extreme winters or had survived their first year of life remained bad even for years. Consequently, predators could kill them more easily than calves in good physical shape (Allen 1974). Wolves, which had considerably increased in response to the abundant prey, kept moose numbers at a relatively low level from 1975 to the early 1980s (Fig. 2.14). After the arrival of disease (highly contagious canine parvovirus) in 1981, wolf population declined by 80 % (Vucetich and Peterson 2013) and remained limited for about a decade (communication R. Peterson, 28 April 2013). Afterwards, moose numbers rapidly increased again. Almost half of the moose living on Isle Royale in

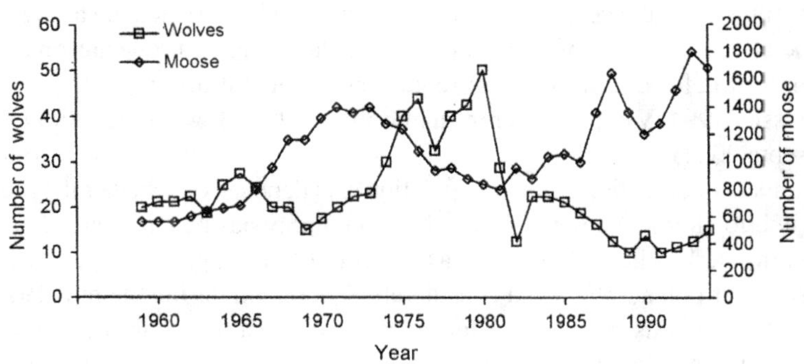

Fig. 2.14 Population dynamics of moose and wolves on Isle Royale. Modified from McLaren and Peterson (1994)

1983 were very likely born in 1981 and 1982 (Peterson et al. 1984). Anyway, moose made up >90 % in the wolves' diet (Pastor et al. 1988). Since then moose population continuously increased with the exception of a temporary decline at the end of the 1980s. Since the beginning of the 1980s wolf numbers have fluctuated at a comparatively low level without a clear response to the growing moose population (cf. Fig. 2.14; McLaren and Peterson 1994).

In the beginning of the present century, moose population decreased from 1,100 in 2001 to 900 in 2003. The decline may possibly be ascribed to warmer temperatures since 1998. Moose do not forage enough in summer if it is too hot. As they have a thick coat and cannot sweat, they stay in the shade or in the water to escape the heat. In addition, molestation by winter ticks increased. Winter ticks obviously profited from the exceptionally warm autumn in 2001. They weakened the physical condition of moose, causing a major mortality across much of North America and also on Isle Royale in the spring of 2002 (Michigan Tech News/Media, 17. May, 2004; communication R. O. Peterson, 28 April, 2013). Weakened moose have also become more vulnerable to wolf attacks, and wolves regained a dominant position as a limiting factor for moose. (Vucetich and Peterson 2013). While wolf numbers reached very low levels in the past 2 years, there are strong indicators of moose population increase (communication R. O. Peterson).

There is some evidence that the social behavior of wolves, which has an effect on the predator-prey relationship, is also influenced by varying climatic conditions. Thus, in severe snow-rich winter, wolf packs chasing moose usually consist of a dozen individuals on average, whereas in milder winter the wolf packs are smaller, usually not exceeding four to five individuals. Formation of larger wolf packs in severe winters has to do with the fact that the sexually mature young wolves are still together with the adults. Large wolf packs again are more effective in killing moose as compared to small wolf packs or – in other words – in milder winters more moose escape fewer wolves. Therefore browse damage is more common in mild winters.

The discussion on the mechanisms regulating moose numbers continues, and the more case studies and models are referred to, the more complex the phenomenon appears. Different regional conditions (e.g., availability of alternative prey, one or several predator species, predator-avoidance strategies of prey animals, plant productivity, nutrient balance, and climate, with snowpack in particular) as well as the post-glacial development of vegetation (e.g., adaptation to browsing pressure by production of defensive substances) require an observant differentiating approach (e.g., Messier 1994; Van Ballenberghe and Ballard 1994; Bowyer et al. 1997; Ripple and Beschta 2004).

The impact of predation on prey population also depends on whether the predators may switch to alternative prey in case their main prey has become rare. Studies on the Nelchina caribou herd (Nelchina Basin, Central Alaska) provide a good example. In this area, caribou numbers had considerably decreased during the mid-1970s to a tenth of the previous population due to three consecutive extreme winters and intensive hunting. Although wolf numbers peaked at this time, the caribou population was able to double by 1981 (Van Ballenberghe 1985). This was explained by the fact that caribou numbers had not fallen below the critical minimum necessary for self maintenance of the herd. This minimum is reached at a ratio of 100 caribou per one wolf. Moreover, as the density of the herd was low, wolf packs encountered caribou in their territories only occasionally and therefore preyed on moose, which were comparatively numerous at that time (Bergerud 1983; Gasaway et al. 1983). If no alternative prey had been available to the wolves, caribou population would probably have fallen below the critical ratio, in particular due to high losses in caribou calves.

Studies on the impact of predators on lemming populations also provide evidence that availability of alternative prey for weasels and other predators has a strong effect during the regression of lemming numbers, besides snowpack conditions and their effect on reproduction (MacLean et al. 1974). Mink predation also depends on availability of alternative prey. American mink (*Mustela vison*), for example, which has been introduced to the upper river Thames, kills fewer rails and grouse if rabbits are available (Ferreras and MacDonald 1999: see also Gerell 1967; Dunstone 1993).

Likewise the interrelationships between lynx and snowshoe hare are by far more complex as is reflected in the growth curves of population dynamics of both species (cf. Fig. 2.12). Just as wolf, also lynx is a non-specialized predator and switches to other prey when snowshoe hares are not available. In case that lynx numbers, controlled by snowshoe hare dynamics, have reached a peak and hares are declining due to shortage of food, lynx will increasingly prey on grouse (Tetraonidae), for example. So, grouse are directly controlled by lynx, whereas lynx numbers follow snowshoe hare cycles, that again depend mainly on the surrounding conditions (climate, forage supply, essential habitat structures, etc.) rather than on predator numbers (Fig. 2.15).

A particular situation has been reported by Bergerud (1984) from Newfoundland, where the original predator–prey ratio changed completely after introduction of arctic snow hares (*Lepus arcticus*). So to speak, the food pyramid became turned upside down. Five native herbivorous mammals (Arctic snow hare, caribou, beaver, eastern meadow mouse = *Microtus pennsylvanicus*, muskrat = *Ondatra zibethicus*) oppose nine carnivores (wolf, red fox, lynx, otter = *Lutra canadensis*, black bear = *Ursus americanus*, stoat = *Mustela ermine*a, American marten = *Martes americanus*, and two bat species). Wolf, which originally was the carnivore preying on caribou, became extinct in 1911. Lynx were very rare at that time and preyed mainly on native arctic snow hare. In 1864 snowshoe hares had been introduced to improve the food resource of the Newfoundlanders. Snowshoe hares reproduced successfully and soon became the main prey of lynx. Consequently, lynx numbers considerably increased. When the snowshoe hare population rapidly declined about 20 years after introduction – which is typical after arrival of an animal species in a new environment – lynx switched back to arctic snow hare and caribou calves and decimated them considerably in the following years. Under the impact of large numbers of lynx and intense hunting, the caribou population which had amounted to approximately 40,000 animals around the turn of the century (nineteenth/twentieth century), decreased to less than 1,000 individuals. After the snowshoe hare population had declined it started fluctuating in a 10 years cycle. Lynx have adjusted to this cycle and switch between snowshoe hare and caribou accordingly (Fig. 2.16).

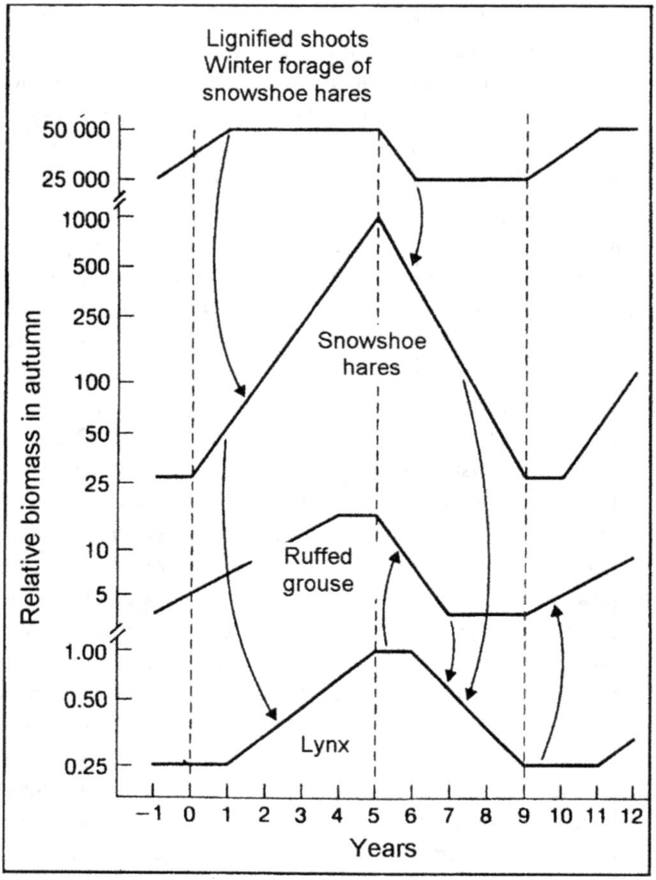

Fig. 2.15 Fluctuations of the major components (relative biomass) of a wildlife cycle in Alberta (Canada) with lynx switching between alternative prey. Modified from Keith (1983)

Arctic snow hare can survive only because it has become so rare. Lynx preys on arctic snow hare when snowshoe hare population is low.

Moose in the Darwin conservation area (former Soviet Union) were in a similar situation as arctic snow hares were in Newfoundland. While wolf numbers ran almost parallel to moose density between 1953 and 1965, wolves sharply increased although moose remained at a very low level during the 1970s. Wolves probably profited from other prey and prevented the recovery of the moose population (Fig. 2.17). In many cases several carnivore species may be competing for prey animals. Consequently, predatory pressure may increase to an extent preventing the prey animals from reaching their natural habitat carrying capacity, determined by forage supply and other vital habitat conditions. Thus, in general the question whether a predator regulates its prey cannot be answered with 'yes' or 'no'. Each case requires careful and reliable differentiating of specific considerations. Sometimes the effects of carnivores are practically negligible while they may be conspicuous in other cases.

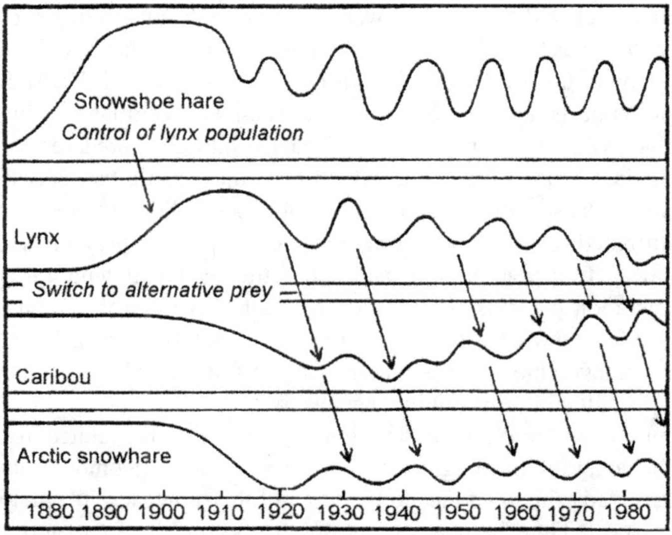

Fig. 2.16 Population dynamics of snowshoe hare (*Lepus americanus*), lynx (*Lynx lynx*), caribou (*Rangifer tarandus*) and Arctic hare (*Lepus arcticus*). The approximately 10 year-cycle of snowshoe hares controls lynx population. When the snowshoe hare density is low lynx preys on caribou and the rare Arctic snow hare. Modified from Bergerud (1989)

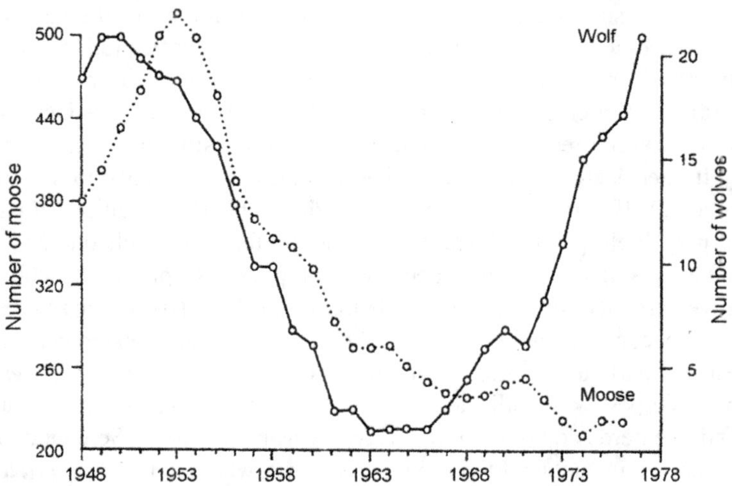

Fig. 2.17 Moose and wolf populations in the Darwin wildlife reserve of the former Soviet Union. Modified after Filonow and Kaleckaja (1985) in Bibikow (1988)

Among other considerations, landscape structures play an important role. Habitat generalists, such as foxes (*Vulpes vulpes*), racoon (*Procyon lutor*) or carrion crow (*Corvus corone*) for example, profit from rich food supply in agricultural landscapes, suburban and urban areas. They also search for food in tree groves and forests where they plunder more nests and kill more nestlings than predators would do that are tied to such habitats (Schneider 2001). In Sweden, for example, the decline of capercaillie (*Tetrao urogallus*) has been ascribed to predatory generalists that have increased due to fragmentation of the original woodlands (Andrén 1992; Hjorth 1994). In the high elevation forests of the northern Black Forest (Southwest Germany), which provide very favorable living conditions to capercaillie, the impact of predators is low compared to marginal areas. Even foxes, which have increased after successful rabies control, have not caused a noticeable decline of capercaillie until present. This may be partly attributed to insufficient availability of alternative prey in the mountain forests far away from agricultural land and settlements. Silvicultural support of spruce regeneration prevents colonization of clear cuts by grasses and herbs that would provide favorable conditions to small rodents (Schroth 1994). Thus, regarding the more or less favorable habitat structures largely shaped by humans, the influence of fox on capercaillie must be put into perspective.

The intensity of browsing/grazing impact may also be related to predator-avoidance behavior of the ungulates. Terrain structures (geomorphology, forest stands interspersed with open meadows, depth of the snowpack) play an important role in this respect. Thus, in the presence of wolves red deer, for example, seem to avoid locations where steep terrain would prevent escape, or where limited visibility prevents detection of predators early enough (Ripple and Beschta 2004). Thus, predation risk is lower in open areas. Damage caused to vegetation may be relatively high in the safer places. Visibility and escape potential may change during succession. After wildfires, for example, aspen stands usually recover by producing stump sprouts and root suckers. These provide high-quality forage in the beginning of secondary succession. Later, however, increasing stem density reduces visibility, and wolves may approach deer without being perceived in good time (Ripple and Larsen 2000; White et al. 2003; Fortin et al. 2005). Thus, kill rate increases and browsing pressure is reduced. Consequently, aspen density continues to increase, visibility further decreases and the habitat becomes less attractive to deer (Ripple and Beschta 2004; Fortin et al. 2005). Although secondary vegetation (grasses and herbs) provide high-quality forage, elk will not prefer burns if chaotic dead stems lying on the forest floor impede escape from wolf packs (Ripple and Beschta 2004).

Moreover, productivity and seasonal change often play an important role (Sinclair 1983) as has become apparent in the East-African grassland, for example, where the rinderpest had particularly affected the herds of wildebeest (*Connochaetes taurinus*). The rinderpest was eradicated in 1964. In Ngorongoro crater with rainfall all year round, the herds do not migrate seasonally as is the case in Serengeti or Mara National Park having a long dry season. In the crater, where lions (*Panthera leo*) and spotted hyenas (*Crocuta crocuta*) kill about a quarter of the wildebeest population per year (Elliott and Cowan 1978), extinction of the rinderpest had no influence on wildebeest numbers. Between 1958 and 1978 wildebeest population did not much vary, ranging between 10,000 and 15,000 animals. This indicates that predators rather than available forage regulated wildebeest numbers in this area.

In Krüger National Park (South Africa), predators appear to have a major effect on wildebeest population. After increased precipitation and resultant better growth of savanna grasses, wildebeests formed smaller herds and grazed where short grasses prevail. Consequently, losses to carnivores increased. Decreasing wildebeest population, however, led to the fallacy that grassland carrying capacity had been exceeded. Therefore, wildlife managers carried out additional culling to adjust wildebeests to the 'right numbers'. However, after culling was stopped, wildebeest decline continued, probably due to the predators' impact (Smuts 1978; Sinclair 1992).

The situation in Serengeti, on he other hand, is completely different from Krüger National Park. Wildebeests numbering several ten thousands and occasionally even several hundred thousands follow the seasonal rainfall regime and move from the south-eastern park area to the northwest and north with short stopovers, and do not return to the starting point before the onset of seasonal rains (Fig. 2.18). In absence

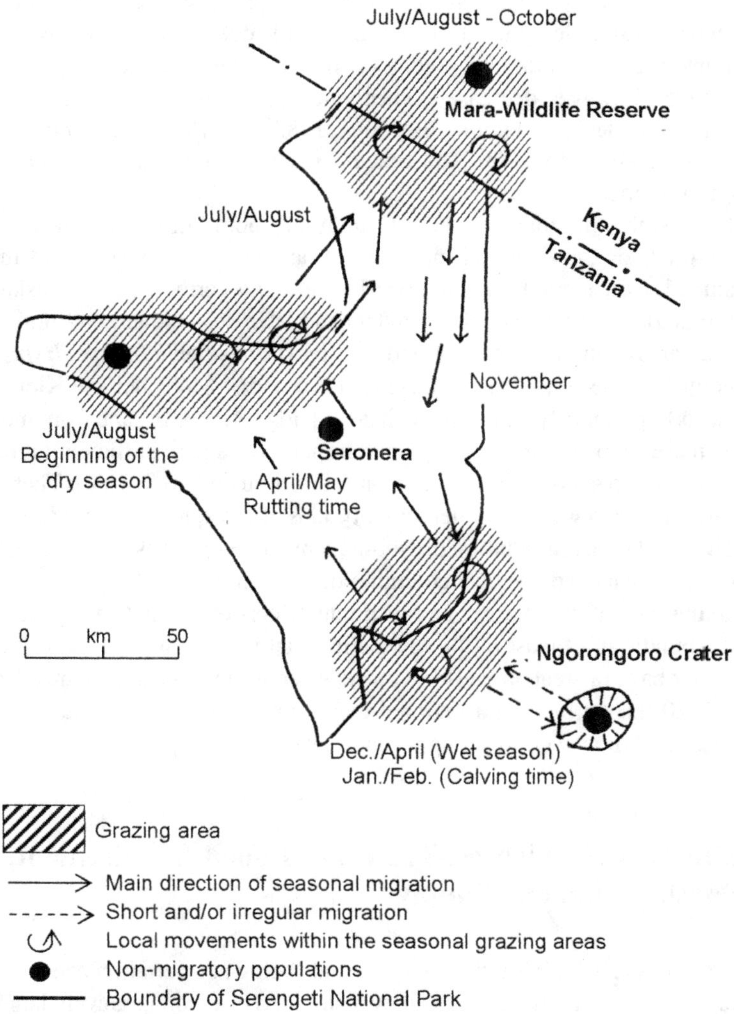

Fig. 2.18 Seasonal migration of wildebeest herds in the Serengeti-Mara area. The herds move regularly beyond the park boundaries. Modified from Leuthold (1977a; see also Sinclair 1995)

of most wildebeest in this area, lions and hyenas rely mainly on topi (*Damaliscus lanatus*), hartebeest (*Alcephalus buselaphus*) and wart hogs (*Phacocoreus aethiopicus*), while they kill only 1 % of wildebeests. Thus, after the rinderpest had become extinct, wildebeest numbers increased five-fold and also African buffalo sharply increased (Sinclair 1979a, b, 1983; Sinclair et al. 1985; Huston 1997). In this case, the seasonal migrations of wildebeest herds relativized the predators' impact (see also Maddock 1979).

Development of the mule deer population (*Odocoileus hemionus*) on Kaibab Plateau located north of Grand Canyon (Arizona) has been mentioned in many textbooks. Between 1906 and 1924 (Rasmussen 1941; Klein 1970) mule deer population erupted from about 4,000 to 100,000 animals. The eruption ran parallel to the decline of mountain lions (*Felis unicolor*) and coyotes (*Canis latrans*), which were killed whenever possible. Thus, the eruption of mule deer appeared to have been the result of insufficient control by predators. In the view of Lauckhart (1961) and Howard (1965), however, decrease of sheep grazing and modification of the habitat by fire were the decisive factors. Between 1889 and 1908 sheep numbers had declined from 200,000 to 5,000. Thus, the decrease of predators has probably been of minor importance.

At closer sight, fluctuations of wild ungulate populations that were readily ascribed to predator–prey relationships often turned out to be more complex than had been assumed. Considerable fluctuations in the Delta caribou herd (Alaska) were ascribed to predation (wolves, bears, eagles) and adverse weather conditions during winter. Yet, harassment by mosquitos and bot flies (*Cephenemia trompe*, *Hypodermia tarandi*) must also be considered (Mörschel 1996, 1999: Mörschel and Klein 1997). These insects continuously attack and irritate grazing caribou. Escape from molesting insects increases expenditure of energy and impairs storage of sufficient fat reserves that caribou need to survive the cold season (cf. moose on Isle Royale). Therefore, it seems plausible that a warm summer with myriads of mosquitos and bot flies may be involved in population crash during the subsequent winter. Wolves and other predators profit from the weakened physical condition of caribou.

Altogether, predation pressure and resultant effects on habitats and biocoenoses vary due a bundle of at least partly interacting factors. The loss of large predators may certainly have far-reaching effects on ecosystem function and balance (see also Estes et al. 2011). As local and regional conditions play an important role in this respect, it is hard, however, to set up a general rule.

2.3 Effects of Herbivores: The Role of Food Requirement, Consumption and Supply

While carnivores rely on individual prey animals being at a certain place at a certain time, forage plants of herbivores are usually distributed over comparatively large areas. Forage usually has a relatively low and much varying nutritional value (Jarman and Sinclair 1979; Senft et al. 1987) compared to the carnivores' prey. The influence of

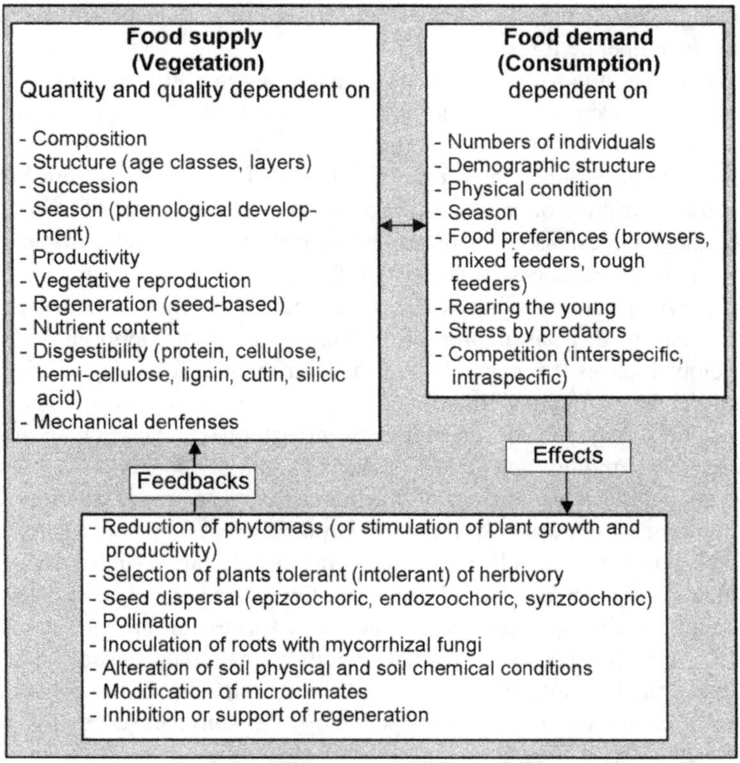

Fig. 2.19 Relationships between forage supply and forage demand of herbivores. Scheme by F.-K. Holtmeier

herbivores on vegetation depends on many factors that are interrelated via food demand and supply (Fig. 2.19). These interrelationships are very complex and require a detailed, well differentiating, approach. Herbivorous mammals and birds will be considered next.

The amount of food required by herbivores depends on the quality of available forage. Moreover, food requirement is influenced by numerous specific physical and physiological characteristics of the herbivore species as well as by animal population density, social and age structure, sex ratio, physical condition and season. Thus, the quantity of consumed phytomass differs. In woodlands, for example, all herbivores together consume less than 10 % of the phytomass only, which is very low compared to other ecosystems such as grasslands. On a worldwide average, herbivores may use between 30 and 50 % or even more of the aboveground phytomass and presumably about 25 % of belowground parts (Coleman et al. 1976; Ingham and Detling 1984; Sala 1988; Chew 1994). Belowground phytomass is used mainly by nematodes and other invertebrates. Even in high-arctic tundra, muskox (*Ovibos moschatus*) consume up to 50 % of the aboveground phytomass (Sect. 3.3.1), which comes nearer to phytomass consumption by mammalian herbivores in tropical grasslands and is much more than consumption of primary production by mammalian herbivores in temperate forests.

Loss of foliage due to herbivory may impair carbon gain and cause high root mortality, which again reduces water and nutrient uptake (Clancy et al. 1995) and possibly impairs competitiveness. On the other hand, reduction of the transpiring plant surface reduces moisture stress. Under certain conditions, severe browsing may stimulate sprouting of new shoots and foliage. Not least, excreta of insects, for example, and their carcasses increase mineralization and thereby availability of nutrients that again enhances productivity (cf. Fig. 2.7).

It is not only the amount of phytomass consumption that matters but also its form when it is consumed. Consumption of the same quantity of energy in the form of fully developed leaves, for example, will have different effects on a tree or a tree stand than feeding on young shoots or eating buds and seeds. Seed consumption or bud clipping may impede or even prevent natural regeneration. Bank voles (*Clethrionomys glareolus*) and yellow-necked field mice (*Apodemus flavicollis*), for example, not only feed on the seeds of beech, maple, fir and spruce but often destroy also plantules of these trees in spring time (Drodz 1966; Ashby 1967; Turcek 1967; Watts 1969; Bäumler and Hohenadl 1980). This impact may result in overaging of forests and radical changes of stand structures. On the other hand, seed dispersal by animals may sustainably influence composition, succession and structure of the plant cover, thereby modifying habitat conditions and biocoenoses (e.g., Laws 1970; Whicker and Detling 1988a, b; Vignolio and Fernández 2010). As to the consequences of phytomass consumption, phenology (season) and succession stage of vegetation after forest fires, grassland fires or clear cutting, for example, play an important role. Even minor effects may have far-reaching consequences to these ecosystems. Thus, herbivores – and also carnivores – mainly influence ecosystems as regulators rather than by transmission of energy (Chew 1994). These effects, however, are hard to be quantified but not less effective than consumption of phytomass by insects for example (see also Janzen 1983). Therefore, information only on the amount of phytomass consumed by insects during a mass-outbreak, for example, does not sufficiently explain ecosystem response to the insect impact.

As to the possible effects of grazing, the vegetation mosaic may be of major importance. It often results from previous grazing. The mosaic usually consists of plant communities characterized by different grazing tolerance. Studies on that issue were carried out for example by Turner and Stratton (1987) on Cumberland Island which is located close to the ocean coast of Georgia (USA).

The vegetation pattern on the extended SW-NE oriented island is closely related to the local topography and varying distance from the ocean shore. Forests are in the higher interior part, while salt meadows occur along the west side and dunes at the east side. Feral horses, cattle and feral pigs were roaming over the entire island until 1974. After removal of cattle and reduction of pigs, about 200 horses and numerous white-tailed deer (*Odocoileus virgianus*) remained on the island. Deer population considerably increased after hunting had stopped and predators had become eradicated. The horses mainly use salt meadows and grassland. They also enter the forest, however, although it does not cover their forage requirement. In contrast to forest vegetation, intensively used salt meadows and grassland rapidly recover from ungulate impact. White-tailed deer have affected plant cover in a similar way. They graze mainly at the upper fringe of the salt meadows, on open grassland and between the dunes. Salt meadows and grassland look like mowed lawn often interspersed with open patches. Deer, however, also browse on broad-leaved trees and shrubs in the island interior. In the oak stands *(Quercus virginiana)*, undergrowth is practically missing due to the deer effects and also because of fire. Grazing impact on the forest will increase if deer become more numerous and forage runs short in the areas outside the forest. In the end, the high grazing tolerance of salt meadows and grassland supporting large numbers of white-tailed deer is the driving factor of forest composition and structure change.

Regular interactions of herbivores and fire play an important role in many forest and grassland ecosystems. Tropical grasslands are particularly prone to fire during the dry season. The development of the open *Hagenia*-forests on the southern slope of Mt. Kenya for example, that are formed by very old and tall trees with diameters (breast height) of 4–9 m has been supported by wild ungulates, in particular by buffalos. These have reduced the forest undergrowth to a herb layer and thus removed most combustible material. As fire, however, is a precondition for natural regeneration of these forests, they have not been able to regenerate for a long time (Lange et al. 1997). On the other hand, forest and grassland fires may considerably influence natural succession because many herbivores may profit from better forage quality, whereas the foraging habitats will become less attractive to the herbivorous mammals in case fires are absent for a longer period of time (e.g., due to fire fighting).

2.3.1 Spatial and Temporal Influences of Foraging Behavior

Large herbivores and ruminants have been assigned to four major types of feeders (Hofmann 1973, 1978, 1989): concentrate feeders, mixed feeders (preferring browse), mixed feeders (preferring grass and bulk), and roughage feeders (grazers). The majority of forest species are small, and most of them are concentrate feeders. Species living in steppes are usually larger and belong to the grazers. In savannas, the proportion of large mammalian herbivores is even greater and consists mainly of bulk and roughage feeders (Cumming 1982).

Depending on the groups of herbivores their use of phytomass varies to a more or less great extent, as not all potential food is available and digestible. A herbivore species usually feeds on one or a few vegetation layers, while others are out of its reach. Roe deer for example use the field layer and shrub layer, whereas most of the tree layer is inaccessible for them. Elephants (*Loxodonta africanus*, Africa; *Elephas maximus*, India) are able to browse low savanna trees, whereas adult giraffes (*Giraffa camelopardalis*) easily reach tree crowns up to five meters above the ground (Photo 2.3). Consequently, there is no competition for food between both large herbivores.

Photo 2.3 Giraffe (*Giraffa camelopardalis*) browsing the lower part of a treetop at about five meters above ground in Tsavo National Park, Kenya. W, Leuthold

Some herbivores species of different taxa feed on the belowground parts of plants such as roots, rhizomes, bulbs, tubers, etc. Parasitically living nematodes, for example, many other invertebrates as well as fossorial rodents and many insect species belong to this group. Grizzly bears (*Ursus arctos horribilis*), other brown bears (*Ursus arctos*) and black bears (*Ursus americanus*) are omnivores using aboveground vegetation (grasses, herbs, berries, pine nuts) as well as bulbs, tubers and roots (Holcroft and Herrero 1984; Kühnlein and Turner 1991; Butler 1992; Mattson and Reinhart 1997). Moreover, they prey on animals ranging in size from moose to insects.

Food selection depends not only on the herbivores' preferences (types of feeders; Hofmann 1973; Coe 1981) but also on availability, palatability and digestibility of food, which again depends on a bundle of factors (Table 2.2). Storage of nutrients and development of secondary compounds as well as physical defenses such as thorns, spines and trichosomes influence forage quality. The stored reserves enable plants to survive extended cold or dry periods and to recover from damage by climatic factors and herbivory by thriving substitute new shoots and foliage. In general, the stored nutrients are easily digestible.

Herbivores usually prefer forage with a narrow C/N ratio. Thus, early successional stages after forest fires, clear cutting or wind throw, for example, are particularly attractive to the herbivores. Plant tissue improving stability (wood) or resistance to

Table 2.2 Availability of nutritive elements for ruminants. Modified from Van Soest (1982)

Components	Availability (%)
Soluble carbohydrates	100
Starch	>90
Organic acids	100
Protein	>90
Cellulose	43–73
Hemicellulose	36–79
Lignin	0
Cutin	0
Silicic acid	0

herbivory such as high amounts of fibre, lignin, cutin, phenols, terpenes or alkaloids (cf. Fig. 2.19) reduce the nutritional value of plants (Van Soest 1982; Coley et al. 1985). Grasses re-sprouting after heavy grazing are usually relatively rich in proteins. At moderate grazing, however, proteins decrease, whereas fibrous tissue increases. Consequently, a thick litter layer accumulates. Its relatively wide C/N ration and the cool and humid conditions, as well as light deficiency under the grass cover, impede mineralization and re-growth of grasses. As a result such areas become less attractive to the grazers (Bakker et al. 1983).

Studies on the impact of red deer (*Cervus elaphus*) on subalpine meadows in the Swiss National Park (Lower Engadine, Grison) provided evidence that the relative proportion of fibrous tissue in grasses and herbs increases more slowly compared to the protein content during the growing season than was observed in undisturbed areas (Holzgang et al. 1996). Studies on the effects of different management practices on the lumbricide fauna on urban lawns showed that the multi-cut English lawn provided much more protein-rich and low-fibre food to the earthworms than lawns mowed less frequently (Keplin 1995; Broll and Keplin 1995). It may be considered as a rule that on intensively grazed pastures the protein (nitrogen) content increases in food plants, whereas on slightly grazed pastures the amount of fibrous tissue (carbon) is comparatively high.

Elephants probably are the most important generalists among herbivores at all. They feed on more than 100 plant species (Bax and Sheldrick 1963) comprising grasses (50 % during rainy season, up to 95 % the other time of the year), leaves, twigs, bast tissue (phloem) of woody shrubs and trees, roots and rhizomes. Marmots (*Marmota* spp.), for example, also are generalists. They however prefer plants and parts of plants rich in carbohydrates and protein (young shoots and leaves, flowers, buds) as well as tubers and roots. Their diet changes seasonally however (Bibikov 1996). Thus, they use almost all plants within their reach. Many other herbivores are selective feeders using only plants with a particularly high nutritional value (e.g., roe deer, *Capreolus capreolus*; impalas, *Aepyceros melampus*; topi, *Damaliscus lunatus*). Common field voles (*Microtus arvalis*) for example rely on certain plants or parts of plants to cover their energy demand. During a controlled feeding experiment it became obvious that about 90 % of the voles that were fed with fresh grass every day died after 8 days due to emaciation. After clover, carrots or food rich in carbohydrates and fat were added, survival rate increased (Pelz and Lauenstein 1989). A few species are extremely specialized such as the Great Panda (*Ailuropoda melanoleuca*) which feeds exclusively on bamboo. The Koala (*Phascolarctos cinereus*) relies on certain *Eucalyptus* species, and mulberry silk moth (*Bombyx mori*) is specialized on mulberry trees. Grey squirrels (*Sciurus carolinensis*; North America, England) feed mainly on nuts and acorns and strip the bark of trees to get to the sugar-rich phloem (Bachmann 1971; Kenward 1989). They also plunder bird nests. Eggs and hatched birds provide them with protein.

Reindeer take up urine-soaked snow to cover their mineral demands (Blüthgen 1942), as is also known from mountain goats (*Oreamnos americanus*), for example. The Inuit make use of this behavior and apply urine into pit falls as bait for caribou. Occasionally, reindeer also feed on lemmings, dead fish, eggs and young birds that are rich in protein (Herre 1955; Kelsall 1968). In addition, they chew intensively on bones and shed antlers on the ground (Jacobi 1931; Herre 1955; Kelsall 1968). Thus, reindeer regain essential vital substances they urgently need. Almost no antlers can be found on the tundra that has not been chewed by reindeer – and/or by small rodents.

Pronghorns (*Antilocapra americana*) and bison (*Bison bison*) often graze on large prairie dog (*Cynomys ludovicianus*) colonies. The grass vegetation, which is kept low growing by these small rodents, is easily digestible and has a much higher nutritional value compared to the grassland outside the prairie dog colonies (Coppock and Detling 1986; Krueger 1986; Whicker and Detling 1988a, b). This is important all the more, as the local conditions may force ungulates to use also hard-digestible plants such as big sagebrush (*Seriphidium tridentatum*). Sagebrush occasionally is an important supplement food. As sagebrush, however, is very rich in terpenes and fibre, mule deer and other ruminants select plants that have comparatively low levels of such toxic compounds. The ruminants vary their consumption of sage in relation to other food plants. When these animals are forced to subsist primarily on sagebrush, high mortality may be the result (Delella-Benedict 1991).

Roe deer have a relatively small rumen. That is why they rely on easily digestible and energy-rich forage (selective browser). On meadows, for example, they nibble flowers and herbs while in forests they take fresh shoots of shrubs and low trees as well as acorns, chestnuts, berries of mountain ash (*Sorbus aucuparia*) and wild service tree (*Sorbus torminalis*), for example (Hofmann 1977, 1978). In case a habitat does not provide sufficient easily digestible and energy-rich browse, roe deer cannot survive (Eisfeld 1975). Feeding only on hay, for example, roe deer would starve to death although the rumen is full of cellulose-rich material. By contrast, cattle and sheep as well as other large herbivores such as horses and elephants (both are non-ruminants) are able to subsist on less digestible forage. As forage remains in the digestive tract for a comparatively long time (e.g., white-tailed deer, 45 h; cattle, 48 h) and due to much gastric juice rich in microorganisms, they are able to digest more than 50 % of the ingested plant tissue (Van Soest 1982; Robbins 1983). On the other hand, bulk or roughage feeders such as the European bison (*Bison bonasus*), for example, do not tolerate very protein-rich and easily digestible forage. Their digestion will collapse after a while due to over-acidification (Hofmann et al. 1998).

The forage preferences of cervids also are clearly related to the nutritional values of the food plants. Studies on forage and rumen content in black-tailed deer and caribou in Alaska showed that the cervids browse the most energy-rich plants during spring and summer until their nutritional values gradually decrease parallel to increasing fibrous tissue. Afterwards, they move to areas where the preferred food plants are still at the early stage of seasonal development (Klein 1962, 1968, 1970). In winter, herbivores have a hard time to find sufficient easily digestible and energy-rich forage. Not all parts of plants have the same quality in this respect. In branches and twigs of birch, for example, secondary compounds enrich towards the end of the grazing season. Some of them act as defenses. Moreover, the proportion of fibrous tissue increases parallel to diameter growth. Thin branches and twigs are easily digestible, but they do not provide much energy and their content in defensive substances is relatively high. Thus, it is important if not even crucial for browsers to find the 'right' twigs (Palo et al. 1992).

The digestive tract of rabbits (*Oryctolagus cuniculus*) and hares (*Lepus* spp.) is comparatively small and contains little gastric juice. Consequently, only less than 10 % of their fiber-rich food (e.g., grass cellulose) can be used efficiently. Therefore their droppings, that are produced in their relatively long blind gut, contain still much plant tissue with protein content of 20–40 %. Moreover, the droppings contain about the fourfold amount of vitamin B1 (so-called vitamin droppings; Scheunert and Zimmermann 1952) compared to normal excreta. Rabbits and hares as well pick up the droppings directly from the anus – young animals from the mother's anus – and ingest them without chewing. In the stomach, the dropping become mixed with the chyme and digested a second time (Leicht 1979). Afterwards, the droppings contain less than 10 % protein (see also Thacker and Brandt 1955; McBee 1971). This uptake of droppings has nothing to do with pathological coprophagy but is physiologically necessary for digestion. Coprophagy prevents severe vitamin deficiency that would otherwise be lethal to the rabbits. Passing of fibrous food through the digestive tract twice has practically the same effect as rumination.

Grzimek and Grzimek (1960a, b) were the first who found that the huge herd of ungulates in the Serengeti neither graze all areas nor use all grass species at the same intensity and that they prefer grasses with higher protein content. Likewise, food selection differs between the ungulate species due to their different digestive systems (Sinclair 1974a, b, 1977; Jarman and Sinclair 1979). Impalas and topis both are selective feeders. Graceful impalas (small body size and a weight of about 50 kg) prefer short green, soft and leaf-rich grasses while rejecting any tall and hard grasses. During the dry season, when fibrous tissue in grasses increase, impalas increasingly browse fresh leaves, flowers and fruits of woody plants. Topis, double the size of impalas, feed preferably on the protein-rich parts of grasses, whereas they do not take fibre-rich stalks and leaf sheaths. Nevertheless, these antelopes are not able to completely use the abundant grass vegetation during the rainy season to the extent that African buffalos (weight 4,000–7,000 kg) can. Buffalos are not choosy and consume large quantities of green grass without being able, however, to completely digest the fibrous matter (Jarman and Sinclair 1979). Digestation efficiency decreases particularly during the dry season when the forage protein content decreases from 8 % to less than 3 % (Sinclair 1977; Botkin et al. 1981). Presumably these large ruminants, profiting most from the high grass quality during the rainy season, store protein reserves on which they rely during the dry season (Jarman and Sinclair 1979).

Forage of zebras (*Equus burchelli*), for example, wildebeests and Thomson gazelles (*Gazella thomsoni*) does not much differ as to the grass species consumed. These ungulates, however, prefer different parts of the food plants. Zebras for example feed on coarse-fibred stalks that they digest easier compared to wildebeest and gazelles, although zebras are non-ruminants. After the grassland has been grazed by zebra, wildebeest can reach the low green stalks and graze them almost to the ground. Then antelopes follow which use the shooting new grasses that are rich in nutrients and growing at high density (Bell 1970).

Innidation of large herbivores in savanna ecosystems by selective feeding and seasonal migrations appears to be less perfect than might be assumed in view of repeated descriptions in literature. Occasionally, a little overlapping of two herbivore species feeding on the same plant species may be enough to strictly limit the food resource for one of the two species. Such interspecific competition occurred for example in the Serengeti during the dry seasons 1967 and 1969. About 1,000 wildebeests – i.e. 7 % of the total population – moved to the riverine grasslands that is the preferred habitat of African buffalos (ca. 2,400 individuals). Their food source, which is limited anyway, became further reduced by the additional grazers. In view of the great absolute population size of wildebeest (about 1.5 million; Sinclair 1979a), buffalos were under serious competitive pressure (Sinclair 1974a, b, 1977).

The ungulates need at least 4–5 % crude protein during dry season to maintain body weight (Sinclair 1974a; Coe 1981). Less protein is reflected in increasing deficiency symptoms. This holds particularly true for selective feeders that rely on heavy concentrate-low roughage diet, such as Kirk's dik-dik (*Madoqua kirki*), sprinbuck (*Antidorcas marsupialis*) and steenbok (*Raphicerus campestris*). At decreasing nutritional value of grasses, gemsbuck (*Oryx gazella*) may only cover its protein demand when additionally feeding on leaves and fruit pods of shrubs and trees (Lovegrove 1993). Even for African elephants, that subsist on a very broad diet, nutrition during dry season may be difficult due to protein deficiency and because of the elephants' relatively inefficient digestive system (Benedict 1936; McCullagh 1969; Wing and Buss 1970). During dry season almost nothing else than leaves of trees and herbs are available (Laws 1970). Thus, during the severe drought 1970/1971 in Tsavo National Park (Kenya) a large number of elephants starved to death although their stomachs were filled with bulk plant material extremely poor in protein (Botkin et al. 1981).

Compared to forage of the African savanna elephant, the food of the smaller forest elephant living in the African tropical rain forests has a higher protein and fat content, whereas the crude fibre content is lower. The diet of the forest elephant consists almost exclusively of leaves, twigs, bark and fruits, while grasses are quantitatively unimportant (Merz 1982). In the rain forest, forage quality does not vary so much seasonally as compared to the savanna. Consequently, in the rain forest, relatively high elephant densities of almost four elephants km^{-2} are possible, without impacting the ecosystem (Laws et al. 1970). By contrast, in forest savanna (Miombo type), more than 0.5 elephant km^{-2}) would already exceed the habitat carrying capacity (Fowler and Smith 1973; Cumming et al. 1997).

Apparently, interspecific competition of large herbivores species in the savanna depends not only on different digestive systems but also on the different anatomy of their mouth and denture. Hippos (*Hippopotamus amphibus*), for example, usually graze in areas of short grass (<15 cm) (Lock 1972) not too far distant (a few kilometres) from permanent water. They grasp the short grass with their sharp-edged horny lips and may keep grass vegetation so low ('hippo-lawns'; Verweij et al. 2006) that other grazers such as the Kob (*Kobus kob*) will be unable to grasp them with their teeth. This may explain the strong increase of waterbuck (*Kobus ellipsiprymnus*) and African buffalo populations in Ruwenzori National Park (previously Queen Elizabeth National Park) after hippos were considerably reduced in 1957. Buffalo population increased sixfold by 1968 (Eltringham 1974). On the other hand, ungulates with slender mouths are able to select the nutrient-rich and easily digestible grasses. They may successfully compete with the square-mouth species and preclude them from the grazing area, as is the case in the Serengeti (Murray and Illius 1996).

2.3.1.1 Influence of the Rainy Seasons

In open savannas, population size of animals as well as quantity and quality of forage depend on the seasonally alternating rainy and dry season (Fig. 2.20). This has already been shown in context with the predator–prey relationships for the long seasonal migrations of wildebeest in the Serengeti-Mara ecosystem (cf. Fig. 2.18). Wildebeest migration extends considerably beyond the Serengeti National Park boundary (e.g., Maddock 1979; Pennycuick 1975). However, also elephants and other ungulates follow the seasonal rains that increase forage supply. During the dry season, elephants, for example, rely on permanent waters, where they come together

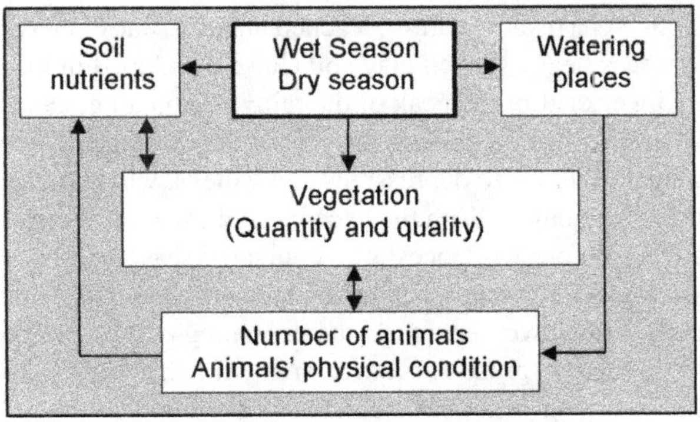

Fig. 2.20 The influence of rainy and dry seasons on the ecological conditions in East African savannas. Modified from Botkin et al. (1981)

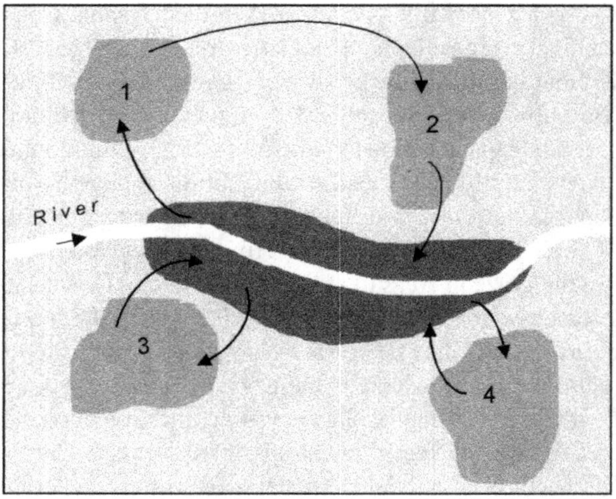

Fig. 2.21 Seasonal habitat use by elephants in Tsavo National Park (dry season habitats dotted). The elephants follow the local rains at the beginning of the wet season. Figures 1–4 show the temporal sequence of herd migrations. Modified from Leuthold (1977b)

in great numbers. At the beginning of the rainy season, the elephants move to areas where rain comes first (Fig. 2.21). They actually seem to 'smell' the rain. As rainfalls vary spatially and temporally, destination and timing of the migrations vary (Leuthold 1977).

In tropical savanna ecosystems, controlled by alternating wet and dry seasons, the animals are important agents that occasionally are even essential for ecosystem function and continued existence. The total biomass of large herbivores and the amounts of precipitation (Coe et al. 1976) are basically linked related to each other. This relationship, however is superimposed by effects of plant chemistry, plant-available nutrients, vegetation structure, food quality (e.g., ratio of structural and assimilating tissue) and availability (Fritz and Duncan 1994).

The distribution of vegetation and their spatial and temporal grazing pattern are closely related to topography and the resultant catena-like distribution patterns of soils (Morrison et al. 1948; Bell 1970; East 1984; Jensen and Belsky 1989; Ben-Shahar 1995; Nellemann et al. 2002). Topography-dependent distribution of soil moisture is the decisive factor as elsewhere in Africa. After rainfall and surface runoff, percolating water moves from convex topography to the adjacent depressions. As a result, shallow, leached and dry sandy soils are typical of the convexities, whereas heavy soaked clay soils have developed in the depressions (Fig. 2.22). Up to 3 weeks after the peak of the rainy season all grazers frequent the convex terrain, where fresh short grasses provide optimal forage. During the dry season, ungulates gradually move to depressions and valleys, where higher soil moisture supports tall grass vegetation. With the exception of the topi, which already move to the moister terrain before wildebeest, the course of seasonal movement is related to the body size of the ungulates. Thompson gazelles, being the smallest ungulate species, leave last for the lower terrain. At the beginning of the rainy season, the smaller ungulates are the first to graze the convex areas, and the larger species follow later (Bell 1970). The foraging pattern of elephants is also linked to terrain characteristics.

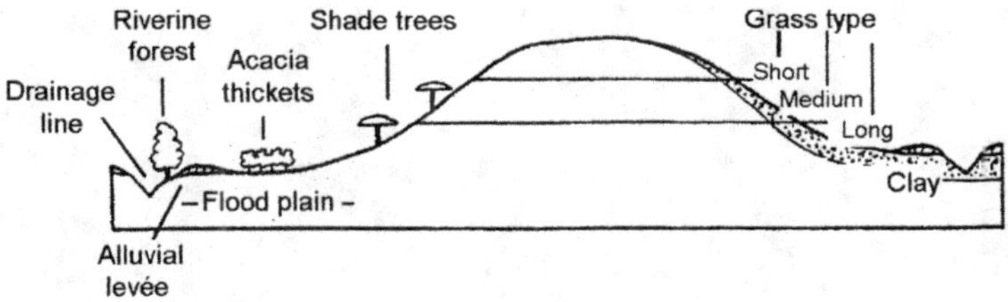

Fig. 2.22 Relationships between local topography, vegetation and soils in the East African savannas (Serengeti). Modified from Bell (1970)

Vesey-Fitzgerald (1960) has described a grazing cycle form in the Rukwa Valley (southwestern Tanzania), which is controlled by alternating flooding and dry conditions. Steep escarpments border this valley on its east and west sides. A shallow salt lake on the valley floor is surrounded by open savanna. The grassland again is enclosed by open tree stands (*Acia* sp. and other tree species). During the rainy season the large herbivores, with the exception of hippos, leave the flooded areas and retreat to higher less frequently flooded terrain. These local 'migrations' reduce the grazing pressure on the lower terrain for a while. Without any drop of rain perennial grasses, rapidly re-sprouting after trampling damage and/or fire during the dry season, provide enough food to grazers in the flooded areas. Some species such as reed buck (*Redunca redunca*), topi and Eland antelope (*Tragelaphus oryx*) usually stay near shade-giving trees, where fresh grasses sprout after fires. There are no water holes. Zebras, buffalos, Puku (*Kobus vardoni*), hippos and elephants, however, prefer open grassland close to watercourses, ponds and lakes (Photo 2.4). Due to this grazing behavior the areas between the tree stands and the open areas at closer proximity to the waters remain almost unused, though also new grasses may sprout there after the fires of the dry season. However, these areas do not increase the carrying capacity for ungulates of the entire grazing area.

In extremely dry years and during the hottest months, when no open water ponds exist anymore and there is no dew-fall during the nights, dense herds of ungulates are found on the alluvial grassland and freshwater swamps. Heavy grazing keeps grass short and stimulates growth. As the plants contain much water, the herbivores may subsist without drinking. In the freshwater swamps, with dominating hippo grass (*Vossia cuspidata*; height 1.5 m), the large and heavy herbivores prepare the way for the lighter ungulates by grazing and trampling down the tall grass (Vesey-Fitzgerald 1960). In a similar way the large herds of topi, together with zebras, Eland antelopes and buffalos, grazing in the drier plain areas during the rainy season, keep the grass short and in a permanently active vegetative state. However, no signs of over-grazing can be found, as it is prevented by seasonal migration of the grazers to other areas.

The semi-permanent freshwater swamps undoubtedly are the richest food resource. They are flooded, however, early at the beginning of the rainy season and stay under water into the dry season. During flooding, hippo grass grows vigorously

Photo 2.4 Elephants and impalas (*right hand-side*) at a water hole. Great Ruaha River, Ruaha National Park, Tanzania. N. Hölzel, September 2010

and reproduces continuously. At least for 8 months grazers are absent, with the exception of hippos. Alluvial grassland dominated by antelope grass (*Echinochloa pyramidalis*) is most widely spread. It often is flooded from the beginning of the rainy season into February. As soon as the first rains fall grass begins to grow vigorously. Flowering and seed formation are completed before the rainy season ends. The grassland remains almost un-grazed for 5 months at least. After seed shedding the stalks die off and nutrients are relocated into the rhizomes. Thus, it becomes relatively unattractive to ungulates. From June onwards savanna fires may occur. After the fires, the grasses immediately shoot up thanks to the stored nutrient reserves. No rain is necessary. During the last 4 months of the dry season the grassland provides optimal food to the ungulates.

In peripheral dry grassland sites, the situation may become exacerbated when ungulate herds in great numbers come in from the flooded areas. The additional grazing may cause severe damage to the plant cover. Without this immigration from flooded areas, the numbers of ungulates would be relatively low because of a limited forage supply during the dry season, and no over-grazing would occur during the rainy season. This is a further example which illustrates the great importance of how biotopes and subunits are spatially and functionally integrated into the overall landscape context.

The grazing areas within the open *Acacia*-stands are formed mainly by annual grasses. They are grazed in the beginning of the rainy season. At this time of year, seedlings grow vigorously and are not very susceptible to trampling and grazing. Before these grazing areas have reached the peak of their annual phenological development, ungulates move to drier areas. Thus, the grasses may produce seeds and parch without being damaged by grazing animals. The ungulates use the dried vegetation again immediately after the end of the rainy season, provided that water is still available. This second grazing, however, does not damage the annual plants because they have completed their life cycle before. Moreover, dry season fires and drying-out of the water ponds make the herds leave this area after a couple of weeks. The grassland surrounding the salt lake is used by hippos only. After forest fires, however, also other ungulates may occasionally graze on the sprouting fresh grass. They, however, prefer the swamp grassland (*Vossia* and *Echinochloa*). Damage by over-grazing never occurs on the lake shore banks influenced by brackish water.

We have to be aware that the studies on inundation were carried out when, as a result of the rinderpest and also after it became under control, forage was abundant over a period of about 70 years and interspecific competition was of minor importance. This changed, however, in the beginning of the 1960s, when the last cases of rinderpest were observed in the Serengeti-plain, in Ngorongoro crater (Tanzania) and in the Mara-Wildlife Reserva (Kenya) (Sinclair 1979a). The populations of wildebeest and African buffalo responded by rapid increase (Fig. 2.23). The numbers of non-ruminant ungulates (e.g., zebra), however, remained stable. This provides evidence that ruminant population growth had to be attributed to the eradication of the pest and not to any other factors, such as decrease of predators, improvement of the food source or decrease of human impact (Sinclair 1979b). At the end of the 1960s wildebeest and buffalo population had more than doubled. Shortage of forage during the dry season and increasing mortality prevented further growth of buffalo population for a while.

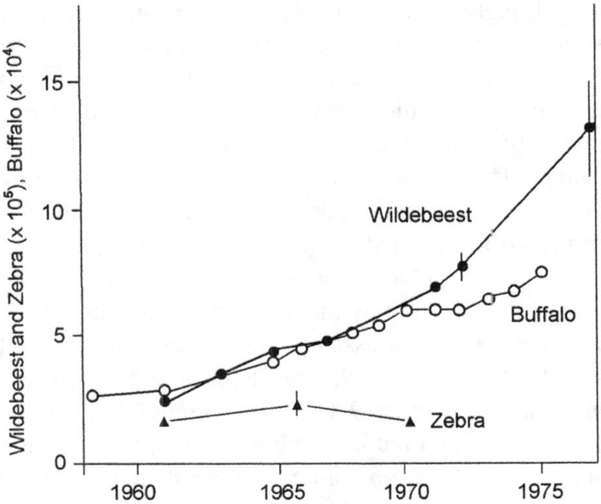

Fig. 2.23 Relationships of wildebeest, buffalo and zebra populations in the Serengeti. Modified from Sinclair et al. (1985); Dublin et al. (1990); Sinclair (1995)

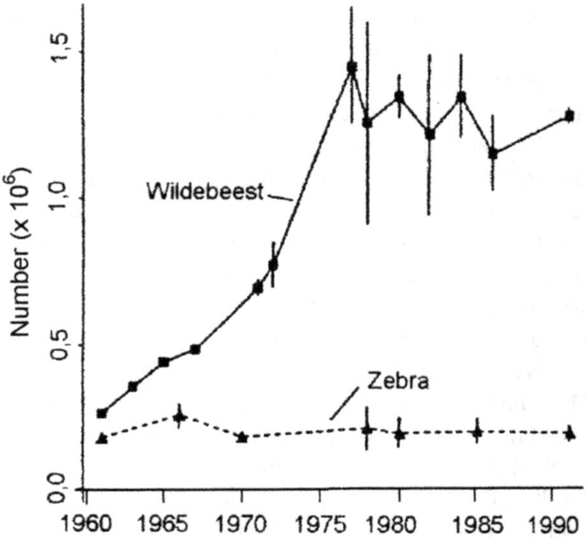

Fig. 2.24 Development of wildebeest and zebra populations in the northern Serengeti. Modified from Sinclair et al. (1985); Dublin et al. (1990); Sinclair (1995)

At the beginning of the following decade, the amount of precipitation considerably increased, and abundant forage was available even during the dry season. In addition, the grassland had expanded, because elephants had destroyed many tree stands (Leuthold 1978). In response to abundant food, wildebeest and buffalo numbers increased again (Sinclair 1979a, b). On the other hand, numbers of hartebeest (*Alcelaphus buselaphus*) and impalas did not increase, probably because wildebeests and buffalos were more successfully competing for food. In the end of the 1970s, wildebeest numbers declined due to shortage of food during the dry season (Fig. 2.24). Since then they stabilized at a level of about 1.3–1.5 million individuals (Dublin et al. 1990; Campbell and Borner 1995).

This had far-reaching consequences to the entire Serengeti ecosystem (Fig. 2.25). The strongly increased wildebeest population extended their grazing area into the shrub and tree savanna (see also Talbot and Talbot 1963; Watson 1967; Pennycuick 1975). The herds influenced vegetation by consumption of phytomass and by changing competition between plants under grazing pressure. First of all, herbivores that did not compete with wildebeests and buffalos for forage profited from this change. In the tree savanna, grazing reduced the amount of combustible material and thus the frequency of dry season fires. As a result, *Acacia*-young growth increased and probably attracted giraffes (*Giraffa camelopardalis*). In any case, female giraffes prefer the foliage of young trees to that of old *Acacia*-trees. The changes of the plant cover brought about by the dominating large herbivores were most conspicuous in the open savanna grassland. The population of topi, hartebeest, warthogs, Grant gazelles (*Gazella granti*) and Thomson gazelles (*Gazella thomsonii*) considerably increased during the 1970s. Oryx (Scimitar horned oryx, *Oryx dammah*) expanded farther west, and buffalos colonized the periphery of the tree savanna, which had been too dry for them before. The populations of grazers (impalas, topis and hartebeests), that were common to the tree savanna before precipitation increased, did

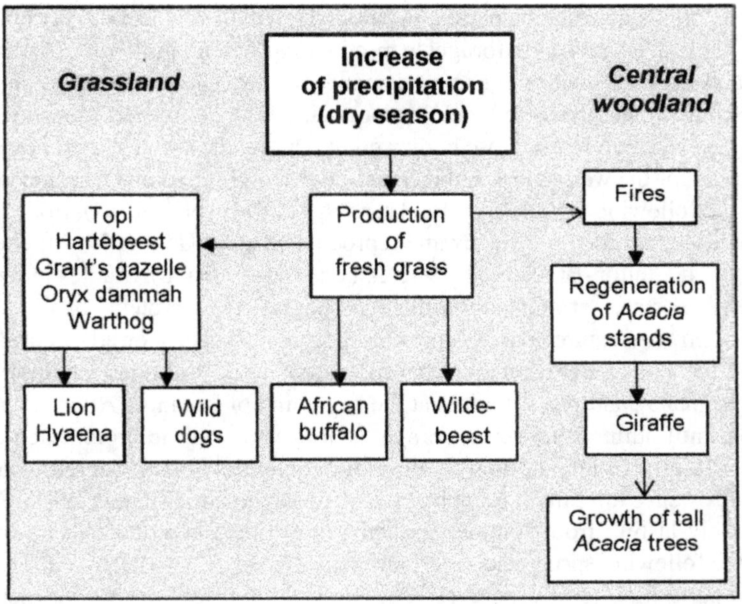

Fig. 2.25 Effects of increasing dry-season precipitation on fauna and vegetation in the Serengeti during the 1970s. *Filled arrows* mean increase, *open arrows* mean decline. Modified from Sinclair (1979a)

not grow because of strong competition with the herds of wildebeests that had increased compared to the 1960s.

Seasonal habitat change due to alternating rainy and dry seasons is also typical of marsh deer (*Blastocerus dichotomus*) in the Brazilian Pantanal wetlands (Tomas and Salis 2000), for example, which is the world's largest inland floodplain. Marsh deer, which appears to be an intermediate concentrate selector (see Hofmann 1973, 1989), prefer leaves of protein-rich easily digestible plants. During the flooding period marsh deer move to higher terrain even though being able to feed on aquatic macrophytes even at >70 cm water depth (Schaller and Vasconcelos 1978). During the dry season, they are found in the lower and wetter areas.

2.3.1.2 Influence of the Thermal Seasons

Many ruminants living in climatic zones with seasonal shortage of forage (winter or dry season) reduce their metabolism and 'adjust' their digestive systems accordingly (Hofmann 1978; National Audubon Society 1992). While in the savannas the amount of forage, the diet spectrum and forage quality vary in response to alternating rainy and dry seasons, outside the tropics the alternating thermal seasons are the controlling factor (e.g., Mraz 1960; Klötzli 1965; Remmert 1973; Ellenberg 1974; Schröder 1977; Hofmann 1978; Hespeler 1992). Likewise, energy demand and protein demand, in particular, are influenced by the thermal seasons. During the gestation period and lactation period, in particular, female reindeer or deer hinds, for example, need protein-rich forage to produce sufficient milk for rearing their calves and to cover their own energy demand. Insufficient summer browsing may considerably impair productivity (Julander et al. 1961). Female cervids stressed by parturition and calve-rearing lose body weight. During the warm season hinds must compensate for the weight loss and increase body weight to as much as they need to survive the following winter when food runs short (Torbit et al. 1985). Body weight also influences fertility, pregnancy and reproductive age (Klein 1970; Mitchell and Brown 1974; Clutton-Brock et al. 1982; Albon et al. 1983a, b). Studies in the Blue Mountains (northeastern Oregon/southern Washington) on protein content of food plants and protein requirement of American elk and mule deer showed – considerable variation of the data disregarded – that forage quality is still relatively high at the end of pregnancy and just sufficient at the beginning of lactation. After the peak of lactation until autumn, however, forage no longer meets the protein demand of deer (Fig. 2.26). In late autumn, the protein content of forage decreases to 7 % (≈the subsistence minimum; French et al. 1955; Mould and Robbins 1981). Protein undersupply lasting from summer through winter must be compensated by deer during the following spring and early summer.

From parturition to next heat, roe deer need the fourfold amount of metabolic energy provided by forage compared to energy required when resting (Hespeler 1992). In winter, basic metabolism energy is very low. The energy demand of roe deer, which were able to store sufficient fat reserves in autumn, ranges only a little above energy needed at rest (Ellenberg 1974). Stress and disturbances, however, raise energy expenditure drastically. Even slow motion increases resting metabolism in roe deer by 45 %, stress due to disturbance by 100 % and escape even by 130–210 % (Wiener 1975). Even in our intensively used cultural landscapes, roe deer, however, appear to have become relatively tolerant to disturbances by human landscape use (agriculture, outdoor recreation, traffic, etc.). Consequently, energy expenditure due to stress might be a little reduced. In richly structured landscape

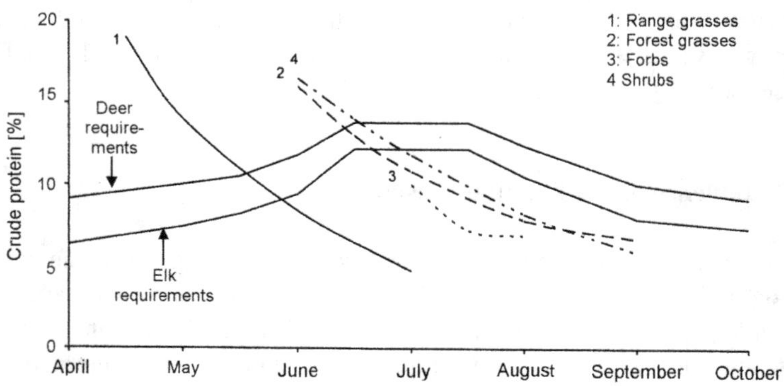

Fig. 2.26 Protein content of different forage plants and protein requirement of American elk (*Cervus canadensis*) and muledeer (*Odocoileus hemionus*) in the Blue Mountains (eastern Oregon). Modified from Irwin et al. (1994)

abundant forage is available almost all year round, in particular along forest edges, tree groves and hedges (Kurt 1991). Even a large cemetery near the city of Münster (Germany), for example, has become a favorite roe deer habitat. It is safe, as hunting is not allowed, and provides ample food, in particular fresh flowers on the graves. Salt licks have recently been installed in a nearby riparian area to lure roe deer away from the cemetery, and an above 2 m high fence is being built to prevent roe deer from coming back. By contrast, poorly structured economic forests usually do not provide sufficient quantities of energy-rich forage. As a result, damage by browsing caused to afforestation and natural regeneration increases.

Red deer, living in habitats free of disturbances, need about five times more energy for movement and maintenance of body temperature compared to energy demand while resting. In mountainous terrain, energy demand for movement may quadruple. Energy expenditure by deer escaping upslope and fighting their way through deep snow may be 60 times higher as at rest (Fig. 2.27). In tourist regions

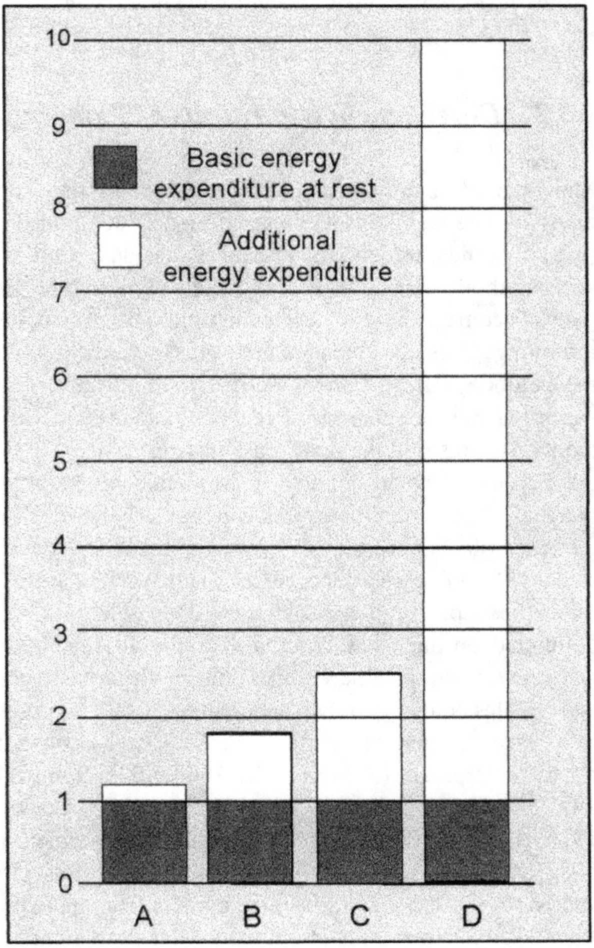

Fig. 2.27 Relative increase of energy expenditure by red deer (*Cervus elaphus*) moving on snow-free and snow-covered ground at different speeds. *A* – Standing, *B* – Walking (1 km on level snow-free ground), *C* – Browsing when walking on snow-covered ground, *D* – Fleeing on snow-covered ground. Modified from Esslinger (1988)

such as the European Alps disturbances by human outdoor activities frequently occur all year round, particularly during winter (downhill and cross-country skiing, etc.). Increased expenditure of energy is particularly critical during this time of the year as the physical condition of red deer may become severely impaired. The animals have to compensate for energy expenditure by forage uptake. Consequently, browsing pressure and damage to vegetation considerably increase (Onderscheka 1986; Reimoser et al. 1987; Esslinger 1988; Petrak 1988; Herbold et al. 1992; Reimoser 2003; Reinhart and Mattson 1990a, b; Holtmeier 2012).

In the Alps, the situation has become exacerbated because red deer habitats are progressively shrinking. Traditional migration routes of red deer from the summer habitats on the mountains to the winter-grazing areas on the valley floors and in the foreland of the Alps are largely blocked by roads, railway, canals, etc. As a result, grazing and browsing pressure has been aggravated in the summer habitats. These are undergoing progressive degradation. Support of red deer by winter feeding on one hand and rigid red deer population control on the other hand are inevitable to prevent severe negative effects on habitats and commercial forests (Onderscheka 1986).

2.3.2 *Development and Effects of Plant Defensive Substances*

Many plants are not grazed or browsed because they contain substances that impair digestation (e.g., lignins, tannins, silific-acid compounds) or have direct toxic effects (e.g., alkaloids, terpenoids, phenol glycosides). Others have mechanical defenses such as thorns, trichosomes or glandular hairs that keep herbivores away. These plants become positively selected in time. The same holds true for plants that tolerate trampling. Thus, the extended matt-grass swards (Leontodo helvetici-Nardetum) on the Feldberg (Black Forest, Germany; Photo 2.5), for example, are the result of sheep grazing for hundreds of years (Wilmanns and Müller 1977). Cattle grazing on alpine pastures pull the matt-grass (*Nardus stricta*) out accidentally and then spit it out because they do not like it. Mountain hikers are usually familiar with such bunches of dead matt-grass that can regularly be found on the pastures.

Other plants respond to browsing impact by development of defensive substances ('induced resistance; Howe and Westley 1993). Mountain birches (*Betula pubescens* ssp. *czerepanovii*), for example, or larches (*Larix decidua*) which became defoliated during mass outbreaks of the autumnal moth or the larch bud moth respectively, produce new foliage during the current growing season. These leaves and needles being not only smaller and less rich in protein than their first generation have a great content of chemical defenses, such as phenols and are not suited as food for the caterpillars (Baltensweiler 1962, 1975; Tenow 1972; Kallio and Lehtonen 1973; Benz 1974; Haukioja and Hakala 1975; Levin 1976; Rhoades and Cates 1976; Baltensweiler et al. 1977; Haukioja and Niemelä 1977, 1979; Haukioja 1980, 1982; Schultz and Baldwin 1982; Haukioja et al. 1983; Werner 1986a, b; Hoogesteger and Karlson 1992). It has become evident from an experiment that mountain birch even responds to application of caterpillar excrement by thriving leaves in the next

Photo 2.5 The vegetation on the Feldberg (Black Forest, Germany) has been grazed for centuries. F.-K. Holtmeier, 29 September 1987

growing season that are inedible for caterpillars of the autumnal moth (Haukioja et al. 1985).

Likewise, chemical defensive substances formed in response to overgrazing by lemmings, microtines and snow hare make the food inedible for these herbivores and contribute to the decline of herbivore density during the population cycles (Haukioja and Hakala 1975; Pease et al. 1979; Haukioja 1980; Laine and Henttonen 1983). In areas with pronounced 10-year snow hare cycles (*Lepus timidus* and *Lepus americanus*, Alaska, Siberia; Wolf 1980; Keith 1983), birches and willows produce more defenses than in Finland, for example, where such cycles do not occur (Keith 1983; Bryant et al. 1989). In feeding trials with free-ranging and captive mountain hares and Alaskan snowshoe hares (Bryant et al. 1989) mountain hares strongly preferred willows with low concentration of phenol glycosides to Alaskan willows (*Salix alaxensis*). Captive mountain hares refused Alaskan willow completely. Snowshoe hares, on the other hand, showed a high tolerance for phenol glycoside in Alaskan willows, and also consumed some Finnish willow species (*Salix caprea*, *Salix phylicifolia*) disdained by mountain hares because of the chemical defenses in these willows. This indicates that snowshoe hares have better adapted than mountain hares to such kind of forage and are able to detoxificate chemical defenses (Brattsen 1979; Bryant et al. 1989). At least, snowshoe hares and willows could coevolve over a longer period of time in the ice-free Pleistocene corridor of Alaska, whereas Finland was covered by the Weichselean continental ice sheet (see also Bowyer et al. 1997) until about 8,000 year BP. Nevertheless, in the feeding trials snowshoe hares also preferred more palatable forage if available.

Secondary plant metabolites also influence litter chemistry and may impair decomposition (Bryant and Chapin III 1986). Although numerous studies have addressed the interactions between herbivores and forage plants (e.g., stimulation of growth, production of defensive substances, mutualistic relationships; see also Owen and Wiegert 1987), explanations are often plausible rather than conclusive, or in other words, there is no lack of new hypotheses but a dearth of experimental evidence.

Selection of food plants often depends on availability of alternatives. Thus, cattle usually avoid mat-grass (*Nardus stricta*) tussocks, while single stalks of this species are eaten when mixed with other grasses and herbs (Nicholson et al. 1970). In many cases, selective grazing of protein-rich and easily digestible plants leads to dominance of less palatable species producing defensive substances. On the other hand, long-lasting grazing pressure must not necessarily support expansion of such species. Occasionally, particularly nutrient-rich and easily digestible plants may also profit and spread if tolerant (Augustine and McNaughton 1998).

2.3.3 Feeding Specialists and Generalists: Their Influence on Species Diversity

The effects of foraging specialists and generalists on composition and development of the plant cover differ. Thus, diversity may either increase or decrease (Harper 1969). In case an area is grazed by generalists, plant species intolerant of grazing are most affected. If these plants are little competitive, they may become eliminated and species richness will probably decrease. By contrast, reduction of the highly competitive species at moderate grazing intensity by generalists may possibly increase species richness, because species suppressed so far may better develop. If grazing is precluded, the highly competitive species will soon prevail afterwards (see also Tansley and Adamson 1925; Klötzli 1977).

Too intensive grazing also leads to decrease of diversity as the herbivores will not find their preferred food plants at sufficient quantities and switch to other species that may be unable to survive under severe grazing pressure. On Schiermonnigkog (western Frisian isles), both a sudden increase as well as decrease of wild rabbit grazing pressure on dune vegegation resulted in loss of species, while moderate grazing led to maximum species diversity (Fig. 2.28; Zeevalking and Fresco 1977). In the Serengeti, McNaughton (1979b, 1983) found the highest species diversity in places moderately grazed by large mammalian herbivores (see also Naveh and Whittaker 1980). It may however be left open whether these findings can be taken for a general rule.

As to the effects of selective grazing it depends on whether the herbivores prefer highly competitive plants or species with low competitive capacity. In the first case, diversity increases, in the latter it declines. Thus, sheep grazing of swards dominated by perennial ryegrass (*Lolium perenne*) and white clover (*Trifolium repens*) increased species diversity (Jones 1933), whereas sheep pastures in the highland of

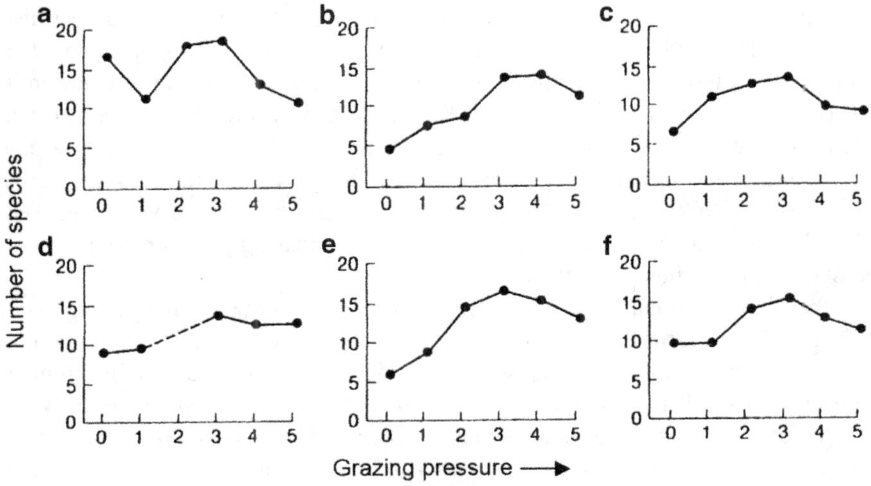

Fig. 2.28 Change of plant species under the influence of different grazing pressure (Scale 1–5) by rabbits on sample sites (1 m^2) on sand dunes (**a–e**) and on average (**f**). Modified from Zeevalking and Fresno (1977)

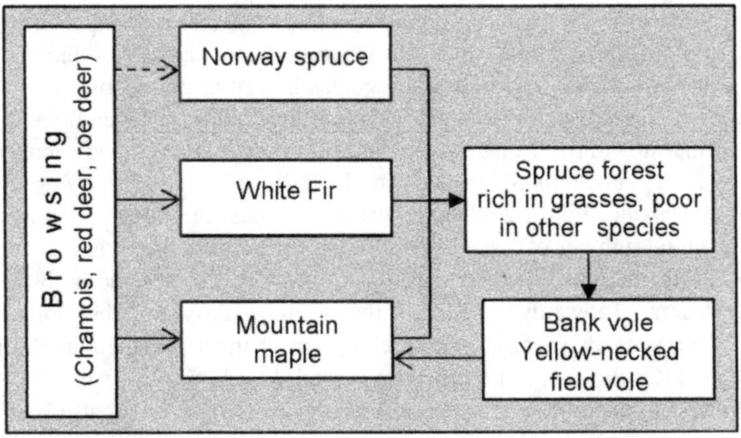

Fig. 2.29 The influence of ungulate browsing (Chamois, red deer and roe deer) in montane mixed forests in Upper Bavaria. *Filled arrow* heads mean increase, *open arrow* heads decrease or impairment. *Solid arrows* mean strong impact, *broken arrows* mean minor influence. Scheme by F.-K. Holtmeier, from different sources

Wales, where grasses prevailed that were largely rejected, such as purple moorgrass (*Molinia caerula*) and red fescue (*Festuca rubra*), benefitted from grazing (Milton 1940, 1947). Under the influence of grazing vegetation mosaics may develop which considerably differ from the surrounding plant cover.

In submontane/high montane mixed forests of Upper Bavaria (Fig. 2.29), having an enormous regeneration potential by nature, as is reflected in great numbers of seedlings and young growth (Schuster 1990), browsing by chamois, red deer and roe deer has impaired natural regeneration. As a result, fir, mountain maple and other additional deciduous tree species decreased, while Norway spruce became dominant (Meister 1969; Bernhart 1988, 1990). Consequently, the mixed forests often develop into stands rich in grasses but poor in other species. Competition between grasses and regeneration of the forest forming tree species gradually increases (e.g., Meister 1969; Gossow 1976). Moreover, voles (bank vole, yellow-necked field vole) profiting from the expansion of grasses, increasingly prevent regrowth of deciduous trees (Schuster 1990.

In the Swiss National Park (Lower Engadine, Grison), grazing pressure on abandoned subalpine pastures by red deer has increased since the 1940s. As a result, not only vegetation structure and composition considerably changed but also the number of species on small plots markedly increased (Schütz et al. 2000). Some species could colonize these patches only because of the red deer impact. Grazing-tolerant plants, in particular, as well as low-growing rosette plants and plants rejected by red deer, benefitted from the grazing pressure. On heavily grazed long-term sample areas, species number increased by 250 % (by 100 % on average), while it slightly declined (by about 6 %) on the slightly grazed patches.

Marmots (*Marmota* spp.) also modify the plant cover at the micro scale by grazing and burrowing. The magnitude of these changes depends on moisture conditions and productivity of the plant communities. Growing marmot impact (*Marmota olympus*) on subalpine meadows in the Olympic Mountains (Washington, USA) caused a slight decline of grasses and species highly preferred by the marmots, whereas ruderal plants and plants that were left untouched considerably increased. Moderate marmot impact, however, reduces dominant species and increases the diversity of the plant community (Del Moral 1984).

In South Dakota (Wind-Cave-National-Park), highest species diversity as well as the highest total number of species were found in recent prairie dog colonies. The high species diversity resulted primarily from invading grasses and herbs. Species diversity decreased towards the edge of the colonies and was higher compared to undisturbed areas. Alterations by prairie dogs were most conspicuous during the first 8 years after the colony became established. Later, species diversity declined again due to the expansion of sagebrush (*Artemisia frigida*). About a half of the grass species vanished again in response to the increased prairie dog impact. Presumably, the growing competition between shrubs and grasses speeded up this process. A similar situation was found in other areas (Coppock and Detling 1986). However, it is also possible that the prairie dogs' preference of grasses caused the decrease, although prairie dogs usually consume also considerable amounts of shrubs and herbs (Koford 1958). Moreover, bison, which often graze on prairie dog colonies, may have contributed to the decline of grasses (Coppock and Detling 1986). Although general statements on the impact of herbivores on vegetation can only be made with special regard to the local conditions and based on long-term studies, also this example shows that herbivores are able to reduce dominant species and encourage less competitive species.

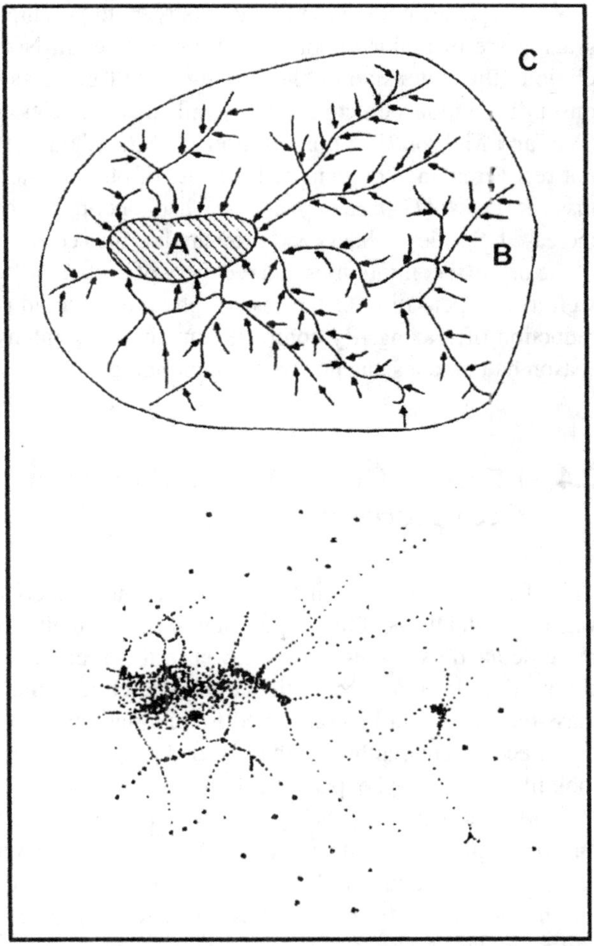

Fig. 2.30 Plan view of a field vole (*Microtus agrestis*) colony (*above*) on a rich meadow. A – Nest area, B – Foraging area, C – Unaffected area. The *arrows* show the direction of aboveground food transport from the foraging area to the subterranean nests. There and along the vole paths nutrients accumulate (*below*). Modified from Leucht (1983)

On permanent mesoxerophytic meadows (Mesobrometum) and rye-grass meadows (Arrhenatheretum) on the 'Schaffhauser Randen' (foot hills of Swiss Jura), microsites were found to be considerably influenced by food selection and excrement of voles (Fig. 2.30). Voles graze on both sides of the vole paths and carry food plants and nesting material into their burrow system. There and along the runways, remains of food, stored food and excrement accumulate. The already crushed plant matter and the mineral-rich excrement accelerate mineralization, even if the organic matter consists of substances hard to decompose. Thus, the mineral nutrients become readily available to plants. Almost homogeneous meadows became converted into a varying vegetation mosaic with considerably higher vegegation and biomass on the patches influenced by voles compared to the vole-free sites. Heterogeneity, species richness and diversity increased in particular on the mesoxerophytic meadows (Leucht 1983).

As to species population and diversity, both grazing intensity and the season of grazing are of major importance. From studies in North Australia it has become obvious that intensive cattle grazing of tall-grass savanna, for example, at the onset of the monsoon considerably influences species composition and production (Ash and McIvor 1998). Even after 2 years of intensive grazing, vegetation had not recovered and compensated the loss of phytomass. The previously dominating tussock grass (*Themedra triandra*) had declined while herbaceous plants had increased. Species richness and diversity had also considerably changed. Preservation of the original communities and recovery of particularly grazing-intolerant plants as well as compensation of the loss of phytomass could only be achieved by drastic reduction of grazing. By contrast, comparatively intensive grazing during the dry season had no consequence worth mentioning.

2.4 Effects of Cyclic Mass-Outbreaks and Other Mass Occurrences of Animals

The effects of defoliation by phyllophagous insects during mass-outbreaks on vegetation and flows of matter have already been briefly addressed (Sect. 2.1.1). A few more detailed examples will be given in this section. Mass-outbreaks occurring at certain time intervals are particularly common in some forest insect species. Insect mass-outbreaks can be considered to be a natural ecological factor inherent to the forest ecosystems, although the insect damage (e.g., defoliation, dying trees) may look like a catastrophe, particularly in a short-time perspective and when economic forests are affected. Mass-occurrences may, however, also be a result of natural hazards such as severe storms, (e.g., 'Vivian', 1990; 'Kyrill', 2007) or big forest fires (e.g., Yellowstone fire, 1988), producing immense amounts of dead wood, and thus, optimal conditions for bark beetles. Moreover, large-scale clear-cutting (e.g., during and after World War II) may promote mass-occurrences of voles and considerable increase of moose. Both voles and moose may seriously affect forest regeneration.

Contrary to the normal local damage caused by 'too high' ungulate populations, defoliation during the peak of mass-outbreaks and its resultant effects usually extend over comparatively wider areas (landscape or even regional scale). Insect 'calamities' may cause a change in the composition of tree species and have lasting effects on forest structures (age classes, vertical layers, stand density, etc.). The insect population dynamics often turn out to be cyclic. Pure forests are usually more affected than mixed forests. Man has often contributed to the development of pure stands, for example, by selective cutting and/or by promoting tree species of particularly high economic value. Thus, most of the extended pure larch forests in the Central Alps, for example, are the result of this practice. Natural succession in many North American forests has completely changed to systematic fire-fighting.

In contrast to the central alpine larch forests, subarctic mountain birch forests (*Betula pubescens* ssp. *czerepanovii*) have developed without similar human support. Both forests, however, are affected by mass-outbreaks of defoliation insects

(*Epirrita autumnata*, *Operophtera brumata*, *Zeiraphera diniana*). Affected trees respond to defoliation through formation of relatively small needles or leaves containing little protein and great amounts of defensive substances causing the decline of the 'noxious' defoliators.

In Central Europe, pure spruce forests were established in order to cover the growing demand for industrial wood. Spruce grows comparatively fast. The spruce forests, however, are particularly vulnerable to windthrow and subsequent bark beetle attack, which may occasionally reach a disastrous magnitude. However, what at first sight seems to be a 'catastrophe' may be a short phase of a long-term cyclicity (Gigon 1983a, b; Gigon and Bolzern 1988).

2.4.1 Autumnal Moth and Winter Moth in Subarctic Birch Forests

Mass-outbreaks of the autumnal moth (*Epirrita autumnata*, Geometridae) occur at intervals of 9–10 years. In the most continental areas of northern Fennoscandia, mass outbreaks are more irregular than in the Scandinavian mountains, where a strong periodicity is typical (e.g., Tenow 1972; Neuvonen et al. 1999; Klemola et al. 2002; Neuvonen et al. 2005). Many square kilometres of birch forests may become defoliated. In Scandinavia, more than a dozen mass-outbreaks have occurred since 1862. Birch forests in the entirety of Norway, in the southern Swedish Scandes, Swedish Lapland and northern Finland were affected (Tenow 1975). Besides climate, man and reindeer, the autumnal moth has probably influenced development and structures of the subarctic birch forests in Northern Europe most sustainably.

Going by all we know, the *Epirrita* mass-outbreak in Finnish Lapland during the mid-1960s was very likely the most disastrous. About 5,000 km^2 of birch forest became completely defoliated (Nikula 1993). In Utsjoki (the northernmost community of Finland), an area of more than 1,300 km^2 was affected. About 50 % of the birches died (Photo 2.6). Those which were less affected gradually recovered by producing stump sprouts. However, birch seedlings were still scarce even 6–7 years after defoliation. Productivity of the birch forests decreased immediately after the moth attack. Consequently, forage for reindeer ran short. Growth rate of surviving birches is relatively low even after a couple of years following defoliation (Nuorteva 1963; Kirchhefer 1996). Thus, it is possible to identify and date mass-outbreaks of the autumnal moth in tree rings.

In Sweden, one of the biggest *Epirrita* mass-outbreaks occurred during the period 1954–1956. Birch forests in the Abisko Valley (Torneträsk area, Sweden) were heavily affected. The birches became completely defoliated and died during the following years. After having consumed the birch leaves, the caterpillars moved to the field layer and fed on the leaves of dwarf birch (*Betula nana*), crowberry (*Empetrum hermaphroditum*), bilberry (*Vaccinium myrtillus*) and cowberry (*Vaccinium vitis idaea*). The dwarf shrubs became temporarily replaced by grasses, in particular by wavy hair-grass (*Deschampsia flexuosa*). Some mushrooms typical of birch forests,

Photo 2.6 Remains of a mountain birch forest (*Betula pubescens* ssp. *czerepanovii*) which was destroyed during a mass outbreak of the autumnal moth (*Epirrita autumnata*) in the mid-1960s. Northwest slope of Jesnalvaara at Kevo, northern Finnish Lapland. F.-K. Holtmeier 6 August 1998

such as rough-stemmed Boletus (*Boletus scaber*), could not be found for many years. The same holds true for some insect species that rely on birch leaves for food. On the other hand, some beetles living on dead wood profited from the dead birches. Secondary 'noxious' insects, such as the large timberworm (*Hylecoetus dermestioides*), attacked the weakened stems and accelerated the decline of the forest stands (Palm 1959; Tenow 1996). Simultaneously, large quantities of nutrients were added to the ecosystem by the caterpillars` excreta and cadavers.

In the birch forests of northern Finland that were affected by the autumnal moth, the herb layer rapidly changed due to better light conditions on the forest floor and an increase of nutrients in the soil (Lehtonen and Yli-Rekola 1979). Differences, however, became apparent dependent on whether the birch forests were of the moist or dry type. In moist forests, the increase of soil nitrogen was probably the main cause of the development of a luxuriant herb layer richer in species compared to the previous forest. Expanding grasses (in particular, *Deschampsia flexuosa* and *Festuca ovina*) and herbaceous plants improved the food source for reindeer. On the other hand, in the dry birch forests, rich in lichens and crowberry, almost no change occurred. In the moist birch forest, however, a retrograde development became obvious after 6–8 years, and not much was left of the secondary lush field layer. Dwarf shrubs (*Vaccinium* spp. and crowberry) and reindeer lichens (*Cladonia* spp.) could expand at the costs of the nutrient-rich grass- and herb-vegetation. Thus, the character of the vegetation became different from that of the birch forest prior to the mass-outbreak (Lehtonen and Yli-Rekola 1979).

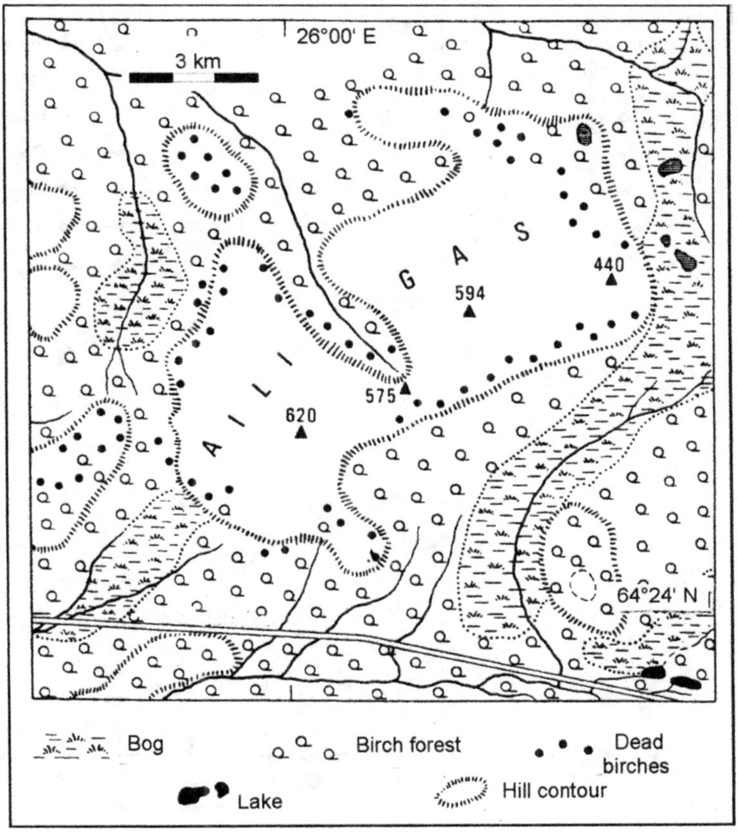

Fig. 2.31 Depression of the upper birch forest limit (*Betula pubescens* spp. *czerepanovii*) following the mass-outbreak of the autumnal moth (*Epirrita autumnata*) on Ailigas (east of Karigasniemi, northern Finnish Lapland) in 1927. Modified from Nuorteva (1963)

In view of the extent of the 'eco-catastrophe', some authors (Kallio and Lehtonen 1973, 1975; Lehtonen and Yli-Rekola 1979) believed that about 1,000 km^2 in northern Finland would turn into treeless 'tundra'. This holds true for areas lacking birch seedlings or where occasional seedlings are destroyed by reindeer or small rodents (Seppäla and Rastas 1980; Heikkinen and Kalliola 1989). That is why earlier *Epirrita* mass-outbreaks were followed by a decline of the upper birch treeline (Fig. 2.31; Kalliola 1941; Nuorteva 1963; Kallio and Lehtonen 1973; Holtmeier 1974, 2009; Holtmeier and Broll 2006), as is evident from numerous rotten birch stumps far above the present birch forest limit (Holtmeier et al. 2003; Holtmeier and Broll 2006). However, even heavily damaged birch forests may occasionally recover, mainly by producing stump sprouts (Vaarama and Valanne 1973; Holtmeier 1974). Birches may thrive as stump sprouts even after decades, and thus, produce new stems (see also Photo 2.25; Holtmeier 1974, 2005b, 2012; Heikkinen and Kalliola 1989). In spring, reindeer are especially drawn towards browsing the tips of

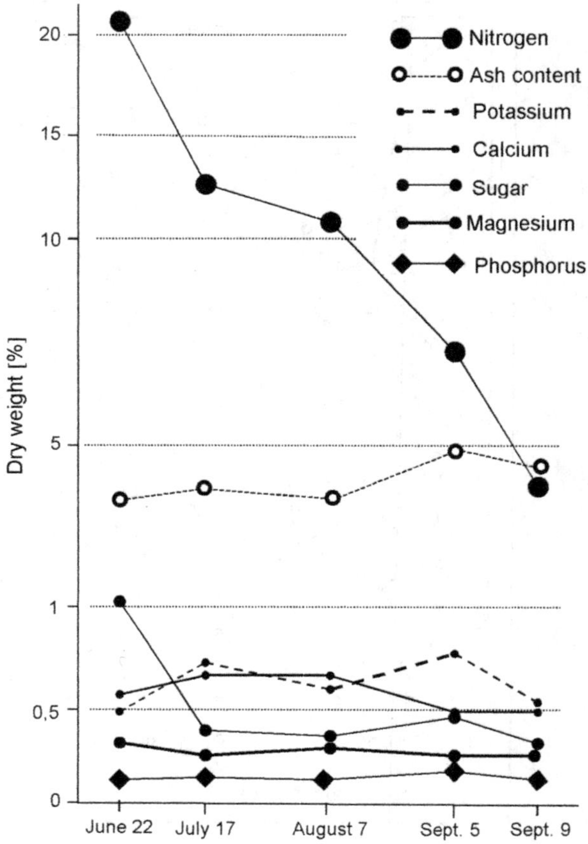

Fig. 2.32 Seasonal change of nutrient content in leaves of mountain birch (*Betula pubescens* spp. *czerepanovii*). Modified from Kallio (1975)

young shoots, as the new leaves and buds are rich in protein (Fig. 2.32). Browsing and even destruction of stump sprouts do not really threaten regeneration of the birches, as new sprouts are regularly produced. However, if these become repeatedly destroyed, vegetative reproduction of birch, and thus, forest recovery from defoliation by *Epirrita* may fail (Lehtonen 1987). This has happened in extended areas in Utsjoki (northernmost Finland), for example, where reindeer numbers are very high. While in the birch forests, particularly in open stands, regeneration from seedlings and also from basal shoots is abundant, seed-based regeneration decreases abruptly above the birch stands at higher elevation and birches in the treeline ecotone, where seedlings are comparatively scarce (Holtmeier et al. 2003). Root rot caused by saprophytic fungi gradually spreading from the decaying root stock into the new shoots may impair recovery of birch (Lehtonen 1987). Infected birch trees respond by excreting compounds that may prevent further expansion of root rot into the new stems (so-called compartmentalization), at least for a while, as has also been observed in many *Epirrita*-damaged areas in northern Finland (Holtmeier 2005b, 2009).

On open windswept terrain, the decline of birch forests due to mass-outbreaks of the autumnal moth may also be detrimental to juniper (*Juniperus communis*)

Photo 2.7 After a mass outbreak of the autumnal moth (*Epirrita autumnata*) at Oadašamjavrik (about 330 m) near Utsjoki (northern Finnish Lapland) the mountain birch forest (*Betula pubescens* ssp. *czerepanovii*) declined. Due to the ensuing decrease of the winter snowpack the common juniper (*Juniperus communis*) was no longer protected by snow and suffered from the consequent adverse climatic conditions. F.-K. Holtmeier, 2003

growing in the understorey. As comparatively little snow accumulates between the birch remains, juniper, previously well protected by the snowpack, becomes exposed and damaged by ice particle abrasion and winter desiccation (Photo 2.7; Kallio and Lehtonen 1973; Holtmeier 2005b).

After *Epirrita* mass-outbreaks, it usually takes many decades from first sprouting out of the stumps until the birch leaf mass will be equal to the leaf mass before the calamity. In Abisko Valley (Sweden), for example, birches attacked by the autumnal moth in 1955 had reached only 75 % of the previous leaf mass 40 years later. Additional 30–40 years are needed for complete compensation, provided no setback occurs (Bylund 1995).

As suggested by Tenow et al. (1995), the long-term development of birch forests is controlled by the combined effects of many interacting factors: climate, fluctuations of *Epirita*, its parasites and predators, and birch response to defoliation (Fig. 2.33). In general, *Epirrita*-mass-outbreaks only occur on old weakened birch stands relatively susceptible to disturbances (e.g., Ruohomäki et al. 1997). Old birches provide optimal conditions for oviposition on the flaky cracked bark and under the lichens covering stems and twigs. The moths obviously prefer such rough and richly-structured surfaces for egg deposition (Tammaru et al. 1995). In any case, overwintering eggs are more numerous on old trees rather than on neighboring young trees (Bylund 1997). Oviposition under lichens, for example, is advantageous to the further development of the insects, as the eggs stick better to the trees and are less exposed to egg-predators and cold winter temperatures.

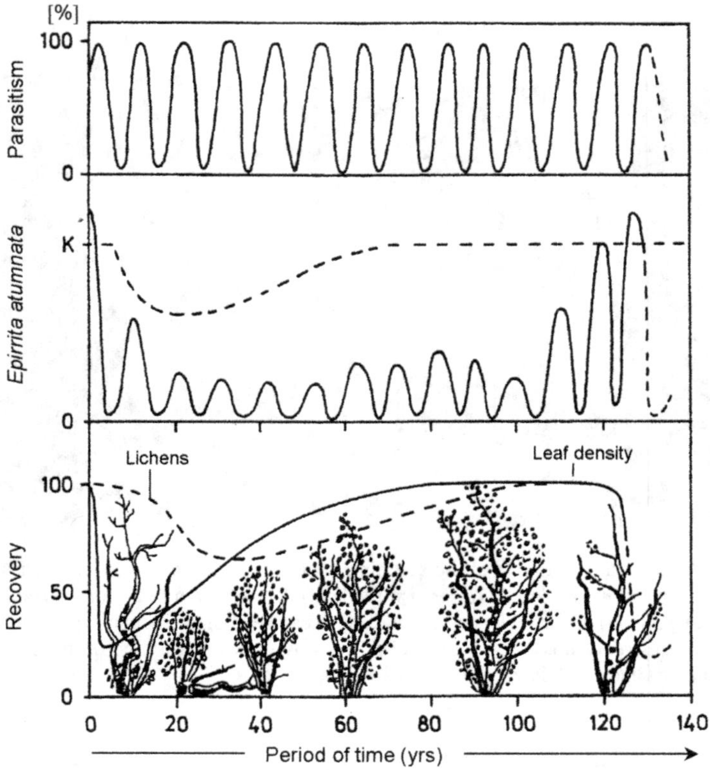

Fig. 2.33 Hypothetical development of larvae parasitism, population density of *Epirrita autumnata* and mountain birch forest after severe defoliation. K – habitat carrying capacity. Modified from Tenow et al. (1995)

After complete defoliation, *Epirrita* population declines due to food shortage and parasites. It will take a couple of years until the population increases again. The dead birch trunks may remain in a vertical position for a decade or even longer (Photo 2.8). They still provide favorable conditions for egg deposition, and the new caterpillars feed on the fresh stump sprouts. And yet, *Epirrita* fluctuations will be moderate during the following decades. This may be explained by the greater vitality of the taller-growing basal shoots and reduced food quality due to defensive substances enriched in the leaves. Moreover, bad weather conditions may occasionally affect the insects. Not least, conditions for oviposition aggravate as the dead birch stems gradually decay. It may take up to 70 years after a mass-outbreak for leaf mass to be as big as before complete defoliation. Food quality has gradually improved, and the increased lichen cover on the birch trunks and twigs proves favorable for egg deposition again. Consequently, *Epirrita* density increases and culminates in a new mass-outbreak. Thus, the next cycle of birch forest development begins.

Photo 2.8 Recovery of mountain birch forests destroyed by defoliation during a mass outbreak of the autumnal moth (*Epirrita autumnata*) in the mid-1960s, Jesnalvaara at about 300 m. F.-K. Holtmeier, 29 August 1996

Obviously, temperatures also strongly influence birch forest recovery. Normally, birches may thrive in leaves a second time after defoliation, provided that they can store sufficient energy reserves in roots and stems during the growing season. Cool and short summers may prevent a second leafing because no 'means for repair' are available. During the '*Epirrita* ecocatastrophe' (Kallio and Lehtonen 1975) in the mid-1960s, the growing season temperatures in northern Finland were extremely unfavorable. The thermal sums (days >5 °C) were the lowest of the twentieth century (Mikola 1971; Kärenlampi 1972; Holtmeier and Broll 2011). Secondary parasites (e.g., borers such as *Hylocoetes dermestioides* and *Agrilus viridis*; Palm 1959) were probably the decisive agents that ultimately caused the collapse of the defoliated birch trees.

The birch timberline has more or less continuously receded from its maximum postglacial position, mainly due to the general cooling until the twentieth century (e.g., Holtmeier and Broll 2006). However, when discussing timberline dynamics, the combined effects of climate, mass-outbreaks of the autumnal moth (and more recently, of the winter moth) and reindeer (browsing, grazing, trampling and resultant soil erosion) must also be considered (Holtmeier 1974; 1995b; Heikkinen and Kalliola 1989). More recently, the animal impacts may even override positive effects of slightly raised temperatures and counteract an advancement of the birch forest to greater elevations (Holtmeier et al. 2003, 2004: Holtmeier and Broll 2006).

The distribution of *Epirrita* damage is closely related to the local climatic conditions (Tenow 1972, 1975; Kallio and Lehtonen 1973, 1975; Holtmeier 1974, 2009). Birch stands at the upper slopes are particularly affected, often up to the timberline,

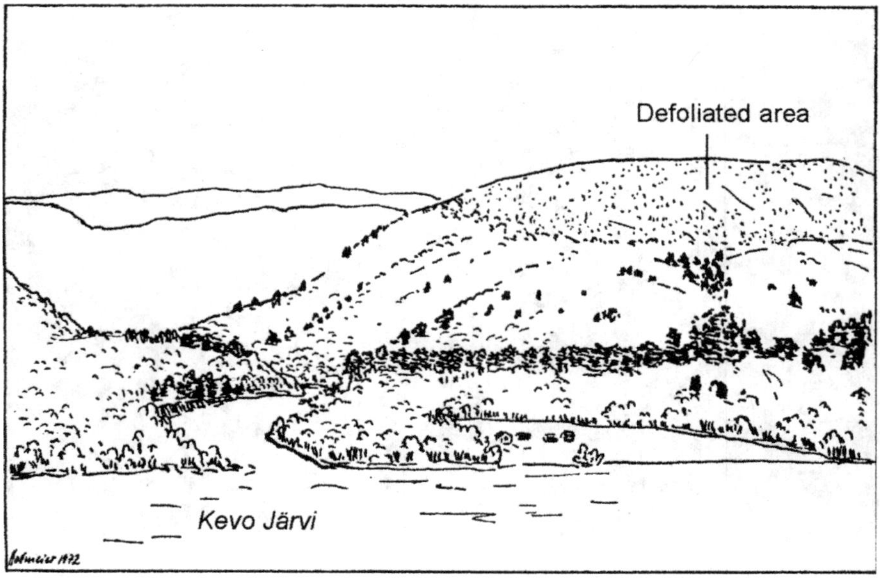

Fig. 2.34 Area of defoliation (*dotted*) by *Epirrita autumnata* on Jesnalvaara (330 m, Kevo area, northern Finnish Lapland). The lower limit of damage is located above the winter cold air layer. Stands of scots pine (*Pinus sylvestris*, dark structures) occur on the valley terraces. Scattered stands are found on the mountain slope. Drawing after a field sketch by the present author on 27 August 1969, from Holtmeier (1974)

whereas those on the lower slopes and on the valley bottom are usually not defoliated (Fig. 2.34 and 2.36). The upper limit of these 'safe' sites mostly coincides with the upper limit of the cold air layers that usually develop during clear nights in winter. On the mountains (~300 m a.s.l.) bordering the Utsjoki Valley (northernmost Finland) and on the upper valley sides of tributary valleys, for example, temperatures may be 15–20 °C higher than on the valley floor (~100 m a.s.l.) (Kallio and Kärenlampi 1971; Kallio and Lehtonen 1973; Neuvonen et al. 1999). At Kevo Research Station, located on a river terrace (~90 m a.s.l.), temperatures often drop below −30 °C. Presumably, such temperatures kill the eggs deposited by *Epirrita* under bark flakes or lichens (*Parmelia olivacea*) (Tenow 1972, 1996; Niemelä 1979; Tenow and Holmgren 1987). In the Abisko area, in 4 out of 12 winters between 1955 and 1962, lethal temperatures were recorded. Between 1984 and 1994, *Epirrita* eggs were killed by too low temperatures in 2 out of 11 winters (Bylund 1999). Embryos are even less frost-tolerant. In the Abisko-Valley, −36 °C is the critical minimum temperature during the diapause. In late winter and spring, however, during embryo development until hatching of the caterpillars, a minimum temperature of −29 °C is lethal already (Nilssen and Tenow 1990). Thus, the extremely low winter temperatures on Finnmarksvidda (northern Norway) that regularly kill the eggs probably prevent mass outbreaks of the autumnal moth in this area (Tenow and Nilssen 1990). According to studies at Kevo (Virtanen et al. 1998), it is unlikely that

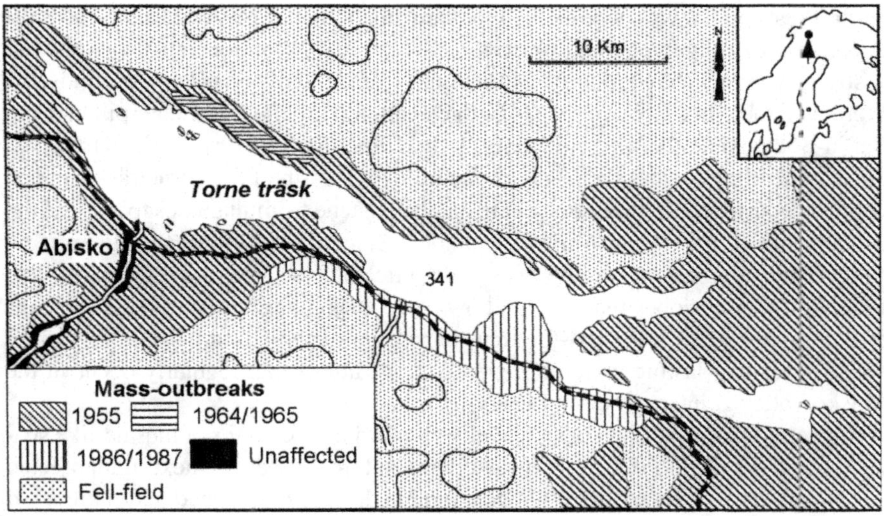

Fig. 2.35 Mountain birch forests affected by several mass outbreaks of *Epirrita autumnata* in the Torneträsk area (Sweden). At lower elevations, birches were not affected, as the very low winter temperatures are lethal to the eggs of the autumnal moth. The *contour lines* above the birch forests are equivalent to the 1,000 m contour line. Modified from Tenow and Bylund (1989)

eggs having survived extreme winter temperatures (often below −36 °C) will be killed by considerably higher frost temperatures in late winter and early spring.

In the northern Torenträsk area (see also Fig. 2.35), influences of the autumnal moth and the polyphagous winter moth (*Operophtera brumata*) occasionally overlap. The winter moth mainly occurs on the west side of the Scandinavian Mountains along the west- and north-coast of Norway, where it initially affects the birch forests at low and medium elevations. In 1964/65 – i.e., 10 years after the presumably biggest mass-outbreak of the autumnal moth – the winter moth irrupted and severely affected the comparatively rich meadow-birch forests (for terminology, see Hämet-Ahti 1963) on the northern side of Torneträsk, characterized by tall single-trunked birches. These meadow-birch forests are favored by the relatively mild maritime air masses from the Norwegian west coast. These regularly cross the mountain saddles and passes (altitude 420–720 m) and influence the north side of Torneträsk. The mass-outbreak of *Operophtera* started in the so-called 'warm slope zone' (Aulitzky 1968; Geiger et al. 1995) and spread from there up- and downhill. Caterpillars of *Epirrita* were also involved in defoliation of the birch trees (Tenow 1996). While the birches on the upper and lower slopes produced new leaves shortly after defoliation, most birches in the 'warm slope zone' died (Sonesson and Hoogesteger 1983). Innumerable dead stems covered the ground, and the original meadow-birch forest has been replaced by open savanna-like birch stands. Even in 1990, the birches produced only 2–3 % of the leaf mass before the mass-outbreak.

The restriction of *Operophtera* damage to the northern side of Torneträsk appears to have various causes. Thus, wind may have blown young caterpillars from infected birch forests on the west side of the mountain chain to its east side. This is the most effective way of winter moth dispersal (Edland 1971), as the female moths are vestigial-winged (having only rudimentary, atrophied wings) and cannot fly. Moreover, there is a forested pass (420 m elevation) that allows the caterpillars from Bardu-Söderdal on the west side to reach the Torneträsk area 'by foot', so to speak. The early leafing birches on the south-facing slopes north of the Torneträsk were an optimal food source for the caterpillars that had hatched simultaneous to the leafing. Birch stands in the Abisko Valley were not affected, probably because there is no comparable connection to the west-side birch forests. Moreover, south- and southwest-facing slopes that would allow early leafing are rare. Not least, the eggs of *Operophtera* are less frost-tolerant than those of the autumnal moth and will not survive long-lasting extreme freezing temperatures that regularly occur in the Abisko Valley at intervals of 3–4 years.

As not all birch stands became equally attacked by the moths, a mosaic-like spatial pattern has developed. More or less damaged tree stands are next to unaffected tree stands, and stands of different age alternate. In case of repeated defoliation, old birch stands that usually suffer most from the insect mass-outbreaks (Bylund 1997) may become completely destroyed or may recover through intense regeneration from stump sprouts (Tenow and Bylund 1989; Tenow et al. 1995). This may also be observed in many locations of the Abisko Valley. Although human use has left its marks on the birch forests, it is not difficult to attribute the usually multi-stemmed mountain birches to various age classes, which result from stump-sprout regeneration following defoliation by the autumnal moth (Emanuelsson 1987; Bylund 1995).

During the last two decades, *Operophtera* has expanded to the northeast into areas where *Epirrita* was previously the dominating outbreak species (Jepsen et al. 2008). Overlapping impact of both *Epirrita* and *Operophtera*, as has been observed in northernmost Finnish Lapland (Holtmeier 2009), northern Norway (Jepsen et al. 2008) and in the Torneträsk area, (Tenow et al. 2005) may become a serious threat to the local subarctic mountain birch forests. In some areas of northeastern Norway, at the eastern distribution limit of *Operophtera brumata*, birch forests became seriously affected during five to six consecutive outbreaks, which started with the autumnal moth and have been followed by the winter moth (Jepsen et al.2008).

In the Nuorgam-Pulmanki area (northern Finnish Lapland), for example, a mass-outbreak of *Operophtera* started in 2004 and continued in 2007 (personal observation, Holtmeier 2009). The eruption was preceded by an outbreak of *Epirrita* which already killed many birches. Birches that had survived one to two consecutive years of defoliation by *Epirrata* seem to be dying now due to attack by the winter moth. In any case, recovery by second leafing and from basal shoots was still extremely rare in 2011 (personal observation). On Koahppeloaivi, for example, a mountain west of Utsjoki village, the birch forest was declining on a horizontal distance of about 2 km (Holtmeier 2009). As *Operophtera* often attacks birch forests 1–2 years after an *Epirrita* mass-outbreak (Hogstad 1997; Tenow et al. 2007), the period of severe stress caused to the birches has been considerably prolonged. The invasion of *Operophtera* might be a result of natural expansion of this species. However, climate warming appears to be the main driving factor (Jepsen et al. 2008). Reduced egg mortality is of particular importance in this respect. However, competition between both moth species for food, and an increase in defensive substances in the host trees as well as predators and parasites might have buffering effects.

2.4.2 *The Larch bud Moth in Central-Alpine Swiss Stone Pine-Larch Forests*

The larch bud moth (*Zeiraphera diniana*) is an additional example of a defoliator that has lasting effects on forest composition, growth and ecology. Similar to the autumnal moth and winter moth, the larch bud moth is also widely distributed. Mass-outbreaks and severe defoliation, however, occur only in the pure larch forests (*Larix decidua*) of the Central Alps and in Siberia (*Larix sibirica*, *Larix dahurica*) at intervals of 8–10 years. In the Alps, outbreaks of the larch bud moth are more spatially restricted compared to the outbreaks of the autumnal moth in northern Europe.

The climate in the central-alpine valleys is moderate-continental, while the Siberian climate is extremely continental. Siberia is considered to be the 'homeland' of the larch bud moth. Defoliation by the larch bud moth, though at irregular intervals, has also occurred in larch plantings in the Pyrenees (*Larix decidua*), England (*Larix decidua*) and Japan (*Larix leptolepis*) (Baltensweiler et al. 1977). The larch forests (*Larix laricina*) in Alaska are also occasionally affected by the larch bud moth (*Zeiraphera* ssp., Werner 1986a, b).

Larch bud moths not only attack larches of every age, but also Swiss stone pine (*Pinus cembra*) and spruce (*Picea abies*). Scots pine (*Pinus sylvestris*) growing at the northern distribution limit of the larch bud moth is also affected (Novak et al. 1989). In Kamchatka, infestation of Siberian dwarf pine (*Pinus pumila*) was observed (Khomentovsky et al. 1997). In the Central Alps, the larch bud moth attacks mainly larch stands at elevations between 500–2,000 m. The uppermost larch stands (1,700–2,000 m) are most seriously affected (Auer 1975; Baltensweiler et al. 1977). In the northern and southern Alps, infestations by the larch bud moth even occurs at elevations below 200 m, at irregular intervals, however, and usually lasting no longer than 1 year. At lower elevation, however, no mass-outbreaks of the larch bud moth have been observed, even though the moth is omnipresent (Graf 1974; Auer 1975).

During the gradation peak, the caterpillars defoliate larch trees completely. Thus, the color of the larch forest canopy turns to reddish-brown in mid-summer and later into greyish-brown before the leaves fall off. Older trees seem to be more susceptible to defoliation than younger ones. Under favorable conditions, larch may produce a second needle generation in August, if unfavorable weather conditions do not prevent this. The situation may become critical when the larches become defoliated during two consecutive summers, or when defoliation overlaps with unfavorable climatic conditions (e.g., drought, early frosts). Infested larch forests often declined after mass-outbreaks of the larch bud moth (Ganzoni 1961; Auer 1961; communication

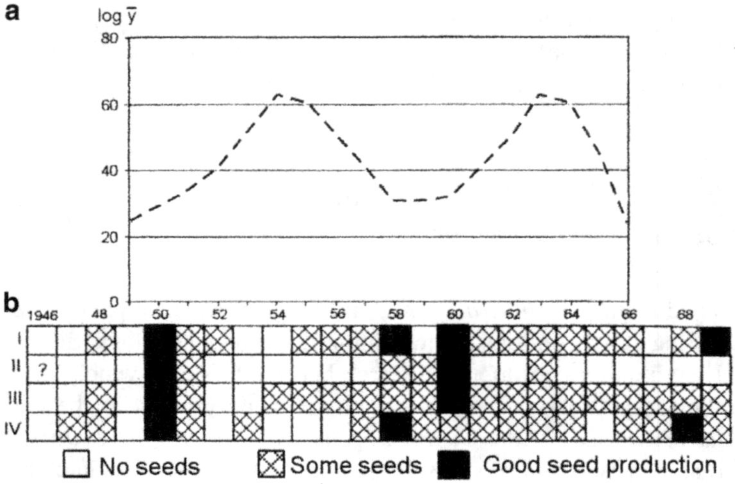

Fig. 2.36 Population cycles (**a**) of the larchbud moth (*Zeiraphera diniana*) in the Upper Engadine (Pontresina and Celerina communities) and seed production (**b**) in Swiss stone pine (*I*), European larch (*II*), Mountain pine (*III*) and Norway spruce (*IV*). Population cycle after Auer (1968). Data on seed production after the internal annual reports of the local forest office. Scheme modified from Holtmeier (1974)

local foresters O. Bisaz and E. Campell). As is the case in the subarctic birch forests destroyed by the autumnal moth and winter moth, the after-effects of defoliation, such as reduced resistance of the larch against climatic influences and secondary parasites, were probably the decisive factors.

Defoliation usually results in reduced radial growth (Auer 1961; Nägeli 1969; Schweingruber 1979; Weber 1995a, b). Abrupt reductions in tree-ring width (conspicuous late wood anomalies) caused by *Zeiraphera diniana* can be distinguished from those that resulted from climatic impact by comparing larch tree ring series with rings of tree species that were not affected by the larch bud moth (e.g., Norway spruce, mature Swiss stone pines). Thus, historical larch bud moth outbreaks may be reconstructed (Rolland et al. 2001).

Moreover, fructification and regeneration of larch are impeded or completely prevented. From comparison of seed years in larch and the other forest-forming tree species in the Upper Engadine (Grisons, Switzerland) with the gradations of larch bud moth (Fig. 2.36), it becomes apparent that, during the low of larch bud moth population and also in the beginning of population increase, larch produced seeds at moderate to high quantities at intervals of several years. However, before the larch bud moth had already culminated and also several years afterwards, seed production completely failed. As the other conifers (*Pinus cembra*, *Picea abies*, *Pinus montana*) produced seeds more regularly, it would appear that the lack of mast years in the larch were related to the mass-outbreak of the larch bud moth (Holtmeier 1974).

In the following, we focus on the effects of the larch bud moth in the Upper Engadine. Pure larch stands are usually found on southern exposures and relative gentle topography, mainly in the footzone of sunexposed slopes just above the valley bottom. This distribution pattern is the result of human forest use that has influenced the mountain forests for centuries (e.g., Campell 1944; Auer 1947; Holtmeier 1967a, b, 1990, 1994, 1995a; Jahn 1968). Under natural conditions, pure larch stands are usually restricted to boulder fans, screes and similar (unstable) sites, where mineral soil is exposed. At advanced successional stages, dense understory vegetation (grasses, dwarf shrubs) prevents anomochorous larch seeds from getting into a suitable seed bed. Thus, light-demanding larch becomes outcompeted by relatively shade-tolerant Swiss stone pine and Norway spruce (<2,000 m). Swiss stone pine will dominate at the climax of natural forest succession.

On terrain where climate and topography allowed pastoral use, Swiss stone pine and Norway spruce, however, were removed, because their dark shade-giving crown impede the growth of pasture plant communities. Thus, open larch forests developed flooded with light. Moreover, Swiss stone is more vulnerable than larch to trampling and browsing. Larch also profited from forest fires, because it is comparatively well protected by its thick cork-like bark. In addition, new needles may flush after a forest fire. By contrast, Swiss stone pine is usually destroyed by fire. Fires were relatively frequent in the past and caused mainly by man, accidentally or intentionally. They were used to remove 'weeds' such as dwarfed juniper (*Juniperus communis*) from pastures and to keep predators away from grazing cattle and sheep (Holtmeier 1974). As a result, larch could spread at the cost of Swiss stone pine through history, as also happened in many other central alpine areas (e.g., Auer 1947; Furrer 1955, 1957; Meyer 1955; Moser 1960; Wendelberger and Hartl 1969; Pitschmann et al. 1970, 1971; Holtmeier 1974, 1987a, b, c, d, 1995a; Mayer 1969). The mass-outbreaks of the larch bud moth have probably increased due to the man-made expansion of pure larch forests. In the Upper Engadine, which is an optimal habitat for the larch bud moth, and in Goms (Canton Valais, Switzerland), which is considered to be a suboptimal habitat, about 60 gradations could be evidenced by dendrological studies (Weber 1995a, b).

After the cease of forest pasture, natural succession towards Swiss stone forest rich in alpine rose (*Rhododenron ferrugneum*) has resumed. In many places, young Swiss stone pines already form a second tree layer in the understorey of the open light-flooded larch stands (cf. Photo 2.13). The larch bud moth, however, often interrupts succession. After having defoliated the larch trees, the caterpillars attack young Swiss stone pines in the understorey (e.g., Bezzola 1989). Due to needle loss and infestation of the young pines by secondary pests (noxious insects) such as bark beetles (e.g., *Pissodes pini*, *Pityogenes bistridentatus*) and others (e.g., *Cinaria cembrae*), the trees may die or develop distorted growth forms. In pure grass-rich (Woolly reedgrass, *Calamagrostis villosa*) larch stands on southern exposures, more than 90 % of Swiss stone pine regeneration was occasionally destroyed (Baltensweiler 1975; Baltensweiler and Rubli 1984). Swiss stone pine is probably threatened more by the larch bud moth than the larch. However, regional differences occur. Thus, in the Piedmont (Italy), for example, the larch bud moth has not played a major role in growth reduction of Swiss stone pine and in succession from larch stands to pine-rich or pine-dominated forests (Nola et al. 2006). In any case, after pure larch stands

Photo 2.9 Larch forest affected by the larchdud moth (*Zeiraphera dininana*) in the 'warm slope zone' on a south-facing slope in the Susauna Valley, Upper Engadine. The trees above and below the 'warm slope' zone were not damaged. F.-K. Holtmeier, October 1972

had developed due to removal of shade-tolerant Swiss stone pines, pastoral use and frequent forest fires, the larch bud moth outbreaks have contributed to the continued existence of pure larch forests.

Similar to infestations by *Epirrita* and *Operophtera*, damage by *Zeiraphera* is closely related to the local climatic conditions (Photo 2.9). Pure larch stands on the valley bottom and the lower slope zone often remain unaffected, while those located above in the 'warm slope zone' become defoliated. However, the reasons for this damage pattern are different. Thus, the eggs of the larch bud moth are extremely frost-resistant and tolerate temperatures of −40 °C (Baltensweiler et al. 1977) or even −50 °C (Bakke 1969). However, temperatures of −40 °C are exceptional in the central alpine valleys. Instead of low temperatures that would kill the eggs, delayed needle flushing of larch within the cold air layers on the valley bottoms and lower slope zones is more likely to be the cause of the lack of mass-outbreaks of the larch bud moth. In the case in which the caterpillars hatch before the needles flush, they will die due to shortage of food. Larch stands at the timberline are often not affected. At elevations >2,000 m, the moths are active into September. As the temperatures usually range below 8 °C at this time of the year, they prevent flying activity as well as oviposition, and thereby population growth (Baltensweiler et al. 1977). Defoliation of *Pinus pumila* in Central Kamchatka (Khomentovsky et al. 1997) during a mass-outbreak (1988–1993) of the larch bud moth can only be explained by survival of the eggs under an insulating snowpack, as in this region, winter temperatures may regularly drop below −50 °C.

While in a short-term perspective, the mass-outbreaks of the larch bud moth appear to be a 'catastrophe', they reveal themselves in the long-term as a factor stabilizing the larch forest ecosystems, as is also the case in the Upper Engadine. Due to the manifold interactions of the trees' different susceptibility and response to defoliation by the larch bud moth, local climate, human impact, forest fires, competition between tree species, and secondary succession, an ecosystem with a 'new quality' has developed, which is characterized by cyclic stability that never existed before. The same holds true for many other herbivorous insects. In the Canadian balsam fir forests (*Abies balsamea*), for example, more than 80 % of the fir trees were killed during a mass-outbreak of the spruce budworm, which initiated drastic changes of stand structures and productivity. In the long-term, however, these calamities support self-maintenance of the balsam fir forests, as young growth profits from elimination of the highly competitive old trees (MacLean 1988).

Although in the long-term larch bud moth mass-outbreaks do not put the existence of the larch forests at risk, they do, however, reduce the esthetic value of the larch forests. In an area where tourism is the most important economic resource, as in the Upper Engadine, for example, this may considerably reduce landscape attractiveness during the summer season. During mass-outbreaks of the larch bud moth, tourists will miss the fresh green foliage in spring and even more the bright autumn colors. People walking through the forests may also feel unpleasant because of caterpillars 'raining' from the trees (Holtmeier 1995a, b).

2.4.3 Noxious Forest Insects in North American Coniferous Forests

Originally, the so-called 'noxious insects' lived in the North American forests without destroying them. Man, however, has disturbed this 'natural balance'. In the forests on the eastern slope of the Cascades in Oregon and Washington, for example, 'noxious insects' and pathogens are the same as 100 years before, with the exception of a few species that were introduced during the twentieth century, such as the white pine blister rust (*Cronartium ribicola*), balsam woolly adelgid (*Adelges piceae*) and larch casebearer (*Coleophora laricella*). However, the spatial and temporal dimensions of the interactions between 'noxious insects', pathogens and their hosts have changed. Before the Europeans arrived, mass-outbreaks and resultant forest damage were spatially restricted and did not last as long as at present, due to the varying mosaic of tree stands of different age and successional stages (composition, age structures). As a result of systematic fire-fighting, however, extended and less structured pure forest replaced the multi-faceted woodland mosaic, and 'noxious insects' and pathogens could rapidly expand over large areas (Hessburg et al. 1994).

In view of the spectacular effects of such mass-outbreaks, the influences of the other forest insects may easily be overlooked, although they are represented by considerably more species than the outbreak species. In the conifer forests of the western United States, for example, every outbreak lepidopter species is confronted by nine species that have levelled out at comparatively low and far less fluctuating densities. Compared to the opportunistic outbreak species (r-strategists), they behave more like K-strategists. They are relatively evenly distributed on trees that also harbor outbreak species. During low population density of the latter, the non-outbreak insects are an alternative food source for predators and parasites of the 'noxious' species (Volker 1978; Munroe 1979). Thus, the non-outbreak insects support the species-rich population of the predators of the larch beetle and probably contribute to the stability of the forest ecosystem (Mason 1987).

Altogether, defoliating outbreak insects as well as bark beetles may severely affect conifer forests at the landscape scale and cause great economic losses in timber harvest. Thus, mass-outbreaks may be considered a 'catastrophe', particularly with regard to the usually long rotation time. In natural non-economic forests, however, mass-outbreaks are an integral factor in the long-term ecosystem dynamics. The ecosystem quality may temporarily change (successional stages), but natural forests will gradually recover from mass-outbreaks of defoliators or bark beetles, as long as the general environmental conditions (e.g. climate) do not change, and human impact, invasion of exotic species and other possible agents, not immanent to the system, do not disturb its long-term balance. In the latter case, prediction of the forest landscape development is difficult, in particular as availability, quantity and quality of data vary.

2.4.3.1 Western Spruce Budworm

The western spruce budworm (*Choristoneura* sp., Tortricidae) is one of 15 lepidopter outbreak species in North America (Furniss and Carolin 1977). The budworm causes severe damage to the coniferous forests in the mountains of western North America. Five of several budworm species (Powell 1980; Carolin 1987) attack fir, Douglas fir, spruce and also larch. One has specialized in pine. Basically, however, the spruce budworm may occur on all conifer species in the west and affect them (Herman 1987). Some *Choristoneura* species are sympatric and can be found in the same forest, but on different tree species. Pure or mixed Douglas fir and fir stands are most affected. The caterpillars feed on male and female pollen or on staminate cones and fresh needles, in particular. Defoliated and discolored tree tops as well as dead apical shoots are symptoms of budworm infestation. Tree mortality will considerably increase in the case that additional adverse factors such as root rot, drought stress and mistletoe affect the trees. Stressed trees are predisposed to bark beetle attacks, for example, by the Douglas fir bark beetle (*Dendroctonus pseudotsugae*) and the fir engraver beetle (*Scolytes ventralis*).

Budworm infestations reduce height and radial growth (narrow, missing or incomplete growth rings). Destruction of the reproductive organs, which happens even at low infestation intensity, will impede regeneration for a long time

(Fellin and Shearer 1968; Dewey 1970; Chrisman et al. 1983), in particular, if the young trees become damaged (Van Sickle 1987) and competition with other vegetation increases. Persistent budworm infestation causes changes in forest structure. Obviously, the spruce budworm caterpillars especially kill small and lesser competitive trees, whereas their impact on tall dominant trees is only marginal (Johnson and Denton 1975; Alfaro et al. 1982; Bousfield and Chase 1982; Hadley and Veblen 1993). In mixed stands, the proportion of fir (*Abies grandis*) decreases due to its higher mortality. In heavily defoliated forest stands, 40–70 % mortality is common.

While infestations by the spruce budworm occurred more sporadically compared to bark beetle attacks in the past, they appear to have become more frequent since the middle of the twentieth century. This may be attributed to human-induced changes in the western economic coniferous forests during this period of time (Carolin 1987; Swetnam and Lynch 1993). On the other hand, the knowledge of spruce budworm calamities has improved in recent times, as more attention has probably been paid to the infestations in general, and because damaged areas have been almost completely monitored by aerial survey since the 1940s (Stipe 1987). In New Mexico, nine regional-scale outbreaks occurred from 1690 to 1989. The most recent outbreak, which began in the late 1970s, was most severe (Swetnam and Lynch 1993).

2.4.3.2 Bark Beetles

Besides herbivorous outbreak insects, there are other non-phyllophagous species, such as bark beetles, for example, that damage trees by boring galleries trough the phloem (vascular tissue). Bark beetles may severely affect the development of trees and have lasting effects on stand dynamics. There are many bark beetle species from different families involved, such as long-horned beetles (Cerambycidae), powder-post beetles (Bostrychidae), jewell beetles (Busprestidae), snout and bark beetles (Curculionidae), and Ips engraver beetles (Scolytinae) (s. also Graf 1971; Novak et al. 1989).

2.4.3.2.1 Mountain Pine Beetle

The mountain pine beetle (*Dendroctonus ponderosae*) is as big as the tip of a match. It is probably the most aggressive of the bark beetle family in North America. At intervals of 20–40 years, a mass-outbreak occurs, destroying extended lodgepole pine forests (*Pinus contorta*). However, in an area extending from New Mexico and Southern California to the Yukon Territory and from the east slope of the Rocky Mountains to the San Rafael Mountains in California, a current epidemic of the mountain pine beetle has been spreading for more than one decade (Logan and Powell 2001; Taylor et al. 2006; Mitton and Ferrenberg 2012). This mass-outbreak, which is the largest in history, has left many millions of dead trees.

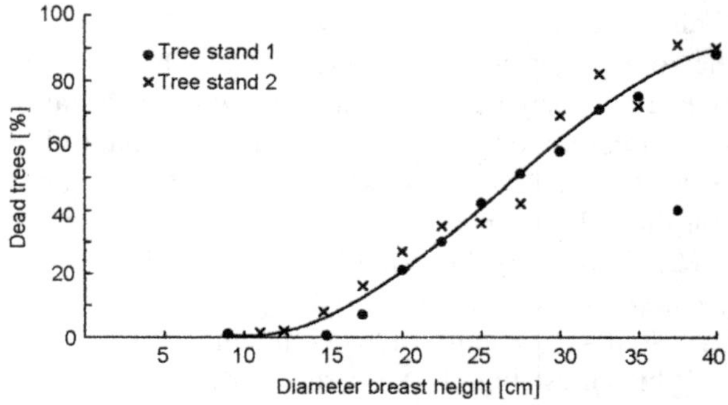

Fig. 2.37 Percentage of lodgepole pine (*Pinus contorta*) killed by mountain pine beetle (*Dendroctonus ponderosae*) per diameter class. From Amman (1977; curve corrected)

The time of infestation depends on when the phloem of some trees is thick enough to allow the beetles excavating galleries for oviposition. After a period of sporadic attacks, the bark beetle population will irrupt and destroy the pines within 1 week. The pines may survive minor attacks, as resin fills the always vertically oriented galleries and kills both eggs and caterpillars (Reid et al. 1967). Summer drought, however, may reduce resin flow, and thus, the ability of pines for defense. If resin does not kill the eggs, the caterpillars will hatch about 2 weeks after egg deposition and feed through the phloem. The horizontal oriented feeding galleries interrupt sapflow (similar to 'girdling'). The mature caterpillars excavate oval cells for pupation. The hatched beetles feed through the bark and, upon reaching the open, attack other trees. Presumably, it is not 'girdling' alone that kills the trees. In addition, fungi (*Ceratostomella* spp. and *Europhium clavigerum*, 'blue stain fungus') may get into the trees during bark beetle oviposition – probably as spores contained in the food. Expanding through the phloem and the water-conducting xylem, they impair sap flow in the pine stems and may also exhaust tree defenses (Berryman et al. 1989.

Bark beetle calamities influence the long-term development of the pine stands. The beetles kill more thick pines than thin ones (Cole and Amman 1969; Figs. 2.37 and 2.39), because trees with a larger diameter have a thicker phloem. The phloem increases parallel to radial growth. During the outbreak peak, the beetles attack the thick pines every year. In the end, only trees with a small diameter survive (Fig. 2.38). As their thin phloem prevents egg deposition, and thus, regeneration of the beetles, the beetle population declines. Studies in western Canada showed that the bark beetles do not attack lodgepole pine stands younger than 50 years. Infestation is more likely in 60–90-year-old trees and regularly occurs in those older than 100 years. However, old tree stands on soils poor in nutrients were not attacked, as they had insufficient diameters (Safranyik et al. 1975).

At higher latitudes and altitudes, thermal deficiency impedes the development of bark beetles. Many of them, still being in a highly susceptible stage of development, succumb to deep frost temperatures during the first or second winter. As a result, infestation intensity is reduced and, compared to lower altitudes and more southern latitudes, more lodgepole pines survive (Amann 1977). However, the situation is changing due to climatic warming (Chap. 7).

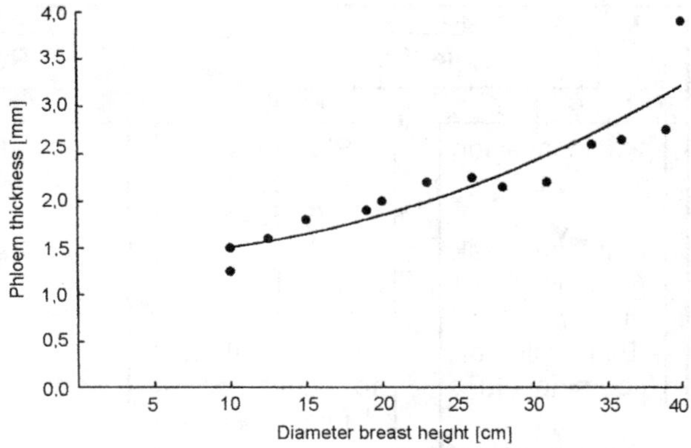

Fig. 2.38 Mean phloem thickness in lodepole pines (*Pinus contorta*) of different diameters. From Amman (1977; curve corrected)

The effects of bark beetle mass-outbreaks may vary considerably, as they are related to the pine stands' composition and stage of development (Fig. 2.39). In lodgepole pine forests, not affected by fire anymore after a pine beetle outbreak, the pines are soon replaced by more shade-tolerant tree species, such as Douglas fir (*Pseudotsuga menziesii*), for example, at lower elevations and by subalpine fir (*Abies lasiocarpa*) at high altitudes. In burned areas, *Pinus contorta*, which is a pioneer species, expands and grows very fast, while seedling growth in *Abies lasiocarpa* and *Picea engelmannii* that became established simultaneously with the pine seedlings lags behind. During mass-outbreaks, the pine beetles eliminate most thick and highly competitive pines, while slim pines and shade-tolerant tree species are not affected and may grow better. After several mass-outbreaks, no pines are left, and in the remaining almost pure fir-spruce stands, pine beetles cannot exist anymore.

In the case of recurrent fires, which are typical of most lodgepole pine forests, stand development is completely different. Due to the large amounts of standing and lying dead wood produced by the beetles, forest fires are very hot. The shade-tolerant tree species will not survive. Even the normally very fire-resistant Douglas fir becomes eliminated. Lodgepole pine, however, regenerates very successfully in burned areas, in particular as its cones only open and release the seeds under the influence of heat. The result is an even-aged lodgepole pine forest that is safe from bark beetle attacks as long as the stem diameters do not yet meet the bark beetles' requirements.

Forests that are only occasionally affected by fire develop a completely different composition and are characterized by highly varying age structures. Until the present, it has not really been understood how lodgepole pine is able to coexist in the long-term with the shade-tolerant tree species. Anyway, in the case in which the old pines are killed by the bark beetles, small gaps are left where pine seedlings may become established. This process may also occur in other places and at a different times.

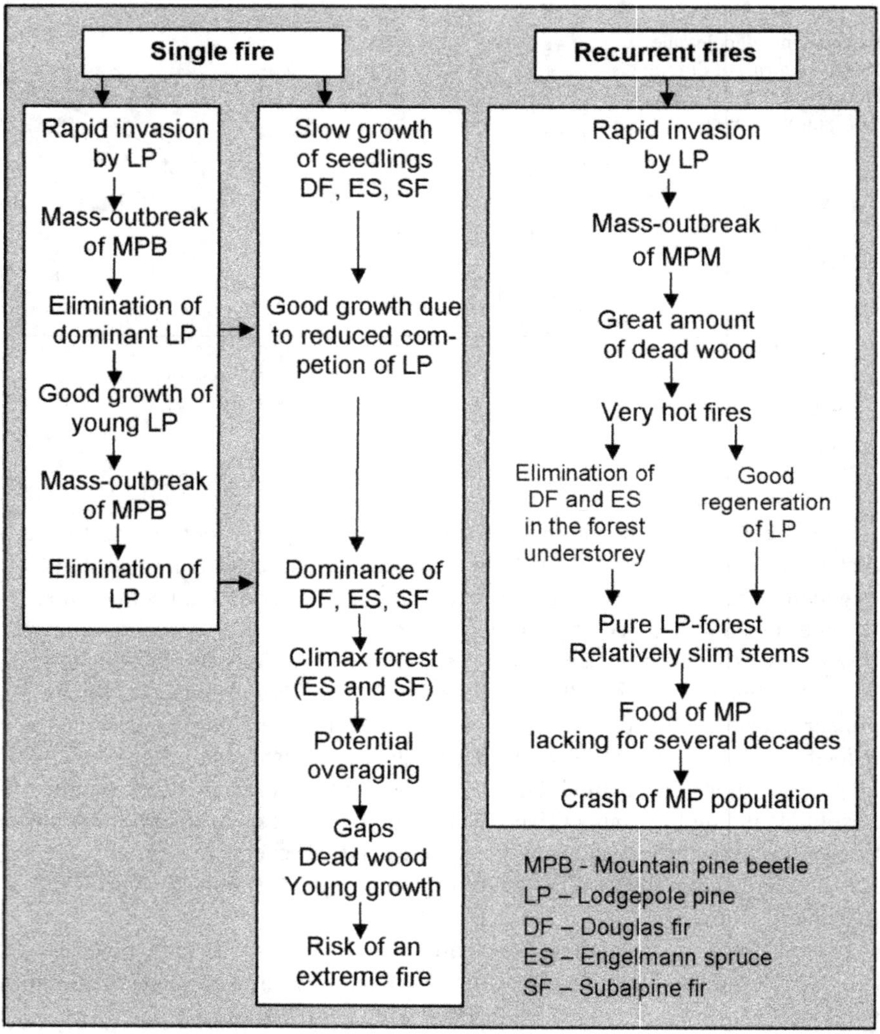

Fig. 2.39 Forest development influenced by a single fire (A) and by several fires (B). Scheme by F.-K. Holtmeier, based on explanations by Amman (1977)

Measures to control mountain pine beetles were usually not very effective or only slowed down the beetle gradation a little. When the mountain pine beetle meets favorable conditions, it is practically impossible to stop the infestation before the caterpillars will have killed all thick pines or man has removed them from the forest (Wickman 1990). So, systematic fire-fighting has increased the risk of pine beetle outbreaks. As forest fires have become less frequent, many tree stems may reach diameters that provide optimal conditions for the reproduction of the mountain pine beetles (Johnson et al. 1994).

2.4.3.2.2 Spruce Beetle

In North America, almost all spruce species are attacked by the spruce beetle (*Dendroctonus rufipennis*), one more representative of the bark beetle family. In the southern and central Rocky Mountains, the Engelmann spruce (*Picea engelmannii*) is particularly threatened. In Canada and Alaska, the white fir (*Picea glauca*) and Sitka spruce (*Picea sitchensis*) are mainly infested. In the Rocky Mountains, *Picea engelmannii* is very common at elevations between 2,700 and 3,000 m. Extreme weather conditions (e.g., drought, wind-breakage) may considerably increase susceptibility of trees to spruce beetle attacks, as was the case on the Kenai Peninsula (Alaska) during the 1970s and 1990s, for example (Berg and De Volder 2000; Berg et al. 2006). Accumulation of combustible materials from fallen trees and invading grass in the understorey increases the risk of severe ground fires (e.g., LaBau and Mead 2001).

The developmental cycle of the spruce beetle usually extends over 2–4 years (semivoltine/trivoltine). Under warm and dry conditions, however, it may even become completed within 1 year (univoltine). First of all, spruce beetles attack fallen trees. Partly uprooted trees having some root contact with the soil remain green and are preferred over totally uprooted and broken trees that probably desiccate and become unsuitable to the spruce beetle earlier (Schmid and Frye 1977). When such trees are missing, the beetles will infest living trees. Similar to the mountain pine beetle, spruce beetles prefers trees with a stem diameter of more than 50 cm. Trees with dead branches in their lower stem section appear to be particularly attractive to spruce beetles. Such trees are more typical of dense stands than in open forests, where the lower branches may grow to great length and often touch the ground. As a result of spruce beetle attack, the average age of the surviving trees, mean stem diameter, stand density and the proportion of spruce within the forest decrease. Surviving spruce trees of medium height, previously suppressed by the very competitive tall trees, may now become a dominant or associated tree species. Seedlings and young growth, profiting from the decline of the upper forest canopy, compete successfully with the herbaceous ground vegetation. However, once the dead stems are removed, young spruces become overwhelmed by the luxuriantly thriving field layer, which benefits from the better light conditions.

In spruce-fir forests in Colorado, 90 % of the trees were spruce and 10 % subalpine fir before a spruce beetle outbreak. After the calamity, the proportion had almost reversed (20 % spruce, 80 % subalpine fir) (Schmid and Hinds 1974; Schmid an Frye 1977). In Alaskan white spruce-paper birch forests (*Picea glauca, Betula papyrifera*), the proportion of spruce declined from 77 to 55 %, while birch increased from 22 to 45 % (Baker and Kemperman 1974).

As suggested by the available observations (Schmid and Hinds 1974; Schmid and Frye 1977), the long-term succession in a spruce-fir stand might proceed as follows (Fig. 2.40). At the beginning, the subalpine fir/spruce ratio is 80–20 %. Both tree species have a breast-height diameter of 8 cm. Young growth of both species is represented in the same proportions as the old trees. This would correspond to the situation in White River National Forest (Colorado) after a severe mass-outbreak of

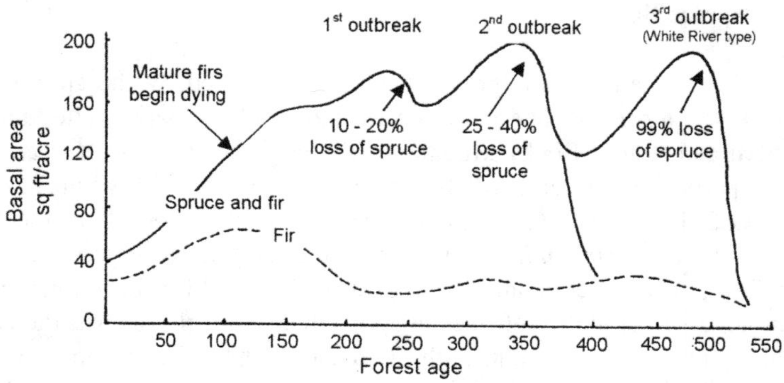

Fig. 2.40 Hypothetical succession in a spruce-fir forest (*Picea engelmannii, Abies lasiocarpa*). Coverage has been used as an indicator of stand structure. Modified from Schmid and Frye (1977)

the spruce beetle during the 1940s. While both spruce and fir are growing rapidly, fir, however, prevails and is even gradually increasing in the upper canopy layer. At an age of 125–175 years, and having reached a stem diameter of 40 cm, firs die for reasons still not well understood. Consequently, young trees no longer suppressed may develop better, and the proportion of fir and spruce in the canopy layer and also in the understorey gradually changes at the cost of fir. The comparatively light-demanding spruces profit more from the gaps in the forest canopy and improved light conditions than the highly shade-tolerant firs. Moreover, fir suffers from browsing much more than spruce. Not least, spruce usually gets much older than fir. The forest gradually reaches its maximum density, which depends on the local site conditions. Subsequently, spruce declines due to recurrent mass-outbreaks of the spruce beetle at periodic intervals. During the first outbreak, 10–20 % of spruce are lost. Afterwards, spruce increases again until the next outbreak, when 25–40 % of spruce are killed by the spruce beetle. Consequently, the spruce beetle population declines. It is possible that a mass-outbreak may occur before the forest has reached its climax stage. A few stands of spruce trees with stem diameters allowing the spruce beetle to complete their life cycles would be sufficient. Although spruce decreases and fir increases, spruce cover will remain 50 %. After the beetle attack, succession towards a one-layered forest is likely. Having reached its maximum density, such a forest is extremely vulnerable to spruce beetle outbreaks. In case of wind throw, the beetles first attack the uprooted trees before they destroy the living ones. The result will be a forest dominated by fir again.

Beetle mass-outbreaks not only influence forest structure and succession, but also the forest animal communities. Elk, mule deer and also voles, for example, profit from the increasing grasses and herbs, whereas fewer spruce seeds are available to red squirrels and pine grosbeaks (*Pinicola enucleator*). On the other hand, after a mass-outbreak, woodpeckers (northern three-toed woodpecker, *Picoides tridactylus*; downy woodpecker, *Dendrocopos pubescens*; hairy woodpecker,

Dendrocopos villosus), which are the most important predators of bark beetles, enjoy an almost inexhaustible food source. During two to three consecutive years, they still benefit from the numerous secondary insects, such as ants, long-horned beetles, and other bark beetles. Afterwards, the number of woodpeckers decreases due to shortage of food. The same holds true for nuthatches (*Sitta* spp.) and brown creeper (*Certia familiaris*). Spruce grouse (*Canachites canadensis*), lacking shelter and spruce buds as winter food, also decline temporarily.

As is evident for the pine beetle in the Colorado Rocky Mountains, spruce beetle populations on the Kenai Peninsula have also been responding to climatic warming. They decreased their development time, switching from a semivoltine life cycle to an univoltine life cycle (Berg et al. 2006).

2.4.3.2.3 Eastern Larch Beetle

The larch beetle (*Dendroctonus simplex*) has specialized in larch and can be found throughout the entire distribution area of American larch or tamarack (*Larix laricina*), which extends from Eastern to Western Canada and into Central Alaska (Johnston 1990; Fowler et al. 1995). It is also a potential risk to native larch species in eastern Siberia and to native and exotic larches in Europe. In Alaska, larch occupies the area between the Brooks Range in the north and the Alaska Range in the south. It is particularly wide-spread in the large river plains of the Yukon, the Koskuvim and Tanana.

Eastern larch beetles usually colonize the phloem of tree stems, exposed roots and larger branches of tamarack. They tunnel through the bark and feed and mate in the bast tissue. The galleries excavated by the beetles interrupt sapflow. Under favorable conditions, larch beetles can produce two complete generations during a summer. High tamarack mortality associated with the larch beetle has occurred throughout eastern North America during the late 1970s and early 1980s (Seybold et al. 2002). In Central Alaska, most tamarack forests became infested, with the exception of the western area of larch distribution (Werner 1986b). In some areas, 50 % of tamarack were killed just within 2 years. The stem diameter of most of the trees (70–99 %) exceeded 10 cm (Seybold et al. 2002). The susceptibility of the trees to beetle attacks was particularly high due to poor site conditions characterized by waterlogged permafrost soils with a shallow and cool active layer. Consequently, growth rates were very small.

In general, mass outbreaks of defoliating insects such as the larch sawfly (*Pristiphora erichsonii*), spruce budworm (*Choristoneura fumiferana*) and the larch casebearer (*Coleophora laricella*) precede larch beetle attacks (Seybold et al. 2002). In the drainage area of the Tanana River, for example, the larch bud moth (*Zeiraphera* sp.) had defoliated extended larch forests in 1975 and 1976, prior to the mass-outbreak of the larch beetle. The larch bud moth attack is reflected in a considerable reduction of growth (Table 2.3; Fig. 2.41). Thus, the severely stressed trees became highly prone to the larch beetle infestation (Werner 1986a, b).

Table 2.3 Growth increment by diameter classes of undamaged larches and damaged larches that died in 1977 in the catchment area of the Tanana river (Alaska). After Werner (1986b)

Diameter class [cm]	Growth increment [mm] in undamaged trees					Growth increment [mm] in damaged trees that died in 1977			
	1974	1975	1976	1977	1978	1974	1975	1976	1977
0–2	1.11	1.01	0.91	0.70	0.82	0.88	0.62	0.52	0.31
>2–4	1.50	1.42	1.32	1.12	1.20	1.08	0.83	0.56	0.29
>4–6	1.92	1.89	1.68	0.91	0.90	1.10	0.92	0.49	0.18
>6–8	0.83	0.79	0.74	0.68	0.81	0.97	0.65	0.32	0.21
>8–10	1.26	1.09	0.93	0.83	1.06	0.81	0.76	0.43	0.15
>10–12	1.14	0.91	0.87	0.69	0.93	0.83	0.77	0.36	0.22
>12–14	0.95	0.86	0.74	0.54	0.71	0.76	0.70	0.21	0.11
>14	0.86	0.74	0.63	0.53	0.52	0.78	0.69	0.32	0.16

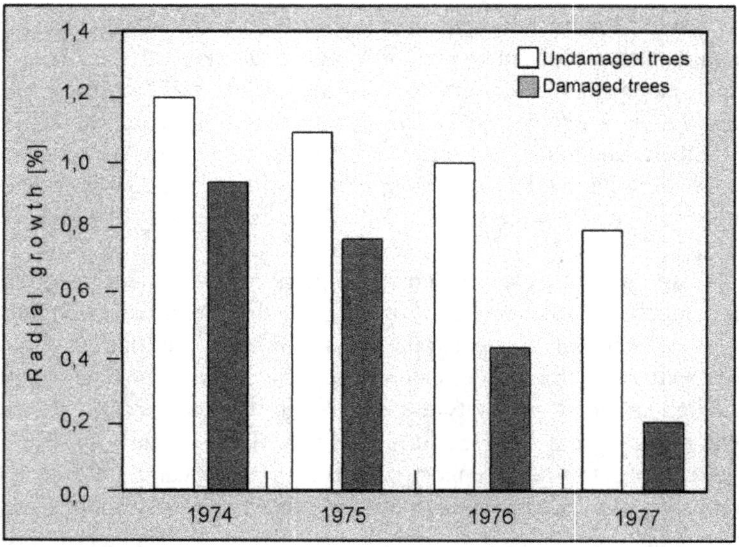

Fig. 2.41 Influence of defoliation by the larchbud moth (*Zeiraphera* spp.) on diameter growth of tamarack (*Larix laricina*) near Fairbanks, Alaska. Modified from Werner (1986)

2.4.4 *African Red-Billed Weaver Bird*

When assessing the impact of animals on landscape ecosystems, both the active role of the animal populations and their passive role and feedbacks under the influence of changing landscape habitat conditions (structure, forage, shelter, landscape use, etc.) must be considered. The latter are often of great importance, as has already been demonstrated in the previous sections. Changing environmental conditions may occasionally result in a drastic increase of species, with far-reaching

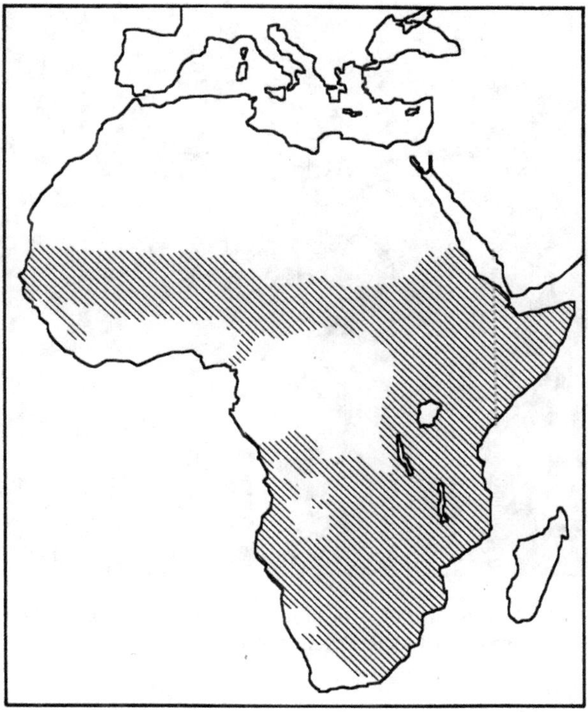

Fig. 2.42 Distribution of the African red-billed weaver bird (*Quelea quelea*). From Magor and Ward (1972)

consequences. The African red-billed weaver bird (*Quelea quelea*), which probably is the most abundant bird species in the world (Fry and Keith 2004), is a particularly spectacular example. This bird, about the same size as a sparrow, lives in the African savannas, where it has become a serious pest. Red-billed weavers cause even more severe damage to cornfields than the gigantic locust swarms that were mentioned in the Bible as one of the ten plagues. Not without reason, people often call the red-billed weaver 'locust bird' (Murton and Westwood 1976) or 'feathered locust' (Markula et al. 2009).

Four geographically-separated species can de distinguished: *Quelea q. quelea*, West Africa; *Quelea q. aethiopica*, Ethiopia, Sudan, northern Somalia; *Quelea q. intermedia*, southern Somalia, Kenya, Tanzania; *Quelea q. lathami*, South Africa (Ward 1971; Dreiser 1993). Weaver birds are spread over the entire region south of the Sahara, with the exception of the rain forests and mountain areas (Fig. 2.42). Weaver birds are probably the most numerous of all granivorous birds. They rely mainly on the seeds of annual grasses. The weavers form huge swarms of some ten thousand or some hundred thousand individuals traveling about in search of food (Photo 2.10). Following the seasonal rains, they may cover distances of many hundred kilometres, or even more than 1,000 km in East Africa (Dreiser 1993). Grass seeds germinate at the onset of the rainy season. Afterwards, they are no longer available. Consequently, the birds move to other areas, where the wet season starts a little earlier and the new grasses have already produced seeds ('early season

Photo 2.10 Swarm of African red-billed weaver birds (*Quelea quelea*) crossing a rice field at Mogaambo (Somalia). R. Peveling, August 1988

migrations' in the sense of Ward (1971)). After 6–8 weeks, the weaver birds return to the area they had left when the wet season started. In this area, they breed. As the fresh grasses have produced seeds in the birds' absence, abundant food is now available in this area (Ward 1965, 1971; Wallin 1990).

Invading weaver birds destroy the grain harvest (corn, millet, wheat and other cereals; GTZ 1987). Such invasions have been known about for a long time. The earliest explorers were already reporting on immense weaver swarms. Since then, the weaver bird populations and the damage they cause have considerably increased parallel to the rapidly growing human population and the expanding area under cultivation (Fig. 2.43). An endless number of water reservoirs have been installed for irrigation of the cornfields. Thus, the weaver birds that originally relied on the seeds of wild grasses now profit from an inexhaustible food source and water available at not too far distances, even during the rainy season. Availability of water was probably the 'bottle neck' for the birds. In agricultural lands weavers can be ten times more abundant compared to natural grassland (Berutti 2000). Overgrazing of the savanna might also have favored re-billed weaver birds, as the proportion of the annual sweet grasses has increased. This has been reported from Kenya (Schmutterer 1965) and southern Somalia (Peveling 1990), at least. In 2009, the drought-tolerant and early maturing sorghum variety 'gadam' was introduced to semiarid eastern Kenya to improve food product diversity and human food security (Wambugu 2011). However, 'gadam' not only gives hope to subsistence farmers but also provides an additional rich food source to

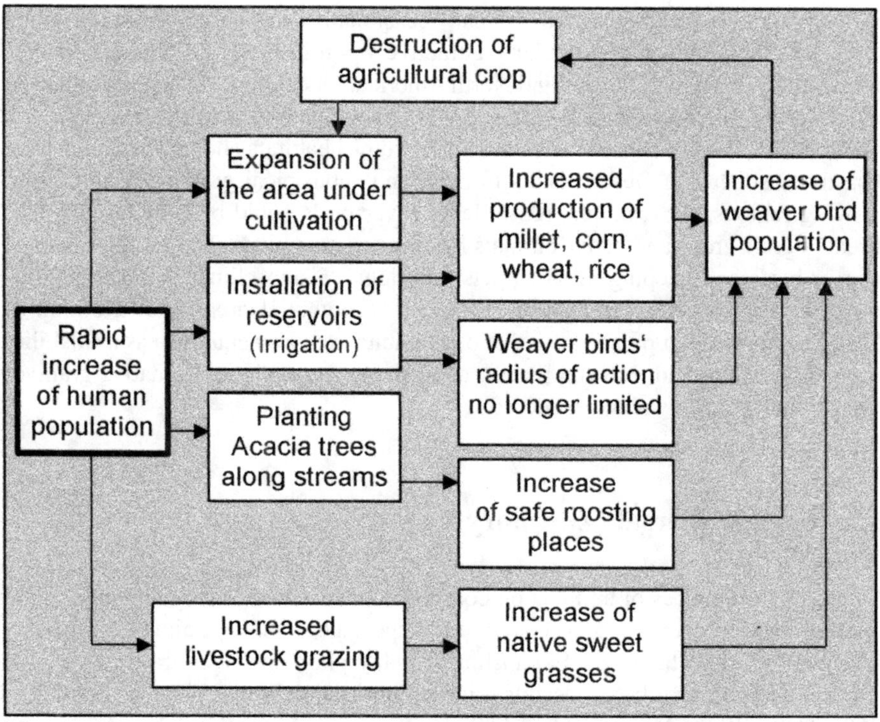

Fig. 2.43 Human population growth as a 'disturbance factor' and its influences on the red-billed weaver birds (*Quelea quelea*) and the environment. Scheme by F.-K. Holtmeier, after different sources

red-billed weaver swarms, which in turn cause a setback to the promising modern agricultural development in Kenya (Esipiu 2013).

Weaver birds nest and breed in large colonies built on thorny shrubs and trees. Reeds (reed fields) and sugar cane plantations, however, are also used. Such colonies may consist of several millions of nests (Ward 1971). Even tall *Acia* trees occasionally break down under the heavy load of birds and their nests (GTZ 1987). Actually, this has also been observed in colonies of social weaver birds (*Philetaris socius*) in the dry regions of South Africa. Social weavers, however, instead of building individual nests, share large coherent nests (up to 5 m in height and 5 m in length), preferably in the crowns of tall trees. Neither natural predators, parasites and diseases nor the extensive control measures (e.g., avicides, fire bombs, flame throwers, explosives, poisoning of water holes) carried out since the 1950s have solved the problem (GTZ 1987; Peveling 1990; Dreiser 1993). Poisoning may also affect other animals. Locust control in the Sahel zone, for example, has been considered a possible cause of an alarming decline in European long-distance migratory birds during the last 25 years (Sánchez-Zapata et al. 2007).

The 'weaver problem' is still peculiar to the sub-Saharan savannas. Situation may change, however, as in many countries people keep weaver birds as pets in birdcages or aviaries from where they can occasionally escape. In the past red-billed weavers – the males are particularly attractive when in breeding plumage – were occasionally brought to Europe and North America, for example. Escapees, however, could not establish viable populations in their new environment (Simberloff et al. 1997; Hinze 2004). Escape or release in suitable habitats, however, could have disastrous effects. In Queensland's tropical and subtropical grasslands and grain-growing regions, for example, red-billed weaver birds could become an immense threat to grain crops and native animals feeding on grass seeds. Therefore, import of weaver birds and keeping them as pets are illegal. Using climate-modelling software Markula et al. (2009) predicted weavers' potential area of distribution in Australia, not without pointing out that other habitat requirements such as availability of food, breeding and roosting places, as well as competition with native granivorous animals must also be considered.

2.5 Seed Dispersal (Zoochory)

Among the influences of animals on composition, structure and dynamics of plant communities, biocoenoses and ecosystems, dispersal of seeds by animals (zoochory) may play an important role. Expansion of many plant species during postglacial times would not have been possible without animals' help (e.g., Müller-Schneider 1948; Davis 1976; Mattes 1978, 1982; Janzen 1983; Frenzel 1983; Johnson and Webb 1989; Farmer 1997). However, in view of the many factors influencing seed dispersal (different behavior of seed-dispersing animal species, food preferences, competition for food, vegetation characteristics, effects of site conditions on germination and seedling performance), general statements on the effects of zoochoric seed dispersal and of seed hoarding on vegetation should be given, if ever, only with great reserve (see also Price and Jenkins 1987).

Most seed dispersers are vertebrates, with the exception of ants. The radius of zoochoric seed dispersal ranges from a few meters to several kilometres. While ants, for example, may disperse seeds within a radius of 70 or some more meters (Müller-Schneider 1977a, b), the Siberian nutcracker (*Nucifraga caryocatactes macrorhyncos*) has been observed carrying seeds of the Siberian stone pine (*Pinus sibirica*) over a distance of 15 km (Reijmers 1959). The Clark's nutcracker (*Nucifraga columbiana*) has been reported to disperse stone pine seeds over even greater distances (up to 25 km; Vander Wall and Balda 1977). Zoochoric seed dispersal may be rather accidental, if, for example, diaspores (seeds with pieces of the pulp, mesocarp) sticking to animals` fur or plumage, beaks or feet, fall off after a more or less long distance of transport. This kind of seed dispersal is called epizoochory. Many animal species hoard seeds in seed caches established in the litter layer and in the topsoil (synzoochory; Vander Wall 1990). In case the inlying seeds are not consumed, they may germinate, favorable conditions provided.

Animals with long fur, such as sheep or Scottish highland cattle (Fischer et al. 1995, 1996; Stender et al. 1997), and not least wild boars having a 'seed collecting' coat with long bristles and thick undercoat, are very efficient dispersers of diaspores. Wild boars, in particular, often travel long distances (Briedemann 1990; Heinken et al. 2002). On Hawaii (Hawaii Volcanoes National Park), for example, feral pigs have been driving the rapid expansion of common velvet grass (*Holcus lanatus*). In Central Europe, wild boars are likely accelerating the expansion of neophytes (Mrotzek et al. 1999).

On Santa Catalina Island (California), where no native large herbivores lived before bison (*Bison bison*) were introduced, these large grazers have contributed to the spread of exotic plant species through both their dung and seeds stuck to their coat (Constible et al. 2005). Cattle, horses and also hares have spread the seeds of exotic plant species with their dung in the flooded grassland pampa (Buenos Aires, Argentina). Many of these plants are weeds competing successfully with the forage species (Vignolio and Fernández 2010).

In tropical forests, primates, for example, feeding on fleshy fruits of trees and shrubs in the understory often are important, though differently effective, seed dispersers (e.g., Lieberman and Lieberman 1986; Zhang and Wang 1995). On Barro Colorado Island (Panama), seed dispersal by white-faced capuchin monkeys (*Cebus capucinus*), for example, has been found to be especially effective, as these primates spend only little time feeding in the tree itself and remove most ingested seeds away from the seed source (Zhang and Wang 1995; Wehncke et al. 2003). Frequent defecation results in a scattered deposition of seeds. Compared to other primate species, post-dispersal seed predation is relatively low (Wehncke et al. 2003).

Fruits and seeds are an important food source for many animal species. Fruits contain much fat and carbohydrates. Seeds are also rich in proteins. The energy content of approximately 35,000 whitebark pine seeds (*Pinus albicaulis*), for example, cached by the Clark's nutcracker in autumn, covers the energy demand of one young bird through winter into spring (Tomback 1982). Some species have specialized in certain kinds of fruits. For others, such as wild boars, foxes, coyotes, badgers or bears, fruits make up only a part of their diet. Most seeds become destroyed when eaten. Thus, red squirrels (*Tamasciurus hudsonicus*), for example, which remove the pulp from juniper berries and eat the seeds, do not contribute to distribution of juniper (Livingston 1972). However, as seeds are usually less digestible than the pulp, they may pass the digestive tract without being damaged. They are deposited with the excrement and, if still viable, seedlings may emerge. Thus, the distribution of many plant species having berries or berry-like fruits (e.g. *Juniperus*, *Taxus*) is supported also by birds, foxes, stone marten (*Martes foina*), wild boars, bears and other carnivores (e.g., Santos et al. 1999). Predators may also disperse plant seeds unintentionally ingested with prey animals (e.g., in their gut) over comparatively large distances. Arctic foxes (*Alopex lagopus*), for example, have been reported to travel at least 10 km per day. Much longer distances are likely, however (Graae et al. 2004). Distribution patterns of endozoochoric plants may be more 'regular' than those of epizoochoric plants, when, for example, seedlings emerge from regularly used latrines or richly-manured sites on alpine pastures. On the other hand, epizoochoric plants may be found at higher density along migration routes, for example.

Seeds contained in the dung of cattle have long been considered to be unimportant, as dung has toxic effects on seedlings of most species. As a result, unvegetated 'dung patches', for example, may develop. These may become gradually recolonized by vegetative propagation of the surrounding vegetation. On pastures in the Mediterranean, however, dung patches are usually occupied by species resistant to toxic dung effects (Malo and Suarez 1995). On the other hand, distribution of *Acacia* seeds (e.g., *Acacia tortilis*, *Acacia nilotica* and *Balanaites wilsonia*) depends exclusively on endozoochoric transport by impalas, kudus, giraffes, elephants, and ostriches (*Struthio camelus*), for example. Elephants also disperse the seed of the Borassus palm (*Borassus aethiopicum*) and many other plant species (Lamprey 1963). In the tropical rain forests of the Congo Basin and the West-African coast, about a third of the tall tree species is dispersed by forest elephants, among them cherry mahogany (*Tieghemella heckelii*), Mubura (*Parinari excelsa*), bush mango (*Irinia gabonensis*) and the Molapa tree (*Gilbertiodendron* sp.) (Beyer 1996). The germination capacity of seeds passing the digestive tract may increase (Lamprey 1963, 1974; Pellew and Southgate 1984; Lieberman and Lieberman 1986; Hoffmann et al. 1989; Miller 1995), or the seeds may be destroyed (Feer 1995). In germination experiments with seeds from elephant dung and seeds taken directly from trees, 50.7 % of the 'dung seeds' germinated compared to only 3 % of the 'tree seeds' (Chapman et al. 1992).

The sparsity of natural regeneration of some forest species with large seeds in the forests of the Ivory Coast (Bia National Park, West Ghana) has been attributed to the disappearance of elephants from this region. At least 30 % of the tree species and 41 % of the individuals of the different species are dispersed by elephants (Alexandre 1978). In the rain forests of Gabon, forest elephants are the only dispersers of 10 species. Elephants prefer big fruits with a number of seeds (Feer 1995).

In northeastern Australia and Papua New Guinea, southern cassowary (*Casuarius casuarius*, a ratide, weight up to 80 kg) is the only disperser of seeds contained in big fleshy fruits of more than 100 rain forest species (Marchant and Higgins 1996; Crome and Bentruppenbaumer 1993).

2.5.1 Ants (Myrmecochory)

Although ants are not specialized granivores in a strict sense, they are important seed dispersers of numerous plant species in many regions. They carry seeds with on-hanging nutrient-rich elaiosomes of some species (e.g., wood anemone, *Anemone nemorosa*); violet, *Viola* sp.; snowdrop, *Galanthus nivalis*; dead nettle, *Lambium* sp.; ramson, *Allium ursinum*) into their nests to feed their larvae. Since the larvae, however, consume the elaisomes only, seedlings may emerge from the remaining seeds. So-called myrmecochory (*Myrmes* = ant) occurs in plants of different taxa. As not only the plants but also the ants profit from myrmecochory, this could be considered a mutualism, at least partially (Handel and Beattie 1990). Ants carrying the seeds away from the seed source partly reduce competitive pressure of the seed plant and other plants on potential seedlings. Moreover, in anthills seeds are safe from seed predators, and, not least, the loose material (substrate) within anthills is favorable for germination, as it is well aerated, consistently moist and rich in nutrients. This holds particularly true for the waste products. Ants partly remove the seeds from the nests after having detached and eaten the elaiosomes. In temperate zones, many myrmecochorous plants have already produced fruits in early spring. As ants will not find dead or living insects at this time of the year, their main food during the warmer seasons, early fruiting may be considered to be an effect of selection (Handel and Beattie 1990).

Seed harvesting by ants is highly selective, but the total amount of harvested seeds is relatively small. In the Mojave Desert, seed removal by black harvester ants (*Veromessor pergandei*) has been estimated to amount to only about 1 % of the total seed crop (Tevis 1958). On the shortgrass plains of north-eastern Colorado, rough harvester ants (*Pogonomyrmex rugosus*) took just about 2 % of the available seeds (Rogers and Lavigne 1974).

On the other hand, species preferred by ants may lose a large proportion of their seed crops to ants (e.g., Buckley 1982 and further literature therein). In Australia, seed dispersal by ants is particularly important. More than 1,500 plant species are myrmecochorous (Berg 1975). On the other hand, consumption of seeds by harvester ants (*Messor* sp.) may impede regeneration of certain tree species, such as Victorian or mountain ash (*Eucalyptus regnans*, *Eucalyptus delegatensis*) for example. About 60 % of the seeds are eaten by ants, with the exception of burned areas (Ashton 1979). Experimental studies on the regeneration of brown stringybark (*Eucalyptus baxteri*) and sheoak (*Casuarina pusilla*) in Australia's southeast (Andersen 1987), however, showed that the number of seedlings increased 15-fold after ants had been removed. Most Eucalypt seedlings, however, died within their first year. Obviously, such long-lived plants depend more on sites favorable for seedling establishment than on the quantity of seeds. By contrast, ephemerals rely directly on the amount of available seeds, and thereby on seed predation by ants. Harvesting ants can be found all over the continent. They are usually omnivores and consume seeds more occasionally, commonly those from woody plants. While outside Australia seeds are harvested almost exclusively by Myrmecinae, within Australia Formicinae and Ponerinae are also involved. Altogether, seed harvesting by ants is so common to all plant communities that it has been considered a keystone factor in Australia's landscape (Andersen 1991).

2.5.2 Mammals

In Central European deciduous and mixed forests, birds consume up to 75 % and squirrels almost 40 % of the annual seed production. Seeds lying on the ground are usually eaten by small mammals. However, in years with good seed production (mast years), seed consumption by granivores does not threaten forest regeneration. On the other hand, in cases of low seed production, small mammals may consume all seeds on the ground, and natural regeneration may outright fail as a result (see Sukachev and Dylis 1964; Castro et al. 1999; Dulamsuren et al. 2013).

In oak forests (*Quercus mongolica* var. *grosseserrata*) of southwestern Japan, even in mast years, no seedlings can be found, as rodents and Asian black bears (*Selenarctos thibetanus japonicus*) eat all acorns before they can germinate (Ida and Nakagoshi 1996). Studies on the influence of herbivorous mammals on the regeneration of encina (*Quercus oleoides*) in Santa Rosa National Park (Costa Rica) showed that only dense oak stands produce sufficient acorns to cover the food demand of the many seed consumers (Agoutis, *Dasyprocta punctata*; variegated squirrel, *Sciurus variegatiodes*; white-throated capuchin, *Cebus capucinus*; spiny pocket mouse, *Liomys salvisi*) so as not to jeopardize the continued existence of the forest. In solitary oaks, however, regeneration fails (Boucher 1981).

In Spanish dry forests, up to 96 % of Scots pine seeds that reached the ground were eaten by rodents and birds (Castro et al. 1999). Seed removal by seed predators was also the main reason for failure of Scots pine seedling emergence in the Mongolian forest steppe. In the forest steppe, 66–92 % of the seeds were lost to seed predators, while 52–76 % were removed at the forest edge (Dulamsuren et al. 2013).

In Californian conifer forests, the losses of seeds to the deer mouse (*Peromyscus maniculatus*) amounts to 70–100 % (Howard and Cole 1967). Similar results were obtained from studies on natural re-forestation of clear-cuts in Douglas-fir forests (*Pseudotsuga menziesii*) close to the coast of British Columbia (Sullivan 1979). At an average medium seed production on clear-cuts in the Cascades, about 5 % of Douglas-fir seeds are lost to mice, birds and chipmunks, whereas the seed losses in western hemlock (*Pseudotsuga heterophylla*) amount to 25 %, and in western red cedar (*Thuja plicata*) to 57 %. As many additional seeds get lost due to other factors, only a few are left for regeneration (Gashwiler 1970). In Rocky Mountain forests, red squirrels (chicarees, *Tamasciurus hudsonicus*) harvest conifer seeds to an extent that seed-based regeneration is possible in mast years only (Finley 1969; Schimpf et al. 1980). Seed predators, however, may feed not only on seeds but also fresh plantules. During seeding experiments in longleaf pine stands (*Pinus palustris*, southwestern Alabama), almost all pine seeds and plantules fell victim to seed predators within 3 months after seeding. Most of them were eaten by mice (58 %) and birds (33 %) (Boyer 1964). In Bavarian mountain forests, bank voles (*Clethrionomys glareolus*) and yellow-necked field mice (*Apodemus flavicollis*) feed mainly on not yet lignified plantules of beech (*Fagus sylvatica*) and mountain maple (*Acer pseudoplatanus*) from spring until summer, and may thereby prevent any natural regeneration (Bäumler and Hohenadl 1980).

In a review, Harmer (1994) comes to the conclusion that successful natural regeneration of oak and beech can only be expected if seed predators are controlled, probably most successfully by their natural enemies. By contrast, Potter (1978), referring to studies in secondary deciduous forests of New Hampshire, holds the view that seed predation by small mammals influences forest structures only during the early successional stages, but not in mature forests.

Many small mammals (mice, voles, hamsters, dormice, squirrels, ground squirrels, etc.) and birds (e.g., nutcrackers, jays, nuthatches, tits, rooks) hoard seeds of

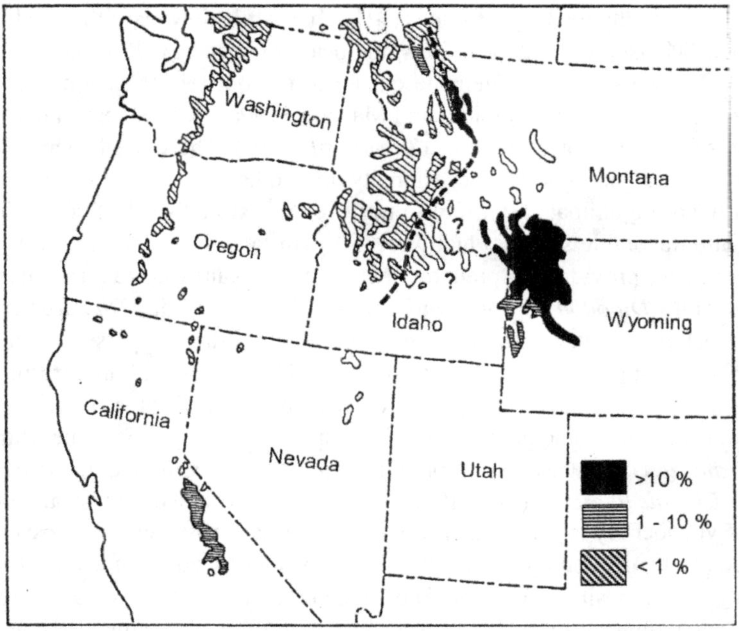

Fig. 2.44 Distribution of whitebark pine (*Pinus albicaulis*) and use of its seeds by bears, differentiated by the percentage (vol. %) of seeds in the bears' droppings. The *broken line* shows the eastern extent of maritime influence. Modified from Mattson and Reinhart (1994)

numerous forest tree- and shrub-species (Vander Wall 1990). Seed predators consume the bulk of the seeds. Squirrels, for example, eat about 90 % of the seeds they cache. Moreover, seed predators often plunder seed caches of other animals. Thus, in the Rocky Mountain whitebark pine (*Pinus albicaulis*) forests, the middens of red squirrels are an important food source of black bears and grizzly bears (Fig. 2.44), in particular. Fully-grown grizzlies are too heavy to climb trees for harvesting pine cones (e.g., Kendall 1983; Kendall and Arno 1990; Mattson and Jonkel 1990; Mattson and Reinhart 1994, 1997).

In Yellowstone National Park and adjacent forest areas, grizzly bears, despite being omnivores, depend first of all on the availability of fleshy berries (*Vaccinium scoparium*, *Vaccinium globulae*, *Shepherdia canadensis*) and on the seeds of whitebark pine (Mattson et al. 1991a). Seed production, however, fluctuates considerably. In years with an abundant seed crop in whitebark pine, grizzly bears practically consume nothing else. As whitebark pine is more common in continental regions, the proportion of pine seeds in bear food is comparatively high compared to more maritime areas (cf. Fig. 2.44). More than 90 % of the whitebark pine seeds consumed by grizzly bears are from red squirrel middens (Mattson et al. 1991a; Mattson and Reinhart 1994, 1997). As red squirrels use the same seed caches for many years, bears retrieve them easily. It has been reported from the Yellowstone area that bears excavate the seed caches year after year or even repeatedly within a single year.

Bears pillage mainly the large middens (>100 m^2), while they leave the smaller ones untouched (Mattson et al. 2001). Thus, the bears save energy. In some areas, seed predation by bears may threaten the existence of squirrels (Mattson and Jonkel 1990; Mattson et al. 2001; Reinhart and Mattson 1990a, b). In Siberia, brown bears excavate seed caches of burunduks (*Tamias sibiricus*). These seed caches contain between 1½ and 2 kg of seeds, occasionally up to 6 kg.

Seed-hoarding animals often fail to retrieve their stored seeds. Seeds that were cached in unfavorable sites gradually rot away. In favorable places, however, they may germinate, provided no predators detect them. Kangaroo rats (Banner-tailed kangaroo rat, *Dipodomys spectabilis*; Ord's kangaroo rat, *Dipodomys ordii*; Merriam' kangaroo rat, *Dipodomys merriami*) are important dispersers of mesquite bush (*Prosopis* spp.; Reynolds 1958; Storer and Usinger 1971; Van der Pijl 1972). In the high-montane Oregon conifer forests, about 50 % of bitterbrush plants (*Purshia tridentata*) emerge from seeds hoarded by the yellow pine chipmunk (*Tamias amoenus*), Townsend's chipmunk (*Tamias townsendii*) and mantled ground squirrel (*Citellus lateralis*) (West 1968). This kind of seed dispersal is called synzoochory. Synzoochory is particularly important in forest ecosystems. In case of abundant seed production, small rodents usually cache more seeds than needed for food. This may have a positive effect on forest regeneration. In other years, the rodents' influence is of minor importance or even negative (Abbot and Quinck 1970). On the other hand, the forest spiny pocket mouse (*Heteromys desmarestianus*), for example, promotes natural regeneration in the tropical lowland forest of Belice (Central America), despite consuming more than 95 % of the hoarded seeds. As the trees may produce abundant seeds per year over a long period, the remaining seeds not used by the rodents are sufficient for intense regeneration (Brewer and Remánek 1999). Jays (*Garrulus glandarius*) and squirrels (*Sciurus vulgaris*) may 'plant' a second tree layer of oak in pine stands and also contribute to expansion of oak beyond closed oak forests.

2.5.3 Birds

As to bird-mediated seed dispersal and seed hoarding, jays and nutcrackers are of particular interest. They are widely distributed over Eurasia and North America, where they (many species) may have a great influence on forests in the temperate zone (Tomback and Linhart 1990).

2.5.3.1 Jays and Nutcrackers

Seed dispersal by jays and nutcrackers has been comparatively well studied. Jays (*Garrulus glandarius*) were originally a forest species (Voous 1962; Goodwin 1976). In the forest, they are far less stressed by competition with other corvids (e.g. carrion crow, *Corvus corone*; rook, *Corvus frugilegus*; magpie, *Pica pica*; jackdaw,

Corvus monedula) than in the open landscape (Bossema 1979). Depending on season and availability, they feed on a multitude of fruits, cereal grains, conifer seeds, acorns, and beech nuts. They also prey on insects, small mammals (mice) and young birds. In Central Europe, acorns of native oak species (*Quercus robur*, *Quercus petraea*) are their main food. They obviously prefer acorns of native oaks to those of the American red oak (*Quercus rubra*) and also to beech nuts (*Fagus sylvatica*). In autumn, each jay caches several thousand acorns.

Jays obviously prefer ripe and undamaged acorns, which have a high germination capacity. They usually cache seeds at a greater distance from the seed trees and put the acorns into the litter or soil at about 4 cm depth, where they are relatively safe from other jays and seed predators, such as pigeons, wild boars, roe deer, squirrels, and mice (Shaw 1968a, b; Bossema 1979). In the upper soil and litter layer, conditions for germination are more favorable than on the surface under the seed trees. Moreover, separately cached acorns are less likely to become infected by fungi than acorns hoarded together in a larger seed cache. As the jays cache acorns usually in open places, seedlings enjoy more favorable growing conditions than in closed tree stands. Not least, they are less affected by leaf-eating caterpillars falling off the tree canopies. Acorn hoarding is advantageous to both jays and oaks. In southern Sweden, young oaks were found to be about ten times as many as beech young growth in spruce forests. This has to be attributed to the jay's clear preference for acorns over beech nuts (Nilsson 1985). Altogether, a true mutualism between jays and oaks is likely. Although jays consume large quantities of acorns and also damage plantules by ripping off the cotyledons, they contribute to maintenance and expansion of oak forests through seed hoarding.

In southern North America, about 50 oak species (e.g., *Quercus palustris*; *Quercus phellos*, *Quercus velutina*) and beech (*Fagus grandifolia*) are spread by the blue jay (*Cyanocitta cristata*) (Darley-Hill and Johnson 1981). Rapid spreading of oaks and beech as well as of hazel, hickory and even chestnut during postglacial times has been possible only with the help of the blue jay (Davis 1976; Johnson and Webb 1989; Farmer 1997), while in Europe, the indigenous jay (*Garrulus glandarius*) was instrumental in bringing about expansion of oak, hazel and beech. Also, Swiss stone pine (*Pinus cembra*) would not have so rapidly re-colonized its preglacial area of distribution without seed dispersal by the European (thick-billed) nutcracker (*Nucifraga c. caryocatactes*), and Siberian stone pine (*Pinus sibirica*) would not have expanded from its refugial areas in the Altai mountains to its present northern limit without the Siberian (slender-billed) nutcracker's help (*Caryocatates c. macrorhyncos*) (Mattes 1978).

Nutcrackers spread the heavy wingless seeds of several pine species (Table 2.4). A true mutualism has developed between the nutcrackers and these pines (Mattes 1978, 1982, 1985; Lanner 1980, 1984, 1990, 1996; Vander Wall and Balda 1977; Tomback 1982, 1983, 1989, 1994a, 1998; Linhart and Tomback 1985; Tomback and Linhart 1990; Tomback and Schuster 1994). Lanner (1996) described this interdependence briefly as '*Tree feeds bird, bird plants tree*'. The seeds of the above-mentioned pines are the main food of nutcrackers. However, similar to jays, they also take insects, cereals, kernels of berries, young mice and other small animals.

Table 2.4 Wingless subalpine pine species with seeds mainly dispersed by nutcrackers at the upper treeline. Modified after different sources from Holtmeier (1993)

Pine species	Nutcracker species	Region
Pinus cembra	*Nucifraga caryocatactes*	Alps, Carpathian Mountains, Northern Mongolia
Pinus sibirica	*Nucifraga caryocatactes*	Siberia, northern Mongolia
Pinus pumila	*Nucifraga caryocatatces*	Northeastern Siberia, Korea, Kamchatka, Japan
Pinus koraiensis	*Nucifraga caryocatactes*	Southeastern Siberia, Eastern Manchuria, Korea, Japan
Pinus albicaulis	*Nucifraga columbiana*	North America
Pinus flexilis	*Nucifraga columbiana*	North America

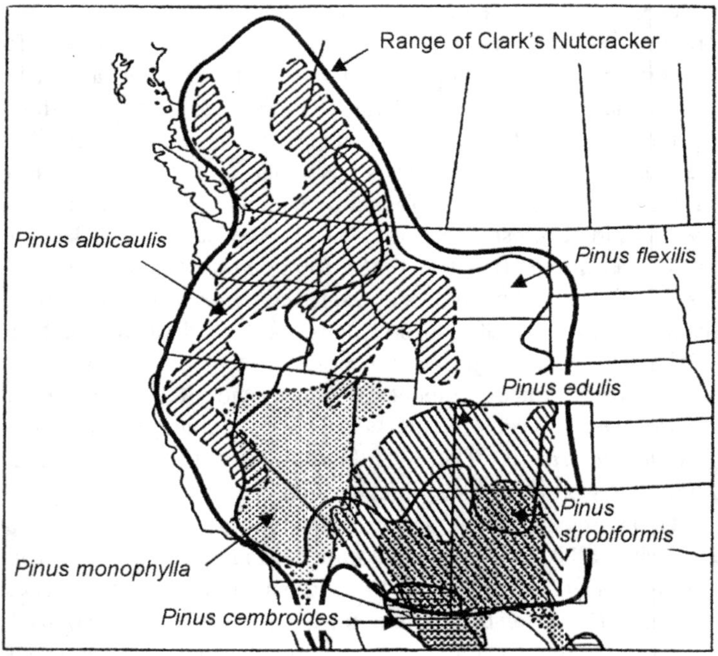

Fig. 2.45 Distribution of the Clark's nutcracker (*Nucifraga columbiana*) and pine species with wingless seeds in North America. Modified from Tomback and Linhart (1994)

In the pinyon-juniper woodlands (*Pinus edulis*, *Pinus monophylla*, *Juniperus utahensis*, *Juniperus osteosperma*) of the southwestern United States (Fig. 2.45), pinyon jay (*Gymnorhinus cyancephalus*) is the most important disperser of pinyon pine. Its seeds ripen around the end of August. Although the cones are still green and resinous at this time, the seeds are well developed and highly nutritive. Pine seed harvest continues throughout autumn. After the cones have dried and the scales have opened, jays pick up the seeds from the cones still attached to the twigs. They harvest only dark brown and fully developed seeds, leaving the light ones, which are usually empty. The jays carry up to 20 seeds in their sublingual pouch to their breeding area over distances of up to 10 km (Lanner 1984). At sunny, and thus, early snow-free places, such as the south-facing side of pinyon stems, the jays put one to two seeds into the upper soil near to the surface (about 2 cm depth) after having loosened it with the beak (Lanner 1996). By late autumn, the jays have established thousands of seed caches in their breeding habitats upon which the sitting birds and their brood rely for food in spring (February–April). Thus, pinyon pine provides ample fat- and protein- rich food to the jays during the period when other food is rare. On the other hand, the seed caches are favorable for germination, as after snow-melt, relatively high soil moisture in the litter layer will last into the dry summer. In case the seeds are not eaten by the jays themselves (or seed predators), it is very likely that the seeds will germinate and new pinyon pines become established. By contrast, seeds that fall off cones still attached to the trees do not have any chance of survival. They will be eaten by rodents or desiccate under the influence of high temperatures before the radicle has found its way into the soil.

Pinyon pine seeds are also the most important food source for small rodents. The latter store seeds in subterranean larders from which seedlings may occasionally emerge. Due to their high olfactory perception, however, rodents are capable of detecting and plundering seed caches of other rodents and jays (Howard and Cole 1967; West 1968; Lanner 1996). Consequently, good regeneration is likely only after abundant seed production, sufficient soil moisture provided (Lanner 1981, 1996).

Other corvids, such as Steller's jay (*Cyanocitta stelleri*), the scrub jay (*Aphelocoma coerulescens*) and Clark's nutcracker (*Nucifraga columbiana*) also scatter-hoard pinyon seeds. In an Arizonan pinyon pine woodland, a flock of about 150 Clark's nutcrackers cached about 1 ton of pine seeds within one autumn (Vander Wall and Balda 1977). This quantity of seeds was estimated to exceed the winter food requirement two to three times. Under such conditions, even great losses to seed predators would not have impeded regeneration of pinyon pine stands.

The seeds of both pinyon pine species were a traditional food source of the Indians living in the Great Basin. There are many cultural peculiarities that developed regarding the use of pinyon pine. Seed coats found in cliff dwellings in Nevada are about 6,000 years old. The Indians not only used seeds but also wood and resin of pinyon pine (Lanner 1981). However, millennia of human use had not seriously threatened the existence of pinyon woodland. This situation, however, totally changed after the Great Basin was colonized by the Europeans. Huge quantities of pinyon pine were used for production of charcoal needed for melting silver ore. Wood was the only energy source in this area in those days. During the silver boom of the 1860s and 1870s, charcoal production was the second most important industry in Nevada after silver mining. When silver mining and the demand for charcoal had ceased, the pinyon pine could re-colonize large parts of its former area of distribution, mainly with the help of jays and Clark's nutcracker. In recent times, however, pinyon pine woodland is being turned into open pastures (Lanner 1984).

The Clark's nutcracker (Photo 2.11; Fig. 2.45) plays an important role at the altitudinal forest limit, where it supports establishment and expansion of limber pine (*Pinus flexilis*) and whitebark pine (*Pinus albicaulis*). The nutcracker also car-

Photo 2.11 Clark's nutcracker (*Nucifraga columbiana*). Rocky Mountains, Colorado. F.-K. Holtmeier August 1974

ries and hoards viable seeds often far beyond the mother trees (Holtmeier 1993, 2012), thus influencing treeline dynamics. Not least, natural reforestation of burned areas would take much longer without the nutcracker seed caching (Tomback 1986, 1989, 1994b; Rebertus et al. 1991). Adults and young nutcrackers were observed to use about 55 % of cached seeds (Tomback 1982, 2001). If each nutcracker out of 25 individuals hoarded about 32,000 pine seeds, approximately 350,000 unused seeds would be left in 95,000 caches. Even if only 1 % of these seeds germinated, regeneration would already be sufficient for maintenance of the pine stands.

On Kamchatka, zoochoric Siberian dwarfed pines (*Pinus pumila*) that developed from seed caches of the Siberian nutcracker (*Nucifraga c.macrorhynchos*) are colonizing fresh lava and volcanic ash blankets (Grishin and Del Moral 1996; Grishin et al. 1996).

Another good example of the complex role of seed hoarding in the function and development of forest ecosystems is the mutualistic relationship of the Swiss stone pine (*Pinus cembra*) and the thick-billed nutcracker (*Nucifraga c.caryoctatctes*) (Photo 2.12; Fig. 2.46) in the subalpine Swiss stone pine - larch forests of the Central Alps. Swiss stone pine and larch form the forest limit and treeline in this region. The nutcracker's role in these forests has been studied in detail in the Engadine (Switzerland) (Campell 1950; Holtmeier 1965, 1967a, b, 1974, 1993, 2009; Mattes 1978, 1982, 1985).

Since the Bronze Age, at the latest, and particularly during the Middle Ages, man has increasingly influenced the high-elevation forests through the grazing of cattle and clear-cutting. High-elevation forests were cleared for pastoral use and to increase the alpine grazing area. In addition, extended forest stands were destroyed by ore-mining, salt works, and charcoal production, especially during the Middle Ages. On all fairly accessible terrain, forest was completely replaced by alpine pastures. The extent of the treeline depression varied regionally and locally. A depression

Photo 2.12 Thick-billed (European) nutcracker (*Nucifraga c. caryocatactes*). Roseg Valley, Upper Engadine. F.-K. Holtmeier, 8 June 2001

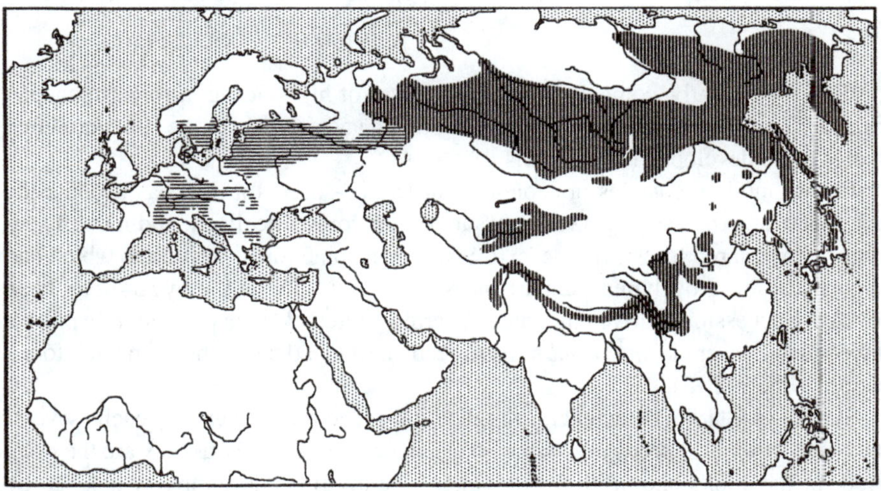

Fig. 2.46 Distribution of the nutcracker in Eurasia (horizontal hatching=*Nucifraga c. caryocatactes*, vertical hatching=*Nucifraga c. macrorhynchos* and other subspecies). Modified from Mattes in Glutz von Blotzheim and Bauer (1993)

of the treeline by 150–300 m compared to its position during the postglacial thermal optimum can be accepted as an average value for the Central Alps (Holtmeier 1974, 1986a, 1994; Burga 1988; Tinner et al. 1996; Carcalliet et al. 1998; Burga and Perret 2001; Kaltenrieder et al. 2005). The remaining forest stands were grazed by

Photo 2.13 Succession in a former larch (*Larix decidua*) pasture forest after grazing ceased. Swiss stone pine (*Pinus cembra*) has already formed a second tree layer under the light larch canopies. North-facing slope of Muottas da Celerina, Upper Engadine. F.-K. Holtmeier, October 1967

cattle. Consequently, composition and structures of high-elevation forests changed considerably. On slopes not too steep for grazing and exposed to the sun, pure larch stands often developed (cf. Sect. 2.4.2).

Due to modern changes in economic structure – mass tourism has become the main base of existence – grazing pressure on the forests declined, and natural succession from larch forests (pioneer) to Swiss stone pine forests (climax), which had been interrupted by human disturbances for many hundreds of years, has been revived. Succession is driven by scatter-hoarding nutcrackers in particular. In places, a second tree layer of *Pinus cembra* has already developed under the open larch forest canopy (Photo 2.13).

Similar to Clark's nutcracker, a single thick-billed nutcracker may cache several tens of thousands of seeds per season (e.g., Mattes 1978, 1982), as they are the main food for the adults and nestlings in winter. The nestlings have already hatched by April. Early brooding guarantees that the young birds will have enough time to learn the special techniques of food-gathering. Thus, they may have already become independent by July when insects are abundant. In autumn, the young nutcrackers establish their own first seed caches (Mattes 1982).

The adults hoard seeds throughout the year, provided seeds are available. When the present author fed nutcrackers with hazel nuts, they cached them immediately in the litter and moss layer nearby. Maybe these were interim caches, from which the inlying seeds were later stored in other places (see also Mattes 1982). As to Swiss

Table 2.5 Seed dispersal by nutcrackers and wind. Modified from Holtmeier (2009)

Seed dispersal by nutcrackers	Wind-mediated seed dispersal
Good quality of seeds harvested by nutcrackers	Variable seed quality, small seeds, low energy content
Distance of seed dispersal dependent on:	Distance of seed dispersal dependent on:
The nutcrackers' radius of action	Height of the seed trees
Attractiveness of sites (terrain structures, plant cover) for seed caching nutcrackers	Weight of the seeds
	Size of seed wings
	Wind velocity
	Weather (dry, humid)
	Terrain structures
	Plant cover
Selection of suitable sites for seed caching	Landing of seeds accidental
Below-ground seed caches invisible to seed predators	Seeds lying on the ground surface easily visible to seed predators
Location of the seed caches relatively favorable for germination and seedling growth	Conditions for germination and seedling growth highly variable
Comparatively favorable moisture and temperature conditions for the seeds cached n the litter layer and in the top soil	Seeds lying on the ground surface exposed to highly varying moisture, high temperatures, and desiccation

stone pine, seed hoarding becomes the most effective agent of dispersal when the seeds begin to ripen.

Usually, both the thick-billed and the slender-billed nutcracker cache more seeds than they need for survival. Unused caches may give rise to pine seedlings and seedling clusters, relatively favorable conditions provided (e.g., Holtmeier 1966, 1967a, b, 2009; Tomback 1978; Mattes 1978, 1982; Hiller et al. 2002). The nutcrackers put the seeds into the litter or topsoil at a depth of 2–4 cm. In general, seed caches in loose or soft soils are larger than those established in hard, compacted soil. Nutcrackers usually cache seeds one by one. However, seed caches containing several seeds are also common. Despite a high retrieval rate (Mattes 1978, 1982) and losses to members of the same species and other seed predators, in particular, redback vole and squirrels, many seeds may germinate. As seeds of Swiss stone pine, as well as those of other subalpine stone pines (Table 2.5), are rich in nutrients, seedlings may rapidly develop.

During succession, larch will gradually be replaced by Swiss stone pine because dense woolly reedgrass (*Calamagrostis villosa*), dwarf shrubs (e.g., *Rhododendron ferrugineum*, *Vaccinium* spp.) and the deep raw humus layer prevent windborne larch seeds from reaching a suitable seed bed. In addition, larch cannot successfully compete with an increasing Swiss stone pine population for light. In the end, larch may establish itself only in places where natural factors, such as avalanches, landslides, and uprooted trees, or special forest management, such as artificial removal of the field layer and raw humus layer, have exposed the mineral soil, or where competition has been reduced by selective cutting of stone pines (Campell 1944;

Photo 2.14 Swiss stone pines (*Pinus cembra*), which have emerged from seed caches of the thick-billed nutcracker (*Nucifraga caryocatactes*) on an abandoned alpine pasture at about 2,300 m. Muottas da Celerina, Upper Engadine. F.-K. Holtmeier, 3 October 1997

Auer 1947; Holtmeier 1990, 1994, 1995a; Bott 1995). On the other hand, young Swiss stone pines in the understorey (cf. Photo 2.13) may be seriously affected by the caterpillars of the larch bud moth.

Nutcrackers cache seeds of Swiss stone pine not only within the mountain forest but even far beyond the present, usually anthropogenic treeline. Thus, they can considerably influence treeline dynamics (Holtmeier 1966, 2009, 2012). When harvesting seeds, nutcrackers may cover horizontal distances of more than 15 km and differences in altitude of up to 700 m (Sutter and Amann 1953; Holtmeier 1966, 1974; Mattes 1978, 1982). Other seed consumers (e.g., woodpeckers, squirrels, voles, mice) do not usually disperse stone pine seeds beyond the forest limit.

While dissemination by wind is usually rather accidental, nutcrackers carefully select the sites for placing seed caches. Site selection for seed caches seems to be influenced by local topography (microtopography), whereas seedling distribution of larch and other anemochorous tree species is also accidental and not related to local landforms in this way. Convex landforms, such as low ridges, spurs, knolls and rocky outcrops appear to be particularly attractive for seed-hoarding nutcrackers (Photo 2.14, Fig. 2.47). The nutcrackers` preference for convex topography is particularly obvious in intensively sculptured terrain, while seed-caching is more irregular on uniform topography (Holtmeier 1993, 2009; Mellmann-Brown 2005; Photo 2.15). There, it seems to be related to surface microstructures (e.g., blocks, rotten stumps or fallen trunks), as was locally observed by the present author on

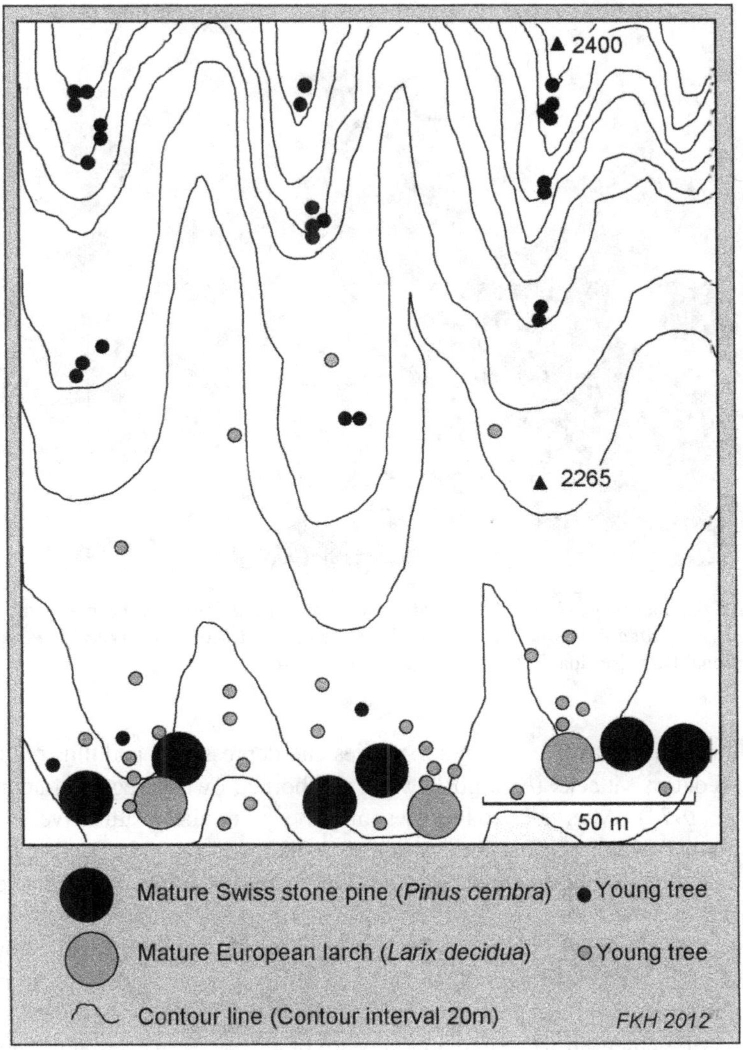

Fig. 2.47 Distribution (schematic) of Swiss stone pine (*Pinus cembra*), European larch (*Larix decidua*) and young growth of both species as related to the distance from the seed trees and to local topography on a mountain slope. Modified from Holtmeier (1993)

wind-swept uniform terrain in the Rocky Mountain treeline ecotone. There, young limberpines (*Pinus flexilis*) became established in close vicinity to bigger stones, providing some shelter from the prevailing winds.

The reasons for the specific site selection are still obscure, however. One may speculate whether convexities act as landmarks that help the nutcrackers to retrieve their seed caches, particularly as the winter snowpack may level the terrain contours (Mattes 1978, 1982). In addition, convex landforms would allow for surveying the

Photo 2.15 Limber pines (*Pinus flexilis*) which have originated from seed caches of the Clark's nutcracker *(Nucifraga columbiana)* at about 3,490 m on Bald Mountain, Wheeler Peak, Great Basin National Park, Nevada. F.-K. Holtmeier, 30 July 1994

surroundings more easily compared to gullies and depressions, and thus reduce the risk of predators' attacks (e.g., goshawk, great horned owl, peregrine falcon, fox; Holtmeier 1974). After all, convexities may be particularly attractive to seed-hoarding nutcrackers, because they would save energy looking for seeds cached at sites with shallow snowpack and early melt-out in spring. Anyway, nutcrackers are able to retrieve their seed caches even under a several decimeter deep snowpack (e.g., Swanberg 1951; Burckhart 1958; Crocq 1974; Mattes 1982). This, however, would be more advantageous within the closed forest with continuous snowpack than in the open wind-swept terrain above which the winter snowpack is rather irregular. Altogether, all these explanations are plausible. However, agglomeration of stone pines on convex sites could just be the result of a higher survival rate of seedlings and saplings. As convexities are usually covered sufficiently with snow but only for a short time, they are relatively favorable for tree establishment, whereas young growth that emerged from seed caches in snow-rich concave locations (gullies, hollows, lee-slopes of convex microtopography) have comparatively little chance of survival because of white snowblight infection (*Phacidium infestans*) or abrasion and breakage by snow slides, snow-creep and settling snow (Holtmeier 2005a, 2009, 2012, for further reading).

However, on wind-exposed topography in dry high-mountain climates, as on Beartooth Plateau (Montana, Wyoming), for example, Mellmann-Brown (2005) found high concentrations of whitebark pine seedlings at the leeward edge of tree groups and also in shallow depressions where snow usually lasts into early summer.

Obviously, wind-shelter and prolonged soil moisture caused by the leeward snowdrifts and snow accumulated in shallow depressions have facilitated seedling establishment from Clark's nutcracker seed caches. On the other hand, when the present author observed hundreds of seed hoarding nutcrackers at treeline on intensely sculptured terrain in the Swiss Alps, only a few of them were caching seeds in convexities. This supports the hypothesis of the nutcracker's preference for convex topography.

Be that as it may, stone pine seedlings and saplings that originated from seed caches on convex landforms enjoy relatively favorable conditions (cf. Table 2.5). Moreover, the germination capacity of seeds hoarded by nutcrackers is comparatively high, as the birds harvest the seeds mainly below the forest limit (Holtmeier 1974; Mattes 1978, 1982). This also holds true for the Clark's nutcracker (Tomback 1982). In addition, nutcrackers prove seed quality by so-called bill-clicking before filling their sublingual pouch (Mattes 1978, 1982). Thus, the nutrient-rich stone pine seeds allow for successful establishment and more rapid growth of seedlings (e.g., Tomback 1978; Keane et al. 1990). This is advantageous to regeneration as the growing season is short within and above the treeline ecotone.

As the open treeline terrain is often exposed to intense direct solar beam radiation, moisture conditions in the underground seed caches are by far more favorable for germination than if the seeds were lying on the surface, as larch seeds do, for example. Dissemination of larch depends on the weather condition (dry or humid), in particular on wind velocity and direction. Hence, it is very irregular, and the number of seeds drastically declines within a short distance from the mother trees (Kuoch 1965; Holtmeier 1974), whereas concentrations of stone pine seedlings and saplings can be found even far above the seed source (cf. Fig. 2.47), depending on the nutcrackers' radius of action (Holtmeier 1974, 1993, 2012; Müterthies 2002). Moreover, on previous alpine pastures, closing grass and dwarfshrub cover impedes light larch seeds from reaching a suitable seed bed. Weight of 1,000 larch seeds is about 90 times less than the weight of 1,000 stone pine seeds. Only during the first years after abandonment of the pastures may greater numbers of larch seeds land and germinate on a few open patches that were created by former cattle grazing and trampling (Holtmeier 1995a). Soon, however, Swiss stone pine, which is usually the climax tree species, will prevail and be a pioneer in natural reforestation of abandoned alpine pastures (Holtmeier 1967a, b). On the other hand, dense stone pine clusters may gradually thin out due to competition and snowblight infection. As the individual seeds hoarded in a seed cache may overly for one or more years, they do usually not germinate simultaneously. Thus, individual trees in a pine cluster are of different age. The taller and older trees causing local turbulences gradually increase snow deposition (Aulitzky 1961). As a result, the younger and smaller pines completely encased in the snowpack are often destroyed by white snowblight (Photo 2.16; Holtmeier 1967b, 1974, 1986a, b).

Not only on abandoned alpine pastures but also on the terminal and lateral moraines of retreating glaciers (e.g., Morteratsch Glacier, Upper Engadine; Photo 2.17), Swiss stone pine is often a pioneer, although larch could be expected to become established first on open mineral substrate. Apparently, however, such

Photo 2.16 Swiss stone pine cluster (*Pinus cembra*) 'planted' by the thick-billed nutcracker (*Caryocatactes caryocatactes*) on the northwest-facing slope of the Upper Engadine main valley at about 2,200 m. The taller and older trees gradually increased snow deposition. They became infected by white snowblight (*Phacidium infestans*) and lost all their needles encased in the winter snowpack. The younger pines were completely destroyed. F.-K. Holtmeier, September 1968

Photo 2.17 Swiss stone pines (*Pinus cembra*) on the 1850-lateral moraine of Morteratsch Glacier, Upper Engadine. The pines originated from seed caching activities of the thick-billed nutcracker (*Nucifraga c. caryocatactes*). F.-K. Holtmeier, 26 July 1988

Photo 2.18 Swiss stone pines (*Pinus cembra*) which emerged from seed caches of the thick-billed nutcracker (*Nucifraga c. caryocatactes*) on a terminal moraine of the Tschierva Glacier, Upper Engadine. F.-K. Holtmeier, 8 June 2001

sites are too dry, particularly close to the moraine crest. While light wind-borne larch seeds may accidentally get into a seed bed, providing sufficient moisture for germination and seedling establishment, nutcrackers hoard pine seeds between blocks or single seeds fall into clefts where fine mineral and organic material accumulates, thus improving conditions for seed germination and seedling growth. On the ice-free valley floor (basal till, in front of the glacier), however, larch is at an advantage, as larger open patches of fine mineral have accumulated between blocks, where larch seeds may land at greater numbers and germinate, profiting from higher soil moisture. Moreover, these locations are possibly less attractive for seed hoarding nutcrackers because of late-lying snowpack, unclear terrain and comparatively great distances from seed trees (Holtmeier 1994, 1995a). However, in case of a more variegated terrain in front of a glacier, concentrations of Swiss stone pines can be found mainly on convexities (hillocks, drumlins, eskers, etc.; Photo 2.18), whereas they are relatively rare on level ground and in concavities.

In the past, nutcrackers were considered to be a competitor for food, as the stone pine seeds were a favorite diet of the mountain people. In addition, nutcrackers often ripped off unripe cones, and therefore were blamed for preventing natural regeneration of stone pine and harvest of sufficient quantities of seed needed for high-altitude afforestation (e.g., Nather 1958). Seed predation by nutcrackers was even considered to be the main cause of Swiss stone pine decline (e.g., Hess 1916). Obviously, there were enough ‘reasons’ to kill nutcrackers whenever possible. Premia were fixed for dead nutcrackers. Some hunters could even make their living from killing nutcrackers (communication O. Bisaz). However, the public attitude has gradually changed. Jenny (cited in Campell 1950), for example, previously wrote: '*Swiss stone pine has stronger enemies than nutcrackers that we also must not shoot, however*'. Nowadays, the nutcracker is completely protected.

Foresters, however, occasionally complain of 'too many' nutcrackers, which cause damage to tree nurseries by digging for recently sown stone pine seeds and pulling plantules out of the ground to get the attached seed. The foresters' aversion to the nutcracker can be easily understood with regard to the absolutely essential maintenance and/or restoration of high-elevation forests up to the potential climatic tree limit. In case of poor cone production, relatively large amounts of viable seeds that are needed to produce seedlings in tree nurseries may be lost to nutcrackers (and other seed predators!). Killing nutcrackers, however, would not solve the problem, in particular because nutcrackers from the surrounding areas would rapidly invade. In any case, natural reforestation of abandoned subalpine pastures and restoration of over-mature forest stands at the present, usually anthropogenic forest limit would be considerably delayed without the nutcracker's help. Hence, the situation at treeline is somewhat different from the Rocky Mountain treeline, for example, because the distance between the closed mountain forest and the potential upper limit of tree growth in the Central Alps is relatively great due to human impact for many hundreds of years. Under such conditions, the nutcracker may be considered to be a keystone species in high-elevation Swiss-stone pine-larch forests and in the treeline ecotone.

By contrast, Kajimoto et al. (1998), who studied dwarf Siberian pine (*Pinus pumila*) in the subalpine zone of Mt. Yumori (Oou Mountains, northern Honshu), came to the result that nutcracker (*Nucifraga c. japonica*) seed hoarding does not essentially support natural pine regeneration, as the nutcrackers cache seeds mainly on open wind-exposed terrain near the seed trees. Seedlings may become established only in case of sufficient summer precipitation.

2.5.3.2 Other Birds

Although endozoochoric dispersal of tree seeds is less effective when compared with seed caching by nutcrackers, seeds defecated by fruit-eating (frugivorous) birds may contribute to the distribution of certain tree species provided that the seeds contained in the animals' droppings have not lost their germination capacity on the way through the digestive tract. This type of seed dispersal may support distribution of plants, such as black cherry (*Prunus serotina*), mountain ash (*Sorbus* spp.) and juniper (*Juniperus* spp.). Black cherry which is native to Eastern North America, was introduced to Europe as an ornamental tree in parks and gardens during the seventeenth century. Since the 1920s it has been used for stabilization of inland dunes, afforestation of heath land, establishment of wind breaks and improvement of soil fertility (e. g., Kowarik 2003). Within closed forest, black cherry reproduces vigorously from root suckers and stump sprouts, whereas seedling density decreases exponentially with increasing distance from the seed source. In Central Europe, about 60 bird species consume black cherry fruits (Turcek 1967).

The spread of mountain ash up to and even beyond the existing tree limit in the Alps, the Scandinavian mountains, Scotland or the dry Himalayas can presumably be ascribed to endozoochoric seed dispersal by fruit-eating birds such as fieldfares (*Turdus pilaris*), ring ouzels (*Turdus torquatus*) (e.g., Mattes et al. 2005), and white-backed thrushes (*Turdus kessleri*, Himalayas; Mauersberger 1995). The same holds true for juniper trees growing on rocky sites at high elevation in the dry regions of the Himalayas and in southern Tibet (Miehe and Miehe 2000). In addition to thrushes, jackdaws (*Corvus monedula*) and choughs (*Pyrrhocorax* spp.) are dispersers of ripe juniper berries (Schickhoff 1993, 2005). However, the seeds are often destroyed or lose their germination capacity when passing through the digestive track. Defecated viable seeds that have reached a suitable seedbed may germinate.

In some species, the germination capacity of undamaged seeds increase by passing through the gut. In bilberry (*Vaccinium myrtillus*), for example, which is dispersed by thrushes (*Turdus* spp.), this 'pretreatment' has turned out to be advantageous as the germination capacity of intact seeds descreases during the fruiting season. Moreover, the germination percentage of excreted seeds was found to be higher than or similar to those of not ingested intact seeds (Honkavaara et al. 2007).

The behavior of the seed dispersers plays an important role with regard to germination success of endozoochorically distributed plant seeds. When using rock outcrops for feeding on berries, the birds also often defecate. Some droppings seeds fall into rock crevices. Shelter from excess direct solar radiation (reduced evaporation and photooxidative stress) and fine mineral and organic matter that accumulates in such places increase soil moisture and nutrients. Thus, germination of still viable seeds and seedling establishment may be facilitated. On pastures in New England, for example, dwarfed juniper (*Juniperus communis* var. *depressa*) can often be found close to big stones. American robin (*Turdus migratorius*), overwintering in small groups in this area, use these stones for a short rest during searches for food. Juniper berries are their main diet at this time of the year. The robins also defecate on the stones. The droppings are washed down and accumulate in cracks and clefts that were caused by freeze-thaw action between the stones and the fine material around them. In such places, seeds are relatively safe from predators (mainly mice) and grazing cattle (trampling, grazing). Moreover, permanent moist conditions (run off from the stone surface, shading, reduced evaporation) support stratification (Livingston 1972).

2.6 Influences of Animals on Soil, Land Surface and Other Site Characteristics

The effects of animals on soils and microtopography have already partly been touched upon in previous sections. Yet, this aspect especially needs to be considered in more detail, as soil is influenced by animals in manifold ways and numerous animal groups are involved. Some are living permanently (e.g., many invertebrates) or during certain development stages in the soil (e.g., some higher insects). Others are living in the soil and on the surface as well (e.g., ants, termites, moles, voles, aardvarks) or exclusively above ground (e.g., ungulates). The abundance of endogenic species even in a small area corresponds to the species richness of a coral reef (Wallwork 1970) – and coral reefs are considered to be the ecosystems richest in species. Very likely, more species are living in the soil than on the ground. Their biomass and influence on soil probably exceed the biomass and impact of the animals living above-ground (Hole 1981).

Apart from selected examples integrated into the general context, a presentation of the manifold, complex and only partly known effects of the real soil fauna (macro-, meso-, micro-fauna) on soil cannot be given within the limitations of this book. Therefore, the reader may refer to the soil-zoological and soil-ecological special literature (e.g., Trolldenier 1971; Edwards and Lofty 1977; Ghilarov 1978; Hole 1981; Topp 1981; Satchell 1983; Gisi et al. 1997; Schrader 1999). Only a few examples will be considered later in this book, among them mound-building termites, which may conspicuously influence landscape physiognomy (Sect. 4.2.1).

Animals influence soil directly by bioturbation, burrowing, digging for roots, bulbs or prey animals, trampling (compaction, erosion) and, not least, by excreta. In addition, animals have indirect effects on soil resulting from their influence on composition and structure of the plant cover, and thereby on microclimates (temperature, humidity), moisture, litter quality and supply and mineralization.

2.6.1 *Effects of Burrowing, Digging and Trampling*

Trampling, burrowing and digging (e.g., for roots or prey animals) often initiate or enforce soil erosion. This is probably the most conspicuous and lasting animal effect on the landscape surface. Bioturbation mainly influences the pedosphere. In addition, it may modify local surface structures. Almost half of the present 800 mammalian genera are involved in bioturbation (Mitchell 1988). Some of them are living permanently or temporarily in the soil (e.g., moles, mulls, pocket gophers, kangaroo rats), while others dig burrows for resting, refuge and breeding (e.g., rabbits, marmots, ground squirrels, prairie dogs, batchers, foxes, warthogs, aardvarks) and/or also for overwintering (e.g., ground squirrels, marmots). Moreover, ungulate herds and even birds may be involved in bioturbation. In southwesten Australia, the males of the superb lyre bird (*Menura novaehollandiae*; Robinson and Frith 1981; Mitchell 1988), for example, create large open display platforms of bare soil. The Australian brush turkey (*Alectura lathami*) living in the tropical and subtropical forests of eastern and northeastern Australia builds large nests of leaves, thin twigs and earth on the ground (Seymour 1992). Other birds, such as Magellanic penguins (*Spheniscus magellanicus*), breed in self-made burrows. Burrowing wedge-tailed shearwaters (*Puffinus pacificus*) on Rottnest Island (Sect. 2.1.2) have been estimated to displace about 210 t soil ha^{-1} (Bancroft et al. 2004). This exceeds the amount of soil excavated and spread on the surface by fossorial mammals.

2.6.1.1 Bioturbation (Soil Mixing, Aeration, Translocation of Soil Material)

While burrows of badgers or foxes are usually not very conspicuous, burrowing rodents, such as prairie dogs, ground squirrels or pocket gophers often have a marked influence on the physiognomy and ecological conditions of their habitats by excavating large quantities of soil and spreading it on the ground surface. The rodents may move many tons of soil per hectare. However, such figures are usually based on data measured or estimated over small areas. They are often hard to compare because different methods were used for calculation, and it is often left open as to whether they refer to wet or dry weight of soil.

Prairie dogs and pocket gophers are probably the most active burrowing rodents in North America. In contrast to pocket gophers, prairie dogs, however, are social animals living in colonies ('towns') usually inhabited by several thousand individuals. In general, they prefer gently sloping terrain (<7°) with deep fertile soils and safe from flooding (Dahlstedt et al. 1981). The size of their subterranean burrows varies in dependence on the local conditions. Their vertical tunnels (diameter 10–15 cm) extend from 1–3 or even 5 m depth (Sheets et al. 1971). There, side tunnels branch off from the vertical tunnels, with the nests in broadenings. The nests are padded with hay. The excavated soil is accumulated on the ground surface, forming a mound similar to a truncated cone (Photo 2.19). The total tunnel system of a typical prairie dog burrow system measures about 15 m in length, and the weight of the excavated soil amounts to 200–250 kg (Whicker and Detling 1988a, 1988b). Thus, in a colony with 100 burrows per hectare (in literature densities of 50–300 burrows are reported) more than 20 t ha^{-1} of excavated loose material might

Photo 2.19 Entrance of a black-tailed prairie dog (*Cynomis ludovicianus*) burrow in the short-grass prairie, Colorado. F.-K. Holtmeier 26 July 1998

be accumulated on the ground. This would correspond to the amount of soil brought to the surface by voles and moles in an oak-beech forest in the Luxembourg Ardennes (Imeson 1976). Just as active are pocket gophers (Geomyidae) and Arctic ground squirrels (*Citellus undulates*, syn. *Spermophilus parryii*).

However, such extrapolations give only a rough idea of the possible magnitude of soil excavated by burrowing rodents and should not be taken for a mean value, as the local conditions such as, substrate, plant cover, population density and fluctuations may vary considerably. In the Wasatch Range (3,000 m, Utah), for example, Ellison (1946) came to the conclusion that 10–40 pocket gophers per hectare excavate about 11–14.5 t ha^{-1}, whereas in the alpine tundra of the Colorado Front Range 4–6 t ha^{-1} may be expected at a density of 11 gophers ha^{-1} (Burns 1979, 1980). Occasionally, however, even 28 t ha^{-1} were estimated (Thorn 1982). Schütz (1998) measured the soil material spread by pocket gophers (*Thomomys talpoides*) on the ground surface of snow glades in ribbon forests on the Colorado Front Range. This material, converted into excavated material per hectare, would amount to 13–30 t. On grazing areas in the Wasatch Mountains, with a pocket gopher density of about 70 individuals per hectare, an amount of excavated soil of 94 t ha^{-1} $year^{-1}$ is not uncommon (Richens 1966). By contrast, in the semi-deserts of the Caspian Lowlands, little sousliks (little ground squirrel, *Citellus pygmaeus*) translocate only about 1.5 t ha^{-1} to the surface (Abaturov 1972). Data on souslik density, however, were not provided. Moles (*Talpa europaea*) excavate between 8 t ha^{-1} in conifer forests and 35 t ha^{-1} on meadows, depending on the local conditions (usage, earthworm density, larvae, bulk density, moisture conditions) (Abaturov 1972; Goszynska and Goszynski 1977). On wooded slopes in Belgium, rabbits at high population density may excavate almost 72 t of soil material per hectare, whereas in a case of low rabbit density, only half a ton is brought to the surface (Voslamber and Veen 1985). Imeson (1976) ascribes the bulk of loose material accumulating on the valley bottoms in the Luxembourg Ardennes to the activities of small burrowing mammals (moles and voles), earthworms and other invertebrates.

The effects of pocket gophers on soils and microgemorphology are no less spectacular than those of prairie dogs, even though gophers are solitary animals. Grinell (1923) was so deeply impressed by their effects on the soils on the semi-arid slope of Sierra Nevada (California) that he compared them in a way to the effects of earthworms that Darwin (1881) had described almost dramatically: '*It may be doubted whether there are many other animals which have played so important a part in the history of the world, as have these lowly organized creatures*'.

Pocket gophers belong taxonomically to three genera (*Thomomys*, *Geomys*, *Cratogeomys*), with 16 species in total (subspecies included). Eight of them belong to the genus *Thomomys*, seven to *Geomys*. The genus *Cratogeomys* is represented by only one species. The distribution area of pocket gophers roughly coincides with the grasslands and shrub steppes of North America (Mielke 1977; Fig. 2.48). Gophers at high densities may severely damage forest plantations by gnawing roots and fresh shoots (cf. Photo 3.17), by girdling the stems of seedlings, and, not least, by intense burrowing (Dingle 1956; Ronco 1967; Hooven 1971; Barnes 1973, 1974). On mountain land (about 2,7000 m) of Arizona, for example, numerous

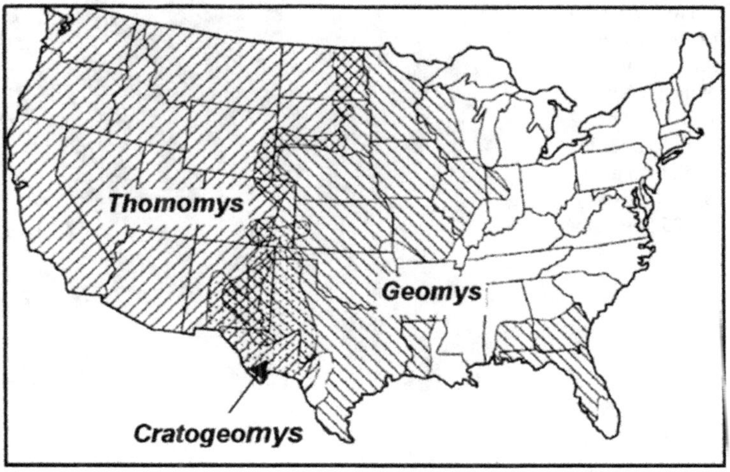

Fig. 2.48 Distribution of the three genera of pocket gophers in the United States. Modified after Teipner et al. (1983)

aspen clones (*Populus tremuloides*) associated with rocky outcrops are surrounded by mountain meadows. These are characterized by deep soils that provide suitable habitat conditions for Botta's pocket gophers (*Thomomys bottae*). The gophers cause serious below-ground injury to aspen root suckers, and thus prevent expansion of aspen into the meadows. The gophers have therefore been considered a 'keystone species' in the mountain meadow ecosystem (Cantor and Whitham 1989).

Moreover, pocket gophers often cause losses of agricultural crops by feeding on above-ground and below-ground parts of useful plants. Not least, their mounds may cover the agricultural plants and also impair machine harvest. In grazing areas, gophers may compete with cattle for food, and they may also reduce cattle's food source by covering vegetation with excavated material (Turner 1973; Andersen and MacMahon 1981; Teipner et al. 1983, ample further references therein; Reichman and Smith 1985; Huntly and Inouye 1988).

In the following paragraphs, the focus is on the northern pocket gopher (*Thomomys talpoides*). Pocket gophers live almost exclusively in the soil. During the snow-free season, they leave their tunnel system with usually just the front part of their body for food uptake (Photo 2.20). In winter, they forage on the ground vegetation under the snowpack.

Each individual has its own burrow with a multiple-branched tunnel system. The tunnels measure 5–6 cm in diameter and reach a total length of several tens of meters. Most tunnels extend to a depth of 10–30 cm below the surface. Some tunnels at a depth of 40–60 cm are connected with the upper tunnels by a vertical link. The nest, padded with dried grass and herbs, and hoarded food are located at the lower level. Loose excavated material is accumulated at the tunnels' exit. Pocket gophers constantly modify their tunnel system. They excavate new tunnels and use the older usually near-surface tunnels to dump excavated mineral and organic (nest)

Photo 2.20 Northern pocket gopher (*Thomomys talpoides*) at burrow entrance in the treeline ecotone on Niwot Ridge at about 3,180 m, Colorado Front Range. The gopher is feeding on alpine avens (*Acomastylis rossii*). H.-U. Schütz, 25 September 1990

material from the lower tunnels (Wilks 1963; Thorn 1978; Teipner et al. 1983), as moles are also known to do (Johannesson-Gross 1996). The tunnel-fillings are usually more compacted than mound material (Turner 1973).

Pocket gophers also construct extended tunnel systems inside the snowpack. There, they are protected from low temperatures and can move with relative safety from their many predators (e.g., owls, red-tailed hawks, goshawks, falcons, coyote, weasels, ermine, foxes, bobcats, badgers, skunks, and others). The snow tunnels also enable the gophers to travel longer distances and to cross barriers, such as streams, paludified terrain, rocky surfaces and block debris, and to colonize new habitats (Vaughan 1963; Hansen and Reid 1973). After snowmelt, the fillings of the snow tunnels form a net of tube-like 'eskers' (winter casts; Photo 2.21) on the ground (Holtmeier 1982, 1987b, 2012; Delella-Benedict 1991; Schütz 1998, 2005). Up to 50 % of a pocket gopher habitat may be covered by 'mounds' and 'eskers' (Buckner 1977; Holtmeier 1982).

In addition, so-called plugs can be found on the ground. Plugs (diameter 5 cm - 6 cm) consist of soil material that pocket gophers push out of the tunnel entrance to forage or to remove excavated material. After having entered the burrow again, the gophers use the plugs to close the tunnel entrance. This habit may help us to find out whether a burrow is still being used. If it is used and one removes the plug, it will not take long until the gopher will close the entrance again. Otherwise, the plugs remain on the ground surface.

When pocket gophers are enjoying favorable habitat conditions and gopher density is high, more than a quarter of their habitat may become covered with excavated loose soil material within as little as a year (e.g., Hooven 1971; Turner 1973). In general, however, only 5–15 % are covered with earth (Grinell 1923; Buechner 1942; Ellison 1946). In some 'ribbon forests' on the Colorado Front Range (cf. Photo 3.16, Fig. 3.15), the present author found 14–16 % of the 'snow glades' covered with loose mound material and tunnel fillings dating from the previous winter (1986/1987), while only a few fresh mounds could be attributed to gopher activity in early summer 1987 (Holtmeier 1987b). However, excavated material may

Photo 2.21 Pocket gopher 'eskers' (snow tunnel fillings) and their remains covering subalpine meadow vegetation on Caribou Hill at about 3,180 m, Colorado Front Range. F.-K. Holtmeier, 10 July 1997

occasionally cover up to 50 % or even more of the glades (cf. Photo 2.21; see also Buckner 1977; Schütz 1998). Vegetation covered with excavated material decomposes more rapidly than litter on the surface (Bowman et al. 1993). In the long-term, carbon content in the topsoil and soil fertility are lowered (Cortinas and Seastedt 1996; Litaor et al. 1996). Some mounds and eskers erode within as little as a few days, while others may continue to exist for several years. An example of extremely long-lived terrain structures created by gophers are the so-called 'Mima mounds', which were mentioned in the introduction. They are perhaps several hundred years old. Mounds and eskers often become covered with fresh material several times (see also Turner 1973). In the long-term, the total snow glade may be influenced by the pocket gophers. No mounds can be found inside the dense tree stands of the 'ribbons' (*Picea engelmannii*, *Abies lasiocarpa*) nor at their edges, which often are formed by shrub willows (*Salix* spp.) and gooseberry currant (*Ribes montigenum*). The same holds true for waterlogged areas. Fillings of snow tunnels, however, which crossed these areas during the winter, may occasionally be found.

Tree seedlings are often affected or even killed when completely buried under excavated material. On the other hand, such open patches occasionally facilitate seedling establishment (e.g., Kallio and Lehtonen 1973; Holtmeier 1974; Butler et al. 2004). Nevertheless, other animals (e.g., American elk, *Cervus canadensis*; mule deer, *Odocoileus hemionus*), climatic injuries, an abbreviated growing season, snow fungus infection (e.g., *Herpotrichia juniperi*; *Phacidium infestans*), lack

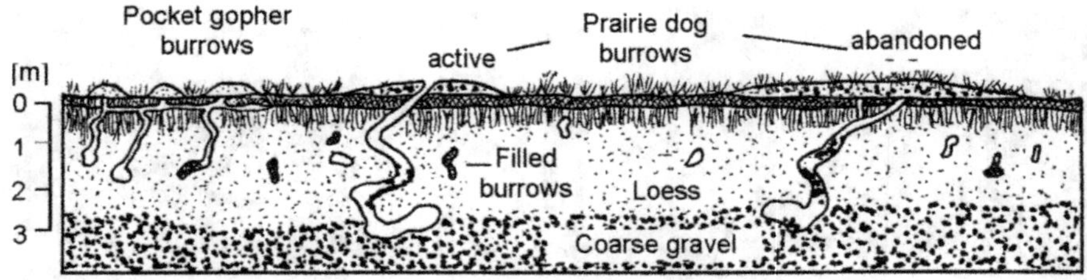

Fig. 2.49 Influence of prairie dogs and pocket gophers on translocation of soil material in the prairie, Colorado. Modified from Thorp (1949)

of moisture and downslope translocation of excavated soil by heavy rainfall and needle ice solifluction (Hall et al. 1999) may have opposite effects. Anyway, the number of open microsites caused by burrowing rodents appears to be insufficient to explain the much higher number of seedlings that have become established above the mountain forest (Butler et al. 2004). Thus, climate warming is suggested as the driving factor.

Burrowing and digging animals influence soil's physical and chemical properties. Normally, fine material and dissolved substances are translocated by water from the soil surface to greater depth, with the exception of semiarid and arid regions, where, under the influence of intense evaporation, soil water with minerals dissolved in it may ascend to the soil surface. Excavated soil material, however, also contains relatively coarse and barely soluble substances from greater depth. In this way, the animals enrich the upper soil horizons with almost unweathered primary material from the substrate (e.g., carbonates, gypsum) and reduce leaching. In the tallgrass prairie of northwestern Minnesota, for example, 'Mima-type mounds' became accumulated on impermeable clay-rich and silty ground overlying a carbonate-enriched horizon at about 60 cm depth. Under the mounds, this horizon is fragmented and the silty mound material is re-enriched in carbonates, which has been ascribed to the burrowing rodents' activities (Ross et al. 1968). In the prairie of Colorado, where chernozem and chestnut-colored soils occur, prairie dog mounds contain much sandy-gravel material, which is buried under a 180–200 cm thick loess cover, whereas the mounds of the gophers that burrow less deeply consist of silty loam (Fig. 2.49; Thorp 1949).

Bioturbation usually has a homogenizing effect on soil material. Occasionally, however, bioturbation may result in stratification (formation of different layers) (see also Hole 1981; Mitchell 1988). In places where ground squirrels or pocket gophers, for example, excavate stone-rich material and spread it on the ground surface (Photos 2.22 and 2.23), the amount of coarse material relatively increases when the finer particles are washed away or blown off (see also Cox et al. 1987). A similar process can be observed in areas with a high density of the giant earthworm (*Lumbricus badensis*), such as in the southern Black Forest (SW-Germany), where moles accumulate fine skeletal material and stones from the subsoil (Stahr 1979) on the soil surface. Due to erosion of the finer particles, coarse material also

Photo 2.22 Columbian ground squirrel (*Spermophilus columbianus*) at its burrow entrance, Glacier National Park, Montana. F.-K. Holtmeier, 1998

Photo 2.23 Soil excavated and spread by ground squirrels on a subalpine meadow. Burrow entrance in the foreground. Glacier National Park, Montana. F.-K. Holtmeier. 1998

enriches relatively. Mole action and selective erosion obviously disguise the effects of bioturbation by this large earthworm species (Lamparski 1985). Normally, the intensive bioturbation by *Lumbcricus badensis* is the main cause of the particularly thick A_h-horizon (humus layer) of the humus-rich brown soils of the southern Black Forest. On grazed plots, earthworm tubes were found down to a depth of 1½ meters (Broll et al. 1996).

By contrast, in the Mount St. Helens area, deposition of tephra on the original soil profile and subsequent coverage with the old soil by pocket gophers and ants, which had survived the eruption (18 May 1980) under a 25 cm deep ash blanket, resulted in distinct layers still recognizable even after a long time (cf. Cox et al. 1987; see also Fig. 2.55). As the old soil, which was brought to the surface after the tephra had cooled, contained original viable mycorrhizal fungi, conditions for plant growth improved (Allen et al. 1984; Andersen and MacMahon 1985; Allen and Grisafulli 1994) and a new organic horizon developed. In western Colorado, pocket gopher mounds were found that consisted of volcanic ash mixed with top soil, whereas eskers and soils devoid of gophers did not contain ash, and so were not affected by bioturbation (Hansen and Morris 1968).

In North- and Northeast-Thailand, Löffler (1996) found bioturbation by termites to be a cause of 'pseudo-stratification'. As the termites convey permanently fine material of weathering conglomeratic sandstone to the surface, coarse gravel and pebbles relatively enrich the lower section of the soil profile, with the sand and other fine soil fractions on top.

On lime stone, moles bring stone fragments to the surface, and thus partly compensate for leaching of calcium (Watt 1974). In mixed forests near Moscow, mole hills contain material from the illuvial horizon located 30–40 cm below the ground surface. This material has been translocated from the topsoil to greater depth by leaching. In the mole hills, the contents of sesquioxides, calcium, magnesium- and phosphate-oxides, as well as the content of exchangeable calcium and magnesium, were considerably higher compared to the A_h horizon of the buried soil profile, whereas the amount of humus and nitrogen enriched in the mole hills was lower than in the A_h horizon (Abaturov 1972). Similar results are reported by other studies on mole hills. Not without reason, the mole has been called a 'zoological base pump' (Graff and Makeshin 1979). Pocket gopher burrowing activities have comparable effects (Mielke 1977; Huntly and Inouye 1988). In the view of Mielke (1977), pocket gophers really control soil chemical properties in the North American prairie. In the Chihuahuan desert (New Mexico), banner-tailed kangaroo rats (*Dipodemus spectabilis*; cf. Sect. 4.2.3), when excavating burrows, break through the hardpan ('caliche', calcium carbonate) close to the surface, and thus allow creosote bush roots access to water and nutrient at greater soil depth (Eldridge et al. 2009).

On hummocky pastures created by the yellow meadow ant (*Lasius flavus*) and occasionally by the common black garden ant (*Lasius niger*) (Schreiber 1969, 1980a; Feldmann 1991; Russow and Heinrich 2001), the physical and chemical properties (e.g., humus content) of the soil of the approximately hemispherical nest hills (height 40 cm - 50 cm) differ from those of the surrounding terrain and from the A_h horizon of the underlying soil. Humus content, for example, is considerably lower in the nest hills. In nests of the yellow meadow ant, increased concentrations of available P and K have been found, while total N, Ca^{2+} and Mg^{2+} were lower than in control plots (Dostál et al. 2005). Soil was homogenized. In the upper 18 cm, soil nutrient concentrations were similar in four consecutive layers, whereas in control plots, nutrients decreased and bulk density dramatically increased (see also Lobry de Bryn and Conacher 1994). Ant mound soils showed a loose structure with particles (0.02–0.1 mm) prevailing. Bulk density was low due to chambers and tunnels (see also Woodell and King 1991).

In nests of harvester ants (*Pogonomyrex barbati*) in the dry North American southwest (Chihuahuan desert, Sonoran desert), concentrations of nitrate, ammonium, phosphorus and potassium are significantly higher than in soils not influenced by ants. The high density of micro-arthropods and protozoa is particularly obvious (Wagner et al. 1977). The ants' activities have caused a high heterogeneity of site conditions and vegetation diversity.

The development of so-called 'krotovina' (from Russian for 'mole hill') has been ascribed to the activities of moles, marmots, prairie dogs, ground squirrels, pocket gophers and other burrowing rodents, which partially fill their burrows with humic or loose mineral material. In light soils, these fillings usually appear as dark spots, whereas they are noticeable as light patches in dark soils (see also Fig. 2.49).

Mounds of the little ground squirrel (souslik, *Citellus pygmaeus*) in the semi-deserts of the northern Caspian Lowlands differ from their surroundings by enriched gypsum, carbonates and calcium- and magnesium-oxides. In these regions especially, where no dissolved substances can ascend from the deep ground water table to the soil surface, ground squirrels, bobaks (*Marmota bobak*) and several invertebrates (ants, termites) are the most important agents of mineral supply from the deeper soil horizons. While readily soluble substances are rapidly washed out afterwards, gypsum and carbonates enrich in the topsoil. The material excavated by sousliks amounts to only 1.4–1.6 t ha^{-1} $year^{-1}$. This is less than that brought to the surface by moles. In the long-term, however, large areas may be covered (Abaturov 1972). The comparatively small amount of soil translocated by sousliks can be ascribed to their foraging behavior. Sousliks graze on the ground surface, while moles search below-ground for earthworms and other invertebrates, and thus have to move comparatively more soil (Graff and Makeshin 1979).

In the deep black soils of the Russian meadow- and forest steppes, the greater mole rat (*Spalax micropthalmus*) plays an important role in bioturbation, along with other burrowing rodents. Similar to pocket gophers and ground squirrels, greater mole rats also fill old unused tunnels with excavated soil. Although most of the excavated material comes from the upper 30 cm of the humus horizon, many light patches of loam rich in calcium carbonates that come from the illuvial horizon below are distributed over the entire humus horizon. The highest calcium contents are found in the tunnel sections next to the nests. In contrast, material that was brought to the surface usually has a lower calcium content than the mounds of ground squirrels and bobaks, of which the burrow systems reach down to much greater depth. Nevertheless, greater mole rats play an important role in the calcium supply of the rooting zone (Zlotin and Khodashova 1980; therein further references). They translocate about double the amount of calcium taken up by the plants per year from the illuvial horizon to the rooting zone. Thereby and by their feces and urine, which are rich in plant-available nitrogen (ammonium- and nitrate ions), they considerably influence the turnover rate in the steppe soils. Zlotin and Khodashova (1980) estimated that, in the steppe, great mole rats inhabiting their colonies for many generations would have 'worked through' the entire topsoil within a period of 2,000–5,000 years.

In the subalpine zone of the eastern Pamir mountains (at about 4,000–4,500 m elevation), long-tailed marmots (*Marmota caudata*), in particular, may completely alter soil profiles (humus content, texture, bulk density, nutrients; see also Tadzhiyev and Odinoshiyev 1987). In all soil types, upward transport of stone- and debris-rich loose material by marmots from lower horizons results in considerable salt, chalk and gypsum enrichment in the upper soil, provided that these substances are not washed out. In some places, Solonchaks may develop. In high-mountain deserts, marmots prevent crust formation on the soil surface, which is typical of terrain not influenced by marmots. A similar effect can be observed in the lowland deserts, as in the Karakum, for example. In places where rodents have broken up the clay crust of takyr soils, shrubs may become established that would otherwise be missing (Walter and Breckle 1986).

Burrowing animals not only move fresh loose mineral material to the upper soil horizons and to the surface but also translocate organic matter to deeper levels. The organic material comes partly from the litter layer, from nests and hoarded forage. Black steppe soils, often characterized by very deep A_h horizons (often >1 m), are probably the most impressive example. Numerous invertebrate groups (earthworms, larvae, etc.), as well as soil-dwelling mammals such as squirrels, marmots, hamsters and others, are involved in extending the humus horizon to greater depth. Bioturbation by pocket gophers, kangaroo rats and gerbils (Gerbillidae), the latter living in the deserts and steps of Asia, has a comparable effect.

Bioturbation and loosening the soil also influence thermal and hygric soil conditions. Bioturbation improves aeration and drainage conditions, particularly in soils rich in clay and/or heavily compacted by grazing. At least locally, as, for example, in rodent colonies, burrowing and digging animals are hardly less important than the permanently (e.g., earthworms, termites) or temporarily soil-dwelling (terricolous) invertebrates. Therefore, bulk density in soils influenced by prairie dogs, ground squirrels, kangaroo rats, moles and mole rats, for example, differs considerably from bulk density in undisturbed soils (Ellison and Aldous 1952; Ross et al. 1968; Laycock and Richardson 1975; Graff and Makeshin 1979; Lee 1986; Moorhead et al. 1988; Mun and Whitford 1990; Heth 1991; Laundre 1993; Cortinas and Seastedt 1996).

Reduced bulk density and the resultant improvement in permeability, aeration and drainage positively influence nutrient turnover. The magnitude of these changes, however, depends on the local conditions (e.g., soil texture and content of skeletal material). This may explain contradictory findings. Soils in Utah and Colorado, for example, influenced by pocket gophers were drier at 5–15 cm depth than undisturbed soils in close vicinity. This was attributed to a higher portion of fine material in the excavated soil, which had silted up ('sealing effect') the soil surface and reduced infiltration (Julander et al. 1959). Contrary to that, on sites intensely influenced by burrowing. infiltration was higher compared to undisturbed terrain (Ellison 1946). The same holds true for steppe soils of Idaho, where meltwater infiltration was considerably higher in places with ground squirrel activities, and therefore lower bulk density than in sites unaffected by the rodents (Laundre 1993).

In the tall-grass prairie of Iowa, soil moisture was higher for many years in places where American badgers (*Taxidea taxus*) in search of food digged for 13-lined ground squirrels (*Citellus tridecimlineatus*) and other soil-dwelling animals. However, soil moisture gradually decreased again parallel to re-colonization of the open patches by vegetation, which took about 20 years (Platt 1975). A loss of the badgers and their digging activities will have lasting effects on ecosystem function and on the distribution pattern of annual plant mosaics (Platt 1975; Elddrige and Whitford 2009).

In general, the tunnel systems of fossorial animals improve aeration. Lower soil moisture around mole tunnels, for example, has been ascribed to aeration by tunnels (Skoczen et al. 1976). However, when pocket gophers, for example, plug their tunnels after returning to their burrows, air circulation will be interrupted (Turner 1973). Relatively high soil moisture was found below the tunnels, probably due to occasional run off on their bottom.

Thermal conditions inside burrows also differ from the surrounding undisturbed terrain as a result of changed physical conditions in the soil (bulk density, aeration, moisture and air humidity) and decomposition of accumulated organic matter. In mounds of the Mount Kenya mole-rat (*Tachyoryctes rex*), for example, in high-elevation valleys of Mount Kenya at about 3,800 m altitude, soil temperatures of 12 °C were measured, whereas normal soil temperatures of the mounds ranged between 5 and 7 °C. The high temperatures inside mole-rat burrows result from decomposition of excreta and already disintegrated plant material. Warmth and food provide optimal conditions for the development of beetle larvae, which were found inside burrows in large numbers (Coe and Foster 1972). Abandoned hills of mole rats are often occupied by groove-toothed rats (*Otomys orests oreste*, *Otomys tropicalis tropicalis*). It may be added that Darlington (1985), referring to her extensive studies in Kenya, holds the opinion that lenticular soil mounds in the savanna were not built by burrowing rodents ('African mole rat, *Tachyoryctes splendens*), but by termites (*Odontotermes* spp.) (see also Martin 1988). This view, however, is contradictory to the observations and hypotheses of other authors (e.g., Cox and Gakahu 1983; Midgley and Musil 1990; Cox at al. 1989; Cox and Scheffer 1991).

Mound-building by gregarious burrowing animals also considerably influences microclimates of the ground surface. The distribution of soil temperatures, in particular, is related to microtopographic structures. In the subalpine belt on Niwot Ridge (Colorado Front Range), for example, the south-facing side of unvegetated gopher mounds is more than 3 °C warmer at 10 cm depth than the north-exposed side. Moreover, the mean temperature of mounds is usually 1–2 °C higher compared to the lower vegetated terrain between them. The diurnal temperature amplitude is also greater on the south-facing mound side than on the northern exposure and on the lower vegetated sites between the mounds. Higher temperatures accelerate decomposition of the older litter layer buried under excavated material (Cortinas and Seastedt 1996). On hummocky pastures in the Swiss Jura, Schreiber (1969) observed that yellow meadow ants prefer the sunny slope of their nest hills during summer and cover it partly with fresh soil. As a result, the nest hills gradually elongate (oval shape).

However, not only microtopography created by burrowing animals but also animal-caused patchy vegetation patterns may cause locally varying soil temperatures. In the steppe, for example, diurnal fluctuations in surface temperature and soil moisture considerably increased in places where the common vole (*Microtus arvalis*) had thinned out the grass vegetation (Zlotin and Khodashova 1980). Changes of plant cover density by prairie dogs, pocket gophers and many other burrowing animals have similar effects (Grinell 1923; Brown and Heske 1990; Hawkins and Nicoletto 1992).

Compared to the effects of ground squirrels, prairie dogs and larger burrowing animals, the influences of the naked mole rats (*Heterocephalus glaber*, Bathyeridae) are less conspicuous (Jarvis 1981). Naked mole rats live in the semideserts of Kenya, Ethiopia and Somalia. Contrary to most naked moles, which usually dig individual tunnels, the East-African moles are 'teamworkers'. About 20–30 individuals usually work together. They use their tunnels to search for fleshy tubers rich in carbohydrates, roots and insects. The tunnel system, usually located close to the soil surface, may be very large. In an area of 10 ha in Kenya, a group of 87 naked mole rats had 'chiselled' a tunnel system of about 3,000 m in total length through the hardened desert soil (Brett, cited in Sherman et al. 1992). These rodents also bring excavated material to the surface, though at lower quantities compared to pocket gophers or termites, for example. In the desert environment, however, the influence of the naked moles on mineralization of the below-ground organic substances appears to be more important. In the tunnels, temperatures range between 30 and 32 °C. Relative air humidity is 90 %. In addition, similarly to termites, naked moles may consume not easily digestible plant material. Completely digested food of hard consistency is excreted in latrines. By contrast, feces consisting of only partly digested food are soft. Young animals, the only female animal in the group and a few other 'non-working' individuals rely on this pre-digested food (Niethammer 1988; Sherman et al. 1992).

Altogether, many effects of burrowing animals on the physical and chemical properties of soil generally correspond to each other. However, they are usually more or less modified by local and regional conditions (e.g., substrate, vegetation, climate, animal species, etc.). Thus, each case requires specific consideration.

On pastures and other cultivated land, burrowing rodents are often considered to be noxious animals enhancing soil erosion and destroying agricultural plants, particularly in the case of eruptive population dynamics. The mass-outbreaks of Brandt's voles in the central Mongolian steppe (Zhong et al. 1999) has already been mentioned in this book. In addition, pikas (*Ochotona curzioniae*) and plateau zokors (*Myospalax baileyi*) on the Tibetan Plateau should be mentioned (Fan et al. 1999). They are controlled by all possible means, partly by rodenticides that are also poisonous to other animal species and humans. More recently, there have been attempts to develop control measures that are less harmful to other members of the biocoenoses and human beings (Fan et al. 1999), or to limit population growth of pest animals through pasture management according to natural control mechanisms (Zhong et al. 1999).

2.6.1.2 Geomorphic Effects

The role of animals as geomorphic agents has been reviewed by Butler (1995/2008). In this section, a few examples of mammals and seabirds that may considerably influence the land surface are considered in more detail.

2.6.1.2.1 Mammals

In most cases, excavated material brought to the surface by burrowing animals is susceptible to erosion by water and wind. The magnitude of erosion, however, depends on local conditions, such as landforms, exposure, distribution of the winter snowpack, precipitation regime, plant cover, and the amount and composition (texture, humus and moisture content) of the loose material.

From studies on surface runoff and translocation of sediments on limestone slopes in the arid Negev Highland (northern Israel), it has become apparent that in the driest areas white-tailed porcupines (*Hystrix indica*) are the only geomorphologically effective animals. Searching for plant bulbs, they break up the compacted and encrusted loess soil surface. The loosened material may become translocated downhill even at low runoff intensity. In more humid areas, desert isopods (*Hemilepistus reaumuri*), the feces of which consist exclusively of mineral material, are more important than white-tailed porcupines in this respect. In the most humid sites, mole rats move twice as much loose material upwards to the ground surface than porcupines and desert isopods in the arid areas (Yair 1974; Yair and Rutin 1981).

In the Karoo (South Africa), the South-African porcupine (*Hystrix austroafricane*) and bat-eared fox (*Otocon megalotis*), which are able to burrow through hardened soil crusts, create microsites with fine loose material on the surface. Higher soil moisture facilitates germination and plant establishment in such microsites (Dean et al. 1999).

Moreover, thaw may cause rapid downslope translocation of loose material. In the Giant Mountains (Czech Republic/Poland), for example, mole hills on grazed slopes eroded under such circumstances (Jonca 1972). The same holds true for the mounds and eskers produced by pocket gophers or soil material excavated by ground squirrels, especially when the loose material accumulated in natural drainage channels. Studies on the Wasatch Plateau (Utah), however, found pocket gophers not to be the primary cause of soil erosion, although they can enforce it. There also was no evidence that pocket gopher tunnels initiate gully erosion. By contrast, loosening of the soil by gophers has been considered to reduce erosion as it increases infiltration, and thus decreases runoff (Ellison 1946). In other regions, however, meltwater flow through the tunnels was found to be an effective erosive agent. As was observed by the present author in the upper subalpine and alpine zone of the Colorado Front Range, abandoned and approximately downhill-oriented burrows were often so intensely washed out that they had collapsed. Small terraces have

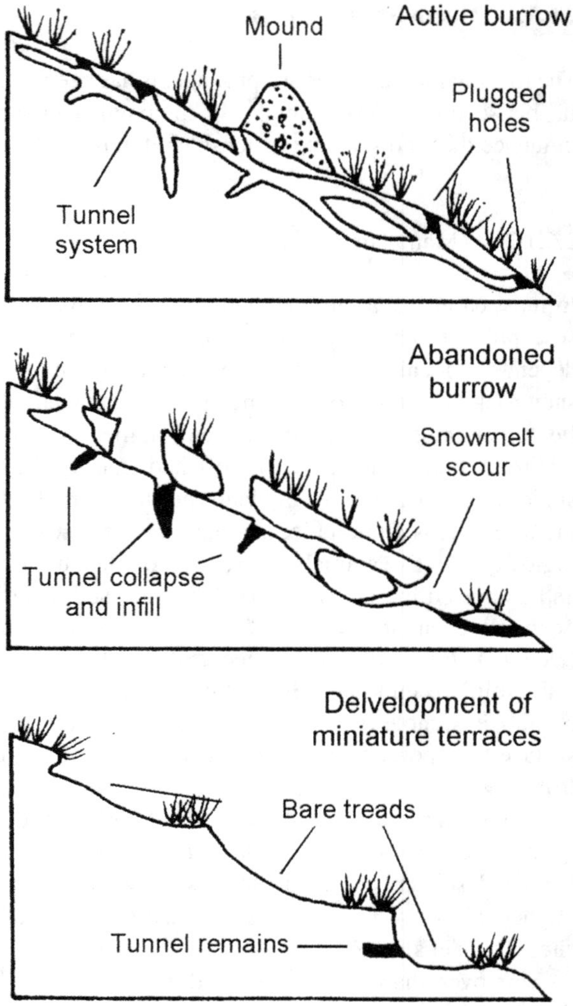

Fig. 2.50 Development of miniature terraces in the alpine zone of the Colorado Front Range under the influence of northern pocket gophers (*Thomomys talpoides*). Modified from Thorn (1978)

remained (Fig. 2.50). These are relatively common on alpine terrain with a closed plant cover (Thorn 1978, 1982). Paradoxically, the dense plant cover that provides abundant forage for the pocket gophers is the cause of enforced biogeomorphic activity. However, in other densely populated colonies of burrowing rodents, such as ground squirrels or marmots, erosion usually starts on terrain undermined by tunnel systems (see also Photo 2.24).

In high-mountains, such biogeomorphic structures are often quite similar to those created by solifluction. On the Quinghai-Xizang Plateau (Tibet), for example, large herds of both wild and domesticated yaks (*Bos mutus*, *Bos grunniens*) grazing on mountain slopes at elevations between 2,800 and 3,200 m have destroyed the closed turf layer. Subsequent needle ice formation has

Photo 2.24 Marmot (*Marmota marmota*) burrows on the south-facing slope of the Languard Valley at about 2500 m, Upper Engadine. F.-K. Holtmeier, 29 September 1998

enhanced erosion (Miehe 1994). Erosion has partly resulted in numerous landslides. Their partly overlapping arch-shaped slip scars give the appearance of 'lobes' (Hall et al. 1999).

Price (1971) presented a detailed study on arctic ground squirrels (*Citellus undulatus*) involved in microgeomorphic processes in the Ruby Range (southwestern Yukon Territory, Alaska). The burrows concentrate on south-exposed terrain characterized by solifluction lobes (Fig. 2.51). Conditions of life are more favorable in such places compared to other exposures. During summer, temperatures are higher, and during winter, relatively greater snow accumulates at the downwind edge (tongue) of solifluction lobes. The snowpack provides shelter from low temperatures and dry winds. Ground squirrels prefer the downwind side of lobes that are between 1 and 3 m high, as the winter snowpack melts earlier compared to deep snow that accumulates downhill (leeward) of the fronts of higher lobes. In such sites, vegetation profits from meltwater supply far into the summer and prospers correspondingly. Soil fertilization by ground squirrels and mixing of mineral and organic substances by bioturbation promote the development of a comparatively rich vegetation, which again provides ample forage for the ground squirrels. In addition, the root mat stabilizes the soil, and thereby the ground squirrels' burrows. Among the burrowing rodents in this area, ground squirrels are the most important soil-influencing agents. They remove the plant cover and bring loose excavated material to the ground surface where it soon erodes. When burrowing, the ground squirrels also translocate loose material from the burrow entrance downhill.

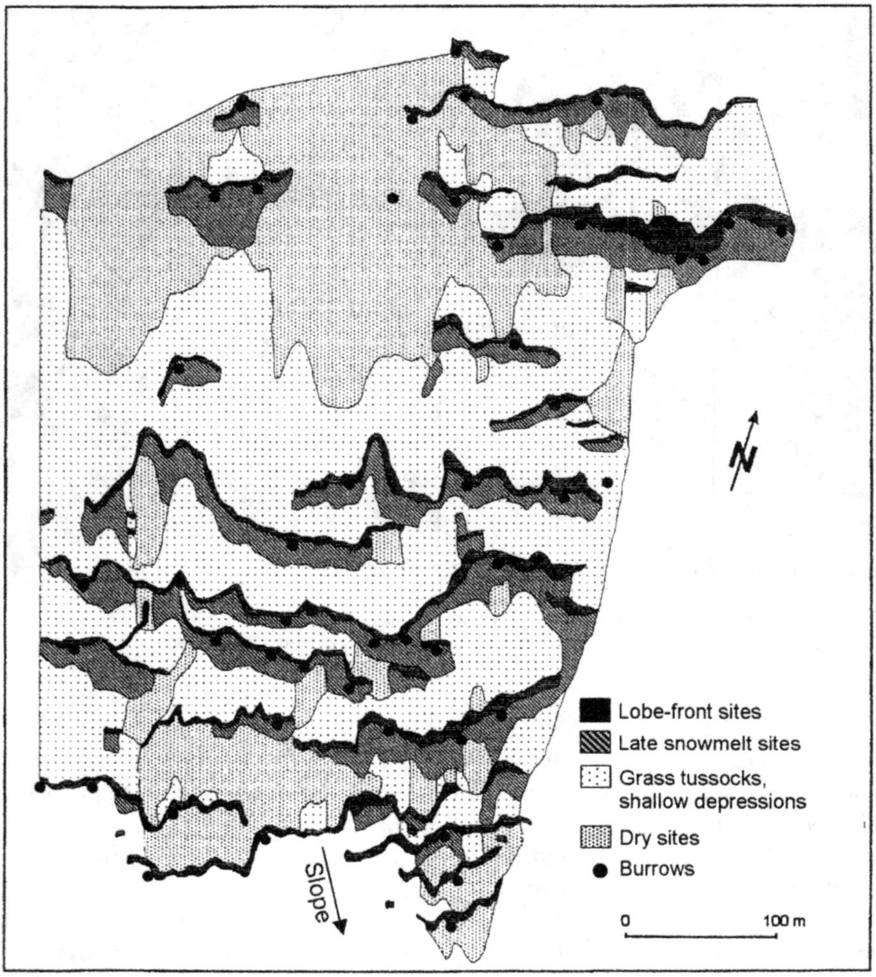

Fig. 2.51 Distribution of burrows of Arctic ground squirrels (*Citellus undulatus*) on a southeast-exposed slope in the Ruby Range, southwestern Yukon Territory, Alaska. Most burrows are located at the wind-protected steep front of solifluction lobes, where snow melts relatively late and ground squirrels enjoy favorable conditions. Modified from Price (1971)

Moreover, they undermine bigger stones that may slide downslope afterwards. Meltwater running down in the tunnels may destabilize the lobes (cf., pocket gophers). These processes, however, interfere with intensive solifluction in this area.

In the Colorado Front range, which is influenced by strong and persistent winds from western directions (Holtmeier 1978, 1996), we also found burrows and pocket gopher activities concentrated in particularly snow-rich sites. In this area, more than 90 % of the burrows are located at the downwind front of high solifluction terraces (Burns 1979), which characterize local topography to a large extent.

Table 2.6 Gross energy of army cutworm moth (*Euxoa auxiliaris*) compared with other bear food (Pitchard and Robbins 1990)

Food item	Kcal/g
Blue berries	4.47
White clover	4.83
Ground squirrels	5.28
Cutthroat trout	5.71
Pinyon pine nuts	6.48
Deer	7.32
Army cutworm moth	7.91

Not only may burrowing animals enforce soil erosion, but also animals that excavate soil when in search of food. Grizzly bears, for example, dig deep holes and shovel big blocks away when digging for bulbs, tubers, subterranean food stores of rodents, or when hunting for marmots, ground squirrels, pikas, and pocket gophers (Mattson et al. 1991a; Craighead et al. 1995). In some Rocky mountain areas, grizzly bears also dig for aggregations of ladybird beetles (*Coccinella*, *Hippodamia*) and adult army cutworm moths (*Euxoa auxiliaris* and others) on steep alpine debris slopes. These invertebrates are a fat-, and thus, very energy-rich diet (Table 2.6) (Pitchard and Robbins 1990; Chapman et al. 1955; Mattson et al. 1991b; French et al. 1994). The adult moths move during summer to high elevations to escape summer heat in the plains. In autumn, they return to the lowlands (Burton et al. 1980).

In present grizzly habitats, as, for example, in Glacier National Park (Montana), such effects of grizzly bears can be found in many places of the forest-alpine tundra ecotone and in the alpine zone. On sufficiently steep terrain, grizzly bears may be a relevant geomorphic agent, moving boulders, stones and large amounts of other excavated material downslope by several meters. Excavations by grizzly bears are usually soon filled with loose material sliding down from the slope above. On the other hand, erosion often starts in such places. Above the treeline in the Canadian Rocky Mountains (Hall et al. 1999), the effects of grizzly bears on soil erosion was found to be even more efficient compared to the influence of voles, ground squirrels and marmots. Grizzly bear diggings in the Alaskan alpine tundra turned out to have lower species richness compared to adjacent mature tundra. The overall richness of mature tundra plant communities combined with bear diggings, however, is higher than species richness of mature tundra alone (Doak and Loso 2003). However, bears' overall impact on soils and also on plant community structure (Doak and Loso 2003) is spatially very restricted. Nevertheless, grizzly bears have possibly played a major role as geomorphic agents before they almost became eradicated by the Europeans (Butler 1992).

In general, loose excavated material on sloping terrain becomes rapidly translocated downhill (Jonca 1972; Imeson 1976), and it is a matter of time and local conditions (substrate, exposure, runoff, frost action, etc.) as to when the geomorphic surface structures caused by burrowing and digging will be levelled. Even on steep slopes, where densely spaced entrances of the burrow systems of marmots encouraged soil erosion (Photo 2.24; see also Bibikov 1996), the animals' geomorphic

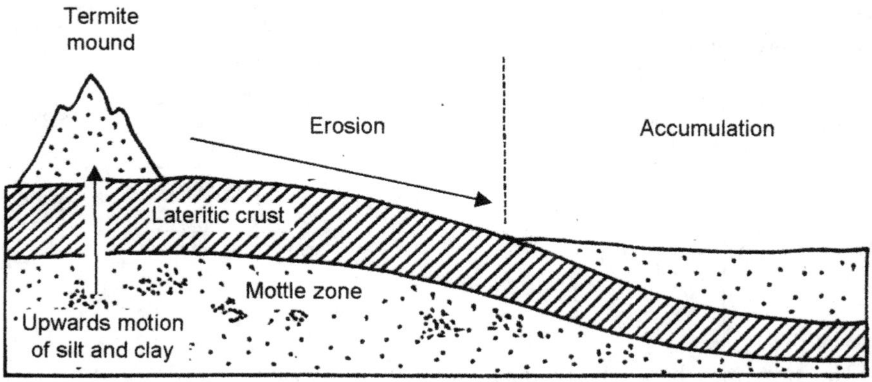

Fig. 2.52 Translocation of loose termite hill material by termites in the Central African Highland. Modified from Tardy and Roquin (1992)

impact can still be seen for a long time. Terracettes resulting from erosion of pocket gopher tunnel systems (cf. Fig. 2.50) on sloping terrain may also exist for several decades and even much longer. On level terrain, zoogenic structures consisting of relatively fine material, such as mole hills or pocket gopher mounds, remain clearly visible for several years, although they usually collapse a little. Under dry conditions, strong winds may remove the loose material. In this context, the very long-lived 'Mima mounds' may be mentioned once again.

In context with the role of animals as geomorphic agents, translocation of loose material from termite hill termites must also be considered (Sect. 4.2.1 for more details). Under certain conditions, loose material may be relocated over much larger distances and in much greater quantities than material excavated by burrowing mammals. Drummond (1888) reminds us of the Greek historian Herodotus (about 484–425 BC), who had called Egypt a 'gift of the Nile', adding that if Herodotus had the present geographical knowledge, he would have attributed the river's sediment load at least partially to the termite activities in the area between Lake Nyasa (Lake Malawi) and Lake Tanganyika (East Africa). Such relocation of termite-hill material has more recently been described by Tardy and Roquin (1992) from the Central African plateau, where termite hills are very common. The plateau is covered by lateritic crusts of varying depth. Termites carry fine material (clay, silt, sand) from the soil horizons underneath the lateritic crust to its surface, where it piles up (see also Tricart 1972). When the termite mounds erode, relocated material usually accumulates downhill upon the laterictic crust (Fig. 2.52).

In general, destruction of the plant cover is the primary cause of erosion initiated or enforced by animals. Besides burrowing and digging, trampling is often an important factor. Trampling by gregarious ungulates, in particular, when staying regularly and for a longer time in an area, is most effective. Damage increases parallel to ungulate numbers. In many regions, erosion as a result of trampling has started after domestic and/or wild ungulates were introduced. New Zealand is a classic example, especially the mountain ranges (e.g., Schweinfurth 1966; Kitching 1995; Sect. 5.1.3.1).

Fig. 2.53 Wildlife paths leading to a drying water hole (W) in Tsavo National Park. *Black dots* represent shrub, tree groves and solitary trees. Drawing by F.-K. Holtmeier after an aerial photo in Leuthold (1977)

Wild ungulates regularly concentrate at water holes and salt licks, for example (Fig. 2.53). Although the soil around the water holes is usually trampled all over, erosion is of minor importance, with the exception of river banks. Salt licks, however, may also be located on steep terrain, where intense trampling by bighorn sheep, for example, or mountain goats and ibexes may considerably enforce slope erosion (Photo 2.25; see also Butler 1993). Erosion often starts from beaten pathways that cover savannas, steppes and tundra, as well as high-mountain grassland and heath, like a net. Rhino pathways, for example, become gradually deepened by surface runoff. However, only on steep terrain and loose substrate is major washout likely (Lock 1972).

Once the plant cover has been destroyed, the litter layer is removed and the topsoil loosened up. The soil beneath, however, is often compacted by permanent trampling. As a result, the ungulate pathways in humid regions are engraved into the ground surface by up to 30 cm. Due to compactation by trampling and reduced water-holding capacity, surface runoff increases and relocates loose material. The effects are particularly spectacular on mountain slopes. In most mountain regions, over-grazing by cattle and sheep has mainly enhanced soil erosion. The influence of wild gregarious ungulates, however, should not be underestimated. Ibex reintroduced to their original habitats in the Upper Engadine (Switzerland), for example, have considerably enhanced local slope denudation on steep southwest-facing mountain slopes (for details, see Sect. 5.2.2).

Photo 2.25 Bighorn sheep (*Ovis canadensis*) at a salt lick in Rocky Mountain National Park, Colorado. Soil erosion initiated by trampling is quite common in such places. F.-K. Holtmeier, 5 March 1990

After destruction and removal of the plant cover, exposed soil may be highly susceptible to deflation. Prairie dogs, for example, remove plants around their burrows systems for a better overview (e.g., Whicker and Detling 1988a, b), thus exposing the soil to precipitation and wind. Bison, which prefer such places for dust wallowing, enforce this effect. Sexually mature males wallow more frequently during the rut (Lott 2002). In the past, when bison were roaming in huge numbers through the prairie (Sect. 3.2.2), this combined action of prairie dogs and bison was probably a major agent of denudation. Bison, as well as other wild or domesticated ungulates, also create dust wallows by themselves. In the dry mountain regions of northern Pakistan or on the Tibetan Plateau, wild and domestic yaks are very important agents in this respect (Photo 2.26). In many places, the terrain is literally dotted with dust wallows. In semideserts, deserts and dry woodland of Eurasia (e.g., Kazakhstan, Turkistan, Uzbekistan, Mongolia), jerboas (Dipodidae), such as the comb-toed jerboa (*Paradipus ctenodactylis*), destroy the plant cover and foster denudation (Walter and Breckle 1986).

In high-mountains and high latitudes, needle ice formation often increases deflation at sites where the soil is exposed (e.g., Holtmeier 1969, 1987a, 1996, 2012; Hall et al. 1999). This probably also applies to dust wallows of yak (communication F. Lehmkul; see also Miehe 1994) and to places where deer, ibex, wild sheep, mountain goats or other wild animals have fragmented or completely destroyed the plant cover (Photo 2.27; see also Photos 5.11 and 5.12). In the afro-alpine zone of Mount Kenya, for example, soils freeze to greater depth at night after hyrax

Photo 2.26 Yaks (*Bos mutus*) in dust-wallows at about 4,300 m in northern Pakistan. Frequent use of the wallows prevents plant cover and makes these places highly susceptible to wind erosion. G. B. Schaller 1974

Photo 2.27 Destruction of subalpine vegetation by American elk (*Cervus canadensis*) and mule deer (*Odocoileus heminonus*) on the wind-facing side of convex topography at about 3,450 m, Colorado Front Range. Needle ice, typically occurring in such disturbed sites, promotes wind erosion. The crescent-like wind scarps clearly reflect strong wind action. F.-K. Holtmeier, 21 August 1977

(*Procavia johnstoni mackinderi*) and groove-toothed rats (*Otomys oreste oreste*) have removed vegetation. Consequently, needle ice formation increases, and on sloping terrain >5°) needle-ice solifluction regularly occurs. After the loose material, uplifted by needle ice, has dried, it becomes relocated by wind, and blow-outs develop (Mahaney and Boyer 1986). The activities of Slogget's rat (ice rat, *Otomys slogetti*) in the mountain land of Lesotho (South Africa; Hall et al. 1999) have similar effects, often resulting in complete destruction of alpine vegetation.

Altogether, the effects of wild animals on denudation are certainly often very spectacular (e.g., Butler 1995), but usually restricted in space, like, for example, severe erosion in the Olympic Mountains (Washington). Erosion has been ascribed, at least partly, to the introduction of mountain goats (*Oreamnos americanus*; Sect. 6.1). However, trampling effects of mountain goats are by far less important than those of cattle and sheep.

2.6.1.2.2 Seabirds (and Other Birds)

Comparatively little is known about the effects of seabird colonies on soil erosion. Many of these bird species are hole breeders which dig nesting holes into the ground. The breeding colonies may reach a considerable size and have long-term effects on geomorphic surface structures. Up to three nesting holes per square meter are not exceptional (Furness 1991). Most studies and observations refer to Subantarctic islands and the Antarctic coastal areas (Müller-Schwarze 1984). The so-called 'mutton birds' (petrels and shearwaters) are one example of nest hole digging seabirds.

In the drainage area of the Kowhai river (Seward Kaikoura Range, east coast the New Zealand South Island), several colonies of the Hutton's shearwater (*Puffinus huttoni*) were found on very steep slopes covered with snow tussock grasses (*Chionocloa pallescens* and *Chionochloa flavescens*) (Evans 1973). The substrate is fine material and possibly of eolian origin. Its maximum depth is greater than elsewhere on the South Island. The densely spaced shearwater burrows are in the loess-like material within the rooting zone of the tussock grasses. About 700 nesting holes were found in a colony in an area of 1,000 m^2. Each burrow measures 1–3 m in length and has a diameter of 10–15 cm. The birds remove the roots from the burrow and pad it with leaves of the nearest tussocks. The excavated material accumulates downhill of the burrow entrance at the front of the grass tussocks and is rapidly removed by water and wind. Once the tussock grass has died off as a result of root damage and recurrent browsing by the shearwaters, the nesting holes collapse and soil erosion takes its course. Chamois, deer and goats (Sect. 5.1.3.1) introduced by the Europeans enforce soil erosion by trampling. When the slopes are riddled with holes, the shearwaters must leave their nesting colonies to establish new ones on undisturbed slope terrain. Thus, it may be safely assumed that permanent nesting colonies do not exist. Also in some areas of the Kaikuora Range, where no nesting colonies currently occur, previous shearwater effects have left their marks in the distribution pattern of vegetation (Evans 1973). However, as to soil conservation in the drainage areas, shearwaters are a minor problem compared to the oversized populations of introduced ungulates.

Stewart Island, located south of New Zealand's South Island, is surrounded by the several small islands called Muttonbird Islands. During the breeding season, many of them are visited by innumerable sooty shearwaters (*Puffinus griseus*, so-called muttonbirds) (Schweinfurth 1966; Fineran 1973). The colonies of sooty shearwaters in New Zealand and those of slender-billed shearwaters in Tasmania have been and still are an important source of human food (Brooke 2004), although they are not everyone's taste.

Obviously, these islands are particularly attractive to these birds because of the deep peat layers that have accumulated under the influence of the cool and humid maritime climate on the granitic bedrock. Muttonbirds dig their nesting holes into the relatively soft peat substrate. They apparently prefer open terrain covered by tussock grass and other low-growing vegetation. However, they also excavate nesting holes at great density in coppice forests next to the sea coast. Densely populated colonies are undermined completely, and a net of unvegetated beaten paths has developed between the nesting holes' entrances. The shearwaters usually nest in the upper 45 cm of the peat layer. This is also the main rooting zone of many trees. As a result, the trees lose their stability and often get into a leaning position or are uprooted by the prevailing strong winds from the west. On the other hand, intense burrowing improves drainage and aeration of the peat. In addition, muttonbirds manure the peaty soil. This is reflected in dense herbaceous vegegation next to the nesting holes. In the so-called muttonbird scrub (*Olearia lyallii*, *Olearia angustifolia*), the soil is completely exposed, and forest plant species that usually cover the forest floor can be found only on peat hummocks, lying tree trunks, and in other places where they are safe from trampling by the birds. In case of permanent impact, the peat surface disintegrates into fine dust-like particles.

The nesting colonies of sooty shearwaters probably enforce soil erosion, particularly on terrain covered with open vegegation. Anyway, no explicit notes can be found in the specific literature on the Muttonbird Islands. More information on soil erosion and destruction of vegegation by seabirds is given for other islands. Naarding (1981), for example, described soil erosion in the 150 Tasmanian breeding colonies of the slender-billed shearwater (*Puffinus tenuirostris*). These colonies consist of about eight million nesting holes at a density of one nest per square meter (Brooke 2004). On Curtis Island (Bass Strait), almost all areas where at least 25 cm thick peat layer built up on the granitic bedrock are completely undermined by muttonbirds (Kirkpatrick et al. 1974). Although intense bioturbation is generally emphasized, no clear evidence of the magnitude of soil erosion is given.

Almost drastic erosion has been reported from subantarctic Marion Island (Prince Edward Island, 46°S, 37°–45°E) (Hall and Williams 1981). About 3.4 million penguins are breeding there. In two colonies, macaroni penguins (*Eudyptes chrysolophus*) have completely removed peat blankets measuring up to 4 m in depth from extended areas (about 100,000 m^2). On an area of 630,000 m^2, severe erosion has been initiated by king penguins (*Aptenodytes patagonicus*). At the coast of the Kerguels Islands of the Falklands and on the opposite treeless Patagonian

Photo 2.28 Magellanic penguins (*Spheniscus magellanicus*) in front of their nesting holes in Los Pinguinos Natural Monumental Monument, Tierra dal Fuego. A. Vogel, March 2002

coast, Magellanic penguins (*Spheniscus magellanicus*) and rockhopper penguins (*Eudyptes crestatus*) breeding in large colonies, have excavated innumerable nesting holes in the loamy and sandy soils (Photo 2.28). Thus, the whole area is pot-holed (Alschner 1980) and soil erosion is enforced by the penguins. In Punta Tombo reserve (Argentina), which is the largest breeding colony of Magellanic penguins in the South American temperate zone, about 50 % of the breeding pairs (about 350,000 at present) nest in underground burrows. On Gough Island (South Atlantic, ca 40°S, 15°W), burrowing by many millions of sea birds is also a major agent in erosion and removal of the peaty soils (Wace 1961).

On New Island (western Falklands), thin-billed prions (*Pachyptila belcheri*) nest mainly in shallow valleys where they usually burrow in peaty soil (Strange 1980). Moreover, the birds nest under over-hanging rocks and beneath the thick cover of low-growing shrub (*Empetrum rubrum*). They also burrow into the fibrous base of tussock grass (*Poa flabellata*). However, no information exists on erosion initiated or enforced by burrowing. This is possibly due to the fact that the island was used for about 150 years by whalers, sealers and penguin oilers as base for their operations. Introduced cattle, sheep and pigs probably had a stronger effect on vegetation and soils than thin-billed prions. On Kidney Island (eastern Falklands), sooty shearwaters (*Puffinus griseus*), white-chinned petrel (*Procellaria aeqinoctioalis*) and Magellanic penguins (*Speniscus magellanicus*) have locally undermined the peat blanket to such an extent that it may collapse at any time (Fig. 2.54).

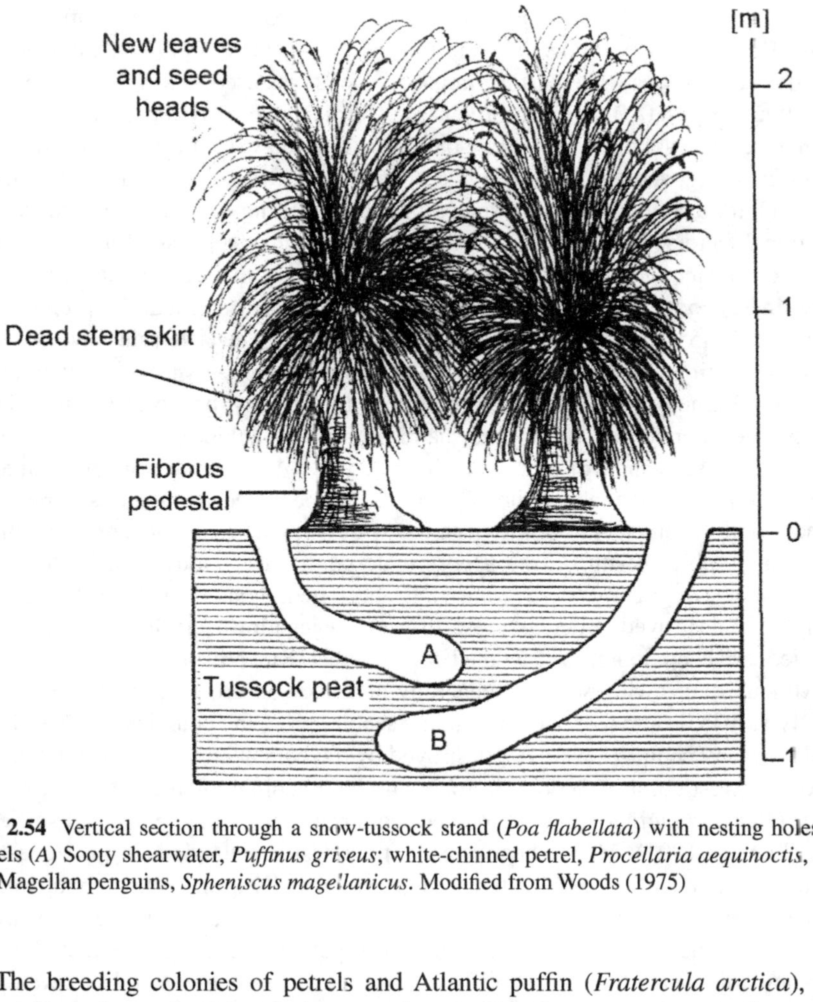

Fig. 2.54 Vertical section through a snow-tussock stand (*Poa flabellata*) with nesting holes of petrels (*A*) Sooty shearwater, *Puffinus griseus*; white-chinned petrel, *Procellaria aequinoctis*, and (*B*) Magellan penguins, *Spheniscus magellanicus*. Modified from Woods (1975)

The breeding colonies of petrels and Atlantic puffin (*Fratercula arctica*), for example, on the northern hemisphere may be considered as counterparts to the southern hemisphere seabird colonies. Atlantic puffins, for example, are breeding in huge numbers on small islands along the coast of the North Atlantic (Kostrzewa 1998). The small Grassholm Island (>9 ha), located off the coast of South Wales, is a typical example. At the end of the nineteenth century, about half a million puffins were living there. Two to three nesting holes per square meter could be found. After the puffins had destroyed the plant cover and completely undermined the terrain, the nesting holes collapsed and erosion increased. At the end of the 1920s, only a few soil pillars with grass vegetation on top gave evidence of the former terrain surface. The puffin population had decreased to about 200 individuals. It seems typical of Atlantic puffins to occupy islands with deep soils overgrown with grass vegetation, to ruin them, and then leave for another place, where they start the same process again. In other words, the birds destroy their own habitat (Lockley 1953). On islands off the west and east coast of Scotland, for example, the destruction of the fescue-turf by puffins resulted in an excessive washout. Consequently, the soil cover slipped off over the cliffs (Darling 1947). Altogether, the puffins play a major role in denudation of islands near the coast. The puffins' impact on soils was used as a welcome 'excuse' for killing about 8,000 puffins on Bleiksøy (Nordland, Norway) in 1940.

In general, nest holes concentrate on terrain covered with substrate that allow for easy burrowing, such as sufficient deep peat layers, deeply weathered soils, loess and sand, for example. However, selection of the breeding sites is also influenced by additional factors, such as sufficient drainage or terrain steepness. Rhinoceros auklets (*Cerorhynca moncerata*) on Protection Island (east end of Juan de Fuca Street between Vancouver Island and the Olympic Peninsula), for example, prefer steep slopes (35°–45°), because starting is easier than from level terrain covered with dense vegetation (Richardson 1961). This possibly also holds true for many other sea birds (Furness 1991; Brooke 2004). Steep topography, however, is particularly susceptible to erosion. On the other hand, rhinoceros auklet colonies can also be found in dense stands of European beach grass (*Ammophila* spp., presumably *Ammophila areanaria*) above the precipices. European beech grass was introduced at the end of the nineteenth century and has widely spread since. Severe denudation, however, combined with slides of loose material, has to be attributed for the most part to intensive grazing by sheep. As a result, breeding sites on steep slopes were completely destroyed, whereas those in grass vegegation near the coast were not affected, as sheep do not feed on the beach grasses (Richardson 1961).

Although burrowing seabirds generally enhance soil erosion, their effects may greatly vary in different areas. Thus, many seabird breeding colonies that have been used for many centuries were not destroyed by the burrowing activities. On the Isle of Rhum (west coast of Scotland), for example, breeding colonies of Manx shearwater (*Puffinus puffinus*) have stabilized the terrain rather than enhanced soil erosion (Furness 1991). Manx shearwaters – they nest on islands along the coast of western Europe and in the Mediterranean area – usually prefer relatively open and steep slopes for breeding, so that enforced denudation might be expected. However, as a result of the high amount of nutrients (nitrogen, phosphate) accumulated in the breeding colonies, a thick turf cover with a dense deep root system has developed that impedes erosion.

Even more complex are the interactions of Antarctic fur seals (*Arctocephalus gazella*) and the biocoenosis on Bird Island (northwest tip of South Georgia). The seals destroy the prevailing tussock grass (*Poa flabellata*) in their colonies The females rest on top of the tussocks to nurse the young seals. In the center of the tussocks, the straws become gradually destroyed without new ones sprouting. In the course of time, the tussocks turn into peat hummocks, bare on top and with laterally down-hanging straws. In the end, the tussocks completely die off, and their peaty base erodes gradually (Bonner 1985). In this way, the habitat of several other animal species is affected. The pipit (*Anthus antarcticus*) and the South-Georgian pintail (*Anas georgica*), for example, lose their breeding sites. The pintail also loses its feeding habitat. Many burrowing species of the much more numerous storm petrels suffer even more from the seals' activities. Their nesting holes in the peat layers collapse under the weight of the seals, and the declining tussock vegegation does not provide sufficient shelter to the smaller species from omnipresent skuas (*Catharacta lonnbergi*). This has a particularly important effect on the population of these petrels, as Bird Island is one of the few islands in this area still free of rats (*Rattus norvegicus*). In many other regions of South Georgia, rats have considerably decimated seabird populations. The impact of fur seals on the island ecosystems will be exacerbated because these marine mammals have drastically increased in number and are more numerous than before their almost complete eradication by the sealers. The increase of the fur seals – as well as of penguins and some other seabird species relying on animal plankton – has been ascribed to the decline of krill-feeding right whales (Laws 1984).

In the tidal estuary of the St. Lawrence River, lesser snow geese (*Chen caerulescens*), which rest there two times a year on their seasonal migrations, appear to enforce erosion of the tidal mud-flats and also of the estuary banks (Dionne 1985). After only a week of intensive grazing by migrating geese during their rest in autumn, the prevailing American bulrush (*Scirpus americanus*) are completely grazed down and cannot impede erosion of the muddy substrate. During low tide, geese dig for the bulrush roots, leaving extended pot-holed areas behind. These holes are only a few centimeters deep and measure about 25 and more centimeters in diameter. Tidal action removes the excavated material. As a result, the surface of the area, which runs dry during ebb, becomes lowered by up to 10 cm yr $^{-1}$. Increasing erosion also affects the rim of the marsh area above the normal tide height. Moreover, snow geese dig for roots, particularly in shallow holes resulting from ice action. Such holes are typical of the rim of the tidal area. Geese gradually enlarging these holes by centimeters foster erosion by tidal currents and wave action. The influence of snow geese is strongest in the lower marsh areas

2.6.2 Influences of Zoogenic Change of Soils on Vegetation

2.6.2.1 Rodents

The indirect influences of animals (bioturbation, burrowing, trampling, nutrient enrichment) on vegetation and ecosystems through change of the pedological conditions is hardly less important than direct animals' effects by consumption of phytomass, selective feeding and seed dispersal. Soil nutrients (nitrogen, potassium and phosphorus) are enriched first of all by excreta, whereas shed antlers, hairs or carcasses, for example, play a minor role and may influence nutrients and vegegation only very locally. As a result of animal-caused change of soil conditions, a variegated vegetation mosaic may develop.

In the tundra near Atkasook (Alaska; McKendrick et al. 1980), for example, the amount of plant available nutrients were many times higher (nitrogen tenfold, phosphorus sevenfold, potassium fivefold) and soil acidity lower (pH 4,2) in places where caribou waste and corpses had rotted away, compared to the surrounding terrain. Vegegation clearly reflected this nutrient enrichment. In total, there were fewer vascular plants, but the coverage by grasses and grass litter was higher than by lichens and dwarf shrubs. Enrichment of nitrogen and phosphorus by the excreta of arctic ground squirrels right next to their burrows has similar effects. Although these rodents travel distances of up to half a kilometre when searching for food, they usually defecate in the immediate vicinity of their burrows. It may take many years, perhaps even hundreds of years, until the effects of nutrient enrichment will become visible in these places.

On the dry steppes of Central Asia, plant productivity was found to be particularly high and vegetation change most obvious in sites where marmots had covered the ground with excavated loose material. High productivity and vegegation change result mainly from nitrogen enrichment and bioturbation improving soil temperature regime and water balance, and thus, mineralization. In addition, due to more favorable thermal and moisture conditions the growing season is longer than outside the area influenced by marmots. Abundant fresh green forage available for a longer period supports expansion of marmot in the dry steppe, which is characterized as such by a very short growing season (Bibikov 1996). In the Arctic tundra of Northern Asia, black-capped marmots (*Marmota camtchatica*), for example, apparently change soil properties and floristic diversity of tundra vegetation (structure, composition) around their burrows through excavated soil, latrines, dumped nesting material and dead marmots that died during hibernation (Semenov et al. 2001).

Bowman et al. (1993) found that pocket gophers increase mineralization and primary productivity in oligotrophic dry alpine grassland on Niwot Ridge (Colorado Front Range) by covering vegetation with excavated material. These alpine grassland communities are short of nitrogen and phosphorus, in particular. Patches characterized by increased productivity are called 'gopher gardens'. Alpine avens (*Acomastylis rossii*) prevails on sites heavily disturbed by gopher burrowing (see also Photo 2.20). The mineral soil exposed by intense gopher activities is prone to deflation (Holtmeier 1996). Bog sedge (*Kobresia myosuroides*), which usually dominates the dry alpine turf, is absent from severely disturbed sites. In the long-term, however, the potential productivity of dry alpine meadows slightly decreases due to rapid mineralization of soil organic matter as a result of the pocket gophers' influence (Burns 1980; Burns and Tonkin 1982; Cortinas and Seastedt 1996; Sherrod and Seastedt 2001).

On Mount St. Helens (Sect. 2.6.1.1), where pocket gophers and ants had already covered about 2 % of the tephra blanket surface with a mixture of 'old soil' material and volcanic ash (Fig. 2.55) 4 months after the eruption (1980), a vegetation became established much richer in species than in the areas without gopher influence. In addition, more seedlings occurred on the 'gopher sites' compared to sterile thepra. Species such as rusty sedge (*Carex subfusca*), for example, profited from bioturbation, as they would not have been able to grow rapidly enough from beneath the tephra cover to its encrusted and very dry surface. By contrast, fireweed (*Chamaenerion angustifolium*) has no problem in this respect, and occurs on gopher-free sites and gopher mounds as well. Both rusty sedge and fireweed are left untouched by pocket gophers. Broadleaf lupine (*Lupinus latifolius*) and western pearly everlasting (*Anaphalis margaritacea*) colonized the

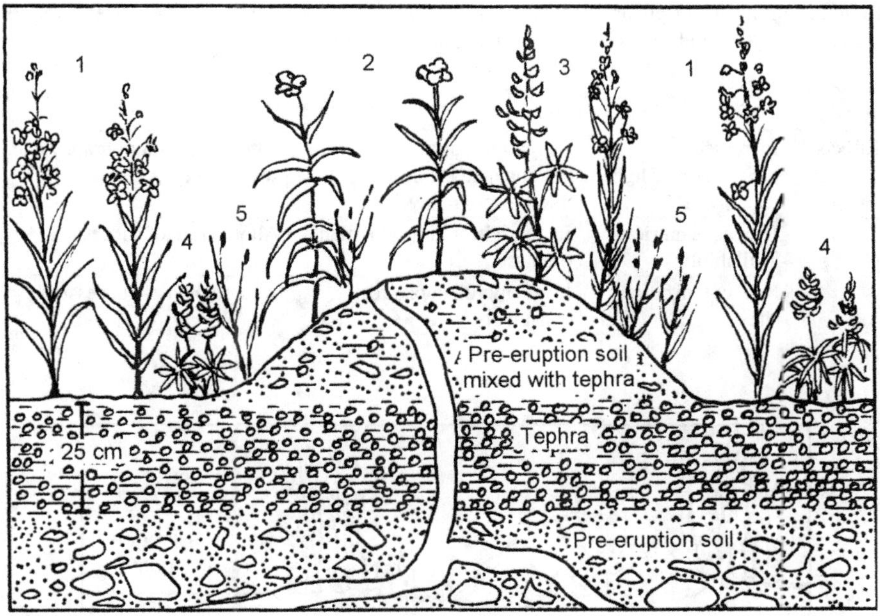

Fig. 2.55 Tephra blanket covered with pre-eruption soil excavated by pocket gophers which survived the eruption of Mount St. Helens on May 18 1980. Site conditions on the mounds differ completely from the tephra surface off the mounds. Vegetation pattern clearly reflects these differences: *1 – Epilobium angustifolium*, *2 – Anaphalis margaritacea*, *3 – Lupinus latifolius*, *4 – Lupinus lepidus*, *5 – Carex subfusca*. Drawing by F.-K. Holtmeier, after several sources

mounds, where they profited from inoculation with mycorrhizal fungi from the old soil buried under tephra. Mycorrhization of their roots was much better than on sites without mound material, and, in addition, fungi spores were more numerous in the mounds (Andersen 1982; Allen et al. 1984; Andersen and MacMahon 1985; Allen and Grisafulli 1994).

Moreover, changed soil chemical properties and soil moisture conditions played a decisive role in vegetation development, as they facilitated both germination and growth of seedlings, The mounds differ from initial soil on pumice by a higher content in organic substances (4 %, in contrast to 0.2 %) and a higher portion of fine soil. Thus, field capacity and moisture of the mound soils are higher compared to pumice. Moreover, their content in essential mineral nutrients is more than tenfold and their cation exchange capacity more than fivefold higher (Andersen and MacMahon 1985) than in the soil on volcanic ashes. Of course, ungulates (e.g., deer) may also disintegrate the tephra crust through trampling, and thus, influence site conditions. The intense mixing of volcanic ash with the original soil material by pocket gophers is much more effective, however.

Wherever pocket gophers occur and change soil's physical and chemical properties, a variegated site mosaic and increased diversity of plant communities have developed (e.g., Turner 1973). Obviously, shade-intolerant species, in particular,

Table 2.7 Influence of plains pocket gophers (*Geomys bursarius*) on their habitat over a variety of spatial and temporal scales. Modified after Huntly and Inouye (1988)

Time	1 week	1 year	50 years
Area	1 m^2	100 m^2	
Effects	Increased penetration of light	Increased spatial heterogeneity (nutrients, micro-topography)	Altered microtopography and soil fertility
	Increased nutrient availability	Increased variability in plant biomass	Altered rate of succession
	New colonization sites	More microhabitats for consumers	Altered path of succession

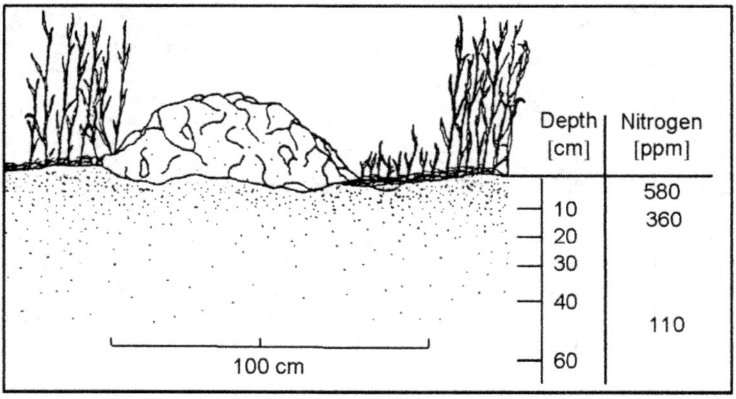

Fig. 2.56 Total nitrogen content at different depths in the prairie at Cedar Creek (Minnesota). Modified from Huntly and Inouye (1988)

profit from gopher disturbances. During long-term studies in old agricultural fields in the prairie of Cedar Creek (Minnesota; Huntly and Inouye 1988), mounds and their immediate vicinity were found to be much richer in species compared to the areas without pocket gopher influences. Changing light conditions and spatially varying soil nitrogen as a result of excavated loose material and consumption of phytomass by the burrowing rodents appeared to have been the decisive factors (Table 2.7). Depending on the spatial and temporal scale of consideration, the development can be even more differentiated. Due to the steep gradient of nitrogen from the soil surface to deeper soil horizons in undisturbed soils (Fig. 2.56; Huntly and Inouye 1988), the content of total nitrogen in the mounds (material from lower soil horizons) is lower than in sites unaffected by pocket gophers. Thus, the vertical distribution of nitrogen is completely different from 'gopher-sites' on Mount St. Helens, where the excavated 'old' soil material on top of the oligotroph volcanic ash (cf. Fig. 2.55) is relatively rich in nitrogen. In the long-term, however, the nitrogen content gradually decreases under the influence of pocket gophers.

Nevertheless, in the prairie of Cedar Creek, the amount of plant-available nitrogen on the mounds and next to them has been found to be a little higher for a while compared to sites without gopher influence, as higher temperatures and better aeration increase mineralization. Moreover, nutrient uptake by the sparse plant cover is comparatively low. But in places with denser plant cover used by pocket gophers (reduced phytomass), the amount of plant-available nitrogen is also temporarily higher compared to undisturbed sites. In the long-term, pocket gopher activities slowed down succession on the study areas at Cedar Creek, as the pioneer plants grew faster at low nitrogen supply and spread more rapidly than secondary plants. Over the 20-year study period, the differences in species composition continuously increased under the influence of pocket gophers (Inouye et al. 1987).

In quaking aspen stands (*Populus tremuloides*) in Utah, the vegetation and site mosaic continuously changed under the influence of pocket gophers during a 4-year study period (McDonough 1974). While some sites were still unvegetated, annuals prevailed at the mound margins and perennials elsewhere. The annual plants, which are relatively rare under undisturbed conditions, benefit from the new open patches permanently created by the gophers. On the other hand, perennial plants such as western coneflower (*Rudbeckia occidentalis*) and mountain brome grass (*Bromus carinatus*), which are able to spread rapidly, also profit from the open patches and may occur at high density within the aspen stands.

On mounds of the Mediterranean pine vole (*Pitymys duodecimcostratus*), which bury the existing vegegation in the subalpine mesoxerophytic grassland (*Festcuca escia* and *Nardus stricta*) of the western Pyrenees (Spain), the species normally dominating the grassland had already decreased after 1 year when buried, while the floristic composition of the vegetation did not change (Gomez-Garcia et al. 1995). As a consequence of reduced competition by the dominant species, diversity increased.

Highveld gerbils (*Tatera brantsii*) in the South-African savanna have comparable effects. While above- and below-ground phytomass considerably decreased as a result of the gerbils' burrowing activities and no differences in species numbers occurred, species diversity and plant evenness increased within the gerbil colonies. The disturbances by gerbils cause a temporally-changing mosaic of severely impacted patches near microsites where the vegetation has recovered. The year-by-year varying intensity of gerbil effects is a precondition for the continued existence of high species diversity (Fig. 2.57; Korn and Korn 1989; Korn 1991).

Under certain conditions, mole hills may also increase species richness, as they favor establishment of less competitive anemochorous species (Watt 1974). On mole hills in the southwestern Spanish Pyrenees (Urbasa-Andia-National Park), for example, not only ruderal and annual plants but also species without mycrorrhiza were more frequent compared to undisturbed grassland. Very likely, the burrowing activities of moles as well as enriched nitrogen in the hills impede mycorrhization. Consequently, the plants without obligate mycorrhiza are at an advantage at such sites compared to mycotroph species (Canals and Sebastía 2000).

Nutrient enrichment and its considerable effects on the composition of vegetation are also characteristic of the possibly zoogenic 'heuweltjies' (low hills, mostly

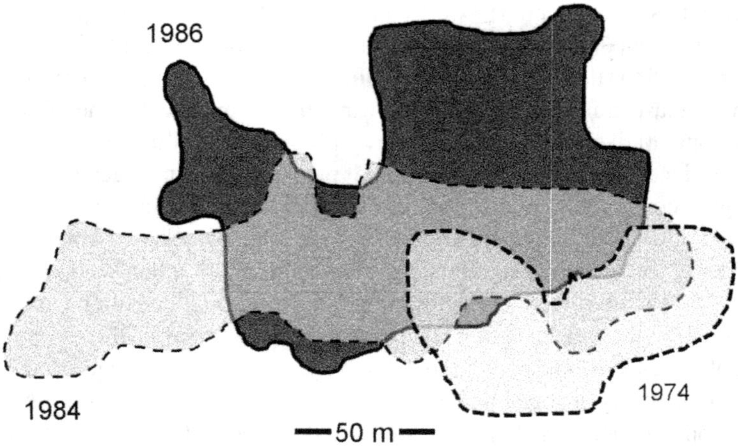

Fig. 2.57 Varying intensity of Highveld gerbil (*Tatera brantsii*) influence on the African savanna during three different years. Modified from Korn (1991)

circular in shape, up to 1 m high and 10–20 m in diameter) in South Africa. Concentrations of calcium, nitrogen and other macro- and micronutrients are higher in the 'heuweltjies', as compared to the soils around them. The same holds true for clay and silt contents (Midgley and Musil 1990). The higher amounts of nitrate result from fast mineralization of the organic substances accumulated by burrowing rodents, ants and/or termites. In the 'heuweltjies', improved aeration and infiltration accelerate mineralization. As a result, plant cover is about 50 % higher than in the off-hill area. Short-lived deciduous plants, in particular, which need much energy to renew foliage regularly, benefit from the relative high amounts of plant available nitrogen and compete successfully with evergreen plants in such places thanks to the effects of burrowing animals. Deciduous species occur on the hills at almost the same frequency and density as the evergreen plants, whereas the latter prevail on undisturbed ground between the hills. As long as burrowing animals are active, rapid turnover rate results in continued nutrient enrichment. Without bioturbation, the above-ground organic matter will no longer be incorporated into the soil and become removed by wind and runoff.

In the Great Karoo (South Africa), 'heuweltjies' act as refuges for certain plant species, such as Scholtzbos (*Pteronia pallens*, Asteraceae), endemic to the Northern and Western Cape Province. The regional mammalian herbivores (Steenbok, *Raphicerus campestris*; common duiker, *Sylvicapra grimmea*; greater Kudu, *Tragelaphus strepsiceros*; Cape hare, *Lepus capensis*) do not feed on *Pteronia pallens*, which spreads from the hills into the former range land. After abandonment by the burrowing animals and subsequent decay of the 'heuweltjies', other plants, such as *Pteronia empetrifolia*, for example, will prevail again. Consequently, the influence of the hills on the surrounding plant communities will gradually decrease (Yeaton and Elser 1990).

Mounds (up to 6 m in diameter) built by Mt. Kenya mole rats (*Tachyoryctes rex*) in the valleys of Mount Kenya cover deep black humus-rich soils, which are well-drained and completely free of stones (Coe and Foster 1972). The mounds are already visible from a distant view because of the dark green color of the shrub (*Alchemilla argyrophylla*) that occupies these sites.

Although burrowing animals often increase local diversity, their effects vary in dependence of the animal species involved. On fallow land in central Germany, for example, continuous burrowing by voles near the soil surface obviously promotes tall-growing zoochoric and anemochoric plants. Common voles (*Microtus arvalis*) also bring diaspores from lower soil horizons (e.g., abandoned agricultural fields, set-aside land) to the surface, and thus contribute to a greater species richness. Creation of open patches, where short-lived species may exist for a while and produce seeds, is more important in this respect than relocation of diaspores (Ryser and Gigon 1985).

Changes caused by burrowing and digging animals do not necessarily increase species richness and diversity, however. In the large colonies (often > several hectares) of tuco-tucos (*Ctenomys talarum*; Ctenomyidae, about 60 species) in the Pampa of southern South America, for example, plant species diversity and/or plant species richness has not changed at sites disturbed by these rodents (bioturbation, excreta, excavated soil). However, although tuco-tucos (like other burrowing rodents) enrich nutrients through their feces and urine (nitrogen, phosphorus, sodium, potassium, magnesia), they modify plant species composition next to their burrows only, where the prevailing herbs and grasses become more abundant compared to undisturbed sites (Malizia et al. 2000). In the central Andes, tuco-tuco effects (different species of *Ctenomys*) on vegetation range from partial local plant damage to complete destruction and disintegration of the plant cover over extensive areas (Werner 1977).

Also, in colonies of coruros (*Spalacopus cyanus*, Octodontidae) in arid coastal Chile and in the Chilean Longitudinal Valley, no significant change in species richness and species diversity was found, although coruro burrowing activities remarkably influence vegegation (Contreras and Gutiérrez 1991). Coruros are social animals living in colonies of 15–30 animals (4–32; Begall 1999; Begall and Gallardo 2000). In search of bulbs and tubers, they construct extensive tunnels, up to 600 m in length (6 cm diameter). The tunnels are located at about 10–15 cm depth below the surface (Reig 1970; Begall and Gallardo 2000). Biomass in tunnelled areas is about 60 % higher compared to undisturbed areas (Contreras and Gutiérrez 1991). Higher biomass is related to bioturbation and presumably also to nutrient enrichment due to mineralization of unused hoarded food. As the food chambers and nests are located mainly between the roots of shrubs, they are not recognizable aboveground. In the southern coruro population studied by Bergall and Gallardo (2000), geophytes (*Dioscorea longipes*) were frequent on older mounds. The coruros probably brought the bulbs to the surface together with excavated material when digging their tunnel systems (communication S. Begall, 22 May 2001).

Modification of vegetation plant cover by burrowing animals is usually most spectacular in open plant formations, such as grassland, semi-deserts and deserts.

In comparatively large colonies, changing impact intensity (fluctuating animal populations, food consumption, subsequent recovery of vegetation) may cause a plant mosaic that is subject to 'rotation'. This may be cyclic, as is the case, for example, on terrain influenced by gerbils (*Tantera brantsii*; cf. Fig. 3.16) or Brandt's vole (*Microtus brandti*). These examples go well with the mosaic-cycle theory (Remmert 1992).

Occasionally, the impact of burrowing animals on soils is still reflected in the distribution pattern of vegetation long after the animals disappear, as can be observed, for example, in abandoned prairie dog colonies in Oklahoma (Wichita Mountains Wildlife Refuge). In the center of the colonies, a concentric arrangement of plant communities has developed, despite continued grazing by cattle, bison, pronghorns, white-tailed deer and red deer (Osborn and Alland 1949). Although the prairie has not been grazed by bison for more than 100 years, their former dust wallows are mirrored in vegetation different from the surrounding grassland (Polley and Collins 1984; Gibson 1989).

2.6.2.2 Wild Boars

Wild boars are a natural disturbance factor influencing forest dynamics (Peterken 1993). They may impair or support tree establishment. In years of poor seed production in oak, wild boars (usually together with other wild ungulates) may almost deplete acorns lying on the ground, and thus impede natural regeneration of oak. Moreover, when 'ploughing up' the soil in search of tubers, bulbs, larvae, voles and even rabbits, they directly destroy seedlings, and thus adversely affect forest regeneration. In addition, due to destruction of the ground vegetation and removal of the litter layer, conditions for life of species that rely on litter get worse. In the deciduous forests of the Smoky Mountains in the southern Appalachians, for example, southern red-back vole (*Clethrionomys gapperi*) and northern shot-tailed shrew (*Blarina brevicauda*) almost became extinct due to wild boars' impact. On the other hand, boars may stimulate sprouting in some tree species, such as beech (*Fagus sylvatica*; Bjerke 1957; *Fagus grandifolia*, Singer et al. 1984), for example, by exposing the roots.

Wild boars churning and exposing the soil favor small-growing therophytes and weeds, in particular. In addition, wild boars can activate seed banks at greater depth that date from previous agricultural use (Milton et al. 1997) or former disturbances (e.g., fires, wind throw). These are reflected by typical indicators such as blackberry (*Rubus* spp.), raspberry (*Rubus idaeus*) and stinging nettle (*Urtica* spp.) (Fischer 1987, 1999). Locally, plant assemblages on set-aside land, for example, have been attributed to activation of old seedbanks by wild boars (Milton et al. 1997). It should be qualified, however, by saying that these findings resulted from very short-term studies. These certainly show that animals have initiated succession but do not allow a long-term perspective.

Wild boars may also bring seeds to the surface that were hoarded by small mammals in subterranean burrows at relatively great depth, where they never would have germinated. In addition, seeds may be activated that are lacking in the present vegetation (Treiber 1997). Seeds of some tree species better germinate in churned soil (Tisdell 1982). Not least, wild boars mix the mineral soil with litter (Jezierski and Myrcha 1975; Singer et al. 1984) and improve soil aeration through their digging activities. Consequently, decomposition of organic matter and supply of plant available nutrients increase. However, under certain conditions, leaching of nutrients (e.g., Ca, P, Zn, Cu, Mg.) is possible (Singer et al. 1984).

Wild boars also change ground vegetation. On relatively nutrient-rich ruderal sites and hay meadows (previous 'Wildäcker' near Groß-Gerau, Germany), for example, soil churning by wild boars increased short-lived species and species richness. As a result, a mosaic of different successional stages on relatively small areas may develop (Simon and Goebel 1999). While species on Californian grassland decreased during the first year after feral pig impact, they increased in the following year and soon exceeded species richness of undisturbed areas (Kotanen 1995). Native grasses colonized the disturbed sites as fast as introduced grasses. This contrasts with Hawaii, where digging and spreading of diaspores by feral pigs resulted in a rapid spread of introduced creeping velvet grass (*Holcus mollis*) at the cost of native grasses (Spatz and Mueller-Dombois 1995). Rooting by wild boars varies annually in dependence on habitat conditions, such as abundance and availability of preferred food and not least from soils. In the Tullgarn Nature Reserve (SSW of Stockholm, Sweden), for example, the rooted area was greater in deciduous forests than in conifer forests and grassland. Moreover, wild boars affected a much larger surface in damp soils compared to dry soils (Welander 2000).

2.6.2.3 Wombats

Three wombat species (marsupials about the same size as a beaver) are native to Australia, the common wombat (*Vomatus ursinus*, three subspecies), the northern hairy-nosed wombat (*Lasiorhinus krefftii*) and the southen hairy-nosed wombat (*Lasiorhinus latifrons*). All three wombat species have suffered from intense hunting, competition with rabbits and cattle for food and – as a non-target species – from poisoning campaigns directed at rabbits.

The common wombat is the most widespread. It (*Vombatus ursinus hirsutus*) nowadays occurs from the southeast tip of Queensland along the coastal regions of New South Wales to the south of Victoria, in Tasmania (*Vombatus ursinus tasmaniensis*) and on Flinders Island northeast of the Tasmanian coast (*Vombatus ursinus ursinus*). The northern hairy-nosed wombat almost went extinct. About 100 individuals are still living in Queensland. The southern hairy-nosed wombat occupies southernmost South Australia and the southeastern tip of West Australia. The Nullarbor Plain adjacent to the southern coast of Australia is their last large refuge, where they can still be found in great numbers (Löffler and Grotz 1995). Wombats are classified as selective grass feeders (mainly fibrous grasses). However, they also feed on shrubs (saltbush, *Atriplex vesicaria*; bluebush, *Maireana sedifolia*).

Photo 2.29 Partly collapsed burrows of the hairy-nosed wombat (*Lasiorhinus latifrons*) in the Nullarbor Plains, Australia. E. Löffler, May 1976

Hairy-nosed wombats impact soils and plant cover in this area quite considerably (Löffler and Margules 1980). Wombat colonies usually consist of 10–30 burrow systems, each having a diameter of 20–30 m. The excavated material towers by about 1 m above the surrounding terrain (Photo 2.29). Farmers and ranchers consider the wombat a noxious animal competing with cattle for food. However, what they say appears not to be true, as wombats feed mainly on hard native grasses that livestock cannot tolerate. Anyway, within fenced 'wombat-proof' places, the plant cover is usually less fragmented than in locations accessible to wombats. Wombats affect their habitat by causing direct damage to the vegetation and, in particular, by burrowing. Farming machinery may be damaged when driving over or breaking into wombat burrows that may also be a hazard to grazing livestock stepping on burrows. Therefore, wombats have become nearly eradicated in agricultural areas. As native animals, wombats are strictly protected, and a special permit is required to kill them if they turn out to be a problem. Culling is relatively inefficient. When a wombat has been shot, a new one will move in from surrounding areas and build a new burrow or repair and extend the older one. Wombat numbers seem to have been increasing more recently.

Although wombats are often blamed for causing extreme soil erosion on farmland and pastures, poor land management practices are probably of major importance. Anyway, destruction of the plant cover and soil excavation by wombats may reach an extent that affected areas can be identified even in satellite imagery (LANDSAT; Löffler and Margules 1980), similar to the impact of reindeer in northernmost Europe (Sect. 3.3.2).

References

Abaturov, B. D. (1972). The role of burrowing animals in the transport of mineral substances in the soil. *Pedobiologia, 12*, 261–266.

Abbot, H. G., & Quinck, T. T. (1970). Ecology of eastern white pine seed caches made by small forest mammals. *Ecology, 51*, 272–278.

Albon, S. D., Guiness, F. E., & Clutton-Brock, T. H. (1983a). The influence of climatic variation on the birth weights of red deer (Cervus elaphus). *Journal of Zoology, 2000*, 295–298.

Albon, S. D., Mitchell, B., & Staines, B. W. (1983b). Fertility and body-weight in female red deer: A density dependent relationship. *Journal of Animal Ecology, 52*, 969–980.

Alcorn, S. M., McGregor, S. E., Butler, G. D., & Krutz, E. B. (1959). Pollination requirements of the Saguaro (Carnegia gigantea). *Cactus and Succulent Journal, 31*, 39–41, Santa Barbara.

Alcorn, S. M., McGregor, S. E., & Olin, G. (1961). Pollination of saguaro cactus by dove, nectar-feeding bats, and honey bees. *Science, 133*, 1594–1595.

Alcorn, S. M., McGregor, S. E., & Olin, G. (1962). Pollination requirements of the organpipe cactus. *Cactus and Succulent Journal of the Cactus and Succulent Society of America, 34*, 134–138, Santa Barbara.

Aldous, S. E., & Krefting, L. W. (1946). The present status of Isle Royale moose. *Transactions of the North American Wildlife Conference, 11*, 296–308.

Alexandre, D.-Y. (1978). Le role disséminateur des élephants en Forêt de Tai, Côte d'Ivoire. *La Terre et la Vie, 32*, 47–72.

Alfaro, R. I., Van Sickle, G. A., Thomson, A. J., & Wegwitz, E. (1982). Tree mortality and radial growth losses caused by western spruce budworm in a Douglas-fir stand in British Columbia. *Canadian Journal of Forest Research, 12*(4), 780–787.

Allen, D. (1974). Of fire, moose and wolves. *Audubon, 74*(1), 39–49.

Allen, M. F. (1991). *The ecology of mycorrhizae*. New York/Port Chester/Melbourne/Sydney: Cambridge University Press.

Allen, M. F., & Grisafulli, C. M. (1994). Recovery of mycorrhizal associations on Mount St. Helens. In *Mount St. Helens: Biological research following the 1980 eruptions. An indexed bibliography and research abstracts (1980–1993)* (General Technical Report PNW-GTR 342). Portland: USDA Forest Service, Pacific Northwest Research Station

Allen, M. F., & MacMahon, J. A. (1988). Direct VA mycorrhizal inoculation of colonizing plants by pocket gophers (Thomomys talpoides) on Mt. St. Helens. *Mycologia, 80*, 754–756.

Allen, S. E., Grimshaw, H. M., & Holdgate, M. W. (1967). Factors affecting the availability of plant nutrients on an Antarctic island. *Journal of Ecology, 55*, 381–396.

Allen, M. F., MacMahon, J. A., & Andersen, D. C. (1984). Reestablishment of endogonaceae on Mount St. Helens: Survival of residuals. *Mycologia, 76*, 1031–1038.

Alschner, G. (1980). *Klipp und klar 100 x (hundertmal) Tierwanderungen*. Zürich: Mannheim Wien.

Amann, G. D. (1977). The role of mountain pine beetle in lodgepole pine ecosystems: Impact on succession. In *Proceedings in Life Sciences, The role of arthropods in forest ecosystems* (pp. 3–18). New York/Heidelberg/Berlin: Springer

Andersen, D. C. (1982). Observations on Thomomys talpoides in the region affected by the eruption of Mount St. Helens. *Journal of Mammalogy, 63*, 652–655.

Andersen, D. C. (1987). Below-ground herbivory in natural communities: A review emphasizing fossorial mammals. *The Quarterly Review of Biology, 62*(3), 261–286.

Andersen, A. N. (1991). Seed harvesting by ants in Australia. In C. R. Huxley & D. F. Citler (Eds.), *Ant-plant interactions* (pp. 521–535). Oxford: Oxford University press.

Andersen, D. C., & MacMahon, J. A. (1981). Population dynamics and bioenergetics of a fossorial herbivore, Thomomy talpoides (Rodentia: Geomyidae), in a spruce-fire sere. *Ecological Monographs, 51*(2), 179–202.

Andersen, D. C., & MacMahon, J. A. (1985). Plant succession following the Mount St. Helens volcanic eruption by a burrowing rodent, Thomomys talpoides. *The American Midland Naturalist, 114*(1), 62–69.

Anderson, J. M., & Coe, M. J. (1974). Decomposition of elephant dung in an arid tropical environment. *Oecologia, 141*, 11–125.

Andersson, M., & Jonasson, S. (1986). Rodent cycles in relation to food resources on an alpine heath. *Oikos, 46*, 93–106.

Andreev, A. V. (1991). Winter habitat segregation in the sexually dimorphic Black-billed capercaillie Tetrao urogalloides. *Ornis Scandinavica, 22*, 287–291.

Andrén, H. (1992). Corvid density and nest predation in relation to forest fragmentation: a landscape perspective. *Ecology, 73*, 794–804.

Area, G., Symes, C. T., & Witkowski, E. T. F. (2013). The birds and the seeds: Opportunistic avian nectarivores enhance reproduction in an endemic montane aloe. *Plant Ecology, 214*, 35–47.

Ash, A. J., & McIvor, J. G. (1998). How season of grazing and herbivore selectivity influence monsoon tall-grass communities of northern Australia. *Journal of Vegetation Science, 9*, 123–132.

Ashby, K. R. (1967). Studies on the ecology of field mice and voles (Apodemus sylvaticus, Clethrionomys glareolus and Microtus agrestis) in Houghall Wood, Durham. *Journal of Zoology, 152*, 389–513.

Ashton, D. H. (1979). Seed harvesting by ants in forests of E. regnans F. Muell, in central Victoria. *Australian Journal of Ecology, 4*, 265–277.

Aston, B. B. (1909). The soils and soil-formers of the subantarctic islands. In C. Chilton (Ed.), *Subantarctic islands of New Zealand* (pp. 745–777). Wellington: Philosophical Institute of Canterbury.

Auer, C. (1947). Untersuchungen über die natürliche Verjüngung der Lärche im Arven-Lärchenwald des Oberengadins. *Mitteilungen der Schweizerischen Anstalt für das forstliche Versuchswesen, 25*, 7–140.

Auer, C. (1961). Ergebniss zwölfjähriger quantitativer Untersuchungen der Populationsbewegungen des Grauen Lärchenwicklers Zeiraphera griseana Hübner (=Guéné) im Oberengadin (1949/60). *Mitteilungen der Schweizerischen Anstalt für das forstliche Versuchswesen, 37*(3), 174–263.

Auer, C. (1975). Ziel und Stand der Forschung über den grauen Lärchenwickler (LW) 1949–1974. *Bündnerwald, 28*(1), 7–32.

Augustine, D. J., & McNaughton, S. J. (1998). Ungulate effects on the functional species composition of plant communities: Herbivore seletivits and plant tolerance. *Journal of Wildlife Management, 62*(4), 1165–1183.

Aulitzky, H. (1961). Über die Windverhältnisse einer zentralalpinen Hangstation in der subalpinen Stufe. *Mitteilungen der forstlichen Bundesversuchsanstalt Mariabrunn, 59*, 209–230.

Aulitzky, H. (1968). Die Lufttemperatureverhältnisse einer zentralalpinen Hanglage. *Archiv für Meteorologie, Geophysik und Bioklimatology, Series B, 16*, 18–69.

Bachmann, P. (1971). Der Eichhörnchenschaden des Jahres 1969 im Kanton Bern. *Bünderwald, 24*(3), 97–98.

Baker, B. H., & Kemperman, J. A. (1974). Spruce beetle effects on a white spruce stand in Alaska. *Journal of Forestry, 72*, 423–425.

Bakke, A. (1969). Extremely low supercooling point in eggs of Zeiraphera diniana (Guenée) (Lepidoptera: Tortricidae). *Norks entomologisk Tidsskrift, 16*, 81–83.

Bakker, J. P., De Leeuw, J., & Van Wieren, S. E. (1983). Micro-patterns in grassland vegetation created and sustained by sheep-grazing. *Vegetatio, 55*, 153–161.

Baltensweiler, W. (1962). Zur Historik der Massenvermehrung des Grauen Lärchenwicklers (Zeiraphera griseana Hübner; Lepidoptera, Tortricidae). *Verhandlungen der Schweizerischen Naturforschenden Gesellschaft, 104*, 121–122.

Baltensweiler, W. (1975). Zur Bedeutung des Grauen Lärchenwicklers (Zeiraphera diniana Gn.) für die Lebensgemeinschaft des Lärchen-Arvenwaldes. *Mitteilungen der Schweizerischen Entomologischen Gesellschaft, 48*(1–2), 5–12.

Baltensweiler, W., & Rubli, D. (1984). *Forstliche Aspekte der Lärchenwickler-Massenvermehrungen im Oberengadin* (Mitteilungen der Eidgenössischen Anstalt für das forstliche Versuchswesen 60, 1). Birmensdorf: Eidgenössische Anstalt für das forstliche Versuchswesen.

Baltensweiler, W., Benz, G., Bovey, P., & Dellucchi, V. (1977). Dynamics of larch-bud moth population. *Annual Review of Entomology, 22*, 79–100.

Bancroft, W. J., Garkaklis, M. J., & Roberts, J. D. (2004). Burrow building in seabird colonies: A soil forming process in island ecosystems. *Pedobiologia, 49*, 149–165.

Barnes, V. G. (1973). *Pocket gophers and reforestations: A problem analysis* (Special Scientific Report on Wildlife 155), USDI Fish and Wildlife Service, Denver Wildlife Research Center, Forest Animal Damage Unit.

Barnes, V. G. (1974). Response of pocket gopher populations to silvicultural practices in central Oregon. In H. C. Black (Ed.), *Proceedings, symposium on wildlife and forest management in the Pacific Northwest 11–12 September 1973, Corvallis, Oregon* (pp. 167–175). Oregon State University, School of Forestry.

Batzli, G. O. (1981). Population and energetics of small mammals in the tundra ecosystem. In L. C. Bliss, O. W. Heal, & J. J. Moore (Eds.), *Tundra ecosystems: A comparative analysis* (The International Biological Program 25, pp. 377–396). Cambridge: Cambridge University Press.

Batzli, G. O., White, R. G., Maclean, S. F., Pitelka, F. A., & Collier, B. D. (1980). The herbivore-based trophic system. In J. Brown, P. C. Miller, L. L. Tieszen, & L. Bunnell (Eds.), *Arctic ecosystems: The coastal tundra at Barrow* (pp. 335–410). Stroudsburg: Alaska.

Bäumler, W., & Hohenadl, W. (1980). Über den Einfluß alpiner Kleinsäuger auf die Verjüngung in einem Bergmischwald der Chiemgauer Alpen. *Forstwissenschaftliches Centralblatt, 99*, 207–221.

Bax, N., & Sheldrick, D. L. (1963). Some preliminary observations on the food of elephants in the Tsavo Royal National Park (east) of Kenya. *East African Wildlife Journal, 1*, 140–153.

Bazeley, D. R., & Jefferies, R. L. (1985). Goose faeces: A source of nitrogen for plant growth in a grazed salt marsh. *Journal of Ecology, 22*, 693–703.

Beck, L. (1989). Lebensraum Buchenwaldboden 1. Bodenfauna und Streuabbau – eine Übersicht. *Verhandlungen der Gesellschaft für Ökologie, 17*, 47–54.

Begall, S. (1999). *Verhaltensökologische und genetische Analysen der Sozial- und Populationsstruktur con Coruros (Spalacopus cyanus, Octodentidae, Rodentia) aus Chile*. Dissertation Universität Essen, Fachberiech 9.

Begall, S., & Gallardo, M. H. (2000). Spalacopus cyanus (Rodentia: Octodontidae): an extremist in tunnel constructing and food storing among subterranean mammals. *Journal of Zoology (London), 251*, 53–60.

Bell, R. H. (1970). The use of the herb layer by grazing ungulates in the Serengeti. In A. Watsons (Ed.), *Animal populations in relation to their food resources* (pp. 111–124). Oxford Blackwell.

Belsky, A. J. (1986). Does herbivory benefit plants? A review of the evidence. *The American Midland Naturalist, 127*(6), 879–892.

Belsky, A. J. (1992). Effects of grazing, competition, disturbance, and fire on species composition and diversity of grassland communities. *Journal of Vegetation Science, 3*, 187–200.

Belsky, A. J., & Canham, C. D. (1994). Forest gaps and isolated forest trees. *BioScience, 44*, 77–84.

Ben-David, M., Bowyer, R. T., Duffy, L. K., Roby, D. D., & Schell, D. M. (1998). Social behavior and ecosystem processes: River otter latrines and nutrient dynamics of terrestrial vegetation. *Ecology, 79*(7), 2567–2571.

Benedict, F. C. (1936). *The physiology of the elephant*. Washington, DC: Carnegie Institution.

Ben-Shahar, R. (1995). Habitat classification in relation to movements and densities of ungulates in a semi-arid savanna. *African Journal of Ecology, 33*, 50–63.

Benz, G. (1974). Negative Rückkoppelung durch Raum- und Nahrungskonkurrenz sowie zyklische Veränderung der Nahrungsgrundlage als Regelprinzip in der Populationsdynamik des Grauen Lärchenwicklers, Zeiraphera diniana (Guenée) (Lep., Tortricidae). *Zeitschrift für Angewandte Entomologie, 76*, 196–228.

Berg, R. J. (1975). Myrmecochorus plants and their dispersal by ants. *Australian Journal of Botany, 23*, 475–508.

Berg, E., & De Volder, A. (2000, April 2–7). *Spruce beetle outbreaks and climate change on the Kenai peninsula, Alaska*. International conference on dendrochronology for the third millennium, Mendoza

Berg, E. E., Henry, J. D., Fastie, C. L., De Volder, A. D., & Matsuoka, S. M. (2006). Spruce beetle outbreaks on the Kenai Peninsula, Alaska, and Kluane National Park and Reserve, Yukon Territory: Relationship to summer temperatures and regional differences in disturbance regimes. *Forest Ecology and Management, 227*, 219–232.

Bergerud, A. T. (1983). The natural population control of caribou. In F. L. Bunnel, D. S. Eastman, & J. M. Peek (Eds.), *Symposium on natural regulation of wildlife populations* (pp. 14–61). Moscow: Forest, Wildlife and Range Experiment Station, University of Idaho.

Bergerud, A. T. (1984). Die Populationsdynamik von Räuber und Beute. *Spektrum der Wissenschaft, 2*, 48–54.

Bernhart, A. (1988). Waldentwicklung, Verjüngung und Wildverbiß im oberbayerischen Bergwald. *Schweizerische Zeitschrift für Forstwesen, 139*(6), 463–484.

Bernhart, A. (1990). Entwicklung der Bestockung im Bergwald Oberbayerns seit 1860. In E. Schuster (Ed.), *Zustand und Gefährdung des Bergwaldes* (Forstwissenschaftliche Forschungen 40, pp. 19–29), Beiheft zum Forstwissenschaftlichen Zentralblatt. Hamburg/Berlin: Parey Verlag.

Berryman, A. A., Dennis, B., Raffa, K. F., & Stenseth, N. C. (1989). Interaction dynamics of bark beetle aggregation and conifer defense rates. *Oikos, 56*, 256–263.

Berutti, A. (2000). The pest status and biology of the red-billed quelea in Bergville-Winterton area of South Africa. In R. A Cheke, L. J. Rosenberg, & M. E. Kieser (Eds.), *Workshop on research priorities for migrant pests of agriculture in southern Africa* (pp. 113–118). Pretoria: Plant Protection Research Institute, 24–26 March 1999. http://icosamp.ecoport.org/archives/mpw/P12.pdf

Beyer, D. N. (1996). Ohne Waldelefanten kein Regenwald. Die segensreiche Wirkung der Urwaldriesen. *WWF Aktuell, 10*.

Bezzola, D. (1989). Zur Verjüngung und Sprossbildung der Arve unter dem Einfluss des Lärchenwicklers. *Schweizerische Zeitschrift für Forstwesen, 140*, 305–312.

Bibikov, D. I. (1996). *Die Murmeltiere der Welt* (Die Neue Brehmbücherei 358, 2nd ed.). Magdeburg/Heidelberg.

Bjerke, S. (1957). Nogle traek af de sydskandinaviske louskoves udujkling gennem de sdiste arhundedret. *Dansk Dendologisk Arssk, 4*, 373–413.

Blakemore, L. C., & Gibbs, H. S. (1968). Effects of gannets on soil at Cape Kidnappers, Hawke's Bay. *New Zealand Journal of Science, 11*, 54–62.

Blumenberg, D. (1986). Telemetrische und endoskopische Untersuchungen zur Soziologie, zur Aktivität und zum Massenwechsel der Feldmaus, Microtus arvalis (Pall.). *Zeitschrift für Angewandte Zoologie, 73*, 301–344.

Blumer, P., & Diemer, M. (1996). The occurrence and consequences of grasshoppers herbivory in an alpine grassland, Swiss Alps. *Arctic, Antarctic, and Alpine Research, 28*(4), 435–440.

Blüthgen, J. (1942). *Die polare Baumgrenze in Lappland* (Veröffentlichungen des Deutschen Wissenschaftlichen Instituts Kopenhagen, Reihe 1: Arktis (10)). Berlin-Zehlendorf: Borntraeger.

Bonner, W. N. (1985). Impact of fur seals on the terrestrial environment at South Georgia. In W. R. Siegfried, P. R. Condy, & R. M. Laws (Eds.), *Antarctic nutrient cycles and food webs* (pp. 641–646). Berlin/Heidelberg: Springer.

Bornemissza, G. F. (1976). The Australian dung beetle project 1965–1975. *Australian Meat Research Commission Review, 30*, 1–30.

Bossema, I. (1979). Jays and oaks: An eco-ethological study of a symbiosis. *Behaviour, 70*, 11–17.

Botkin, D. B., Mellilo, J. M., & Wu, L. S.-Y. (1981). How ecosystem processes are linked to mammal population dynamics. In C.-W. Fowler & T. D. Smith (Eds.), *Dynamics of large mammal populations* (pp. 373–387). New York: Wiley.

Bott, G. (1995). Wirtschaftliche Bedeutung der Arve im Obernegadin. *Bündnerwald, 6*, 56–60.

Boucher, D. H. (1981). Seed predation by mammals and forest dominance by Quercus oleoides in a tropical lowland. *Oecologiea, 49*, 409–414.

Bousfield, W., & Chase, A. (1982). *Progress report on spruce budworm damage in the Wall Creek Area, Beaverhed National Park* (USDA Forest Service, Forest Insect and Disease Management, State and Private Forestry, Northern Region Report, pp. 82–12).

Bowman, W. D., Theodose, T. A., Schardt, J. C., & Conant, R. T. (1993). Constraints of nutrient availability on primary production in two alpine tundra communities. *Ecology, 74*(1), 2085–2097.

Bowyer, R. T., Van Ballenberghe, V., & Kie, J. G. (1997). The role of moose in landscape processes: Effects of biogeography, population dynamics and predation. In J. A. Bisonette (Ed.), *Wildlife and landscape ecology. Effects of pattern and scale* (pp. 265–287). New York: Springer.

Boyd, W. L., Staley, J. T., & Boyd, J. W. (1966). Ecology of microorganisms of Antarctica. In J. F. C. Tedrow (Ed.), *Antarctic soils and soil forming processes* (Antarctic Research Series 8, pp. 129–159).

Boye, P. (1996). Die Rolle von Säugetieren in mitteleuropäischen Ökosystemen. *Schriftenreihe für Landschaftspflege und Naturschutz, 46*, 11–18.

Boyer, W. D. (1964). Longleaf pine and seed predators in southwestern Alabama. *Journal of Forestry, 62*, 481–484.

Brandner, T. A., Peterson, R. D., & Risenhoover, K. L. (1990). Balsam fir on Isle Royale: Effects of moose herbivory and population density. *Ecology, 71*(1), 155–164.

Brattsen, L. B. (1979). Biochemical defense mechanisms in herbivores against plant allelochemics. In G. A. Rosenthall & D. H. Janzen (Eds.), *Herbvores: Their interaction with secondary plant metabolites* (pp. 199–270). New York: Academic Press.

Braun, H. M. (1973). Primary production in the Serengeti: Purpose, methods and some results of the research. *Annales Université d'Abidjan, Serie E (Ècologie), 6*(2), 171–188.

Brewer, S. W., & Remánek, M. (1999). Small rodents as significant seed dispersers of tree seeds in a Neotropical forest. *Journal of Vegetation Science, 10*, 165–174.

Briedemann, R. (1990). *Das Schwarzwild*. Berlin: VEB Deutscher Landwirtschaftsverlag.

Broll, G., & Keplin, B. (1995). Ecological studies on urban lawns. In H. Sukopp, M. Numata, & A. Huber (Eds.), *Urban ecology as the basis of urban planning* (pp. 71–82). Berlin/Heidelberg/New York: Springer.

Broll, G., Brauckmann, H.-J., & Schreiber, F.-F. (1996). Changes of soil-ecological conditions due to long-term fallow, Bernau. In *9th ISCO Conference, Bonn, Excursion Guide Post Conference Tour South* (pp. 100–103).

Brooke, M. D. L. (2004). *Albatrosses and petrels across the world*. Oxford: Oxford University Press.

Brown, J. H., & Heske, E. J. (1990). Control of a desert-grassland transition by a keystone rodent guild. *Science, 250*, 1705–1707.

Bryant, J. P. (1981). The regulation of snowshoe hare feeding behavior during winter by plant antiherbivory chemistry. In: S. K. Meyer & C. D. MacInnes (Eds.), *Proceedings of the First International Lagomorph Congress* (pp. 720–731). Guelph.

Bryant, J. P., & Chapin, F. S., III. (1986). Browsing-woody plant interactions during boreal forest succession. In K. Van Cleve, F. S. Chapin III, P. W. Flanagan, et al. (Eds.), *Forest ecosystems in the Alaskan taiga* (pp. 213–225). New York: Springer.

Bryant, J. P., Chapin, F. S., III, & Klein, D. R. (1983). Carbon/nutrient balance of boreal plants in relation to herbivory. *Oikos, 40*, 357–368.

Bryant, J. B., Wieland, G. D., Clausen, T., & Kuropat, P. (1985). Interactions of snowshoe hare and feltleaf willow in Alaska. *Ecology, 66*, 1564–1573.

Bryant, J. P., Tahvanainen, J., Sulkinoja, M., Jukkunen-Tiito, R., Reichart, P., & Green, T. (1989). Biogeographic evidence for the evolution of chemical defense by boreal birch and willow against mammalian browsing. *The American Midland Naturalist, 134*(1), 20–34.

Buckley, R. C. (1982). Ant-plant interactions: A world review. In R. C. Buckley (Ed.), *Ant-plant interactions in Australia* (pp. 111–141). The Hague: Dr. W. Junk Publishers.

Buckner, D. L. (1977). *Ribbon-forest development and maintenance in the Central Rocky Mountains of Colorado*. Dissertation University of Colorado, Boulder.

Buechner, H. K. (1942). Interrelationships between the pocket gopher and land use. *Journal of Mammalogy, 23*(3), 346–348.

Buechner, H. K., & Dawkins, H. C. (1961). Vegetation change induced by elephants and fire in Murchison Falls National Park, Uganda. *Ecology, 42*(4), 752–766.

Burckhart, D. (1958). Vom Vorratsanlegen des Tannenhähers. *Bündnerwald, 11*, 102–144.

Burga, C. A. (1988). Swiss vegetation history during the last 18,000 years. *New Phytologist, 110*, 581–602.

Burga, C. A., & Perret, R. (2001). Monitoring of eastern and southern Swiss alpine timberline ecotones. In C. A. Burga & A. Kratochwil (Eds.), *Biomonitoring: General and applied aspects in regional and global scale* (pp. 179–194). Dordrecht: Kluwer Academic Publishers.

Burger, A. E., Lindeboom, H. J., & Williams, A. J. (1978). The mineral and energy contributions of guano of selected species of birds to the Marion Island terrestrial ecosystems. *South African Journal of Antarctic Research, 8*, 59–70.

Burns, S. F. (1979). The northern pocket gopher (Thomomys talpoides). A major geomorphic agent on the alpine tundra. *Journal of Colorado-Wyoming Academy of Sciences, 27*, 395–410.

Burns, S. F. (1980). *Alpine soil distribution and development, Indian Peaks, Colorado Front Range*. Dissertation, University of Colorado, Boulder.

Burns, S. F., & Tonkin, P. J. (1982). Soil-geomorphic models and the spatial distribution and development of alpine soils. In C. E. Thorn (Ed.), *Space and time in geomorphology* (pp. 25–43). London: Allen & Unwin.

Burton, R. L., Starks, K. J., & Peters, D. C. (1980). *The army cutworm* (Bulletin B-749). Stillwater: Agricultural Experiment Station, Division or Agriculture, Oklahoma State University.

Butler, D. R. (1992). The grizzly bear as an erosional agent in mountainous terrain. *Zeitschrift für Geomorphologie, Neue Folge, 36*(2), 179–189.

Butler, D. R. (1993). The impact of mountain goat migration on unconsolidated slopes in Glacier National Park, Montana. *Geographical Bulletin, 35*, 98–106.

Butler, D. R. (1995). *Zoogeomorphology: Animals as geomorphic agents* (2nd ed., 2008). Cambridge: Cambridge University Press.

Butler, D. R., Malanson, G. P., & Resler, L. M. (2004). Turf-banked terrace treads and risers, turf exfoliation and possible relationships with advancing treeline. *Catena, 58*, 259–5274.

Bylund, H. (1995). *Long-term interactions between the autumnal moth and mountain birch: The role of resources, competitors, natural enemies, and weather*. Dissertation, Swedish University of Agricultural Sciences, Uppsala.

Bylund, H. (1997). Stand age-structure influence in a low population peak of Epirrita autumnata. *Ecography, 20*, 319–329.

Bylund, H. (1999). Climate and the population dynamics of two insect outbreak species in the north. *Ecological Bulletins, 47*, 54–62.

Cambefort, Y., & Hanski, I. (1991). Dung beetle population biology. In I. Hanski & Y. Cambefort (Eds.), *Dung beetle ecology* (pp. 36–50). Princeton, New Jersey: Princeton University Press.

Campbell, K., & Borner, M. (1995). Population trends and distribution of Serengeti herbivores: Implications for management. In A. R. E. Sinclair & P. Arcese (Eds.), *Serengeti II*. Chicago/London: University of Chicago Press.

Campbell, I. B., & Clairidge, G. G. C. (1966). A sequence of soils from a penguin rookery, Inexpressible Island, Antarctic. *New Zealand Journal of Science, 9*, 361–372.

Campell, E. (1944). *Der Wald des Oberngadins im Wandel der Zeiten*. Festschrift zur 124. Jahresversammlung der Schweizerischen Naturforschenden Gesellschaft (pp. 93–112).

Campell, E. (1950). Der Tannen- oder Nußhäher und die Arvenverbreitung. *Bündnerwald, 4*(1), 3–7.

Canals, R.-M., & Sebastía, M.-T. (2000). Soil nutrient fluxes and vegetation changes on molehills. *Journal of Vegetation Science, 11*, 23–30.

Cantor, L. F., & Whitham, T. G. (1989). Importance of belowground herbivory: Pocket gophers my limit aspen to rock outcrop refugia. *Ecology, 70*(4), 962–970.

Carcalliet, C., Talon, B., & Barbero, M. (1998). Pinus cembra et incendies pendant l'Holocène, 300 m au-dessus de la limite actuelle des arbres dans le massif de la Vanoise (Alps dur nord-ouest). *Écologie, 29*, 277–282.

Carey, A. B., Kershner, J., Biswell, B., & Dominguez de Toledo, K. (1999). Ecological scale and forest development: Squirrels, dietary fungi, and vascular plants in managed and unmanaged forests. *Wildlife Monographs, 142*, 1–71.

Cargill, S. M., & Jefferies, R. L. (1984). The effects of grazing by lesser snow geese on the vegetation of a subarctic salt marsh. *Journal of Applied Ecology, 21*, 669–686.
Carl, E. A. (1971). Population control in arctic ground squirrels. *Ecology, 52*, 395–413.
Carolin, V. M. (1987). Life history and behavior. In M. H. Brookes, R. W. Campbell, J. J Colbert, R. G. Mitchell, & R. W. Stark (Eds.), *Western spruce budworm* (Technical Bulletin 1694, pp. 29–42). USDA Forest Service, Cooperative State Research Service. Washinton, DC: U. S. Government Printing Office.
Castro, J., Gómez, J. M., García, D., Zamora, R., & Hódar, J. A. (1999). Seed production and dispersal in relict pine forests in southern Spain. *Plant Ecology, 145*, 115–123.
Caughley, G. (1970). Eruption of ungulate populations, with emphasis on Himalayan Tahr in New Zealand. *Ecology, 51*(1), 53–72.
Caughley, G. (1983). Dynamics of large mammals and their relevance to culling. In N. Owen-Smith (Ed.), *Management of large mammals in African conservation areas* (pp. 115–126). Pretoria: HAUM Publishers.
Challies, C. N. (1975). Feral pigs (Sus scrofa) on Auckland Island: Status and effects on vegetation and nesting sea birds. *New Zealand Journal of Zoology, 2*(4), 479–490.
Chapin, S. F., III. (1980). The mineral nutrition of wild plants. *Annual Review of Ecological Systems, 11*, 233–260.
Chapin, F. S., III, Barsdale, R. J., & Barél, D. (1978). Phosphorus cycling in Alaskan coastal tundra. A hypothesis for the regulation of nutrient cycling. *Oikos, 31*, 189–199.
Chapman, J. A., Romer, J. I., & Stark, J. (1955). Ladybird beetles and army cutworm adults as food for grizzly bears in Montana. *Ecology, 36*, 156–158.
Chapman, L. J., Chapman, C. A., & Wrangham, R. W. (1992). Balanites milsoniana: Elephant dependent dispersal. *Tropical Ecology, 8*, 275–283.
Chew, R. M. (1994). Consumers as regulators of ecosystems: An alternative to energetics. *The Ohio Journal of Science, 74*, 359–370, Ohio Academy of Science, Columbus.
Choquenot, D. (1991). Density-dependent growth, body conditions, and demography in feral donkeys: Testing the food hypothesis. *Ecology, 72*, 805–813.
Chrisman, A. B., Blake, G. M., & Shearer, R. C. (1983). *Effect of western spruce budworm on Douglas-fir cone production in western Montana* (Research Paper INT-308). Ogden: USDA Forest Service, Intermountain Forest and Range Experimental Station.
Christiansen, E. (1983). Fluctuations in some small rodent populations in Norway 1971–1979. *Holarctic Ecology, 6*, 24–31.
Christiansen, F. B., & Fenchel, T. M. (1977). *Theories of populations in biological communities* (Ecological Studies 20). Berlin/New York: Springer.
Clancy, K. M., Wagner, M. R., & Reich, P. B. (1995). Ecophysiology and insect herbivory. In W. K. Smith & T. M. Hinckley (Eds.), *Ecophysiology in coniferous forests* (pp. 125–180). San Diego/London: Academic Press.
Clutton-Brock, T. H., Guiness, F. E., & Albon, S. D. (1982). *Red deer: Behavior, and ecology of two sexes*. Chicago: University of Chicago Press.
Clutton-Brock, T. H., Ajor, M., Albon, S. D., & Guiness, F. E. (1987). Early development and population dynamics in red deer I: Density-dependent effects on juvenile survival. *Journal of Ecology, 56*, 53–67.
Coe, M. J. (1981). Large herbivores and food quality. In J. A. Lee, S. McNeill, & I. H. Rorison (Eds.), *Nitrogen as an ecological factor* (pp. 345–386). Oxford: Blackwell.
Coe, M. J., & Foster, J. B. (1972). The mammals of the northern slopes of Mt. Kenya. *Journal of East African Natural History Society and Nature Museum, 131*, 1–18.
Coe, M. J., Cumming, D. H., & Phillipson, J. (1976). Biomass and production of large African herbivores in relation to rainfall and primary production. *Oecologia, 22*, 341–354.
Cole, W. E., & Amman, G. D. (1969). *Mountain pine beetle infestations in relation to lodgepole pine diameters* (USDA Forest Service Research Note INT-95). Forest Service, Intermountain Forest and Range Experiment Station, Ogden, UT.
Coleman, D. C., Andrews, R., Ellis, J. E., & Singh, J. S. (1976). Energy flow and partitioning in selected managed and natural ecosystems. *Agrosystems, 3*, 45–54. Forest Service, Intermountain Forest and Range Experiment Station, Ogden, UT.

Coley, P. D., Bryant, J. P., & Chapin, F. S., III. (1985). Resource availability and plant antiherbivore defense. *Science, 230*, 895–899.

Collins, N. M., & Wood, T. G. (1984). Termites and atmospheric gas production. *Science, 224*, 84–85.

Constible, J. M., Sweitzer, R. A., Van Vuren, D. H., Schuyler, P. T., & Knapp, D. A. (2005). Dispersal of non-native plants by introduced bison in an island ecosystem. *Biological Invasions, 4*, 2–24.

Contreras, L. C., & Gutiérrez, J. R. (1991). Effects of subterranean herbivorous rodent Spalacopus cyanus in herbaceous vegetation in arid costal Chile. *Oecologia (Berlin), 87*, 106–109.

Coppock, D. L., & Detling, J. K. (1986). Alteration of bison and black-tailed prairie dog grazing interaction by prescribed burning. *Journal of Wildlife Management, 50*, 452–455.

Cortinas, M. R., & Seastedt, T. T. R. (1996). Short- and long-term effects of gophers (Thomomys talpoides) on organic matter dynamics in alpine tundra. *Pedobiologia, 40*, 162–170.

Costantino, R. F., Desharnais, R. A., Cushing, J. M., & Dennis, B. (1991). Chaotic dynamics in an insect population. *Science, 275*, 389–391.

Cox, G. W., & Gakahu, C. G. (1983). Mima mounds in the Kenya highlands: Significance for the Dalquest-Scheffer hypothesis. *Oecologica, 57*, 68–84.

Cox, G. W., & Scheffer, V. B. (1991). Pocket gophers and Mima terrain in North America. *Natural Areas Journal, 11*(4), 193–198.

Cox, G. W., Gakahu, C. G., & Allen, D. W. (1987). Small-stone content of Mima mounds of the Columbia Plateau and Rocky Mountains regions: Implication for mound origin. *Great Basin Naturalist, 47*(4), 609–619.

Craighead, J. J., Summer, J. S., & Mitchell, J. A. (1995). *The grizzly bear population in the Yellowstone ecosystem, 1959–1992*. Washington, DC: Island Press.

Crocker, R. E. (1966). *Das Meer – der größte Lebensraum*. Hamburg/Berlin: Verlag Paul Parey.

Crocq, C. (1974). Notes complémentaires sur la identication due Casse-noix Nucifraga caryocatatces dans les Alpes francaises. *Alauda, 42*, 3950.

Crome, F. J. H., & Bentruppenbaumer, J. (1993). Special people, a special animal and a special vision: The first step to restoring a fragmented tropical landscape. In D. A. Saunders, R. J. Hobbs, & P. R. Ehrlich (Eds.), *Nature conservation 3. The reconstruction of ecosystems* (pp. 267–279). Chipping Norton: Surrey Beatty & Sons Pty Ltd.

Croxall, J. P. (1984). Seabirds. In R. M. Laws (Ed.), *Arctic ecology* (pp. 553–616). London: Academic.

Cumming, D. H. M. (1982). The influence of large herbivores on savanna structure in Africa. In B. J. Huntley & B. Walker (Eds.), *Ecology of tropical savannas* (Ecological Studies 42, pp. 217–245). Berlin/New York/Heidelberg: Springer.

Cumming, D. H., Fenton, M. B., Rautenbach, I. L., Taylor, R. D., Cumming, G. S., Cumming, M. S., Dunlop, J. M., Ford, A. G., Hovorka, M. D., Johnston, D. S., Kalcounis, M., Mahlangu, Z., & Portfors, C. V. R. (1997). Elephants, woodlands and biodiversity in southern Africa. *South African Journal of Science, 93*, 231–236.

Dahlstedt, K. J., Staher-Blair, S., Worcester, B. K., & Klukas, R. (1981). Application of remote sensing to prairie dog management. *Journal of Range Management, 34*, 218–223.

Darley-Hill, S., & Johnson, W. C. (1981). Acorn dispersal by the Blue Jay (Cyanocitta cristata). *Oecologia, 50*, 231–232.

Darling, F. (1947). *Natural history in the Highlands and islands*. London: Collins.

Darlington, J. P. E. C. (1985). Lenticular soil mounds in the Kenya highlands. *Oecologia, 66*, 116–121.

Darwin, C. R. (1881). *The formation of vegetable mould and earthworms: With observations on their habits*. London: John Murray.

Davis, M. B. (1976). Pleistocene biogeography of temperate deciduous forest. *Geoscience and Man, 13*, 13–26.

Dean, W. R. J., Milton, S. J., & Jeltsch, F. (1999). Large trees, fertile islands, and birds in arid savanna. *Journal of Arid Environments, 41*, 61–78.

Dearing, D. (2001). Plant-herbivore interactions. In W. D. Bowman & T. R. Seastedt (Eds.), *Structure and function of an alpine ecosystem* (pp. 266–282). Oxford: Oxford University Press.
Del Moral, R. (1984). The impact of the Olympic Marmot on subalpine vegetation structure. *American Journal of Botany, 71*(9), 1228–1236.
Delella-Benedict, A. D. (1991). *A Sierra Club naturalist's guide: The southern Rocky Mountain regions of southern Wyoming, Colorado and northern New Mexico*. San Francisco: Sierra Club Books.
Delwiche, C. C., & Wijler, J. (1956). Non-symbiotic nitrogen fixing in soil. *Plant Soil, 17*, 113–129.
Detling, J. K., Dyer, M. I., Procter-Gregg, C., & Winn, D. T. (1980). Plant herbivore interactions: Examination of potential effects of bison saliva on regrowth of Bouteloa gracilis (H.B.K.) Lag. *Oecologia, 45*, 26–31.
Dewey, J. E. (1970). Damage to Douglas-fir cones by the budworm, Choristoneura occidentalis. *Journal of Economic Entomology, 63*(6), 1804–1806.
Dighton, J. (1978). Effects of synthetic lime aphid honey on populations of soil organisms. *Soil Biology and Biochemistry, 10*, 369–376.
Dingle, R. W. (1956). Pocket gophers as a cause of mortality in eastern Washington pine plantations. *Journal of Forestry, 54*(12), 832–835.
Dionne, J.-C. (1985). Tidal marsh erosion by geese, St. Lawrence estuary, Québec. *Géographie physicale et Quarternaire, 39*(1), 99–105.
Doak, D. F., & Loso, M. G. (2003). Effects of grizzly bear digging on alpine plant community structure. *Arctic, Antarctic, and Alpine Research, 35*(4), 421–428.
Dostál, P., Březnova, M., Kozličkova, V., Herben, T., & Kovář, P. (2005). Ant-induced soil modification and its effects on plant below-ground biomass. *Pedobiologia, 49*, 127–137.
Doube, B. M., Macqueen, A., Ridsdill, T. J., & Weir, T. A. (1991). Native and introduced dung beetles in Australia. In I. Hanski & Y. Cambefort (Eds.), *Dung beetle ecology* (pp. 255–278). Princeton: Princeton University Press.
Dreiser, C. (1993). *Mapping and monitoring of Quelea habitats in East Africa* (Berliner Geographische Studien 37), Berlin.
Drodz, A. (1966). Food habits and food supply of rodents in beech forest. *Acta Theriologica, 11*, 363–384.
Drummond, H. (1888). *Tropical Africa*. London: Hodder & Stoughton.
Dublin, H. T., Sinclair, A. R. E., & McGlade, J. (1990). Elephants and fire as causes of multiple stable states in the Serengeti-Mara woodlands. *Journal of Animal Ecology, 59*, 1147–1164.
Dulamsuren, C., Hauck, M., & Leuschner, C. (2013). Seedling emergence and establishment of Pinus sylvestris in the Mongolian forest-steppe ecotone. *Plant Ecology, 214*, 139–142.
Dunstone, N. (1993). *The mink*. London: T & A.D. Poyser Ltd.
Dyer, M. I., & Bokhari, U. G. (1976). Plant animal interactions: Studies on the effects of grasshopper grazing in blue grama grass. *Ecology, 57*, 762–772.
East, R. (1984). Rainfall, soil nutrient status and biomass of large African savanna mammals. *African Journal of Ecology, 22*, 245–270.
Edland, T. (1971). Wind dispersal of the winter moth larvae Operophtera brumata L. (Lep. Geometridae) and its relevance to control measures. *Norsk Entomologisk Tidskrift, 18*, 103–105.
Edwards, C. A., & Lofty, J. R. (1977). *Biology of earthworms* (2nd ed.). New York: Wiley.
Eisfeld, D. (1975). Der Eiweiß- und Energiebedarf des Rehes (Capreolus L.), diskutiert anhand von Laboruntersuchungen. *Verhandlungen der Gesellschaft für Ökologie, Erlangen, 1974*, 129–139.
Eldridge, D. J., Whitford, W. G., & Duval, B. D. (2009). Animal disturbances promote shrub maintenance in a desertified grassland. *Journal of Ecology, 97*, 1302–1310.
Ellenberg, H. (1973). Ziele und Stand der Ökosystemforschung. In H. Ellenberg (Ed.), *Ökosystemforschung* (pp. 1–31). Berlin/Heidelberg/New York: Springer.
Ellenberg, H. (1974). *Beiträge zur Ökologie des Rehes (Capreolus caprolus L. 1758). Daten aus den Stammhamer Versuchsgehegen*. Dissertation, Universität Kiel.

Ellenberg, H. (1978) Zur Populationsökologie des Rehes (Capreolus capreolus L., Cervidae) in Mitteleuropa. *Spixiana München,* Suppl 2, 1–121.

Elliott, J. O., & Cowan, I. M. (1978). Territoriality, density, and prey of the lion in Ngorongoro crater, Tanzania. *Canadian Journal of Zoology, 56*, 26–34.

Ellison, L. (1946). The pocket gopher in relation to soil erosion on mountain range. *Ecology, 27*(2), 101–114.

Ellison, L., & Aldous, C. M. (1952). Influence of pocket gophers on vegetation of subalpine grassland in Central Utah. *Ecology, 33*(2), 177–186.

Eltringham, S. K. (1974). Changes in the large mammal community of Mweya Pensinsula, Ruwenzori National Park, Uganda, following the removal of hippopotamus. *Journal of Applied Ecology, 11*, 855–866.

Emanuelsson, U. (1987). Human influence on vegetation in the Torneträsk area during the last three centuries. *Ecological Bulletins, 38*, 95–111.

Esipiu, I. (2013, August 7). *Climate-smart Kenyan crop hits a setback – Hungry birds*. Thomson Reuters Foundation website.

Esslinger, H. (1988). *Ski Heil! ARD Sendereihe Globus*. Beiheft 1 zur Sendung vom 10. Januar 1988.

Estes, J. A., Terborgh, J., Brashares, J. S., Power, M. E., Berger, J., Bond, W. J., Carpenter, S. R., Essington, T. E., Holt, R. D., Jackson, J. B., Marquis, R. J., Oksanen, L., Oksanen, T., Paine, R. T., Pikitch, E. K., Ripple, W. J., Sandin, S. A., Scheffer, M., Schoener, T. W., Shurin, J. B., Sinclair, A. R. E., Soulé, M. E., Virtanen, R., & Wardle, D. A. (2011). Trophic downgrading of planet earth. *Science, 333*, 301–306.

Eurola, S., & Hakala, A. V. K. (1977). The bird cliff vegetation of Svalbard. *Aquilo Series Botany, 15*, 1–18.

Evans, G. R. (1973). Hutton's shearwaters initiating local soil erosion in the Seward Kaikoura range. *New Zealand Journal of Sciences, 16*, 637–642.

Everson, I. (1977). *The living resources of the Southern Ocean*. Rome: UN Development Program, Food and Agricultural Organization.

Fabian, P. (1991). Klima und Wald. Perspektiven für die Zukunft. *Forstwissenschaftliches Zentralblatt, 110*, 286–304.

Faegri, K., & Van der Pijl, L. (1971). *Principles of pollination ecology*. Oxford: Pergamon Press.

Fan, N., Zhou, W., Wei, W., Wang, Q., & Jiang, Y. (1999). Rodent pest management in the Quinhai-Tibet alpine meadow ecosystem. In G. R. Singleton, L. A. Hunds, H. Leir, & Z. Zhang (Eds.), *Ecologically-based management of rodent pests* (pp. 285–304). Canberra: Australian Center for International Agricultural Research.

Farmer, R. E. (1997). *Seed ecophysiology of temperate and boreal zone forest trees*. Delray Beach: St. Lucie Press.

Feer, F. (1995). Morphology of fruits dispersed by African forest elephant. *African Journal of Ecology, 33*, 279–284.

Feinsinger, P. (1983). Coevolution and pollination. In D. F. Futuyama & M. Slatkins (Eds.), *Coevolution* (pp. 283–310). Sunderland: Sinauer Associates.

Feldmann, R. (1991). Buckelweiden – Buckelraine – Buckelwälder: Biogene Keinreliefbildung in der Kulturlandschaft. *Natur und Museum, 121*(7), 204–210.

Fellin, D. G., & Shearer, R. (1968). Spruce-budworm larvae damage western larch cones and seeds. *Journal of Forestry, 66*(7), 568–570.

Ferrar, P., & Watson, J. A. L. (1970). Termites associated with dung in Australia. *Journal of the Australian Entomological Society, 9*, 100–102.

Ferreras, P., & MacDonald, D. W. (1999). The impact of American mink Mustela vison on water birds in the upper Thames. *Journal of Applied Ecology, 36*, 701–708.

Fiedler, L. A. (1988). Rodent pest problems and management in eastern Africa. *FAO Plant Production Bulletin, 36*, 125–134.

Fineran, A. A. (1973). A botanical survey of seven Mutton-bird islands, southwest Stewart Island. *Journal of the Royal Society of New Zealand, 3*(4), 475–526.

Finley, R. B. (1969). Cone caches and middens of Tamiasciurus in the Rocky Mountains. In J. K. Jones (Ed.), *Contributions in mammalogy* (University of Kansas Natural History Museum, Miscellaneous Publications 51, pp. 233–273).

Fischer, A. (1987). *Untersuchungen zur Populataionsdynamik am Beginn von Sekundärsukzessionen. Die Bedeutung von Samenbank und Sameniederschlag für die Wiederbesiedlung vegetationsfreier Flächen in Wald- und Grünlandgesellschaften*. Dissertationes Botanicae 110, Berlin/Stuttgart.

Fischer, A. (1999). Der Einfluß des Schalenwildes auf die Bodenvegetation. In M. Suda, A. Fischer, W. Hockenjos, F. Straubinger, G. Meister, & G. Sperber (Eds.), *Waldökosystem und Schalenwild* (pp. 23–44). Rothenburg o. d. Tauber: Ökologischer Jagdverein Bayern e.V.

Fischer, S. F., Poschlod, P., & Beinlich, B. (1995). Die Bedeutung der Wanderschäferei für den Artenschutz zwischen isolierten Schaftriften. *Beiheft zu den Veröffentlichungen zur Naturschutz und Landschaftspflege, Baden-Württemberg, 83*, 229–256.

Fischer, S. F., Poschlod, P., & Beinlich, B. (1996). Experimental studies on the dispersal of plants and animals on sheep in calcareous grasslands. *Journal of Applied Ecology, 33*, 1206–1222.

Fitzgerald, B. M. (1981). Predatory birds and mammals. In L. C. Bliss, O. W. Heal, & J. J. Moore (Eds.), *Tundra ecosystems: A comparative analysis* (The Third International Biological Program 25, pp. 485–508). Cambridge: Cambridge University Press.

Flook, D. R. (1962). Range relationships of some ungulates native to Banff and Jasper National Parks, Alberta. In J. D. Criso (Ed.), *Grazing in terrestrial and marine environments* (pp. 119–128). Oxford: Blackwell.

Fogel, R., & Trappe, J. M. (1978). Fungus consumption (mycophagy) by small animals. *Northwest Science, 52*(1), 1–31.

Forman, R. T. T. (1995). *Land mosaics. The ecology of landscapes and regions*. Cambridge: Cambridge University Press.

Forman, R. T. T., & Godron, M. (1981). Patches and structural components for a landscape ecology. *BioScience, 31*(10), 733–740.

Fortin, D., Beyer, H. L., Boyce, M. S., Smith, D. W., Duchesne, T., & Mao, J. S. (2005). Wolves influence elk movements: Behavior shapes a trophic cascade in Yellowstone National Park. *Ecology, 86*(5), 1320–1330.

Fowler, C. W., & Smith, T. (1973). Characterizing stable populations, an application to the African Elephant population. *Journal of Wildlife Management, 37*(4), 513–523.

Fowler, D. P., Park, J. S., & Loo-Dinkins, J. (1995). Larix laricina – silvics and genetics. In W. C. Schmidt & K. J. McDonald (Compilers). *Ecology and management of Larix forests: A look ahead.* Proceedings of an International Symposium, Whitefisch, 5–9 October 1992 (USDA Forest Service, Intermountain Research Station, General Technical Report GTR-INT 319, pp. 54–57).

Frank, F. (1953). Zur Entstehung übernormaler Populationsdichten im Massenwechsel der Feldmaus. *Zoologisches Jahrbuch, 81*, 610–624.

Frank, F. (1954). Untersuchungen über den Zusammenbruch von Feldmausplagen (Microtus arvalis Pallas). *Zoologisches Jahrbuch, 82*, 95–136.

Frank, F. (1986). Regulierung der Nagetiere und des Maulwurfs im Garten und in der Land- und Forstwirtschaft. *Tierärztliche Wochenschrift, 93*(5), 194–198.

Frank, D. A., & Groffman, P. M. (1998). Ungulate vs. landscape control of soil C and N processes in grasslands of Yellowstone National Park. *Ecology, 79*(7), 2229–2241.

French, C. M., McEwen, L. C., Magruder, N. D., et al. (1955). *Nutritional requirement of white-tailed deer for growth and antler development* (Bulletin 600). University Park: Pennsylvanian Agriculture Experiment Station.

French, S. P., French, M. G., & Knight, R. R. (1994). Grizzly bear use of army cutworm moths in the Yellowstone ecosystem. *International Conference on Bearch Research and Management, 9*(1), 389–399.

Frenzel, B. (1983). Die Vegetationsgeschichte Süddeutschlands im Eiszeitalter. In H. Müller-Beck (Ed.), *Urgeschichte in Baden-Württemberg* (pp. 91–166). Stuttgart: Theiss.

Fritz, H., & Duncan, P. (1994). On the carrying capacity of large ungulates of African savanna ecosystems. *Proceedings Royal Society London B, 256*, 77–82.

Fry, C. H., & Keith, S. (2004). *The birds of Africa.* London: Christopher Helm.

Furness, R. W. (1991). The occurrence of burrow-nesting among birds and its influence on soil fertility and stability. *Symposium of the Zoological Society London, 63*, 53–67.

Furniss, R. L., & Carolin, V. M. (1977). *Western forest insects* (USDA Forest Service Miscellaneous Publications 1339).

Furrer, E. (1955). Probleme um den Rückgang der Arve in den Schweizer Alpen. *Mitteilungen der Schweizerischen Anstalt für das forstliche Versuchswesen, 31*(3), 669–705.

Furrer, E. (1957). *Das schweizerische Arvenareal in pflanzengeographischer und forstgeschichtlicher Sicht* (Berichte des Geobotanischen Forschungsinstitutes Rübel in Zürich, pp. 16–23). Bern: Verlag Hans Huber.

Ganzoni, R. Z. (1961). Die Waldungen des Oberengadins. *Schweizerische Zeitschrift für Forstwesen*, 40–44, 77–81

Gasaway, W. C., Stephenson, R. O., Davis, J. L., Shepherd, P. E. K., & Burris, O. E. (1983). Interrelations of wolves, prey, and man in interior Alaska. *Wildlife Monographs, 84*, 1–50.

Gashwiler, J. S. (1970). Further study of conifer seed survival on a Western Oregon clearcut. *Ecology, 51*, 849–854.

Geiger, R., Aron, R. H., & Toddhunter, P. (1995). *The climate near the ground* (5th ed.). Braunschweig: Vieweg-Verlag.

Gerell, R. (1967). Food selection in relation to habitat in mink (Mustela vison Schreber) in Sweden. *Oikos, 18*, 233–246.

Gerken, B., Kriedemann, K., & Grupe, M. (1992). Dynamik im Rotbuchenwald durch Eisbruch und Vogelkolonien – ein Beitrag zum Verständnis der Verlichtungsdynamik im mitteleuropäischen Wald. *Laufener Seminarbeiträge, 2*, 71–79.

Gessaman, J. A., & MacMahon, J. A. (1984). Mammals in ecosystems: Their effects on the composition and production of vegegation. *Acta Zoologica Fennica, 172*, 11–18.

Ghilarov, M. S. (1978). Bodenwirbellose als Indikatoren des Bodenhaushaltes und von bodenbildenden Prozessen. *Pedobiologia, 18*, 300–309.

Gibson, D. J. (1989). Effects of animal disturbance on tallgrass prairie vegetation. *American Midland Naturalist, 121*, 144–154.

Gigon, A. (1983a). Über das biologische Gleichgewicht und seine Beziehungen zur ökologischen Stabilität. *Berichte des Geobotanischen Institutes ETH, Stiftung Rübel, 50*, 149–177.

Gigon, A. (1983b). Typen ökologischer Stabilität und Instabilität mit Beispielen aus Waldökosystemen. In H. Mayer (Ed.), *Urwald symposium, Wien 1982* (pp. 23–24). Wien: IUFRO Gruppe Urwald, Waldbau-Institut, Universität für Bodenkultur.

Gigon, A., & Bolzern, H. (1988). Was ist biologisches Gleichgewicht? *Aus Forschung und Medizin, 3*(1), 18–28.

Gisi, U., Schenker, R., Schulin, R., Stadelmann, F. X., & Sticher, H. (1997). *Bodenökologie* (2nd ed.). Stuttgart/New York: Thieme.

Gomez-Garcia, D., Borghi, C. E., & Giannoni, S. M. (1995). Vegetation differences cause by pine vole mound building in the subalpine plant communities in the Spanish Pyrenees. *Vegetatio, 117*, 61–67.

Goodwin, D. (1976). *The crows of the world*. Ithaca: Cornell University Press.

Goretzky, J. (1999, June 16). Einfluß von Beutegreifern auf die Feldhasenpopulation. In Bundesminister für Ernährung, Landwirtschaft und Forsten (Ed.), *Feldhasensymposium, Informationsveranstaltung am*, Bonn.

Gossow, H. (1976). *Wildökologie*. München: BLV Verlagsgesellschaft.

Goszynska, W., & Goszynski, J. (1977). Effect of the burrowing activities of the common vole and the mole on the soil and vegetation of the biocoenoses of cultivated fields. *Acta Theriologica, 22*(10), 181–190.

Gould, M. S., & Swingland, I. R. (1980). The tortoise and the goat: Interactions on Aldabra Island. *Biological Conservation, 17*, 267–279.

Graae, B. J., Pagh, S., & Bruun, H. H. (2004). An experimental evaluation of the Arctic fox as a seed disperser. *Arctic, Antarctic, and Alpine Research, 36*(4), 468–473.

Graf, J. (1971). *Tierbestimmungsbuch*. München: J. F. Lehmanns Verlag.

Graf, E. (1974). Zur Biologie und Gradologie des Grauen Lärchenwicklers, Z. diniana Gn. (Lep. Tortricidae) im schweizerischen Mittelland. *Zeitschrift für angewandte Entomologie 76*: 233–251, 347–379.

Graff, O., & Makeshin, F. (1979). Der Einfluß der Fauna auf die Stoffverlagerung sowie die Homogenität und die Durchlässigkeit von Böden. *Zeitschrift für Pflanzenernährung und Bodenkunde, 142*, 476–491.

Grinell, J. (1923). The burrowing rodents of California as agents in soil formation. *Journal of Mammalogy, 4*(3), 137–149.

Grishin, S. Y., & Del Moral, R. (1996). Dynamics of forests after catastrophic eruptions of Kamchatka volcanoes. In I. M. Turner, C. H. Diong, S. S. Lim, & P. K. L. Ng (Eds.), *Biodiversity and the dynamics of ecosystems* (DIWPA Series 1).

Grishin, S. Y., Del Moral, R., Krestov, P. V., & Verkholat, V. P. (1996). Succession following the catastrophic eruption of Ksudach volcano (Kamcharka, 1907). *Vegetatio, 127*, 129–153.

Gruttke, H. (1994). Investigations on the ecology of Laemostenus terricola (Coleoptera, Carabidae) in agricultural landscape. In K. Desender, M. Dufrêne, M. Lorau, M. L. Luff, & J.-P. Maelfait (Eds.), *Carabid beetles: Ecology and evolution* (pp. 145–151). Dordrecht: Kluwer Academic.

Grzimek, M., & Grzimek, B. (1960a). Census of plains animals in the Serengeti National Park, Tanganyika. *Journal of Wildlife Management, 24*, 27–37.

Grzimek, M., & Grzimek, B. (1960b). A study of the game of Serengeti plains. *Zeitschrift für Säugetierkunde (Sonderheft), 25*, 1–61.

GTZ. (1987). *The ecology and control of the red-billed weaver bird (Quelea quelea L.) in Northeast Nigeria*. Eschborn: Deutsche Gesellschaft für Technische Zusammenarbeit.

Haberd, D. J. (1962). Some observations on natural clones of Festuca ovina L. *New Phytologist, 61*, 85–100.

Haberd, D. J. (1967). Observations on natural clones of Holcus mollis. *New Phytologist, 66*, 401–408.

Hadley, K. S., & Veblen, T. T. (1993). Stand response to western spruce budworm and Douglas-fir bark beetle outbreaks, Colorado Front Range. *Canadian Journal of Forest Research, 23*, 479–491.

Hall, K. J., & Williams, A. J. (1981). Animals as agents of erosion at subantarctic Marion Island. *South African Journal of Antarctic Research, 10–11*, 18–24.

Hall, K., Boelhouwers, J., & Driscoll, K. (1999). Animals as erosion agents in the alpine zone: Some data and observations from Canada, Lesotho, and Tibet. *Arctic, Antarctic, and Alpine Research, 31*(4), 436–446.

Hämet-Ahti, L. (1963). Zonation of the mountain birch forest in northernmost Fennoscandia. *Annales Botanici Societatis Zoologicae Botaniocae Fennicae 'Vanamo', 34*(4), 1–127.

Handel, S. N., & Beattie, A. J. (1990, October). Verbreitung von Pflanzensamen durch Ameisen. *Spektrum der Wissenschaft*, pp. 150–156

Hansen, R. M., & Morris, M. J. (1968). Movement of rocks by northern pocket gophers. *Journal of Mammalogy, 49*, 391.399.

Hansen, R. M., & Reid, V. H. (1973). Distribution and adaptation of pocket gophers. In G. T. Turner, R. M. Hansen, V. H. Reid, H. P. Tietjen, & A. L. Ward (Eds.), *Pocket gophers and Colorado mountain rangeland* (Colorado State University Fort Collins, Bulletin 554S, pp. 1–19). Fort Collins: Colorado State University.

Hansen, H. L., Krefting, L. W., & Kurmis, V. (1973). *The forest of Isle Royale in relation to fire history and wildlife* (Technical bulletin 294). St. Paul: Minnesota Agricultural Experiment Station.

Harmer, R. (1994). Natural regeneration of broadleaved trees in Britain: II Seed production and predation. *Forestry, 67*(4), 275–286.

Harper, J. L. (1969). The role of predation in vegetational diversity. In *Diversity and stability in ecological systems. Brookhaven Symposium in Biology 22*, pp. 48–62.

Hartmann, H. (1980). Beitrag zur Kenntnis der Pflanzengesellschaften Spitzbergens. *Phytocoenologia, 8*(1), 65–147.

Hatton, J. C., & Smart, N. O. E. (1984). The effect of long-term exclusion of large herbivores on soil nutrient status in Murchison Falls National Park, Uganda. *African Journal of Ecology, 22*, 23–30.
Haukioja, E. (1980). On the role of plant defences in the fluctuation of herbivore populations. *Oikos, 5*, 203–213.
Haukioja, E. (1982). Inducible defences of white birch to a geometrid defoliator, Epirrita autumnata. In *Proceedings 5th International Symposium Insect-Plant Relationships* (pp. 199–203). Wageningen: Pudoc.
Haukioja, E., & Hakala, T. (1975). Herbivore cycles and periodic outbreaks. Formulation of a general hypothesis. *Reports from the Kevo Subarctic Research Station, 12*, 1–9.
Haukioja, E., & Niemelä, P. (1977). Retarded growth of a geometrid larva after mechanical damage to leaves of its host tree. *Annales Zoologici Fennici, 14*, 48–52.
Haukioja, E., & Niemelä, P. (1979). Birch leaves as a resource for herbivores: Seasonal occurrence of increased resistance in foliage after mechanical damage of adjacent leaves. *Oecologia, 39*, 151–159.
Haukioja, E., Kapiainen, K., Niemelä, P., & Tuomi, J. (1983). Plant availability hypothesis and other explanations of herbivore cycles: Complementary or exclusive alternatives. *Oikos, 40*, 419–432.
Haukioja, E., Suomala, J., & Neuvonen, S. (1985). Long-term inducible resistance in birch foliage: Triggering cues and efficiency on a defoliator. *Oecologia, 65*, 363–365.
Hawkins, L., & Nicoletto, P. F. (1992). Kangaroo rat burrows structure the spatial organization of ground dwelling animals in a semiarid grassland. *Journal of Arid Environments, 23*, 199–208.
Hayes, J. P., Gross, S. P., & McIntire, P. W. (1986). Seasonal variation in mycophagy by the western red-backed vole, Clethrionomys californicus in southern Oregon. *Northwest Science, 60*(4), 250–257.
Heikkinen, R. K., & Kalliola, R. J. (1989). *Vegetation types and maps of Kevo nature reserve, northernmost Finland* (Kevo Notes 8). Turku: Kevo Subarctic Research Institute.
Heinken, T., Hanspach, H., Rautnitschka, D., & Schaumann, F. (2002). Dispersal of vascular plants by four species of wild mammals in a deciduous forest in NE Germany. *Phytocoenologia, 32*(4), 627–643.
Heinrich, B., & Bartholomew, G. A. (1980). Afrikanische Kotkäfer. *Spektrum der Wissenschaft, 1*, 61–67.
Henttonen, H., & Wallgren, H. (2001). Rodent dynamics and communities in the birch forest zone of northern Fennoscandia. In F. E. Wielgolaksi (Ed.), *Nordic mountain birch ecosystems* (Man and the Biosphere Series 27, pp. 261–278). New York/London: The Parthenom Publishing Group.
Herbold, H., Suchentrunk, F., Wagner, S., & Willing, R. (1992). Einfluss anthropogener Störreize auf die Herzfrequenz von Rotwild (Cervus elaphus) und Rehwild (Capreolus capreolus). *Zeitschrift für Jadgwissenschaft, 38*, 145–159.
Herman, R. K. (1987). Description of the host species. In M. H. Brookes, R. W. Campbell, J. J. Colbert, R. G. Mitchell, & R. W. Stark (Eds.), *Western spruce budworm* (USDA Forest Service, Cooperative State Research Service, Technical Bulletin 1694, pp. 43–56). Washington, DC: U.S. Forest Service, Cooperative State Research Service.
Herre, W. (1955). *Das Ren als Haustier. Eine zoologische Monographie*. Leipzig: Akademische Verlagsgesellschaft.
Hespeler, B. (1992). *Rehwild heute. Lebensraum, Jagd und Pflege (4. Auflage)*. München/Wien/Zürich: BLV.
Hess, A. (1916). Der Tannenhäher in forstwirtschaftlicher Bedeutung. *Schweizerische Zeitschrift für Forstwesen, 67*(9/10).
Hessburg, P. F., Mitchell, R. G., & Filip, G. M. (1994). *Historical and current roles of insects and pathogens in eastern Oregon and Washington forested landscapes*. (USDA Forest Service, Pacific Northwest Research Station, General Technical Reopt PNW GTR-327) Portland.
Heth, G. (1991). The environmental impact of subterranean mole rats (Spalax ehrenbergii) and their burrows. *Symposium of the Zoological Society London, 63*, 265–280.

Hickie, P. F. (1936). Isle Royale moose studies. *Proceedings of the North American Wildlife Conference, 1*, 396–399.
Hilderbrand, G. V., Hanley, T. A., Robbins, C. T., & Schwarz, C. C. (1999). Role of brown bears (Ursus arctos) in the flow of marine into a terrestrial ecosystem. *Oecologia, 121*, 545–550.
Hiller, B., Müterthies, A., Holtmeier, F.-K., & Broll, G. (2002). Investigations on spatial heterogeneity of humus forms and natural regeneration of larch (Larix decidua Mill.) and Swiss stone pine (Pinus cembra L.) in an alpine timberline ecotone (Upper Engadine), Central Alps, Switzerland). *Geographica Helvetica, 2*, 81–90.
Hinz, W. (1976). Zur Ökologie der Tundra Zentralspitzbergens. *Norsk Polar Institut Skrifter, 163*, 1–15.
Hinze, I. (2004). *Weaver massacre*. Bird Channel.com/bird/latest bird-news/news-archive/bird-.
Hjorth, I. (1994). *Tjädern. En skogsfågel*. Jönköping: Skogsstyrelsen.
Hodges, K. E., Krebs, C. J., & Sinclair, A. R. E. (1999). Snowshoe hares demography during a cyclic population low. *Journal of Animal Ecology, 68*, 581–594.
Hoffmann, R. T., Cowling, R. M., Douie, C., & Pierce, S. M. (1989). Seed predation and germination of Acacia eriloba in the Kuiseb River Valley, Namib Desert. *South African Journal of Botany, 78*, 443–457.
Hofmann, R. (1973). *The ruminant stomach* (East African monographs, biology 2). Nairobi: East African Literature Bureau.
Hofmann, R. (1977). Wildbiologische Erkenntnisse, ein Hilfsmittel zur Minderung der Waldschäden. *Allgemeine Forstzeitung, 32*(5), 111–115.
Hofmann, R. (1978). Die Ernährung des Rehwildes im Jahresablauf. *Allgemeine Forstzeitung, 33*(4), 1279–1284.
Hofmann, R. R. (1989). Evolutionary steps of ecophysiological adaptation and diversification of ruminants: A comparative view of their digestive system. *Oecologica, 78*, 443–457.
Hofmann, R., Cornelius, R., Linder, R., & Scheibe, K. (1998). Wiedererstehen des postglazialen Großsäugerspektrums als landschaftsgestaltendes Multi-Spezies-System: ein Pilotversuch zur "Redynamisierung" halboffener Landschaften. *Schriftenreihe für Landschaftspflege und Naturschutz, 56*, 301–315.
Hogstad, O. (1997). Population fluctuations of Epirrita autumnata (Bkh.) and Operophtera brumata (L.) (Lep. Geometridae) during 25 years and habitat distribution of their larvae during a mass-outbreak in a subalpine birch forests in Central Norway. *Fauna Norvegica Series B, 44*, 1–10.
Hohmann, U. (1994). Status specific habitat use in the common buzzard Buteo buteo. In B.-U. Meyburg & R. D. Chancellor (Eds.), *Raptor conservation today* (pp. 359–369). Berlin: The Banks.
Holcroft, A. C., & Herrero, S. (1984). Grizzly bear digging sites for Hedysarum sulphurescens roots on southwestern Alberta. *Canadian Journal of Zoology, 62*, 2571–2595.
Hole, F. (1981). Effects of animals on soil. *Geoderma, 25*, 75–112.
Holtmeier, F.-K. (1965). *Die Waldgrenze im Oberengadin in ihrer physiognomischen und ökologischen Differenzierung*. Dissertation, Mathematisch-Naturwissenschaftliche Fakultät der Rheinischen Friedrich-Wilhelms-Universität, Bonn.
Holtmeier, F.-K. (1966). Die ökologische Funktion des Tannenhähers im Zirben-Lärchenwald und an der Waldgrenze im Oberengadin. *Journal für Ornithologie, 4*, 337–345.
Holtmeier, F.-K. (1967a). Die Verbreitung der Holzarten im Oberengadin unter dem Einfluß des Menschen und des Lokalklimas. *Erdkunde, 21*(4), 249–258.
Holtmeier, F.-K. (1967b). Zur natürlichen Wiederbewaldung aufgelassener Alpweiden im Overengadin. *Wetter und Leben, 10*, 195–200.
Holtmeier, F.-K. (1969). Die Landschaft von Pontresina im Luftbild. *Erdkunde, 23*(2), 133–142.
Holtmeier, F.-K. (1974). *Geoökologische Beobachtungen und Studien an der subarktischen und alpinen Waldgrenze in vergleichender Sicht (nördliches Fennoskandien/Zentralalpen)* (Erdwissenschaftliche Forschung VIII). Wiesbaden: Franz Steiner Verlag.
Holtmeier, F.-K. (1978). Die bodennahen Winde in den Hochlagen der Indian Peaks Section, Colorado Front Range, USA. *Münstersche Geographische Arbeiten, 3*, 3–47.

Holtmeier, F.-K. (1982). "Ribbon-forest" und "Hecken". Streifenförmige Verbreitungsmuster des Baumwuchses an der oberen Waldgrenze in den Rocky Mountains. *Erdkunde, 36*, 142–153.

Holtmeier, F.-K. (1986a). Die obere Waldgrenze unter dem Einfuß von Klima und Mensch. *Abhandlungen aus dem Museum für Naturkunde, 48*(2), 395–412.

Holtmeier, F.-K. (1986b). Ökologische Studien in Lappland und Südfinnland. *Gesellschaft zur Förderung der Westfälischen Wilhelms-Universität, 1984/85*, 22–26.

Holtmeier, F.-K. (1987a). Biber und Steinwild. Ökologische Aspekte der Wiederansiedlung von Tieren in ihren ehemaligen Lebensräumen. *Münstersche Geographische Arbeiten, 26*, 99–117.

Holtmeier, F.-K. (1987b). Beobachtungen und Untersuchungen über den Ausaperungsverlauf und einige Folgewirkungen in "ribbon-forests" and der oberen Waldgrenze in der Front Range, Colorado. *Phytocoenologia, 15*(3), 373–396.

Holtmeier, F.-K. (1987c). Human impacts on high altitude forests and upper timberline with special reference to middle latitudes. In T. Fujimori & M. Kimura (Eds.), *Human impacts and management of mountain forests*. Proceedings of an International Workshop of the IUFRO project group P.1.07-00, held in Japan 5–13 September 1987, pp. 9–20. Forestry and Forest Products Research Institute, Tsukaba Norin Kenkyu Danchi-Nai, Ibaraki.

Holtmeier, F.-K. (1987d). Der Baumwuchs als klimaökologischer Faktor an der oberen Waldgrenze. *Münstersche Geographische Arbeiten, 27*, 145–151.

Holtmeier, F.-K. (1990). Disturbance and management in larch-cembra pine forests in Europe. In W. C. Schmidt (Compiler), *Proceeding of the Symposium on Whitebark Pine Ecosystems: Ecology and Management of a High-mountain Resource* (USDA Forest Service, Intermountain Research Station, General Technical Report INT-270, pp. 25–36), Ogden, UT.

Holtmeier, F.-K. (1993). Der Einfluß der generativen und vegetativen Verjüngung auf das Verbreitungsmuster der Bäume und die ökologische Dynamik im Waldgrenzbereich. Beobachtungen in Hochgebirgen Norderamerikas und den Alpen. *Geoökodynamik, 14*, 153–182.

Holtmeier, F.-K. (1994). Introduction to the Upper Engadine and its forests. In W. Schmidt & F.-K. Holtmeier (Compilers). *Proceedings of the International Workshop on Subalpine Stone Pines and their Environment: The Status of Our Knowledge*, St. Moritz, 5–11 September 1992 (USDA Forest Service, Intermountain Research Station, General Technical Report INT-GTR-309, pp. 9–17).

Holtmeier, F.-K. (1995a). European larch (Larix decidua MILL = Larix europaea DE CANDOLLE) in Middle Europe with special reference to the Central Alps. In W. C. Schmidt & K. J. McDonald (Compilers). *Proceedings Symposium on Ecology and Management of Larix Forests: A Look Ahead*, Whitefish, 5–9 October 1992 (USDA Forest Service, Intermountain Research Station, General Technical Report INT-319, pp. 41–49).

Holtmeier, F.-K. (1995b). Waldgrenzen und Klimaschwankungen. Ökologische Aspekte eines vieldiskutierten Phänomens. *Geoökodynamik, 15*, 1–24.

Holtmeier, F.-K. (1996). Die Wirkungen des Windes in der subalpinen und alpinen Stufe der Front Range, Colorado, U.S.A. *Arbeiten aus dem Institut für Landschaftsökologie, Westfälische Wilhelms-Universität, Münster, 1*, 19–45.

Holtmeier, F.-K. (2005a). Relocation of snow and its effects in the treeline ecotone – With special regard to the Rocky Mountains, the Alps and Northern Europe. *Die Erde, 136*(4), 343–373.

Holtmeier, F.-K. (2005b). Change in the timberline ecotone in northern Finnish Lapland during the last thirty years. *Reports from the Kevo Subarctic Research Station, 23*, 97–113.

Holtmeier, F.-K. (2009). *Mountain timberlines. Ecology, patchiness, and dynamics* (Advances in Global Change Research 36). Dordrecht: Springer Science+Business Media B. V.

Holtmeier, F.-K. (2012). Impact of wild herbivorous mammals and birds on the altitudinal and northern treeline ecotones. *LandscapeOnline, 30*, 1–28. doi:10.3097/LO.201230.

Holtmeier, F.-K., & Broll, G. (2006). Radiocarbon-dated peat and wood remains from the Finnish Subarctic – Evidence of treeline and landscape history. *The Holocene, 16*(5), 743–751.

Holtmeier, F.-K., & Broll, G. (2011). Response of scots pine (Pinus sylvestris) to warming climate at its altitudinal limit in northernmost subarctic Finland. *Arctic, 64*(3), 269–280.

Holtmeier, F.-K., Broll, G., Müterthies, A., & Anschlag, K. (2003). Regeneration of trees in the treeline ecotone: Northern Finnish Lapland. *Fennia, 181*(2), 103–128.

Holtmeier, F.-K., Broll, G., & Anschlag, K. (2004). Winderosion und ihre Folgen im Waldgrenzbereich und in der alpinen Stude einiger nordfinnischer Fjelle. *GeoÖko, 25*, 204–224.

Holzgang, O., Achermann G., & Gigon, A. (1996). *Productivity and usage by red deer (Cervus elaphus, L.) of two subalpine grasslands in the Swiss National Park* (pp. 35–45). Zurich: Bulletin Geobotanisches Institut ETH.

Honkavaara, J., Siitari, H., Saloranta, V., & Viitala, J. (2007). Avian seed ingestion changes germination patterns of bilberry, Vaccinium myrtillus. *Annales Botanici Fennici, 44*, 8–17.

Hoogesteger, J., & Karlsson, P. S. (1992). Effects of defoliation on radial growth and photosynthesis in the mountain birch (Betula pubescens ssp. tortuosa). *Functional Ecology, 6*, 217–323.

Hooven, E. F. (1971). Pocket gopher damage on ponderosa pine plantations in southwestern Oregon. *Journal of Wildlife Management, 35*(2), 346–353.

Höper, J. (1997). *Stoffausträge aus Böden forstlich bewirtschafteter Standorte im Wassereinzugsgebiet Hohe Ward, Münster*. Diploma thesis, Institut für Lanschaftökologie, Westfälische Wilhelms-Universität Münster.

Houston, D. B., & Stevens, V. (1988). Resource limitation in mountain goats: A test by experimental cropping. *Canadian Journal of Zoology, 66*, 228–238.

Howard, W. E. (1965). Interactions of behavior, ecology and genetics of introduced mammals. In H. G. Bakers & G. L. Stebbins (Eds.), *The genetics of colonizing species* (pp. 147–150). New York: Academic.

Howard, W. R., & Cole, R. E. (1967). Olfaction in seed detection by deer mice. *Journal of Mammalogy, 48*, 147–150.

Howe, H. F., & Westley, L. C. (1993). *Anpassung und Ausbeutung: Wechselbeziehungen zwischen Pflanzen und Tieren*. Heidelberg/Berlin/Oxford: Spektrum Akademischer Verlag.

Huntly, N., & Inouye, R. (1988). Pocket gophers in ecosystems: Patterns and mechanisms. *BioScience, 38*(11), 786–793.

Huston, N. A. (1997). Hidden treatments in ecological experiments: Re-evaluating the ecosystem function of biodiversity. *Oecologica, 110*, 449–460.

Ida, H., & Nakagoshi, N. (1996). Gnawing damage by rodents to the seedlings of Fagus crenata and Quercus mongolica var. grosseserrata in a temperate Sasa grassland-deciduous forest series in southwestern Japan. *Ecological Research, 11*, 97–113.

Imeson, A. C. (1976). Some effects of burrowing animals on slope processes in the Luxembourg Ardennes. Pt. 1: Excavation of animal burrows in experimental plots. *Geografiska Annaler, 58A*, 115–125.

Ingham, R. E., & Detling J. K. (1984). Plant-herbivore interactions in a North American mixed-grass prairies. III. Soil nematode populations and root biomass on cynomis ludovicianus colonies and adjacent uncolonized areas. *Oecologica, 63*, 307–313.

Inouye, R. S., Huntly, N. ., Tilman, D., Tester, J. R., Sillwell, M. A., & Zinnel, K. C. (1987). Old-field succession on a Minnesota sand plain. *Ecology, 68*, 12–26.

Jacob, J., & Tkadlec, E. (2010). Rodent outbreaks in other regions: A search for generalities. In G. R. Singleton, S. R. Belmain, P. R. Brown, & B. Hardy (Eds.), *Rodent outbreaks: Ecology and impacts* (pp. 207–223). Los Baños: International Rice Research Institute (IRRI).

Jacobi, A. (1931). *Das Rentier, eine zoologische Monographie der Gattung Rangifer* (Zoologischer Anzeiger 96). Leipzig: Akademische Verlagsgesellschaft.

Jahn, E. (1968). Beobachtungen zum Maissenwechsl von Forstinsekten im Hochgebirge. *Verhandlungen der Deutschen Zoologischen Gesellschaft Innsbruck, 32*, 734–740.

Janzen, D. H. (1983). Food webs: Who eats what, why, how, and with what effects in a tropical forest. In F. B. Golley (Ed.), *Tropical forest ecosystems* (pp. 167–182). Amsterdam/Oxford/New York: Elsevier Scientific Publishing.

Jarman, P. J., & Sinclair, A. R. E. (1979). Feeding strategy and the pattern of resource partitioning in ungulates. In A. R. E. Sinclair & M. Norton-Griffiths (Eds.), *Sergengeti: Dynamics of an ecosystem* (pp. 130–163). Chicago: University of Chicago Press.

Järvinen, A. (1987). Microtine cycles and plant production: What is cause and effect? *Oikos, 49*(3), 352–357.

Jarvis, J. U. M. (1981). Eusociality in a mammal: Cooperative breeding in naked mole-rat colonies. *Science, 212*(4494), 571–573.

Jensen, C. L., & Belsky, A. J. (1989). Grassland homogeneity in Tsavo National Park (West), Kenya. *African Journal of Ecology, 27*, 35–44.

Jepsen, J. U., Hagen, S. B., Ims, R. A., & Yoccoz, N. G. (2008). Climate change and outbreaks of the geometrids Operophtera brumata and Epirrita autumnata in subarctic birch forest: Evidence of a recent outbreak range expansion. *Journal of Animal Ecology, 77*, 257–264.

Jezierski, W., & Myrcha, A. (1975). Food requirements of a wild boar population. *Polish Ecological Studies, 1*, 61–83.

Johannesson-Gross, K. (1996). Qualitative Indizien zum Vorkommen des Maulwurfs (Talpa europaea): Seine gewöhnlichen und außergewöhnlichen Aktivitätsspuren. *Schriftenreihe für Landschaftspflege und Naturschutz, 46*, 77–83.

Johnson, P. C., & Denton, R. E. (1975). *Outbreaks of the western spruce budworm in the American northern Rocky Mountain area from 1922 through 1971* (General Technical Report INT-20). Ogden: USDA Forest Service, Intermountain Forest and Range Experimental Station.

Johnson, W. C., & Webb, T. (1989). The role of blue jay (Cyanocitta cristata) in the postglacial dispersal of fagaceous trees in Eastern North America. *Journal of Biogeography, 16*, 561–571.

Johnson, C. G., Clausnitzer, R. R., Mehringer, P. J., & Oliver, C. D. (1994). *Biotic and abiotic processes of eastside ecosystems: The effects of management on plant and community ecology, and on stand and landscape vegetation dynamics* (USDA Forest Service, Pacific Northwest Research Station, General Technical Report PNW-GTR 322).

Johnston, W. F. (1990). Laric laricina (Du Roi) K. Koch. Tamarack. In R. M. Burns & H. Honkala (Technical Coordinators), *Silvics of North America, Vol. 1. Conifers* (Agricultural Handbook 654). Washington, DC: USDA Forest Service.

Joly, Y., Frenot, Y., & Vernon, P. (1987). Environmental modifications of a subantarctic peat-bog by the wandering albatross (Diomedea exulans): A preliminary study. *Polar Biology, 8*, 61–72.

Jonca, E. (1972). Winter denudation of molehills in mountainous areas. *Acta Theriologica, 17*(31), 407–417.

Jones, M. G. (1933). Grassland management and its influence on the sward. *Empire Journal of Experimental Agriculture, 1*, 43–57.

Julander, O., Low, J. B., & Morris, O. W. (1959). Pocket gophers on seeded Utah mountain range. *Journal of Range Management, 12*(5), 2119–2224.

Julander, O., Robinette, W. L., & Jones, A. D. (1961). Relation of summer range condition to mule deer herd productivity. *Journal of Wildlife Management, 25*, 54–60.

Kajimoto, T., Onodera, H., Daimaru, H., & Seki, T. (1998). Seedling establishment of subalpine stone pine (Pinus pumila) by nutcracker (Nucifraga) and seed dispersal on Mt. Yumori, Northern Japan. *Arctic and Alpine Research, 30*(4), 408–417.

Kalela, O. (1957). Regulation of reproduction rate in subarctic populations of the vole Clethrionomys rufocanus (Sund). *Annales Academiae Scientarum Fenniae, Ser. A, IV, Biologica, 34*, 1–60.

Kalela, O. (1962). On the fluctuations in the numbers of arctic and boreal small rodents as a problem of production biology. *Annales Academiae Scientarum Fenniae, Series A, IV. Biologica, 66*, 1–38.

Kallio, P. (1975). Kevo, Finland. In T. Roswall & O. W. Heal (Eds.), *Structure and function of tundra ecosystems* (Ecological Bulletins 20, pp. 193–223). Stockholm: Swedish Natural Science Resource Council.

Kallio, P., & Kärenlampi, L. (1971). A review of the stage reached in the Kevo IBP in 1970. In O. W. Heal (Ed.), *IBP Tundra Biome* (Working meeting on analysis of ecosystems, Kevo, Finland, Sept. 1970, pp. 79–89). London: Tundra Biome Steering Committee.

Kallio, P., & Lehtonen, J. (1973). Birch forest damage caused by Oporinia autumnata (BKH. in 1965–1966 in Utsjoki, N-Finland. *Reports from the Kevo Subarctic Research Station, 10*, 55–59.

Kallio, P., & Lehtonen, J. (1975). On the ecocatastrophe of birch forests caused by Oporinia autumnata (Bkh.) and the problem of reforestation. In F. E. Wielgolasky (Ed.), *Fennoscandian tundra ecosystems* (Ecological Studies 17, pp. 174–180). Berlin/New York: Springer.

Kalliola, R. (1941). Tunturimittari (Oporinia autumnata) subalpiniisten koivikoiden tuholainen. *Luonnon Ystävä 45*: 45–60 (cited in Nuorteva 1963).

Kaltenrieder, P., Tinner, W., & Amnmann, B. (2005). Zur Langzeitökologie des Lärchen-Arvengürtels in den südlichen Walliser Alpen. *Botanica Helvetica, 115*, 137–154.

Kärenlampi, L. (1972). On the relationships of Scots pine annual ring width and some climatic variables at the Kevo Subarctic Research Station. *Reports from the Kevo Subarctic Research Station, 9*, 78–81.

Kausrud, K. L., Mysterud, A., Steen, H., Vik, J. O., Østbye, E., Cazelles, B., Framstad, E., Eikeset, A. M., Mysterud, I., Solhøy, T., & Stenseth, N. C. (2008). Linking climate change to lemming cycles. *Nature, 456*, 93–97.

Keane, R., E., Arno, S. F., Brown, J. K., & Tomback, D. F. (1990). Simulating disturbances and conifer succession in whitebark pine forests. In W. C. Schmidt (Ed.), *Proceedings of the symposium on whitebark pine ecosystems: Ecology and management of a high-mountain resource, Bozeman, 29–31 March* (General Technical Report INT 270, pp. 274–288). USDA Forest Service, Intermountain Research Station.

Keith, L. B. (1983). Role of food in hare population cycles. *Oikos, 40*, 385–395.

Keith, L. B., Cary, J. R., Rongstad, O. J., & Brittingham, M. C. (1984). Demography and ecology of a declining snowshoe hare population. *Wildlife Monographs, 90*, 1–43.

Kelsall, J. A. (1968). *The migratory barren-ground caribou of Canada* (Canadian Wildlife Service Monographs 3). Minister of Indian Affairs and Northern Development, Ottawa.

Kendall, K. C. (1983). Use of pine nuts by grizzly and black bears in the Yellowstone area. *International Conference on Bear Research and Management, 5*, 166–173.

Kendall, K. C., & Arno, S. F. (1990). Whitebark pine – An important but endangered wildlife resource. In W. C. Schmidt (Ed.), *Proceedings of the Symposium on Whitebark Pine Ecosystems: Ecology and Management of a High Mountain Resource*, Bozeman, 29–31 March (USDA Forest Service, Intermountain Research Station, General Technical Report INT 270, pp. 264–273).

Kenward, R. E. (1989). Bark-stripping by grey squirrels in Britain and North America: Why does the damage differ? In R. J. Putnam (Ed.), *Mammals and pests* (pp. 144–154). London/New York: Chapman and Hall.

Keplin, B. (1995). *Untersuchungen zur Bodenfauna städtischer Grünflächen unter dem Einfluß verschiedener Pflegemaßnahmen*. Dissertation Mathematisch-Naturwissenschaftliche Fakultät, Westfälische Wilhelms-Universität, Arbeitsbereich Landschaftsökologie, Münster.

Khomentovsky, P. A., Baltenseiler, W., Efremova, L. S., Pavlenko, T. V., & Marycheva, E. M. (1997). The first record of an outbreak of the larchbud moth, Zeiraphera diniana Gn. (Lep., Tortricidae) on an evergreen conifer host (Pinus pumila [Pall.] Regel) in north-eastern Asia. *Journal of Applied Entomology, 121*, 1–7.

Kielland, K., Bryant, J. P., & Ruess, R. W. (1997). Moose herbivory and carbon turnover of early successional stands in interior Alaska. *Oikos, 80*, 25–30.

Kingston, T. J., & Coe, M. (1977). The biology of a giant dung beetle (Heliocopris dilloni) (Coleoptera Scarabeidae). *Journal of Zoology (London), 181*, 243–263.

Kirchhefer, A. (1996). A dendrochronological study on the effect of climate, site, and insect outbreaks on the growth of Betula pubescens coll. in northern Norway. In B. Frenzel (Ed.), *Holocene treeline oscillations, dendrochronology and palaeoclimate* (Special issue: ESF project European palaeoclimate and man 13, pp. 93–106). Stuttgart: G. Fischer.

Kirkpatrick, J. B., Massey, J. S., & Parssons, R. F. (1974). Natural history of Curtis Island, Bass Strait. 2. Soils and vegetation. *Papers and Proceedings of the Royal Society of Tasmania, 107*, 131–144.

Kitching, R. M. (Ed.). (1995). *The handbook of New Zealand animals*. Oxford University Press New Zealand: Auckland/Oxford/Melbourne/New York.

Klaus, S., & Andreev, A. V. (2001). Raufußhuhn als "Landschaftsgärtner": Steinauerhühner in Nordostsibirien. *Der Falke, 4*, 132–137.

Klein, D. R. (1962). Rumen contents analysis as in index of range quality. *Transactions of the North American Wildlife Conference, 27*, 15–164.
Klein, D. R. (1968). The introduction, increase and crash of reindeer on St. Matthew Island. *Journal of Wildlife Management, 32*, 350–367.
Klein, D. R. (1970). Food selection by North American deer and their response to over-utilization of preferred plant species. In A. Watson (Ed.), *Animal populations in relation to their food resources* (pp. 25–46). Oxford: Blackwell.
Klekowsky, R. Z., & Opalinski, K. W. (1986). Matter and energy flow in Spitsbergen ornithogenic tundra. *Polar Research 4*(New Series), 187–197.
Klemola, T., Tanhuanpää, M., Korpimäki, E., & Ruohomäki, K. (2002). Specialists and generalists natural enemies as an explanation for geographical gradients in population cycles of northern herbivores. *Oikos, 99*, 83–94.
Klimetzek, D., & Wellestein, G. (1978). Assimilatentzug und Zuwachsminderung an Forstpflanzen durch Baumläuse (Lachnidae) unter dem Einfluß von Waldameisen (Formicidae). *Forstwissenschaftliches Centralblatt, 97*, 1–12.
Klötzli, F. (1965). *Quantität und Qualität der Rehäsung*. Zürich: Veröffentlichungen des Geobotanischen Instituts ETH, Stiftung Rübel, 38.
Klötzli, F. (1977). *Wild und Vieh im Gebirgsland Äthiopiens. Einfluß der Nutzung auf die Stabilitat des Graslandes* (Berichte über das Internationale Symposium der Vereinigung für Vegetationskunde, Vegetation und Fauna (Rinteln, 12. – 24. 4. 1976), pp. 499–512). Vaduz.
Koford, C. B. (1958). *Prairie dogs, whitefaces, and blue grass* (Wildlife monographs, 3, pp. 1–78).
Korn, H. (1991). Small mammals and the mosaic-cycle concept of ecosystems. In H. Remmert (Ed.), *The mosaic cycle concept of ecosystems* (Ecological Studies 85, pp. 106–131). Berlin/Heidelberg: Springer.
Korn, H., & Korn, U. (1989). The effect of gerbils (Tatera brantsii) on primary production and plant species composition in a southern African savanna. *Oecologia, 79*, 271–278.
Koshinka, T. V. (1970). On population dynamics of the lemming. *Fauna Ekologia Grusinoc, 9*, 11–16 (in Russian, English summary).
Kostrzewa, R. (1998). *Die Alken des Nordatlantiks*. Wiesbaden: Aula-Verlag.
Kotanen, P. M. (1995). Responses of vegetation to a changing regime of disturbance effects of feral pigs in a Californian coastal prairie. *Ecography, 18*, 190–199.
Kotter, M. M. (1981a). *Interrelationships of tassel-eared squirrels, ponderosa pine, and hypogeous mycorrhizal fungi*. M.S. thesis, Ohio State University, Columbia.
Kotter, M. M. (1981b). Interrelationships of tassel-eared squirrels. *Mycologia, 76*(4), 760–763.
Kotter, M. M., & Farentinos, R. C. (1984a). Formation of ponderosa pine ectomycorrhizae after inoculation with feces of tassel-eared squirrels. *Mycologia, 76*(4), 760–763.
Kotter, M. M., & Farentinos, R. C. (1984b). Tassel-eared squirrels Sciurus aberti ferreus as spore dispersal agents of hypogeous mycorrhizal fungi. *Journal of Mammalogy, 65*, 684–687.
Kowarik, I. (2003). *Biologische Invasionen: Neophyten und Neozoen in Mitteleuropa*. Stuttgart: Verlag Eugen Ulmer.
Krebs, C. J., & Myers, J. H. (1974). Population cycles in small mammals. *Advances in Ecological Research, 8*, 267–399.
Krefting, L. W. (1974). *The ecology of Isle Royal moose with special reference to the habitat* (Agricultural Experiment Station, State University of Minnesota, Technical Bulletin 297, Forestry Series 15). Minneapolis.
Krueger, K. (1986). Feeding relationships among bison, pronghorn, and prairie dogs: An experimental analysis. *Ecology, 67*, 760–770.
Krüger, U. (1999). Das niederländische Beispiel: Die Oostvaardersplassen - ein Vogelschutzgebiet mit Großherbivoren als Landschaftsgestalter. *Natur und Landschaft, 74*(10), 428–435.
Kühnlein, H., & Turner, N. (1991). *Traditional food plants of Canadian indigenous peoples: Nutrition, botany and use*. Philadelphia: Gordon and Breach.
Kuoch, R. (1965). Der Samenanfall 1962/63 an der oberen Fichtenwaldgrenze im Sertigtal. *Mitteilungen der Schweizerischen Anstalt für das forstliche Versuchswesen, 41*, 63–85.
Kurt, F. (1982). Das comeback. *Natur, 11*, 83–91.

Kurt, F. (1991). *Das Reh in der Kulturlandschaft*. Hamburg/Berlin: Paul Parey Verlag.

LaBau, V. J., & Mead, B. R. (2001, December). Phytomass associated with the spruce beetle (Dendroctonus rufipennis Kirby) epidemic on Kenai Peninsula and other Alaska Studies. In *Proceedings of the Alaska Reforestation Council May 24, 2000 Workshop, University of Alaska*. Agricultural and Forestry Experiment Station, University of Alaska Fairbanks, Miscellaneous Publication 2001–2. Fairbanks.

Laine, K., & Henttonen, H. (1983). The role of plant production in microtine cycles in northern Fennoscandia. *Oikos, 40*, 407–418.

Lamont, B. B., Ralph, C., & Christensen, P. E. S. (1985). Mycophagous marsupials as dispersal agents for ectomycorrhiza fungi on eucalyptus calophylla and grastrolobium bilobum. *The New Phytologist, 101*, 651–656.

Lamparski, F. (1985). *Der Einfluß der Regenwürmer Lumbricus badensis auf Waldböden im Südschwarzwald* (Freiburger Bodenkundliche Arbeiten 15). Freiburg: Inst. f. Bodenkunde u. Waldernährungslehre, Univ.

Lamprey, H. F. (1963). Ecological separation of the large mammal species in the Tarangire Game Reserve, Tanganyika. *East African Wildlife Journal, 1*, 63–92.

Lamprey, H. F. (1974). Notes on the dispersal and germination of some tree seeds through the agency of mammals and birds. *East African Wildlife Journal, 5*, 179–180.

Lange, S., Bussmann, R. W., & Beck, E. (1997). Stand structure and regeneration of the subalpine Hagenia abyssinica forests of Mt. Kenya. *Botanica Acta, 110*, 473–480.

Lanner, R. M. (1980). Avian seed dispersal as a factor in the ecology and evolution of limber and whitebark pines. In *Proceedings of the sixth American forest biology workshop*, University of Alberta, Edmonton (pp. 16–48).

Lanner, R. M. (1981). *The pinon pine, a natural and cultural history*. Reno: University of Nevada Press.

Lanner, R. M. (1984). *Trees of the Great Basin*. Reno: University of Nevada Press.

Lanner, R. M. (1990). Biology, taxonomy, evolution and geography of the stone pines of the world. In W. Schmidt & K. J. McDonald (Compilers). *Proceedings – Symposium on whitebark pine ecosystems: Ecology and management of a high-mountain resource* (General Technical Report INT 270, pp. 14–24). Ogden: USDA Forest Service, Intermountain Research Station.

Lanner, R. (1996). *Made for each other. A symbiosis of birds and pines*. Oxford/New York: Oxford University Press.

Lauckhart, J. B. (1961). *Predator management*. 41. Annual Conference, Western Association of the State Game and Fish Commission, New Mexico.

Laundre, J. W. (1993). Effects of small mammal burrows on water infiltration in a cool desert environment. *Oecologia, 94*, 43–48.

Laws, R. M. (1968). Interactions between elephant and hippopotamus populations and their environment. *East African Agricultural and Forestry Journal, 33*(Special Issue), 140–147.

Laws, R. M. (1970). Elephants as agents of habitat and landscape change in East Africa. *Oikos, 21*, 1–15.

Laws, R. M. (1984). Seals. In R. M. Laws (Ed.), *Antarctic ecology* (pp. 621–715). London: Academic Press.

Laws, R. M. (1985). The ecology of the Southern Ocean. *American Scientist, 73*, 26–40.

Laws, R. M., Parker, I. S. C., & Johnstone, R. C. B. (1970). Elephants and habitats in North Bunyoro, Uganda. *East African Wildlife Journal, 8*, 162–180.

Laycock, W. A., & Richardson, B. Z. (1975). Long-term effect of pocket gopher control on vegetation and soils in a subalpine grassland. *Journal of Range Management, 28*(6), 458–462.

Leamy, M. L., & Blakemore, L. C. (1960). The peats of the Auckland Islands. *New Zealand Journal of Agricultural Research, 3*, 526–546.

Lee, J. A. (1986). Origin of mounds under creosote bush (Larrea tridentata) on terraces of the Salt River, Arizona. *Journal of the Arizona-Nevada Academy of Science, 21*, 23–27.

Lee, K. E., & Wood, T. G. (1971). *Termites and soil*. London/New York: Academic.

Lehtonen, J. (1987). Recovery and development of birch forests damaged by Epirrita autumnata in Utsjoki area. *Reports from the Kevo Subarctic Research Station, 20*, 35–39.
Lehtonen, J., & Yli-Rekola, M. (1979). Field and ground layer vegetation in birch forests after Oporinia damage. *Reports from the Kevo Subarctic Research Station, 15*, 27–32.
Leicht, W. H. (1979). *Tiere der offenen Kulturlandschaft, Teil 2, Feldhamster, Feldmaus* (Ethologie einheimischer Säugetiere 2). Heidelberg: Quelle und Mayer.
Leirs, H., Verhagen, R., Verhayen, W., Mwanjabe, P. M., & Mbise, T. (1996). Forecasting rodent outbreaks in Africa: An ecological basis for Mastomyx control in Tanzania. *Journal of Applied Ecology, 33*, 937–943.
Leser, H. (1976). *Landschaftsökologie* (1st ed.). Stuttgart: Eugen Ulmer Verlag.
Leser, H. (1986). Problems of biotic compartments in geoecosystems. In: H. Richter & G. Schönfelder (Eds.). *Landscape synthesis part I; Geoecological foundations* (Kongressbericht Martin-Luther-Universität Halle-Wittenberg WB 1986/16 (Q 16), pp. 43–51). Halle/Saale.
Leser, H. (1997). *Landschaftökologie* (3rd ed.). Halle/Saale: Eugen Ulmer Verlag.
Leucht, A. (1983). *Einfluß der Feldmaus, Microtus arvalis (Pall.) auf die floristische Zusammensetzung von Wiesen-Ökosystemen*. Zürich: Veröffentlichungen des Geobotanischen Institutes ETH, Stiftung Rübel 79.
Leuthold, W. (1977). Spatial organization and strategy of habitat utilization of elephants in Tsavo National Park, Kenya. *Zeitschrift fur Saugetierkunde, 42*(6), 358–379.
Leuthold, W. (1978). Kann Afrikas Großwild überleben? *Vierteljahresschrift der Naturforschenden Gesellschaft Zürich, 123*, 333–340.
Levin, D. A. (1976). The mechanical defenses of plants to pathogens and herbivores. *Annual Review of Ecology and Systematics, 7*, 121–159.
Lieberman, M., & Lieberman, D. (1986). An experimental study of seed ingestion and germination in a plant-animal assemblage in Ghana. *Journal of Tropical Ecology, 2*, 113–126.
Linhart, Y. B., & Tomback, D. F. (1985). Seed dispersal by Clark's nutcracker causes multitrunk growth forms in pines. *Oecologia, 67*, 107–110.
Litaor, M. I., Mancinelli, R., & Halfpenny, J. C. (1996). The influence of pocket gophers on the status of nutrients in alpine soils. *Geoderma, 70*, 37–48.
Livingston, R. B. (1972). Influence of birds, stones and soil on the establishment of pasture juniper, Juniperus communis and red cedar, J. virginiana, in New England pastures. *Ecology, 52*, 1141–1147.
Llewellyn, M. (1972). The effects of the lime aphid Eucaliterus filiae L. (Aphidae) on the growth of lime Tilia x vulgaris Hayne. I. Energy requirements of the aphid populations. *Journal of Applied Ecology, 9*, 261–282.
Lobry de Bryn, L. A., & Conacher, A. J. (1994). The bioturbation activity of ants in agricultural and naturally vegetated habitats in semi-arid environments. *Australia Journal of Soil Research, 32*, 555–570.
Lock, J. M. (1972). The effects of hippopotamus grazing on grasslands. *Journal of Ecology, 60*, 445–467.
Lockley, R. M. (1953). *Puffins*. London: J. M. Denton and Sons.
Löffler, E. (1996). *Bioturbation als Faktor der Reliefgenese in den wechselfeuchten Tropen*. Leipzig: Klett-Perthes.
Löffler, E., & Grotz, R. (1995). *Australien*. Darmstadt: Wissenschaftliche Buchgesellschaft.
Löffler, E., & Margules, C. (1980). Wombats detected from space. *Remote Sensing of Environment, 9*, 47–56.
Logan, J. A., & Powell, J. A. (2001). Ghost forests, global warming, and the mountain pine beetle (Coleoptera: Scolytidae). *American Entomologist, 47*, 160–173.
Lott, D. F. (2002). *American bison. A natural history*. Berkeley/Los Angeles: University of California Press.
Lovegrove, B. (1993). *The living deserts of Southern Africa*. Vlaeberg: Fernwood Press.
MacLean, D. A. (1988). Effects of spruce budworm outbreaks on vegetation, structure, and succession of balsam fir forests on Cape Breton Island, Canada. In M. J. A. Werger, H. J. During, & J. T. A. Verhoeven (Eds.), *Plant form and vegetation structure* (pp. 253–261). The Hague: SPB Academic Publishing b.v.

MacLean, S. F., Fitzgerald, B. M., & Pitelka, F. A. (1974). Population cycles in arctic lemmings: Winter reproduction and predation by weasels. *Arctic and Alpine Research, 6*, 1–12.

MacMahon, J. A. (1981). Successional processes: Comparison among biomes with special reference to probable role of and influence on animals. In H. Shugart, D. Botkin, & P. West (Eds.), *Forests succession: Concept and application* (pp. 277–304). New York/Heidelberg/Berlin: Springer.

Maddock, L. (1979). The 'migration' and grazing succession. In A. R. E. Sinclair & M. Northon-Griffiths (Eds.), *Serengeti, dynamics of an ecosystem* (pp. 104–129). Chicago/London: University of Chicago Press.

Madge, D. S. (1965). Leaf fall and litter disappearance in a tropical rain forest. *Pedobiologia, 5*, 273–288.

Magor, J. I., & Ward, P. (1972). *Illustrated descriptions, distribution maps and bibliography of the species Quelea (Weaver bird: Proceidae)* (Tropical Pest Bulletin 1). London: Center for Overseas Pest Research.

Mahaney, W. C., & Boyer, M. G. (1986). Appendix: Small herbivores and their influence on landforms origins in the Mount Kenya afroalpine belt. *Mountain Research and Development, 6*(3), 256–260.

Maher, W. J. (1967). Predation by weasels on a winter population of lemmings, Banks Island, Northwest Territories. *Canadian Field Naturalist, 81*, 248–250.

Maher, J. W. (1970). The pomarine jaeger as a brown lemming predator in northern Alaska. *Wilson Bulletin, 82*, 130–157.

Malajczuk, N., Trappe, J. M., & Molina, R. (1987). Interrelationships among some ectomycorrhizal trees, hypogeous fungi and small mammals: Western Australian and northwestern American parallels. *Australian Journal of Ecology, 12*, 53–55.

Malizia, A. I., Kittlein, M. J., & Busch, C. (2000). Influence of the subterranean herbivorous rodent Ctenomys talarum on vegetation and soil. *Zeitschrift fur Saugetierkunde, 65*, 172–182.

Malloch, D. W., Pirochinski, K. H., & Raven, P. (1980). Ecological and evolutionary significance of mycorrhizal symbiosis in vascular plants (a review). *Proceedings of the National Academy of Sciences of the USA, 77*, 2113–2118.

Malo, J. E., & Suarez, F. (1995). Establishment of pasture species on cattle dung: The role of endozoochorous seeds. *Journal of Vegetation Science, 6*, 169–174.

Marchant, S., & Higgins, P. J. (1996). *Handbook of Australian, New Zealand and Antarctic birds: Vol. 1. Ratites to ducks*. Melbourne: Oxford University Press.

Markula, A., Jones, M. H., & Csurhes, S. (2009). *Redbilled quelea*. Brisbane: The State of Queensland, Department of Employment, Economic Development and Innovation, Brisbane.

Martin, G. H. (1988). Mima mounds in Kenya – A case of mistaken identity. *African Journal of Ecology, 26*, 127–1233.

Maser, Z., & Maser, C. (1987). Notes on mycophagy of the yellow-pine chipmunk (Eutamias amoenus) in northeastern Oregon. *The Murrelet, 68*, 24–27.

Maser, C., Trappe, J. M., & Nussbaum, R. A. (1978a). Fungal-mammal interrelationships with emphasis on Oregon forests. *Ecology, 59*(4), 799–809.

Maser, C., Trappe, J. M., & Ure, D. C. (1978b). *Implications of small mammal mycophagy to the management of western coniferous forests*. Transactions 43rd North American Wildlife and Natural Resources Conference, Washington, DC.

Maser, Z., Maser, C., & Trappe, J. M. (1985). Food habits of the northern flying squirrel (Claucomys sabrinus) in Oregon. *Canadian Journal of Zoology, 63*, 1086–1088.

Maser, C., Maser, Z., Witt, J. W., & Hunt, G. (1986). The northern flying squirrel: A mycophagist in southwestern Oregon. *Canadian Journal of Zoology, 64*, 2086–2089.

Mason, R. R. (1987). Nonoutbreak species of forest lepidoptera. In P. Barbosa & J. Schultz (Eds.), *Insect outbreaks* (pp. 31–57). San Diego: Academic.

Mattes, H. (1978). Der Tannenhäher im Engadin. Studien zur Ökologie und Funktion im Arvenwald. *Münstersche Geographische Arbeiten, 2*, 1–87.

Mattes, H. (1982). Die Lebensgemeinschaft von Arve und Tannenhäher. *Eidgenössische Anstalt für das forstliche Versuchswesen, Berichte 24*, 1–74.

Mattes, H. (1985). The role of animals in cembran pine forest regeneration. In H. Turner & W. Tranquillini (Eds.), *Establishment and tending of subalpine forest: Research and management. Proceedings on the 3rd IUFRO workshop P.1.07-00 1984* (Eidgenössische Anstalt für das forstliche Versuchswesen, Berichte 270, pp. 197–205). Birmensdorf.

Mattes, H., Maurizio, R., & Bürkli, W. (2005). *Die Vogelwelt im Oberengadin, Bergell und Puschlav*. Sempach: Schweizerische Vogelwarte.

Mattson, D. J., & Jonkel, C. (1990). Stone pines and bears. In W. C. Schmidt (Ed.), *Proceedings of the symposium on whitebark pine ecosystems: Ecology and management of a high mountain resource*, Bozeman, 29–31 March (USDA Forest Service, Intermountain Research Station, General Technical Report INT 270, pp. 223–236).

Mattson, D. J., & Reinhart, D. P. (1994). Bear use of whitebark pine seeds in North America. In W. C. Schmidt & F.-K. Holtmeier (Compilers). *Proceedings of the international workshop on subalpine stone pines and their environment: The status of knowledge*, St. Moritz, 5–11 September 1992 (USDA Forest Service Intermountain Research Station, General Technical Report INT-GTR 309, 212–220).

Mattson, D. J., & Reinhart, D. P. (1997). Excavation of red squirrel middens by grizzly bears in the whitebark zone. *Journal of Applied Ecology, 34*, 926–940.

Mattson, D. J., Blanchard, B. M., & Knight, R. R. (1991a). Food habits of Yellowstone grizzly bears, 1977–1987. *Canadian Journal of Zoology, 69*, 1619–1629.

Mattson, D. J., Gillin, C. M., Bensson, S. A., & Knight, R. (1991b). Bear feeding activity at alpine insect aggregation sites in the Yellowstone ecosystem. *Canadian Journal of Zoology, 69*, 2430–2435.

Mattson, D. J., Kendall, K. C., & Reinhart, D. P. (2001). Whitebark pine, grizzly bears, and red squirrels. In D. F. Tomback, S. F. Arno, & R. E. Keane (Eds.), *Whitebark pine communities. Ecology and restoration* (pp. 121–136). Washington, DC/Covelo/London: Island Press.

Mauersberger, G. (1995). *Vögel. Urania Tierreich*. Leipzig/Jena/Berlin: Urania-Verlag.

Mayer, H. (1969). *Wälder des Ostalpenraumes. Standort, Aubau und waldbauliche Bedeutung der wichtigsten Waldgesellschaften in den Ostalpen samt Vorland*. Stuttgart: Gustav Fischer Verlag.

McBee, R. H. (1971). Significance of intestinal microflora in herbivory. *Annual Review of Ecology and Systematics, 2*, 165–176.

McCullagh, K. G. (1969). The growth and nutrition of African elephants. I. Seasonal variations in rate of growth and the urinary excretion of hydroxyproline. *East African Wildlife Journal, 7*, 85–90.

McDonough, W. T. (1974). Revegetation of gopher mounds on aspen range in Utah. *Great Basin Naturalist, 34*(4), 267–275.

McGregor, S. E., Alcorn, S. M., & Olin, G. (1962). Pollination and pollinating insects of the saguaro. *Ecology, 43*, 259–267.

McKendrick, J. D., Batzli, G. O., Everett, K. R., & Swanson, J. C. (1980). Some effects of mammalian herbivores and fertilization on tundra soils and vegetation. *Arctic and Alpine Research, 12*(4), 565–578.

McLaren, B. E., & Peterson, R. O. (1994). Wolves, moose, and tree rings on Isle Royale. *Science, 266*, 155–158.

McNaughton, S. J. (1979a). Grazing as an optimization process: Grass-ungulate relationships in the Serengeti. *American Naturalist, 113*, 691–703.

McNaughton, S. J. (1979b) Grassland-herbivore dynamics. In A. R. E. Sinclair & M. Norton-Griffiths (Eds.), *Dynamics of ecosystems* (pp. 46–81). Chicago/London: University of Chicago Press.

McNaughton, S. J. (1985). Ecology of grazing ecosystems: The Serengeti. *Ecological Monographs, 55*(3), 260–294.

Mech, L. D., McRoberts, R. E., Peterson, R. O., & Page, R. E. (1987). Relationship of deer and moose populations to previous winter's snow. *Journal of Animal Ecology, 56*, 6115–6627.

Meister, G. (1969). Ziele und Ergebnisse forstlicher Planung im oberbayerischen Hochgebirge. *Forstwissenschaftliches Centralblatt, 88*, 97–130.

Mellmann-Brown, S. (2005). Regeneration of whitebark pine in the timberline ecotone of the Beartooth Plateau, U.S.A.; Spatial distribution and responsible agents. In G. Broll & B. Keplin (Eds.), *Mountain ecosystems. Studies in treeline ecology* (pp. 97–115). Berlin/Heidelberg/New York: Springer.

Merz, G. (1982). *Untersuchungen über Lebensraum und Verhalten des Afrikanischen Waldelefanten im Tai-Nationalpark der Republik Elfenbeinküste unter dem Einfluß der regionalen Entwicklung*. Dissertation Naturwissenschaftlich-Mathematische Gesamtfakultät, Universität Heidelberg.

Messier, F. (1994). Ungulate population models with predation: A case study with the North American moose. *Ecology, 75*(2), 478–488.

Meyer, A. (1955). Frühere Verbreitung der Holzarten und einstige Waldgrenze im Kanton Wallis. *Mitteilungen der Schweizerischen Anstalt für das forstliche Versuchswesen, 31*(3), 563–668.

Midgley, G. F., & Musil, C. F. (1990). Substrate effects of zoogenic soil mounds on vegetation composition in the Worcester-Robertson valley, Cape Province. *South African Journal of Botany, 52*(2), 158–166.

Miehe, G. (1994). Recent climatic changes in subhumid to arid regions of High Asia (Karakorum, Kunlun, Himalaya, Tibet) as indicted by vegetation features. In D. Zhen, Q. Thang, & Y. Pan (Eds.), *Proceedings of international symposium on the Karakorum and Kunlun Mountains* (pp. 333–346). Beijing: China Meteorological Press.

Miehe, G., & Miehe, S. (2000). Comparative high mountain research on the treeline ecotone under human impact. *Erdkunde, 54*, 34–50.

Mielke, H. (1977). Mound building by pocket gophers (Geomyidae): Their impact on soils and vegetation in North America. *Journal of Biogeography, 4*, 171–180.

Mikola, P. (1971). Reflexion of climatic fluctuation in the forestry practices of Northern Finland. *Reports from the Kevo Subarctic Research Station, 8*, 116–121.

Miller, M. F. (1995). Acacia seed survival, seed germination and seedling growth following pod consumption by large herbivores and seed chewing rodents. *African Journal of Ecology, 33*, 194–210.

Milton, W. E. J. (1940). The effect of manuring, grazing and liming on the yield, botanical and chemical composition of natural hill pastures. *Journal of Ecology, 28*, 326–356.

Milton, W. E. J. (1947). The composition of natural hill pastures, under controlled and free grazing, cutting and manuring. *Welsh Journal of Agriculture, 14*, 182–195.

Milton, S. J., Dean, W. R. J., & Klotz, S. (1997). Effects of small-scale animal disturbances on plant assemblages of set-aside land in central Germany. *Journal of Vegetation Science, 8*, 45–54.

Mitchell, P. (1988). The influence of vegetation, animals and micro-organisms on soil processes. In H. Viles (Ed.), *Biogeomorphology* (pp. 43–82). Oxford: Basil Blackwell.

Mitchell, B., & Brown, D. (1974). The effect of age and body size on fertility in female red deer, cervus elaphus. *International Congress of Game Biologists, 10*, 437–450.

Mitton, J. B., & Ferrenberg, S. M. (2012). Mountain pine beetle develops an unprecedented summer generation in response to climatic warming. *The American Naturalist 1979*(5), 9 pp. doi:10.1086/665007.

Möller, I. (2000). Pflanzensoziologische und vegetationskundliche Studien in Nordwestspitzbergen. *Mitteilungen der Geographischen Gesellschaft Hamburg 90*. Stuttgart: Franz Steiner Verlag.

Moorhead, D. L., Fisher, F. M., & Whitford, W. C. (1988). Cover of spring annuals on nitrogen-rich Kangaroo rat mounds in a Chihuahuan desert grassland. *The American Midland Naturalist, 120*(2), 443–447.

Morrison, C. G. T., Hoyle, A. C., & Hope-Simpson, J. F. (1948). Tropical soil vegetation catenas and mosaics. *Journal of Ecology, 36*, 1–84.

Mörschel, F. M. (1996). *Effects of weather and parasitic insects on summer ecology of caribou of the Delta Herd*. M.S. Thesis, University of Alaska, Fairbanks.

Mörschel, F. M. (1999). Dynamische Prozesse in Huftierpopulationen am Beispiel von Karibus in Alaska und ihre Bedeutung für Mitteleuropa. In B. Gerken & M. Görner (Eds.), *Europäische Landschaftsentwicklung mit großen Weidetieren. Geschichte, Modelle und Perspektiven* (Natur und Landschaft 3, pp. 193–197). Höxter/Jena: Huxaria Druck und Verlag.

Mörschel, F. M., & Klein, D. R. (1997). Effects of weather and parasitic insects on behavior and group dynamics of caribou of the Delta Herd, Alaska. *Canadian Journal of Zoology, 75*, 1659–1670.

Moser, L. (1960). Verbreitung und Bedeutung der Zirbe im italienischen Alpengebiet. *Jahrbuch des Vereins zum Schutze der Alpenpflanzen und –tiere, 25*, 1–6.

Mosimann, T. (1984). *Landschaftsökologische Komplexanalyse*. Stuttgart: Franz Steiner Verlag.

Mould, E. D., & Robbins, C. T. (1981). Nitrogen metabolism in elk. *Journal of Wildlife Management, 45*, 323–334.

Mraz, K. (1960). *Zusammenhang zwischen Wildbestand und Waldgesellschaft*. Biosoziologie, Berichte über das Internationale Symposium in Stolzenau 78, Weser, pp. 331–340.

Mrotzek, R., Halder, M., & Schmidt, W. (1999). Die Bedeutung von Wildschäden für die Diasporenausbreitung von Phanerogamen. *Verhandlungen der Gesellschaft für Ökologie, 29*, 437–443.

Müller-Schneider, P. (1948). Untersuchungen über die endozoochore Samenverbreitung durch Weidetiere im Schweizerischen Nationalpark. *Ergebnisse der Untersuchungen im Schweizerischen Nationalpark, 2*, 1–13.

Müller-Schneider, P. (1977a). *Über die Rolle der Tiere bei der Samenverbreitung*. Berichte über das Internationale Symposium der Internationalen Vereinigung für Vegetations-kunde, Vegetation und Fauna, Rinteln, 12–15 April 1976, pp. 119–130.

Müller-Schneider, P. (1977b). *Verbreitungsbiologie (Diasporologie) der Blütenpflanzen*. Zürich: Veröffentlichungen des Geobotanischen Institutes ETH, Stiftung Rübel 61.

Müller-Schwarze, D. (1984). *The behavior of penguins*. Albany: State University of New York Press.

Mun, H. T., & Whitford, W. G. (1990). Factors affecting animal plant assemblages on banner-tailed kangaroo rat mounds. *Journal of Arid Environments, 18*, 165–173.

Munroe, E. (1979). Lepidoptera. In H. V. Danks (Ed.), *Canada and its insect fauna* (Memoirs of the entomological society of Canada, Vol. 108, pp. 427–481). Ottawa: Entomological Society of Canada.

Müri, H. (1999a). Witterung, Fortpflanzung und Bestandesdichte beim Reh. *Zeitschrift für Jagdwissenschaft, 45*, 1–8.

Müri, H. (1999b). Veränderungen im Dispersal von Rehen in einer stark frequentierten Landschaft. *Zeitschrift für Ökologie und Naturschutz, 8*, 41–51.

Murie, A. (1934). *The moose of Isle Royale* (University of Michigan, Museum of Zoology, Miscellaneous Publications 25). Ann Arbor: University of Michigan Press.

Murray, M. G., & Illius, A. W. (1996). Multispecies grazing in the Serengeti. In J. Hodgson & A. W. Illius (Eds.), *The ecology and management of grazing systems* (pp. 247–272). Wallingford: CAB International.

Murton, R. K., & Westwood, N. J. (1976). Bird as a pest. In T. H. Coaker (Ed.), *Applied Biology I* (pp. 1–181). London: Academic Press.

Müterthies, A. (2002). *Struktur und Dynamik der oberen Grenze des Lärchen-Arvenwaldes im Bereich aufgelassener Alpweiden im Oberengadin*. Arbeiten aus dem Institut für Landschaftsökologie 11, Westfälische Wildhelms-Universität, Münster.

Myrberget, S. (1973). Geographical synchronisms of cycles of small rodents in Sweden. *Oikos, 24*, 224–240.

Myrberget, S. (1986). A note on cycles in plant production and microtine populations. *Oikos, 46*, 264–267.

Myrcha, A., & Tatur, A. (1991). Ecological role of the current and abandoned penguin rookeries in the land environment of the maritime Antarctic. *Polish Polar Research, 12*(1), 3–24.

Myrcha, A., Pietr, S. J., & Tatur, A. (1985). The role of penguin rookeries in nutrient cycles at Admiralty Bay, King George Island. In W. R. Siegfried, P. R. Condy, & R. M. Laws (Eds.), *Antarctic nutrient cycles and food webs* (pp. 156–162). Berlin/New York: Springer.

Naarding, J. A. (1981). *Study on the short-tailed Shearwater (Puffinus tenuirostris) in Tasmania* (Tasmanian Wildlife Division, Technical Report 81 (3)). Hobart: National Parks and Wildlife Service.

Nägeli, W. (1969). Waldgrenze und Kampfzone in den Alpen. *Hespa-Mitteilungen, 19*(1), 1–44.

Nather, H. (1958). Zur Keimung der Zirbensamen. *Centralblatt für das gesamte Forstwesen, 75*(1), 61–70.

National Audubon Society. (1992). *Beringia natural history* (Notebook Series). Anchorage: National Audubon Society.

Naveh, Z., & Whittaker, R. H. (1980). Structural and floristic diversity of shrublands and woodlands in northern Israel and other Mediterranean areas. *Vegetatio, 41*, 171–190.

Nellemann, C., Moe, S. R., & Rutina, L. P. (2002). Links between terrain characteristics and forage pattern of elephants (Loxodonta africana) in northern Botswana. *Journal of Tropical Ecology, 18*, 835–844.

Neuvonen, S., Niemelä, P., & Virtanen, T. (1999). Climate change and insect outbreaks in boreal forests: The role of winter temperatures. *Ecological Bulletin, 47*, 63–67.

Neuvonen, S., Bylund, H., & Tømmervik, H. (2005). Forest defoliation risks in birch forests by insects under different climate and land use scenarios in northern Europe. In F. E. Wielgolaski (Ed.), *Plant ecology, herbivory, and human impact in Nordic mountain birch forests* (pp. 125–138). Berlin/Heidelberg: Springer.

Newey, S., Willebrand, T., Haydon, D. T., Dahl, F., Aebischer, N. J., Smith, A. A., & Thirgood, J. (2007). Do mountain hare populations cycle? *Oikos, 116*, 1447–1557.

Nicholson, J. A., Paterson, I. S., & Currie, A. (1970). A study of vegetational dynamics: Selection by sheep and cattle in Nardus pasture. In A. Watson (Ed.), *Animal population in relation to their food resources* (Symposium of the British Ecological Society, pp. 129–143). Oxford: Blackwell.

Niemelä, P. (1979). Topographical delimitation of Oporinia-damages: Experimental evidence of the effect of winter temperature. *Reports from the Kevo Subarctic Research Station, 15*, 27–32.

Niethammer, J. (1988). *Sandgräber. Grzimeks Enzyklopädie, Vol. 3, Säugetiere* (pp. 310–313). München: Kindler Verlag.

Nikula, A. (1993). Animals as pests in Finnish Lapland. In R. Jalkanen, A. T. Aaalto, & M.-L. Lahti (Eds.), *Forest pathological research in northern forests with a special reference to abiotic stress factors. Extended SNS meeting in forest pathology in Lapland, Finland, 3–7 August 1992* (Finnish Forest Research Institute Research Papers 451, pp. 22–29). Rovanieni.

Nilssen, A., & Tenow, O. (1990). Diapause embryo growth and supercooling capacity of Epirrita autumnata eggs from northern Fennoscandia. *Entomologia Experimentalis et Applicata, 57*, 39–55.

Nilsson, S. G. (1985). Ecological and evolutionary interactions between reproduction of beech Fagus silvatica and seed eating animals. *Oikos, 44*, 152–164.

Nola, P., Morales, M., Motta, R., & Villalba, R. (2006). The role of larch budmoth (Zeiraphera diniana Gn.) on forest succession in a larch (Larix decidua Mill.) and Swiss stone pine (Pinus cembra L.) stand in the Susa Valley (Piedmont, Italy). *Trees, 20*, 371–382.

Novak, V., Hrozinka, F., & Stary, B. (1989). *Atlas schädlicher Forstinsekten* (translated and revised by K. Rack). Stuttgart: Ferdinand Enke Verlag.

Nuorteva, P. (1963). The influence of Oporinia autumnata (Bkh.) (Lep., Geometridae) on the timberline in subarctic conditions. *Annales Entomologici Fennici, 29*, 270–277.

Odening, K. (1984). *Antarktische Tierwelt. Einführung in die Biologie der Antarktis*. Leipzig/Jena/Berlin: Urania Verlag.

Odum, E. (1971). *Fundamentals of ecology* (3rd ed.). Philadelphia/London/Toronto: Saunders.

Odum, E. (1983). *Grundlagen der Ökologie in 2 Bänden* (2nd ed.). Stuttgart/New York: Thieme.

Odum, E. (1991). *Prinzipien der Ökologie. Lebensräume, Stoffkreisläufe, Wachstumsgrenzen.* Heidelberg: Verlag Spektrum der Wissenschaft.

Oksanen, L., & Oksanen, T. (1981). Lemmings (Lemmus lemmus) and grey-sided voles (Clethrionomys rufocanus) in interaction with their resources and predators in Finnmarksvidda, northern Norway. *Reports from the Kevo Subarctic Research Station, 17*, 7–31.

Onderscheka, K. (1986). Auswirkungen der Umweltveränderungen auf das Rot- und Gamswild. *Alpine Umweltprobleme, Ergebnisse des Forschungsprojektes Achenkirch, 19*, 31–35.
Osborn, B., & Alland, P. F. (1949). Vegetation of an abandoned prairie-dog town in tall grass prairie. *Ecology, 30*, 322–332.
Owen, D. F., & Wiegert, R. G. (1976). Do consumers maximize plant fitness? *Oikos, 38*, 488–492.
Owen, D. F., & Wiegert, R. G. (1981). Grasses and grazers: Is there a mutualism? *Oikos, 38*, 258–259.
Owen, D. F., & Wiegert, R. G. (1987). Leaf eating as a mutualism. In P. Barbosa & J. C. Schultz (Eds.), *Insect outbreaks* (pp. 81–95). London: Academic Press.
Owen-Smith, N. (1982). Factors influencing the consumption of tropical savannas. In B. J. Huntley & B. H. Walker (Eds.), *Ecology of tropical savannas* (Ecological Studies 42, pp. 359–404). Berlin/New York: Springer.
Palm, T. (1959). Följdverkningar ave fjällbjörkmätarens härjning I Abiskodalen 1854–1956. En koleopterisk undersökning somrarna 1958 och 1959. *Entomologisk Tidsskrifte, 80*, 120–136.
Palo, R. T., Bergström, R., & Danell, K. (1992). Digestibility, distribution of phenols, and fiber at different twig diameters of birch in winter. Implication for browsers. *Oikos, 65*, 450–454.
Parker, I. S. C. (1983). The Tsavo story: An ecological case history. In N. Owen-Smith (Ed.), *Management of large mammals in African conservation areas* (pp. 37–49). Pretoria: HAUM.
Pastor, J., Naiman, R. J., Dewey, B., & McInnes, P. (1988). Moose, microbes, and the boreal forest. *BioScience, 38*, 770–777.
Pease, J. L., Vowles, R. H., & Keith, L. B. (1979). Interactions of snowshoe hares and woody vegetation. *Journal of Wildlife Management, 43*, 43–60.
Pellew, R. A. P., & Southgate, B. J. (1984). The parasitism of Acacia tortilis in the Serengeti. *African Journal of Ecology, 22*, 73–75.
Pelz, H.-J. & Lauenstein, G. (1989, August 22–29). *Nutritional deficiencies as cause for changes in density in Microtus arvalis (Pallas) on grassland (Abstract)*. 5th International Theriological Congress, Rome, pp. 549–550.
Pennycuick, L. (1975). Movements of the migratory wildebeest population in the Serengeti area between 1960 and 1973. *East African Wildlife Journal, 13*, 65–87.
Perry, D. A. (1994). *Forest ecosystems*. London/Baltimore: Johns Hopkins University Press.
Petelle, M. (1980). Aphids and melizitose. A test of Owen's 1978 hypothesis. *Oikos, 35*, 127–128.
Petelle, M. (1984). Aphid sugars and soil nitrogen. *Soil Biology and Biochemistry, 16*(3), 203–206.
Peterken, G. (1993). *Woodland conservation and management* (2nd ed.). London: Chapman & Hall.
Peterson, R. O., Page, R. E., & Dodge, K. M. (1984). Wolves, moose, and the allometry of population cycles. *Science, 224*, 1350–1352.
Petrak, M. (1988). Skilanglauf und Rothirsch (Cervus elaphus Linné, 1758) in der Eifel. *Bilanz des Konzeptes zur Lebensraumberuhigung im Wintersportgebiet Monschau. Zeitschrift für Jagdwissenschaft, 34*, 104–114.
Petrak, M. (1993). Beäsung als Standotfaktor unter besonderer Berücksichtigung des Rehwildes. In R. R. Hofmann & Z. Pielowski (Eds.), *Rehwild Symposium Gießen* (Schriftenreihe des Arbeitskreises Wildbiologie an der Justus Liebig Universität Gießen e.V. 21, pp. 105–116).
Peveling, R. (1990). Nutzungsabhängige Zusammensetzung und Verbreitung südsomalischer Grasgesellschaften und ihre Einfluss auf das Schadenspotential von Quelea quela (L.). *Geomethodica, 15*, 107–139.
Pitchard, G. T., & Robbins, C. T. (1990). Digestive and metabolic efficiencies of grizzly and black bears. *Canadian Journal of Zoology, 68*, 1645–1651.
Pitelka, F. A. (1957). Some characteristics of microtine cycles in the Arctic. In H. Hansen (Ed.), *Arctic Biology, Eighteenth Annual Biological Colloquium Oregon* (pp. 73–88). Corvallis.
Pitschmann, H., Resigl, H., Schiechtl, H.-M., & Stern, R. (1970). *Karte der aktuellen Vegetation von Tirol 1: 100 000. I. Teil, Blatt 6, Innsbruck-Stubaier Alpen*. Documents pour la Végétation des Alpes. IX., Grenoble, pp. 7–34.

Pitschmann, H., Reisigle, H., Schiechl, H.-M., & Stern, R. (1971). *Karte der aktuellen Vegetation cvon Tirol 1: 100 000. II. Teil, Blatt 7, Zillertaler und Tuxer Alpen*. Documents pour la Végétation des Alpes. X., Grenoble, pp. 109–132.

Platt, W. J. (1975). The colonization and formation of equilibrium plant species associations on badger disturbances in a tall-grass prairie. *Ecological Monographs, 45*, 285–305.

Polley, H. W., & Collins, S. L. (1984). Relationships of vegetation and environment in buffalo wallows. *American Midland Naturalist, 112*, 178–186.

Potter, G. L. (1978). The effect of small mammals in forest ecosystem structure and function. In D. P. Snyder (Ed.), *Population of small mammals under natural conditions* (Special publication 5, pp. 181–191). Pittsburgh: Pymatung Laboratory of Ecology.

Powell, J. A. (1980). *Nomenclature of nearctic conifer-feeding Chorsitoneura (Lepidoptera: Tortricidae): historical review and present status* (USDA Forest Service, Pacific Northwest Forest and Range Experimental Station, Portland, General Technical Report PNW-100).

Price, L. W. (1971). Geomorphic effect of the Arctic ground squirrel in an alpine environment. *Geografiska Annaler, 53A*, 100–106.

Price, M. V., & Jenkins, S. H. (1987). Rodents as seed consumers and dispersers. In D. Murray (Ed.), *Seed dispersal* (pp. 191–235). Sydney: Academic Press.

Rabotnov, T. A. (1995). *Phytozönologie*. Stuttgart: Ulmer.

Rasmussen, D. I. (1941). Biotic communities of Kaibab Plateau, Arizona. *Ecological Monographs, 11*, 229–275.

Rasmussen, R. A., & Khalil, M. A. K. (1986). Atmospheric trace gases: Trends and distribution over the last decades. *Science, 232*, 1623–1627.

Raupach, M. R., Kirby, J. M., Barrett, D. J., & Briggs, P. R. (2001). *Balances of water, carbon, nitrogen and phosphorous in Australian landscapes: (1) Project description and results*. Canberra: CSIRO Land and Water.

Reardon, P. O., Leineweber, C. L., & Merril, L. B. (1972). The effect of bovine saliva on grass. *Journal of Animal Science, 34*, 897–898.

Reardon, P. O., Leineweber, C. L., & Merril, L. B. (1974). Response of sideoats grama to animal saliva and thiamine. *Journal of Range Management, 27*, 400–401.

Rebertus, A. J., Burns, B. R., & Veblen, T. T. (1991). Stand dynamics of Pinus flexilis dominated subalpine forests in the Colorado Front Range. *Journal of Vegetation Science, 2*, 445–458.

Reichman, O. J., & Smith, S. C. (1985). Impact of pocket gopher burrows on overlaying vegetation. *Journal of Mammalogy, 66*, 720–725.

Reid, R. W., Whitney, H. S., & Watson, J. A. (1967). Reactions of lodgepole pines to attack by Dendroctonus ponderosae Hopkins and the blue stain fungi. *Canadian Journal of Botany, 45*, 1115–1126.

Reig, O. A. (1970). Ecological notes on the fossorial octodont rodent Spalacopus cyanus (Molina). *Journal of Mammalogy, 51*, 592–602.

Reijmers, F. F. (1959). Birds of the cedar pine forests of South-Central Siberia and their role in the life of cedar pine. *Truddy Bioligischeskokogo Instituta, Sibirskogo Otdelenie Akad, Nauk, USSR 5*, 121–166.

Reimoser, F. (2003). Steering the impacts of ungulates on temperate forests. *Journal of Nature Conservation, 10*, 243–252.

Reimoser, F., Mayer, H., Holzinger, A., & Zandl, J. (1987). Einfluß von Sommer- und Wintertourismus auf Waldschäden durch Schalenwild im Angertal (Bad Gastein). *Centralblatt für das gesamte Forstwesen, 104*, 95–118.

Reinhart, D. P., & Mattson. (1990a). Schalenwild und Wintersport. *Laufener Seminarbeiträge, 6*, 39–45.

Reinhart, D. P., & Mattson, D. J. (1990b). Red squirrels in the whitebark zone. In W. C. Schmidt (Ed.), *Proceedings of the symposium on whitebark pine ecosystems: Ecology and management of a high-mountain resource, Bozeman, 29–31 March, 1989* (General Technical Report INT-270, pp. 256–263). Ogden: USDA Forest Service, Intermountain Research Station.

Remmert, H. (1968). Über die Bedeutung volkreicher Meeresvogelkolonien und pflanzenfressender Landtiere für die Tundra Spitzbergens. *Veröffentlichungen des Instituts für Meeresforschung, Bremerhaven, 11*, 47–60.

Remmert, H. (1973). Über die Bedeutung warmblütiger Pflanzenfresser für den Energiefluß in terrestrischen Ökosystemen. *Journal für Ornithologie, 114*, 227–249.
Remmert, H. (1980a). *Ökologie*. Berlin/Heidelberg/New York: Springer.
Remmert, H. (1980b). *Arctic animal ecology*. Berlin/Heidelberg/New York: Springer.
Remmert, H. (1992). Das Mosaikzyklus-Konzept und seine Bedeutung für den Naturschutz -eine Übersicht. *Laufener Seminarbeiträge, 2/92*, 45–57.
Reynolds, H. G. (1958). Merriam kangaroo rat (Dipodomys merriami Mearns) on the grazing land of southern Arizona. *Ecological Monographs, 28*, 111–127.
Rhoades, D. F. (1983). Herbivore population dynamics and plant chemistry. In R. F. Denno & M. S. McClure (Eds.), *Variable plants and herbivores in natural and managed systems* (pp. 155–220). New York: Academic.
Rhoades, D. F., & Cates, R. (1976). Towards a general theory of plant antiherbivore chemistry. In J. W. Wallace & R. L. Mansell (Eds.), *Biochemical interactions between plants and insects* (Recent Advances in Phytochemistry 10). New York: Plenum Press.
Richardson, F. (1961). Breeding colony of the rhinoceros auklet on Protection Island, Washington. *Condor, 63*, 456–473.
Richens, V. B. (1966). Notes on the digging activity of the northern pocket gopher. *Journal of Mammalogy, 47*(3), 531–533.
Riney, T. (1964). *The impact of introductions of large herbivores on tropical environment* (New Series 4, pp. 261–273). Gland, Switzerland: IUCN Publications.
Ripple, W. J., & Beschta, R. L. (2004). Wolves and the ecology of fear: Can predation risk structure ecosystems? *BioScience, 54*(8), 755–766.
Ripple, W. J., & Larsen, E. J. (2000). Historic aspen recruitment, elk, and wolves in northern Yellowstone National Park, USA. *Biological Conservation, 95*, 361–370.
Robbins, C. T. (1983). *Wildlife feeding and nutrition*. New York: Academic Press.
Robinson, F. N., & Frith, H. J. (1981). The superb lyre bird Menura novahollandiae at Tindinbilla A. C. T. *Emu, 81*, 145–157.
Rogers, L. E., & Lavigne, R. J. (1974). Environmental effects of western harvester ants on the shortgrass plains ecosystems. *Environmental Entomology, 3*(6), 994–997.
Rolland, C., Baltensweiler, W., & Petitcolas, V. (2001). The potential for using larix decidua rind widths in reconstructions of larchbud moth (Zeiraphera diniana) outbreak history: Dendrochronological estimates compared with insect surveys. *Trees, 15*, 414–424.
Romme, W. H., & Despain, D. G. (1990). Die Yellowstone-Brände: Feuer als Lebensspender. *Spektrum der Wissenschaft, 1*, 114–123.
Ronco, F. (1967). *Lessons from artificial regeneration studies in a cutover beetle-killed spruce stand in western Colorado* (USDA Forest Service, Research Note RM 90). Fort Collins, CO: Rocky Mountain Forest and Range Experiment Station.
Ross, B. A., Tester, J. R., & Breckenridge, W. J. (1968). Ecology of mima-type mounds in northwestern Minnesota. *Ecology, 49*, 172–177.
Rudolph, J. (1994). Anomalous methane. *Nature, 368*, 19–20.
Ruohomäki, K., Virtanen, T., Kaitaniemi, P., & Tammaru, T. (1997). Old mountain birches at high altitudes are prone to outbreaks of Epirrita autumnata (Lepidoptera: Geometridae). *Environmental Entomology, 26*, 1096–1104.
Russow, F., & Heinrich, J. (2001). Jungholozäne Überprägungen von quartären Deckschichten und Böden des Mittelgebirgsraumes durch biomechanische Prozesse in Mitteleuropa. *Geoöko, 22*, 37–290.
Ryser, P., & Gigon, A. (1985). Influence of seed bank and small mammals on the floristic composition of limestone grassland (Mesobrometum) on northern Switzerland. *Berichte des Geobotanischen Instituts ETH, Stiftung Rübel, Zürich, 52*, 41–52.
Safranyik, L., Shrimpton, D. M., & Whitney, H. S. (1975). An interpretation of the interaction between lodpole pine, the mountain pine beetle and its associate Blue stain fungi in western Canada. In D. M. Baumgartner (Ed.), *Management of lodgepole pine ecosystems, symposium proceedings* (pp. 406–428). Pullman: Washington State University Cooperative Extension Service.

Sala, O. E. (1988). The effect of herbivory on vegetation structure. In M. J. A. Werger, P. J. M. Van der Aart, H. J. During, & J. T. A. Verhoeven (Eds.), *Plant form and vegetation structure* (pp. 317–330). The Hague: SPB Academic Publishing.

Sánchez-Zapata, J. A., Donázar, J. A., Deldgado, A., Forero, M. G., Ceballos, O., & Hiraldo, F. (2007). Desert locust outbreaks in the Sahel: Resource competition, predation and ecological effects of pest control. *Journal of Applied Ecology, 44*, 323–329.

Santos, T., Telleria, J. L., & Virgos, E. (1999). Dispersal of Spanish juniper Juniperus thurifera by birds and mammals in a fragmented landscape. *Ecography, 22*, 193–201.

Satchell, J. E. (Ed.). (1983). *Earthworm ecology from Darwin to vermiculture*. London/New York: Chapman and Hall.

Schaffer, W. M., & Tamarin, R. H. (1973). Changing reproductive rates and population cycles in lemmings and voles. *Evolution, 27*, 111–124.

Schaller, G. B., & Vasconcelos, J. M. C. (1978). A marsh deer census in Brazil. *Oryx, 14*, 345–351.

Scheunert, T. A., & Zimmermann, K. (1952). Bakterielle Symbiose im Blinddarm und Koprophagie beim Kaninchen. *Archiv für Tierernährung, 2*(4), 217–222.

Schickhoff, U. (1993). *Das Kaghan-Tal im Westhimalaya (Pakistan)* (Bonner Geographische Abhandlungen 87). Bonn: Ferdinand Dümmlers Verlag.

Schickhoff, U. (2005). The upper timberline in the Himalayas, Hindu Kush and Karakorum; a review of geographical and ecological aspects. In G. Broll & B. Keplin (Eds.), *Mountain ecosystems. Studies in treeline ecology* (pp. 275–354). Berlin/Heidelberg/New York: Springer.

Schimpf, D. J., Henderson, J. A., & MacMahon, J. A. (1980). Some aspects of succession in the spruce-fir zone of northern Utah. *The Great Basin Naturalist, 40*(1), 1–26.

Schmid, J. M., & Frye, R. H. (1977). *Spruce beetle in the Rockies* (General Technical Report RM-49). Fort Collins: USDA Forest Service, Rocky Mountain Forest and Range Experimental Station.

Schmid, J. M., & Hinds, T. E. (1974). *Development of spruce-fir stands following spruce beetle outbreaks* (Research Paper RM-131). Fort Collins: USDA Forest Service, Rocky Mountain Forest and Range Experimental Station.

Schmutterer, H. (1965). Großschädling Blutschnabelweber. Eine neue Plage für die afrikanische Landwirtschaft. *Kosmos, 11*, 469–4577.

Schneider, M. F. (2001). Habitat loss, fragmentation and predator impact: Spatial implications for prey conservation. *Journal of Applied Ecology, 38*, 720–735.

Schönwiese, C. D. (1992). *Klima im Wandel - Tatsachen, Irrtümer, Risiken*. Stuttgart: Deutsche Verlagsanstalt.

Schrader, S. (1999). Bodentiere als "ecosystem engineers" in Agroökosytemen. *Wasser und Boden, 51*(12), 48–51.

Schreiber, K.-F. (1969). Beobachtungen über die Entstehung von "Buckelweiden" auf den Hochflächen des Schweizer Jura. *Erdkunde, 23*(4), 280–290.

Schreiber, K.-F. (1980a). Brachflächen in der Kulturlandschaft. Daten und Dokumente zum Umweltschutz. *Sonderreihe Umwelttagung, Universität Hohenheim, 30*, 61–93.

Schreiber, K.-F. (1980b). Entstehung von Ökosystemen und ihre Beeinflussung durch menschliche Eingriffe. In Der Minister für Umwelt, Raumplanung und Bauwesen Saarland (Ed.), *Eine Welt darin zu leben* (pp. 34–51). Saarbrücken: Der Minister für Umwelt, Raumordnung und Bauwesen Saarland.

Schreiber, K.-F. (1989). The history of landscape ecology in Europe. In I. S. Zonneveld & R. T. T. Forman (Eds.), *Changing landscapes: An ecological perspective* (pp. 22–33). Berlin/Heidelberg/London/Paris/Tokyo: Springer.

Schröder, W. (1977). Räumliche Verteilung und Nahrungsauswahl von Gams und Rotwild im Hochgebirge. *Forstwissenschaftliches Centralblatt, 96*, 94–99.

Schroth, K.-E. (1994). *Zum Lebensraum des Auerhuhns (Tetrao urogallus L.) im Nordschwarzwald. Eine Analyse der Kaltenbronner Auerhuhnhabittate und deren Veränderungen seit Beginn der geregelten Forstwirtschaft (1843–1993)*. Mitteilungen der forstlichen Versuchs- und Forschungsanstalt Baden-Württemberg 178, Abteilung Landespflege, Freiburg.

Schultz, A. M. (1972). A study of an ecosystem: The arctic tundra. In G. M. Van Dyne (Ed.), *The ecosystem concept in natural resource management* (pp. 77–93). New York: Academic Press.
Schultz, J. (1988). *Die Ökozonen der Erde*. Stuttgart: Ulmer.
Schultz, J. (2000). *Handbuch der Ökozonen*. Stuttgart: Ulmer.
Schultz, J. C., & Baldwin, I. T. (1982). Oak leaf quality decline in response to defoliation by gypsy moth larvae. *Science, 217*, 149–151.
Schuster, E. (1990). Exkurisonsbericht Zustand und Gefährdung der Bergwälder. In E. Schuster (Ed.), *Zustand und Gefährdung des Bergwaldes* (Forstwissenschaftliche Forschungen 40, Beiheft zum Forstwissenschaftlichen Centralblatt, pp. 11–18).
Schütz, H.-U. (1998). *Untersuchungen zur Ökologie von "ribbon forests" der Colorado Front Range (Rocky Mountains, U.S.A.) unter besonderer Berücksichtigung von Thomomys talpoides (Geomyidae)*. Dissertation Mathematisch-Naturwissenschaftliche Fakultät, Westfälische Wilhelms-Universität, Münster.
Schütz, H.-U. (2005). Pocket gopher – Actor under the stage. Studies on Niwot Ridge, Colorado Front Range, U.S.A. In G. Broll & B. Keplin (Eds.), *Mountain ecosystems. Studies in treeline ecology* (pp. 153–180). Berlin/Heidelberg: Springer.
Schütz, H., Seiler, W., & Rennnenberg, H. (1990). Soil and land-use related sources and sinks of methane in the context of the global methane budget. In A. F. Bowman (Ed.), *Soils and the greenhouse effect* (pp. 269–288). Chichester: Wiley.
Schütz, M., Wildi, O., Achermann, G., Krüsi, B. O., & Nievergelt, B. (2000). Predicting the development of subalpine grassland in the Swiss National Park: How to build a succession model based on data from long-term permanent plots. *Nationalparkforschung in der Schweiz, 89*, 207–235.
Schweinfurth, U. (1966). *Neuseeland, Beobachtungen und Studien zur Pflanzengeographie und Ökologie der antipodischen Inselgruppe* (Bonner Geographische Abhandlungen 36). Bonn: Ferdinand Dümmlers Verlag.
Schweingruber, F. H. (1979). Auswirkungen des Lärchenwicklerbefalls auf die Jahrringstruktur der Lärche. *Schweizerische Zeitschrift für Forstwesen, 130*(2), 1071–1093.
Semenov, Y., Ramousse, R., & Le Berre, M. (2001). Impact of black-capped marmot (Marmota camtchatica bungei) on floristic diversity of Arctic tundra in northern Siberia. *Arctic, Antarctic, and Alpine Research, 33*(2), 204–210.
Senft, R. L., Coughenour, M. B., Bailey, D. W., Rittenhouse, L. R., Sala, O. E., & Swift, D. M. (1987). Large herbivores foraging and ecological hierarchies. *BioScience, 37*(11), 789–799.
Seppäla, M., & Rastas, J. (1980). Vegetation map of northernmost Finland with special reference to subarctic forest limits and natural hazards. *Fennia, 158*, 41–61.
Seybold, S. J., Albers, M. A., & Katovic, S. A. (2002). *Eastern larch beetle* (Forest Insect and Disease Leaflet 175). Washington, DC: USDA Forest Service.
Seymour, R. S. (1992). Der natürliche Brutschrank des Buschhuhns. *Spektrum der Wissenschaft*, Februar 1992, pp. 60–67.
Shaw, M. W. (1968a). Factors affecting the natural regeneration of sessile oak (Quercus petraea) in North Wales. I. A preliminary study of acorn production, viability and losses. *Journal of Ecology, 56*, 565–668.
Shaw, M. W. (1968b). Factors affecting the natural regeneration of sessile oak (Quercus petraea) in North Wales. II. Acorn losses and germination under field conditions. *Journal of Ecology, 56*, 647–660.
Sheets, R. G., Linder, R. L., & Dahlgren, R. B. (1971). Burrow system of prairie dogs in South Dakota. *Journal of Mammalogy, 52*, 451–453.
Sherman, P. W., Jarvis, J. U. M., & Braude, S. H. (1992). Die enge Gemeinschaft der Nacktmulle. *Spektrum der Wissenschaft*, Oktober 1992, pp. 90–98.
Sherrod, S. K., & Seastedt, T. R. (2001). Effects of the northern pocket gopher (Thomomys talpoides) on alpine soil characteristics, Niwot Ridge, CO. *Biogeochemistry, 55*, 195–218.
Shigo, A. L. (1967). Succession of organisms in discoloration and decay of wood. In J. A. Romberger & P. Mikola (Eds.), *International review of forest research* (Vol. 2, pp. 239–299). New York: Academic.

Sicuro, F. L., & Olivlira, L. F. B. (2002). Coexistence of peccaries and feral hogs in the Brazilian Pantanal Wetland: An ecomorphological view. *Journal of Mammalogy, 83*(1), 207–217.

Simberloff, D., Schmitz, D. C., & Brown, T. C. (1997). *Strangers in paradise: Impact and management of nonindigenous species in Florida*. Washington, DC: Island Press.

Simon, O., & Goebel, W. (1999). Zum Einfluss des Wildschweins (Sus scrofa) auf die Vegetation und Bodenfauna einer Heidelandschaft. In B. Gerken & M. Görner (Eds.), *Europäische Landschaftsentwicklung mit großen Weidetieren. Geschichte, Modelle, Perspektiven* (Natur und Kulturlandschaft 3, pp. 172–177). Höxter/Jena: Huxaria Druck und Verlag.

Sinclair, A. R. E. (1974a). The resource limitation of trophic levels in tropical grassland ecosystems. *Journal of Animal Ecology, 44*, 497–520.

Sinclair, A. R. E. (1974b). The natural regulation of buffalo population in East Africa. IV. The food supply as a regulating factor, and competition. *East African Wildlife Journal, 12*, 291–312.

Sinclair, A. R. E. (1977). *The African buffalo. A study of resource limitation of populations*. Chicago/London: University of Chicago Press.

Sinclair, A. R. E. (1979a). Dynamics of the Serengeti ecosystem, process and pattern. In A. R. E. Sinclair & M. Norton-Griffith (Eds.), *Dynamics of an ecosystem* (pp. 1–30) Chicago/London: University of Chicago Press.

Sinclair, A. R. E. (1979b). The eruption of ruminants. In A. R. E. Sinclair & M. Norton-Griffith (Eds.), *Dynamics of an ecosystem* (pp. 82–103). Chicago/London: University of Chicago Press.

Sinclair, A. R. E. (1983). The function of distance movements in vertebrates. In A. R. E. Sinclair & P. J. Greenwood (Eds.), *The ecology of animal movement* (pp. 240–258). Oxford: Clarendon Press.

Sinclair, A. R. E. (1985). Does interspecific competition shape the African ungulate community? *Journal of Animal Ecology, 54*, 899–918.

Sinclair, A. R. E. (1992). The adaptation of African ungulates and their effects on community function. In F. Boulière (Ed.), *Tropical savannas* (Ecosystems of the World 13, pp. 401–436). Amsterdam: Elsevier.

Sinclair, A. R. E. (1995). Serengeti past and present. In A. R. E. Sinclair & P. Arcese (Eds.), *Serengeti* (pp. 3–30). Chicago/London: University of Chicago Press.

Sinclair, A. R. E., Dublin, H., & Borner, M. (1985). Population regulation of Serengeti wildebeest: At test of the food hypothesis. *Oecologia, 65*, 266–268.

Singer, F. J., Swank, W. T., & Clebsch, E. E. C. (1984). Effects of wild pig rooting in a deciduous forest. *Journal of Wildlife Management, 48*(2), 464–473.

Skoczen, S., Nagawiecka, H., & Boron, K. (1976). The influence of mole tunnels on soil moisture on pastures. *Acta Theriologica, 21*(38), 543–548.

Slavik, A. D., & Janke, R. A. (1987). The vascular flora of Isla Royale National Park. *Michigan Botanist, 26*, 91–134.

Smith, V. R. (1976). The effect of burrowing species of Procellariidae on the nutrient status of inland tussock grassland on Marion Island. *Journal of South African Botany, 42*(2), 265–277.

Smith, J. M. N., Krebs, C. J., Sinclair, A. R. E., & Boonstra, R. (1988). Population biology of snowshoe hares. II. Interactions with winter food plants. *Journal of Animal Ecology, 57*, 269–286.

Smuts, G. I. (1978). Interrelations between predators, prey and their environment. *BioScience, 28*, 316-320, Chicago/London: BioScience.

Sonesson, M., & Hoogesteger, J. (1983). Recent tree-line dynamics (Betula pubescens Erh. ssp. tortuosa Ledeb. Nyman) in northern Sweden. *Nordicana, 47*, 47–54.

Spatz, G., & Mueller-Dombois, D. (1995). Successional patterns after pig digging in grassland communities on Mauna Loa, Hawaii. *Phytocoenologia, 3*, 346–373.

Stahr, K. (1979). *Die Bedeutung periglazialer Deckschichten für die Bodenbildung und Standortseigenschaften im Südschwarzwald* (Freiburger Bodenkundliche Abhandlungen 9). Freiburg im Breisgau: Institut für Bodenkunde und Waldernährungslehre der Albert-Ludwigs-Universität.

Stary, B. (1990). *Atlas der schädlichen Forstinsekten (Deutsche Ausgabe, überarbeitet und ergänzt von G. Benz)*. Stuttgart: Ferdinand Enke Verlag.

Steiner, H. (1999). *Der Mäusebussard (Buteo buteo) als Indikator für Struktur und Bodennutzung des ländlichen Raumes: Produktivität im heterogenen Habitat, Einfluss von Nahrung und Witterung und Vergleich zum Habicht (Accipiter gentilis)*. Stapfia 62.

Stender, S., Poschlod, P., Vauk-Henzelt, E., & Dernedde, T. (1997). Die Ausbreitung durch Galloway-Rinder. *Verhandlungen der Gesellschaft für Ökologie, 27*, 173–180.

Stipe, L. E. (1987). History. In M. H. Brookes, R. W. Campbell, J. J. Colbert, R. G. Mitchell, & R. W. Stark (Eds.), *Western spruce budworm* (Technical Bulletin 1694, pp. 1–16). Washington, DC: USDA Forest Service, Cooperative State Research Service.

Storer, T. I., & Usinger, R. L. (1971). *Sierra Nevada natural history*. Berkeley/Los Angeles/ London: University of California Press.

Strandgaard, S. (1982). *Factors affecting the moose population in Sweden during the 20th century with special attention to silviculture* (Report 8). Uppsala: Swedish University of Agricultural Sciences, Department of Wildlife Ecology.

Strange, I. J. (1980). The thin-billed prion, Pachyptila belcheri, at New Island, Falkland Islands. *Le Gerfaut, 70*, 411–445.

Sukachev, V., & Dylis, N. (1964). *Fundamentals of forest biogeocoenology* (translated from Russian). Edinburgh/London: Oliver & Boyd.

Sullivan, T. P. (1979). Repopulation of clear-cut habitat and conifer seed predation by deer mice. *Journal of Wildlife Management, 43*(4), 861–871.

Summerhayes, V. S., & Elton, C. S. (1928). Further contributions to the ecology of Spitzbergen. *Journal of Ecology, 16*, 193–268.

Sutter, E., & Amann, F. (1953). Wie weit fliegen vorratssammelnde Tannenhäher? *Ornithologischer Beobachter, 50*, 89–90.

Swanberg, P. O. (1951). Food storage, territory and song in the Thickbilled Nutcracker. In 10th *International Ornithological Congress*, Uppsala 1950 (pp. 545–554).

Swank, W., Waide, J. B., Crossley, D. A., & Todd, R. L. (1981). Insect defoliation enhances nitrate export from forest ecosystems. *Oecologia, 51*, 297–299.

Sweanor, P. Y., & Sandegren, F. (1989). Winter-range philopatry of seasonally migratory moose. *Journal of Applied Ecology, 26*, 25–33.

Swetnam, T. W., & Lynch, A. M. (1993). Multi centural, regional-scale patterns of western spruce budworm outbreaks. *Ecological Monographs, 63*(4), 399–424.

Tadzhiyev, U., & Odinoshiyev, A. (1987). Influence of marmots on soil cover of the eastern Pamir. *Soviet Soil Science, 2*, 22–30.

Talbot, L. M., & Talbot, M. H. (1963). The wildebeest in western Masailand, East Africa. *Wildlife Monographs, 12*, 1–88.

Tammaru, T., Kaitaniemi, P., & Ruohomäki, K. (1995). Oviposition choices of Epirrita autumnata (Lepidoptera: Geometridae) in relation to its eruptive dynamics. *Oikos, 74*, 296–304.

Tansley, A. G., & Adamson, R. S. (1925). Studies on vegetation of the English chalk. III. The chalk grasslands of the Hampshire-Sussex border. *Journal of Ecology, 13*, 117–123.

Tardy, Y., & Roquin, C. (1992). Geochemistry and evolution of lateritic landscapes. In I. P. Martini & W. Chesworth (Eds.), *Weathering, soils and paleosols* (pp. 407–443). Amsterdam/London/ New York/Toronto: Elsevier.

Tast, J., & Kalela, O. (1971). Comparison between rodent cycles and plant production in Finnis Lapland. *Annales Academiae Scientarum Fennicae, Series A (IV), 186*, 1–14.

Tatur, A. (1989). Ornithogenic soils of the maritime Antarctic. *Polish Polar Research, 10*(4), 481–532.

Tatur, A., & Myrcha, A. (1984). Ornithogenic soils on King George Island, South Shetland Islands (Maritime Antarctic Zone). *Polish Polar Research, 5*, 31–60.

Taylor, S. W., Carroll, A. L., Alfaro, R. I., & Safranyik, L. (2006). Forest, climate and mountain pine beetle outbreak dynamics in western Canada. In L. Safranyik & B. Wilson (Eds.), *The mountain pine beetle: A synthesis of biology, management and impacts in lodgepole pine* (pp. 67–94). Victoria: Natural Resources Canada, Canadian Forest Service.

Tedrov, J. C. F., & Ugolini, F. C. (1966). Antarctic soils. In J. C. F. Tedrov (Ed.), *Antarctic soils and soil forming processes* (American Geophysical Union Research Series 8, pp. 161–177). Washington, DC: American Geophysical Society.

Teipner, C. L., Garton, E. O., & Nelson, L., Jr. (1983). *Pocket gophers in forest ecosystems* (General Technical Report INT-154). Ogden: USDA Forest Service, Intermountain Forest and Range Experiment Station.

Tenow, O. (1972). *The outbreaks of Oporinia autumnata BKH. and Operophtera spp. (Lep., Geometridae) in the Scandinavian mountain chain and northern Finland 1862–1968* (Zoologisk Bidrag fran Uppsala, Supplement 2, pp. 1–107). Stockholm: Almquist & Wiksells.

Tenow, O. (1975). Topographical dependence of an outbreak of Oporinia autumnata Bkh. (Lep. Geometridae) in a mountain birch forest in northern Sweden. *ZOON, 3*, 85–110.

Tenow, O. (1996). Hazards to a mountain birch forest – Abisko in perspective. *Ecological Bulletins, 45*, 104–114.

Tenow, O., & Holmgren, B. (1987). Low winter temperatures and an outbreak of Epirrita autumnata along a valley of Finnmarksvidda, the "cold pole" of Fennoscandia. In H. Alexandersson & B. Holmgren (Eds.), *Climatological extremes in the mountains* (UNGI Rapport 65, pp. 203–216). Uppsala: Uppsala University.

Tenow, O., & Bylund, H. (1989). A survey of winter cold in the mountain birch/Epirrita autumnata system. *Memoranda Societatis pro Fauna et Flora Fennica, 65*, 67–72.

Tenow, O., & Nilssen, A. (1990). Egg cold hardiness and topoclimatic limitations to outbreaks of Epirrita autumnata in northern Fennoscandia. *Journal of Applied Ecology, 27*, 723–734.

Tenow, O., Bylund, H., & Strong, D. R. (1995). Multiple mechanisms in cycles of the autumnal moth on mountain birch. In H. Bylund (Ed.), *Long-term interactions between the autumnal moth on mountain birch: The roles of resources. Competition, natural enemies and weather.* Dissertation Sveriges Landbruksuniversitet, Uppsala.

Tenow, O., Bylund, H., Nilssen, A. C., & Karlsson, P. S. (2005). Long-term influence of herbivores on northern mountain birch forests. *Ecological Studies, 180*, 165–180.

Tenow, O., Nilssen, A., Bylund, H., & Hogstad, O. (2007). Wave and synchrony in Epirrita autumnata/Operophtera brumata outbreaks. I. Lagged synchrony: Regionally, locally and among species. *Journal of Animal Ecology, 76*, 258–268.

Tevis, L. (1958). Interrelations between the harvester ant Vermessor perangadei (Mayr) and some desert ephemerals. *Ecology, 39*, 695–704.

Thacker, E. J., & Brandt, S. (1955). Coprophagy in the rabbit. *Journal of Nutrition, 55*, 375–385.

Thannheiser, D., & Wüthrich, C. (1999). Flora und Vegetation am St. Jonsfjord (Spitzbergen). *Norden, 13*, 291–301.

Thompson, D. Q. (1955). The role of food and cover on population fluctuations of the brown lemming at Point Barrow, Alaska. *Transactions of the Twentieth North American Wildlife Conference, 20*, 166–176.

Thorn, C. (1978). A preliminary assessment of the geomorphic role of pocket gophers in the alpine zone of the Colorado Front Range. *Geografiska Annaler, 60A*(3–4), 181–187.

Thorn, C. E. (1982). Gopher disturbance: Its variability by Braun-Blanquet vegetation units in the Niwot Ridge alpine tundra, Colorado Front Range, USA. *Arctic and Alpine Research, 14*(1), 45–51.

Thorne, E. T., Dean, R. E., & Hepworth, W. G. (1976). Nutrition during gestation to successful reproduction in elk. *Journal of Wildlife Management, 40*, 330–335.

Thorp, J. (1949). Effects of certain animals that live in soils. *Scientific Monthly, 68*, 180–191.

Tinner, W., Ammann, B., & Germann, P. (1996). Treeline fluctuations recorded for 12 500 years by soil profiles, pollen, and plant macrofossils in the Central Swiss Alps. *Arctic and Alpine Research, 28*, 131–147.

Tisdell, C. A. (1982). *Wild pigs: Environmental pest or economic resource?* Sydney/Oxford/New York/Toronto/Paris/Frankfurt: Pergamon Press.

Tomas, W. M., & Salis, S. M. (2000). Diet of the marsh deer (Blastocerus dichotomus) in the Pantanal wetland, Brazil. *Studies on Neotropical Fauna and Environment, 35*, 165–172.

Tomback, D. F. (1978). Foraging strategies of Clark's nutcracker. *The Living Bird, 16*, 123–163.

Tomback, D. F. (1982). Dispersal of whitebark pine seeds by Clark's nutcracker: A mutualism hypothesis. *Journal of Animal Ecology, 51*, 451–467.

Tomback, D. F. (1983). Nutcrackers and pines: Coevolution or coadaptation? In M. H. Nitecki (Ed.), *Coevolution* (pp. 179–223). Chicago: University of Chicago Press.
Tomback, D. F. (1986). Post-fire regeneration of krummhol whitebark pine: A consequence of nutcracker seed caching. *Madrono, 33*(2), 100–110.
Tomback, D. F. (1989). The broken cycle: Fire, birds and whitebark pine. In T. Walsh (Ed.), *Wilderness and wildfire* (Vol. 50, pp. 14–17). Missoula: Wilderness Institute Montana Forestry and Range Experimental Station, School of Forestry, University of Montana, Miscellaneous Publications.
Tomback, D.-F. (1994a). Genetic population structures and growthform distribution in bird-dispersed pines. In W. Schmidt & F.-K. Holtmeier (Compilers), *Proceedings – International workshop on subalpine stone pines and their environment: The status of our knowledge*, St. Moritz, 5–11 September 1992 (USDA Forest Service, Intermountain Research Station, General Technical Report INT-GTR-309, pp. 43–50).
Tomback, D. F. (1994b). Effects of seed dispersal by Clark' nutcracker on early postfire regeneration of whitebark pine. In W. Schmidt & F.-K. Holtmeier (Compilers), *Proceedings – International workshop on subalpine stone pines and their environment: The status of our knowledge*, St. Moritz, 5–11 September 1992 (USDA Forest Service, Intermountain Research Station, General Technical Report INT-GTR-309, pp. 193–198).
Tomback, D. F. (1998). Clark's nutcracker. In A. Poole & F. Gill (Eds.), *The birds of North America*, No. 331. Philadelphia: The Academy of Natural Sciences.
Tomback, D. F. (2001). Clark's nutcracker: Agent of regeneration. In D. F. Tomback, S. F. Arno, & R. E. Keane (Eds.), *Whitebark pine communities. Ecology and restoration* (pp. 88–104). Washington/Covelo/London: Island Press.
Tomback, D. F., & Linhart, Y. B. (1990). The evolution of bird-dispersed pines. *Evolutionary Ecology, 4*, 185–219.
Tomback, D. F., & Schuster, W. F. (1994). Genetic population structure and growth form distribution in bird-dispersed pines. In W. C. Schmidt & F. -K. Holtmeier (Compilers), *Proceedings of the international workshop on subalpine stone pines and their environment: The status of our knowledge, St. Moritz, 5–11 September, 1992* (General Technical Report INT-GTR-309, pp. 43–50). USDA Forest Service, Intermountain Research Station.
Topp, W. (1981). *Biologie und Bodenorganismen*. Heidelberg: Quelle & Meyer.
Torbit, S. C., Carpenter, L. H., Swift, D. M., & Alldredge, A. W. (1985). Differential loss of fat and protein by mule deer during winter. *Journal of Wildlife Management, 49*, 80–85.
Treiber, R. (1997). Vegetationsdynamik unter dem Einfluß des Wildschweins (Sus scrofa L.) amd Beispiel bodensaurer Trockenrasen der elsässischen Harth. *Zeitschrift für Ökologie und Naturschutz, 6*, 83–95.
Tricart, J. (1972). *The landforms of the humid tropics, forests and savannas*. London: Longman.
Trolldenier, G. (1971). *Bodenbiologie. Die Bodenorganismen im Haushalt der Natur*. Frankfurt/Stuttgart: Franckh'sche Verlagshandlung.
Tscherko, D., Bölter, M., Beyer, L., Chen, J., Elster, J., Kandeler, E., Kuhn, D., & Blume, H.-P. (2003). Biomass and enzyme activity of two soil transects at King George Island, Maritime Antarctic. *Arctic, Antarctic, and Alpine Research, 35*(1), 34–47.
Turcek, F. (1967). *Ökologische Beziehungen der Säugetiere und Gehölze*. Bratislava: Vydavateltvo Slovenskij akademic vied.
Turner, G. T. (1973). Effects of pocket gophers on the range. In G. T. Turner, R. M. Hansen, V. H. Reid, H. P. Tietjen, & A. L. Ward (Eds.), *Pocket gophers and Colorado mountain rangeland* (Bulletin 554, pp. 51–61). Fort Collins: Colorado State University Experimental Station.
Turner, M. G., & Stratton, S. P. (1987). Fire, grazing, and the landscape heterogeneity of a Georgia barrier island. In M. G. Turner (Ed.), *Landscape heterogeneity and disturbance* (Ecological Studies 64, pp. 85–101). New York: Springer.
Vaarama, A., & Valanne, T. (1973). On the taxonomy, biology and origin of Betula tortuosa Ledeb. *Reports from the Kevo Subarctic Research Station, 10*, 70–84.
Van Ballenberghe, V. (1985). Wolf predation on caribou: The Nelchina herd case history. *Journal of Wildlife Management, 49*(3), 711–720.

Van Ballenberghe, V., & Ballard, W. B. (1994). Limitation and regulation of moose population: The role of predation. *Canadian Journal of Zoology, 72*, 2071–2077.
Van der Pijl, L. (1972). *Principles of dispersal in higher plants*. Berlin: Springer.
Van Sickle, G. A. (1987). Host responses. In M. H. Brookes, R. W. Campbell, J. J. Colbert, R. G. Mitchell, & R. W. Stark (Eds.), *Western spruce budworm* (Technical Bulletin 1694, pp. 57–70). Washington, DC: USDA Forest Service, Cooperative State Research Service.
Van Soest, P. J. (1982). *Nutritional ecology of the ruminant*. Corvallis: O & O Books.
Vander Wall, S. B. (1990). *Food hoarding in animals*. Chicago/London: The University of Chicago Press.
Vander Wall, S. B., & Balda, R. P. (1977). Coadaptations of the Clark's nutcracker and Pinyon pine for efficient harvest and dispersal. *Ecological Monographs, 47*, 22–37.
Vaughan, T. A. (1963). Movements made by two species of pocket gophers. *The American Midland Naturalist, 69*(2), 367–377.
Verme, L. J. (1969). Reproductive patterns of white-tailed deer related to nutritional plane. *Journal of Wildlife Management, 33*, 881–887.
Verweij, R. J., Verrelst, J., Loth, P. L., Heitkönig, I. M. A., & Brunsting, M. H. (2006). Grazing lawns contribute to the subsistence of mesoherbivores on dystrophic savannas. *Oikos, 114*, 108–116.
Vesey-Fitzgerald, D. F. (1960). Grazing succession among East African game animals. *Journal of Mammalogy, 41*(2), 161–172.
Vignolio, O. R., & Fernández, O. N. (2010). Cattle dung as a vector of spreadings seeds of exotic species in the Flooding Pampa grasslands (Buenos Aires, Argentina). *Annales Botanici Fennici, 47*, 14–22.
Virtanen, T., Neuvonen, S., & Nikula, A. (1998). Modelling topoclimatic patterns of egg mortality of Epirrita autumnata (Lepidoptera: Geometridae) with Geographical Information Systems: Predictions of current and warmer climate scenarios. *Journal of Applied Ecology, 32*, 1–12.
Vitala, J. (1977). Social organization in cyclic subarctic populations of the voles Clethrionomys rufocanus (Sund.) and Microtus agrestis (L.). *Annales Zoologici Fennici, 14*, 53–93.
Vogel, L. (1985). Das Schilf reguliert seine Schädlinge selbst, eine "Monokultur" als strukturierter Lebensraum. forschung – Mitteilungen DFG 2, pp. 26–28.
Vogel, A. (1988). Angelica archangelica-Hochstaudenfluren auf Island - natürlich oder anthropogen? *Flora, 180*, 19–29.
Volker, K. C. (1978). *Ecology of parasites and predators of the Douglas fir tussock moth in the Pacific Northwest*. Dissertation, University of Idaho, Moscow.
Voous, K. H. (1962). *Die Vogelwelt Europas und ihre Verbreitung*. Hamburg/Berlin: Parey.
Voslamber, B., & Veen, A. W. L. (1985). Digging by badgers and rabbits on some wooded slopes in Belgium. *Earth Surface Processes and Landforms, 10*, 799–882.
Vucetich, J. A., & Peterson, R. O. (2013). *Ecological studies of wolves on Isle Royale*. Annual Report 2012–2013. www.isleroyalewolf.org
Wace, N. M. (1961). The vegetation of Gough Island. *Ecological Monographs, 31*, 337–367.
Wagner, D., Brown, M. J. F., & Gordon, D. M. (1977). Harvester and nests, soil biota and soil chemistry. *Oecologia, 112*, 232–236.
Wallin, D. O. (1990). *Habitat dynamics for an African weaver-bird: The red-billed Quelea (Quelea quelea)*. Dissertation, University of Virginia, USA.
Wallwork, J. A. (1970). *Ecology of soil animals*. New York: McGraw-Hill.
Walter, H. (1977). *Vegetationszonen und Klima* (3rd ed.). Stuttgart: Ulmer.
Walter, H., & Breckle, S. (1986). *Ökologie der Erde 3. Spezielle Ökologie der gemäßigten und arktischen Zonen Euro-Nordasiens*. Stuttgart: Gustav Fischer Verlag.
Walter, H., & Breckle, S. W. (1991). *Ökologie der Erde 4. Gemäßigte und arktische Zonen außerhalb Euro-Nordasiens*. Stuttgart: Gustav Fischer.
Wambugu, S. W. (2011). *Increasing sorghum utilization and marketability through food product diversification. End of Project Report, Period covered June 2009–August 2011*. Nairobi: Kenya Industrial Research and Development Institute (KIRDI).

Ward, P. (1965). Feeding ecology of the black-faced dioch Quelea quelea in Nigeria. *The Ibis, 107*(2), 173–214.

Ward, P. (1971). The migration patterns of Quelea quelea in Africa. *The Ibis, 113*(3), 276–297.

Waterhouse, D. F. (1974). The biological control of dung. *Scientific American, 230*, 100–109.

Watson, R. M. (1967). *The population ecology of wildebeest (Connochaetes taurinus albojubatus Thomas) in the Serengeti*. Ph.D. thesis, Cambridge University.

Watt, A. S. (1974). Senescence and rejuvenation in ungrazed chalk grassland (Grassland B) in Breckland: The significance of litter and moles. *Journal of Applied Ecology, 11*(3), 1157–1171.

Watts, C. (1969). The regulation of wood mouse (Apodemus syvaticus) number in Whytham Woods, Berkshire. *Journal of Animal Ecology, 38*, 283–304.

Weber, U. M. (1995a). *Jahrringe als Ausdruck von Insektengradationen in den schweizerischen Zentralalpen und der Front Range, Colorado*. Dissertation, Philosophisch-Naturwissenschaftliche Fakkultät, Universität Basel

Weber, U. M. (1995b). Der Lärchenwickler im Spiegel der Jahrhunderte. *Bündnerwald, 6*, 61–71.

Wehncke, E. V., Hubbel, S. P., Foster, R. B., & Dalling, J. W. (2003). Seed dispersal patterns produced by white-faced monkeys: Implications for the dispersal limitation of neotropical tree species. *Journal of Ecology, 91*, 677–685.

Welander, J. (2000). Spatial and temporal dynamics of wild boar (Sus scrofa) rooting in a mosaic landscape. *Journal of Zoology (London), 252*, 263–271.

Wendelberger, G., & Hartl, H. (1969). Untesuchungen im Brandgebiet aletschwald. *Schweizerische Zeitschrift für Forstwesen, 120*(9), 453–475.

Werner, D. J. (1977). Vegetationsveränderungen in der argentinischen Puna unter dem Einfluß von Bodenwühlern der Gattung Ctenomys Blainville. In R. Tüxen (Ed.), *Vegetation und Fauna, Bericht über das Internationale Symposium der Internationalen Vereinigung für Vegetationskunde* (pp. 433–449). Rinteln

Werner, R. A. (1986a). Association of plants and phytopathogenous insects in the Alaska Taiga. *Ecological Studies, 57*, 205–212.

Werner, R. A. (1986b). *The eastern larch beetle in Alaska* (Research Paper PNW-357). Portland: USDA Forest Service, Pacific Northwest Research Station.

Weseloh, D. V., & Brown, R. T. (1971). Plant distribution within a heron rookery. *The American Midland Naturalist, 86*(1), 57–64.

West, N. E. (1968). Rodent-influenced establishment of ponderosa pine and bitterbrush seedlings in central Oregon. *Ecology, 49*, 1009–1011.

Westrich, P. (1989). *Die Wildbienen Baden-Württembergs*. Stuttgart: Eugen Ulmer Verlag.

Whicker, A. D., & Detling, J. K. (1988a). Ecological consequences of prairie dog disturbances. *BioScience, 38*(11), 778–785.

Whicker, A. D., & Detling, J. K. (1988b). Modification of vegetation structure and ecosystem processes by North American mammals. In M. J. A. Werger et al. (Eds.), *Plant form and vegetation structure – Adaption, plasticity and relation to herbivory. International symposium on vegetation structure* (Vol. 1, pp. 301–316). The Hague: SPB Academic Publishing.

White, C. A., Feller, M. C., & Bayley, S. (2003). Predation risk and the functional response of elk-aspen herbivory. *Forest Ecology and Management, 181*, 77–97.

Whittaker, R. J., & Jones, S. H. (1994). The role of frugivorous bats and birds in rebuilding of a tropical forest ecosystem, Krakatau, Indonesia. *Journal of Biogeography, 21*, 245–258.

Wickman, B. E. (1990). *The battle against bark beetles in Crater Lake National Park: 1925–1934* (General Technical Report PNW.GTR 259). Portland, OR: USDA Forest Service, Pacific Northwest Research Station.

Wiener, J. G. (1975). Nutrient cycles, nutrient limitations, and vertebrate populations. *Biologist, 57*(3), 104–124.

Wilks, B. J. (1963). Some aspects of the ecology of population dynamics of the pocket gopher (Geomys bursarius) in southern Texas. *Texas Journal of Science, 15*, 241–283.

Wilmanns, O., & Müller, K. (1977). *Zum Einfluß der Schaf- und Ziegenbeweidung auf die Vegetation im Schwarzwald*. Bericht über das Internationale Symposium der Internationalen Vereinigung für Vegetationskunde, Vegetation und Fauna, 12–15 April 1976, Rinteln, pp. 465–479.

Wing, L. D., & Buss, I. O. (1970). Elephants and forests. *Wildlife Monographs, 19*, 37–92.

Wolf, J. O. (1980). The role of habitat patchiness in the population dynamics of snowshoe hare. *Ecological Monographs, 50*, 111–129.

Woodell, S. R. J., & King, T. J. (1991). The influence of mound-building ants on British lowland vegetation. In C. R. Huxley & D. F. Cutler (Eds.), *Ant-plant interactions* (pp. 521–535). Oxford: Oxford University Press.

Woodmansee, R. G. (1978). Additions and losses of nitrogen in grassland ecosystems. *BioScience, 28*(7), 448–553.

Woods, R. W. (1975). *The birds of the Falkland Islands*. Wiltshire: Compton Press.

Wormell, P. (1976). The shearwaters of Rhum. *Scottish Birds, 9*, 103–118.

Wüthrich, C. (1992). Stofftransporte Meer-Land. Vogelklifftundren und ornithogene Böden. *Stuttgarter Geographische Studien, 117*, 177–192.

Wüthrich, C. (1994). *Die biologische Aktivität arktischer Böden mit spezieller Berücksichtigung ornithogen eutrophierter Gebiete (Spitzbergen und Finnmark)* (Basler Beiträge zur Physiogeographie 17). Basel: Wepf Verlag.

Wüthrich, C., & Thannheiser, D. (1997). Die Ökosysteme der polaren und subpolaren Zonen. In H. Leser (Ed.), *Handbuch des Geographieunterrichts* (Vol. 11, pp. 200–214). Köln: Aulis Verlag.

Yair, A. (1974). Report on present day geomorphological processes. Abhandlungen der Akademie der Wissenschaften in Göttingen. *Mathematisch-Physikalische Klasse 29*, 403–417.

Yair, A., & Rutin, J. (1981). Some aspects of the regional variation in the amount of available sediment eroded by isopods and porcupines, northern Negev, Israel. *Earth Surface Processes and Landforms, 6*, 221–234.

Yeaton, R. I., & Elser, K. J. (1990). The dynamics of a succulent Karoo vegetation. A study of species association and recruitment. *Vegetatio, 88*, 103–113.

Zeevalking, H. J., & Fresco, L. F. M. (1977). Rabbit grazing and diversity in a dune area. *Vegetatio, 35*, 193–196.

Zhang, S. Y., & Wang, L. X. (1995). Fruit consumption and seed dispersal of Ziziphus cinnamomum (Rhamnaceae) by two sympatric primates (Cebus apella and Ateles paniscus) in French Guiana. *Biotropica, 27*, 397–401.

Zhong, W., Wang, M., & Wan, X. (1999). Ecological management of Brandt's vole (Microtus brandti) in Inner Mongolia, China. In G. R. Singleton, L. A. Hinds, H. Leirs, & Z. Zhang (Eds.), *Ecologically-based management of rodents pests* (pp. 199–214). Canberra: Australian Center for International Agricultural Research.

Zimmermann, P. R., Greenberg, J. P., Waniga, S. O., & Crutzen, P. J. (1992). Termites. A potentially large source of atmospheric methane, carbon dioxide, and molecular hydrogen. *Science, 218*, 563–565.

Zlotin, R. I. (1971). Invertebrate animals as a factor of the biological turnover. In Organismes du sol et production primaire. IV. Colloquium Pedobologiae, Annales de Zoologie, Ecologie Animale, pp. 455–462.

Zlotin, R. I., & Khodashova, K. S. (1980). *The role of animals in biological cycling of forest-steppe ecosystems* (English edition by French, N. R., translated by Lewus, W. & Grant, W. E.). Stroudsburg: Dowden, Hutchison & Ross.

Chapter 3
Selected Landscapes Under the Influence of Wild Herbivorous Mammals

3.1 African Savannas

African savannas belong to the largest biomes in the world. In the more humid regions savanna is covered by open woodland (tree savanna). In the drier regions open woodland gives way to grassland savanna. Both woodland and grassland savanna show a varying landscape mosaic dependent on landforms and local moisture conditions. The enormous populations of large mammalian herbivores are one of the most conspicuous traits of the African continent. No other continent is comparable in this respect. In South America, most native large herbivores did not survive the Pleistocene extinction (McNaughton et al. 1993). In Australia, large grazers disappeared 40.000 or more years ago. The potential reasons of extinction – adverse climatic influences, habitat alteration and rapidly growing hunting pressure – are still debated (e.g., Gibbson 2004; Koch and Barnosky 2006). Leaf-cutting ants and termites have occupied the functional niche of the original mammalian grazers as primary consumers in South American and Australian savannas (McNaughton et al. 1993).

3.1.1 Large Ungulate Herds

Africa has 91 ungulate species, whereas only 21 live in South America, for example. Only a few of them, such as white-tailed deer (*Odocoileus virgianus*), pampas deer (*Ozoateros bezoarticus*) and marsh deer (*Blastocerus dichotomus*) live in small populations in the savanna (Ojasti 1992). Large herbivore biomass in African grass- and tree-savannas amounts to 280 kg ha^{-1} and thus is almost ten times the biomass of the large herbivores that lived in the North American prairie prior to colonization by Europeans. This difference will turn out even more spectacular when comparing the biomass of large African herbivores with the biomass of the large herbivores of the northern temperate woodland. The biomass of moose (7 kg ha^{-1}) on Isle Royale, for example, amounts to only a fortieth of the biomass of herbivorous mammals in

F.-K. Holtmeier, *Animals' Influence on the Landscape and Ecological Importance*,
DOI 10.1007/978-94-017-9294-3_3

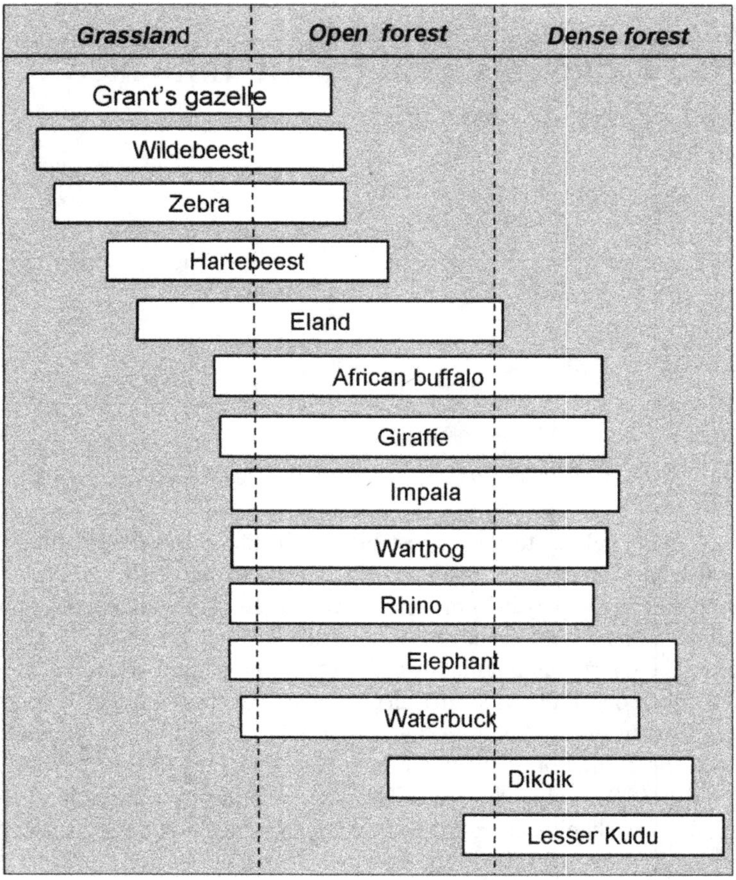

Fig. 3.1 Habitat preferences of the prevailing large herbivores in the Tangarire-Masai area (Tanzania) (Modified from Lamprey 1963)

Murchison-Falls National Park, although moose density is the worldwide highest (Botkin et al. 1981). Forty-four large mammalian herbivores live in Africa, 31 one of them (out of 24 genera) in the savannas. Most of them are solipeds (one-hoofed, Perissodactyla) and cloven-hoofed (Artiodactyla), and elephants (Owen-Smith 1982). During a long evolution they occupied their functional and spatial niches mainly by their different preferences for food (Fig. 3.1) and different capability to utilize it (cf. Table 2.2; Lamprey 1963; Sinclair 1974a, b, 1977; Hirst 1975; Jarman and Sinclair 1979; McNaughton 1983a, b).

The following report of a credible eye-witness from 1929 (cited in Alschner 1980) may give an idea of the formerly huge size of the ungulate herds in East Africa.

> Just in front of us was the first row. And what a row! At least 10 million zebras and wildebeest covered the terrain for kilometres and kilometres. I am fully aware that this statement sounds exaggerated. However, if I had not calculated for hours to get an idea of the approximate

unbelievable size of the herds I would not think of writing down may approximations. The front of the huge herd was about 16 km wide. We could overlook the slopes of Ngorongoro for about 45 km: and as far as the eye could see, it saw animals and animals again.

Anyway, the size of the ungulate herds is still impressive also at present. In the Serengeti (14,750 km^2, without Mara-Wildlife Reserve; see also Fig. 2.18) live about 1.3 million wildebeests, 200,000 zebras and about 440,000 Thomson gazelles, other numerous ungulates disregarded (Sinclair 1995). According to figures provided by UNESCO World Heritage Centre (2014), populations have increased to about two million wildebeest, 300,000 zebras and 900,000 Thomson gazelles. Already before humans appeared on the stage, the large ungulates living in groups or herds had sustainably altered the so-called climatic zonal vegetation in their area of distribution. Without the impact of mammalian herbivores, the grasslands would look completely different, even if the influence of herbivory played a minor role compared to the effects of geomorphology, soil-chemical properties and soil moisture conditions (Cole 1986; Belsky 1995). In other words: without considering the influence of the herbivores we cannot understand these ecosystems (MacMahon 1981; McNaughton et al. 1988).

The herbivores directly modify physiognomy, composition and structure of the plant cover. Moreover, they influence nutrient cycles (turnover rates) and nutrient balance to a varying extent in dependence from the geographical position (climate, soils) of the grazing areas and from the spatial-temporal grazing behavior of the different species (Wing and Buss 1970; Botkin et al. 1981). In grass- and woodland savannas, with precipitation of 740–1,200 mm, storage of nutrients in the plants and herbivores (consumption of green phytomass, protein-rich excreta, accelerated mineralization, immediate uptake of the available nutrients by the vegetation) are likely to prevent loss of nutrients by leaching (Botkin et al. 1981). In the end, the different nutrient requirements and the seasonally varying use of vegetation allows the coexistence of many herbivorous mammal species (e.g., Ferrar and Walker 1974; Hirst 1975; Jarman and Sinclair 1979; McNaughton et al. 1988), provided that no other factors have opposite effects.

Dependent on the varying soil fertility, the effects of herbivory differs. More than a half of grass- and bush-savanna extends over very old usually crystalline land surfaces, where soils are dry and poor in nutrients (Scholes and Hall 1996). In areas characterized by basic intrusive and effusive rocks, soils are more fertile (Jones 1973). Soils that developed on carbonate-rich sediments or on alluvial and aeolean material have an intermediate position. This holds also true for net primary production. The nitrogen content (~1 %) of the grass vegetation on the old infertile land surfaces ranges below the minimum content required for efficient digestation of the ruminants. Protein contents of at least 4–5 % are needed for maintaining ungulate body weight (Sinclair 1974a, b). Accordingly, the total biomass of mammals is low, and large mammals (>200 kg) prevail. In the fertile areas, however, nitrogen content ranges above the subsistence minimum all year round. Thus, these areas carry a great number of different grazers, and small species (<100 kg) are comparatively numerous. In tropical grass-, bush- and tree savannas that have developed on fertile soils, the herbivores consume about 10 % of net primary production, whereas they use about 80 % of net primary production in comparable vegegation types on unfertile soils (Scholes and Walker 1993).

McNaughton et al. (1988) have described three different turn-over types characteristic of the Serengeti. Serengeti (1,220–2,000 m), located in the north of Tanzania, is a sloping plain, interspersed with mountain ranges, inselbergs and valleys. The southeastern part is covered with grass savanna, which gradually merges in the west and north into tree- and bush savanna in response to the increased amount of precipitation. The consumption of aboveground phytomass increases from less than 50 % in the northern Serengeti to above 80 % in the southeastern region.

In areas intensively grazed throughout the whole growing season, as in the south and southeast of Serengeti, rapid turnover is typical. Most plants are eaten before their fibrous tissue increases due to aging and their nutritional value declines. With the excreta of the grazers the soil is supplied with many nutrients already available to the plants (e.g. ammonium) or rapidly decomposable (e.g. urea). In addition, the herds prevent larger amounts of standing dead plants by trampling them down. The dead matter becomes disintegrated and more easily decomposable chemically. Thus, the herbivorous mammals considerably accelerate turnover not only on the southern and southeastern Serengeti plains but also on the hills in more rain-loaden areas, where herds concentrate during the rainy season. In the areas where fire and herbivory are unimportant, turnover is very slow because the proportion of not easily decomposable plant material (stems, structural tissue) increases and is larger compared to the intensively grazed areas.

In other regions, with periodical grazing and fires outside growing season (during dormancy), the nutrients that accumulated during the growing season in differently nutrient-rich and decomposable plant tissue, are returned to the soil with the ash and herbivore excreta. Moreover, dead plant material becomes mixed with soil by trampling. In the end, it depends on quality and availability of forage in the different grazing areas which factor is most important for nutrient cycling. In case of high food quality, turnover rate is influenced mainly by the excreta of the herbivores, whereas at poor quality the ashes and mechanical disintegration of the plant material by trampling are the decisive factors. In such 'pulsating cycles' nutrient loss may be higher than in the other turnover types, because the reflux of nutrients to the soil with the excreta of the large herbivores is not synchronous with vegetation nutrient requirement. Moreover, some basic nutrients are lost due to fires, and, in addition, rapidly moving herds may 'export' nutrients (McNaughton 1985).

Although man has considerably reduced the populations of large herbivores directly by persecution (hunting and poaching) and indirectly by partly radical alteration of the landscape, large herbivores strongly influence their habitats as ever by their effects on composition and structure of the vegetation and on soil (Laws 1970). This holds particularly true for the national parks and wildlife reserves in East-, Central- and South Africa (Norton-Griffith 1979; Perry 1994) with their huge populations of large herbivores, that considerably exceed their original numbers. Exclosures give an idea of the extent of the herbivores' impact. After exclusion, conspicuous changes occurred in vegetation composition and structure, biomass, diversity as well as in soil-physical and soil-chemical conditions (e.g., Laws 1970; Lock 1972; Edroma 1981; Hatton and Smart 1984; Smart et al. 1985; Belsky 1992, 1995).

In Murchison Falls National Park (Uganda), for example, areas that were not grazed by large herbivores for 24 years became re-colonized by *Acacia* trees, while half of the grass species vanished. Moisture conditions improved, primary decomposers (termites, ants etc.) became abundant, and soil nutrient conditions more favorable. The content of organic substances in the soil considerably increased due to *Acacia*-litter fall and nitrogen fixation by mycorrhiza. This is also reflected in the now darker coloring of the top soil of the previously reddish-brown loamy soils. Humus enrichment can be found down to one meter depth. Due to the increased content of organic matter soils may accumulate more plant-available nutrients. The amount of extractable potassium increased more than five-fold, the amount of calcium and magnesia two- to three-fold, while pH rose from 6.0 to 7.4. Parallel with organic matter the total amount of nitrogen increased by 50 %. Although no data on the quantity of mineral nitrogen are available, a sufficient supply of the mineral soil with ammonium and nitrate is likely, as is suggested by the total amount of nitrogen and the conditions for mineralization. These changes, however, resulting from complete exclusion of large herbivores, are unlikely under natural conditions.

3.1.2 Elephants and Other Pachyderms

Originally, elephants were spread almost all over the African continent, although their distribution area was not continuous (Laws 1970). Moreover, local elephant populations were not as large as they are in national parks at present (Photo 3.1; Cumming et al. 1997). In East Africa, elephants can be found up to 4,000 m elevation. In the alpine belt of Mount Kenya and the Virunga volcanoes, for example, elephants seriously impact the stands of the up to 8 m high giant rosette plants (*Senecio keniodendron*), which become gradually replaced by grassland (Mulkey et al. 1984). Climatic change and human impact, in particular (hunting and poaching for meat and ivory), as well as habitat loss and fragmentation have very much reduced elephant habitats. Consequently, elephant populations have drastically increased in the remained habitats and often exceed the carrying capacity (Buechner and Dawkins 1961; Buss 1961; Brooks and Buss 1962; Lamprey et al. 1967; Norton-Griffith 1979; Owen-Smith 1983; Dublin et al. 1990; Dublin 1995; Cumming et al. 1997). The main cause, however, is the extreme growth of human population. In Zimbabwe, for example, where less than half a million people lived around the turn of the nineteenth to the twentieth century, present population ranges around 12 million. In Kenya – almost 30 million people in the 1990s – population has increased to >43 million in 2012. Situation is similar in other regions of Africa, although population growth often has become retarded because of bad economic conditions, insufficient health care, warfare and Aids.

Elephants influence their environment in many ways. Effects on vegetation and on biodiversity are profound (e.g., Ben-Shahar 1993; Van de Vijver et al. 1999;

Photo 3.1 Elephant herd in Tsavo National Park, Kenya. W. Leuthold

Nellemann et al. 2002), especially at high elephant densities in confined areas. This depends mainly on their large body size, immense forage demand (about 100 kg day^{-1} at an average) (e.g., Laws 1970), and great physical strength. 'Devastations' by elephants are particularly spectacular and often reported, in particular when conflicting with landscape use by man. About 70 % of the range of African elephant are outside protected areas (Blanc et al. 2007). Elephant numbers are increasing in some areas and have become a real 'pest'. In Uganda, for example their number has doubled since 1996, and most of them are living outside the reserves. Obviously, increased elephant population is not only due to better protection but results also from elephants which escaped the war-like conflicts in the neighbored Democratic Republic of the Congo (Wasswa and Frentzen 2013). Elephants occasionally devastate agricultural land and destroy ripening crops (e.g., maize, millet, sorghum, beans, ground nuts, melons, coffee, cotton), mainly at the end of the wet season (Hoare 1999). They raid settlement and food stores and also destroy watertanks and irrigation systems. Often, people that try to defend their property against marauding elephants are killed. Great efforts are made to mitigate these conflicts for the benefit of both people and elephants (e.g. Hoare 2000; Jackson et al. 2008). The main problem is the overlapping of spatial distribution of elephants and people. Overlap might be reduced by using chemical deterrents (Sitati et al. 2005; Sitati and Walpole 2006), for example, or by providing mineral licks to attract

Photo 3.2 Baobab tree (*Adansonia digitata*) perforated by elephants to get to the soft fibrous wood in the stem interior. Ruaha National Park, Tanzania. N. Hölzel, September 2010

elephants and thus divert them from cultivated land. Elephants usually frequent sodium-rich soils (e.g., termite mounds) and water to meet their mineral demand (Holdo et al. 2002). Reduction of elephants by culling does not always really meet the complex 'elephant problem' (e.g., Jackson et al. 2008). As elephants densities are usually highest across artificial water holes during the dry season, making such areas inaccessible to elephants or not opening the water holes during the dry season, have been suggested to reduce local elephant overabundance and the risk of human-elephant conflicts in the adjacent areas (Chamaillé et al. 2007). On the other hand, farmers near elephant refuges suffered from marauding elephants also in the past, though to a lesser extent (e.g., Bell 1984; Hawkes 1991).

Elephants often strip the bark of the trees and pull off the bast until sapflow is interrupted. They destroy the thick stems of baobab trees (*Adansonia digitata*) with their tusks to get to the soft fibrous wood in the stem interior on which they feed during the dry season (Leuthold 1978; Photo 3.2). Elephants browse small trees when available, even those that are smaller than one meter ('seedlings') and snap the tops off taller trees (e.g. Dublin 1995; Birkett 2002). They also overthrow large trees (Photos 3.3 and 3.4) to browse leaves and twigs that otherwise were out of reach. Moreover, elephants under stress occasionally destroy trees. Stress often results from high population pressure, shortage of food and competition (Murray 1976). Breakage of trees may result in multi-stemmed growth forms with limited

Photo 3.3 Tree stand (*Acacia* sp.) devastated by elephants. Ruaha National Park, Tanzania. N. Hölzel, September 2010

Photo 3.4 Elephant destroying trees in Amboseli National Park, Kenya. G. B. Schaller

vertical growth (Holdo 2003). In addition, elephants can prevent recruitment in tree species they prefer for food. On the other hand, elephants may promote height growth of young trees by uprooting tall trees and creating gaps in the woodland canopy (e.g., Childes and Walker 1987). Thinning out tree stands by elephants also encourages expansion of grass vegetation and thus improves the food source of other herbivores. In other cases, however, elephants have turned wood- and bushland into almost unvegetated surfaces (Laws 1970; Hatton and Smart 1984; Van Wijngaarden 1985; Cumming et al. 1997). Murchison-Falls-National-Park in Uganda (Buechner and Dawkins 1961; Laws 1970), Tsavo-National-Park in Kenya (Parker 1983; Belsky 1992) and the southern Zambesi area in Zimbabwe (Cumming et al. 1997) are characteristic examples. The total elephant population in northern Botswana and adjacent Zimbabwe and Namibia probably is the largest in Africa. It has been estimated at about 231,000 (Blanc et al. 2007). In Chobe-National-Park (Botswana), *Acacia* is drastically declining under the influence of the numerous elephants and other herbivores, and, once again, the decisive factor is shrinkage of habitats due to human impact (Perry 1994; see also Tchamba 1995). Since 1900, the elephant population has increased from a few thousand to an estimated 41,000 at present, exceeding the carrying capacity (25,000) by far. Nevertheless, trophy hunting will be completely prohibited from 2014 onwards.

On the other hand, elephants may positively influence the habitats of other animals. In swamp savannas, for example, elephants often make new food sources accessible to smaller herbivores by trampling down the almost impenetrable several meters high vegetation. In this way they also encourage the growth of highly nutritive grasses, as has been described for example from the swamp savanna around Rukwa Lake in Tanzania and from the Amboseli-Basin in Kenya (Darling 1960; Vesey-Fitzgerald 1960; Western 1989). After elephants have grazed and trampled down the high swamp grasses, buffalos use the new sprouting plants. As a result, open patches with short grasses develop that are preferred by topi (grazing succession). In addition, uprooting of trees by elephants improves aeration of the soil, which is particularly important on clay-rich not well-drained substrate as in the mopane woodland (*Colospherum mopane*), for example. Though this is only a local effect, larger areas may be influenced in this way in the long-term (Darling 1960). During the dry season elephants excavate water holes of which also other animals benefit.

In the end of the 1960s and in the 1970s, the development in Tsavo National Park became the focus of public attention, when, in view of serious 'devastations', rigid culling of elephants was subjected to controversial discussions. After all, culling was carried out with the objective to prevent further decline of the woodland (Laws 1970; Parker 1983). It was a big problem, however, as the state of knowledge of how to manage elephants and their habitats was still insufficient those days and not a sound basis for such far-reaching decisions. Thus, decisions had to be made based mainly on assumptions and plausibility considerations.

Looking back on the Tsavo landscape history as far as possible, the 'elephant problem' turns out to be relatively young. The first Europeans travelling through the region at the end of the nineteenth century came across comparatively few elephants. Probably, elephants were killed in great numbers at that time already by the native people for meat and ivory, in particular, because of the strong demand for ivory on the world market (Parker 1983). During the peak of ivory trade elephant hunters had to travel very long distances (ca. 1,000 km) from the coastal region of Mombasa to the land interior, up to the east side of Lake Rudolph to get large tusks at quantities worth the effort. At these occasions they regularly crossed the area of today's Tsavo National Park without catching sight of elephants. At least there are no reports on elephants, which should be expected if hunting were profitable in this region located relatively close to the coast (Wing and Buss 1970; Parker 1983). In the end, only few elephants lived east of Lake Rudolph, while they were numerous in the area of present Tsavo National Park. Despite safari trophy hunting and poaching, elephant populations increased until the park became established. In 1957, a paramilitary operation terminated poaching, at least for the time being. Afterwards, elephant density rapidly increased, particularly due to invasion from the densely populated and intensively used agricultural surroundings of the park (Laws 1970). According to estimates of reliable authorities (cited in Parker 1983) roughly estimated 48,000–50,000 elephants lived in the park and adjacent areas.

When the park became established, the dense and high *Commiphora-Acacia* wood- and bushland, interspersed with solitary baobabs (*Adansonia digitata*) and other tall trees, almost prevented wildlife observation by safari tourists. The park management tried repeatedly to thin out the tree stands and bushes using fire. These efforts, however, failed because there was no easily inflammable and sufficiently dense grass cover. Even the fires caused by flying sparks of the railway commuting between Nairobi and Mombasa for about half a century (1900 until after 1945) had local effects only. Most of the park turned out to be 'fireproof' (Parker 1983). Until the end of the 1950s, however, elephants had thinned or destroyed the woodland, in particular along the rivers and artificial dams, and converted more than 12,000 km^2 of woodland into open grassland (Ngene et al. 2011). As a result, grassland fires became more frequent. Elephant culling was considered to be an effective preventive measure. However, when drastic elephant culling met worldwide furious opposition of animal- and nature-conservationists, it was terminated (Kurt 1977). Degradation of vegegation continued, followed by erosion and drying-up. Periods of drought, occurring in Tsavo National Park at intervals of about 10 years, resulted in further deterioration of the food source. Conditions were a little better only along the Galana River and other waters for a while. Then, however, many elephants died due to starvation because no forage was available anymore within a reachable distance away from waters. Between 6,000 and 7,000 elephants as well as most rhinos and smaller herbivores fell victim to the extreme drought in 1970/1971 (Corfield 1973; Leuthold 1978; Cobb 1980), particularly in the eastern section of the park, where the amount of precipitation is low anyway (Cobb 1980). Installation of artificial water holes retarded elephant decline for a while but could not stop it. After the drought, additional losses came along with intensive poaching. Elephant numbers in Tsavo decreased from estimated 30,000 (40,000) to about 6,000 (Leuthold 1995, 1996). During the 1980s, even more elephants were killed by poachers than died during the drought of 1970/1971 (Leuthold 1995, 1996; Coe 1999). Since the early 1990s, elephant numbers have gradually grown again

Table 3.1 Ungulate species counted along a road on Galana Ranch (Tsavo National Park, Kenya) in 1963 and 1981–1982

Species	1963	1981/1982	Losses/increase (%)
Browsers			
Rhino (*Ceratotherium simum*)	1.3	0	−100
Dikdik (*Madoqua* sp.)	48.7	0.5	−99
Lesser kudu (*Tragelaphus imberbis*)	10.4	0.9	−91
Gerenuk (*Litocranius walleri*)	7.9	1.5	−81
Giraffe (*Giraffa camelopardalis*)	6.7	4.1	−39
Grazers			
Eland antelope (*Tragelaphus oryx*)	1.6	3.3	+44
Intermediate herbivores			
Water buck (*Kobus ellipsiprymnus*)	2.2	2.1	−5
Warthog (*Phacochoreus aethiopicus*)	3.5	5.3	+51
Grant's gazelle (*Gazella granti*)	5.8	22.3	+284
Oryx (*Oryx dammah*)	9.6	37.4	+290
Zebra (*Equus burchelli*)	2.3	23.6	+926

The *figures* represent the percentage of the total number a species of all species observed (Modified from Parker 1983)

due to tighter conservation and protection. In 2008, 11,733 elephants were counted by aerial survey (Omondi et al. 2008) and 12,537 elephants in 2011 (Ngene et al. 2013). About 70 % were observed inside the protected area and about 30 % outside. The 2009-drought claimed in particular young and aged elephants as well as many other small animals. Poaching is still a major threat to elephants.

Due to degradation of the tree savanna by elephants, herbivores previously characteristic of wood- and bushland have declined, whereas grassland herbivore species have considerably increased (Parker 1983; Table 3.1), as can also be observed in other regions (Cumming et al. 1997). After the wood- and bushland destruction in Tsavo National Park had culminated during the 1960s and 1970s, the plant cover started to recover. Many tree species that were destroyed by elephants, such as *Acacia tortilis*; *Lannea* spp., *Commiphora* spp. and *Boscia coricacea*, have expanded again (Leuthold 1995; Photos 3.5 and 3.6). Their development, however, is impeded by fires that now regularly occur after the grasses increased due to the elephants' impact. Under more humid conditions the development of a tree savanna has been considered to be possible, and elephant populations could also recover if protected from poaching, as a cyclic model suggests (Cobb 1980). However, a fire-controlled 'climax vegetation' is also possible for an indefinitely long time (Parker 1983). The results of simulation models suggested (Dublin et al. 1990) that fires alone would be sufficient to transform wood- and bushland into open grassland, while afterwards elephants together with giraffes and smaller ungulates would prevent tree establishment (Fig. 3.2). It also has been hypothesized that changes like those that occurred in Tsavo National Park during the 1960s and 1970s represent a snapshot of a long-term natural cyclic process that is characterized by the manifold

Photo 3.5 Vegetational aspect in Tsavo National Park near the park entrance at Manyani. W. Leuthold, 21 August 1970

Photo 3.6 The same place as in Photo 2.10 almost 30 years later. Numerous trees (Species of *Commiphora* and *Boswellia*) have established themselves after the drastic decline of the elephant population. Dense grass and shrub vegetation hides the abandoned diagonal road in the center of the picture. The tall baobab tree (*Adansonia digitata*, *right hand-side*) is missing. W. Leuthold, 29 July 1997

interactions between elephants and their environment (Ford 1966; Cobb 1980; Caughley 1976, 1983). These cycles are supposed to cover approximately 150–200 years. Such a cycle might have taken a course as follows (Fig. 3.3): bush- and tree-stands provide suitable living conditions to the tsetse fly (*Glossina* spp.),

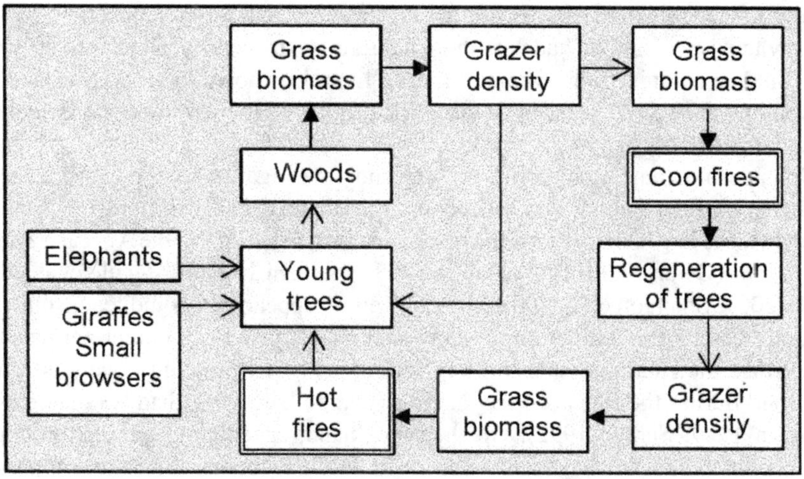

Fig. 3.2 The influence of ungulates, elephants and fire on vegetation in the Serengeti (Modified from Belsky 1989)

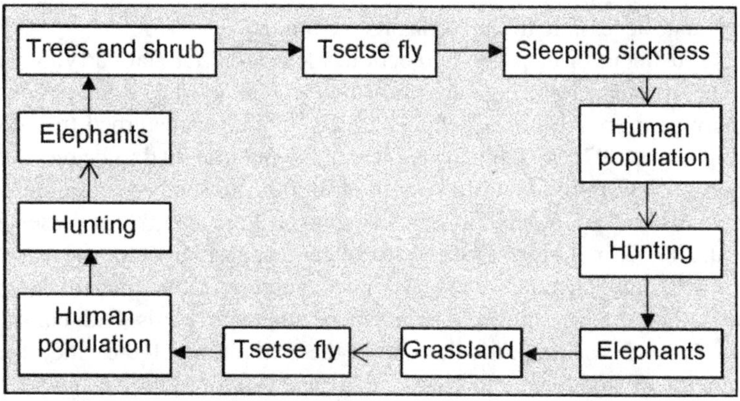

Fig. 3.3 Hypothetical cycles of environmental changes induced by tsetse flies, humans and elephants. *Filled arrows* mean increase, *open arrows* mean decrease (Scheme by F.-K. Holtmeier from various sources)

which transmits the sleeping sickness (*Trypanosomiasis*) to humans and cattle. Sleeping sickness took a heavy toll on human life, and large areas became depopulated. As hunting pressure decreases, elephant population grows. Tree- and bush savanna turn into grassland again, which is re-colonized by native people hunting elephants. Tree vegetation recovers, and conditions for the tsetse fly improve. Consequently, native people population declines again and the next cycle begins. This hypothesis is based on the assumption that such cycles were spatially limited and did not cause extensive losses of species. In addition, refugia must have existed from where species that had left their habitats destroyed by elephants survived and could re-colonize their former habitats. In reality, however, conditions are not so well 'balanced'. At least in South Africa, none of the assumptions is realistic (Cumming et al. 1997).

Considering the innumerable reports and investigations on the influence of elephants on their habitat and biocoenoses in context, one will hardly hesitate to concede these largest herbivores the role of a keystone species in the African savanna ecosystems (Owen-Smith 1988; Dublin 1995; Shoshani 1993; Mosegulo et al. 2002; Holdo 2007; Shannon et al. 2008, 2011). Their disappearance could lead, similar to a 'domino effect', to extinction of many smaller herbivores and other animals that would lose their habitats due to increasing density of the plant cover, as it also happened during the Pleistocene when many large herbivores had become extinct (Western 1989; Schüle 1992). On the other hand, speculating on such potential effects one should not ignore that the large wildlife conservation areas represent only a small fraction of the original living space and harbor by far too many elephants. If the existence of the elephants were really threatened, then over-utilization and progressive destruction of their food sources would be the most important factors. In Zimbabwe, for example, elephant numbers increase by 5 % per year. Because of human population growth and land use conversion, the possibilities to make new or former habitats accessible to the elephants are limited.

Thus, adjustment of elephant numbers to the present habitat carrying capacity seems to be the only promising way to ensure elephant existence and to conserve a biodiversity as it probably was peculiar to the original undisturbed tree savannas (Cumming et al. 1997). In Zimbabwe, elephant population had to be reduced from more than 60,000 individuals to less than 30,000 to reach a tolerable density (five elephants km^{-2}). However, it is estimated to be almost 85,000 at present (2006; Blanc et al. 2007). Similarly to Tsavo National Park, opposition of the public seriously interferes with implementation of such control measures (Cumming et al. 1997).

Likewise, in the Masai-Mara area where ungulate herds had considerably increased (wildebeest in particular) after the extinction of the rinderpest and transformation of woodland into grassland, a natural reforestation is not very likely at the present elephant population (1,300). Elephant impact has to be considered in combination with the seasonal migration of wildebeest (see also Fig. 2.18) (Dublin 1995). More accidentally, grazing wildebeests destroy young trees. Elephants rely on the leaves, particularly during the dry season (Fig. 3.4), because the foliage contains much more protein (13–33 %) than the grasses usually preferred by elephants (e.g., *Themeda triandra* and *Cynodon dactylon*, protein content about 5–8 %; Douglas et al. 1964; Sinclair 1975). As a result, elephants browse the survived hardwoods. The situation is aggravated by poaching in the surroundings of the wildlife reserve. As poaching prevents elephants from following their traditional migration routes, elephant population in the Masai-Mara area has reached extreme density. In the northern Serengeti, which borders the Masai-Mara wildlife reserve in the south, the surviving elephants (400–500) decreased between 1970 and 1986 by about 80 %, mainly due to poaching and migration of surviving individuals to the

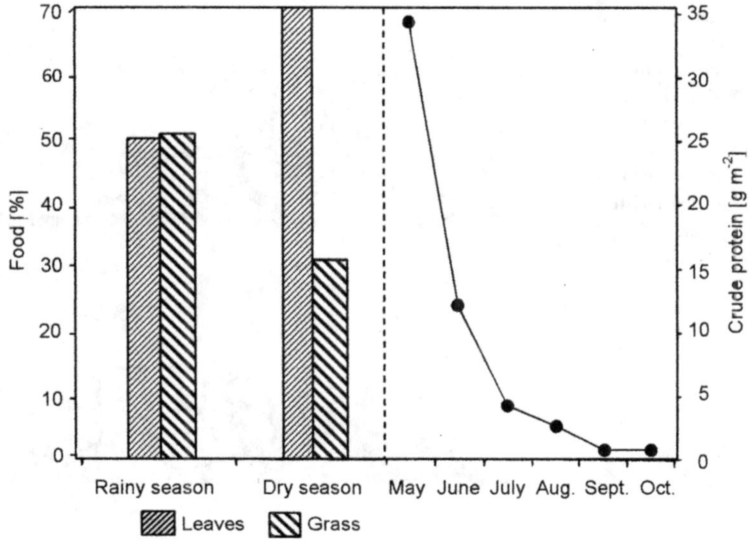

Fig. 3.4 Available protein (g m^{-2}) of tall grasses during season (*right*) and amount of grasses and leaves in the forage during wet and dry seasons (*left*) (Modified from Dublin et al. 1990)

safer Mara wildlife reserve (Dublin 1995). Since then, intense regeneration of *Acacia* (*A. clavigera*, *A. gerrardii*) and other woody plants has occurred (see also Tsavo National Park).

In Addo National Park (Cape Province, South Africa), the vegetation, which is rich in endemic species, is increasingly becoming destroyed by pastoral use. The first settlers already decimated elephants, because such animals caused serious damage to agricultural fields and plantations. Thus, only 16 elephants were living in this area about 80 years ago. After the park became established, elephant numbers rapidly increased up to 400 (ca. four elephant km^{-2}) at present. This density is considered to be equivalent to the tolerable maximum. The first signs of over-utilization, however, became already apparent when about 60 elephants were living there (Pentzhorn et al. 1974). Similar to other African National Parks, the park management is in a dilemma; to guarantee long-term preservation of the elephant habitat and, on the other hand, to conserve the great variety and diversity of the unique vegetation (Johnson et al. 1999). Destruction of the vegegation cannot be ascribed to elephants only but is also due to excessive cattle and goats, in particular. Goats appear to be even more destructive than elephants. Regeneration and spread of the tree-shaped succulents (*Portulacaria afra*), for example, which spreads mainly by layering (formation of adventitious roots) of their long branches contacting the ground at their tips, is seriously affected by goats, whereas elephants browse the tree canopies ‘from the top’ only. Thus, the trees though being coppiced, may continue to propagate by layering (Fig. 3.5). Nevertheless, conservation of the vegegation structures and plant diversity seems to be impossible at the present elephant impact

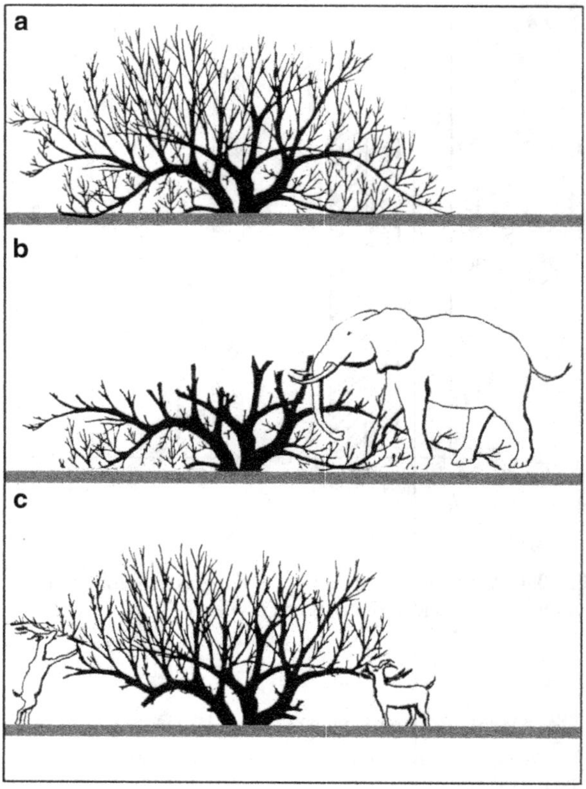

Fig. 3.5 Growth forms and layering (formation of adventitious roots) in elephant bush (*Portulacaria afra*). (**a**) undisturbed, (**b**) disturbed by foraging elephants, (**c**) browsed by goats (From Stuart-Hill (1992), elephant and goats added by F.-K. Holtmeier)

(Moolman and Cowling 1994; Johnson et al. 1999). Efforts to reduce the elephant population, however, is conflicting with the massive interest of tourism, which is an important currency source – and, of course, tourists want to see many elephants. Proposals have been made to enlarge Addo National Park by incorporation of a former national park located in the northwest. An enlarged Addo would provide a suitable habitat with sufficient forage for about 500 elephants (Kerley and Boshoff 1997). Successful preservation of the threatened vegegation, however, will only be possible if sufficiently large park areas are completely protected from elephants.

In view of observations in Lake Manyara National Park (northern Tanzania), the impact of elephants on woody vegetation has to be seen in perspective, and once again it becomes apparent that a generalization and transfer of locally obtained results to other areas is always problematic. Poaching reduced elephant density in Lake Manyara National park between 1987 and 1991 from six elephants km^{-2} to less than one elephant km^{-2}. Parallel to elephant decline *Acacia tortilis* and other tree-and shrub species invaded. This development seemed to confirm the so-called 'cycle hypothesis' (Caughley 1976). However, closer examination and evaluation of oral tradition showed that the decrease of elephant numbers had not been the primary cause of shrub invasion, as in the northern park area shrub encroachment already began a few years prior to elephant decline, and in the southern part even about 10 years earlier. In both areas, shrub invasion had coincided with the outbreak of an anthrax pandemic (*Bazillus anthracis*). The epidemic decimated in particular the numerically dominating impalas (*Aepyceros melampus*). In contrast to typical grazers, impalas feed not only on grasses and herbs but also on *Acacia* pods and seedlings. Giraffes browse tree foliage several meters above ground (cf. Photo 2.3) and do not impede establishment of *Acacia* seedlings. As matters stand, the coincidence of the anthrax epidemic and phases of *Acacia* regeneration probably was the main cause of shrub invasion in Manyara National Park. The introduction of the rinderpest to East Africa at the end of the nineteenth century had a comparable effect. At least, it looks very much as if even-aged *Acacia* and baobab trees established themselves when the rinderpest had reduced the grazing impact. In Manyara National Park, fluctuations of ungulate populations probably have temporarily reduced grazing pressure on *Acacia* seedlings, as is reflected in about four age classes of *Acacia* which became established during the last 200 years. Such local ungulate crashes due to epidemics may be considered to be narrow time-slots that allow seedling establishment and growth. It would appear that also occurrences of old even-aged *Acacia* stands can be attributed to short-term periods of reduced or lacking grazing impact (Prins and Van der Jeugd 1993). At present, severe grazing pressure impairs regeneration of *Acacia* considerably.

In Chobe National Park (Botswana), *Acacia* woodland was also able to temporarily recover from grazing impact after elephant and ungulate populations had crashed due to the rinderpest pandemic and dry-out of water holes during the end of the nineteenth century. When the herds had recovered, probably due to more favourable humid conditions, the woodland declined once again (Lewin 1986). Experimental studies (Moe et al. 2009) in the elephant-transformed Chobe riparian woodland showed that seedling mortality is closely related to abundant local impalas rather than to any other browsers. Impala considerably increased after the tall dense woodland changed into shrubland. They feed on tree seedlings mainly during the dry season. The studies also support the theory that the older closed-canopy woodlands became established during episodic time windows favorable to seedling survival, as it has also been suggested for Manyara National Park (Prins and Van der Jeugd 1993; see above). Theoretically, restoration of the woodland might be achieved by reducing seedling predators as well as elephants, which, however, affect the taller trees. Anyway, this is hardly practicable at the landscape scale (Moe et al. 2009). In the ecological view there is no reason to reduce elephants and ungulates. The woodland would turn into a shrubland again as it probably prevailed during the nineteenth century (Campbell 1990). However, economic and social reasons for elephant control cannot be disregarded (Skarpe et al. 2004). Thus, a 'permissible elephant range' could be defined, for example, and elephant groups leaving this area would be destroyed. The conflict between elephants and human land-use activities would be mitigated, and local economy would profit from legal harvesting of 'surplus elephants' (Skarpe et al. 2004).

Compared to the effects of elephants and of the large herds of wildebeests and antelopes, hippos influence their habitat more locally, because they rely on water

Photo 3.7 Hippos (*Hippopotamus amphibus*) grazing low grass vegetation near a stream in the dry season. Lake Manze, Selous Game Reserve, Tanzania. N. Hölzel, September 2010

within reach. More detailed studies on the impact of hippos on landscape have been carried out in Ruwenzori National Park and Murchison Falls National Park in Uganda (Laws 1968; Lock 1972). After the parks were established, hippo populations considerably increased, and overgrazing by hippos appeared to become a threat to the environment.

During the dry season hippos usually stay in close proximity of water courses, lakes and other stagnant waters (Photo 3.7). During the wet season, however, they expand their radius of action and frequent also temporary waters and mud holes in the savanna. They graze there during the night before returning to rivers and lakes during the day. Hippos usually defecate in the waters and thereby considerably increase eutrophication. A hippo consumes about 18 kg (dry weight) per day. Apparently, there is no comparable backflow of nutrients from the waters to the savanna (Laws 1968). Hippos may graze several kilometres away from water presumably always using the same trails. Consequently, beaten paths develop lacking almost any vegetation, with the exception of a small ribbon of grasses between the footsteps. Due to compaction and reduced infiltration, surface runoff during heavy rains follows these trails. On steeper terrain, the runoff may erode real drainage channels.

Hippos usually feed only on grasses not higher than 15 cm (cf. Photo 3.7). Therefore, some areas are overgrazed while others are left untouched, and a locally varying plant mosaic develops (e.g., Eltringham 1999). Hippos pull out tufts of grasses with roots inclusive. Thus, rooting and less palatable grasses such as the

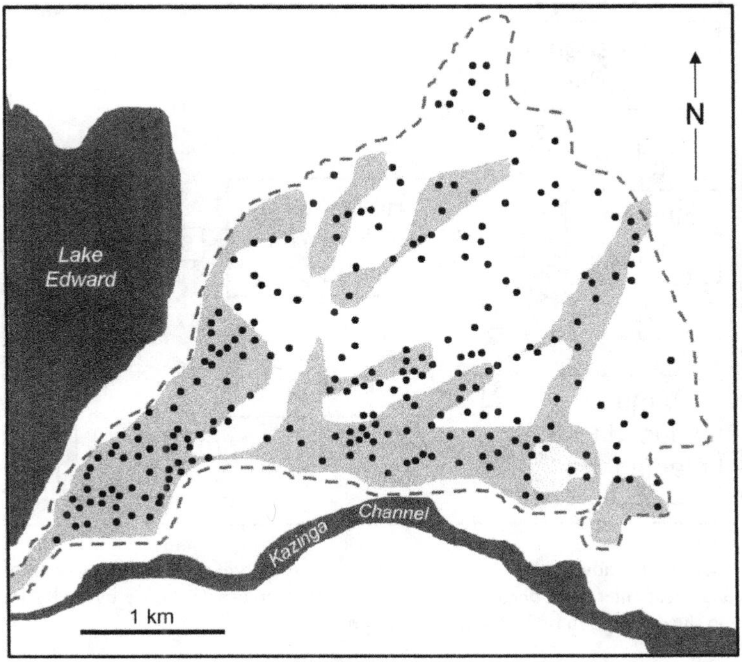

Fig. 3.6 Distribution (*dots*) of warthogs (*Phascochoreus aethiopicus*) in 'mosaic grassland' (*light-gray*) east of Lake Edward, Ruwenzori National Park, Uganda. One dot represents one sighting of any numbers of warthogs. The *broken line* encloses the study area (Modified from Lock 1972)

tussock grass (*Sporobolus pyramidales*), which expands vegetatively, may become predominant under high hippo grazing pressure. If hippos are excluded, it will soon be replaced by other grasses. On intensely grazed areas, surface runoff and erosion increase and herbivores other than hippos decline (Laws 1968).

Warthogs (*Phaocoreus aethiopicus*), however, apparently benefit frcm hippos grazing as is reflected in conspicuously high warthog density on the 'mosaic grassland' (Fig. 3.6). Warthogs digging for rhizomes of *Alternantherc pungens* (Amaranthaceae), for example, during the dry season, very likely contribute to continued existence of unvegetated patches (Lock 1972). While elephants usually transform wood- and shrubland into grassland, hippos are more likely to promote the establishment of shrub vegegation, in particular near waters. They reduce the amount of combustible grass vegegation and thereby the frequency of grassland fires that would prevent shrub expansion.

Around Lake Edward (Uganda) more than 6,000 of about 14,000 hippos living there in 1957 were culled between 1958 and 1966. Great changes in vegetation that occurred after culling and also enclosure of hippos provide clear evidence that these large grazers had considerably contributed to conservation of the varied vegetation

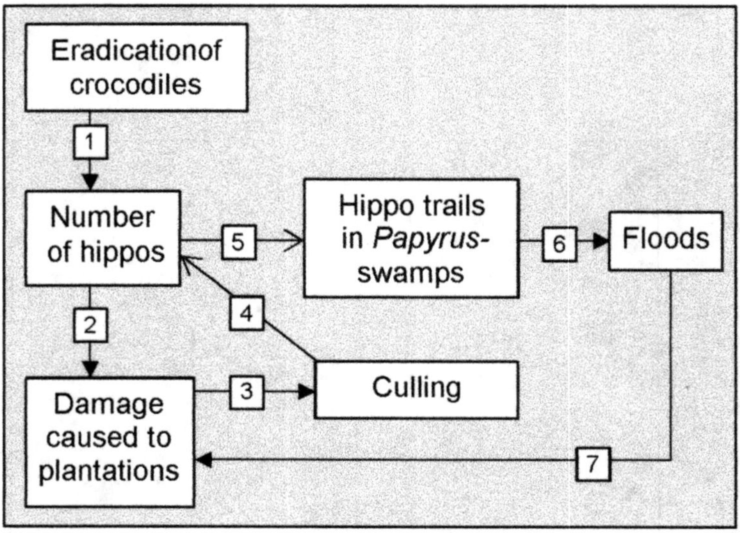

Fig. 3.7 Effects of eradication of crocodiles (*Crocodilus niloticus*) in *Papyrus*-swamps. *Filled arrow* heads mean increase, *open arrow* heads mean decrease (Scheme by F.-K. Holtmeier according to the description by Milne and Milne 1960)

mosaic or even were its main cause. After reduction of hippos, however, the other grazers rapidly increased, whereas warthog numbers considerably declined (Laws 1968). Lock (1972) pointed out that hippos, keeping the grass vegegation short, improve the possibilities of wildlife watching, an aspect that should not be disregarded in a national park highly frequented by safari-tourists.

In papyrus swamps, the paths trampled by hippos improve drainage conditions (Darling 1960). In case of hippo decline, however, these paths become overgrown and flooding will increase. Milne and Milne (1960) reported a similar development triggered by eradication of crocodiles (*Crocodylus niloticus*). Crocodiles were hunted because of their skins. When crocodile predation on young hippos ceased, hippo population rapidly increased. As hippos, however, caused severe damage to farms and plantations along the rivers, one tried to solve the problem by culling and promptly got another one – flooding (Fig. 3.7).

Not least, rhinos may lastingly influence their habitats. Due to grazing by square-lipped rhino (*Ceratotherium simum*), for example, grazing-tolerant short grasses expand at the costs of taller grasses, herbs and shrub. As rhinos are less dependent on open water than hippos they influence vegetation over larger areas. In South Africa, extinction of rhinos and resultant vegetation change was followed by disappearance of several smaller ungulate species, without any influence of hunting (Owen-Smith 1983, 1987). By contrast, black (hooked-lipped) rhinos (*Diceros bicornis*) are browsers and feed on the leaves of bushes. In case black rhinos have gone and no other similarly effective herbivores replace them, impenetrable thickets will develop that most medium-sized ungulates cannot use (Owen-Smith 1983, 1987).

In addition to the browsing impact, rhino-dung enriches soil nutrients. The dung also contains plant seeds that may germinate. Moreover, rhinos create wallows that, after filled with water, provide favorable life conditions to other animals such as frogs and insects, which feed in them and where they complete their life cycles. Not least, like elephants and hippos, rhinos open pathways through dense thickets and make new food sources accessible to smaller herbivores.

As to the management of large herbivores in savanna ecosystems, the management objectives must be clearly defined. The goals may be protection of viable populations of endangered animals or conservation of original vegetation (e.g., woodland or grassland), or both. Management has to be adjusted accordingly. If it is preservation or restoration of woodland savanna, for example, the browsing pressure has to be reduced to a level that allows natural tree establishment. In this case, the combined effect of the main browsers must be considered. In Sweetwaters Game Reserve (Kenya), for example, with an elephant population of 100, a black rhino population of 25 and 150 giraffes, the habitat is already being threatened (Birkett 2002). It has been supposed that at the present population increase of both elephant and rhinos 32 % of *Acacia depranolobium* trees will be lost in 7 years. To prevent woodland savanna decline, 50 elephants or 75 giraffes and surplus rhinos (exceeding a population of 25) have to be removed (Birkett 2002).

3.1.3 The Influence of Poaching

As poaching has recently assumed unprecedented proportions, a few additional remarks on its effects on animals and ecosystems may complete the previous sections. International ivory trade has been banned by the Convention on International Trade in Endangered Species of Wild Fauna and Flora (CITES). However, unregulated and illegal international ivory trade has continued, and still there is a demand for raw ivory, particularly on domestic markets (Stiles 2004). While elephant poaching does not seem to have increased in South Africa as of the beginning of the present century, the situation in Central Africa (Kenya, Zambia and Zimbabwe) has become aggravated (Stiles 2004).

In the Serengeti-Mara area, for example, as well as in other African national parks and wildlife reserves, poaching has considerably increased (Arcese et al. 1995; Campbell and Hofer 1995; Sinclair 1995; Martin and Vigne 2003). This development is directly related to the political and demographic change in Africa. Thus one might say it has been triggered from 'outside' the parks. Increasing density of human population (incomers, population growth by about 3 %) and expanding colonization immediately west of the park boundary are the main driving factors. After the border between Tanzania and Kenya were closed tourists stayed away, and the income of the park government considerably decreased. Consequently, anti-poaching patrols had to be reduced. Poaching caused severe damage to wildlife and to the ecosystem as a whole. However, not all species were equally affected. Wildebeest and zebra populations remained relatively stable, although they are

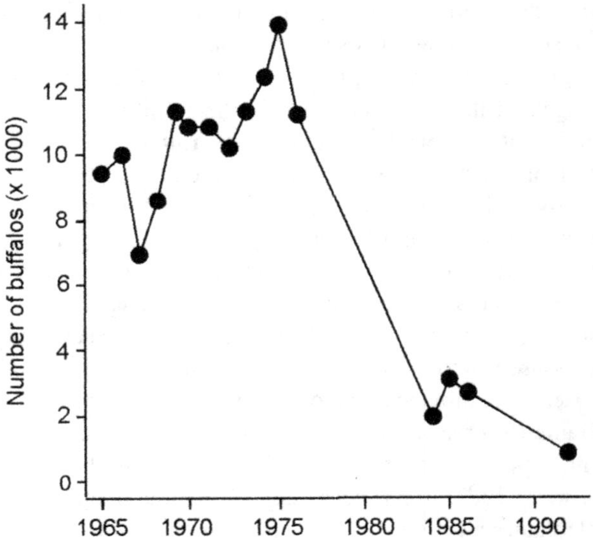

Fig. 3.8 The development of buffalo population in the northern Serengeti (Sinclair 1995)

poached in particular for meat (Magombe and Campbell 1989; Campbell and Borner 1995). Giraffes, waterbucks, reedbucks, topis, impalas and warthogs also suffered seriously from poaching, at least locally. Rhinos and elephants, however, are most severely affected. In 1980 already, the black rhino was almost eradicated for its horn. Elephant numbers declined by 80 % after the closure of the park boundary. Only when ivory trade became proscribed worldwide was poaching no longer worth the effort and collapsed abruptly. Buffalo population, that had reached a peak in 1976, became also almost completely eradicated by poachers (Fig. 3.8; Dublin et al. 1990). As buffalos do not go on long-range seasonal migrations as is typical of wildebeest, the decline of these large grazers brought about a considerable change in the spectrum of plants (Sinclair 1995).

In Amboseli National Park (Kenya), the 2009-drought had far-reaching impact on the park ecosystem. All wildlife, wildebeests, zebras, buffalos, hippos and elephants, were heavily affected. Wildebeest, for example, decreased by more than 95 %. Cattle lost 81 %, and sheep and goat decreased by 64 % (African Conservation Centre 2012). Increase of predator impact on smaller herbivores and livestock is likely. Due to the drastic decline of livestock, native people will become dependent on other natural resources such as charcoal production, bush meat and trophy hunting, and poaching for both ivory and rhino horn.

The ban on ivory trade (1989) had mitigated poaching pressure on elephants. Countless tusks confiscated from poachers or taken from elephants that died due to old age, for example, were kept in safe depots of African gamekeepers. These tusks, however, were so to speak 'dead capital'. To capitalize on the hoarded tusks, some African countries successfully tried to get an exception from the ivory trade ban. In 2008, Zimbabwe, Botswana, South Africa and Namibia were allowed to sell

108 t of ivory. Legalization of international ivory trade at limited quota had been expected to reduce the demand for illegal ivory. Considering the total African savanna elephant population of 350,000–472,000 individuals (Blanc et al. 2007), tusks from elephants that died a natural death or were culled in human-elephant conflicts were seemingly sufficient for supplying the global market with good quality ivory at reasonable prices (Stiles 2004).

Theoretically, international legal ivory trade can be better controlled than the domestic markets (ivory carvers selling name seals, jewellery boxes, pendants, letter openers, etc.). However, the controversial discussion on whether ivory trade should be legalized to a certain extent seems to have become outdated by the most recent development. As a result of legal ivory sale, the demand for more ivory promptly increased, and an unprecedented slaughter of elephants is now going on. In China and some other countries of Southeast Asia, prospering economy and improving living standard have fuelled ivory demand. Consequently, ivory prices have shot up to almost US\$2,000 kg^{-1}.

In Africa, poaching has been 'militarized', and ivory has become an important financial source of terror gangs and numerous local war-like conflicts. Between 2002 and 2010 the population of the African forest elephant declined by 62 %. Poaching is blamed for being a main cause (Maisels et al. 2013). The Janaweed (Dafur, Sudan) and the Lord's Resistance Army (Uganda) for example, are killing elephants at large scale. Also, the Uganda military, the Congolese Army and South Sudan's military are involved in elephant poaching and dealing with ivory (New York Times 2012a, b). Warlords of the Islamic Shabab send their warriors to Kenya for elephant hunting. Real massacres on elephants have recently been reported from Garamba National Park (Democratic Republic of Congo) (New York Times 2012a). In Bouba Ndjida National Park (Cameroon), mounted warriors carrying Kalashnikovs killed 350 of the park's elephants in early 2012 (Knaup and Puhl 2012). Tusks are exchanged for weapons and ammunition (New York Times 2012b; Knaup and Puhl 2012). In Hwange National Park (Zimbabwe), poachers recently poisoned about 90 elephants with industrial cyanide, which is normally used in gold mining. As the poachers put cyanide into water holes and on mineral licks (NBC News, 25 September 2013) not only elephants were killed.

Ivory trade is controlled by international mafia-like organized structures which supply the Asian market (e.g., China, Thailand, Malaysia, Laos, Vietnam) with elephant tusks. The criminals operate all over the continent from the east to the west coast. Poaching has particularly increased in elephant-rich areas, where Chinese workers are constructing roads. Thus, one might suspect that there possibly is a connection. Game wards, policemen and custom officers are usually not able to efficiently control poaching and ivory deals, and, being ill-paid, they often are corruptible. To fight against ivory poaching, many countries (e.g., Kenya, Gabon, the Philippines, France, United States, and even China as the latest) have crushed or burned their stockpiles of confiscated ivory during the last few years. Presumably, they have been acting on public opinion rather than on scientific facts and without having a clear idea of the ecological constraints and possible side effects on the native people and economy. Anyway, illegal ivory trade has partly collapsed, at least for a while. In the long-term, however, these actions may have an opposite effect, as increasing shortage of ivory will keep on forcing up prices and thus fuel poaching.

In view of the African confused political situation, bad economy and impossibility to enforce legal and efficient control of poaching and ivory trade, the future of African elephants seems rather gloomy. As elephants have a keystone function in the savanna ecosystems, unpredictable fundamental changes are likely, exceeding a possible influence of climate change by far.

Rhino is even more threatened than elephant. Rhino horn, however, does not consist of ivory but of keratin, the same as in human hares or nails. Rhino horn is mainly used as a remedy for nearly everything (e.g., malaria, epilepsy, cancer, as aphrodisiac, etc.). In China and Southeast Asia, however, the demand for rhino horn has also been growing in recent times. South Africa's rhino population is most affected as it has 70–80 % of the global population. In South Africa, rhino kills have risen from 13 in 2007 to 455 in 2012 (BBC 2012; Cota-Larson 2012). In the Yemen, rhino horn is used to make the handles of the traditional curved daggers (so-called jambiyas). Although nowadays these dagger handles are usually made of the horns of water buffalo, those made of rhino horn are still most appreciated (Martin and Vigne 2003). Trade with rhino horns is very lucrative. As to the end of September 2013, poachers have killed more than 700 rhinos in South Africa, exceeding the annual record of 668 in 2012 (ABC-News, 20 October 2013). Rhino horn is sold at prices up to 65,000 $ kg^{-1} on the black market (e.g., BBC 2012; Focus Online 2012), which is more than the price of gold. Poaching for rhino horns is hard to fight against for the same reasons that prevent effective control of elephant poaching; political chaos, lax law enforcement, corruption and people's poverty. Although rhinos' effects on the savanna environment are less conspicuous and of minor importance compared to those of elephants, continued decline of the rhino population will leave its mark on the ecological interrelationships between animals, plants and soils in the savanna landscapes.

3.2 Grasslands Outside the Tropics

Although the biomass of herbivorous mammals living in the temperate grasslands was small compared to the biomass of the huge herds of African ungulates, they have had a marked influence on the natural grasslands of Eurasia and North America right through the nineteenth century. The situation in the North American west, however, differs as the shrub steppes were only little affected by native ungulates such as American elk, mule deer and pronghorn (Heady 1968; Daubenmire 1970; Galbraight and Anderson 1971; Franklin and Dyrness 1973). In contrast to the Great Plains, bison never played a role in this region (Galbraight and Anderson 1971). Grazing pressure considerably increased when enormous numbers of cattle and sheep of the Europeans occupied the west. During the last two centuries the natural grasslands were converted into agricultural fields and pasture land – the North American prairie and the Hungarian Puszta after the middle of the nineteenth century, the steppes of Central Asia during the twentieth century (Billings 1970).

3.2.1 Eurasian Steppes

The history of the ungulate herds in the natural grasslands outside the tropics is a 'history of extermination'. Originally, the steppes and desert-steppes harbored many millions of antelopes (e.g., saiga, *Saiga tartarica*; wild horses, *Equus przewalskii*, Asian half ass, *Equus hemionus*; kiang, onager, kulan, dschiggetai; Denzau and Denzau 1999) and wild camels (*Camelus ferus*). These ungulate populations have drastically declined due to intensive persecution and loss of habitats resulting from expansion of agriculture and pastoral use. The comparatively small remaining populations often are near extinction, with the exception of eastern Mongolia where large herds of many thousand gazelles (*Procapra gutturosa*) can still be found (Mallon and Jiang 2009; Olson et al. 2009). Such big ungulate herds are rather exceptional outside south- and east-African savannas.

The history of the saiga antelope differs a little from the development of the other ungulates living in steppes and deserts (Bannikov et al. 1967; Sokolov 1974). Saiga antelopes were originally distributed over a huge area, reaching from Poland over the northern Caucasus (Kalmyk Steppe), Kazakhstan, northern Uzbekistan and southwestern Mongolia to Xinijang (China). The herds migrated seasonally over distances of 200–400 km between their southern overwintering habitats and the summer grazing area in the north (Alschner 1980). However, not all saigas migrated seasonally. In Mongolia, for example, seasonal migrations are unknown. East of the Volga river, a minor number of saigas live there all year round, while most of them move to areas with little snow in winter (Sokolov 1974). Although physiology and behavior of the saigas are well adapted to the extreme climatic conditions of the steppe natural environment, periods of severe drought and extremely snow-rich winters often cause severe losses. However, saiga population may rapidly recover due to their high reproductive capacity (Chan et al. 1995).

Excessive hunting rather than natural climatic extremes had brought saiga antelopes close to extinction. In 1930, the population at rest ranged around 1,000 individuals (Sokolov 1974). Hunting for saiga has a long tradition. It dates back to the mid-Pleistocence. Saigas were hunted mainly for their skins and meat. Since about 200 years at the latest, trading with saiga horns increased hunting pressure on saiga considerably. In the eighteenth century already great drives (battues) were organized and saiga horns sold at large quantities and high prices to China (Pallas 1777, cited in Bannikov et al. 1967).

In the former Soviet Union, extensive protective measures and careful wildlife management by the government supported saiga population which had grown to more than one million animals over a period of 20 years (1950–1970). Since the Soviet Union declined, however, poaching and also increasing intensive land use have become again a serious threat to saigas. Moreover, there is a great demand for the horns as constituents of Far East medicine (Sokolov 1974; Chan et al. 1995). Still as ever, export of saiga horns into East-Asian countries is an important source of foreign currency. In particular, opening the Russian-Chinese border has revived the trade with saiga horns on the black market (Milner-Guland 1994; Milner-Guland et al. 1995; Luschekina and Struchkov 2001). In some areas, as in the Kalmyk steppe, for example, economic development, improvement of traffic infrastructure, construction of irrigation systems and industrial plants, as well as cattle breeding together with extensive fencing adversely affect saiga populations. In those areas, where traditional migration routes became blocked, relatively large herds now concentrate around the water reservoirs. Overgrazing and its negative effects, such as reduced fertility and increasing mortality of saiga antelopes are the results. Moreover, soil erosion is increasing locally (Chan et al. 1995).

As saiga horns have run short due to high demand, hunting pressure on other cornigerous (horn-bearing) herbivores such as Tibetan antelope (Tschiru, *Panatholops hodgsoni*) and various gazelle species (Tibetan gazelle, *Procapra picticaudata*; Mongolian gazelle; Zeren, *Procapra subguttorosa*; Przewalski's gazelle, *Procapra przewalskii*) has considerably increased. Still during the 1930s huge herds of zeren were living in Mongolia (Andrews 1932, cited in Schaller 1995). Only a few populations have survived until present. In Central Mongolia, more than 4,000 zeren were killed within 1 month in 1979 ('Dainly China', cited in Schaller 1995). In the Tibetan Highland (China), Tibetan antelopes have been under increasing poaching pressure since the late 1980s mainly because of their wool (so-called shatoosh), although the species is considered an endangered species and protected. Shatoosh is even finer and more expensive than the wool of Cashmere goat. The wool is exported to Kashmir to be woven into shawls that are very popular in solvent high society. The underfur of four to five tschirus is needed to produce one of these luxury shawls. About 2,000 tschirus are annually killed by poachers. Tschirus were estimated to have declined by one million during the last 100 years (Schaller 2003). Though reliable information is missing, it is likely that plant cover has changed.

The Eurasian steppes are occupied by vast numbers of small rodents. European souslik (*Citellus citellus*), bobak (*Marmota bobak*) and Brandt's vole (*Microtus brandti*) may strongly influence soils and vegetation, whereas local disturbance by the Siberian marmot (tabargan marmot, *Marmota sibirica*) appears to have only little influence on regional plant richness (Sasaki and Yoshihara 2013). In Mongolia, vegetation cycles have been observed that are controlled by Brandt's vole (Remmert 1988, 1992). The vole colonies are so densely populated that almost no vegetation becomes established. When the food source is depleted, voles leave these places. Due to intensive bioturbation by voles, the soil is well drained and aerated. In addition, voles have enriched nutrients. After a while, a relatively species-rich plant cover develops, which becomes gradually replaced by luxuriantly growing forage plants. After plant productivity decreases due to grazing by large herbivorous mammals, voles recolonize these locations, and a new cycle begins. Although voles occur on only 2 % of the total area, about 40 % of the total Mongolian steppe is influenced by voles in this way. Obviously, the excessively grazed areas, in particular, provide optimal habitat conditions for Brandt's vole (Zhong et al. 1999; Fig. 3.9). However, if these habitats no longer meet the requirements of Brandt's voles, Mongolian gerbils (*Meriones unguiculatus*) will occupy these locations. Moreover, comb-toed jerobas (*Paradipus ctenodactylis*), which destroy the plant cover, also encourage denudation (Walter and Breckle 1986).

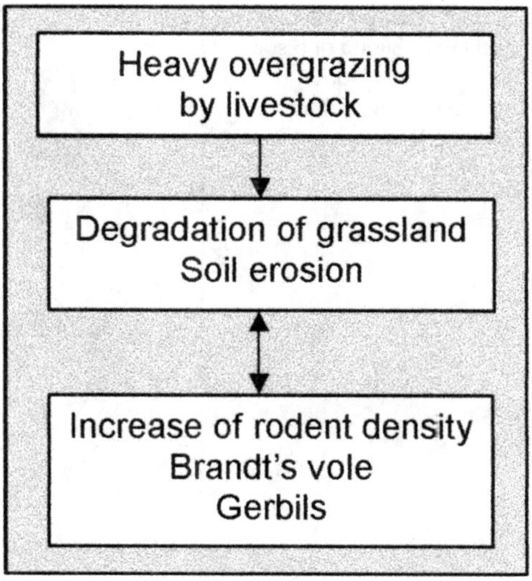

Fig. 3.9 Relationships between overgrazing and rodent density in the central Mongolian Steppe (Modified from Zhong et al. 1999)

3.2.2 North American Prairies

Prior to the arrival of the Europeans, the North American prairie harbored about 50–60 million bison (*Bison bison*) (Roe 1970; Haugen and Shult 1972). Seton (1929) estimated even 75 million. One must be aware, however, that these estimations are based on observations made during rutting time in summer, when the buffalos formed herds of up to 100,000 individuals. In winter, they roamed about in small groups. According to estimations based on the presumable carrying capacity of the former prairie grazing areas, the total bison population probably did not exceed 30 million (e.g., McHugh 1972). The bison herds were really shot up and became almost extirpated during the 1870s (Fig. 3.10; Soper 1941; Campbell et al. 1994). These days, bison occupy less than one percent of the original range (Gates et al. 2010). The destruction of the bison, however, cannot be ascribed to these massacres alone (Isenberg 2000). It was also before the 'big slaughter', that bison numbers heavily fluctuated, mainly due to the droughts regularly occurring in the semi-arid west. In the early nineteenth century, a third of the calves were killed by wolves which were still numerous at this time (about one million). In addition, severe snow storms, prairie fires and also competition with other grazers, such as the many horses of the Indians, for example, influenced bison population dynamics. An estimated 400,000–900,000 Indian ponies grazed on the western prairie, and about two million wild horses were living in the southern grassland region. The horses presumably consumed as much grass as two million bison.

Phases of bison decline were very likely induced by periods of high precipitation, resultant abundant forage, and subsequent mild winters. Bison responded to these

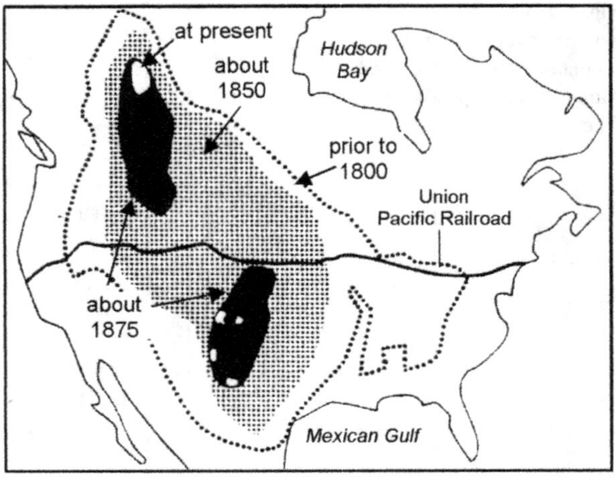

Fig. 3.10 Historical and present distribution of bison (*Bison bison*) in North America (Modified after Ziswiler 1965, cited in Illies 1971)

favorable conditions by successful reproduction. Consequently, their numbers considerably increased and exceeded the carrying capacity of the grazing areas during following less favorable years. In addition, Indian buffalo hunters often killed by far more animals than needed to cover their needs for subsistence – which contrasts with the romantically idealized picture of the wild native living in harmony with his environment. Anyway, bison populations regularly recovered from these impacts, in particular as the grass vegegation, well adapted to the semi-arid conditions, rapidly resprouted. Not last, cattle and sheep of the European immigrants competed with bison and other native grazers for food. In many areas, such as the valleys of the Platte river and Arkansas river, cattle considerably impaired the food resources of bison and drove them out of their traditional grazing grounds that they had used for thousands of years. Nevertheless, the white 'bison slaughters' are responsible for the complete crash of the total bison population from which bison would never have recovered on their own. The fate of pronghorns (*Antilocapra americana*) was not better. Their population, estimated at about 40 million decreased to about 30,000 at the beginning of the twentieth century (Knapp 1965). Since then pronghorn population has increased again.

During the last 30 years of the last century, the bison population increased from 30,000 to an estimated 400,000–500,000 individuals which live on around 6,400 farms in the US and Canada (Gates et al. 2010; Craine 2013). About 10 % of them are owned by the state and live on fenced grassland. Fewer than 20.000 are still running wild (Gates and Aune 2008; Wildlife Conservation Society 2014). The present bison population is genetically impoverished to a high degree, as it descends from only about 500 individuals that survived the nineteenth-century massacres. Genetic impoverishment is also common, by the way, in free-roaming European bison (*Bison bonasus*) in Poland and in some areas of the former Soviet Union. They are descendants of only 13 individuals of 54 zoo bison. The increase of American bison on private land has to be attributed to the strong demand for bison meat rather than to conservation efforts. Bison meat is not only very tasty but also has lower fat and cholesterol contents than beef. Nutrition-conscious Americans appreciate it very much and do not hesitate to pay the double price. As rearing bison is not more expensive than cattle breeding, bison meat production has turned out to be very profitable.

Since the last glaciation until the arrival of the white man, bison markedly influenced the north American prairie. This is particularly apparent in view of the changes that occurred after they disappeared. Although millions of cattle now graze on original bison land, plant species characteristic of the tall-grass prairie could expand also into the western grassland on a broad front. In the southern plains mesquite-bush (*Prosopis glandulosa*) has spread, and in the Canadian grassland aspen groves (*Populus tremuloides*) have become established in great numbers. Probably there was no competition between the native ungulates (bison, pronghorn, American elk, mule deer), as their habitat preferences and feeding habits differ. Bison and pronghorns could coexist in the original prairie because of their different foraging behavior (Haugen and Shult 1972). Pronghorns and mule deer, both of comparatively small size and being concentrate feeders (cf., Geist 1974), do not compete very much, as mule deer prefers shrub- and woodland, while pronghorns occupy the open grassland. Because of their large body size, bison are not able to forage in mule deer habitats (Wydeven and Dahlgren 1985). The habitats of elk and mule deer overlap particularly during summer. Elk, however feed mainly on grasses while mule deer prefer more protein-rich food such as herbs, leaves, buds and young shoots of shrubs and trees. Nevertheless, both ungulate species compete occasionally as seems to be reflected in the opposite course of population development of elk and pronghorn in Wind Cave National Park (South Dakota) in the early 1960s (Lovaas 1973). It is possible, however, that also other factors were involved (Wydeven and Dahlgren 1985). In Yellowstone National Park, on the other hand, where between 18,000 and 20,000 elk are living in the northern part, they do not limit pronghorn population (Yellowstone National Park 1997).

Compared to the large bison herds in the short- and mixed grass prairies, bison density was low in the tallgrass prairie (Shaw and Lee 1995). As far as we know, bison did not seasonally migrate as an immense compact herd – so to speak the animals pushing each other forward – as it is suggested by many reports on the 'Old West' and also in some text books (e.g. Sedlag 1995). Instead, smaller herds roved around 'nomadically' through the prairie searching for forage (e.g., Hanson 1984; Shaw and Lee 1995). The spatial grazing pattern was very likely similar to that in the present bison grazing areas. Bison obviously prefer locations occupied by prairie dogs and burned areas where they find high quality forage plants (Coppock and Detling 1986). The combined spatial and temporal effects of bison grazing and prairie fires cause a mosaic-like landscape pattern ('mosaic cycles'; cf. Remmert 1988). Grazing on burns reduces the grazing pressure on surrounding unburned areas. There, much combustible material accumulates increasing the risk of fire. In the long-term, these recurrent processes cause a varying vegetation mosaic and a higher species and landscape diversity as well (Vinton et al. 1993; Hamilton 1996; Hartnett et al. 1996; Fuhlendorf and Engle 2004).

In the tallgrass prairie, grasses, over half a meter high, such as big bluestem (*Andropogon gerardii*), Indian grass (*Sorgastrum nutans*) and switch grass (*Panicum virgatum*) prevail. The inflorescences of big bluestem may even reach two metres in height. Dry leaves remain at the stalks until they become ripped off and trampled down by bison or cattle and/or destroyed by prairie fires. In case much dead matter has accumulated, primary production will be adversely affected (Knapp and Seastedt 1986, there further literature). Therefore, grazing by bison or other ungulates may promote lower growing herbaceous plants (cf. the effects of elephants in swamp savannas). These not only profit from better light and thermal conditions but also from better moisture supply due to the reduced transpiring surface of grass vegetation (White and Brown 1972; Parton and Risser 1980; McNaughton 1983b; Archer and Detling 1986). Studies in Wind Cave National Park showed that in places where bison urinate the nitrogen content in little bluestem (*Schizachrium scoparium*) and common meadow-grass (*Poa pratensis*, imported from Germany) increases. Although the 'urine patches' make only 2 % of the total study area, they produce 7 % of the biomass and 14 % of the total nitrogen contained in the grasses eaten by the large herbivores (bison, elk, mule deer, pronghorns). Moreover, growth of common meadow-grass starts earlier and ends later compared to other sites (Day and Detling 1990). Early onset of growth may also be promoted by higher temperatures. Towards the end of the growing season, herbaceous plants benefit from the increased soil moisture. These combined effects of bison grazing on site conditions not only increase production, phytomass and plant vitality (Fahnestock and Knapp 1994) but also cause a higher heterogeneity of the grassland plant communities (Collins and Steinauer 1998). Moderate grazing may increase production up to 50 % (Sims and Singh 1971; Frank and McNaughton 1993). In long-term field experiments in native tallgrass prairie (Konza prairie, northeastern Kansas) grazing by bison or cattle may reverse loss of species diversity caused by frequent burning (Collins et al. 1998). There is no doubt, that bison had sustainably influenced the natural grassland ecosystems prior to the arrival of the Europeans. Not much is reported, however, about the other large herbivores, although they were also involved.

Anyway, the above-mentioned studies in Wind Cave National Park (South Dakota), provide interesting insight into alterations of vegetation, soils and turnover rates in natural grassland ecosystems induced by prairie dogs and enforced by other herbivores. These investigations focused on the 'grazing association' of prairie dogs, bison and pronghorns. Prior to colonization by the Europeans, prairie dog colonies extended over an estimated area of 49 million hectares. About 5 billion prairie dogs presumably lived in North America at the end of the nineteenth century (Costello 1970). As the ranchers considered prairie dogs to be competing with livestock for food, they killed them whenever and wherever possible. Shooting prairie dogs was considered a 'sport'. The present population of prairie dogs amounts to only about 2 % of the original population (Summer and Linder 1978). The influence of prairie dog colonies on soils, vegetation and other animals have important implications for the North American grassland ecosystem. The black-footed ferret (*Mustela nigripes*), for example, which is one of America's most endangered mammals, subsist mainly on prairie dogs, making up to >90 % of its diet (e.g., Clark 1986). Thus, with regard to landscape and nature conservation it would be wise to reduce popular 'recreational' prairie-dog shooting, with the exception of areas where conflicts with private landowners are inevitable (e.g., Pauli and Buskirk 2007).

Prairie dog colonies are areas of permanent disturbance by grazing and burrowing (Carlson and White 1984). Moreover, bison, pronghorns and also elk graze on prairie dog colonies during the growing season (Coppock et al. 1983b; Wydeven and Dahlgren 1985; Krueger 1986; Whicker and Detling 1988a, b) because the vegetation is richer in protein and more easily digestible than in the surrounding area (Coppock et al. 1983a; Krueger 1986). Consequently, consumption of aboveground phytomass is very high. In the grass-dominated areas of the colonies, prairie dogs and wild ungulates consume between 60 and 80 % of the primary production (Whicker and Detling 1988a).

The high protein concentration has various reasons. First, after the areas have been grazed, new grasses shoot up in great numbers. This compensates for nitrogen loss due to aging of the plants. In addition, the grasses in prairie dog colonies take up more nitrogen than those in the surroundings. Not least, the animals' excreta supply the soil with great amounts of already plant available nitrogen (ammonium and nitrate ions; Coppock et al. 1983b; Whicker and Detling 1988a). In this respect, conditions are similar to those of the mounds of banner-tailed kangaroo rat (*Dipodomys spectabilis*; Sect. 4.2.3) (Moorhead et al. 1988). After exclusion of prairie dogs and bison, nitrogen content in grass vegegation decreases slightly but significantly because of the wider C/N ratio of grasses, which get older than before. Moreover, nitrogen content in grasses decreases due to lacking nitrogen supply by excreta, and reduced nitrogen uptake by the ungrazed grasses (e.g., Cid et al. 1993). Thus, a limited number of prairie dog colonies evenly spread across the grassland and continuous but not too intensive grazing could possibly contribute to an improvement of this food resource for large herbivores. In case of too high grazing pressure, however, by prairie dogs, large herds of bison and pronghorn, prairie dogs will give up their colonies and occupy new areas, if possible. Prairie dog colonies are also regularly frequented by bison to take a dust bath in the fine mineral soil excavated and spread on the surface by the prairie dogs. In this way, bison produce shallow bowl-like shaped depressions ('dust wallows') that may reach several metres in diameter in the course of time. The dust wallows are produced by the bulls during the rutting season in spring and summer. Later, the wallows are also used by the females. As bison usually wallow at the same place, the wallows remain without vegetation for a long time and become susceptible to wind erosion. In frequently used wallows, soil may become compacted and temporarily soaked (saturated). In this case, bison give the wallows up and produce new ones in more suitable places (Polley and Collins 1984; Collins and Glenn 1988). Compaction is also common on heavily trampled bison trails. These effects, together with bioturbation by prairie dogs, influence soil moisture conditions, vegegation, microclimates and flows of matter and energy in the food webs of belowground consuments (earthworms, nematodes, and others; Ingham and Detling 1984; Archer and Detling 1986).

3.3 Tundra

Like the herds of large mammalian herbivores in the grasslands outside the tropics, reindeer and caribou herds in the tundra and subarctic and boreal forests can only be compared with some reservation to the huge herds of grazers in the African savannas, as far as the numbers of individuals and their influence on their environment are concerned. Occasionally, however, old sources give evidence of very large reindeer and caribou herds in the past. Pallas (1777, cited in Bannikov et al. 1967 and in Sedlag 1995), for example, reported reindeer passing without interruption for about three subsequent days his camp site near the Anadyr river (flows into the Bering Sea). Contemporary witnesses reported of 'inexhaustible' or 'countless millions' of caribou in Canada, and considered them to be the arctic/subarctic counterpart of the many millions of bison in the North American grassland (Kelsall 1968).

Reindeer and muskox are the only large herbivorous mammals adapted to the extreme conditions of life in the Arctic, although extreme events such as exceptionally cold summers and severe storms may cause considerable losses. Muskox, however, is more restricted to certain habitats than reindeer. As has become apparent from studies on various arctic islands, caribou do not compete with muskox for food. Muskox graze sedges and grasses in lower areas with shallow snowpack and on slopes covered with low-growing willows (*Salix arctica*) (Klein and Bay 1990), while caribou use drier slopes and areas with varied vegegation structures. Perhaps reindeer have pushed out muskox from their preferred grazing sites (White et al. 1981, further literature therein).

3.3.1 *Muskox*

Muskoxen (Photo 3.8) live in small groups widely dispersed across the tundra. Natural populations can be found in western and northern Alaska, in northern Canada and on its the Arctic islands. Muskoxen also occur in the north and northeast of Greenland. They were introduced to Nunavik Island (Bering Sea) as well as to northern Europe (Norway, Sweden) and to Russia (Crawford 2014), An introduction to Svalbard failed (Myrberget 1987). Population density usually does not exceed a few individuals per square kilometre. Thus, their influence on the tundra ecosystem was considered to be very low (e.g., Bliss 1986). Studies on Sverdrup-Pass in central Ellesmere Island (Canadian High Arctic), however, have come to a different result (Raillard and Svoboda 2000). Along streams and rivulets and in the footzone of slopes, where seepage comes out and soil is usually poorly drained or soaked, relatively closed sedge communities (particularly *Carex aquatilis* var. *stans*, *Carex membranacea*) and cotton grass (*Eriophorum angustifolium* ssp. *triste*) prevail, while the major part of the area is covered with patchy tundra vegetation. The sedges are the most important food for muskox in winter, as it is also the case in other Arctic regions (Larter and Nagy 2001). In central Ellesmere Island, muskoxen move to the drier slopes with little and usually patchy snowpack when temperature drops

Photo 3.8 Muskox (*Ovibos moschatus*) in the Arctic tundra near Lake Hazen, Ellesmere Island Canada. Cotton grass is in the foreground. G. Broll, July 2008

below –20 °C, as sedges in the lower sites are inaccessible under deep and hardened snow at such conditions. On terrain with shallow snowpack, muskoxen may expose vegetation by scraping it snow-free. Energy expenditure, however, is higher than energy gain from grazing the nutrient-poor forage (Fig. 3.11). As soon as temperature increases and the snowcover softens, muskoxen move to the lower terrain with more easily accessible energy-rich sedge communities, despite the comparatively deep snowpack. Although sedge communities cover only a third of the study area, muskox spent more than 80 % of total grazing time in these locations. Muskox density occasionally increased up to 25 individuals per square kilometre. They grazed roughly 50 % of aboveground phytomass per year. This is more than is consumed by large herbivores in temperate and in many tropical grasslands. Muskoxen grazing stimulates the growth of the sedges. In addition, their excreta enhance productivity. It is this fertilization that enables the sedges to tolerate intensive grazing and to compensate for the great loss of aboveground tissue. *Carex aquatilis* which is highly tolerant of grazing and trampling (high amount of well-protected rhizomes and root suckers) is at an advantage compared to the associated plants and becomes gradually dominant (Tolvanen and Henry 2000). Thus, on Sverdrup Pass, which is located between large ice fields, a luxuriant and very productive oasis has developed, that would probably not continue to exist without the muskoxen's influence (Henry and Svoboda 1993). While muskoxen are a keystone-like factor in this area, caribou do not play a comparable role.

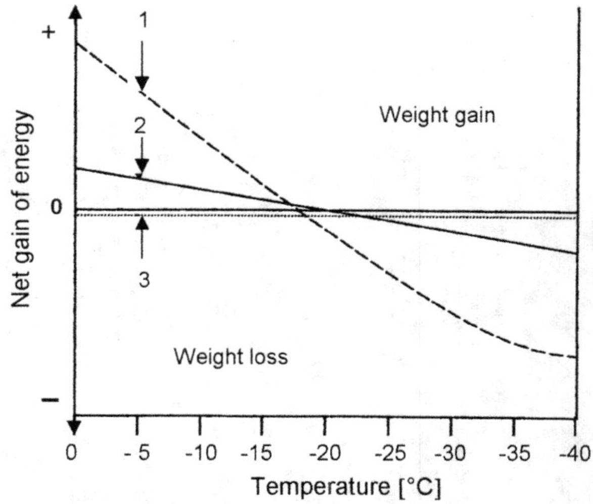

Fig. 3.11 Winter energy balance of muskox (*Ovibos moschatus*) in dependence on grazed vegetation and temperatures. *1* Sedges (*Carex aquatilis* var. *stans*), *2* Willows (*Salix arctica*), *3* Vegetation on dry slopes (*Dryas integrifolia*, *Salix arctica*) (Modified from Raillard and Svoboda 1989)

3.3.2 Reindeer and Caribou

In contrast to muskoxen, reindeer and caribou graze extensively under natural conditions, as has become apparent from studies in North America, for example. When foraging, they are almost permanently in motion, grazing here and there for a few moments, and then run relatively long distances before next feeding. Both reindeer and caribou usually forage on low plants. When browsing birch and willow leaves they also take them beyond a hand's breadth above ground. In case only protein-poor food is available, as ruminants they are able to reduce excretion of nitrogen and to return it partly as urea to their rumen and intestinal microorganisms (Russel and Martell 1984). However, they select the most palatable and easily digestible parts of plants, if possible. This also applies to lichens (Skuncke 1969). This intensive foraging, however, requires large areas, where these cervids do not have to compete for food with less 'careful' grazing livestock (cattle, sheep, goats). Kelsall (1968) reported that in the Canadian arctic tundra and forest-tundra that are not so much influenced by man compared to the fell heaths in northern Europe, the plant cover hardly showed any damage even after large caribou herds had passed through. There, however, are also contradictory observations (Bee and Hall 1956). On Svalbard, for example, summer foraging by reindeer (*Rangifer tarandus platyrhyncus*) in high density has been found to suppress flowering and thus recruitment of mountain avens (*Dryas octopetala*), which is the vascular plant species with the highest-percentage cover. The impact varies depending on reindeer foraging pressure (Cooper and Wookey 2003).

Populations of wild reindeer have markedly decreased compared to the past mainly due to intense hunting pressure (e.g., Kalliola 1939; Haapasaari 1988; communication National Audubon Society 1992). Nevertheless, the influence of reindeer and caribou on turnover rates should not be underestimated (e.g., Remmert 1980). This applies particularly to the semi-wild or domestic reindeer (Herre 1955) that have replaced wild reindeer in many regions. Reindeer husbandry is the most important economic resource of many ethnic groups (Sámi, Samojedes or Nenets, Chukchi, and others) from Lapland in the west to the Bering Sea in the east. Although old cultural contacts exist between Eurasia and North America, reindeer husbandry was not introduced to the American Arctic until the end of the nineteenth century, when Sámi and Chukchi reindeer herders and their reindeer were sent to Alaska to familiarize the Inuit with reindeer breeding and economy.

In winter, one reindeer consumes 3–5 kg of lichens per day (Richardson and Young 1977). Reindeer prefer *Cladonia* species, particularly *Cladonia stellaris* (Hollemann and Luick 1977; Oksanen 1978; Helle and Aspi 1983; Oksanen et al. 1995; Ossenbrink 1996; Väre et al. 1996). Over-grazing of winter pastures is common in all reindeer herding areas. On Chukchi Peninsula, for example, lichen cover became so seriously affected during the 1960s that reindeer numbers declined from 100,000 to 45,000 individuals. Afterwards, lichen vegegation began to recover from this impact (communication National Audubon Society 1992). A similar development was observed on a couple of islands near the Alaskan coast, as on the Pribilov Islands, St. Paul, St. George, and Nunivak, for example. In 1920, about 100 reindeer were introduced to Nunivak. During the following 20 years their population increased to 22,000 individuals. After reindeer had depleted the lichen vegetation they heavily affected shrubs and bushes. Finally, in the end of the 1950s, the Nunivak reindeer population collapsed. Only about 5,000 individuals survived. On Seward and Baldwin Peninsulas, introduced reindeer also failed (Herre 1955). The situation was different on Southampton Island (Northwest Territories, northern Hudson Bay). After native caribou became extinct, caribou were re-introduced to the island. In 1967, about 50 caribou were living there again, and 24 years later reindeer population had increased to 13,700 individuals (1-year-old and older). Lichen vegetation decreased parallel to the increase of reindeer numbers, particularly on windswept terrain with little or no snow in winter, whereas plants in more protected sites were not affected. Increased hunting was considered to be the only way to prevent over-grazing (Quellet et al. 1993). On the Russian Kolguyev Island, located in the southeastern Barents Sea, about 1,000 reindeer recently died due to exhaustion, which has been ascribed to bad weather conditions and inaccessibility of forage plants buried under frozen snow. Shortage of food, however, as a result of over-grazing the Arctic tundra and wrong herding practice have probably played a major role. Although the tolerable maximum reindeer population is restricted to only 5,000 reindeer, about 8,000 were grazing on the island (Narjan-Mar; dpa 15 April 2014).

In the area on both sides of the Rivière George, which flows into Ungava Bay (Labrador), caribou populations considerably increased during the twentieth century despite predation by wolves. Consequently, vegegation in the caribou summer-grazing areas considerably changed. Lichen cover and dwarf birch (*Betula glandulosa*) became particularly affected (Manseau et al. 1996). Compared to the un-grazed areas, productivity of the summer pastures decreased by 50 % (Crête et al. 1996). The phytomass of lichens and dwarf birch declined to <50 g m^{-2}, i.e., below the carrying capacity for caribou (see also Batzli et al. 1980). It has been assumed that due to continued intensive grazing pressure, dwarf birches might not be able to store sufficient reserves for the next growing season and to produce a leaf mass that would compensate for the leaf mass consumed by caribou. It seemed possible, however, that forage depletion would slow down or even stop further population growth of caribou (Crête et al. 1996).

In many areas of northern Europe, populations of semi-domestic reindeer also considerably exceed the carrying capacity of the subarctic dwarf shrub-lichen heath (Holtmeier 1974; Heikkinen and Kalliola 1989; Evans 1995). In some areas, as in northern Finnmark (Norway) between Tana- and Laksefjord, for example, the impact of reindeer on dwarf shrub-lichen heath and resultant soil erosion are so conspicuous that they can be identified even in satellite images taken at an elevation of 750 km above ground (Evans 1995; further references therein).

Reindeer originally were an integral component of northern forests and fell tundra ecosystems. Population density had been fluctuating mainly due to the influence of natural physical and biological factors such as availability of forage, climate (e.g., strong winters), diseases, and also large predators. Mortality of new-born calves probably was the most important factor regulating reindeer densities (Skogland 1985; Kojola et al. 1992). For centuries, reindeer have been the main base of existence of the Laplanders without over-using natural vegetation. The situation changed in modern times. In many areas, reindeer numbers have considerably increased (Holtmeier 1974, 2002; Heikkinen and Kalliola 1989; Evans 1995; Väre et al. 1996; Helle and Kojola 2006).

In northern Norway (Troms County), extreme reindeer densities amounted to 25–100 individuals km^{-2}, before reindeer numbers had considerably decreased due to governmental control measures. The present reindeer population is probably below the tolerable carrying capacity of the grazing areas (Kashulina et al. 1997; further references therein). In the continental interior as, for example, on Finnmarksvidda and in large areas in Finnish Lapland, reindeer density is usually <1 reindeer km^{-2} with the exception of the communities of Kittilä, Enontetiö and Utsjoki, where the density is 5–10 reindeer km^{-2} (Oksanen et al. 1995).

In northern Finland, situation became critical first due to closure of the state borders between Finland and its neighboring countries in the nineteenth century (Norway in 1853, Sweden in 1889) that terminated traditional seasonal migrations of reindeer to their distant summer grazing areas at the coast. Reduced mortality due to supplementary winter-feeding with hay and imported lichens (since 1974; Helle and Kojola 1993; Burgess 1999), easier surveying of the herds (off-road vehicles, snow mobiles; Pelto 1973), better veterinarian care (e.g., improved parasite control) and almost complete extinction of large predators (Väre et al. 1996) mainly by the reindeer herders (e.g., Väisänen 1998) accelerated increase of reindeer numbers (Fig. 3.12). Reindeer population has more than doubled since the mid-1970s (Kumpula and Nieminen 1992; Oksanen et al. 1995; Burgess 1999; Helle and Kojola 2006). Reindeer numbers peaked in the 1990s (5–10 reindeer km^{-2}, Oksanen et al. 1995; see also Danell et al. 1999; Suominen and Olofsson 2000). As also other people than Sámi (such as EU citizens residing permanently in the reindeer

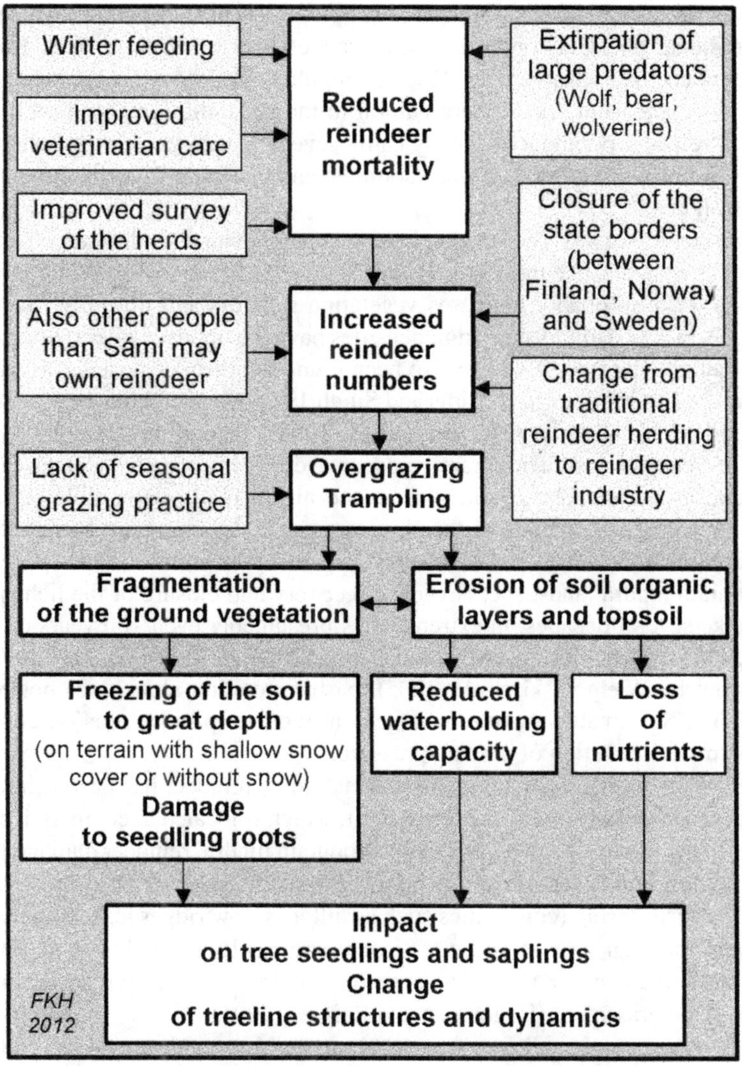

Fig. 3.12 Causes of oversized reindeer population and its impact on mountain birch forest and treeline in northernmost subarctic Finland (Scheme by F.-K. Holtmeier, modified from Holtmeier 2012)

husbandry area) can own reindeer (www.mmm.fi Ministry of Agriculture and Forestry, 22.09.2011), their density is much higher in Finland as compared to Norway or Sweden. In addition, the lack of seasonal grazing practice has enforced excessive summer grazing pressure (Käyhkö and Pellikka 1994) exceeding the natural carrying capacity of the grazing areas. When the demand for reindeer meat was low after the Chernobyl nuclear disaster (1986), reindeer were not slaughtered. Consequently their numbers culminated and their impact on both summer and winter ranges, where lichens were already depleted, became even more critical. As a result, reindeer physical condition and reproduction were affected (e.g., Kumpula and Nieminen 1992; Kojola et al. 1995; Kumpula et al. 1998, 2002; Colpaert et al. 2003). Thus, although reindeer are natural to the subarctic forests and fell-tundra oversized reindeer populations locally cause severe disturbances (e.g., Evans 1995, 1996; Holtmeier 1974, 2009, 2012; Suominen and Olofsson 2000; Holtmeier et al. 2003; Broll et al. 2007).

In northern Norway, lichen-dominated subalpine and alpine vegetation has decreased by 85 % since the 1970s (NORUT 1999; Löffler 2000). While moderate reindeer grazing probably increases vegetation richness and diversity (Suominen and Olofsson 2000), reindeer at high densities have the opposite effect. Such effects of different grazing intensity have also been found in other ecosystems, as has been shown in the last sections (e.g., Sims and Singh 1971; McNaughton 1979, 1983a, b; Frank and McNaughton 1993; Krüsi et al. 1995). In northern Finland, damage caused to vegegation by temporary reindeer overgrazing are of the same order of magnitude as the damage resultant from the nickel industry on Kola Peninsula. Without overgrazing by reindeer, the physiognomy of the dwarf shrub - lichen heath would completely change and white-gray to green-gray colors (lichens) prevail in the mountain tundra landscape. However, recovery and closure of the lichen cover would take several decades, in extreme sites presumably even 100 years or longer because of the very slow growth of lichens (Pégau 1970, 1975; Ahti 1977). As pointed out by Wielgolaski et al. (2005), pasture rotation alone does not prevent overgrazing. Sustainable pasture management requires pasture rotation combined with controlled reduction of grazing pressure.

Destruction of the lichen- and moss-vegetation by reindeer has far-reaching consequences, particularly above the forest limit, which is located at comparatively low altitude. In the treeline ecotone and in the mountain tundra, reindeer damage is most spectacular on windswept terrain, where snowpack is shallow or even missing in winter and early spring (convexities such as hillocks, low ridges, eskers, remains of lateral and terminal moraines, etc.). Lichens are easily accessible in such places (Helle and Tarvainen 1984). During summer reindeer frequent windy locations to escape insect harassment (Hagemoen and Reimers 2004).

Tree seedlings and saplings often are destroyed by reindeer grazing and trampling directly, or indirectly by the effects of soil erosion initiated by reindeer in such places (Photos 3.9 and 3.10; Holtmeier 1974, 2012; Haapasaari 1988; Evans 1995; Kashulina et al. 1997; Holtmeier et al. 2003; Holtmeier and Broll 2006, 2011; Uhlig and Zink 2006; Broll et al. 2007). S*terocaulon paschale* often prevails while reindeer lichens (*Cladonia* spp.) are less important (see also Heikkinen and Kalliola 1989). A few species, however, such as sheep fescue (*Festuca ovina*), three-leaved rush (*Juncus trifidus*), Bigelow's sedge (*Carex bigelowii*) and Lapland reed grass (*Calamagrostis lapponica*) apparently benefit from reindeer grazing (Kalliola 1939; Oksanen 1978).

A closed lichen cover prevents soil from rapid drying out. The water content of lichens may be threefold to tenfold as high as their dry weight (Kershaw and MacFarlane 1980). Under lichens, soil moisture remains relatively high even during

Photo 3.9 Wind-eroded convex topography on an ENE-trending mountain spur on the north-facing slope of Koahppeloaivi, northern Utsjoki, Finnish Lapland (Photo from 250 to 300 m above ground, view to the south). The main wind direction is northwest (from the *right-hand side*). After destruction of the dwarf shrub- lichen cover by reindeer trampling, strong winds have removed the organic layer and eroded the upper topsoil. F.-K. Holtmeier, 5 August 1998

lasting drought periods (see also Larsen and Kershaw 1976). Dry lichen cover (Skuncke 1969; Holtmeier et al. 2003; Kumpula et al. 2011), however, as well as a few millimetres thick 'biological crusts formed by algae, fungi and mosses (Belnap 2001; Belnap et al. 2001) are particularly susceptible to trampling.

Rapidly draining sandy and coarse substrates (up to 2 cm, Seppäla 2004; Butler et al. 2009) as well as dry peat layers covering the ground are particularly vulnerable to wind erosion (Holtmeier 1974, 1979; Ukkola 1995; Holtmeier et al. 2004; Holtmeier and Broll 2010). Once the litter layer is removed, microsite conditions become more critical (Holtmeier et al. 2003, 2004). Supply of fresh litter comes to an end. The old organic matter rapidly decomposes (warmer conditions), and wind may also erode the topsoil. Consequently, water holding capacity of the soil and nutrients decrease (Table 3.2; Holtmeier et al. 2003, 2004; Holtmeier and Broll 2005). Although nutrients contained in urine and pellets may have a mitigating effect, they are unlikely to compensate completely for nutrient loss in such places. Rapidly drying shallow peat layers are not only destroyed by trampling but also by reindeer wallowing (Photo 3.11).

As exposed soil freezes to great depth in such places, damage to fine seedling and dwarf shrub roots is likely. Roots are much less frost tolerant than plant tissue projecting above the snowpack (e.g., Larcher 1980; Coleman et al. 1992;

Photo 3.10 Detail of the wind-eroded topography (327 m) on Koahppeloaivi (cf. photo 2.14). The wind-scarps mark the downwind migrating front of wind erosion. Right alongside the wind- scarps are remains of the former B-horizon. Soil further windward has already eroded down to the BC-horizon. Under the thin dwarf shrub vegetation (*right*) the original topsoil still exists. F.-K. Holtmeier, 26 August 1997

Table 3.2 Soil moisture (topmost 6 cm) at different microsites in the birch-treeline ecotone, northernmost Finnish Lapland

Soil moisture (vol. %)			
Site	Mean (±SE)	Minimum	Maximum
Wind-eroded site (a)	8.7 (0.2)	0.5	23.6
Lichen heath (a)	12.2 (0.3)	bdl	66.0
Willow scrub (b)	65.6 (0.9)	1.5	57.3
Sedge mire (b)	77.4 (0.6)	32.4	87.6

Number of measurements: (a) n=500; (b) n=300; bdl=below detection limit (Data from Anschlag et al. 2008)

Sutinen et al. 1997; Ryyppö et al. 1998; Repo et al. 2001). In exposed fine soils, frequent frost-heave and needle-ice formation may push occasional tree seedlings out of the ground (Holtmeier 1974, 2012). Moreover, loose mineral soil particles uplifted by needle ice and covering the soil surface after needle ice has melted are particularly prone to deflation (cf. Photos 3.9 and 3.10). These effects and drought stress, typical of sandy substrate, usually override possible positive effects of relatively

Photo 3.11 Wind-eroded shallow peaty layer on top of a low ridge on the northwest-exposed slope of Rodjanoaivi (northern Finnish Lapland) at 377 m. Wind erosion has been triggered by reindeer which fragmented the lichen cover by trampling and wallowing. F.-K. Holtmeier, 25 August 2003

high soil temperatures during summer (Holtmeier et al. 2004; Holtmeier 2009). Soil moisture and nutrient deficiency as well as severe microclimates impede or even prevent mountain birch (*Betula pubescens* ssp. *czerepanovii*) and other trees (e.g., Scots pine) from colonizing such eroded sites (Holtmeier et al. 2003, 2004; Broll et al. 2007; Anschlag et al. 2008). Frost damage in dwarf shrubs increases.

Heavy reindeer impact probably prevents also climatically-driven expansion of deciduous dwarf shrubs such as downy birch (*Betula nana*) and greyleaf willow (*Salix glauca*) into such locations (den Herder et al. 2008; Ost and Pedersen 2008; Zöckler et al. 2008; Olofsson et al. 2009). Continued fragmentation and intense browsing of willow stands by reindeer stands destroy an important habitat (food source, shelter) of grouse population (e.g., den Herder et al. 2008; Henden et al. 2011; Ehrich et al. 2012).

In our study areas in northern Finnish Lapland (e.g., Holtmeier et al. 2003, 2004; Holtmeier and Broll 2006; Broll et al. 2007; Anschlag et al. 2008), the area of open patches with mineral substrate exposed by reindeer and subsequent erosion in the treeline ecotone and above is still expanding (see also Käyhkö 2007). Similar observations have been made in the southern Swedish Scandes (Kullman 2005) and in northern Norway (Evans 1995). There, large-scale landscape degradation combined with overgrazing by reindeer has caused a depression of the altitudinal belts and birch forest limit together with soil erosion (Löffler 2005).

3.4 Woodlands

Although in forests, of which structure and composition has often been influenced by man over centuries or even thousands of years, consumption of phytomass by herbivorous mammals is of minor importance, it may have far-reaching and lasting effects due to the different foraging behavior. There is a wealth of scientific literature on this issue in Europe and North America. The role of mammalian herbivores in forest ecosystem will be highlighted in the following section by way of examples.

3.4.1 Central and Western Europe

In Middle Europe, densities of wild ungulate populations were possibly about ten times lower in the past than at present (Widmann 1991; Danilkin 1996). In many areas, increasing human impact (hunting, livestock grazing, fragmentation of habitats) brought native wild ungulates close to extinction. They could recover as hunting pressure decreased, due to the bubonic plague ('Black Death'; fourteenth century), for example, or during feudal times when sovereigns and nobility reserved exclusive hunting rights. The populations of big game were supported by winter feeding, and predators were killed whenever possible. As a result, excessive damage to field crops and pastures caused by over-abundant red deer occasionally led to war-like conflicts between aristocrats and farmers (Schröder et al. 1984). After the revolutions (1789–1799, 1848), hunting rights passed principally over to landowners. Consequently, hunting pressure on game considerably increased. Deer, for example, almost became extinct in Central Europe. Lately, however, wild ungulate populations not only rapidly recovered but have grown to ever-greater numbers due to fundamental reform of hunting rights and wildlife management (Wotschikowsky et al. 2010).

In Europe, the populations of red deer, for example, have been higher in recent decades than ever before (e.g., Gill 1990; Bradshaw et al. 2003) even though appropriate habitats have been continuously shrinking due to the increasing impact of multiple landscape use (agriculture, tourism, outdoor recreation, residential and natural resource development) and artificial barriers that preclude deer from moving seasonally between different habitats (e.g., Gill 1990; Bradshaw et al. 2003). Red deer has lost about 90 % of its original habitats (Gill 1990). In Germany, red deer density in about 70 % of the present habitats (agricultural land inclusive) ranges below 25 deer 1,000 ha^{-1} (Kinser et al. 2010). Local densities of >100 deer 1,000 ha^{-1} have been estimated in some forest habitats, however. In the remained habitats, winter feeding may compensate for scarcity of food and reduce deer mortality. Winter feeding however simulates a habitat carrying capacity that does not really exist (Richter 1988).

Roe deer has benefited in particular from the varied mosaic of agricultural landscape (Table 3.3). The mosaic of agricultural fields, fallow land, forests and tree groves provides favorable habitats (sufficient shelter and ample food supply; e.g., Kurt 1991). In England, red deer range nearly doubled (Dolman et al. 2010).

Table 3.3 Roe deer densities in different European countries during the 1980s

Country	Roe deer (1,000 ha^{-1}) in the total area	Roe deer (1,000 ha^{-1}) in the forested area
Austria	119.0	270.3
Denmark	32.6	269.2
West Germany	68.5	228.5
Great Britain	23.1	200.0
Switzerland	29.3	116.5
The Netherlands	6.1	75.8
Italy	17.3	71.4
Poland	16.0	59.2
Romania	12.2	45.2
Jugoslavia	13.7	40.2
Norway	5.0	20.8
Sweden	8.9	15.6
Finland	1.0	1.5

After Danilkin (1996)

Deer densities >10 km^{-2} are common these days in temperate zones (e.g., Fuller and Gill 2001; Russell et al. 2001). In many areas, current red deer populations exceed the tolerable carrying capacity of their fragmented habitats (Côté et al. 2004). However, deer impact varies locally due to different habitat qualities, hunting regulations and land ownership (private or public ground). In northern Europe, moose and also roe deer have also increased to numbers probably greater than ever before during postglacial times (Persson et al. 2000).

Browsing and grazing by great numbers of both red deer and roe deer are conflicting in particular with forestry. Cervids differ in their diet. Red deer is an intermediate grazer that causes forest damage mainly by bark stripping, fraying and browsing trees, whereas the choosy roe deer, for example, takes mainly energy-rich buds, and fresh leaves and shoots that are palatable and easily digestible (for review see Gill 1992). The extent of forest damage increases with ungulate population density, shortage of forage, frequency of disturbances, and susceptibility of tree species to browsing. White fir (*Abies alba*) and mountain maple (*Acer pseudoplatanus*), for example, are highly susceptible, while Norway spruce (*Picea abies*) is comparatively tolerant.

In general, forests are resilient to moderate grazing by wild ungulates. Moderate grazing may even increase plant species diversity and facilitate expansion of forest tree species (Risenhoover and Maass 1987; Krüsi et al. 1995). Increasing browsing pressure, however, may impede natural regeneration and reduce tree species richness as well as diversity. Browsing is particularly critical to seedlings and saplings. As long as ample forage is available, ungulates browse their favorite species. After forage is almost depleted also other species will be browsed (e.g., Motta 1996). Damage caused by browsing in winter affects evergreen trees more than broadleaved, whereas broadleaved trees suffer more from summer browsing (Eiberle 1978). Moreover, the understorey vegetation is particularly vulnerable to browsing (Fischer 1999; Gill and Beardall 2001; Rooney and Waller 2003; Côté et al. 2004). Resultant alteration of habitat structures and plant communities may affect other mammals (e.g., voles), avian communities, and invertebrates (e.g., Mlot 1991; Wagner 1993; Gill 1992, 2000; Väre et al. 1995; Kienast et al. 1998; Suominen 1999; Suominen et al. 1999; McShea and Rappole 2000; Fuller 2001; Perrins and Overall 2001; Rooney and Waller 2003; den Herder et al. 2004).

At high density, red deer and other wild ungulates may also locally initiate soil erosion by trampling (Packer 1963; Holtmeier 2012; Holtmeier et al. 2003, 2004; Beschta and Ripple 2010). Compacted soils are particularly susceptible to erosion during high-intensity summer rains. On the other hand, exposing the mineral soil ungulates may create niches suitable for tree regeneration (e.g., Miles and Kinnaird 1979; Mitchell and Kirby 1990). Nevertheless, browsing and trampling may considerably reduce the chance of undisturbed tree-seedling growth and survival in such places. Urination and defecation influence decomposition, nutrient cycles, and mycorrhiza (e.g., Holland and Detling 1990; Holland et al. 1992; Molvar et al. 1993; Hobbs 1996; Persson et al. 2000). Reliable data are rare, however (Persson et al. 2000).

Red deer (*Cervus elaphus*) and roe deer (*Capreolus capreolus*) have become a serious problem particularly for many mountain forests in Central Europe (Fig. 3.13; see also Sect. 6.1). Red deer damage trees mainly by bark-stripping, browsing buds, fresh shoots and twigs, while roe deer injure trees usually by browsing. Fiber-rich food – ungulates need it for digestation – makes up to about 30 % in red deer and about 70 % of the total food intake in roe deer (Schauer 1982). Damage caused by rubbing (to clean the velvet), fraying or banging (during rutting time) the antlers against trees play a minor role (reviews in Kraft et al. 1990). From oversea countries, where European cervids were introduced as, for example, to Nahuel Huapi National Park in Argentina, also severe damage by browsing and bark-stripping and negative effects on the native vegegation have been reported (Relva and Kitzberger 2000). The dramatic effects of introduced red deer and other ungulates on vegetation, fauna and soil in New Zealand will be considered in a special section (5.1.3).

In Switzerland and also other Alpine countries, damage by wild ungulates (red deer, roe deer, chamois and ibex) have often become intolerable with regard to maintaining the protective functions of the mountain forests (prevention of avalanche and soil erosion) (cf. Fig. 3.13; Bodenmann and Eiberle 1967; Holtmeier 1969a, b, 1987a; Schröder 1972; Mayer 1973, 1975; Kurt 1977; Zeltner 1979; Pfister et al. 1987; Voser 1987; Bernhart 1988, 1990; Liss 1988; Moser 1988; Schlund 1988; Moser et al. 1989; Rotenhan 1990; Schuster; 1990; Brändli 1995; Ott et al. 1997). Red deer (and also chamois) affect high-altitude reforestation more than natural young growth in the treeline ecotone (e.g., Bavier 1976; Schönenberger et al. 1990; Senn 1999). Losses of more than 70 % of young trees have been reported (Löw and Mettin 1977). This impact must be ascribed not only to oversized ungulate populations but also to the slow tree growth at high altitude, especially in the treeline ecotone (Holtmeier 1974, 2009, 2012; Kennel 1998). During snow-rich winters

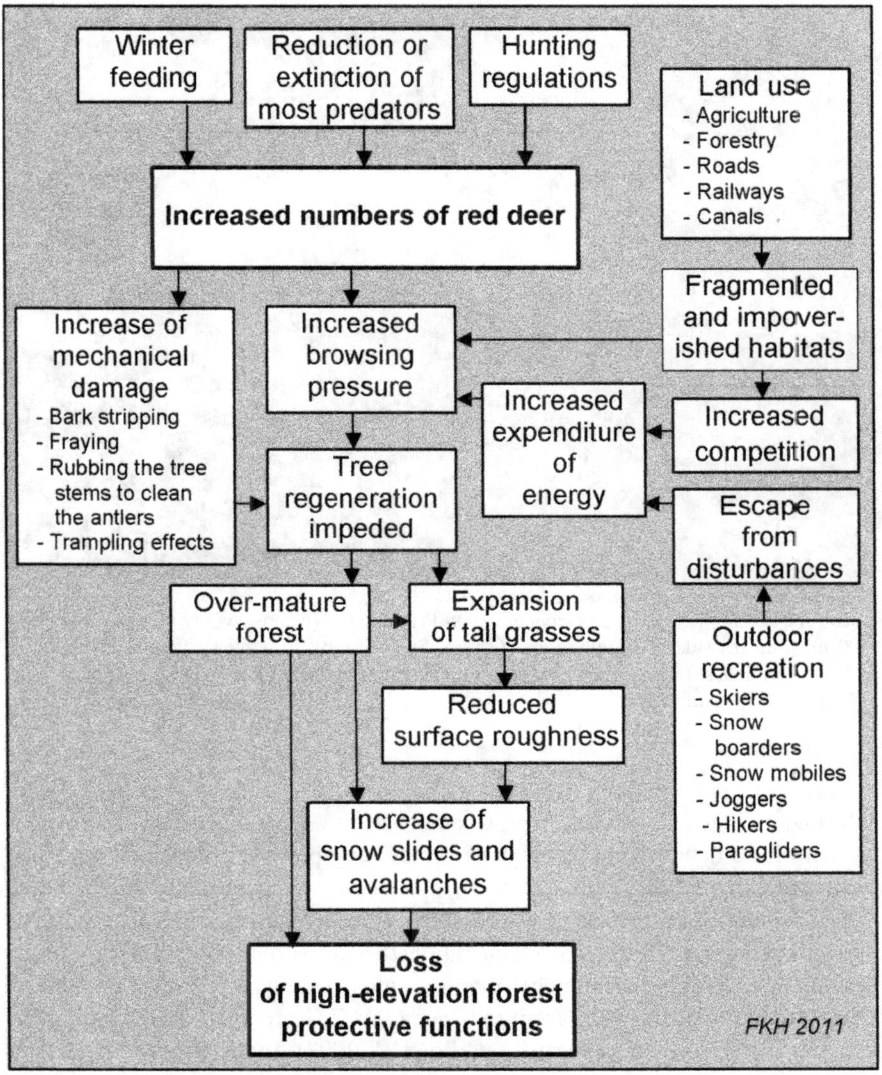

Fig. 3.13 Causes and consequences of over-sized red deer populations and their impact on the protective functions of high-elevation forests (Scheme by F.-K. Holtmeier)

red deer, if not supported by supplemental feeding within the mountain forest at lower elevation, frequent the treeline ecotone and the alpine zone where strong winds remove the snow from exposed terrain thus making food more accessible. Foraging red deer prefer sun-exposed slopes. Seedlings and young growth that became established in such sites may be damaged or even destroyed by browsing and trampling. Very windy conditions, however, make red deer retreat to the shelter given by upper forest stands (Schmidt 1993a), where they may affect young growth

Photo 3.12 Chamois (*Rupicapra rupicapra*) on Pru dal vent (means 'windy meadow') at about 2,200 m, southern side of Bernina Pass (Grison, Switzerland). In winter, chamois frequent this place from which strong northern winds remove the snow and make forage easily accessible. F.-K. Holtmeier, March 1970

in the understorey and in forest gaps. Chamois also frequent wind-exposed topography with little or no winter-snow, where they may prevent establishment of tree seedlings (Photo 3.12).

In the Veneto Prealps (NE Italy), for example, oversized red deer population causing severe damage to tree regeneration (beech, spruce, fir) adversely affect 'near-nature' forest management, that is almost exclusively based on natural regeneration. Fir suffers most from red deer. Only 7 % are unaffected. Culling has been considered to be necessary (Caudillo et al. 2002). In the Scottish Highlands (Cairngorm Mountains), large red deer populations prevented almost any regeneration of Scots pine (*Pinus sylvestris*) until the 1950s (Watt and Jones 1948). Pine could spread again after grazing by reed deer and sheep, and burning as well, had ceased (French et al. 1997). Modeling the impact of grazing and trampling impact of sheep, cattle, red deer, and smaller herbivores on heather (*Calluna vulgaris*) in upland Scotland showed that removal of sheep grazing may bring about intensified use by red deer impeding recolonization by heather. Thus, if the management objective is to stop heather loss, red deer population growth will have to be controlled adequately (Albon et al. 2007).

Also, inside many mountain forests grazed by cattle, damage caused to the trees must be attributed for the most part to red deer (e.g., Liss 1988, 1990; Burschel et al. 1990; Rösch 1992; Suda 1990; Photo 3.13). In mixed mountain forests of the eastern

Photo 3.13 Heavily browsed Norway spruce (*Picea abies*) in the subalpine forest (Swiss stone pine, European larch, Norway spruce) above St. Moritz, Upper Engadine. The winter has protected the basal part from being browsed. F.-K. Holtmeier, 1998

Bavarian Alps, for example, no natural regeneration can be expected in the forests even after termination of cattle grazing, as long as ungulate numbers are not drastically reduced (Liss 1988, 1990; Burschel et al. 1990; Suda 1990; Rösch 1992). Regeneration of ash (*Fraxinus excelsior*), mountain maple (*Acer pseudoplatanus*) and white fir (*Abies alba*) is failing almost completely. Providing fields with forage to take away grazing pressure from the forest can only be a supporting measure. Winter feeding is also unlikely to facilitate natural regeneration of forest trees. Studies in the Karwendel Mountains (Achenkirch, Tyrol) showed that natural regeneration of all tree species improved, even without fencing, after drastic reduction of red deer and roe deer and decrease of forest pasture – parallel with more careful methods of timber harvesting and construction of forest access roads (Schwab and Messner 1987).

Maintenance of mountain forests and high-altitude afforestation will only be possible if wild ungulate numbers are considerably and persistently reduced (e.g., Bavier 1976; Schönenberger 1986; Pfister et al. 1987). However, reduction of deer numbers by shooting alone is not always an appropriate and the only measure, as damage degree depends not only on deer density but also on habitat preconditions, such as attractiveness of the forest for deer (forage, shelter), forest age structure and composition, regeneration, vulnerability of tree species, and, not least, intensity of disturbances by human activities (timber harvesting, outdoor recreation, etc.). Despite rigorous culling significant damage can still take place. As red deer habitats are progressively shrinking due to increasing disturbances from outdoor recreation (cf. Fig. 3.13), habitats free of disturbances must be preserved. In the Angerbach Valley (Badgastein/Salzburg, Austria), for example, the total red deer habitat has shrunk by about 80 % of its original size after the area was opened for ski tourism. The winter-grazing area, originally comprising 31 % of the deer habitat, was reduced to 6 %. Depending on the weather conditions, only parts of this remained area can be grazed. When one-and-a half meter deep snowpack, for example, covers this terrain, forage is accessible only on 0.5 % of the winter-grazing area.

As red deer concentrate in the relatively small forest stands still undisturbed by ski tourists, they may cause severe damage, although their numbers were reduced (Reimoser et al. 1987; Reimoser 1999). In snow-rich climates, deep winter snowpack makes foraging difficult and increases energy requirement. Additional stress due to frequent disturbances by human activities and constant movement to escape disturbances increases energy expenditure (c.f. Fig. 2.27). Increased energy needs have to be compensated by forage uptake. As a result, grazing and browsing impacts in the remaining habitats considerably increase (Fig. 3.13; Onderscheka 1986; Esslinger 1988; Petrak 1988; Herbold et al. 1992).

Not least, many high elevation forests are particularly sensitive to deer damage because they have been grazed by cattle for centuries and often were already overaged before red deer density increased (Holtmeier 1967a, b). In many high elevation spruce, larch and stone pine forests, suppression of tree regeneration by red deer has locally supported the expansion of tall grasses such as woolly reed grass (*Calamagrostis villosa*). As a result of reduced surface roughness, snow slides can become more frequent, affecting in turn occasional young growth of trees and development of sound stand structures (e.g., Gampe 1989; Ott et al. 1997).

On the other hand, grazing by red deer may also have 'positive' effects, such as conservation of plant diversity, for example. Having said that, we must add that present plant diversity is usually not natural but the result of mountain landscape cultural history. Conservation of plant diversity, may require 'sufficient' grazing impact, as has been suggested by studies of plant succession of up to 80-years-old permanent plots and by other research in the Swiss National Park (Lower Engadine; Krüsi et al. 1995). The grazing/browsing pressure by the very large red deer and chamois populations were even considered to be too low rather than too high for maintenance of the existing forest stands in proportion to open areas and preservation of the resultant plant diversity. On some subalpine pastures with good forage quality, these wild ungulates have prevented natural reforestation, although these areas were not grazed for more than 50 years. On the other hand, under too intense grazing, fertile meadows and pastures, comparatively poor in species and characterized by high-quality grasses and herbs, have turned into species-rich rough meadows. Because of the bad forage quality these are less attractive to red deer and become gradually invaded by shrubs. In the long-term, a return of forest is likely. Due to red

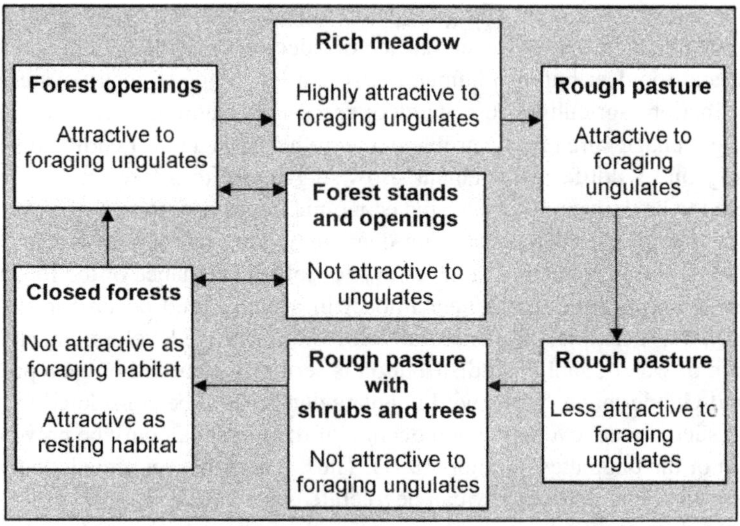

Fig. 3.14 Hypothetical development of forests and open areas under the influence of ungulates in the Swiss National Park (Lower Engadine) after park foundation (1914) and cease of utilization (Modified from Krüsi et al. 1995)

deer impact, natural reforestation takes place much more slowly than was expected (Schütz et al. 2000).

The further development might process as follows: The forest stands serve as resting place to the wild ungulates, which prevent tree regeneration. Consequently, the forest stands over-age and gaps develop. Under certain conditions, these may gradually turn again into preferred foraging places due to rich nutrient supply (Fig. 3.14). Such a cycle, however, would require a sufficient ungulate density. According to a simulation model for the time period 1900–2000, which, contrary to previous models, also considers the influences of microtopography and grazing intensity, floral diversity on subalpine meadows in the Swiss National Park will culminate around 2200 AD., i.e. halfway of succession from grassland to pine forest (Risch et al. 2001). The speed of succession, however, depends on topography and will be relatively low on convex terrain and slopes with nutrient loss and on moderately to intensively grazed level terrain. Contrary to that, more rapid change is likely in concave, little to moderately grazed, sites where nutrients accumulate.

Red deer impact on natural reforestation of abandoned subalpine pastures up to the potential tree limit is likely to keep these areas open for a while. Species diversity would be higher as if closed mountain forests covered the former grazing areas (e.g., Grabherr and Pauli 1994; Theurillat et al. 1998; Luckman and Kavanagh 1998; Theurillat and Guisan 2001; Tinner and Kaltenrieder 2005; Malanson et al. 2007; Sundquist et al. 2008; Holtmeier 2009, 2012; Nagy and Grabherr 2009; Pauchard et al. 2009).

Almost everywhere in the European cultural landscape, red deer control appears to be inevitable. Control, however, has to consider not only the habitat demands of red deer and roe deer but also human impact on landscape (forestry, water supply and distribution, agricultural use, outdoor recreation, hunting, etc.). Definition of 'browsing damage' already is a problem as is reflected in a never ending discussion in forestry and wildlife management. Browsing injury to afforestation is usually considered to be damage when causing economic losses. Contrary to that, browsing effects in a near-natural forest are no damage as long as they do not impede or prevent natural tree regeneration necessary for self-maintenance of the forest.

If regeneration fails due to deer impact in a typical red beech forest (*Fagus sylvatica*), for example, or if natural reforestation of redevelopment areas for protection of water supply and distribution is seriously impeded or even prevented (Suda and Gundermann 1994), wild ungulate density will be considered to be 'too high'. In such a case, overaging and decline of the forest can only be prevented by reduction of the ungulates (Guthörl 1991). Thus, it depends very much on the local situation which ungulate density can be tolerated (Sect. 6.2).

The red deer problem, which is partly an effect of too high red deer density in separated habitats, could be mitigated, if more open grazing areas outside the forests and migration routes were available. Damage caused to the forests and agricultural fields might probably decrease, provided that the relatively lower red deer numbers (in a larger area) would be under strict control. Better genetic exchange between isolated habitats would be an additional positive effect.

3.4.2 *Northern Europe*

The influence of the few cervid species on boreal forests does by far not reach the order of magnitude that would be comparable to the impact of about 13 antelope species on their savanna habitats; and yet, it should not be underestimated. In the subarctic birch forests and in the northern boreal forest of Eurasia it is mainly reindeer that may seriously impair young growth (Holtmeier 1974, 2012; Mattson 1995). Reindeer may cause deformed growth or destroy tree seedlings and saplings completely (Holtmeier and Broll 2011; Holtmeier 2012). When we studied the response of Scots pine to warming climate at its altitudinal limit in northernmost Finland (Holtmeier and Broll 2011) we found that >80 % (total 231) of young pines were more or less deformed by climatic injury, reindeer and moose. Locally, mountain hare had severely damaged low-growing prostrate pines on wind-exposed terrain in the treeline ecotone (Holtmeier 2005, 2012). All young pines less than 150 cm high are at a constant risk of being affected by reindeer.

Reindeer damage seedlings and small trees mainly when pawing through the snowpack for lichens (e.g., Skuncke 1969). Moreover, they often cause severe damage to saplings when cleaning their antlers (Holtmeier and Broll 2011). Such mechanical damage reduces resistance of young trees against climatic injuries and biotic influences (e.g., wound parasites). Kujala (1950), for example, supposed that reindeer grazing on snow-rich terrain often promote lethal damage to young pines by pine canker (*Dasyscypha fuscosanguinea*). This fungus gets into the young trees through the wounded bark. In an outlier of the boreal pine forests located in the subarctic birch forest zone of northern Utsjoki (Finland, 60° N), Holtmeier (1974) found about 60–90 % of pine seedlings and saplings damaged by reindeer. This damage rate appears to be very high if compared to observations by other authors (e.g. Arnborg 1955; 36 %). It corresponds, however, roughly to the findings by Andreev (1954; cited in Hustich 1966), who observed reindeer damage in almost 75 % of young pines. Damage of this order of magnitude typically occur in areas with dense (and even-aged) pine regeneration, while multi-layered pine stands are less affected (Helle 1966). Contrary to moose, reindeer do not browse pine.

On the other hand, reindeer may improve pine regeneration in dry pine forests by destroying dense lichen mats that often prevent pine seeds from reaching a suitable seed bed. Moreover, the removal of star-tipped reindeer lichen (*Cladonia stellaris*) facilitates pine seedling establishment, as the lichens produce toxic substances that impair the development of pine mycorrhiza (Brown and Mikola 1974). Reindeer may also impede spread of snow blight (*Phacidium infestans*) by compacting the snowpack. Thinning or removal of dense ground vegetation by reindeer may also facilitate the early phase of birch seedling establishment (reduced competition; Lehtonen 1987; Suominen and Olofsson 2000; Holtmeier et al. 2003), while birch seedlings are usually rare or even absent in wind-eroded treeline sites. Mechanical damage caused to seedlings may however override these positive effects.

Continuous reindeer grazing modifies the forest field layer. In coniferous forests (with *Pinus sylvestris* prevailing) grazed by reindeer and ungrazed for 30–50 years in northern Finland and on Kola Peninsula (Väre et al. 1995, 1996), mosses (in particular *Dicranum*-species and *Pleurozium schreberi*) have increased and oligotrophic conditions changed into mesotrophic conditions in the grazed forest.

Intense reindeer grazing and also voles feeding on birch seedlings have been among the factors preventing recovery of mountain birch forests after defoliation during regional (local) episodic mass-outbreaks of the autumnal moth (*Epirrita autumnata*) and the winter moth (*Operophtera brumata*) (e.g., Lehtonen and Heikkinen 1995; Oksanen et al. 1995; Holtmeier 2002, 2005, 2009; Holtmeier et al. 2003; Cairns and Moen 2004; Holtmeier and Broll 2006). The combined effects of lasting high reindeer density and recurrent outbreaks of leaf-eating insect probably are the greatest threat to treeline birch forests (Holtmeier et al. 2003, 2004; Lempa et al. 2005; Helle 2005; Neuvonen and Wielgolaski 2005). In many areas, the Holocene decline of birch treeline that is usually attributed to the general cooling ('Little Ice Age') was very likely enforced by the impact of reindeer and the autumnal moth (Holtmeier and Broll 2006). At present, reindeer impact often prevents climatically-driven re-colonization of formerly forested terrain in the present treeline ecotone in northernmost Finland (Holtmeier et al. 2003; Holtmeier and Broll 2011).

Moose (*Alces alces*) is spread all over Fennoscandia from low elevations up to the treeline and above. During the 1920s and again after World War II, moose were close to extinction. Populations, however, regularly recovered. Moose numbers have considerably increased since the 1960s into the late 1970s (e.g., Nygren 1987; Bergström and Hjeljord 1987; Bergström and Willebrand 1992; Persson et al. 2000; Hörnberg 2001; Lavsund et al. 2003; Kouki et al. 2004). The main reasons are hunting practices (trophy hunting), lack of predators, decrease of pastoral use and, in particular, clear felling that replaced selective felling during the 1950s. Luxuriant secondary vegetation (grasses, herbs, and deciduous trees) that followed clear cutting and land abandonment as well as recent afforestation has considerably improved the food source of moose. In addition, nitrogen fertilization of forest has probably increased energy-rich forage (e.g., Strandgaard 1982).

In many areas, high moose density is problematic. Vehicle collisions, for example, with moose have considerably increased. During the 1990s, about 5,000 accidents were officially registered in Sweden. Additional 5,000 unreported collisions were likely, however (Seiler 2005). In many areas of Sweden and Finland, moose have become a serious problem to forestry. Though preferring mountain ash (*Sorbus aucuparia*), trembling aspen (*Populus tremula*) and different willow species (*Salix* spp.) moose also forage on birch (*Betula pubescens*, *Betula pendula*; Wam and Hjeljord 2010) and Scots pine (*Pinus sylvestris*). Moose impact on pine and birch is strongest in areas where the preferred deciduous tree species occur at high densities (Heikkilä and Härkönen 1996). In winter, moose feed on pine, in particular (e.g., Bergström and Hjeljord 1987; Histøl and Hjeljord 1993; Heikkilä and Härkönen 1996; Löyttyniemi 1985; Nikula 1992; Stöcklin and Körner 1999) because of easy availability and high digestibility of young pine tissue (Stolter et al. 2009). Afforestation with pine and also birch is most seriously affected (Löyttyniemi 1985; Lavsund 1987; Niemelä et al 1989; Nikula 1992; Heikkilä and Härkönen 1996).

In southern Sweden (Dalarna Province), damage by moose has considerably increased in intensively used economic forests (*Pinus syslvetris*, *Picea abies*, *Betula pubescens*, *Betula pendula*) located in moose winter habitats (Lavsund 1987). During long and snow-rich winters (snow depth >70 cm) damage increases and is strongest in comparatively small areas with shallow snowpack. Moose impact is reflected in disturbed growth forms in most trees that have started projecting beyond the winter snowpack (40–50 cm). Pines less than 250 cm in height are under the constant risk of being affected by moose (Bjørneraas et al. 2012).

As many other large ungulates, moose also influence other animals by modifying their habitat. In old-growth boreal forests of eastern Karelia and southern Finland, large moose populations are likely to be eliminating trembling aspen. Trembling aspen is one of the most important micro-habitats for several threatened species (Esseen et al. 1997; Kouki et al. 2004; De Chantal et al. 2009).

During the last three decades moose have been invading in great numbers northernmost Finland (Utsjoki) (http://riistaweb.riista.fi/riistatiedot/riistatietohaku.mhtml, 16 December 2010). Moose are coming from northern Norway across the Tenojoki (=Tana River; state border between Finland and Norway). Moose numbers seem to be still increasing. Obviously, this area has been very attractive for moose for several reasons. First, there is extensive pine afforestation (*Pinus sylvestris*) on the fluvio-glacial terraces (valley fillings) of the Tenojoki and the Utsjoki, a southern tributary of the Tenojoki. Secondly, natural regeneration of pine has considerably increased within the old pine forests on the river terraces and lower slopes during the second half of the twentieth century (Holtmeier 1974, 2012; Holtmeier and Broll 2011). Thirdly, pastures and meadows (hay making, silage) on the river terraces provide additional food during summer. As pine is a preferred winter diet, moose seriously affect Scots pine afforestation and natural pine regeneration. Pine is more vulnerable to moose impact than mountain birch, willow or mountain ash. It is possible that moose have already delayed the climatically-driven advance of Scots pine beyond the current altitudinal pine limit (Holtmeier and Broll 2011). In addition to forest damage, traffic accidents (car, train) with moose involved have considerably increased. However, this situation is also due to the increasing traffic volume.

3.4.3 North America

Many North American forest are exposed to excessive browsing pressure. Only a few examples from Newfoundland, Anticosti Island (Gulf of St. Lawrence, Canada) and several national parks (Banff and Jasper National Parks, Alberta; Isle Royale, Lake Superior, Michigan; Apostle Islands, Lake Superior, Wisconsin; Denali National Park, Alaska; Yellowstone National Park, Wyoming/Montana; Rocky Mountain National Park and adjacent Southern Rocky Mountains, Colorado) as well as from the Blue Mountains (northeastern Oregon) and Devils Tower National Monument (Wyoming) are presented in the following paragraphs.

In general, the boreal balsam fir - paper birch forests (*Abies balsamea*, *Betula papyrifera*) of central Newfoundland rapidly recover from disturbances such as clear cut, for example. Moose is not native there, but was introduced in the early twentieth century as wild game animals. Moose not only impair regeneration of birch but also severely damage young growth of balsam fir by browsing and by uprooting (Bergerud and Manuel 1968). Thus, they considerably delay natural forest succession and thereby destroy the food resource of other herbivores such as snowshoe hare and beaver. Both species rely on birch as food in this area. In Newfoundland, moose has become a pest animal. About 200,000 moose have been estimated to be living there. Though moose are a tourist attraction, the local people are concerned about their growing impact on the forests and also on private ground. When the park government decided to drastically reduce moose population, the hunters' lobby in particular objected, as only high numbers of moose guarantee good hunting success. Moreover, the outfitters feared considerable income loss (Neue Osnabrücker Zeitung, 8 April, 2011). On Anticosti Island (Gulf of St. Lawrence, Québec, Canada), where about 220 white-tailed deer were introduced in the late 1890s, high deer densities (about 20 deer km^{-2}) caused a shift from native balsam fir toward spruce forest (white spruce prevailing) in the long-term (Potvin et al. 2003; Tremblay et al. 2007). After canopy disturbance (clear cut), selective browsing has been the principal cause preventing recruitment of the preferred species (Tremblay et al. 2007).

Competition of for forage will usually have an adverse effect on the less competitive species. In Banff and Jasper National Parks (Alberta, Canada; Flook 1962) and in Banff and Jasper National Parks (Alberta), where American elk and beaver are competing for food, beaver is the one which suffers. Elk also severely damage young growth of balsam fir by browsing and uprooting. Balsam fir is an important winter food of elk (Bergerud and Manuel 1968). In the long-term, as a result of paper birch and balsam fir decline, white spruce (*Picea glauca*) might become the dominant species. As moose does not feed on this tree species, natural decrease of moose numbers is likely.

The development of vegetation on Isle Royale (Lake Superior, USA; cf. Sect. 2.2.2) is an additional example for the lasting impact of wild ungulates on (near-natural) forest ecosystems that have remained relatively untouched by man (Snyder and Janke 1976). Moose arrived on the island in the beginning of the twentieth century (Allen 1974). In the early 1930s, moose numbers (4–10 moose km^{-2}) already exceeded the habitat carrying capacity (Murie 1934; Hickie 1936). In 1933/1934 moose population crashed due to malnutrition and fluctuated considerably during the following decades (Krefting 1974). Some severe winters claimed many moose, in particular young calves (Mech et al. 1987). After forest fires, secondary vegetation improved browsing conditions. Predation by wolves that had invaded Isle Royale between 1948 and 1950 did not prevent regular recovery of the moose population (Hansen et al. 1973).

Under the influence of large numbers of moose the forest became more open and even-aged. A luxuriant field layer developed, while the spectrum of species did not much change. The composition of tree species, however, became severely altered. As in Newfoundland, white spruce (*Picea glauca*) spread at the costs of balsam fir (*Picea balsamea*) of which sustainable regeneration was impeded by moose. White spruce gradually invades the intensely browsed and grazed areas. Later, when these trees produce seeds, spruce groups establish themselves around the mother trees and gradually expand. In the beginning of succession, selectively feeding moose may increase the diversity of the forest communities, which, however, decreases again due to growing spruce populations. Parallel to this change, the distribution of nitrogen change and influences vegetation development. Limited nitrogen supply impedes plant growth and regeneration after the browsing/grazing impact has ceased (Pastor et al. 1988).

Paper birch suffered less from moose damage, probably because the trees were already high enough to be not seriously affected when moose arrived at the island. By contrast, mountain ash (*Sorbus americana*) which is preferred by moose as well as yew (*Taxus canadensis*) in the shrub layer have almost disappeared, even in only slightly browsed/grazed areas. Highbush cranberry (*Viburnum pauciflorum*) and skunk currant (*Ribes glandulosum*) are no longer able to reach full size. In some areas, moose population crashed during the mid-1970s until the early 1980s (see also Fig. 2.14). As a result, balsam fir stands could temporarily regenerate and thus partly compensate losses due to overaging. In other places, however, balsam is gradually declining due to intensive browsing by moose, and a development of fir-dominated forests, typical of the previous landscape before moose arrived, is prevented.

Due to the high browsing pressure, balsam fir is also unable to recolonize the large burn areas in the central part of the island. The forest fires were caused for the most part by man. Normally, typical pioneer trees such as paper birch (*Betula papyrifera*) and quaking aspen (*Populus tremuloides*) invade the burn areas. After about 100 years they become outcompeted by balsam fir and other shade-tolerant tree species (Brandner et al. 1990).

In the aspen-spruce forests (*Populus tremuloides*, *Picea glauca*) of Denali National Park (Alaska) up to 75 % of quaking aspen and admixed willows (*Salix bebbiana*) were severely damaged by bark stripping. Moose will switch to bark in case protein-rich forage is rare. Resultant tree mortality increases parallel with moose density. Bark stripping and browse damage accelerate succession towards white spruce forests (Miquelle and Van Ballenberghe 1989). On the other hand, browsing the terminal shoots stimulates production of new leaves, that again are preferred by moose (Bowyer and Bowyer 1997). Moose influence their habitats not only by consumption of green phytomass but also by their droppings and urination. As the nitrogen content of droppings and urine is higher than in fresh leaves ingested by moose, mineralization and productivity increase. At the northern treeline in Denali National Park, mineralization was significantly higher in places frequented by moose than in other treeline sites. Moose has even been considered a keystone factor in this area (Molvar et al. 1993).

Different ungulate species live in Banff and Jasper National Parks: moose, American elk, mule deer, mountain goats and bighorn sheep (*Ovis canadensis*). About 100 years ago, the numbers of mountain goats, moose and mule deer exceeded American elk population by far. Only a small elk herd was left of the large elk population that occupied the parks prior to 1880. The possible causes of elk decline are not well known. The most plausible explanation is that many elk died during several strong winters. Conditions became more favorable for the ungulates again, when forest fires became more frequent – probably in the wake of railroad construction and human colonization Secondary vegetation on the burned areas improved the food source. Consequently, moose, mule deer, bighorn sheep and elk, in particular, considerably increased in numbers again. In addition, release of 235 elk from Yellowstone National Park between 1918 and 1920 has supported elk population growth. Nowadays about 3,200 elk live (during summer) in Banff National Park, and about 1,300 in Jasper National Park. As a result, a growing elk population increasingly competes with mule deer, bighorn sheep and moose for food. Bighorn sheep are most affected, as they usually forage on south-facing slopes in the subalpine zone that also attract many elk, mainly during winter and spring time. There is usually little snow in winter, and grasses sprout first in these places. Grazing during spring time, in particular, strongly affects the plant cover and the topsoil. Thus, forage runs short, and elk and bighorn increasingly compete for food. During the extremely snow-rich winter of 1948, food competition was very likely the main cause of high mortality in bighorn sheep and elk. In view of these negative effects, the park management began already during the 1940s to reduce elk population. Subsequently, grass vegetation recovered. The aspen stands, however, remained in bad condition and still did not tolerate continuous use by elk.

Meanwhile the general situation has changed due to systematic control of forest fires. Secondary vegetation that originally colonized burn areas and was an important food source is now missing (see also Peterson 1983 for Kenai Wildlife Refuge, Alaska). Spruce is colonizing the previous burns. Without fire, spruce, which is highly competitive, may become the dominating tree species. Consequently, the food source will deteriorate, and a decline of elk, mule deer and bighorn sheep is likely (see also Flook 1962).

In contrast to bighorn sheep and mule deer, moose were not only affected directly by food competition but also indirectly by the effects of elk on beaver habitats. During summer, moose feed primarily on abundant aquatic and marsh plants, whereas beavers use softwoods, and aspen in particular. Due to additional severe browse by elk, regeneration (mainly from root suckers) of aspen and other softwoods largely failed and could not compensate for the losses of trees felled by beavers. Thus, forage ran short for the ungulates and also for beavers. After depletion of the food source, beavers left their ponds. When beaver dams collapsed and the ponds dried out, moose were deprived of their favorite food source.

In the Rocky Mountain National Park (Colorado) and also in other parts of the southern Rocky Mountains, strong browsing impact of elk on the extended willow shrub communities (mainly *Salix brachycarpa* and *Salix planifolia*) is very common. Willows make up to 20 % of elk food (Stevens 1980). Aspen also exhibit severe damage, mainly by barkstripping (Photos 3.14 and 3.15). During the 1930s already, shortage of winter forage had become obvious. It was due to the considerably increased elk and mule deer populations (Musselman 1969). The park government tried to mitigate or even terminate the browsing impact, for example by incorporation of about 18 km^2 of former cattle pastures into the park. The enlargement of the park area, however, reduced deer impact only for a short while, and in the following years the cervid numbers exceeded again the carrying capacity of the winter-grazing areas. Thus, reduction of elk and mule deer by culling appeared to be the only alternative. First culling was carried out in 1941 (Wright 1992). At present 3,000–4,000 elk occupy the park area, but this number is, as always, too high.

In Yellowstone National Park, the impact on willows and aspen in the elk winter-grazing areas is even stronger (e.g., Chadde and Kay 1996). For more than 80 years there has been concern about too large numbers of elk, which are considered to be the main cause of the decline in white-tailed deer and beavers as well as of the soil erosion that has considerably increased since the turn of the nineteenth/twentieth century. Drastic reduction of elk population by culling and capturing followed, culminating during the 1960s. Afterwards, however, the park management changed its mind and since 1968 'self regulation' was considered to possibly be the most appropriate 'ecological' factor regulating elk population (Cole 1971; Houston 1971, 1976; Yellowstone National Park 1997).

From numerous studies that examined the hypothesis of balanced interactions of cervids and their habitats of northern Yellowstone, it has become apparent that the conspicuous changes in the park area have to be attributed to a bundle of factors and cannot, if ever, be ascribed to elk alone. (e.g., Singer et al. 1994; Singer 1996). In any case, no causal relationships could be evidenced, neither between high elk density

Photo 3.14 Aspen (*Populus tremuloides*) affected by American elk (*Cervus canadensis*) near Beaver Meadow, Rocky Mountain National Park, Colorado. The tree stems are scarred over up to about two meters above ground. F.-K. Holtmeier, 5 March 2000

and increase of erosion (Engstrom et al. 1991) nor between elk numbers and the decline of other cervids and pronghorns. The decline of willow stands is partly attributed to the drier (10–20 mm precipitation less) and a little warmer (0.5–1.0 °C) climate compared to the nineteenth century (Balling et al. 1992). The drastic decline of the beaver population since about 90 years due to food competition with elk probably is also involved (e.g., Houston 1982; Singer et al. 1994; Chadde and Kay 1996). After the collapse of beaver dams, streams and rivers incised more rapidly. As a result, the ground water table sunk and site conditions for willows deteriorated to an extent that willows became unable to resist the increasing browsing pressure. In the end, however, the severe impact of elk on willow and aspen stands is apparent and cannot be 'discussed away'. Growth of softwoods, their productivity and regeneration (catkins form only on undamaged shoots) as well as their capability of resprouting after browsing damage are considerably impeded (Singer et al. 1994).

Advocates of the 'self-regulation hypothesis' feel confirmed by these studies in their opinion that there is no 'overpopulation' of elk and other ungulates

Photo 3.15 Exclosure at Beaver Meadow, Rocky Mountain National Park, Colorado Front Range. While aspen inside the fenced area have vigorously reproduced from root suckers, elk (*Cervus canadensis*) and muledeer (*Odocoileus hemionus*) have suppressed any reproduction of aspen outside the exclosure. F.-K. Holtmeier 24 July 1987

(Yellowstone National Park 1997). For corroboration they refer to the assumption that the original elk population – i.e., prior to the arrival of the Europeans (1882) – was comparatively high (Schullery and Whittlesey 1992) and very likely in balance with the natural environment (Despain et al. 1986). This seems to be supported by the recent development. About 20 years ago (1988) approximately 25,000 elk were still living in Yellowstone all-year round. During summer, additional 7,000 moved into the park from the surrounding area. After the big forest fire in 1988, the elk population declined drastically due to shortage of food and because many elk that had left the park during and immediately after the forest fire were killed by hunters (Singer et al. 1989; Singer and Harter 1996). Very few animals died in the fire. Only about 1 % of the total elk population was lost. Losses were even smaller among the other animals (Singer et al. 1989). The animal populations gradually recovered. They profited from the secondary vegetation that colonized the burned areas and provided ample high-quality forage. Since 1991 elk numbers have ranged between 16,000 and 20,000. Winter mortality of very young and very old individuals increased parallel to the growing elk population and is about 20 %. Malnutrition seems to be the main cause. In the period 1987–1990, 31 % of the new calves on average fell victim to grizzly bears, black bears, and coyotes. This proportion has increased due to wolves which have been re-introduced to Yellowstone since 1990.

The opponents of this more than 30 years old 'field experiment' suspect the total collapse of the ecosystem (Beetle 1974; Chase 1986) and that it is only a matter of time until this will happen. In any case, Yellowstone is not a closed system, and the present elk density cannot be attributed to its carrying capacity alone, but is also influenced by land use outside the park borders. Moreover, many deer come into the park to escape massive hunting pressure in the surrounding areas. Prediction of the future ecological situation in the park is impossible, as not only the interactions of elk and willows have to be considered but the complete biocoenoses in relation to the varying Yellowstone landscape. The lack of long-term studies, in particular, is a big problem, which results from an inappropriate park policy. As the objectives of the park management changed repeatedly – one time priority was given to elk, another time to grizzly bears or to reintroduction of wolves into the park – no sound budget for long-term research projects was held out (Wright 1992).

Thus, when the first wolves were re-introduced to Yellowstone in 1990, it was an experiment of which the outcome was uncertain. However, the new predators were expected to contribute effectively to natural (ecological) regulation of the elk population. As the decline of woody browse species began after wolves had been eradicated in the 1920s (Ripple and Beschta 2003, 2004), it seemed plausible that re-introduced wolves would reduce elk numbers, and softwoods would recover from browsing pressure – provided that wolves would build up a persistent vital population. The experiment seems to have turned out well. After reintroduction of wolves, recovery of until then suppressed browse plants (willows, aspen) was locally observed (Vucetich et al. 2002; Smith et al. 2003; Ripple and Beschta 2003, 2004; Vucetich et al. 2005; Beschta and Ripple 2010; Estes et al. 2011). In between, wolves, obviously enjoying favorable habitat conditions, have rapidly increased in numbers. In 2011, about 10 wolf packs (about 100 wolves in total) were living in Yellowstone. Elk numbers have decreased by about 60–70 % (Garrot et al. 2009) almost parallel to growing wolf numbers.

In view of this seemingly simple 'mechanism' one may be inclined to ascribe elk decline mainly to wolf predation. Ripple and Beschta (2004) have considered wolf a keystone species in the Yellowstone ecosystem. However, natural elk population control may be more complex, as other large predators (Black bear, *Ursus americanus*; Grizzly bear, *Ursus arctos*; mountain lion, *Felis concolor*) and, in particular, climate (drought, severe snow-rich winters) are also involved (Kauffman et al. 2010). Moreover, wolves also prey on bison living in great numbers in Yellowstone. Thus, the effectiveness of wolf predation as the sole regulating factor might be overestimated. On the other hand, wolf predation is influencing elk population permanently, whereas periods of extreme drought or snow-rich winters usually occur at irregular intervals. They may however have lasting effects.

In 2009, a comprehensive presentation of 16 years of integrated field studies on the ecology of large mammals in Central Yellowstone was published (Garrot et al. 2009, eds.) showing the great complexity of predator-ungulate relationships. More recently, new findings have been published on linkage of elk population to the decline of native trout (cutthroat, *Oncorhyncus clarkii*) after introduction of American lake trout (*Salvelinus namaycush*) and resultant increasing predator pressure by bears on young elk (Middleton et al. 2013). Lake trout, illegally introduced to Yellowstone Lake, has almost outcompeted cutthroat, as the habitat requirements of both species largely overlap and lake trout prey on small cutthroats. Until the end of the 1980s, cutthroats, when moving upward in the shallow tributary streams of Yellowstone Lake for oviposition, had been an important energy-rich food source of the grizzlies. Lake trout, by contrast, spawn in the lake itself. Parallel to the decline of cutthroats, bears increasingly switched to young migratory elk as alternative prey, mainly near the cutthroats' spawning grounds. Consequently, elk calves decreased by 4–16 %, which means a decrease in elk population growth by 2–11 %. This change in the food web might have lasting consequences for species interactions in Yellowstone National Park.

Probably, the combined effects of predation by wolves and bears and hunting, together with extreme winter weather events, will keep elk numbers at a lower level and relieve especially softwoods from sever browsing pressure in the future. In any case, predation risk may influence elks' foraging behavior and thus vegetation structures. Aspen still is hardly available, whereas willows recovered more rapidly. Actually, all of the Yellowstone beavers subsist mainly on willow and aquatic plants. Occasionally, they also use lodgepole pine. Aspen within the beavers' reach have long been cut, and cottonwood plays a minor role as beaver food (Smith and Tyers 2008).

As to the assessment of the impact of wild ungulates, one must be aware that the ecosystems have been altered by man to a varying degree and that human influences have often overruled the original regulating mechanisms. This does not mean, however, that the influence of large herbivores would be unimportant under natural conditions. Observations in forests that have remained relatively untouched as compared to the forest in Central Europe, for example, show that plant communities (number of species, diversity), succession and thus age structures have often been and still are influenced by the large mammalian herbivores in a manifold way.

Thus, elk and mule deer, for example, contribute to the continued existence of the spatial pattern of so-called 'ribbon forests', that locally occur in the Rocky Mountain treeline ecotone (cf. Sect. 2.6.1.1). 'Ribbon forests' are usually located downwind of extended alpine areas from where strong permanent winds relocate much snow to the leeward slope below. A 'ribbon forest' usually consists of longitudinal, relatively compact tree stands ('ribbons') alternating with almost treeless glades ('snow glades'). These structures are oriented parallel to the prevailing direction (NW, W, SW) of the permanent downslope winds (Billings 1969; Buckner 1977; Holtmeier 1978, 1982, 1987b, 1996). The tree stands, measuring up to a hundred or more metres in length and between 10 m –50 m in width, are mainly formed by Engelmann spruce (*Picea engelmannii*) and subalpine fir (*Abies lasiocarpa*). The glades, covered with grasses and herbs, are up to three times wider than the ribbons (Fig. 3.15; Photo 3.16).

There is some evidence that shelter giving solifluction terraces and similar landforms oriented almost perpendicular to the prevailing wind direction and to the contour lines have favored the establishment of the initial tree stands. Growing taller, these trees acted as a 'snow fence' that gradually increased accumulation of snow at their leeward edge and thus duration of the winter snowpack. Late-lying snow curtails the growing season and provides favorable conditions for brown felt-blight (*Herpotricia juniperi*). Brown felt-blight is a parasitic fungus which

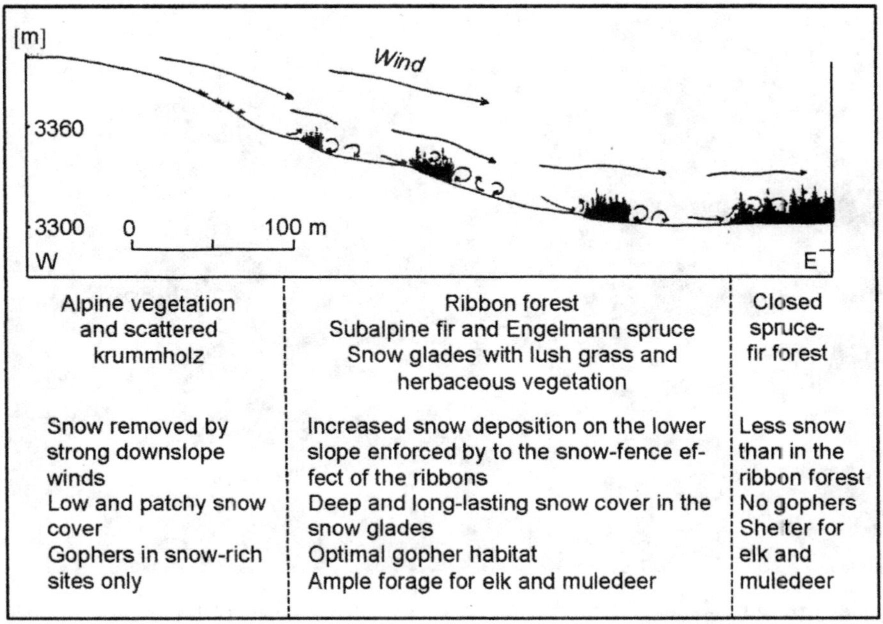

Fig. 3.15 Vertical section through a 'ribbon forest' (schematic, ca. three-fold over-heightened) on the eastern slope of Mount Audubon, Colorado Front Range (Modified from Holtmeier 1987d)

attacks the needles of evergreen conifers buried in long-lasting snow and may prevent the establishment of new trees. Consequently, an almost treeless 'snow glade' develops. As the snow-fence-effect gradually decreases with distance from the upwind 'ribbon', a next 'ribbon' can developed at a distance roughly corresponding to three times the height of the tree stands. A 'ribbon forest' may consist of five to six or even more 'ribbons' and 'snow glades' before merging into the closed downwind forest.

The upwind parts of the glades are often covered with snow until mid-July. The latest snow patches melt in the beginning of August. Thus, soil moisture is high compared to less snow-rich terrain and supports almost luxuriant subalpine meadows (Holtmeier 1987b): These provide optimal forage to elk and mule deer (Fig. 3.16). Their droppings, that are found everywhere in the glades, improve nutrient supply to the plants. Young spruces and firs may establish themselves only on small convexities (less snow), along the wind-facing side of a ribbon and within small gaps between ribbon sections where wind-funnelling removes the snow ('blow outs').

Occasional slow-growing seedlings and saplings usually succumb after a couple of years or even after decades to the brown snow felt fungus or are destroyed by elk and deer. In addition, northern pocket gophers (*Thomomys talpoides*, Geomyidae) enjoying optimal conditions of life within the snow glades, where they are protected from cold winter temperatures and their many predators, damage young trees under the snowpack by girdling stems and roots (Photo 3.17). Pocket gophers are burrowing

Photo 3.16 Ribbon forests (3,350–3,430 m) on a north-facing slope of Niwot Ridge, Colorado Front Range. Between the tree ribbons almost treeless snow-glades have developed, where snow accumulation is enforced by the 'snow-fence effect' of the ribbons. Snow drifts may linger into early summer. The almost treeless glades provide favorable habitat conditions for pocket gophers (*Thomomys talpoides*), American elk (*Cervus canadensis*) and mule deer (*Odocoileus hemionus*). The combined effects of these animals are among the factors keeping the snow glades treeless (cf. Fig. 3.16). F.-K. Holtmeier, 13 July 1997

rodents that spend most time of their life in the ground. Gophers accumulate excavated soil on the surface ('mounds', snow tunnel fillings). In the Colorado Front Range, the present author found up to 50 % of pocket gopher habitats covered with excavated material (Holtmeier 1982; see also Buckner 1977).

Young trees may be pushed out of the soil by the gophers' burrowing activity or become buried by excavated soil (see also Teipner et al. 1983, further references therein). On the other hand, such open patches may facilitate seedlings establishment (Butler et al. 2004; Holtmeier 2012). Nevertheless, elk, mule deer, climatic injuries, short growing seasons, snow fungus infection, moisture stress in summer and downslope translocation of excavated soil by heavy rainfall and needle ice solifluction (Hall et al. 1999) may ultimately kill tree seedlings in such sites. Besides the effects of long-lasting winter snowpack the impact of elk and pocket gophers have presumably contributed to persistence of the ribbon forests spatial pattern without major changes for several hundreds or even thousands of years (e.g., Buckner 1977; Holtmeier 1987b).

Beals et al. (1960) presented an interesting study on the influence of white-tailed deer on the vegetation of the Apostle Islands – a group of 22 small islands in Lake Superior (Wisconsin). Actually, prevailing yellow birch (*Betula lutea*) and/or

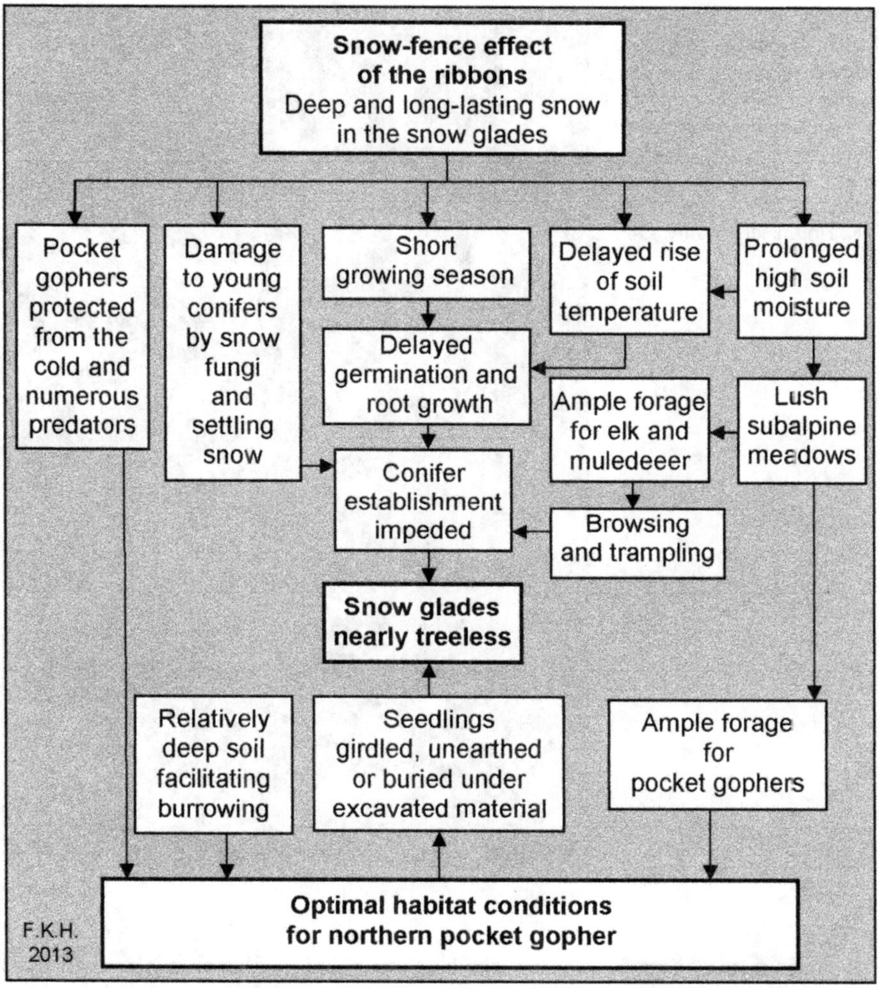

Fig. 3.16 Interactions of biotic and physical factors in a 'ribbon forest' ecosystem, according to the present author's observations in the Colorado Front Range (Scheme by F.-K. Holtmeier, from Holtmeier 2009)

white cedar (*Thuja occidentalis*) provide deer with abundant food. However, deer prefer Canadian yew (*Taxus canadensis*), which is growing in the forest understorey. Even minor browsing causes marked changes. Thus, yew declines first. As browsing pressure increases, cedar also declines and deer rely mainly on the more tolerant deciduous trees that may recover from browsing by producing basal sprouts. Under heavy browsing pressure, however, these also decline. In the long-term, growing deer population will destroy its food source.

Remarkably, some of the islands do not harbor deer, although nutrient-rich forage is abundant. Apart from that, deer could have easily reached these islands as

Photo 3.17 Young spruce (*Picea engelmannii*) girdled by pocket gopher (*Thomomys talpoides*) under the winter snowpack in a snow glade (about 3,340 m) of a ribbon forest at Jenny Lake, Colorado Front Range. F.-K. Holtmeier, 25 July 1989

they are located not far distant from the lake shore and neighboring islands that are occupied by deer. It is possible that on the still deer-free islands essential habitat conditions for deer such as forest openings, for example, are lacking. In any case, after clearings were created artificially, deer promptly invaded. As the forests on some of these islands are still in a near-natural state, the authors (Beals et al. 1960) have suggested preventing deer invasion. It would already be sufficient to simply not clear these forests.

In the Blue Mountains (northeastern Oregon), foraging and bark stripping by great numbers of mule deer and elk are important factors in dieback and death of Pacific yew (*Taxus brevifolia*), which is a favorite browse of these ungulates. The continued existence of yew in this area might be threatened (Parks et al. 1998).

In Devils Tower National Monument (Black Hills, USA), native deciduous woodlands appear to be dying under heavy browsing by white-tailed and black-tailed deer. Supplemental foraging on agricultural land probably reduces fluctuations of the deer population. As the woodlands patches are relatively small, fencing them may be an alternative to culling deer (Merrill et al. 2003).

3.4.4 *Large Herbivores and the Woodland Hypothesis*

It has been discussed repeatedly whether wild large herbivores have also influenced the original Central European vegetation and landscape (Tubbs 1986; Beutler 1992; Geiser 1992; Schüle 1992; Bunzel-Drüke et al. 1994; Geissen 1996; Bunzel-Drüke 1997; Hofmann et al. 1998) more than one should expect in view of the 'woodland hypothesis', which means that Central Europe once was completely covered with dense forests (Wilmanns 1973; Ellenberg 1982; Kaule 1991; Pott and Hüppe 1991; May 1993; Otto 1994; Pott 1995, 1997, 1998; Zoller and Haas 1995; Fischer 1999).

Referring to forest development during prehistoric and historic times, the advocates of historical large herbivores' influence speculate that closed forests could only develop after man had eradicated most large herbivores. The argumentation sounds plausible and understandable. There is evidence that during the Neolithic Age (about 5,000 BP) pastoral use of forest, practiced since the beginning of agriculture, and cattle herding, brought about structural changes and also decay of forest. In many areas, forest became replaced by scattered shrub vegegation, open pastures and heaths (Ellenberg 1978; Burrichter et al. 1980; Pott and Hüppe 1991; Pott 1998)

Nowadays, closed forests will continue to exist only if domestic ungulates are excluded. The same holds true for over-sized wild ungulates. At the northern rim of the prairie in southwestern Canada (Campbell et al. 1994), aspen have been expanding during the twentieth century. Aspen parkland forms a transition zone between the prairie in the south and the boreal woodland in the north. Expansion of aspen started immediately after bison was extirpated and elk populations had decreased. Thus, a causal relationship is not unlikely. Both ungulate species had prevented establishment of aspen by browsing the root suckers or destroying them by trampling. Regeneration from root suckers is typical of aspen as a seral species after wild fires. These assumptions are supported by direct observations of elk herds impacting aspen stands, and vegetative reproduction by root suckers in particular (cf. Photo 3.14; Blyth and Hudson 1987; see also Bartos et al. 1994). In today's Elk Island National Park (Alberta), for example, an aspen forest which had developed after bison and other large herbivores had declined, was destroyed by fire in 1895. After re-introduction of bison and increase of elk numbers, the grassland could expand, whereas after culling the ungulates during the 1930s and 1950s, aspen considerably increased (Blyth and Hudson 1987).

What would be more plausible than ascribing a comparable effect on vegetation to the large ungulates (mammoth, straight-tusked elephant, Merck's rhinoceros, aurox, European bison, European wild horse; e.g., Remmert and Zell 1984; Beutler and Schilling 1991; Beutler 1992, 1996, 1997; see also Billings 1970 for North America and Owen-Smith 1987 for the worldwide view), which lived in great numbers in Central Europe during the last glacial period and in the early Holocene? Their influence on natural vegetation was probably not less significant than has been the impact of domestic and wild ungulates during historic times. In Central Europe, roe deer, red deer, chamois, ibex, wild boar, and moose (in eastern and northern Europe), are the only ungulates which have survived up to the present. Roe deer is the only omnipresent species.

It may be safely assumed that ungulate species native to Central Europe have occupied their ecological niches by food preferences and modes of behavior in a similar way as has been described of the wild ungulates in the African savannas (Vesey-Fitzgerald 1960; Bell 1970; Ferrar and Walker 1974; Hirst 1975; Jarman and Sinclair 1979; McNaughton 1988) or of the prairie in North America (e.g., Haugen and Shult 1972; Wydeven and Dahlgren 1985). However, inferring the situation in Holocene Europe from the present African savannas must be considered with some reserve (see also May 1993). Nevertheless, it is tempting to presume that the grazing/browsing impact of the comparatively many herbivore species in Central and western Europe had already created an open parkland-like landscape before humans and their domestic ungulates became of major importance. The ancient parkland was probably characterized by a variegated mosaic of open pastures, tree groves and shrubs.

The parkland provided suitable habitats to non-forest species. Buzzards (*Buteo buteo*), for example, relying on open hunting grounds, had probably not survived in a landscape completely covered with dense forests. Some thermophilic reptile species needing sunny and warm places would also not have been able to exist under a closed forest canopy (Glandt and Geiger 1990). Not least, closed forest would have been hostile to some songbird and butterfly species. The hypothesis, that only the alpine zone, coastal areas, flood plains, a few bogs and dunes as well as avalanche pathways and terrain affected by rockfall were originally devoid of forest, can hardly be supported (see also Hofmann et al. 1998), as the ungulates would have practically been restricted to such locations. This, however, seems very unlikely. About the population densities of prehistoric large herbivores, we can only speculate (e.g., Beutler 1997). In any case, they must have occurred in relatively large numbers to preserve an open parkland. Reliable data are missing, however.

Extinction of the original large herbivores would have considerably supported the expansion of closed forests covering entire Central Europe – provided that the herbivores really had a strong influence. After eradication of the wild ungulate herds, the domestic grazers adopted their functional role. During historic time, the direct impact of cattle as well as cutting leaves for hay making and littering stables and cow sheds have sustainably influenced woodland structures and composition. Natural regeneration, in particular, was impaired or prevented by the domestic ungulates. In many areas, an open parkland developed, of which the ecological dynamics and succession were completely controlled by the interactions of domestic grazers and vegetation (Burrichter et al. 1980; Harding and Rose 1986; Pott and Hüppe 1991; Van Wieren 1991; Pott 1993, 1996, 1998; Bakker et al. 2003). The historical woodland pastures of north-western Germany may serve as an example. Palatable tree species such as oak (*Quercus* spp.) could only grow to full size if mantled by thorny shrubs (*Prunus spinosa, Crataegus laevigata, Rosa canina, Rhamnus catharticus*) giving shelter from grazing cattle. As tree cover increased, shade-intolerant shrubs became outcompeted, and recruitment of light-demanding trees largely failed. In the end, senescent tree stands collapsed and gave way to open grassland patches (e.g., Vera 2000). These became sucessively recolonized by light-demanding shrubs and trees, with the exception of regularly used cattle paths and sites exposed to soil erosion. As succession varied spatially and temporally, a highly divers site mosaic developed. It is not unlikely (e.g., Bakker et al. 2003) that comparable interactions also existed between large native grazers and vegetation during ancient times.

Changes in species spectrum and composition that occurred in the remained woodland-pastures after the decrease of grazing may support the 'woodland hypothesis'. On the other hand, it also is apparent that the 'natural landscape' cannot be 'reconstructed' without including the ungulates' impact (see also Scherzinger 1996). In this context we refer once again to the observations on the influence of red deer and chamois on vegetation dynamics in the Swiss National Park (Krüsi et al. 1995) that reflect a 'forest-openland-forest cycle' (cf. Fig. 3.14) similar to the 'forest-mosaic-cycle' (Remmert 1988 and later).

Although it will probably be impossible to reliably assess the influence of previous ungulate populations on vegetation development (Bradshaw and Mitchell 1999), the large herbivores very likely prevented natural forestation of open burn and windthrow areas, at least for a while (Cornelius and Hofmann 1999). This may be backed up by observations on vegetation development within and outside exclosures on the western Olympic Peninsula (Washington), for example. The long-term dynamics of non-economic forests are mainly controlled by occasional but very effective extreme events such as forest fires or severe storms, whereas deer and mule deer influence secondary succession and tree regeneration after fire or windthrow (Woodward et al. 1994). Similar conditions were found in mixed-conifer forests of the Oregon Blue Mountains (Riggs et al. 2000). Riggs et al. (2000) consider the large herbivores to be 'chronic disturbance factors' which, in the view of the authors, have a keystone function as far as their effects on nutrient conditions, productivity and vegetation structures are concerned. However, the ungulates do not hold up succession towards closed forests.

In the end, also in Central Europe, large herbivores could not prevent the development of dense forests (Schmidt 1999), particularly because they probably were by far less numerous than are wild ungulates at present (Widmann 1991). Thus, Schreiber (2000) believed that the large herbivores played only a minor role as far as their impact on the landscape environment is concerned. In the view of Wilmanns (2005) wild ungulates only created sunlit patches rather than open woodland or parkland. In his detailed monograph on grazing-ecology and forest history Vera (2000), however, objects to the alleged general validity of the 'woodland hypothesis', which is largely based on circular reasoning. To substantiate his objection he puts forward strong arguments in favor of a substantially more differentiating consideration of vegetation history including the impact of the large herbivores. In any case, missing grass pollen cannot be taken for evidence of previous closed woodlands, as it is exactly grazing that prevents or restricts flowering of grasses. Thus, it is not surprising that considerably more tree pollen than grass pollen were produced on open terrain grazed by cattle all year round (Groenmann-Van Wateringe 1993). In the opinion of Mitchell (2005), increased large herbivore populations influenced tree species composition, but did not open the forests canopies in the primeval European landscape (see also Mitchell 1990; Mitchell and Cole 1998). Gaps caused by fire and windthrow (Bradshaw and Hannon 2004), however, probably were maintained by large herbivores, at least for a while.

References

African Conservation Centre. (2012). *News: The impact of the 2009 drought on wildlife, livestock and tourism: recommendations for ecosystem restoration.* Nairobi: ACC.

Ahti, T. (1977). Lichens of the boreal coniferous zone. In M. R. D. Seaward (Ed.), *Lichen ecology* (pp. 145–181). London: Academic Press.

Albon, S. D., Brewer, M. J., O'Brien, S., Nolan, A. J., & Cope, D. (2007). Quantifying the grazing impacts associated with different herbivores on rangeland. *Journal of Applied Ecology, 44*, 1176–1187.

Allen, D. (1974). Of fire, moose and wolves. *Audubon, 74*(1), 39–49.

Alschner, G. (1980). *Klipp und klar 100 x (hundertmal) Tierwanderungen*. Zürich: Mannheim Wien.

Andreev, V. N. (1954). Extension of arboreous plants into the tundra (original in Russian). *Botanical Journal*, Moscow.

Anschlag, K., Broll, G., & Holtmeier, F.-K. (2008). Mountain birch seedlings in the treeline ecotone, Subarctic Finland: Variation in above- and below-ground growth depending on microtopography. *Arctic, Antarctic, and Alpine Research, 40*(4), 609–616.

Arcese, P., Hando, J., & Campbell, K. (1995). Historical and present-day-anti-poaching efforts in Serengeti. In A. R. E. Sinclair & P. Arcese (Eds.), *Serengeti II* (pp. 506–533). Chicago/London: University of Chicago Press.

Archer, S., & Detling, J. K. (1986). Evaluation of potential herbivore mediation of plant water status in a North American mixed-grass prairie. *Oikos, 47*, 287–291.

Arnborg, T. (1955). Skador pa tallplantor inom ett starkt renbetat omrade i Hede socken. *Norrlands Skogvardsförbunds Tidskrift, 4*, 5–66. Skogvardsförbunds.

Bakker, E. S., Olff, H., Vandenberghe, C., de Maeyer, K., Smit, R., Gleichman, J. M., & Vera F. W. M. (2003). In E. S. Bakker (Ed.), *Herbivores as mediators of their environment: The impact of large and small species on vegetation dynamics*. PhD-thesis, Wageningen University, Wageningen.

Balling, R. C., Meyer, G. A., & Wells, S. G. (1992). Climate change in Yellowstone National Park: The drought related risk of wildfire increasing? *Climatic Change, 22*, 35–45.

Bannikov, A. G., Zhirnov, L. V., Lebedeva, L. S., & Fandeev, A. A. (1967). Main Administration of Hunting and Reserves at the Council of Ministers of the RSFSR, Astrachan Laboratory for the Biology of the Saiga (translated from Russian). Jerusalem. Jerusalem Program for Scientific Studies.

Bartos, D. L., Brown, J. K., & Booth, G. D. (1994). Twelve years biomass response on aspen communities following fire. *Journal of Range Management, 47*, 79–83.

Batzli, G. O., White, R. G., Maclean, S. F., Pitelka, F. A., & Collier, B. D. (1980). The herbivore-based trophic system. In J. Brown, P. C. Miller, L. L. Tieszen, & L. Bunnell (Eds.), *Arctic ecosystems: The coastal tundra at Barrow* (pp. 335–410). Stroudsburg: Alaska.

Bavier, G. (1976). Graubündens Waldwirtschaft. Ein Vergleich und Ausblick. *Bündner Wald, 29*(6), 220–248.

BBC. (2012, October 16). *BBC News Africa*. London: BBC.

Beals, E. W., Cottam, G., & Vogl, R. J. (1960). Influence of deer on vegetation of the Apostle Islands, Wisconsin. *Journal of Wildlife Management, 24*(1), 68–80.

Bee, J. W., & Hall, E. R. (1956). *Mammals of northern Alaska on the Arctic Slope* (Miscellaneous Publications Museum of Natural History 8). Kansas: University of Kansas.

Beetle, A. A. (1974). The zootic disclimax concept. *Journal of Range Management, 27*, 30–32.

Bell, R. H. (1970). The use of the herb layer by grazing ungulates in the Serengeti. In A. Watsons (Ed.), *Animal populations in relation to their food resources* (pp. 111–124). Oxford: Blackwell.

Bell, R. H. V. (1984). The man-animal interface: an assessment of crop damage and wildlife control. In R. H. V. Bell & E. Meshane-Caluzi (Eds.), *Conservation and wildlife management in Africa* (pp. 387–416). Malawi: US Peace Corps.

Belnap, J. (2001). Biological crusts and wind erosion. In J. Belnap & O. Lange (Eds.), *Biological crusts: Structure, function and management* (Ecological Studies 150, pp. 340–347). New York: Springer.

Belnap, J., Büdel, B., & Lange, O. (2001). Biological crusts: Characteristics and distribution. In J. Belnap & O. Lange (Eds.), *Biological crusts: Structure, function and management* (Ecological Studies 150, pp. 1–30). New York: Springer.

Belsky, A. J. (1989). Landscape pattern in a semi-arid ecosystem in East Africa. *Journal of Arid Environments, 17*, 265–270.

Belsky, A. J. (1992). Effects of grazing, competition, disturbance, and fire on species composition and diversity of grassland communities. *Journal of Vegetation Science, 3*, 187–200.

Belsky, A. J. (1995). Spatial and temporal landscape pattern in arid and semi-arid African savannas. In L. Hansson, L. Fahrig, & G. Merriam (Eds.), *Mosaic landscapes and ecological processes* (pp. 31–56). London: Chapman & Hall.

Ben-Shahar, R. (1993). Patterns of elephant damage to vegetation in Northwestern Botswana. *Biological Conservation, 65*, 249–256.

Bergerud, A. T., & Manuel, F. (1968). Moose damage to balsam fir-white birch forests in central Newfoundland. *Journal of Wildlife Management, 32*(4), 729–746.

Bergström, R., & Hjeljord, O. (1987). Moose and vegetation interactions in northwestern Europe and Poland. *Swedish Wildlife Research Supplement, 1*, 213–228.

Bergström, R., & Willebrand, T. (1992). Research and statistics, tools in game management. In R. Bergström, H. Huldt, & U. Nilsson (Eds.), *Swedish game, biology and management* (pp. 96–109). Uppsala: Almquist & Wicksell.

Bernhart, A. (1988). Waldentwicklung, Verjüngung und Wildverbiß im oberbayerischen Bergwald. *Schweizerische Zeitschrift für Forstwesen, 139*(6), 463–484.

Bernhart, A. (1990). Entwicklung der Bestockung im Bergwald Oberbayerns seit 1860. In E. Schuster (Ed.), *Zustand und Gefährdung des Bergwaldes* (Forstwissenschaftliche Forschungen 40, pp. 19–29), Beiheft zum Forstwissenschaftlichen Zentralblatt.

Beschta, R. L., & Ripple, W. J. (2010). Recovering riparian plant communities with wolves in Northern Yellowstone, U.S.A. *Restoration Ecology, 18*(3), 380–389.

Beutler, A. (1992). Die Großtierfauna Mitteleuropas und ihr Einfluß auf die Landschaft. *Landschaftsökologie, Weihenstephan, 6*, 49–69.

Beutler, A. (1996). Die Großtierfauna Mitteleuropas und ihr Einfluß auf die Landschaft. In B. Gerken & C. Meyer (Eds.), *Wo lebten Pflanzen und Tiere in der Naturlandschaft und der frühen Kulturlandschaft Europas?* (Natur- und Kulturlandschaft 1, pp. 107–112). Höxter/Jena: Huxaria Druck und Verlag.

Beutler, A. (1997). Das Weidelandschaftsmodell: Versuch einer Rekonstruktion der natürlichen Landschaft. In: B. Gerken & C. Meyer (Eds.), *Vom Waldesinnsaum zur Hecke Geschichte, Situation und Perspektiven eines Natur-Lebensraum-Gefüges* (Natur und Kulturlandschaft 2, pp. 194–196). Höxter/Jena: Huxaria Druck und Verlag.

Beutler, A., & Schilling, D. (1991). Säugetiere. In G. Kaule (Ed.), *Arten und Biotopschutz* (2nd ed., pp. 198–205). Stuttgart: Eugen Ulmer Verlag.

Billings, W. D. (1969). Vegetational pattern near alpine timberline as affected by fire-snow drift interactions. *Vegetatio, 19*, 192–207.

Billings, W. D. (1970). *Plants, man, and the ecosystem* (Fundamentals of Botany Series, 2nd ed.). Belmont: Wadsworth.

Birkett, A. (2002). The impact of giraffe, rhino and elephant on the habitat of a black rhino sanctuary in Kenya. *African Journal of Ecology, 40*, 276–282.

Bjørneraas, K., Herfindal, I., Solberg, E. J., Sæther, B. E., Van Moorter, B., & Rolandsen, C. M. (2012). Habitat quality influences population distribution, individual space use, and functional responses in habitat selection by a large herbivore. *Oecologia, 168*, 231–243.

Blanc, J. J., Barnes, R. F. W., Craig, G. C., Dublin, H. T., Thouless, C. R., Douglas-Hamilton, I., & Hart, J. A. (2007). *African Elephant Status Report 2007: An update from the African Elephant Database* (Occasional Paper of the IUCN Species Survival Commission, no. 33). Gland: IUCN/Species Survival Commission African Elephant Specialist Group.

Bliss, L. C. (1986). Arctic ecosystems: Their structure, function and herbivore carrying capacity. In O. Gudmundsson (Ed.), *Grazing research at northern latitudes* (pp. 5–25). New York: Plenum Press.

Blyth, C. B., & Hudson, R. J. (1987). *A plan for management of vegetation and ungulates, Elk Island National Park*. Alberta: Canadian Parks Service.

Bodenmann, A., & Eiberle, K. (1967). Über die Auswirkungen des Verbisses der Gemse im Aletschwald. *Schweizerische Zeitschrift für Forstwesen, 7*, 461–470.

Botkin, D. B., Mellilo, J. M., & Wu, L. S.-Y. (1981). How ecosystem processes are linked to mammal population dynamics. In C.-W. Fowler & T. D. Smith (Eds.), *Dynamics of large mammal populations* (pp. 373–387). New York: Wiley.

Bowyer, J. W., & Bowyer, R. T. (1997). Effects of previous browsing on the selection of willow stems by Alaskan moose. *Alces, 33*, 11–18.

Bradshaw, R. H., & Hannon, G. (2004). The Holocene structure of north-west European forest induced from palaeoecological data. In O. Honnay, K. Verheyen, B. Bossuyt, & M. Hermy (Eds.), *Forest biodiversity: Lessons from history for conservation* (pp. 11–25). Oxford: CAC International.

Bradshaw, R., & Mitchell, F. J. G. (1999). The paleoecological approach to reconstructing former grazing-vegetation interactions. *Forest Ecology Management, 120*, 3–12.

Bradshaw, R. H. W., Hannon, G. E., & Lister, A. M. (2003). A long-term perspective on ungulate-vegetation interactions. *Forest Ecology and Management, 181*, 267–280.

Brändli, U.-B. (1995). Zur Verjüngung und Wildschadensituation im Gebirgswald der Schweiz. Regionale Ergebnisse des ersten Landesforstinventars (LFI), 1983/85. *Schweizerische Zeitschrift für Forstwesen, 146*(5), 355–378.

Brandner, T. A., Peterson, R. D., & Risenhoover, K. L. (1990). Balsam fir on Isle Royale: Effects of moose herbivory and population density. *Ecology, 71*(1), 155–164.

Broll, G., Holtmeier, F.-K., Anschlag, K., Brauckmann, H.-J., Wald, S., & Drees, B. (2007). Landscape mosaic in the treeline ecotone on Mt Rodjanoaivi, Finnish Subarctic. *Fennia, 185*(2), 89–105.

Brooks, A. C., & Buss, I. O. (1962). Past and present status of the elephant in Uganda. *Journal of Wildlife Management, 26*(1), 38–50.

Brown, R., & Mikola, P. (1974). The influence of fruticose lichens upon the mycorrhizae and seedling growth of forest trees. *Acta Forestalia Fennica, 141*, 1–23.

Buckner, D. L. (1977). *Ribbon-forest development and maintenance in the Central Rocky Mountains of Colorado*. Dissertation University of Colorado, Boulder.

Buechner, H. K., & Dawkins, H. C. (1961). Vegetation change induced by elephants and fire in Murchison Falls National Park, Uganda. *Ecology, 42*(4), 752–766.

Bunzel-Drüke, M. (1997). Großherbivore und Naturlandschaft. *Schriftenreihe für Landschaftspflege und Naturschutz, 54*, 109–128.

Bunzel-Drüke, M., Drüke, J., & Vierhaus, H. (1994). Quaternary Park, Überlegungen zu Wald, Mensch und Megafauna. *ABU Info, 17/18*(1), 4–38.

Burgess, P. (1999). *Human environmental interactions in Upper Lapland, Finland* (Arctic Centre Reports 27). Arctic Centre, University of Lapland, Rovaniemi.

Burrichter, E., Pott, R., Raus, T., & Wittig, R. (1980). Die Hudelandschaft "Borkener Paradies" im Emstal Bei Meppen. *Abhandlungen aus dem Museum für Naturkunde, 42*(4), 1–69.

Burschel, P., Binder, F. E., Kateb, H., & Moser, R. (1990). Erkenntnisse zur Walderneuerung in den Bayerischen Alpen. *Beiheft zum Forstwissenschaftlcihen Centralblatt, 40*, 40–49.

Buss, I. O. (1961). Some observations on food habits and behavior of the African elephant. *Journal of Wildlife Management, 25*(2), 131–148.

Butler, D. R., Malanson, G. P., & Resler, L. M. (2004). Turf-banked terrace treads and risers, turf exfoliation and possible relationships with advancing treeline. *Catena, 58*, 259–5274.

Butler, D. R., Malanson, G. P., Resler, L. M., Walsh, S. J., Wilkerson, F. D., Schmid, G. L., & Sawyer, C. F. (2009). Geomorphic patterns and processes at alpine treeline. In D. R. Butler, G. P. Malanson, S. J. Walsh, & D. B. Fagre (Eds.), *The changing alpine treeline: The example of Glacier National Park, Montana, USA. Developments in earth surface processes* (Vol. 12, pp. 63–84). Amsterdam/Boston/Heidelberg: Elsevier.

Cairns, D. M., & Moen, J. (2004). Herbivory influences treelines. *Journal of Ecology, 92*(6), 1019–1024.

Campbell, A. C. (1990). History of elephants in Botswana. In P. Handcock (Ed.), *The future of Botswana's elephants. Proceedings of a workshop organized by the Kalahari Conservation Society in conjunction with the Department of Wildlife and National Parks* (pp. 5–15), Gaborone.

Campbell, K., & Borner, M. (1995). Population trends and distribution of Serengeti herbivores: Implications for management. In A. R. E. Sinclair & P. Arcese (Eds.), *Serengeti II*. Chicago/London: University of Chicago Press.

Campbell, K., & Hofer, H. (1995). People and wildlife: Spatial dynamics and zones of interaction. In A. R. E. Sinclair & P. Arcese (Eds.), *Serengeti II* (pp. 534–570). Chicago/London: University of Chicago Press.

Campbell, C., Campbell, I. D., Blyth, C. B., & McAndrews, H. H. (1994). Bison extirpation may have caused aspen expansion in western Canada. *Ecography, 17*(4), 360–362.

Carlson, D. C., & White, E. M. (1984). Effect of prairie dogs on mound soils. *Soil Science Society of America Journal, 51*, 389–393.

Caudillo, G., De Battisti, R., Colpi, C., Vazzola, C., & Da Ronch, F. (2002). Ungulate damage and silviculture in the Cansiglio Forest (Veneto Prealps, NE Italy). *Journal of Nature Conservation, 10*, 233–241.

Caughley, G. (1976). The elephant problem – An alternative hypothesis. *East African Wildlife Journal, 14*, 265–283.

Caughley, G. (1983). Dynamics of large mammals and their relevance to culling. In N. Owen-Smith (Ed.), *Management of large mammals in African conservation areas* (pp. 115–126). Pretoria: HAUM.

Chadde, S. W., & Kay, C. E. (1996). Tall-willow communities on Yellowstone's northern range: A test of the 'natural regeneration' paradigm. In: F. J. Singer (Ed.), *Effects of grazing by wild ungulates in Yellowstone National Park* (Technical Report NPS/NTYELL/NRTR 96-01, pp. 165–183). Denver: National Park Service, Natural Resource Information Division.

Chamaillé, S., Valeix, M., & Hervé, F. (2007). Managing heterogeneity in elephant distribution: Interactions between elephant population densities and surface-water availability. *Journal of Applied Ecology, 44*, 625–633.

Chan, S., Maksimuk, A. V., & Zhrinov, L. V. (1995). From steppe to store: The trade in Saigas antelope horn. In *Deer production* (Royal Society of New Zealand Bulletin 22, pp. 23–36) Wellington: Royal Society of New Zealand.

Chase, A. (1986). *Playing God in Yellowstone: The destruction of America's first national park.* San Diego/New York/London: Atlantic Monthly Press/Harcourt Brace & Company.

Childes, S. M., & Walker, B. H. (1987). Ecology and dynamics of the woody vegetation on the Kalahari sands in Hwange National Park. *Vegetatio, 72*, 111–128.

Cid, M. S., Detling, J. K., & Brizuela, M. A. (1993). Vegetational response of a mixed-grass prairie site following exclusion of prairie dogs and bison. *Journal of Range Management, 44*, 100–1051.

Clark, T. (1986). Some guidelines for management of the black-footed ferret. *Great Basin Naturalist Memoirs, 8*, 160–168.

Cobb, S. (1980). Tsavo, the first thirty years. *Swara, 3*, 12–15.

Coe, M. (1999). Introduction. In M. Coe, N. McWilliam, G. Stone, & M. Packer (Eds.), *Mkomazi: The ecology, biodiversity and conservation of a Tanzanian savanna* (pp. 5–13). London: Royal Geographical Society.

Cole, G. C. (1971). An ecological rationale for the natural and artificial regulation of native ungulates in parks. *Transactions of the American Wildlife Conference, 36*, 1417–1425.

Cole, M. M. (1996). *The savannas: Biogeography and botany*. London: Academic Press.

Coleman, M. D., Hinckley, T. M., McNaughton, G., & Smit, B. A. (1992). Root cold hardiness and native distribution of subalpine conifers. *Canadian Journal of Forest Research, 22*, 932–938.

Collins, S. L., & Glenn, S. M. (1988). Disturbance and community structure in North American prairies. In H. J. During, M. J. A. Werger, & J. A. Willems (Eds.), *Diversity and pattern in plant communities* (pp. 131–143). The Hague: SPB Academic Publishers.

Collins, S. L., & Steinauer, E. M. (1998). Disturbance, diversity and species interactions in tall-grass prairie. In A. K. Knapp, J. M. Briggs, D. C. Hartner, & S. L. Collins (Eds.), *Grassland dynamics* (pp. 140–156). Oxford: Blackwell.

Collins, S. L., Knapp, A. K., Briggs, J. M., Blair, J. M., & Steinauer, E. M. (1998). Modulation of diversity by grazing and mowing in native tallgrass prairie. *Science, 280*, 745–747.

Colpaert, A., Kumpula, J., & Nieminen, M. (2003). Reindeer pasture biomass assessment using satellite remote sensing. *Arctic, 56*(2), 147–158.

Cooper, E. J., & Wookey, P. A. (2003). Floral herbivory of Dryas octopetala by Svalbard reindeer. *Arctic, Antarctic, and Alpine Research, 35*(3), 369–376.

Coppock, D. L., & Detling, J. K. (1986). Alteration of bison and black-tailed prairie dog grazing interactions by prescribed burning. *Journal of Wildlife Management, 50*, 452–455.

Coppock, D. L., Detling, J. K., Ellis, J. E., & Dyer, M. I. (1983a). Plant herbivore interactions in a North American mixed-grass prairie I. Effects of black-tailed prairie dogs on intraseasonal aboveground biomass and nutrient dynamics and plant diversity. *Oecologia, 56*, 1–9.

Coppock, D. L., Ellis, J. E., Detling, J. K., & Dyer, M. I. (1983b). Plant herbivore interactions in a North American mixed-grass prairie II. Responses of bison to modification of vegetation by prairie dogs. *Oecologia, 50*, 10–15.

Corfield, T. F. (1973). Elephant mortality in Tsavon National Park, Kenya. *East African Wildlife Journal, 11*, 339–368.

Cornelius, R., & Hofmann, R. R. (1999). Huftiergemeinschaften und Vegetation – Entstehung, Zerstörung und schrittweise Rekonstruktion eines natürlichen Gefüges. In B. Gerken & M. Görner (Eds.), *Europäische Landchaftsentwicklung mit großen Weidetieren. Geschichte, Modelle und Perspektiven. Natur und Kulturlandschaft* (Vol. 3, pp. 382–389). Höxter/Jena: Huxaria Druck und Verlag.

Costello, D. F. (1970). *The world of the prairie dogs*. Philadelphia/New York: Lippincott.

Cota-Larson, R. (2012, November 20). *South Africa: 455 rhinos killed in 829 days*. Annamiticus, facebook.

Côté, S. D., Rooney, T. P., Tremblay, J.-P., Dussault, C., & Waller, D. M. (2004). Ecological impacts of deer overabundance. *Annual Review of Ecology, Evolution, and Systematics, 35*, 113–147.

Craine, J. M. (2013). Long-term climate sensitivity of grazers performance: A croo-site study. *PLoS ONE*. doi:10.1371/journal.pone.0067065.

Crawford, R. M. M. (2014). *Tundra-taiga biology. Human, plant, and animal survival in the Arctic*. Oxford: Oxford University Press.

Crête, M., Couturier, S., Hearn, B. J., & Chubbs, T. E. (1996). *Relative contribution of decreased productivity and survival to recent changes in the demographic trends in the Rivière George caribou herd* (pp. 27–36), Special Issue 9, Rangifer.

Cumming, D. H., Fenton, M. B., Rautenbach, I. L., Taylor, R. D., Cumming, G. S., Cumming, M. S., Dunlop, J. M., Ford, A. G., Hovorka, M. D., Johnston, D. S., Kalcounis, M., Mahlangu, Z., & Portfors, C. V. R. (1997). Elephants, woodlands and biodiversity in southern Africa. *South African Journal of Science, 93*, 231–236.

Danell, Ö., Staatland, H., & Nieminen, M. (1999). Rennäring i Nordvest-Europa. In H. K. Dahle, Ö. Danell, E. Gaare, & M. Nieminen (Eds.), *Reindrfit i Nordvest-Europa i 1998- biologiske muligheter og begrensinger* (pp. 19–30). København: Nordisk ministerråd.

Danilkin, A. (1996). *Behavioural ecology of Siberian and European roe deer*. London/Glasgow/Weinheim/New York/Tokyo/Melbourne/Madras: Chapman & Hall.

Darling, F. F. (1960). *Wildlife in an African territory*. London/New York/Toronto: Oxford University Press.

Daubenmire, R. (1970). *Steppe vegetation of Washington* (Technical Bulletin 1962). Pullman: Washington Agricultural Experiment Station.

Day, T. A., & Detling, J. K. (1990). Grassland patch dynamics and herbivore grazing preference following urine deposition. *Ecology, 71*(1), 180–188.

De Chantal, M., Lindberg, H., & Kallonen, S. (2009). The condition and survival of Populus tremula and other deciduous saplings in a moose winter-foraging are in southern Finland. *Annales Botanici Fennnici, 46*, 280–290.

den Herder, M., Virtanen, R., & Roininen, H. (2004). Effects of reindeer browsing on tundra willow and its associated insect herbivores. *Journal of Applied Ecology, 41*, 870–879.

den Herder, M., Virtanen, R., & Roininen, H. (2008). Reindeer herbivory reduces willow growth and grouse forage in a forest-tundra ecotone. *Basic and Applied Ecology, 9*, 324–331.

Denzau, G., & Denzau, H. (1999). *Wildesel*. Stuttgart: Jahn Torbecke Verlag.

Despain, D., Houston, D., Meagher, M., & Schullery, P. (1986). *Wildlife in transition: Man and nature on Yellowstone's northern range*. Boulder: Roberts Rinehart.

Dolman, P., Fuller, R., Gill, R., Hooton, D., & Tabor, R. (2010). Escalating ecological impacts of deer in lowland woodland. *British Wildlife, 10*, 242–254.

Douglas, H. W., Drysdale, V. M., & Glover, P. E. (1964). The chemical composition of Themedra triandra and Cynodon dactylon. *East African Wildlife Journal, 2*, 82–121.

Dublin, H. T. (1995). Vegetation dynamics in the Serengeti-Mara ecosystem: The role of elephants, fire and other factors. In A. R. E. Sinclair & P. Arcese (Eds.), *Serengeti II* (pp. 71–90). Chicago/London: University of Chicago Press.

Dublin, H. T., Sinclair, A. R. E., & McGlade, J. (1990). Elephants and fire as causes of multiple stable states in the Serengeti-Mara woodlands. *Journal of Animal Ecology, 59*, 1147–1164.

Edroma, E. L. (1981). The role of grazing in maintaining high species composition in Imperata grassland in Rwenzori National Park, Uganda. *African Journal of Ecology, 19*, 215–233.

Ehrich, D., Henden, J.-A., Ims, R. A., Doronina, L., Killengren, S. T., Lecomte, N., Pokrovsky, I. G., Skogstad, G., Sokolov, A. A., Sokolov, V. A., & Yoccoz, N. (2012). The importance of willow thickets for ptarmigan and hares in shrub tundra: The more the better. *Oecologia, 168*(1), 141–151.

Eiberle, K. (1978). The consequential effects of simulated deer browsing damage on the development of young forest trees. *Schweizerische Zeitschrift für Forstwesen, 129*, 757–763.

Ellenberg, H. (1978) Zur Populationsökologie des Rehes (Capreolus capreolus L., Cervidae) in Mitteleuropa. *Spixiana München*, Suppl 2, 1–121.

Ellenberg, H. (1982). *Vegetation Mitteleuropas mit den Alpen* (3rd ed.). Stuttgart: Verlag Eugen Ulmer.

Eltringham, S. K. (1999). *The hippos. Natural history and conservation*. London/San Diego: Academic.

Engstrom, D. R., Whitlock, C., Fritz, S. C., & Wright, H. E. (1991). Recent environmental changes inferred from the sediments of small lakes in Yellowstone's northern range. *Journal of Paleoclimatology, 5*, 139–174.

Esseen, P.-A., Ehnström, B., Ericson, L., & Sjöberg, K. (1997). Boreal forests. *Ecological Bulletins, 46*, 16–47.

Esslinger, H. (1988). *Ski Heil! ARD Sendereihe Globus*. Beiheft 1 zur Sendung vom 10. Januar 1988.

Estes, J. A., Terborgh, J., Brashares, J. S., Power, M. E., Berger, J., Bond, W. J., Carpenter, S. R., Essington, T. E., Holt, R. D., Jackson, J. B., Marquis, R. J., Oksanen, L., Oksanen, T., Paine, R. T., Pikitch, E. K., Ripple, W. J., Sandin, S. A., Scheffer, M., Schoener, T. W., Shurin, J. B., Sinclair, A. R. E., Soulé, M. E., Virtanen, R., & Wardle, D. A. (2011). Trophic downgrading of Planet Earth. *Science, 333*, 301–306.

Evans, R. (1995). *Impacts of reindeer grazing in soils and vegetation in Finnmark* (NORUT infomasjonsteknologi. Rapport 2003/02-94). Tromsö: Northern Research Institute.

Evans, R. (1996). Some impacts of overgrazing by reindeer in Finnmark, Norway. *Rangifer, 16*, 3–19.

Fahnestock, J. T., & Knapp, A. K. (1994). Plant response to selective grazing by bison: Interactions between light, herbivory and waterstress. *Vegetatio, 115*, 123–131.

Ferrar, A. A., & Walker, B. H. (1974). An analysis of herbivore/habitat relationships in Kyle National Park, Rhodesia. *Journal of South African Wildlife Management Association, 4*(3), 137–147.

Fischer, A. (1999). Der Einfluß des Schalenwildes auf die Bodenvegetation. In M. Suda, A. Fischer, W. Hockenjos, F. Straubinger, G. Meister, & G. Sperber (Eds.), *Waldökosystem und Schalenwild* (pp. 23–44). Rothenburg o. d. Tauber: Ökologischer Jagdverein Bayern e.V.

Flook, D. R. (1962). Range relationships of some ungulates native to Banff and Jasper National Parks, Alberta. In J. D. Criso (Ed.), *Grazing in terrestrial and marine environments* (pp. 119–128). Oxford: Blackwell.

FOCUS Online. (2012, August 6). Wilderer haben es auf Zoo Nashörner abgesehen. FOCUS online. www.focus.de/...r-haben-es-auf-zoo-nashoerner-abgesehen_aid_794528.htm

Ford, J. (1966). The role of elephants in controlling the distribution of tsetse flies. *Bulletin of the International Union for Conservation of Nature, 19*, 6.
Frank, D. A., & McNaughton, J. (1993). Evidence for the promotion of aboveground grassland production by native large herbivores in Yellowstone National Park. *Oecologia, 96*, 157–161.
Franklin, J. F., & Dyrness, C. T. (1973). *Natural vegetation of Oregon and Washington* (USDA Forests Service, General Technical Report PNW 8). Portland: Oregon State University Press.
French, D. D., Miller, G. R., & Cummins, R. P. (1997). Recent development of high-altitude Pinus sylvestris in the northern Cairngorm Mountains, Scotland. *Biology and Conservation, 79*, 133–144.
Fuhlendorf, S. D., & Engle, D. M. (2004). Application of the fire-grazing interaction to restore a shifting mosaic on tallgrass prairie. *Journal of Ecology, 41*, 604–614.
Fuller, R. J. (2001). Response of woodland birds to increasing numbers of deer: A review of evidence and mechanisms. *Forestry, 74*, 289–298.
Fuller, R. J. A., & Gill, R. M. A. (2001). Ecological impacts of increasing numbers of deer in British woodland. *Forestry, 74*, 193–199.
Galbraight, W. A., & Anderson, E. W. (1971). Grazing history of the Northwest. *Range Management, 24*(1), 6–12.
Gampe, S. (1989). *Über Sanierungsaufforstungen in den Bayerischen Alpen* (Mitteilungen aus der Staatsforstverwaltung Bayerns 44). Regensburg: Staatsforstverwaltung Bayerns.
Garrot, R. A., Ehite, P. J., Becker, M. S., & Gower, C. N. (2009). Apparent competition and regulation in a wolf-ungulate system: Interactions of life history characteristics, climate, and landscape attributes. In R. A. Garrott, P. J. White, & F. G. R. Watson (Eds.), *The ecology of large mammals in central Yellowstone* (pp. 520–540). Amsterdam/Boston/Heidelberg/London/New York/Oxford/Paris/San Diego/San Francisco/Singapore/Sydney/Toronto: Academic.
Gates, C., & Aune, K. (2008). Bison bison. In *IUCN Red List of Threatened Species*. Version 2013.2, www.iucnredlist.org. Accessed 19 Jan 2014.
Gates, C. C., Freese, C. H., Gogan, P. J. P., & Kotzman, M. (Eds. and comps.). (2010). *American bison. Status and conservation guidelines*. Gland: IUCN.
Geiser, R. (1992). Auch ohne Homo sapiens wäre Mitteleuropa von Natur aus eine halboffene Weidelandschaft. *Laufener Seminarbeiträge, 2*, 22–34.
Geissen, H.-P. (1996). *Welche Natur wollen wir? Naturschutz auf naturgeschichtlicher Grundlage* (Bibliothek Natur und Wissenschaft 9). Solingen: Verlag Natur und Wissenschaft.
Geist, V. (1974). On the relationship of social evolution and ecology in ungulates. *American Zoologist, 14*, 205–220.
Gibbson, R. (2004). Examining the extinction of the Pleistocene megafauna. *Stanford Undergraduate Research Journal, 3*, 222–227.
Gill, R. M. A. (1990). *Monitoring the status of European and North American cervids* (GEMS Information Series 8). Nairobi: Global Environmental Monitoring System, United Nations Environment Programme.
Gill, R. M. A. (1992). A review of damage by mammals in north temperate forests: 3. Impact on trees and forests. *Forestry, 65*(4), 363–388.
Gill, R. (2000, August). *The impact of deer on woodland biodiversity* (pp. 1–6). Edinburgh: Forestry Commission.
Gill, R. M. A., & Beardall, V. (2001). The impact of deer on woodlands: The effects of browsing and seed dispersal on vegetation structures and composition. *Forestry 74*(3), Special Issue: 209–218.
Glandt, D., & Geiger, A. (1990). Reptilienschutz in Nordrhein-Westfalen. Situation – Forschungsstand – Probleme – Maßnahmen. *NZ/NRW Seminarbericht, 9*, 9–13.
Grabherr, G., & Pauli, H. (1994). Climate effects on mountain plants. *Nature, 369*, 448.
Groenmann-Van Wateringe, W. (1993). The effects of grazing on the pollen production of grasses. *Vegetation History and Archaeobotany, 2*, 157–162.
Guthörl, V. (1991). Rehwildverbiß und Waldvegetation. *Allgemeine Forstzeitung, 4*, 175–177.
Haapasaari, M. (1988). The oligotrophic heath vegetation of northern Fennscandia and its zonation. *Acta Botanica Fennica, 135*, 1–219.
Hagemoen, R. I. M., & Reimers, E. (2004). Reindeer summer activity pattern in relation to weather and insect harassment. *Journal of Ecology, 71*, 883–892.

Hall, K., Boelhouwers, J., & Driscoll, K. (1999). Animals as erosion agents in the alpine zone: Some data and observations from Canada, Lesotho, and Tibet. *Arctic, Antarctic, and Alpine Research, 31*(4), 436–446.

Hamilton, R. G. (1996). Using fire and bison to restore a functional tallgrass prairie landscape. In *Transactions 61st North American wildlife and natural resources conference* (pp. 208–214). Oklahoma City.

Hansen, H. L., Krefting, L. W., & Kurmis, V. (1973). *The forest of Isle Royale in relation to fire history and wildlife* (Minnesota Agricultural Experiment Station, Technical Bulletin 294). St. Paul: Agricultural Experiment Station, University of Minnesota.

Hanson, J. R. (1984). Bison ecology in the northern plains and a reconstruction of bison patterns for North Dakota region. *Plains Anthropologist, 29*(104), 93–113.

Harding, P. T., & Rose, F. (1986). *Pasture-woodland in lowland Britain*. Huntingdon: Institute of Terrestrial Ecology.

Hartnett, D. C., Hickman, K. R., & Fischer, W. L. E. (1996). Effects of bison grazing, fire, and topography on floristic diversity in tallgrass prairie. *Journal of Range Management, 49*, 413–420. Baltimore: American Society of Range Management.

Hatton, J. C., & Smart, N. O. E. (1984). The effect of long-term exclusion of large herbivores on soil nutrient status in Murchison Falls National Park, Uganda. *African Journal of Ecology, 22*, 23–30.

Haugen, A. D., & Shult, M. J. (1972). Approximate pre-white-man animal influences and relationships in prairie natural areas. *Proceedings of the Midwest Prairie Conference, 2*, 17–19.

Hawkes, R. K. (1991). *Crop and livestock losses to wild animals in the Bulilima-Mangwe natural resources management project area* (Center for Applied Social Sciences Report). Harare: University of Zimbabwe.

Heady, H. F. (1968). Grassland response to changing animal species. *Journal of Soil and Water Conservation, 23*(5), 173–176.

Heikkilä, R., & Härkönen, S. (1996). Moose browsing in young Scots pine stands in relation to forest management. *Forest Ecology Management, 88*, 179–186.

Heikkinen, R. K., & Kalliola, R. J. (1989). *Vegetation types and maps of Kevo nature reserve, northernmost Finland* (Kevo Notes 8). Turku: Kevo Subarctic Research Institute of the University of Turku.

Helle, T. (1966). An investigation of reindeer husbandry ib Finland. *Fennia, 95*(4), 65.

Helle, T. (2005). Mountain birch forests and reindeer husbandry. In F. F. Wielgolaski (Ed.), *Nordic mountain birch ecosystems* (Man and the Biosphere Series 27, pp. 279–291). New York/London: UNESCO and The Parthenon Publishing Group.

Helle, T., & Aspi, J. (1983). Effects of winter grazing by reindeer on vegetation. *Oikos, 40*, 337–343.

Helle, T., & Kojola, I. (1993). Reproduction and mortality of Finnish semi-domesticated reindeer in relation to density and management strategies. *Arctic, 46*(72), 681–688.

Helle, T., & Kojola, I. (2006). Population trends of semi-domesticated reindeer in Fennoscandia – Evaluation of explanations. In B. C. Forbs, M. Bölter, L. Müller-Wille, J. Hukkinen, F. Müller, N. Gunslay, & Y. Konstantinov (Eds.), *Reindeer management in northernmost Europe: Linking practical and scientific knowledge in social-ecological systems* (pp. 319–339). Berlin: Springer.

Helle, T., & Tarvainen, I. (1984). Determination of the winter digging period of semi-domestic reindeer in relation to snow conditions and food resources. *Reports from the Kevo Subarctic Research Station, 19*, 49–56.

Henden, J.-A., Ims, R. A., Yoccoz, N. G., & Killengreen, S. T. (2011). Declining willow ptarmigan populations: The role of habitat structure and community dynamics. *Basic and Applied Ecology, 12*, 413–422.

Henry, G. H. R., & Svoboda, J. (1993). Composition of grazed and non-grazed high Arctic sedge meadows. In J. Svoboda & B. Freedman (Eds.), *Ecology of a polar oasis: Alexandra Fjord, Ellesmere Island, Canada* (pp. 193–194). Toronto: Campus University.

Herbold, H., Suchentrunk, F., Wagner, S., & Willing, R. (1992). Einfluss anthropogener Störreize auf die Herzfrequenz von Rotwild (Cervus elaphus) und Rehwild (Capreolus capreolus). *Zeitschrift für Jadgwissenschaft, 38*, 145–159.

Herre, W. (1955). *Das Ren als Haustier. Eine zoologische Monographie*. Leipzig: Akademische Verlagsgesellschaft.
Hickie, P. F. (1936). Isle Royale moose studies. *Proceedings of the North American Wildlife Conference, 1*, 396–399.
Hirst, S. M. (1975). Ungulate-habitat relationships in a South African woodland/savanna ecosystem. *Wildlife Monographs, 44*, 5–69.
Histøl, T., & Hjeljord, O. (1993). Winter feeding strategies of migrating and nonmigrating moose. *Canadian Journal of Zoology, 71*, 1421–1428.
Hoare, R. (1999). Determinants of human-elephant conflict in a land-use mosaic. *Journal of Applied Ecology, 36*, 689–700.
Hoare, R. (2000). African elephants and humans in conflict: The outlook for co-existence. *Oryx, 34*(1), 34–38.
Hobbs, N. T. (1996). Modification of ecosystems by ungulates. *Journal of Wildlife Management, 60*(4), 695–713.
Hofmann, R., Cornelius, R., Linder, R., & Scheibe, K. (1998). Wiedererstehen des postglazialen Großsäugerspektrums als landschaftsgestaltendes Multi-Spezies-System: ein Pilotversuch zur "Redynamisierung" halboffener Landschaften. *Schriftenreihe für Landschaftspflege und Naturschutz, 56*, 301–315.
Holdo, R. M. (2003). Woody plant damage by African elephants in relation to leaf nutrients in western Zimbabwe. *Journal of Tropical Ecology, 19*, 189–196.
Holdo, R. M. (2007). Elephant, fire, and frost can determine community structure and composition in Kalahari woodlands. *Ecological Applications, 17*(4), 558–568.
Holdo, R. M., Dudley, J. P., & McDowell, L. R. (2002). Geophagy in the African elephant in relation to availability of dietary sodium. *Journal of Mammalogy, 83*, 652–664.
Holland, E. A., & Detling, J. K. (1990). Plant response to herbivory and below-ground nitrogen cycling. *Ecology, 71*, 1040–1049.
Holland, E. A., Parton, W. J., Detling, J. K., & Coppock, D. L. (1992). Physiological response of plant populations to herbivory and their consequences for ecosystem nutrient flow. *American Naturalist, 140*, 685–706.
Hollemann, D. F., & Luick, J. R. (1977). Lichen species preferred by reindeer. *Canadian Journal of Zoology, 55*, 1368–1369.
Holtmeier, F.-K. (1967a). Die Verbreitung der Holzarten im Oberengadin unter dem Einfluß des Menschen und des Lokalklimas. *Erdkunde, 21*(4), 249–258.
Holtmeier, F.-K. (1967b). Zur natürlichen Wiederbewaldung aufgelassener Alpweiden im Overengadin. *Wetter und Leben, 10*, 195–200.
Holtmeier, F.-K. (1969a). Das Steinwild in der Landschaft von Pontresina. *Natur und Museum, 99*(1), 15–24.
Holtmeier, F.-K. (1969b). Die Landschaft von Pontresina im Luftbild. *Erdkunde, 23*(2), 133–142.
Holtmeier, F.-K. (1974). *Geoökologische Beobachtungen und Studien an der subarktischen und alpinen Waldgrenze in vergleichender Sicht (nördliches Fennoskandien/Zentralalpen)* (Erdwissenschaftliche Forschung VIII). Wiesbaden: Franz Steiner Verlag.
Holtmeier, F.-K. (1978). Die bodennahen Winde in den Hochlagen der Indian Peaks Section, Colorado Front Range, USA. *Münstersche Geographische Arbeiten, 3*, 3–47.
Holtmeier, F.-K. (1979). Die polare Waldgrenze (forest-tundra-ecotone) in geoökologischer Sicht. In L. Müller-Wille & H. Schroeder-Lanz (Eds.), *Kanada und das Nordpolargebiet* (Trierer Geographische Arbeiten, Sonderheft 2, pp. 230–252). Trier: Geographische Gesellschaft Trier/ Fachgruppe Geographie der Universität Tier.
Holtmeier, F.-K. (1982). "Ribbon-forest" und "Hecken". Streifenförmige Verbreitungsmuster des Baumwuchses an der oberen Waldgrenze in den Rocky Mountains. *Erdkunde, 36*, 142–153.
Holtmeier, F.-K. (1987a). Biber und Steinwild. Ökologische Aspekte der Wiederansiedlung von Tieren in ihren ehemaligen Lebensräumen. *Münstersche Geographische Arbeiten, 26*, 99–117.
Holtmeier, F.-K. (1987b). Beobachtungen und Untersuchungen über den Ausaperungsverlauf und einige Folgewirkungen in "ribbon-forests" and der oberen Waldgrenze in der Front Range, Colorado. *Phytocoenologia, 15*(3), 373–396.

Holtmeier, F.-K. (1987c). Der Baumwuchs als klimaökologischer Faktor an der oberen Waldgrenze. *Münstersche Geographische Arbeiten, 27*, 145–151.

Holtmeier, F.-K. (1996). Die Wirkungen des Windes in der subalpinen und alpinen Stufe der Front Range, Colorado, U.S.A. *Arbeiten aus dem Institut für Landschaftsökologie, Westfälische Wilhelms-Universität, Münster, 1*, 19–45.

Holtmeier, F.-K. (2002). *Tiere in der Landschaft. Einfluss und ökologische Bedeutung*. Stuttgart: UTB 8230.

Holtmeier, F.-K. (2005). Change in the timberline ecotone in northern Finnish Lapland during the last thirty years. *Reports from the Kevo Subarctic Research Station, 23*, 97–113.

Holtmeier, F.-K. (2009). *Mountain timberlines. Ecology, patchiness, and dynamics* (Advances in Global Change Research 36). Dordrecht: Springer Science+Business Media B. V.

Holtmeier, F.-K. (2012). Impact of wild herbivorous mammals and birds on the altitudinal and northern treeline ecotones. *LandscapeOnline, 30*, 1–28. doi:10.3097/LO.201230.

Holtmeier, F.-K., & Broll, G. (2005). Sensitivity and response of northern hemisphere altitudinal and polar treelines to environmental change at landscape and local scales. *Global Ecology and Biogeography, 14*, 395–410.

Holtmeier, F.-K., & Broll, G. (2006). Radiocarbon-dated peat and wood remains from the Finnish Subarctic – Evidence of treeline and landscape history. *The Holocene, 16*(5), 743–751.

Holtmeier, F.-K., & Broll, G. (2010). Wind as an ecological agent at treelines in North America, the Alps, and the European Subarctic. *Physical Geography, 31*(3), 203–233.

Holtmeier, F.-K., & Broll, G. (2011). Response of scots pine (Pinus sylvestris) to warming climate at its altitudinal limit in northernmost subarctic Finland. *Arctic, 64*(3), 269–280.

Holtmeier, F.-K., Broll, G., Müterthies, A., & Anschlag, K. (2003). Regeneration of trees in the treeline ecotone: Northern Finnish Lapland. *Fennia, 181*(2), 103–128.

Holtmeier, F.-K., Broll, G., & Anschlag, K. (2004). Winderosion und ihre Folgen im Waldgrenzbereich und in der alpinen Stude einiger nordfinnischer Fjelle. *GeoÖko, 25*, 204–224.

Hörnberg, S. (2001). Changes in population density of moose (Alces alces) and damage to forest in Sweden. *Forest Ecology and Management, 149*, 141–151.

Houston, D. B. (1971). Ecosystems of natural parks. *Science, 172*, 648–651.

Houston, D. B. (1976). The status of research on ungulates in northern Yellowstone Park. In *Research in the parks. Transactions of National Parks Centennial Symposium, 1971* (National Park Service Syposium Series 1, pp. 11–27). Washington, DC: U.S. Government Printing Office.

Houston, D. B. (1982). *The northern Yellowstone elk-ecology and management*. New York: Macmillan.

Hustich, I. (1966). On the forest-tundra and the northern tree-lines. *Reports from the Kevo Subarctic Research Station, 3*, 7–47.

Illies, J. (1971). *Einführung in die Tiergeographie*. Stuttgart: Fischer.

Ingham, R. E., & Detling, J. K. (1984). Plant-herbivore interactions in a North American mixed-grass prairies. III. Soil nematode populations and root biomass on cynomis ludovicianus colonies and adjacent uncolonized areas. *Oecologica, 63*, 307–313.

Isenberg, A. C. (2000). *The destruction of the bison. An environmental history, 1750–1920*. Cambridge: Cambridge University Press.

Jackson, T. P., Mosojane, S., Ferreira, S. M., & Van Aarde, R. J. (2008). Solutions for elephant Loxodonta Africana crop raiding in northern Botswana: Moving away from symptomatic approaches. *Oryx, 42*(1), 83–91.

Jarman, P. J., & Sinclair, A. R. E. (1979). Feeding strategy and the pattern of resource partitioning in ungulates. In A. R. E. Sinclair & M. Norton-Griffiths (Eds.), *Sergengeti: Dynamics of an ecosystem* (pp. 130–163). Chicago: University of Chicago Press.

Johnson, C. F., Cowling, R. M., & Phillipson, P. B. (1999). The flora of the Addo Elephant National Park, South Africa: Are threatened species vulnerable to elephant damage? *Biodiversity and Conservation, 8*, 1447–1456.

Jones, M. J. (1973). The organic matter content of the savanna soils of West Africa. *Journal of Soil Science, 24*, 42–53.

Kalliola, R. (1939). Pflanzensoziologische Untersuchungen in der alpinen Stufee Finnisch-Lapplands. *Annales Botanici Societatis Zoologicae-Botanicae Fennicae 'Vanamo', 13*(2), 1–326.

Kashulina, G., Reinmann, C., Finne, T. E., Halleraker, J. H., Äyräs, M., & Chekushin, V. A. (1997). The state of the ecosystems in the Central Barents Region: Scale, factors and mechanisms of disturbance. *The Science of Total Environment, 206*, 203–225.

Kauffman, M. J., Brodie, J. F., & Jules, E. S. (2010). Are wolves saving Yellowstone's aspen? A landscape-level test of a behaviorally mediated trophic cascade. *Ecology, 91*(9), 2742–2755.

Kaule, G. (1991). *Arten- und Biotopschutz* (2nd ed.). Stuttgart: Verlag Eugen Ulmer.

Käyhkö, J. (2007). Aeolian blowout dynamics in subarctic Lapland based on decadal levelling investigations. *Geografiska Annaler, 89A*(1), 65–81.

Käyhkö, J., & Pellikka, P. (1994). Remote sensing of the impact of reindeer grazing on vegetation in northern Fennoscandia using SPOT XS data. *Polar Research, 13*, 115–124.

Kelsall, J. A. (1968). *The migratory barren-ground caribou of Canada* (Canadian Wildlife Service Monographs 3). Ottawa: Minister of Indian Affairs and Northern Development.

Kennel, E. (1998). Was kann das Vegetationsgutachten zum nachhaltigen Management eines waldverträglichen Schalenwildbestandes leisten? Vorschlag zur Bewertung von Verbissbefunden. *Berichte ANL, 22*, 51–58.

Kerley, G. I. H., & Boshoff, A. (1997). *A proposal for a greater Addo National Park and natural conservation development opportunity* (Terrestrial Ecology Unit Report 17). Port Elizabeth: University of Port Elizabeth.

Kershaw, K. A., & MacFarlane, J. D. (1980). Physiological-environmental interactions in lichens. X. Light as an ecological factor. *New Phytologist, 84*, 687–702.

Kienast, F., Wildi, O., & Brzeziecki, B. (1998). Potential impacts of climate change on species richness on mountain forests – An ecological risk assessment. *Biological Conservation, 83*, 291–305.

Kinser, A., Koope, K., & Freiherr von Münchhausen, H. (2010). Die Rotwildverbreitung in Deutschland. *AFZ-Der Wald, 5*, 32–34.

Klein, D. R., & Bay, C. (1990). Foraging dynamics of muskoxen in Peary Land, northern Greenland. *Holarctic Ecology, 13*, 269–280.

Knapp, R. (1965). *Die Vegetation von Nord- und Mittelamerika und der Hawaii-Inseln*. Stuttgart: Gustav Fischer-Verlag.

Knapp, A. K., & Seastedt, T. R. (1986). Detritus accumulation limits productivity of tallgrass prairie. *BioScience, 36*, 662–668.

Knaup, H., & Puhl, J. (2012). Bringt sie zur Strecke. *DER SPIEGEL, 37*, 102–103.

Koch, P. L., & Barnosky, A. D. (2006). Late Quaternary extinction: State of the debate. *Annual Review of Ecology, Evolution, and Systematics, 37*, 215–250.

Kojola, I., Helle, T., Aikio, P., & Niskanen, M. (1992, October 6–8). Carrying capacity of reindeer ranges in northern Lapland. In E. Tikkanen, M. Varmola, & T. Katermaa (Eds.), *Extended abstracts*. Symposium on the state of the environmental monitoring in Northern Fennoscandinavia and Kola Peninsula, Rovaniemi (pp. 288–289).

Kojola, I., Helle, T., Niskanen, M., & Aikio, P. (1995). Effects of lichen biomass on winter diet, body mass and reproduction of semi-domesticated reindeer Rangifer t. tarandus in Finland. *Wildlife Biology, 1*(1), 33–38.

Kouki, J., Arnold, K., & Martikainen, P. (2004). Long-term persistence of aspen – A key host for many threatened species – Is endangered in old-growth conservation areas in Finland. *Journal for Nature Conservation, 12*, 41–52.

Kraft, U., Kräusslich, H., Grauvgl, A., Bogner, H., & Alps, H. (1990). Ethologische Untersuchungen zum Baumschälen durch Rinder. *Landwirtschaftliches Jahrbuch, 67*(3), 251–310.

Krefting, L. W. (1974). *The ecology of Isle Royal moose with special reference to the habitat* (Agricultural Experiment Station, State University of Minnesota, Technical Bulletin 297, Forestry Series 15), Minneapolis.

Krueger, K. (1986). Feeding relationships among bison, pronghorn, and prairie dogs: An experimental analysis. *Ecology, 67*, 760–770.

Krüsi, B. O., Schütz, M., Wildi, O., & Grämiger, H. (1995). Huftiere, Vegetationsdynamik und botanische Vielfalt im Schweizerischen Nationalpark. *Cratschla/Mitteilungen aus dem Schweizerischen Nationalpark 3*(2), 1–25, Zernez.

Kujala, V. (1950). Über Kleinpilze der Koniferen in Finland. Ascomycetes, Fungi imperfecti, Uredinales. *Communicationes Instituti Forestalis Fenniae, 38*(4), 124pp.

Kullman, L. (2005). Wind-conditioned 20th century decline of birch treeline vegetation in the Swedish Scandes. *Arctic, 58*(3), 286–294.

Kumpula, J., & Nieminen, M. (1992, October 6–8). The inventory of reindeer pastures in Finland by satellite imageries. In E. Tikkanen, M. Varmola, & T. Katermaa (Eds.), *Extended abstracts: Symposium on the state of the environment and environmental monitoring in Northern Fennoscandia and Kola Peninsula*, Rovaniemi, pp. 294–296.

Kumpula, J., Colpaert, A., & Nieminen, M. (1998). Reproduction and productivity of semidomesticated reindeer in northern Finland. *Canadian Journal of Zoology, 76*, 269–277.

Kumpula, J., Stark, S., & Holand, Ø. (2011). Seasonal grazing effects by semi-domesticated reindeer on subarctic mountain birch forests. *Polar Biology, 34*, 441–453.

Kurt, F. (1977). *Wildtiere in der Kulturlandschaft*. Erlenbach/Zürich/Stuttgart: Eugen Rentsch Verlag.

Kurt, F. (1991). *Das Reh in der Kulturlandschaft*. Hamburg/Berlin: Verlag Paul Parey.

Lamprey, H. F. (1963). Ecological separation of the large mammal species in the Tarangire Game Reserve, Tanganyika. *East African Wildlife Journal, 1*, 63–92.

Lamprey, H. F., Gover, P. E., Turner, M. I. M., & Bells, R. H. V. (1967). Invasion of the Serengeti National Park by elephants. *East African Wildlife Journal, 5*, 151–166.

Larcher, W. (1980). *Ökologie der Pflanzen auf pyhsiologischer Grundlage* (3rd ed.). Stuttgart: Verlag Eugen Ulmer.

Larsen, D. W., & Kershaw, K. A. (1976). Studies on lichen-dominated systems. XVII: Morphological control of evaporation in lichens. *Canadian Journal of Botany, 54*, 2061–2073.

Larter, N. C., & Nagy, J. A. (2001). Variation between snow conditions at Peary caribou and muskox feeding sites and elsewhere in foraging habitats on Banks Island in the Canadian High Arctic. *Arctic, Antarctic, and Alpine Research, 33*(2), 123–130.

Lavsund, S. (1987). Moose relationships to forestry in Finland, Norway and Sweden. *Swedish Wildlife Research Supplement, 1*, 229–246.

Lavsund, S., Nygren, T., & Solberg, E. J. (2003). Status of moose populations and challenges to moose management in Fennoscandia. *Alces, 39*, 109–130.

Laws, R. M. (1968). Interactions between elephant and hippopotamus populations and their environment. *East African Agricultural and Forestry Journal, 33*(Special Issue), 140–147.

Laws, R. M. (1970). Elephants as agents of habitat and landscape change in East Africa. *Oikos, 21*, 1–15.

Lehtonen, J. (1987). Recovery and development of birch forests damaged by Epirrita autumnata in Utsjoki area. *Reports from the Kevo Subarctic Research Station, 20*, 35–39.

Lehtonen, J., & Heikkinen, R. K. (1995). On the recovery of mountain birch after Epirrita damage in Finnish Lapland, with a particular emphasis on reindeer grazing. *Ecoscience, 2*, 349–356.

Lempa, K., Neuvonen, S., & Tømmervik, H. (2005). Effects of reindeer grazing on pastures – A necessary basis for sustainable reindeer herding. In F. Wielgolaski (Ed.), P. S. Karlsson, S. Neuvonen, & D. Thannheiser (Ed. board). *Plant Ecology, herbivory, and human impact in Nordic mountain birch forests* (pp. 157–164). Berlin/Heidelberg: Springer.

Leuthold, W. (1978). Kann Afrikas Großwild überleben? *Vierteljahresschrift der Naturforschenden Gesellschaft Zürich, 123*, 333–340.

Leuthold, W. (1995). Langfristige Veränderungen der Vegetation im Tsavo-National Park, Kenya. *Vierteljahresschrift der Naturforschenden Gesellschaft Zürich, 140*(4), 163–1971.

Leuthold, W. (1996). Recovery of woody vegetation in Tsavo National Park, Kenya. *African Journal of Ecology, 34*, 101–112.

Lewin, R. (1986). In ecology, change brings stability. *Science, 234*, 1071–1073.

Liss, B.-M. (1988). *Versuche zur Waldweide - der Einfluß von Weidevieh und Wild auf Verjüngung, Bodenvegetation und Boden im Bergmischwald der ostbayerischen Alpen* (Forstliche Forschungsberichte München 87). München: Verlag Frank.

Liss, B.-M. (1990). Beweidungseffekte im Bergwaldgebiet. Ergebnisse aus fünfjährigen Untersuchungen zur Waldweide unter besonderer Berücksichtigung des Wildverbisses. *Beiheft zum Forstwissenschaftlichen Centralblatt, 40*, 50–65.

Lock, J. M. (1972). The effects of hippopotamus grazing on grasslands. *Journal of Ecology, 60*, 445–467.

Löffler, J. (2000). High mountain ecosystems and landscape degradation in Northern Norway. *Mountain Research and Development, 20*(4), 358–363.

Löffler, J. (2005). Reindeer grazing impact in arctic-alpine landscapes in western Greenland, and central and northern Norway. *Geoöko, 26*, 1–18.

Löyttyniemi, K. (1985). On repeated browsing of Scots pine saplings by moose Alces alces. *Silva Fennica, 19*, 387–391.

Lovaas, A. L. (1973). A cooperative elktrapping program in Wind Cave National Park. *Wildlife Society Bulletin, 1*, 93–100.

Löw, H., & Mettin, C. (1977). Der Hochlagenwald im Werdenfelser Land. *Forstwissenschaftliches Centralblatt, 96*, 108–120.

Luckman, B. H., & Kavanagh, T. A. (1998). Documenting the effects of recent climate change at treeline in the Canadian Rockies. In M. Beniston & J. K. Innes (Eds.), *Impact of climate variability on forests* (pp. 121–144). Heidelberg: Springer.

Luschekina, A., & Struchkov, A. (2001). The saiga antelope in Europe: Once again on the brink? *The Open Country, 3*, 11–24.

MacMahon, J. A. (1981). Successional processes: Comparison among biomes with special reference to probable role of and influence on animals. In H. Shugart, D. Botkin, & P. West (Eds.), *Forests succession: Concept and application* (pp. 277–304). New York/Heidelberg/Berlin: Springer.

Magombe, J. K., & Campbell, K. (1989). *Poaching survey* (Serengeti Wildlife Research Center Bi-annual Report 1988–89).

Maisels, F., Strindberg, S., Blake, S., Wittemyer, G., Hart, J., Williamson, E. A., Aba'a, R., Abitsi, G., Ambahe, R. D., Amsini, F., Bokkabana, P. C., Hicks, T. C., Bayogo, R. E., & Warren, Y. (2013). Devastation decline of forest elephants in Central Africa. *PLoS One, 8*(3), e5949. doi:10.1371/journal.pone.0059469.

Malanson, G. P., Butler, D. R., Fagre, D. B., Walsh, S. J., Tomback, D. F., & Daniels, L. D. (2007). Alpine treeline of western North America: Linking organism-to landscape dynamics. *Physical Geography, 28*, 378–396.

Mallon, D. P., & Jiang, Z. (2009). Grazers on the plains: Challenges and prospects for large herbivores in Central Asia. *Journal of Applied Ecology, 46*, 516–519.

Manseau, M., Huot, J., & Crête, M. (1996). Effects of summer grazing by caribou on composition and productivity of vegetation: community and landscape level. *Journal of Ecology, 84*, 503–513.

Martin, E., & Vigne, L. (2003). Trade in rhino horn from eastern Africa to Yemen. *Pachyderm, 34*, 75–87.

Mattson, J. (1995). Human impact on the timberline in the North of Europe. *Prace Geograficzne, 98*, 41–44.

May, T. (1993). Beeinflussten Großsäuger die Waldvegetation der pleistozänen Warmzeiten Mitteleuropas? Ein Diskussionsbeitrag. *Natur und Museum, 123*(6), 157–170.

Mayer, H. (1973). Möglichkeiten und Grenzen der Schalenwildhege im Gebirgswald. *Zeitschrift des Schweizerischen Forstvereins, Beiheft, 52*, 90–118.

Mayer, H. (1975). Der Einfluß des Schalenwildes auf die Verjüngung und Erhaltung von Naturwaldreservaten. *Forstwissenschaftliches Centralblatt, 94*, 209–222.

McHugh, T. (1972). *The time of the buffalo*. New York.

McNaughton, S. J. (1979) Grassland-herbivore dynamics. In A. R. E. Sinclair & M. Norton-Griffiths (Eds.), *Dynamics of ecosystems* (pp. 46–81). Chicago: University of Chicago Press.

McNaughton, S. J. (1983a). Serengeti grassland ecology: The role of composite environmental factors and contingency in community organization. *Ecological Monographs, 53*(3), 291–320.

McNaughton, S. J. (1983b). Compensatory plant growth as response to herbivory. *Oikos, 40*, 329–336.

McNaughton, S. J. (1985). Ecology of grazing ecosystems: The Serengeti. *Ecological Monographs, 55*(3), 260–294.

McNaughton, S. J. (1988). Mineral nutrition and spatial concentration of African ungulates. *Nature, 334*, 343–345.

McNaughton, S. J., Riess, R. W., & Seagle, S. W. (1988). Large mammals and process dynamics in African ecosystems. *BioScience, 38*, 794–800.

McNaughton, S. J., Sala, O. E., & Oesterheld, M. (1993). Comparative ecology of African and South American arid to subhumid ecosystems. In P. Goldblatt (Ed.), *Biological relationships between Africa and South America* (pp. 548–567). New Haven: Yale University Press.

McShea, W. J., & Rappole, J. H. (2000). Managing the abundance and diversity of breeding bird populations through manipulation of deer populations. *Conservation Biology, 12*, 1161–1170.

Mech, L. D., McRoberts, R. E., Peterson, R. O., & Page, R. E. (1987). Relationship of deer and moose populations to previous winter's snow. *Journal of Animal Ecology, 56*, 6115–6627.

Merrill, E. H., Beyer, H. L., Jones, G. P., & McDaniel, G. W. (2003). Deciduous woodland conservation under heavy deer browsing on Devils Tower National Monument, USA. *Journal for Nature Conservation, 10*, 221–232.

Middleton, A., Morrison, T. A., Fortin, J. K., Robbins, C. T., Proffitt, K. M., White, P. J., McWhirter, D. E., Koel, T. M., Brimeyer, D. G., Fairbanks, W. S., & Kauffman, M. J. (2013). Grizzly bear predation links the loss of native trout to the demography of migratory elk in Yellowstone. *Proceedings of the Royal Society B 280*(1762).

Miles, J., & Kinnaird, J. W. (1979). Grazing: with particular reference to birch, Juniper and cots Pine in the Scottish Highlands. *Scottish Forestry, 33*, 280–289.

Milner-Guland, E. J. (1994). A population model for the management of the saiga antelope. *Journal of Applied Ecology, 31*, 25–39.

Milner-Guland, E. J., Bekenov, A. B., & Grachov, Y. A. (1995). The real threat to saiga antelopes. *Nature, 377*, 488–489.

Milne, L. J., & Milne, M. (1960). *The balance of nature*. New York: Alfred A. Knopf.

Miquelle, D. G., & Van Ballenberghe, V. (1989). Impact of bark stripping by moose on aspen-spruce communities. *Journal of Wildlife Management, 53*(3), 557–568.

Mitchell, F. J. G. (1990). The impact of grazing and human disturbance on the dynamics of woodland in S. W. Ireland. *Journal of Vegetation Science, 1*, 245–254.

Mitchell, F. J. G. (2005). How open were European primeval forests? Hypothesis testing using palaeoecological data. *Journal of Ecology, 93*, 168–177.

Mitchell, F. J. G., & Cole, E. (1998). Reconstruction of long-term successional dynamics of temperature woodland in Bialowieza Forest, Poland. *Journal of Ecology, 86*, 1042–1061.

Mitchell, F., & Kirby, K. (1990). The impact of large herbivores on the conservation of semi-natural woodlands in the British uplands. *Forestry, 63*, 333–353.

Mlot, C. (1991). The trouble with deer. *BioScience, 41*, 754–755.

Moe, S. R., Rutina, L. P., Hytteborn, H., & du Toit, J. T. (2009). What controls woodland regeneration after elephants have killed big trees? *Journal of Applied Ecology, 46*, 223–230.

Molvar, E. M., Bowyer, R. T., & Van Ballenberghe, V. (1993). Moose herbivory, browse quality, and nutrient cycling in an Alaskan treeline community. *Oecologia, 94*, 472–479.

Moolman, H. J., & Cowling, R. M. (1994). The impact of elephant grazing on the endemic flora of South African succulent thicket. *Biological Conservation, 88*, 53–61.

Moorhead, D. L., Fisher, F. M., & Whitford, W. C. (1988). Cover of spring annuals on nitrogen-rich Kangaroo rat mounds in a Chihuahuan desert grassland. *The American Midland Naturalist, 120*(2), 443–447.

Mosegulo, D. K., Moe, S. R., Ringrose, S., & Nellemann, C. (2002). Vegetation changes during a 36-year period in northern Chobe National Park, Botswana. *African Journal of Ecology, 40*, 232–240.

Moser, O. (1988). Verbiß durch Schalenwild - tragbar oder waldzerstörend? *Österreichische Forstzeitung, 11*, 51–52.

Moser, O., Widhalm, H., & Hochbichler, E. (1989). Schutzwaldsanierung und Wildbewirtschaftung in Österreich auf neuen Wegen? *Allgemeine Forstzeitung, 9–10*, 242–244.

Motta, R. (1996). Impact of wild ungulates on forest regeneration and the composition of mountain forests in the Western Italian Alps. *Forest Ecology and Management, 88*, 93–98.
Mulkey, S. S., Smith, A. P., & Young, T. P. (1984). Predation by elephants on Senecio keniodendron (Compositae) in the alpine zone of Mount Kenya. *Biotropica, 16*(3), 246–248.
Murie, A. (1934). *The moose of Isle Royale* (University of Michigan, Museum of Zoology, Miscellaneous Publications 25). Ann Arbor: University of Michigan Press.
Murray, N. (1976). *The love of elephants*. London: Octopus Books.
Musselman, L. K. (1969). *Rocky Mountain National Park, 1915–1965: An administrative history*. Dissertation, University of Denver.
Myrberget, S. (1987). Introduction of mammals and birds in Norway including Svalbard. *Meddelser fran Norstk Viltforskning, 3*, 1–26.
Nagy, L., & Grabherr, G. (2009). *The biology of alpine habitats* (The Biology of Habitats Series). Oxford/New York: Oxford University Press.
National Audubon Society. (1992). *Beringia natural history* (Notebook Series). Anchorage: National Audubon Society.
Nellemann, C., Moe, S. R., & Rutina, L. P. (2002). Links between terrain characteristics and forage pattern of elephants (Loxodonta africana) in northern Botswana. *Journal of Tropical Ecology, 18*, 835–844.
Neuvonen, S., & Wielgolaski, F. E. (2005). Herbivory in northern birch forests. In F. E. Wielgolaski (Ed.), *Plant ecology, herbivory, and human impact in Nordic mountain birch forests* (pp. 183–189). Berlin/Heidelberg: Springer.
New York Times. (2012a, September 8). Elephant slaughter. *New York Times*. www.nytimes.com/2012/09/opinion/sunday/elephant-slaughter.html?_r=0
New York Times. (2012b, November 17). Elephants dying in epic frenzy as ivory fuels wars and profits. *New York Times*. www.nytimes.com/2012/09/04/world/africa/africas-elephants-are-being-slaughteres-in poachin-frenzy-html?pagewanted=all&_r=0
Ngene, S., Ihwagi, F., Nzisa, M., Mukeka, J., Njumbi, S., & Omodi, P. (2011). *Total aerial census of elephants and other large mammals in the Tsavo-Mkomazi ecosystem*. Nairobi: Kenya Wildlife Service.
Ngene, S., Njumbi, S., Nzisa, M., Kimitei, K., Mukeka, J., Muya, S., Ihwagi, F., & Osmondi, P. (2013). Status and trends of the elephant population in the Tsavo-Mkomazi ecosystem. *Pachyderm, 53*, 38–50.
Niemelä, P., Hagman, M., & Lehtilä, K. (1989). Relationships between Pinus sylvestris L. origin and browsing preference by moose in Finland. *Scandinavian Journal of Forest Research, 4*, 239–246.
Nikula, A. (1992). Animals as forest pests in Finnish-Lapland. *Finnish Forest Research Institute, Research papers, 451*, 22–29.
Norton-Griffith, M. (1979). The influence of grazing, browsing, and fire on the vegetation dynamics of the Serengeti. In A. R. E. Sinclair & M. Norton-Griffith (Eds.), *Serengeti. Dynamics of an ecosystem* (pp. 310–352). Chicago/London: University of Chicago Press.
Norut. (1999). *Endringer I vegetasjondekket på Finnmarksvidda*. http://itek.norut.no/vegetasjon/10fi/10fi.htm.Norut
Nygren, T. (1987). The history of moose in Finland. *Swedish Wildlife Research*, (Suppl. 1), 49–54.
Ojasti, J. (1992). Ungulates and large rodents of South America. In F. Bourlière (Ed.), *Tropical savannas* (Ecosystem of the World 13, pp. 427–439). Amsterdam: Elsevier.
Oksanen, L. (1978). Lichen grounds of Finnmarksvidda, northern Norway, in relation to summer and winter grazing by reindeer. *Reports from the Kevo Subarctic Research Station, 14*, 64–71.
Oksanen, L., Moen, J., & Helle, T. (1995). Timberline patterns in northernmost Fennoscandia. Relative importance of climate and grazing. *Acta Botanica Fennica, 153*, 93–107.
Olofsson, J., Oksanen, L., Callaghan, T., Hulme, P. E., Oksanen, T., & Suominen, O. (2009). Herbivores inhibit climate-driven shrub expansion on the tundra. *Global Change Biology, 15*, 2681–2693.

Olson, K. A., Mueller, T., Bolortsetseg, S., Leimgruber, P., Fangan, W. F., & Fuller, T. K. (2009). A mega-herd of more than 200 000 Mongolian gazelles Procapra gutturosa: A consequence of habitat quality. *Oryx, 43*, 149–153.

Omondi, O., Bitok, E. K., Mukeka, J., Mayienda, R. M., & Litoroh, M. (2008). *Total aerial count of elephants and other large mammal species of Tsavo/Mkomazi ecosystem*. Nairobi: Kenya Wildlife Service.

Onderscheka, K. (1986). Auswirkungen der Umweltveränderungen auf das Rot-,Reh- und Gamswild. *Alpine Umweltprobleme, Ergebnisse des Forschungsprojektes Achenkirch, Bd. A 98*, 31–35).

Ossenbrink, R. (1996). *Standortuntersuchungen in einer Fjellheide im nördlichen Finnisch-Lappland unter besonderer Berücksichtigung des Einlusses der Rentiere auf die Vegetation*. Diploma thesis, Institut für Landschaftsökologie, Westfälische Wilhelms-Universität, Münster.

Ost, E., & Pedersen, C. (2008). Opposing plant community responses to warming with and without herbivores. *Proceedings of the National Academy of the Science USA, 105*(34), 12353–12358.

Ott, E., Frehner, M., Frey, H.-U., & Lüscher, P. (1997). *Gebirgsnadelwälder. Ein praxisorientierter Leitfaden für eine standortgerechte Waldbehandlung*. Bern/Stuttgart/Wien: Verlag Paul Haupt.

Otto, H.-J. (1994). *Waldökologie*. Stuttgart: Verlag Eugen Ulmer.

Owen-Smith, N. (1982). Factors influencing the consumption of tropical savannas. In B. J. Huntley & B. H. Walker (Eds.), *Ecology of tropical savannas* (Ecological Studies 42, pp. 539–404). Berlin/New York: Springer.

Owen-Smith, N. (Ed.). (1983). *Management of large mammals in African conservation areas*. Pretoria: HAUM.

Owen-Smith, N. (1987). Pleistocene extinctions of the pivotal role of megaherbivores. *Paleobiology, 13*(3), 351–362.

Owen-Smith, R. N. (1988). *Megaherbivores: The influence of very large body size on ecology* (Cambridge Studies on Ecology). Cambridge/New York: Cambridge University Press.

Packer, P. E. (1963). Soil stability requirement for the Gallatin elk winter range. *Journal of Wildlife Management, 27*, 401–410.

Parker, I. S. C. (1983). The Tsavo story: An ecological case history. In N. Owen-Smith (Ed.), *Management of large mammals in African conservation areas* (pp. 37–49). Pretoria: HAUM.

Parks, C. G., Bednar, L., & Thiedemann, A. R. (1998). Browsing ungulates – An important consideration in dieback and mortality of Pacific yea (Taxus brevifolia) in a northeastern Oregon stand. *Northwest Science, 72*(3), 190–197.

Parton, W. J., & Risser, P. G. (1980). Impact of management practices on the tallgrass prairie. *Oecologia, 46*, 223–234.

Pastor, J., Naiman, R. J., Dewey, B., & McInnes, P. (1988). Moose, microbes, and the boreal forest. *BioScience, 38*, 770–777.

Pauchard, A., Kueffer, C., Dietz, H., Daehler, C. C., Alexander, J., Edwards, P. J., Arevalo, J. R., Cavieres, L. A., Guisan, A., Haider, S., Jacobs, G., McDougall, K., Millar, C. I., Naylor, B. J., Parks, C. G., Rew, L. J., & Seidl, T. (2009). Ain't no mountains high enough: Plant invasions reaching new elevations. *Frontiers in Ecology and the Environment, 7*(9), 479–486.

Pauli, J. N., & Buskirk, S. W. (2007). Risk-disturbance overrides density dependence in a hunted colonial rodent, the black-tailed prairie dog Cynomys ludovicianus. *Journal of Applied Ecology, 44*, 1219–1230.

Pégau, R. E. (1970). Succession in two exclosures near Uvalakleet, Alaska. *The Canadian Field Naturalist, 84*, 175–177.

Pégau, R. E. (1975). *Analysis of the Nelchina caribou range*. Biological Papers, University of Alaska, Special Report 1, pp. 316–323.

Pelto, P. (1973). *The snowmobile revolution: Technology and social change in the Arctic*. Menlo Park, California: Cummings Publishing Co.

Pentzhorn, B. L., Robertse, P. J., & Olivier, M. C. (1974). The influence of the African elephant on vegetation of the Addo Elephant National Park. *Koedoe, 17*, 137–158.

Perrins, C. M., & Overall, R. (2001). Effect of increasing numbers of deer on bird populations in Wytham Woods, central England. *Forestry, 74*, 299–309.

Perry, D. A. (1994). *Forest ecosystems*. London/Baltimore: Johns Hopkins University Press.

Persson, I.-L., Danell, K., & Bergström, R. (2000). Disturbance by large herbivores in boreal forests with special reference to moose. *Annales Zoologici Fennici, 37*, 251–263.

Peterson, R. O. (1983). Wolf management and harvest patterns on the Kenai Wildlife Refuge, Alaska. In L. N. Carbyn (Ed.), *Wolves in Canada and Alaska. Proceedings of the Wolf Symposium, Edmonton, 12–14 May 1981* (Canadian Wildlife Service, Report Series 45, pp. 96–99).

Petrak, M. (1988). Skilanglauf und Rothirsch (Cervus elaphus Linné, 1758) in der Eifel. *Bilanz des Konzeptes zur Lebensraumberuhigung im Wintersportgebiet Monschau. Zeitschrift für Jagdwissenschaft, 34*, 104–114.

Pfister, F., Walther, H., Erni, V., & Candrian, M. (1987). *Walderhaltung und Schutzaufgaben im Berggebiet*. Eidgenössische Anstalt für das forstliche Versuchswesen (EAFV), Berichte 294: Birmensdorf.

Polley, H. W., & Collins, S. L. (1984). Relationships of vegetation and environment in buffalo wallows. *American Midland Naturalist, 112*, 178–186.

Pott, R. (1993). *Farbatlas Waldlandschaften*. Stuttgart: Verlag Eugen Ulmer.

Pott, R. (1995). The origin of grassland plant species and grassland communities in Central Europe. *Fitosciologica, 29*, 7–32.

Pott, R. (1996). *Biotoptypen, schützenswerte Lebensräume Deutschlands und angrenzender Regionen*. Stuttgart: Verlag Eugen Ulmer.

Pott, R. (1997). Von der Urlandschaft zur Kulturlandschaft – Entwicklung und Gestaltung mitteleuropäischer Kulturlandschaften durch den Menschen. *Verhandlungen der Gesellschaft für Ökologie, 27*, 5–26.

Pott, R. (1998). Effects of human interference on the landscape with special reference to the role of grazing livestock. In M. F. Wallis de Vries, J. P. Bakker, & S. H. Van Wieren (Eds.), *Grazing and conservation management* (pp. 107–134). Dordrecht: Kluwer Academic Publishers.

Pott, R., & Hüppe, R. (1991). *Die Hudelandschaften Nordwestdeutschlands* (Abhandlungen aus dem Westfälischen Museum für Naturkunde 53 (1/2)). Münster: LWL-Museum für Naturkunde.

Potvin, F., Beaupré, P., & Laprise, G. (2003). The eradication of balsam fir stands by white-tailed deer on Anticosti Island, Québec: A 150 year process. *Ecoscience, 10*, 448–495.

Prins, H. H. T., & Van der Jeugd, H. P. (1993). Herbivore population crashes and woodland structure in East Africa. *Journal of Ecology, 81*, 305–314.

Quellet, J. P., Heard, D. C., & Boutin, S. (1993). Range impacts following the introduction of caribou on Southampton Island (Northwest Territories, Canada. *Arctic and Alpine Research, 25*(2), 136–141.

Raillard, M., & Svoboda, J. (1989). Muskox winter feeding strategies at Sverdrup pass, Ellesmere Island, N. W. T. *The Muskox, 37*, 86–92.

Raillard, M., & Svoboda, J. (2000). High grazing impact, selectivity, and local density of Muskoxen in central Ellesmere Island, Canadian High Arctic. *Arctic, Antarctic, and Alpine Research, 32*(3), 278–285.

Reimoser, F. (1999). Schalenwild und Wintersport. *Laufener Seminarbeiträge, 6*, 39–45.

Reimoser, F., Mayer, H., Holzinger, A., & Zandl, J. (1987). Einfluß von Sommer- und Wintertourismus auf Waldschäden durch Schalenwild im Angertal (Bad Gastein). *Centralblatt für das gesamte Forstwesen, 104*, 95–118.

Relva, M. A., & Kitzberger, T. (2000, April 2–7). A tree rind-based reconstruction of fraying and bark stripping damage by exotic deer in northern Patagonia, Argentine. In *Conference on dendrochronology for the third millennium*, Mendoza (p. 261).

Remmert, H. (1980). *Ökologie*. Berlin/Heidelberg/New York: Springer.

Remmert, H. (1988). Gleichgewicht durch Katastrophen. *Aus Forschung und Medizin, 3*(1), 7–17.

Remmert, H. (1992). Das Mosaikzyklus-Konzept und seine Bedeutung für den Naturschutz -eine Übersicht. *Laufener Seminarbeiträge, 2/92*, 45–57.

Remmert, H., & Zell, R. A. (1984). Tiere der Urzeit: Ausgestorben oder ausgerottet? *Bild der Wissenschaft, 21*(9), 50–51.

Repo, T., Nilsson, J.-E., Rikala, R., Ryyppö, A., & Sutinen, M.-L. (2001). Cold hardiness of Scots pine (Pinus sylverstris). In F. J. Bigras & S. J. Colombo (Eds.), *Conifer cold hardiness* (Tree Physiology 1, pp. 463–493). Dordrecht/Boston/London: Kluwer.

Richardson, D. H. S., & Young, C. M. (1977). Lichens and vertebrates. In M. R. Seaward (Ed.), *Lichen ecology* (pp. 121–144). London: Academic Press.

Richter, H. (1988). Rotwild und Biotokapazität. *Allgemeine Forstzeitung, 44*, 1184–1185.

Riggs, R. A., Tiedemann, A. R., Cook, J. G., Ballard, T. M., Eggerton, P. J., Vara, M., Krueger, W. C., Hall, F. C., Bryant, L. D., Irwin, L. L., & Delucto, T. (2000). *Modification of mixed-conifer forest by ruminant herbivores in the Blue Mountains ecological province* (Research Paper PNW-RP 570). Portland: USDA Forest Service, Pacific Northwest Research Station.

Ripple, W. J., & Beschta, R. K. (2003). Wolf reintroduction, predation risk, and cottonwood recovery in Yellowstone National Park. *Forest Ecology and Management, 184*, 299–313.

Ripple, W. J., & Beschta, R. L. (2004). Wolves and the ecology of fear: Can predation risk structure ecosystems? *BioScience, 54*(8), 755–766.

Risch, A. C., Krüsi, B. O., Grämiger, H., & Schütz, M. (2001). Spatially specific simulation of the long-term development of a subalpine pasture in the Swiss National Park. *Bulletin des Geobotanischen Instituts ETH, 67*, 27–40.

Risenhoover, K. L., & Maass, S. A. (1987). The influence of moose on the composition and structure of Isle Royale forests. *Canadian Journal of Forest Research, 17*, 357–364.

Roe, F. G. (1970). *The north American buffalo* (2nd ed.). Toronto: University of Toronto Press.

Rooney, T. P., & Waller, D. M. (2003). Direct and indirect effects of white-tailed deer in forest ecosystems. *Forest Ecology and Management, 181*, 165–176.

Rösch, K. (1992). *Einfluß der Beweidung auf die Vegetation des Bergwaldes. Nationalpark Berchtesgaden* (Forschungsbericht 26). Berchtesgaden: Nationalparkverwaltung Berchtesgaden.

Rotenhan, S. Frh. von. (1990). Stellungnahme zur Wald und Schalenwild: Aus der Sicht der Arbeitsgemeinschaft Naturgemäße Waldwirtschaft. *Allgemeine Forstzeitung, 4*, 102–103.

Russel, D. E., & Martell, A. M. (1984). Winter range ecology of caribou (Rangifer tarandus). In R. Olson, R. Hastings, and F. Geddes (Eds.), *Northern ecology and resource management. Memorial essays honouring Don Gill* (pp. 117–124). Edmonton: University of Alberta Press.

Russell, F. L., Zippin, D. B., & Fowler, N. L. (2001). Effects of white-tailed deer (Odocoileus virgianus) on plants, plant populations and communities: A review. *American Midland Naturalist, 146*, 1–26.

Ryyppö, A., Repo, T., & Vappaavuori, E. (1998). Development of freezing tolerance in roots and shoots of scots pine seedlings at non-freezing temperatures. *Canadian Journal of Forest Research, 28*, 557V–565V.

Sasaki, T., & Yoshihara, Y. (2013). Local-scale disturbance by Siberian marmots has little influence on regional plant richness in a Mongolian grassland. *Plant Ecology, 214*, 29–34.

Schaller, G. B. (1995). *Der letzte Panda*. Rowohlt: Reinbek bei Hamburg.

Schaller, G. B. (2003). Sterben für die Mode. Warum die Tschirus gejagt werden. *National Geographic, 9*, 148–149.

Schauer, T. (1982). *Die Belastung des Bergwaldes durch Schalenwild, Waldweide und Naturschutz* (Laufener Seminarberichte 9/82). Laufen: Bayerische Akademie für Naturschutz und Lanschaftspflege (ANL).

Scherzinger, W. (1996). *Naturschutz im Wald, Qualitätsziele einer dynamischen Waldentwicklung*. Stuttgart: Ulmer.

Schlund, G. (1988). Rotwild and Biotopkapazität. In *AFZ-Diskussion: Ökosystemgerechte Jagd - Stellungnahme zu Paul Müller in AFZ Nr. 27–28* (pp. 761–772).

Schmidt, K. (1993a). Winter ecology of nonmigratory Alpine red deer. *Oecologia, 95*, 226–233.

Schmidt, M. (1993b). Naturschutzmanagement – Nowendigkeit oder Übereifer? In Landesamt für Umweltschutz Baden Württemberg (Ed.), *Landschaftspflege - quo vadis?* (pp. 138–142). Karlsruhe.

Schmidt, O. (1999). *Stellungnahme zur "Megaherbivorentheorie" aus forstlicher Sicht*. Duplicated manuscript, Bayerische Landesanstalt für Wald- und Forstwirtschaft Freiburg.

Scholes, R. J., & Hall, D. O. (1996). The carbon budget of tropical savannas, woodlands and grasslands. In A. I. Breymeyer, D. O. Hall, J. M. Mellilo, & G. I. Agren (Eds.), *Global change effects on coniferous forests and grasslands* (pp. 69–100). London: Wiley.

Scholes, R. J., & Walker, B. H. (1993). *An African savanna: Synthesis of the Nylsvley study*. Cambridge: Cambridge University Press.

Schönenberger, W. (1986). Wiederaufforstung eines Bergwaldes. *Neue Zürcher Zeitung, 38*, 35–36.
Schönenberger, W., Frey, W. & Leuenberger, F. (1990). Ökologie und Technik der Aufforstung im Gebirge. Anregungen für die Praxis. Eidgenössische Anstalt für das forstliche Versuchswesen, Berichte 325. Teufen. F. Flück-Wirth
Schröder, W. (1972). *Funktion und Situation von großen Pfanzenfressern in Waldökosystemen Mitteleuropas*. Gesellschaft für Ökologie, Tagung Gießen 1972, Tagungsbericht Belastung und Belastbarkeit von Ökosystemen (pp. 55–88).
Schreiber, K.- F. (2000). Überlegungen zum Einfluss der Großwildfauna auf die Landschaft im Holozän. Rundgespräch der Kommission für Ökologie 18, *Entwicklung der Umwelt seit der letzten Eiszeit* (pp. 77–89). München: Dr. Friedrich Pfeil.
Schröder, W., Wotschikowski, U., & Georgii, B. (1984). Besser schonen durch klügeres Schießen. *Natur, 12*, 52–71.
Schüle, W. (1992). Vegetation, megaherbivores. Man and climate in the Quaternary and the genesis of closed forest. In J. G. Goldammer (Ed.), *Tropical forests in transition* (pp. 45–56). Basel: Birkhäuser Verlag.
Schullery, P., & Whittlesey, L. (1992). The documentary record of wolves and related wildlife species in the Yellowstone Park area prior to 1882. In J. D. Varley & G. Brewster (Eds.), *Wolves for Yellowstone? A report to the United States Congress, Vol. 4, Research and analysis* (pp. 1–173). Yellowstone: Yellowstone National Park Service.
Schuster, E. (1990). Exkurisonsbericht Zustand und Gefährdung der Bergwälder. In E. Schuster (Ed.), *Zustand und Gefährdung des Bergwaldes* (Forstwissenschaftliche Forschungen 40, Beiheft zum Forstwissenschaftlichen Centralblatt, pp. 11–18). Heidelberg: Springer
Schütz, M., Wildi, O., Achermann, G., Krüsi, B. O., & Nievergelt, B. (2000). Predicting the development of subalpine grassland in the Swiss National Park: How to build a succession model based on data from long-term permanent plots. *Nationalparkforschung in der Schweiz, 89*, 207–235.
Schwab, P., & Messner, L. (1987). Vorschläge zur Schalenwildregulierung. Ergebnisse des Forschungs- und Vesuchsprojektes "Alpine Umweltgestaltung" des Fonds für Umweltstudien (FUST) in Achenkirch. Tirol/Bonn. *Allgemeine Forstzeitung, 11*, 247–249.
Sedlag, U. (1995). *Tiergeographie*. Leipzig/Jena/Berlin: Urania Tierreich.
Seiler, A. (2005). Predicting locations of moose-vehicle collisions in Sweden. *Journal of Applied Ecology, 42*, 371–382.
Senn, J. (1999). Tree mortality caused by Greminiella abietina in a subalpine afforestation in the Central Alps and its relationship with duration of snow cover. *European Journal of Forest Pathology, 29*, 65–74.
Seppäla, M. (2004). *Wind as a geomorphic agent in cold climates* (Studies in Polar Research). Cambridge: Cambridge University Press.
Seton, E. (1929). *Lives of game animals*. Doubleday: New York to Garden City.
Shannon, G., Druce, D. J., Page, B. R., Eckhardt, H. C., Grant, R., & Slotov, R. (2008). The utilization of large savanna trees by elephant in southern Kruger 475 National Park. *Journal of Tropical Ecology, 24*, 281–289.
Shannon, G., Thaker, M., Vanak, A. T., Page, B. R., Grant, R., & Slotov, R. (2011). Relative impacts of elephant and fire on large trees in a savanna ecosystem. *Ecosystems, 14*, 1372–1381.
Shaw, J. H., & Lee, M. (1995). *Ecological interpretation of historical accounts of bison and fires on the southern plains with emphasis on tallgrass prairie* (Final Report to the Nature Conservation). Stillwater: Oklahoma State University.
Shoshani, J. (1993). Elephants the super keystone species. *Swara, 16*(2), 25–29.
Sims, P. L., & Singh, J. S. (1971). Herbage dynamics and net primary production in certain ungrazed and grazed grasslands in North America. In N. R. French (Ed.), *Preliminary analysis of structure and function in grasslands* (Range Science Department Scientific Series 10, pp. 56–124). Fort Collins: Colorado State University.
Sinclair, A. R. E. (1974a). The resource limitation of trophic levels in tropical grassland ecosystems. *Journal of Animal Ecology, 44*, 497–520.

Sinclair, A. R. E. (1974b). The natural regulation of buffalo population in East Africa. IV. The food supply as a regulating factor, and competition. *East African Wildlife Journal, 12*, 291–312.
Sinclair, A. R. E. (1975). The resource limitation of trophic levels in tropical grassland ecosystems. *Journal of Animal Ecology, 44*, 497–520.
Sinclair, A. R. E. (1977). *The African buffalo. A study of resource limitation of populations*. Chicago/London: University of Chicago Press.
Sinclair, A. R. E. (1995). Serengeti past and present. In A. R. E. Sinclair & P. Arcese (Eds.), *Serengeti* (pp. 3–30). Chicago/London: University of Chicago Press.
Singer, F. J. (Ed.). (1996). *Effects of grazing wild ungulates in Yellowstone National Park* (Technical Report NPS/NRYELL/NTR/96-01). Denver: USDI National Park Service, Natural Research Program Center, Natural Resource Information Division.
Singer, F. J., & Harter, M. K. (1996). Comparative effects of elk herbivory and the fires of 1988 on grassland in northern Yellowstone National Park. In F. J. Singer (Ed.), *Effects of grazing by wild ungulates in Yellowstone National Park* (Technical Report NPS/NRYELL/NRTR/96-01). Denver: USDI National Park Service, Natural Research Program Center, Natural Information Division.
Singer, F. J., Schreier, W., Oppenheim, J., & Garton, O. (1989). Drought, fire and large mammals. *BioScience, 39*, 716–722.
Singer, F. J., Mack, L. C., & Cates, R. C. (1994). Ungulate herbivory of willows on Yellowstone's northern winter range. *Journal of Range Management, 47*, 435–443.
Sitati, N. W., & Walpole, M. J. (2006). Assessing farm-based measures for mitigating human-elephant conflict in Transmara District, Kenya. *Oryx, 40*, 279–286.
Sitati, N. W., Walpole, M. J., & Leader-Williams, N. (2005). Factors affecting susceptibility of farms to crop raiding by African elephants: Using a predictive model to mitigate conflicts. *Journal of Applied Ecology, 42*, 1175–1182.
Skarpe, C., Aarrestad, P. A., Andreassen, H. P., Dhillion, S. S., Dimakatso, T., du Toit, J., Halley, D. J., Hytteborn, H., Makhabu, S., Mari, M., Marokane, W., Massunga, G., Modise, D., Moe, S. R., Mojapohko, R., Mosugelo, D., Motsumi, S., Neo-Mahupelneng, G., Ramotadima, M., Rutima, L., Sechele, L., Sejoe, T. B., Stokke, S., Swenson, J. E., Taolo, C., Vandewalle, M., & Wegge, P. (2004). The return of the giants: Ecological effects of an increasing elephant population. *Ambio, 33*, 276–282.
Skogland, T. (1985). The effect of density-dependent resource limitations on the demography of wild reindeer. *Journal of Animal Ecology, 54*, 359–374.
Skuncke, F. (1969). *Reindeer ecology and management in Sweden* (Biological Papers University of Alaska 8), Fairbanks.
Smart, N. O. E., Hatton, J. C., & Spence, D. H. N. (1985). The effect of long-term exclusion of large herbivores on vegetation on Murchison Falls National Park, Uganda. *Biological Conservation, 33*, 229–245.
Smith, D. W., & Tyers, D. B. (2008). The beavers of Yellowstone. *Yellowstone Science, 16*, 4–15.
Smith, D. W., Peterson, R. O., & Houston, D. B. (2003). Yellowstone after wolves. *BioScience, 53*, 330–340.
Snyder, J. D., & Janke, R. A. (1976). Impact of moose browsing on boreal-type forests of Isle Royale National Park. *The American Midland Naturalist, 95*(1), 79–95.
Sokolov, V. E. (1974). Saiga tartarica. *Mammalian Species, 38*, 1–4.
Soper, J. D. (1941). History, range, and home life of the northern bison. *Ecological Monographs, 11*, 347–412.
Stevens, D. R. (1980). *The deer and elk of Rocky Mountain National Park: a ten-year-study*. Est Park: National Park Service, Rocky Mountain National Park, ROMO-N-13.
Stiles, D. (2004). The ivory trade and elephant conservation. *Environmental Conservation, 31*(4), 309–321.
Stolter, C., Niemelä, P., Ball, J. P., Julkunen-Tiito, R., Vanhatalo, A., Danell, K., Varvikko, T., & Ganzhorn, J. U. (2009). Comparison of plant secondary metabolites and digestibility of three different boreal coniferous trees. *Basic and Applied Ecology, 10*, 19–26.
Stöcklin, J., & Körner, C. (1999). Recruitment and mortality of Pinus sylvestris near the nordic treeline: The role of climate change and herbivory. In A. Hogaard, J. P. Ball, K. Danell, &

T. V. Callaghan (Eds.), *Animal response to global change in the north* (Ecological bulletins 47, pp. 168–177). Copenhagen: Munksgaard International Publishers.

Strandgaard, S. (1982). *Factors affecting the moose population in Sweden during the 20th century with special attention to silviculture* (Report 8). Uppsala: Swedish University of Agricultural Sciences, Department of Wildlife Ecology.

Stuart-Hill, G. C. (1992). Effects of elephants and goats on the Kaffarian succulent thicket on the eastern Cape, South Africa. *Journal of Applied Ecology, 29*, 699–710.

Suda, M. (1990). Die Entwicklung der Schalenwildbestände im bayerischen Alpenraum seit Anfang des 19. Jahrhunderts. *Beiheft zum Forstwissenschaftlicheh Centralblatt, 40*, 30–39.

Suda, M., & Gundermann, E. (1994). *Auswirkungen und monetäre Bewertung von Wildschäden im Bereich wasserwirtschaftlicher Sanierungsflächen des Bayerischen Alpenraumes* (Forstliche Forschungsberichte München 143). München: Frank.

Summer, C. A., & Linder, R. L. (1978). Food habits of the black-tailed prairie dog in western South Dakota. *Journal of Range Management, 31*, 134–136.

Sundquist, M. K., Björk, R. G., & Molau, U. (2008). Establishment of boreal forest species in alpine dwarf-shrub heath in Subarctic Sweden. *Plant Ecology & Diversity, 1*(1), 65–75.

Suominen, O. (1999). Impact of cervid browsing and grazing on the terrestrial gastropod fauna in the boreal forests of Fennoscandia. *Ecography, 22*, 651–658.

Suominen, O., & Olofsson, J. (2000). Impacts of semi-domesticated reindeer on structure of tundra and forest communities in Fennoscandia: A review. *Annales Zoologici Fennici, 37*, 233–249.

Suominen, O., Danell, K., & Bergström, R. (1999). Moose, trees, and ground-living invertebrates: Indirect interactions in Swedish pine forests. *Oikos, 84*, 215–226.

Sutinen, M.-L., Ritari, A., Holapppa, T., & Kujala, K. (1997, June 10–12). Seasonal changes in soil temperature and in the frost hardiness of Scots pine under subarctic conditions. In I. K. Islander, E. A. Wright, J. K. Radke, B. S. Shmatt, O. H. Groenvelt, & L. B. Hinzman (Eds.), *International symposium on physics, chemistry and ecology of seasonally frozen soils*, Fairbanks (Cold Regions Research and Engineering Laboratory Special Report 97-10, pp. 513–517). Hanover.

Tchamba, M. N. (1995). The impact of elephant browsing on the vegetation in Waza National Park, Cameroon. *African Journal of Ecology, 33*, 184–193.

Teipner, C. L., Garton, E. O., & Nelson, L., Jr. (1983). *Pocket gophers in forest ecosystems* (General Technical Report INT-154). Ogden: USDA Forest Service, Intermountain Forest and Range Experiment Station.

Theurillat, J.-P., & Guisan, A. (2001). Potential impact of climate change on vegetation in the European Alps: A review. *Climate Change, 50*, 77–109.

Theurillat, J.-P., Felber, F., Geissler, P., Gobat, J.-M., Fierz, M., Fischlin, A., Küpfer, P., Schussel, A., Velutti, C., & Zhao, G.-F. (1998). Sensitivity of plant and ecosystems of the Alps to climate change. In P. Cebon, U. Dahinden, H. C. Davies, D. Imboden, & C. C. Jaeger (Eds.), *Views from the Alps: Regional perspectives on climate change* (pp. 225–308). Cambridge, MA: MIT Press.

Tinner, W., & Kaltenrieder, P. (2005). Rapid responses of high-mountain vegetation to early Holocene environmental changes in the Swiss Alps. *Journal of Ecology, 93*, 936–947.

Tolvanen, A., & Henry, G. H. R. (2000). Population structure of three dominant sedges and muskox herbivory in the High Arctic. *Arctic, Antarctic, and Alpine Research, 32*(4), 449–455.

Tremblay, J.-P., Huot, J., & Potvin, F. (2007). Density-related effects of deer browsing in the regeneration dynamics of boreal forests. *Journal of Applied Ecology, 44*, 552–562.

Tubbs, C. R. (1986). *The new forest*. London: Collins.

Uhlig, C., & Zink, A. (2006). Changes in organic horizon soil properties due to reindeer herding and changing management. In B. B. Forbes, M. Bölter, L. Müller-Wille, J. Hukkinen, F. Müller, N. Gunslay, & Y. Konstantinov (Eds.), *Reindeer management in northernmost Europe: Linking practical and scientifique knowledge in social-ecological systems* (Ecological studies 184, pp. 245–264). Berlin: Springer.

Ukkola, R. (1995). Trampling tolerance of plants and ground cover in Finnish Lapland, with an example from the Pyhätunturi National Park. In O. Heikkinen, B. Obrebska-Starkel, & S. Tuhkanen (Eds.), *Environmental aspects of the timberline in Finland and in the Polish Carpathians* (Prace Geograficzne-Zeszyt 98, pp. 91–110). Kraków: Uniwersitet Jagiellonski.

UNESCO World Heritage Centre. (2014). Serengeti National Park.
Väisänen, R. A. (1998). Current Research trends in mountain biodiversity in Europe. *Prineos, 151/152*, 131–156, Jaca.
Van de Vijver, C. A. D. M., Foley, C. A., & Olff, H. (1999). Changes in the woody component of an East African savanna during 25 years. *Journal of Tropical Ecology, 15*, 545–564.
Van Wieren, S. E. (1991). The management of populations of large mammals. In I. F. Spellenberg, G. M. Goldsmith, & M. G. Moser (Eds.), *The scientific management of temperate communities for conservation* (pp. 103–127). Oxford: Blackwell Scientific.
Van Wijngaarden, W. (1985). *Elephants-trees-grazers. Relationships between climate, soil, vegetation and larger herbivores in a semi-arid savanna ecosystem (Tsavo, Kenya).* Dissertation Landgebouwhogeschool te Wageningen, ITC Puclication 4.
Väre, H., Ohtonen, R., & Oksanen, J. (1995). Effect of reindeer grazing on understorey vegetation in dry Pinus sylvestris forests. *Journal of Vegetation Science, 6*, 523–530.
Väre, H., Ohtonen, R., & Mikkola, K. (1996). The effect and extent of heavy grazing by reindeer in oligotrophic pine heaths in northern Fennoscandia. In H. Väre (Ed.), *Relationship between spatial heterogeneity in vegetation communities and in belowground ecosystems* (Acta Universitatis Ouluensis A 275, pp. 1–14). Oulu: University of Oulu.
Vera, F. W. M. (2000). *Grazing ecology and forest history*. Wallingford/Oxford/New York: CABI Publishing.
Vesey-Fitzgerald, D. F. (1960). Grazing succession among East African game animals. *Journal of Mammalogy, 41*(2), 161–172.
Vinton, M. A., Hartnett, D. C., Finck, E. J., & Briggs, J. M. (1993). Interactive effects of fire, bison (Bison bison) and plant community composition in tallgrass prairie. *The American Midland Naturalist, 129*, 10–18.
Voser, P. (1987). Einfluß hoher Rothirschbestände auf die Vegetation im Unterengadin und im Münstertal, Kanto Graubünden. *Ergebnisse der Untersuchungen im schweizerischen Nationalpark, XVI*, 143–220.
Vucetich, J. A., Peterson, R. O., & Schaefer, O. L. (2002). The effect of prey and predator densities on wolf predation. *Ecology, 83*, 3003–3012.
Vucetich, J. A., Smith, D., & Strahler, D. R. (2005). Influence of harvest, climate and wolf predation on Yellowstone elk, 1961–2004. *Oikos, 111*, 259–270.
Wagner, H. (1993). "Natur total" – Biotoppflege oder Sukzession? In Landesamt für Umweltschutz Baden Württemberg (Ed.), *Kolloquium zur Standortsbestimmung und Entwicklung der Landschaftspflege, Tagungsbericht, 38–52* (Landesamt für Umweltschutz Baden Württemberg, pp. 38–53). Karlsruhe: Landesamt für Umweltschutz Baden Württemberg.
Walter, H., & Breckle, S. (1986). *Ökologie der Erde 3. Spezielle Ökologie der gemäßigten und arktischen Zonen Euro-Nordasiens*. Stuttgart: Fischer.
Wam, H. K., & Hjeljord, O. (2010). Moose summer and winter diets along a large scale gradient of forage availability in southern Norway. *European Journal of Wildlife Research, 56*, 745–755.
Wasswa, H., & Frentzen, C. (2013, March 5). Elephanten-Population wird in Uganda zur Plage. Die Welt. Hamburg/Berlin/Dresden: Axel Springer
Watt, A. S., & Jones, E. W. (1948). The ecology of the Cairngorms. Part 1. The environment and the altitudinal zonation. *Journal of Ecology, 36*(2), 283–304.
Western, D. (1989). The ecological role of elephants in Africa. *Pachyderm, 12*, 42–45.
Whicker, A. D., & Detling, J. K. (1988a). Ecological consequences of prairie dog disturbances. *BioScience, 38*(11), 778–785.
Whicker, A. D., & Detling, J. K. (1988b). Modification of vegetation structure and ecosystem processes by North American mammals. In M. J. A. Werger et al. (Eds.), *Plant form and vegetation structure – Adaption, plasticity and relation to herbivory. International symposium on vegetation structure* (Vol. 1, pp. 301–316). The Hague: SPB Academic Publishing.
White, L. M., & Brown, J. H. (1972). Nitrogen fertilization and clipping effects on green needlegrass (Stipa viridula Trin.): II Evaporation, water use efficiency, and nitrogen recovery. *Agronomy Journal, 64*, 487–490.

White, R. G., Bunnel, F. L., Gaare, E., Skogland, T., & Hubert, B. (1981). Ungulates on arctic ranges. In L. Bliss, O. W. Heal, & J. J. Moorhead (Eds.), *Tundra ecosystems: A comparative analysis* (The international biological programme 25, pp. 397–483). Cambridge: Cambridge University Press.

Widmann, P. (1991). Zur Synökologie von Wild und Waldvegetation. *Allgemeine Forstzeitung, 8*, 382–384.

Wielgolaski, F. E., Karlsson, P. S., Neuvonen, S., Thannheiser, D., Tømmervik, H., & Gautestad, A. O. (2005). The Nordic mountain birch ecosystem – Challenges to sustainable management. In F. E. Wielgolaski (Ed.), *Plant ecology, herbivory, and human impact in Nordic mountain birch forests* (Ecological Studies 180, pp. 344–356). Berlin/New York: Springer.

Wildlife Conservation Society. (2014). *Bison*. www.wcs.org/saving-wildlife/hoofedmammals/bison apex

Wilmanns, O. (1973). *Ökologische Planzensoziologie* (4th ed.). Heidelberg: Quelle & Meyer.

Wilmanns, O. (2005). Kältezeitliche Reliktpflanzen der Schwäbischen Alb: aktualistische Überlegungen zur prähistorischen Landschaft. Hoppea, Denkschrift der Regensburger Botanischen Gesellschaft 66 (Schönfelder-Festschrift), pp. 447–468.

Wing, L. D., & Buss, I. O. (1970). Elephants and forests. *Wildlife Monographs, 19*, 37–92.

Woodward, A., Schreiner, E. G., Houston, D. B., & Moorhead, B. B. (1994). Ungulate-forest relationships in Olympic National Park: Retrospective exclosure studies. *Northwest Science, 68*(2), 97–110.

Wotschikowsky, U., Simon, O., Elmauer, K., & Herzig, S. (2010). *Leitbild Rotwild. Wege für ein forschrittliches Management* (2nd actualized ed.). Hamburg: Deutsche Wildtier Stiftung.

Wright, R. G. (1992). *Wildlife research and management in the national parks*. Urbana/Chicago: University of Illinois.

Wydeven, A. P., & Dahlgren, R. B. (1985). Ungulate habitat relationships in Wind Cave National Park. *Journal of Wildlife Management, 49*, 805–813.

Yellowstone National Park. (1997). *Yellowstone's northern range complexity and change in a wild land ecosystem*. Mammoth Hot Springs: National Park Service.

Zeltner, J. (1979). Verarmung der Baumartenmischung durch übersetzte Rehwildbestände. *Schweizerische Zeitschrift für Forstwesen, 130*, 81–84.

Zhong, W., Wang, M., & Wan, X. (1999). Ecological management of Brandt's vole (Microtus brandti) in Inner Mongolia, China. In G. R. Singleton, L. A. Hinds, H. Leirs, & Z. Zhang (Eds.), *Ecologically-based management of rodents pests* (pp. 199–214). Canberra: Australian Center for International Agricultural Research.

Zöckler, C., Liles, L., Fish, L., Wolf, A., Rees, G., & Danks, F. (2008). Potential impact of climate change and reindeer density on tundra indicator species in the Barents Sea region. *Climate Change, 87*, 119–130.

Zoller, H., & Haas, J. N. (1995). War Mitteleuropa ursprünglich eine halboffene Weidelandschaft oder von geschlossenem Wald bedeckt? *Schweizerische Zeitschrift für Forstwesen, 146*(5), 321–354.

Chapter 4
Keystone Species

4.1 The Term 'Keystone Species'

So-called keystone species are of particular interest from the view of landscape ecology and biocoenology (e.g., Müller 1977; Remmert 1980; Odum 1991). The term was coined by Paine (1969). A few years before (Paine 1966), this author had already described the function of keystone species in the intertidal zone, without calling these species explicitly keystone species, however. This term is associated with the idea of the 'last stone' that has to be inserted into the apex of an arch to give it stability.

It has been demonstrated in the previous chapters of this book that many vertebrate species (mammals as well as birds), for example, that play a minor role in biomass production and flows of matter and energy, may, however, have important effects on habitat structure, species composition and biogeochemical processes through physical disturbances, predator-prey relationships, competition and interactions with other species. This undoubtedly holds true, for example, for prairie dogs, ground squirrels, pocket gophers and kangaroo rats, all of them influencing vegetation and soils, in particular. By burrowing, food hoarding (seeds, plant material) and deposition of feces and urine in their burrow systems, they create microhabitats that completely differ from undisturbed places. Numerous fungi, algae, microarthropods and also larger animals profit from the organic matter (Seastedt et al. 1986, Hawkins and Nicoletto 1992). Mites, collemboles and nematodes are more frequent in mounds compared to the off-mound sites, and the invertebrates are a source of food for predators, such as lizards, for example. Rattle snakes and other vertebrates use abandoned burrows for shelter. Burrows still in use, food chambers and defecating places of blind mole rats (*Spalax ehrenbergii*) provide habitats for many invertebrate species (Heth 1991). In Cedar Creek Natural History Area (Minnesota), Huntly and Inouye (1988) found a close relationship between the density of pocket gopher mounds and the abundance of grasshoppers (mainly *Melanoplus* sp.), for example, which partly resulted from favourable conditions for oviposition on the mounds and on abundant forage supply by vegetation in sites disturbed by the gophers.

F.-K. Holtmeier, *Animals' Influence on the Landscape and Ecological Importance*,
DOI 10.1007/978-94-017-9294-3_4

African elephants, beavers, moose and musk oxen, as well as nutcrackers, may have lasting effects on their entire habitats, ecosystem stability, structure, function and dynamics. If these species were missing, the ecosystem character would completely change. The same applies to Southern cassowary (*Casuarius casuarius*; Sect. 2.5) living in the forests of northeastern Australia and Papua New Guinea. The increasing clearing and fragmentation of the forests in the coastal lowland and high losses of cassowaries due to traffic accidents, hunting and, not least, wild pigs, which destroy cassowary nests and kill the young birds (Crome and Bentruppenbaumer 1993), will bring about profound change in these forest ecosystems (Bennett 1999), as the spread of many tree species relying on seed dispersal by the cassowary will fail. On the other hand, as the cassowary has lost much of its original habitat, it is increasingly invading towns, where this ratite is considered a pest.

Only, a few species even create their habitat by themselves, such as, for example, termites and beavers. These species and their function cannot be replaced by other species. The role of nutcrackers, for example, in the Swiss stone pine – larch forests (Holtmeier 1966, 2009, 2012) could not be fulfilled by the other seed dispersers, such as squirrels, woodpeckers, mice or voles living in the same ecosystem. When a beaver population declines, their habitat and its biocoenosis will undergo fundamental change.

While many animal species (mammals, birds, invertebrates) are acting as 'ecosystem engineers', the term 'keystone species' in its proper sense should be restricted to cases where ecosystem stability, structure, function and dynamics are determined by only one species (Müller 1977). In so doing, the numbers of 'keystone species' will be considerably reduced. Three examples of keystone species will be considered in more detail in the following sections.

4.2 Three Selected Examples of Keystone Species' Influence on the Landscape

4.2.1 Termites

Termites are colonial insects, like ants. Some termite groups may considerably influence landscape character and ecological conditions. Numerous genera of termites are distributed over South and North America, Africa, Asia and Australia. Most of them live in the tropics. However, termites can also be found outside the tropics up to 48° N and 45° S (Harris 1961; Lal 1987; Goudie 1988). In the tropics, termites represent up to 95 % of all terricolous insects (Eggleton et al. 1996; Watt et al. 1997). Africa, harboring about 700 species, is the continent richest in termites. The family of the Termitidae represents about 75 % of the so-called higher termite species, among them the subfamily of hypogaeic Macrotermitinae. Their comparatively large above-ground mounds are typical of many tropical landscapes (Lee and Wood 1971). The mounds may be up to 9 m high and 20–30 m in diameter (e.g., in Africa *Macrotermes bellicosus*, *Macrothermes subhyalinus*,

Macrotermes falciger, *Odontothermes* spp., *Bellicositermes bellicosus* and in Australia *Nasutitermes triodiae*, *Amitermis meridionales*; Goudie 1988). About 200 mounds of Macotermitinae per square kilometer were found in East African (Tanzania) miombo woodlands (Jones 1990). The mounds may cover as much as 30 % of the ground surface (Wood and Sands 1978).

Troll (1936), founder of landscape ecology, presented his idea of landscape as an association of ecotopes (Sect. 2.1; see also Troll 1963) referring to East African savannas (Iringa Highland and northern Tanzania) characterized by termite mounds. The conspicuous pattern of sharply contoured hygrophyllous tree and shrub islands in mesophytic grassland had repeatedly been reported by earlier travellers and researchers (references in Troll 1963). The patchy distribution pattern is related to differences in soil conditions, which are caused by the termites' activities. Troll considered a similar landscape patchiness related to the effects of leaf-cutting ants (*Atta* spp.) in the New World savannas to be a 'biocoenotic convergence' to the palaeotropical savannas. To emphasize the regular coincidence of tree and bush islands with termite mounds, he introduced the term 'termite savanna' into scientific literature.

However, in many savanna areas, termite mounds are without vegetation, as, for example, the above 6 m high mounds of *Nasutitermes triodiae* (Nasutitermitinae) or *Amitermis meridionales* (Amitermitinae) in Northern Australia (Mertens 1961; Lee and Wood 1971; Photo 4.1). In the southeastern Serengeti, even low termite mounds (up to 1 m high) are covered with grass vegetation which reflects the particular local edaphic conditions and differs from the surrounding savanna vegetation. The termite mounds consisting of very salt-rich material are covered by salt-tolerant grasses such as *Digitaria macroblephora* (Poaceae) and Bermuda grass (*Cyonodon dactylon*), while Andropogon (*Andropogon greenwayi*) and red grass (*Themedra triandra*) occupy the other sites. After the mounds have decayed and salt has been leached, salt-intolerant grasses may invade these sites (Sinclair 1979).

Termite mounds covered with and often completely hidden under dense shrub and tree vegetation occur in both dry and seasonally flooded savannas. In both savanna types, the physical and chemical properties of the soil of the termite mounds provide more favourable growing conditions to trees compared to the surrounding grasslands with scattered trees. By constructing passages through the mounds, termites improve aeration and drainage (e.g., Pomeroy 1976; Lal 1987). Better aerated and well-drained soils are decisive factors for tree and shrub vegetation, and the termites themselves benefit from the resultant constant moderate moisture conditions. Termite mounds are key habitats for a number of small mammal species (Flemming and Loveridge 2003), probably because the structural diversity of resources in termitaria is higher compared to open savanna (Okullo et al. 2013).

In dry savannas, soil moisture in mounds is higher than in the off-mound area. Flooded savannas are completely waterlogged during the rainy season. They are characterized by black clay-rich soils ('Mbuga soils'). In the dry season, their surface becomes encrusted and disintegrates into polygons. The termite mounds, however, that project above the flooding level are not affected and provide similarly suitable conditions for trees and shrubs as dry savanna termite mounds. In a way,

Photo 4.1 Unvegetated termite hill in the Northern Territory, Australia. E. Löffler, 15 September 1998

conditions can be compared to seasonally irrigated paddy fields in northeastern Thailand (Khorat Plateau), for example. In this area, large termite mounds (about seven to ten mounds per hectare) covered with trees locally occur on terrain a little above the irrigated paddy fields. When working in the paddy fields, the farmers usually visit these tree stands to escape the heat, at least for a short while. Moreover, they build shelter huts on the termite mounds that also provide shelter from heavy rains (Löffler 1977; 1996).

Mound-building termites bring loose material from several meters depth to the surface. In mounds of *Bellicosithermes bellicosus*, material was even found that came from 12 m depth (Boyer 1975). In small mounds, the weight of excavated material ranges between 10,000 and 45,000 kg ha^{-1}, depending on the abundance of the particular termite species. The weight of large mounds of *Macrotermes* in Africa may even reach 100–240 t ha^{-1} (Lee and Wood 1971; Wood and Sands 1978). The texture of the mound material largely corresponds to the below-ground material. However, it is often richer in clay than the buried topsoil (Hesse 1955)

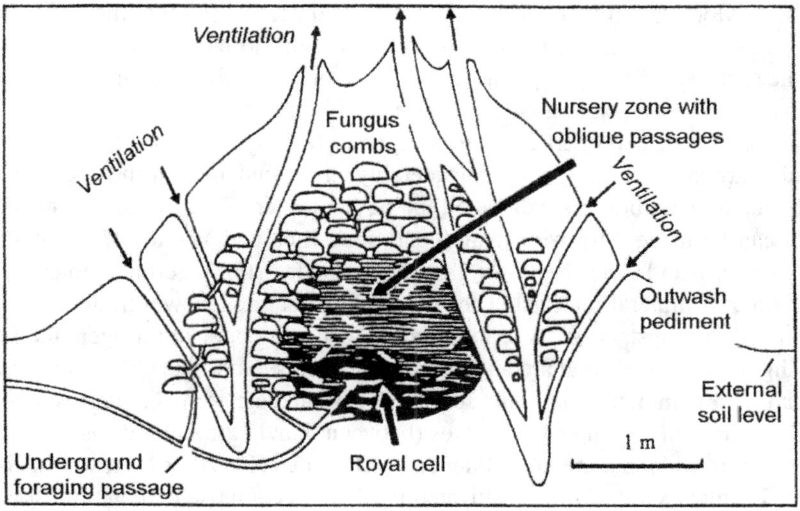

Fig. 4.1 Vertical section through a termite hill (*Macrotermes subhyalinus*) (Modified from Darlington 1984, in Goudie (1988))

and the soils around the mounds. Obviously, termites prefer clay-rich material (Lee and Wood 1971; Grassé 1984; Goudie 1988; Lepage et al. 1998; Konate et al. 1998). Mound material becomes stabilized by secretions (saliva and also feces) (Lee and Wood 1971; Gumnior and Thiemeier 2003). Because of its high clay content, mound material is used in some areas for construction of mud huts (Cosar 1934; Perry 1994).

Termites convey litter and also fresh plant material into their burrows, where it is safe from the frequent grassland fires (Josens 1992). In addition, termites (Macrotermitinae) cultivate cellulose-decomposing fungi (so-called 'fungus gardens') in their nests (Fig. 4.1). For construction of fungus combs, the termites exclusively use their feces, which consist only of little decomposed plant remains (Wood 1988). As these fungi (Termitomyces; this genus includes all Basidiomycetes living in symbiosis with the Marcrotermitinae; Heim 1963; Sands 1969) decompose cellulose, lignin is left (Grassè and Noirot 1958; Butler and Buckerfield 1979). Inside the termitaria, the portion of organic matter is usually greater than outside and in the surrounding soils (e.g., Stoops 1964; Lee and Wood 1971; Perry 1994). Microbiological activity becomes stimulated, as the fungi transform fresh termite feces characterized by a relatively wide C/N-ratio into a substrate with a comparatively narrow C/N-ratio, on which the termites may feed again (Matsumoto 1976; Collins 1977, 1981a, b; Abbadie and Lepage 1989).

Numerous passages from both the above-ground and subterranean parts of the termitaria give access to the fungus gardens. The tunnels, locally broadened for storage of food, not only serve for food transport but also have a ventilation effect keeping temperature constant and humidity relatively high (Lal 1987; see also

Fig. 4.1). Moreover, they increase infiltration and percolation. On the other hand, the fine excavated material brought from deeper soil horizons to the surface may seal the surface. This may explain why roots often follow the well-drained and aerated passages that reach to great depth (Lee and Wood 1971; Goudie 1988).

As the chemical properties of termite soils (acidity, content of nitrogen and organic carbon, C/N-ratio) vary considerably in dependence on the regional and local general conditions as well as on the particular termite species, general statements can be made only with many conditions (Lee and Wood 1971; Lal 1987; Lobry de Bryn and Conacher 1990; López-Hernández 2001). Termite mounds consist mainly of mineral material. The carton-like structures, however, are relatively rich in organic substances. As termite food is generally poor in nitrogen, the fungi growing on the stored food are an important protein source. The C/N-ratio in East-African termite mounds and in the soil at their immediate periphery corresponds to the C/N within the termites themselves (between 5 and 12), and thus, is much narrower than in soils that are not influenced by termites (10–20 or even wider; Jones 1990). Termites and the fungi cultivated by them decompose carbon almost completely. Numerous microorganisms in their digestive system enable the termites to digest even heavy decomposable substances.

In almost all studies on termite mounds, the calcium content has been found to be high compared to the off-mound sites. The opinions on the causes of the high content of exchangeable cations differ, however. Translocation of calcium-rich material from great depth and transport of water from high groundwater table by the termites themselves, residues of evaporation of ground water sucked upwards in the 'ventilation chimneys', and other factors have been considered to be involved. In the view of Lee and Wood (1971), the high content of magnesium and potassium they found in Australian termite mounds comes from the plant tissue consumed by the termites (see also Trapnell et al. 1976; Wood 1988).

In East Africa, termitaria are often used for crop cultivation, because of their favourable physical and chemical soil properties (Troll 1936; Milne 1947; Pullan 1979). Pendleton (1941, 1942) mentioned better growth of cultivated plants on termite mounds in Thailand. However, he also reported many cases in which the yield of crop cultivation on levelled termite mounds was relatively low. Comparable observations were made on areas under shifting (slash-and-burn) cultivation in Sri Lanka (Joachim and Kandiah 1940). The nutrient content of the topsoil of decayed or levelled termite mounds in Nigeria was also found to be lower than in the surrounding soils (Nye 1955). Growth of maize was bad on termite mounds. However, Nyamapfene (1986) suggested to manure unproductive soils with termite-mound material in case mineral fertilizer or dung are not available. In a subsistence economy, termite- mound material is widely used for soil amendment (Watson 1977; Logan 1992) as an alternative to chemical fertilizers (Oliveira and Paiva 1985; Bishoa Menea and Boloy 1995). As to the effect of termites in tropical agrosystems, there is still a great need for research (Black and Okwakol 1997).

The hypothesis that nutrient enrichment in termite mounds is a result of stored plant litter and its decomposition by fungi is widely accepted. It is indirectly supported by the low content in organic carbon and by the low number of bacteria and

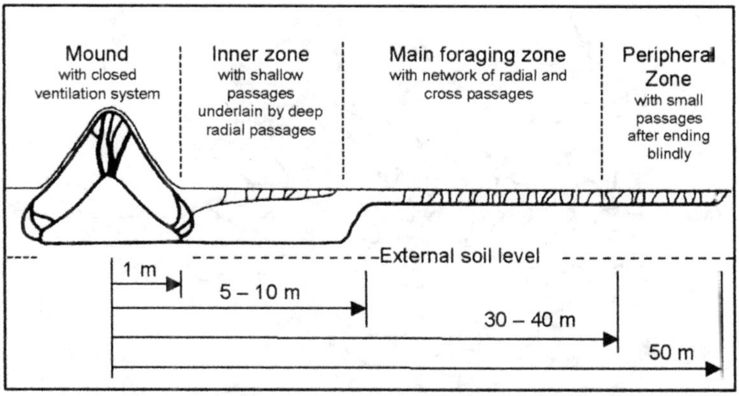

Fig. 4.2 Vertical section (not to scale) through a termite mound ecosystem of *Macrotermes michaelseni* near Kaijiado, Kenya (Modified from Darlington (1982))

fungi in the off-mound sites (Arshad 1982; Keya et al. 1982; Jones 1990). The low carbon content, however, does not only result from the termites' effects, but also from the frequent seasonal grassland fires. The trees and copses on the termitaria are affected at their periphery only. The fires are probably the main cause of the distinct limit between them and the open grassland (Troll 1936; Pullan 1979). Nevertheless, this spectacular physiognomical contrast should not lead to the assumption that the termite mounds proper are 'self-regulating ecological units'. They certainly form distinct biocoenoses within the grassland (Troll 1936), but their dense subterranean passage and gallery system usually extends up to 50 m away from the termite mound proper (Figs. 4.2 and 4.3; Arshad 1982; Darlington 1982). Thus, the termite population of a termitaria influence soils of a much larger area than might be assumed at first sight. Enrichment of nutrients in the mounds contrasts with impoverishment of the surrounding soils, where the termites collect the litter to carry it into the mounds. The passage systems extending from the individual mounds into all directions appear to be the main reason (competition for litter) for the great distances between large termite mounds (Jones 1990). Termites defend their territory, which spans their foraging area (Darlington 1982).

The life span of termitaria ranges from several decades to centuries (Watson 1967; Darlington 1985). Abandoned termitaria are often colonized and enlarged by other termite species (Coaton 1962; Trapnell et al. 1976; Pullan 1979). Termite mounds covered by dense copses are well protected from erosion (e.g., Brown 1972), in particular by run-off. If, however, the copses become thinned out and fragmented by elephants, other large herbivores or humans, grasses will invade. Consequently, the vulnerability of the tree stands to grassland fires will increase. After decline of the tree cover, the mounds are exposed to erosion, particularly at the onset of the wet season before sprouting of the grass vegegation. Elephants may destroy termite mounds when breaking them up with their tusks to eat the mineral-rich soil or when using the mounds to scrub their skin (see also Leuthold

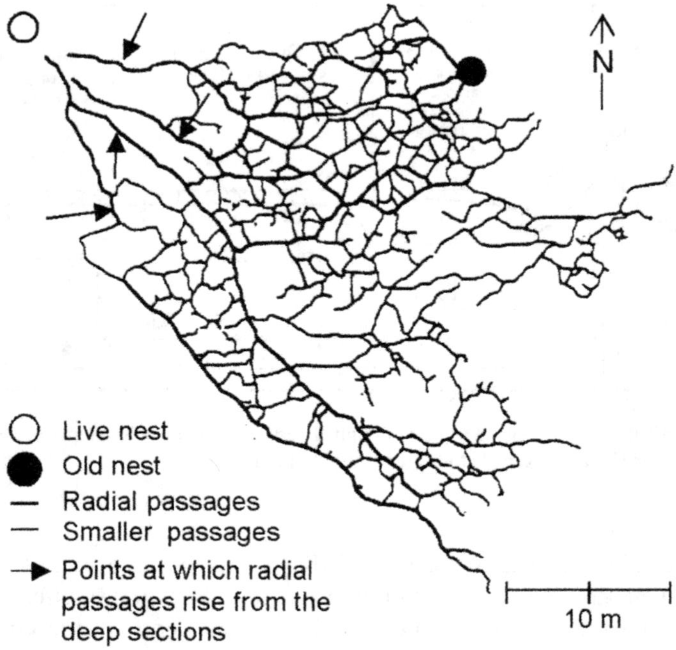

Fig. 4.3 Plan view of subterranean galleries (shallow galleries omitted) of *Macrotermes michaelseni* near Kajiado, Kenya (Modified from Darlington (1982))

1977a, b). Aardvarks (*Orycteropus afer*) dig in the mounds for termites, which they prefer for food. Other wild animals use the mounds as salt licks and also as lookout (antelopes, buffalos, lions, and hyenas; Brown 1972). They may enhance erosion by trampling. The grassland fires prevent re-colonization of the eroded mounds by trees and shrubs (Pullan 1979).

The material of decaying termite mounds, which is relatively rich in clay, largely accumulates on the adjacent terrain. Especially around the large mounds of fungi-cultivating species a more or less broad footzone develops. In dry savannas, such footzones usually remain unvegetated, whereas under humid conditions, they will gradually become colonized by plants. This vegetation is attractive to herbivores, which cause increasing trampling damage. As a result, characteristic vegetation patterns develop.

Such patterns have been studied in detail on the Loita plains (southern Kenya) (Glover et al. 1964), for example. The Loita plains (1,700–1,900 m) are covered with open grass savanna dissected by the tributaries of the Mara-river system. The grass savanna is characterized by circular or elliptically elongated vegetation structures that become obvious, however, only in the plan view (Fig. 4.4). The pattern is related to comparatively low termite mounds (*Odontotermes* spp.), which have caused varying pedological conditions at a fine-scale. The circular vegetation structures (6–10 cm in diameter) cover the gentle interfluves between the drainage

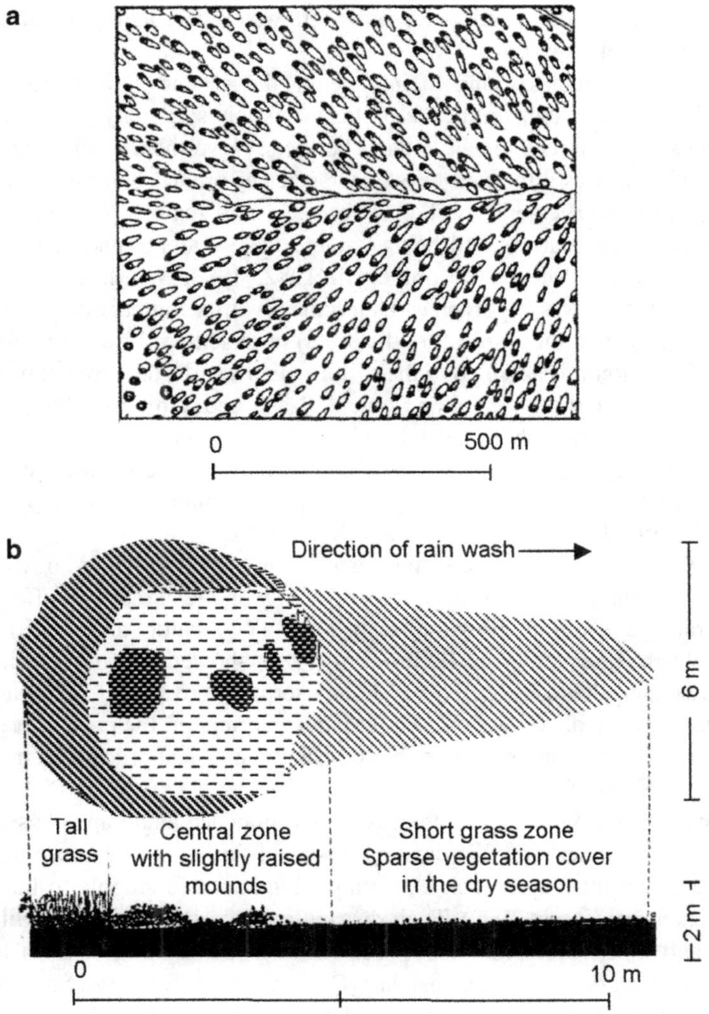

Fig. 4.4 Distribution of vegetation pattern related to low termite mounds (*Odontotermes* sp.) on sloping ground (Loita plains, Kenya). (**a**) Ground plan, (**b**) Plan view and vertical section of a single geomorphic pattern (Modified from Glover et al. (1964))

channels. On steeper terrain sloping towards the drainage channels, the structures assume an elliptical ground plan elongated downslope due to the direction of surface runoff and erosion, i.e., more or less perpendicular to the contour lines. In the plan view, the pattern is similar to a peacock's feather fan (cf. Fig. 4.4). Aardvarks dig for termites in the mounds located in the center of the vegetation structures. Surface runoff relocates the excavated material downhill, where it forms the lower edge of the elliptical structures. The rapidly hardening and little permeable relocated material is colonized by a short and, during the dry season, very sparse grass vegegation. On the disturbed sites, however, water percolating to great depth promotes luxuriant plant growth.

Many termite species living mainly below the ground surface do not build mounds. That is why their influence on landscape physiognomy is usually not as conspicuous as the effects of the mound-builders. Nevertheless, subterranean termites are often just as important keystone species as mound-building termites and influence soil properties and the biocoenoses. In arid regions, they are the main decomposers of dead above-and below-ground plant material and of the dung of herbivores (Buxton 1981; Whitford et al. 1982, 1988; Whitford 1991, 1999). Particularly in areas where polyphagous termites (with a comparatively broad food spectrum) exist, as in the Karoo (South Africa) or in semiarid Australia, for example, only little carbon is supplied to the soil. In other words, the more numerous the termites, the lower the soil carbon content (Nash and Whitford 1995; Whitford 1999). However, as the subterranean termites, similar to the mound-builders, 'wallpaper' their passages, galleries and nests with feces, they locally enrich carbon and nutrients. Diversity and heterogeneity of the plant communities may possibly be ascribed first of all to this behaviour (Whitford 1999).

Subterranean termites may also considerably influence the hydrology of desert ecosystems through their influence on soil porosity and resultant side effects (infiltration, drainage, aeration, growth of plant roots) (Wood and Sands 1978). Studies in the Chihuahuan desert (e.g., Elkins et al. 1986) showed that in areas lacking a shrub cover (creosotebush, *Larrea tridentata*), termites are a very significant factor for the maintenance of soil water characteristics that support the present vegetation. Under creosote bush canopies, other biological and physical processes appeared to override the effects of termites' activities.

Subterranean termites may also occasionally influence landscape physiognomy. So-called 'fairy circles' (mythic dancing places of fairies) are one example. They occur in a partly interrupted belt extending from southern Angola to the Orange River (Becker and Getzin 2000). Their distribution is restricted to areas with sandy substrate and annual precipitation between 50 and 100 mm. 'Fairy circles' are open unvegetated or sparsely vegetated patches (Coaton 1958; Coaton and Sheasby 1972; Lovegrove 1993; Moll 1994; Becker and Getzin 2000) embedded in short and sparse grass vegetation (Photo 4.2). They measure 5–8 m in diameter. A ring of tightly packed tussocks (Silky bushman grass, *Stipa uniplumis*) usually forms the outer rim of the 'fairy circles'. These tussocks are noticeably higher than the surrounding grassland, which largely consists of the same species. Many attempts have been made to explain the origin of the 'fairy circles'. Becker and Getzin (2000; see also Becker 2001) have presented an interesting study on 'fairy circles' in Kaokoland (Northwest Namibia) which gives a plausible explanation of the influence of 'fairy circles' on the vegetation mosaic and local landscape physiognomy. Harvester termites (*Hodotermes mossambicus*) appear to play a decisive role. In a few days, the termites harvest all grass within a radius of a few meters around the entrances of their subterranean galleries extending (up to 270 m; Becker 2001) from the colony center. By consuming large amounts of grass, the termites may strongly compete with cattle for forage, as is also known from harvester ants (*Pogonomyrex occiden-*

Photo 4.2 Fairy circles in northwestern Kaokoland, Pre-Namib, Africa. T. Becker

talis) in western North America, which also create unvegetated patches similar to 'fairy circles' (Sharp and Barr 1960). Anyway, the most important predators of termites in Africa, such as aardvarks (*Orycteropus afer*) and bat-eared foxes (*Otocyon megalotis*), are persistently persecuted by the ranchers: aardvarks because they undermine protective fences (cattle), bat-eared foxes because they allegedly prey on cattle (Coaton and Sheasby 1972).

In years with average rainfall, termites only harvest grasses next to the brood chambers, which are located in the center of the colonies. During dry periods, when the growth of grasses is considerably reduced, they expand their gallery system, and new 'fairy circles' may develop around the galleries' entrances. Simultaneously, vegetation may recover on the abandoned 'circles'. In case the food source improves for the longer time due to more favourable moisture conditions, the termites give up the distant supply galleries, and grass will gradually recolonize the bare circular patches.

The development of the outer fringe of the 'fairy circles' formed by higher and more compact grass tussocks appears to be dependent on several interacting factors. On the one hand, competition for water and nutrients is reduced compared to the surrounding grassland, as vegetation is lacking within the 'fairy circles'. In addition, loose mineral and organic matter removed by wind from the unvegetated patch is accumulated within the relatively high tussock grasses at the circles' margin. Relocation of loose material by wind may also be the cause of the often slightly concave form of the 'fairy circles'. Wind-eroded material accumulated at the tussock-covered periphery enriches nutrients. Not least, moisture conditions might be more favourable compared to the bare center of the 'fairy circles' due to increased humus content (comment of the present author). However, the discussion on the origin of 'fairy circles' is ongoing (e.g., Tschinkel 2012; Jürgens 2013). In the view of Tschinkel (2012), for example, a 'fairy circle' pattern results from self-organization of plants under arid conditions. Obviously, the riddle of formation of 'fairy circles' has not yet been solved. Those who reject the 'termite-hypothesis' are calling for more experimental studies.

As has become evident from studies in the northern Chihuahuan desert (Whitford et al. 1982, 1988), artificial removal of termites was followed by marked changes in ecosystem structure and processes. On intensively grazed pastures, decomposition of the great amount of dung was considerably retarded, and productivity decreased. Sprinkling experiments (Elkins et al. 1986) showed higher runoff and heavier erosion on the areas sparsely covered with perennial vegetation and free of termites compared to the areas occupied by termites. After the subterranean galleries had collapsed, bulk density increased. In areas without creosotebush – i.e., in about two thirds of the total area – termites considerably influenced water balance and other vegetation than creosotebush by increasing infiltration and water-holding capacity of the topsoil. Similar relationships were found in areas with mounds of *Odontotermes* at the Ivory Coast (Konate et al. 1998).

To close this section, a short look at the so-called 'soil-feeding' termites is given. This group is richest in species and most widely distributed in African tropical rainforests, where they influence soils profoundly. Obviously, the low amount of organic matter in mineral soil is a sufficient food source for these termites. In forest soils of different age and degree of disturbance in Cameroon, influenced by *Cubitermes fungifaber*, acidity was low, while organic carbon, moisture and the relative clay content increased compared to the mineral parent material. In African rain forests, the huge termite biomass is probably one of the most important sources of the great soil microhabitat heterogeneity (Donovan et al. 2001). Altogether, the function of termites as keystone species is evident. As has become apparent from the foregoing examples, every single case, however, requires a well differentiated consideration and assessment with respect to the regional and local peculiarities (e.g., climate, vegetation, soils, disturbances, etc.).

This also holds true for the impact of termites on human economy and health, which may go beyond the scope of landscape ecology, however. Termites are usually considered to be noxious animals. Most termite species, however, are harmless. For example, only 20 of the 120 termite species of Nigeria cause damage to crops and buildings. A few are transmitters of diseases. On the other hand, termite mound material can be used for fertilization and construction (high clay content), and termites are also a protein- and energy-rich food for humans, particularly when other food runs short. Humans prefer winged termites, although in some regions, 'queens', 'soldiers' and 'workers' are also eaten. In many countries, termite mounds are privately owned, and only family members have the legal right to harvest the termites for food (Logan 1992).

4.2.2 Beavers

The beaver is originally an animal of the northern hemisphere (Eurasia, North America), where it occurs from the subtropics to the rim of the Arctic. Beavers may radically change and manage their riparian habitat for their own benefit (Holtmeier 1987; Naiman et al. 1994; Butler 1995; Huntly 1995; Pollock et al. 1995; Guernell 1998; Collen and Gibson 2001; Donkor and Fryxell 1999; Rosell et al. 2005). Not without reason, the beaver has been called a 'master builder in the animal kingdom' or an 'upstream engineer'. The beaver is actually an epitome of a 'keystone species'. Very likely, more publications exist about beavers and their keystone function than on the effects of other animals on the landscape ecosystems (abundant references e.g., in: Kämpfer 1967; Djoshkin and Safonov 1972; Stocker 1985; Medin and Torquemada 1988; Naiman et al. 1986, 1988, 1993; Naiman 1988; Butler and Malanson 1994; Butler 1995; Zahner 1997; Leary 2012), perhaps with the exception of termites.

The North American beaver (*Castor Canadensis*, Photo 4.3) differs from the European (Eurasian) beaver (*Castor fiber*) morphologically, anatomically and genetically. Both species are subdivided into many subspecies: 24 of the American beaver (Hall and Kelson 1959; Baker and Hill 2003) and, historically, eight subspecies of the European beaver (Durka et al. 2005; Halley 2011). Recently, Durka et al. (2005) reported that phylogenetic analysis (mitochondrial DNA) revealed only two ESU units (Evolutionarily Significant Units), the western and the eastern haplogroup.

Before the beaver became almost completely extinct over much of its range, it was practically spread all over Europe, with the exception of Ireland, Iceland and Nowaja Semlja. In North America, beaver also occurred almost everywhere along

Photo 4.3 Beaver (*Castor canadensis*). Guanella Pass, Colorado Front Range. F.-K. Holtmeier 1989

streams and lakes over most of the continent from the Mackenzie River delta south to northern Mexico. Relic populations survived in some regions of North America, Russia, southern Norway, eastern Germany and southern France. Beaver was re-introduced to many of its original habitats (Sect. 5.2.1). Beaver is not originally native to the southern hemisphere. However, at the end of the 1940s, Canadian beavers (*Castor canadensis*) were introduced to Tierra del Fuego where they rapidly multiplied (Godoy 1963; Dietrich 1985; Vásquez 2002; Anderson et al. 2012).

Beavers live in streams and in bodies of standing water, provided that these allow the beavers to move freely, including in the winter. Beavers do not occupy little streams, which usually dry in the summer, nor water bodies that freeze down to bottom in winter. Moreover, they do not settle very broad and rapidly flowing rivers (Djoshkin and Safonov 1972; Slough and Sadleir 1977; Howard and Larson 1985; Beier and Barret 1987; Hartman 1996). Beavers will usually not impound streams that are deep enough to meet their ecological demands anyway. In mountainous terrain, beavers prefer streams of secondary and higher order (e.g., Naiman et al. 1986; Butler and Malanson 1994; Snodgrass 1997). Sustained water supply, forage availability and river channel characteristics are the most important ecological factors for beaver occupancy. In lowlands, beavers colonize steep river banks and lake shores. Contrary to the common view, beavers are not put off by stony substrate (Photo 4.4).

Photo 4.4 Beavers have started to cut this big poplar (Diameter 70 cm) at Blackfoot River, Montana. F.-K. Holtmeier, 27 February 1990

Subterranean burrows are probably the most common type of dwelling. As beavers carefully adjust their burrow systems to the local situation, the burrows may differ considerably. Occasionally, whole river banks are undermined by a labyrinth of passages, nests, storage chambers and emergency exits. Thus, subterranean beaver burrows are similar to those of muskrat (Sect. 5.1.4.1).

Vegetation is usually most affected, apart from occasional collapsing river banks and lake shores (Butler and Malanson 1994). Beavers feed almost exclusively on aquatic plants and riparian vegetation. Beavers' diet comprises about 300 species. Practically, there is almost no plant in a beaver habitat which is not used (Djoshkin and Safonov 1972). Nevertheless, beavers prefer some plant species, while they eat others only in case of forage shortage.

The beavers' diet varies seasonally. During summer, they feed mainly on grasses, herbs, aquatic plants, ferns and fresh leaves of woody plants. In autumn, they switch to twigs, bark and bast, mainly of softwoods such as aspen, willows and birches. Beavers remove the bark of birch and eat the bast, which is easily digestible. Perhaps the broad food spectrum prevents enrichment of secondary compounds that might have toxic effects (e.g., Freeland and Kanzen 1974). Feeding experiments simulating an unbalanced diet resulted in noticeable malnutrition and symptoms of deficiency (O'Brien 1938). Contrary to that, beavers, which were released into the Eifel area (western Germany), subsisted exclusively on an uprooted red beech (*Fagus sylvatica*) for 6 weeks without negative effects on their physical condition (Schneider and Schulte 1985). At least, no adverse effects were reflected in their outward appearance. In case of food shortage, beavers obviously are not very choosy and even feed on the bark of conifers and sage brush (*Artemisia* spp.) (Djoshkin and Safonov 1972; Jenkins 1975, 1979; Smith et al. 1996; Persico and Meyer 2009).

Willows and aspen are not only an important beaver food but also represent the main dam building material (e.g., Baker and Hill 2003). Nevertheless, beavers also cut conifers for construction of lodges and dams (MacDonald 1956; by contrast Wilsson 1971). Beavers usually prefer trees of medium size for forage and construction wood. Occasionally, they fell trees measuring more than 50 cm in diameter (Photo 4.5). Wilsson (1971) reported on the felling of a birch that had a diameter of more than 100 cm. This, however, might be exceptional. When beavers have cut a large tree, they will only use leaves, thin twigs and parts of the bark from the upper stem section for food, whereas the stem proper, with its coarse and apparently not very palatable bark, is left untouched. By contrast, small trees are utilized completely. Altogether, aspen and willows are used disproportionately to their frequency.

Beavers influence terrestrial vegetation only within a 45–60 m wide strip bordering streams and lakes. On broad valley bottoms, however, the total riparian zone may be altered. Only exceptionally can beavers be found at a greater distance from the waters (e.g., Jenkins 1980; Schneider 1985; Johnston and Naiman 1987; Barnes and Dibble 1988). Perhaps beavers prefer to stay relatively close to their ponds to avoid predator attacks (Baskin 2011). Particularly in mountain valleys, the beavers' radius of action is often limited by steep valley sides, and also by roads and railroad on the valley floor (see also Photos 4.8 and 4.9). Steep topography may hinder beavers from moving to neighboring catchment areas.

Photo 4.5 Beaver lodge at the stone-rich shore of Lake MacDonald, Montana. The beavers obviously preferred this relatively steep shore section to the sandy level shore in the background. F.-K. Holtmeier, 27 February 1990

Beaver is the only animal capable of felling complete tree stands. Beavers use almost all trees next to streams, lakes and ponds, whereas tree stands at a greater distance are less affected. Rutherford (1964) reported that in Colorado, beavers had literally 'mowed down' complete aspen forests over the course of only one summer. That is why the influence of beaver on vegetation is disproportionately great compared to the effect of other herbivorous mammals. Each of the six beavers living in a beaver colony in Minnesota harvested double the phytomass per hectare as does an ungulate herd in the Serengeti (Johnston and Naiman 1990a).

Under the influence of beavers, tree density and composition of woody species are particularly subject to change. Next to streams or lakes, willows and alders gradually replace aspen, which beavers prefer for food. In the drainage area of the Truckee River (California), beavers were released between 1938 and 1946. About 40 years later, beavers had occupied all habitats, even those that were largely lacking preferred forage. In these marginal habitats, quaking aspen (*Populus tremuloides* and *Populus trichocarpa*) considerably declined. Apparently, this process could not be stopped without beaver control. On the other hand, willow stands survived despite severe temporary impact (Beier and Barret 1987). Occasionally, conifers such as white spruce (*Picea glauca*) and balsam fir (*Abies balsamea*) expand after aspen have decreased (Johnston and Naiman 1990a). Relieved from competition, conifers may become the dominating tree species within such a short time that they prevent establishment of shrubs. In many cases, aspen cut by beavers recover by intensive production of root suckers.

The beaver's role as a true 'landscape organizer', however, begins with dam building. Dams are built in streams that are not too broad and not flowing too fast.

Photo 4.6 Beaver ponds and beaver meadows in the Bow River Valley, Alberta, Canada. The beavers have completely modified the riparian ecosystem. F.-K. Holtmeier, August 1972

Dam building changes water courses, usually of secondary or higher order (e.g., Naiman et al. 1986; Butler and Malanson 1994: Snodgrass 1997), into a row of 'stair-case-like' arranged ponds. As beavers live in colonies, pond systems may be of considerable size (Photo 4.6, see also Photo 4.12). Beavers usually build their lodges in the central part of the ponds (Photo 4.7), where water is deep enough not to freeze through to the ground in winter. The lodges consist of branches and twigs of the felled trees and are sealed with mud (see also Photos 4.4 and 4.9).

Beavers usually build dams where streams are narrow. The noise of running turbulent water has been supposed to have a stimulating effect (Wilsson 1971; Hogdon and Lancia 1983; Heidecke 1985). However, beavers also construct dams where water is flowing slowly and quietly. The dams are always optimally adjusted to stream flow. Beavers use tree stems, branches, twigs and, in case of gravel-rich streams, stones for dam construction. Blocks in the streambeds and trees next to the riverbanks are often incorporated into dam construction. Even garbage such as old tires, useless jerry cans, and wooden boards are used as construction material (personal observation). Mosses, reed and mud are used for sealing.

Number and extent of dams depends on beaver population size and on the local possibilities of utilization by beavers. A colony can be considered a 'family unit' that occupies a discrete territory (e.g., Hogdon and Lancia 1983; Hay 2009). A beaver family usually consists of the parents and 1-year-old and 2-year-old beavers. Between two and eight young are whelped per year. Thus, 10 beavers may be expected to be living in a colony on average. Under less favorable conditions, the

Photo 4.7 Beaver lodge built in the middle of a beaver pond at about 3,000 m on Guanella Pass, Colorado Front Range. Mud used for sealing the lodge is clearly discernible under the wooden pieces. F.-K. Holtmeier, 27 August 1990

Photo 4.8 Aspen (*Populus tremuloides*) that were cut by beavers fell on Highway 62 northwest of Placerville, Colorado. The beaver ponds are located between the road and the valley side at the *left hand-side*. They are not visible on the photo F.-K. Holtmeier, 4 August 1997

Photo 4.9 Beaver lodge located further down on the European Highway 4 near Skuleberget (Ångermanland, Sweden). F.-K. Holtmeier, 22 August 2001

colonies are usually smaller. After food has become depleted within reach, beavers move upstream and build new dams.

Between 8.6 and 16 dams (mean 10.6) were counted per kilometre stream length in southern Québec. On the Kabetogama peninsula in northern Minnesota, 2.0–3.9 dams were found per kilometre (Naiman et al. 1988). In the Colorado Rocky Mountains (e.g., on Guanella Pass near Georgetown and in the Rocky Mountain National Park), the present author found 10 or more dams per kilometre. A similar dam density has been reported by Butler and Malanson (1994) from Waterton National Park (Rocky Mountains, Montana). Woo and Waddington (1990) calculated a dam density of more than 14 dams per kilometre in northern Ontario. On level terrain, as, for example, in the Mississippi swamp area or along the Russian lowland rivers, beaver dams may be several hundred meters in length. Only such dams will allow impoundment of sufficiently deep ponds. Along the comparatively small mountain rivers and streams, beaver dams usually measure only a few meters or decameters in length, but often several meters in height.

The size and shape of beaver pond systems are closely related to topography. Commonly, conditions are more favorable in more gentle sections of mountain streams where they meander on relatively broad floodplains (Guernell 1998; Holtmeier 1999, 2002; Butler 2012). On flood plains, beaver ponds are comparatively large, often covering several hectares (e.g., Collen and Gibson 2001). Thus, impoundment by beavers may create extended floodplain wetlands. These are usually bordered by river terraces or valley sides (see also Photos 4.6 and 4.12). By contrast, in the upper sections of the high-mountain streams, where space for beaver ponds is often limited by steep local topography, ponds are usually relatively long and narrow. Slow plant growth and relatively small size of riparian softwood forests limit beaver population in such stream channel pond systems.

However, beavers are able to adjust pond building to a certain extent to the given local conditions, even in close proximity to human infrastructure (e.g., roads). In the Colorado Rocky Mountains (NW of Placerville), for example, the present author found small beaver ponds in a narrow, only 20–40 m wide brook between a steep slope on one side and a highway on the other. The dams were 2–5 m long and about 2 m high. Several ponds were lined up like pearls on a string over a distance of more than 100 m. The beavers had been clever enough to incorporate the lie of the highway as an 'artificial valley side', so to speak, into their dam building 'concept'. Thus, impoundment was possible by construction of very short dams, which saved the beavers' energy. Obviously, the beavers do not feel bothered by road traffic. On the other hand, felled aspen that occasionally fall on the road are a risk for car traffic (Photo 4.8). In Sweden, the present author found a beaver pond system bordering the very busy European Highway 4 near Skuleberget (Ångermanland). A large beaver lodge is located at the rim of the pond only about 20 m distant from this highway (Photo 4.9), and, as in the beaver habitat near Placerville, the relatively high noise level in this place does not worry the beavers. There are many other comparable examples (Sect. 5.2.1). The remarkable great adaptability of beavers to particular local conditions, which is independent from species and race, has often been mentioned in literature (e.g., Pilleri 1960: Richard 1983; Holtmeier 1987; Roller 2001). To get to forage places at a greater distance from the ponds, beavers regularly use the same trails. The beavers actively deepen and widen them to real 'channels', which they also use for floating material they need for dam building and construction of lodges.

In beaver ponds, flow velocity is very low. The ponds often look like stagnant water. Beavers are permanently busy adjusting the dams to varying water levels and velocity of flow. Much water evaporates. In case of subsiding water, level beavers heighten the dams to reduce discharge; in case of high water, they create outlets for better drainage. Surplus water runs over the dam. Well-maintained dams are almost impermeable. Dams may leak where beavers are behind with repair, as can often be observed in older dams. Occasionally, water seeps through at the dam base. Altogether, water flows slowly from the upstream ponds to the downstream impoundments (Woo and Waddington 1990). In case of high water, new drainage channels develop downstream of the beaver dams a little above the original stream level. These channels usually fall dry in case of low run-off.

Impoundment alters the ecosystem character of streams and the riparian habitat in manifold ways. However, with regard to the great variety of local conditions general statements can be made only with many conditions. This also holds true for the next figure (Fig. 4.5), which shows interactions of ecological factors in a beaver pond ecosystem. The figure is based on the present author's observations in the Rocky Mountains and publications referring to this region (Rutherford 1964; Holtmeier 1986, 1987). Rutherford (1964) classifies such a beaver-pond ecosystem characterized by aspen (*Populus tremuloides*) and willows (*Salix* spp.) as the 'aspen-willow type', which is typical of the mountains, and compares it with the 'plains

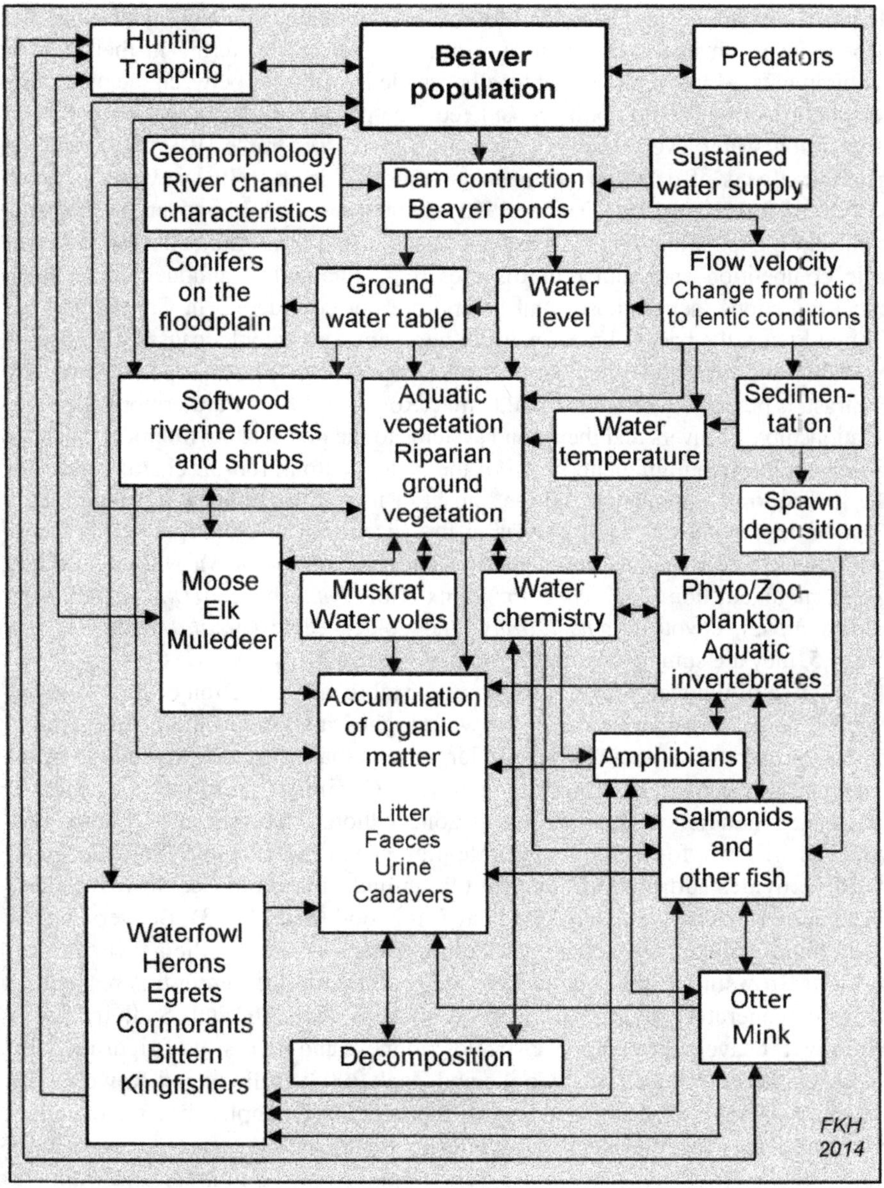

Fig. 4.5 Interrelationships of ecological factors in a beaver pond ecosystem. *Filled arrow* heads mean increase, *open arrow* heads mean decrease or impairment (Scheme by F.-K. Holtmeier, Modified from Holtmeier (1987))

cottonwood riverbottom type' formed by cottonwood (*Populus sargentii*) and willows along the prairie streams.

In order to guarantee sufficient clarity, only the most important components of a beaver-pond ecosystem and their mutual interrelationships are considered in Fig. 4.5.

Trout, for example, stands for the total fish population, although changes in the ecosystem result in a succession of various fish species differing in their habitat requirements. Moreover, regrettably, all possible interactions between the many factors and processes could not be considered. The positive effects of beaver ponds on trout at low water levels during dry periods (Cox 1940; Grasse 1949, 1951) and the negative effects of dams on upstream migration of trout to its spawning grounds (Johnson 1927; Reid 1951; Denney 1952) are not presented. Moreover, mechanisms controlling population dynamics as well as social and territorial behavior, intraspecific competition and numerous diseases (e.g., tularemia, salmonella infections, cocci infections, helminthoses and others) are not considered. In Canada and the United States, the bacterium *Francisella tularenis*, for example, causes high beaver mortality and occasional decline of beaver populations. The European beaver, by contrast, is practically resistant to tularemia (Romašov 1992). Furthermore, the close relationship of beavers and their pond systems to the physical environment, such as, for example, stream gradient, width of the valley bottom, run-off characteristics of the stream to be impounded and the impact of fire on the primary vegetation (e.g., *Populus tremuloides* as seral species in fire succession) are disregarded. Potential predators are bear (*Ursus americanus*, *Ursus arctos horribilis*), wolverine (*Gulo gulo*), mountain lion (*Felis concolor*), lynx (*Lynx lynx),* bobcat (*Lynx rufus*), wolf (*Canis lupus*), coyote (*Canis latrans*) and golden eagle (*Aquila chrysaetos*). In Fig. 4.5, they are summed up under 'Natural predators'.

The rise of the water level, ground water table and reduction of flow velocity cause changes in vegetation. On the waterlogged margins of the ponds, swamp plants spread and soon cover a much larger area than was occupied by the original riparian vegetation before impoundment. Conifers next to the pond die due to oxygen deficiency in the rooting zone (Photo 4.10, see also Photo 4.6), whereas willows, for example, tolerate high ground water table. The new vegetation provides additional forage for elk, muledeer and moose. However, these large herbivores compete with the beaver for food (Sect. 3.4.3). Beavers, on the other hand, reduce soft woods, particularly aspen, which are their favorite winter diet. Browsing of the root suckers plays an important role in this respect, as aspen regenerates mainly from root suckers (Slough and Sadleir 1977). Moreover, beavers may increase species richness and diversity of riparian habitats (e.g., Wright et al. 2002; Cooke and Zack 2008; Bulluck and Rowe 2013). Beaver ponds attract waterfowl and wild game. For example, they are excellent roosting places for migratory and wintering waterfowl (Arner and Hepp 1989). So, valleys settled by beavers are very popular among hunters and trappers. Beaver ponds are also a favorable habitat for muskrats (*Ondatra zibethica*) (Grasse 1951; Beard 1953; Stegeman 1954; Rutherford 1955; Neff 1957; Arner et al. 1969). These occasionally live in beaver lodges. In addition, Canadian river otter (*Lutra canadensis*), European river otter (*Lutra lutra*) and mink (*Mustela vison*) benefit from beaver ponds. Beaver ponds also give shelter to wild animals from wild fires. Not least, beaver ponds may harbor many fish species (lotic and lentic; Collen and Gibson 2001). Thus, beaver ponds are appreciated by anglers as excellent fishing grounds.

Photo 4.10 Dead spruce trees (*Picea engelmannii*) in a beaver meadow in Kawuneeche Valley at about 2,750 m. The conifers died off due to the high ground water table and resulting oxygen deficiency. Willows (*Salix* sp.) tolerate the wet conditions. Rocky Mountain National Park, Colorado. F.-K. Holtmeier, 28 July 1989

In the beaver ponds, sedimentation and temperature increase. According to J. Sedell (cited in Bergstrom 1985), a beaver pond covering only 0.1 % of the Fish Creek (Cascades, Oregon) produces 6–8 % of the relatively small Coho salmon (*Oncorhynucus kisutch*) in the total river catchment area. Young salmon that were in the pond increased their weight sixfold within 4 month due to abundant food supply and slightly increased temperature. In the ponds, trout also grow faster at first, compared to trout in rapidly flowing and cold water. Rising temperature, however, and increasing decomposition of organic matter reduce oxygen content (see also Harthun 2000). Trouts do not tolerate temperatures above 22 °C. In the course of time, spawning conditions also get worse because fine suspended mineral and organic material is increasingly accumulating on the originally gravel-rich stream bottom, upon which trouts rely for spawning (Cox 1940; Denney 1952; Gard 1961). So, new ponds are generally more favorable for trout reproduction than older ones (Salyer 1935; Rutherford 1955), and in the course of time, trout may be replaced by other fish species (Reid 1951).

Forage supply and available space for impounding running water are the main factors limiting beaver population growth (Yeager and Rutherford 1957; Rutherford 1964; Johnston and Naiman 1990a), whereas predators presumably 'skim off' surplus individuals. Hunting and trapping, the main cause of near-extinction of beaver in the past, is now an instrument of beaver management and control.

As the physical (topography, climate, stream flow characteristics) and the biological conditions (predator species and numbers, competition for food with cervids) may vary considerably, the system of interactions presented in Fig. 4.5, which is based on observations in the Rocky Mountains, can be transferred to other areas with reserves only. Thus, for example, while beaver ponds in the Rocky Mountains have a positive effect on trout population, at least during the first years after dam building, the situation is different at the natural distribution limits of trout (Reid 1951). However, even in the drainage area of a single mountain stream, the relative importance of the factors differs much. Alteration of vegetation, for example, is probably more important for deer in the winter grazing habitats at lower elevation than in the summer grazing areas (Rutherford 1955).

Beaver ponds change more or less with the passage of time in dependence on the particular local conditions, such as size (area, depth), age, succession of vegegation, flow conditions, magnitude of sedimentation of mineral and organic matter, and development of the beaver population. Data published in literature are seldom comparable because of different methods and databases. Anyway, they may give an idea of possible variation.

In a beaver pond system more than 700 m in length and consisting of 22 dams in Mission Creek (Washington State) almost 4,000 m^3 of fine material (silt) accumulated (Table 4.1). According to Naiman et al. (1988), upstream of small dams having a volume of 8–18 m^3, the amount of accumulated sediments ranges from 2,000 to 6,500 m^3. Naiman et al. (1986) roughly estimated the total accumulation of sediments in the beaver ponds of the almost undisturbed drainage area of the Matamek River (total water area 7.7 million km^2) in Québec. Taking 10 dams per stream kilometre and 1,000 m^3 sediment per pond as a basis, the authors came to a total accumulation of 3.6 million m^3. This amount would have been sufficient to cover the total river bottom with a sediment layer of 42 cm. Smith (1980) estimated that 90 % of the sediment load carried by the streams he studied accumulated in the beaver ponds there. On Guanella Pass in the Colorado Rocky Mountains, the present author found the sediment layers of beaver ponds exposed after the dam had collapsed, probably due to an extreme run-off event. The sediments consisted of sand alternating with thin layers of silt and organic (peaty) material (Photo 4.11) that had accumulated on the original gravelly stream bottom The sediments measured 120 cm in depth. The peaty layers formed as a result of anaerobic conditions that reduced decomposition.

Extended beaver pond complexes are characterized by a more or less variegated mosaic of habitats at different successional stages (ecotopes in the sense of Tansley 1939 or Troll 1970, 'patch bodies' sensu Johnston and Naiman 1987). A few long-term studies (Kabetogama Peninsula, northern Minnesota, 46 years; Johnston and Naiman 1990b; Adirondack State Park, New York, 40 years; Remillard et al. 1987) provide insight into the development of such beaver pond complexes. On the Kabetogama Peninsula, for example, the number of dams had increased more than tenfold, and the area influenced by beavers had increased 13-fold between 1940 and 1986 (Fig. 4.6). This was ascribed to favorable forage supply – aspen had considerably expanded due to large forest fires around the turn of the century – and to the

Table 4.1 Size of beaver dams and accumulation of fine material (silt) in the beaver ponds, lower section of the eastern headwater of Mission Creek, Washington State

No.	Distance [m] from the first dam	Dimensions [m] of the dams		Fine material [m^3] accumulated in the ponds
		Length	Height	
1	0	4.6	0.61	157.2
2	21.3	4.9	0.61	47.8
3	58.8	18.3	1.07	336.5
4	98.4	4.6	0.30	55.5
5	144.8	3.7	0.45	34.0
6	157.3	7.6	1.07	44.0
7	245.4	15.8	1.52	312.7
8	194.2	9.8	1.07	540.5
9	280.4	9.1	1.21	141.7
10	313.9	8.5	1.07	129.4
11	341.4	8.2	1.40	156.8
12	365.8	16.8	1.22	300.5
13	401.1	20.4	1.37	58.7
14	427.6	10.1	0.91	195.1
15	437.4	8.5	0.30	36.9
16	442.0	7.6	0.30	15.5
17	479.5	33.5	0.61	531.3
18	502.9	27.4	0.91	348.5
19	528.8	36.6	0.30	222.8
20	542.5	20.7	0.61	132.6
21	559.3	5.5	0.91	31.2
22	621.8	2.4	1.22	93.0
Total		248.7		391.9

Data from Scheffer (1938), after Holtmeier (1987)

many streams appropriate for beavers. While the distribution of ponds was more spotty in the beginning, density increased after 1961. When the beavers had completely occupied the area in 1986, they began to extend and improve the old ponds. Apparently, beaver settled the optimal locations first (see also Howard and Larson 1985). After having depleted the local food source, they moved to the less favorable places. In other areas, however, beavers who had left ponds that they had occupied for a long time, came back after 1 or 2 years (Hall 1971).

Beaver dams may persist for decades or even centuries (Butler and Malanson 1994); provided that sufficient food is available, the beavers maintain their dams and will not leave the area or decline (diseases, hunting). Beavers are absolutely able to use all aspen within their reach. After the food source is depleted by some time or another, beavers begin to build new dams, usually upstream and at a short distance from the old ones. In abandoned areas, succession begins with grasses and shrubs and may lead again to aspen or conifer stands.

Photo 4.11 Organic (peaty) and mineral (sandy to loamy) layers that accumulated in a beaver pond system at about 3,120 m have become eroded down to the former gravelly stream bed of South Clear Creek. Both the peaty layers and the mineral sediments measure about 70 cm in depth each. F.-K. Holtmeier, 23 July 1992

Cyclic changes of biotopes caused by the complex effects of beavers, riparian vegetation and hydrogeomorphic processes were also observed by Gill (1972) in the Mackenzie River Delta (Northwest Territories, Canada). As a result of helicoidal (spiral-like) stream flow, meandering streams are characterized by outer concave banks (undercut banks) and inner convex banks where sand and gravel bars (point bars) accumulate. Longitudinal arcuate depressions (meander scroll depressions) often follow behind the point bars (towards the stream bank). These depressions, although seldom affected by spring floods, are usually permanently filled with water from precipitation, run-off and melting snow (Gill 1972). These alluvial sites are occupied by balsam poplars (*Populus balsamea*) that provide an excellent food source for beavers. On slightly higher terrain, white spruce (*Picea glauca*) become established. As in the river delta, rapid flow, particularly during snowmelt, prevents almost any dam building, and thus, beavers usually burrow in the river banks. At the lower rim of meander scrolls, however, they dam up the water remaining from inundation and snowmelt. After the water is deep enough, beavers start building lodges in the ponds. Permafrost and much fine clay, accumulated during high floods, seal the ground and make drainage impossible. Over the course of time, poplars, which do not tolerate being permanently flooded, die and become replaced by alder (*Alnus crispa*) and willows (e.g., *Salix alaxensis*, *Salix arbusculoides*, *Salix richardsonii*). As the place is no longer attractive for beavers, they move to other locations. After recovery of the poplar stands, beavers may return. Such spatially and temporally

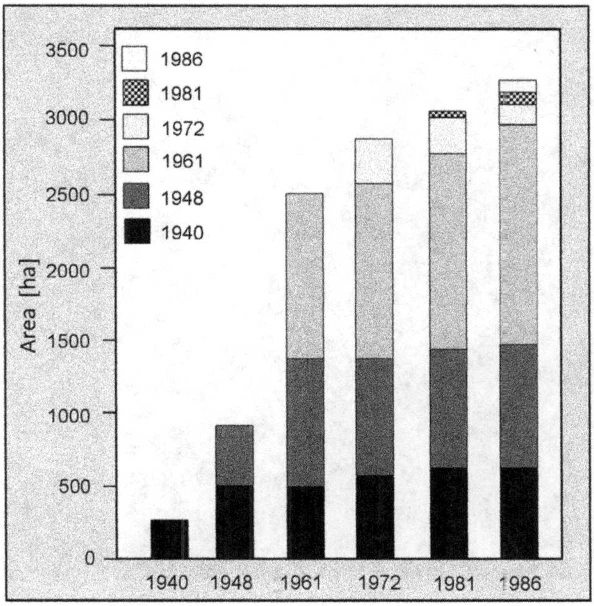

Fig. 4.6 Cumulative pond area by age class (Modified from Johnston and Naiman (1990b))

staggered cycle-like successions cause a high ecological and landscape diversity, as long as no severe disturbances occur. Under natural conditions, the more or less frequent forest fires played an important role, as they promoted regeneration and expansion of balsam poplar (e.g., Rutherford 1955: Slough and Sadleir 1977). In many areas of North America, aspen and poplars have become replaced by conifers as a result of systematic fire control. In addition, enormously increased consumption of root suckers by beavers, which have come back after almost complete eradication, impedes natural regeneration of the softwoods.

Beaver has very likely occupied all suitable habitats in the Old and the New World during the Holocene and influenced them as described above with spatially and temporally varying intensity. In Central Europe, landscape patterns related to former beaver effects have become almost completely distorted by hydrological works (river control), drainage systems, removal of riparian forests, fish farming, and conversion of floodplains into arable land and pastures. Gradually, and in a few areas only, re-introduced beavers have again started influencing riparian and lake shore landscapes (e.g., Harthun 1999) (see also Sect. 5.2.1). In Russia and North America, however, signs of former beaver influence on landscape can be partly traced back far into the past. Certainly, beaver populations had also shrunk to a minimum in these regions, but efforts were made comparatively early to bring beavers back to their original habitats, and the still existing beavers have managed to build up viable populations. That is why, in those regions, beavers' influence on landscape has been more continuous compared to Central Europe. This also holds true for many areas in the Rocky Mountains.

Photo 4.12 Upper section of Kawuneeche Valley (about 2,750 m, cf. Photo 4.10) is completely occupied by a beaver meadow at different successional stages. F.-K. Holtmeier, 27 August 1989

Illustrated by the uppermost catchment area of the Colorado River (Rocky Mountain National Park), Ives (1942) showed that large parts of the floodplain landscape mosaic characterized by countless meanders, cutoffs, oxbows, paludified areas, and moist meadows (Photo 4.12) have to be attributed to ancient and present beaver activities rather than being the result of infilling of glacial ponds and lakes. Recent beaver effects, however, cover a comparatively short period of time, as the beaver populations that survived intense hunting and trapping during the nineteenth century did not begin to recover before the 1930s. It is possible that Ives (1942) as well as Warren (1926) overshot the mark a little when explaining the development of the comparatively broad flood plain through centuries of beaver action alone (see also Rutherford 1964). On the other hand, Ruedemann and Schoonmaker (1938) had also ascribed many almost level floodplain sections in northern North America to infilling of fine sediments increased by beaver ponds.

The development of abandoned pond systems largely depends on the susceptibility of the under-ground to erosion. In case of no or little erosion, grasses and sedges will invade the drained ponds. It usually takes only a few months until a closed meadow will have developed (Neff 1957), with the possible exception of erosion channels cut into still fresh sediments. These channels, however, do not usually reach the gravelly original stream bottom or the bedrock. The groundwater table remains relatively high. Such 'beaver meadows' may persist for many decades, if not invaded by willow shrub or forest trees. The meadows are an attractive food source for cervids and other large herbivores (cf. Fig. 4.5), which may prevent establishment of forest stands. In North America, these natural meadows and their nutrient-rich soils were considered to have facilitated early settlement by European pioneers (Schott 1934; Ruedemann and Schoonmaker 1938).

In cases of less resistant substrate, deep erosion channels rapidly develop (see also Photo 4.11), and great quantities of sediments are translocated downstream. The ground water table often drops to such a great depth that willows die off due to moisture deficiency and grass vegetation gets increasingly scattered. Consequently, erosion increases. It will take many years until a closed plant cover will have become established again (Neff 1957; Rutherford 1964). Thus, the way succession goes on depends on the local conditions in the floodplain, and space- and time-staggered cycle-like developments may occur every time.

4.2.3 Kangaroo Rats

The function of several terrestrial species as 'keystone species' has also become evident from experiments. In the Chihuahuan desert (southern Arizona; Brown and Heske 1990) and in New-Mexico (Fields et al. 1999), for example, the role of kangaroo rats in the steppe ecosystem was studied. Alternatively, all rodents, the native kangaroo rat species (Banner-tailed kangaroo rat, *Dipodomys spectabilis*; Ord's kangaroo rat, *Dipodomys ordii*; Merriam's kangaroo rat, *Dipodomys merriamii*) and only the biggest of them (*Dipodemus spectabilis*) were excluded from different sampling sites. Twelve years later, the original shrub steppe had turned into grass steppe. Selective consumption of large seeds and effects on soil conditions by the kangaroo rats were the decisive factors. However, the location of the study area between two vegetation formations apparently had favored the change. A similar effect can usually be observed in the ectone between grassland and desert the development of which usually tends towards desert under grazing impact (MacMahon and Wagner 1985). On sample areas, from which only horses and cattle had been excluded, changes were nowhere near as spectacular as on the areas from which kangaroo rats had been removed. This means that the selective consumption of large seeds and effects on soil conditions by the three taxonomically and ecologically closely related kangaroo rat species were the driving factors. Brown and Heske (1990) called them a 'keystone species guild', as the removal of banner-tailed kangaroo rats alone, which significantly influenced the distribution of the other rodents, did not satisfactorily explain the vegetation change. On the other hand, some other rodents that had remained after removal of the kangaroo rats could not prevent the change from shrub steppe to grass steppe.

Studies on the impact of the banner-tailed kangaroo rat on plant community structure and species dominance in the transition zone from shortgrass steppe to desert steppe (north of Albuquerque, New Mexico) corroborate the role of kangaroo rats as a 'keystone species'. Two areas were compared, one covered with blue grama grass (*Bouteloa gracilis*), which prevails on the shortgrass steppe, and one covered with black grama grass (*Bouteloa eriopoda*), which is dominant in the desert steppe.

On both sample areas, black grama covered the margin of the kangaroo rat mounds, whereas the coverage by blue grama was comparatively low on all mounds and at their margins. Obviously, the response of blue grama grass to disturbances is independent from its position within the surrounding vegetation (*Bouteloa gracilis*). On the other hand, the coverage by black grama grass is much higher on the microsites at the mound rims that are dominated by blue grama grass, compared to the undisturbed sites. The coverage by perennial grasses and the high amount of herbs, shrubs and succulents on the mound margins, which is relatively low compared to sites dominated by blue grama, indicates that vegetation around disturbed sites plays a major role in the dynamics initiated by the kangaroo rats.

References

Abbadie, L., & Lepage, M. (1989). The role of subterranean fungus comb chambers (Isoptera, Macroterminae) in soil nitrogen cycling in a preforest savanna (Côte d'Ivoire). *Soil Biology and Biochemistry, 21*(8), 1067–1071.

Anderson, C. B., Soto, N., Cabello, J. L., Pastur, G. M., Lencinas, M. V., Wallem, P. K., Atúnez, D., & Davis, E. (2012). Castor canadensis Kuhl (North American Beaver). Building effective alliances between research and management to mitigate the impacts of an invasive ecosystem engineer: Lessons from the study and control of Castor canadensis in the Fuegian Archipelago. In R. A. Francis (Ed.), *A handbook of global freshwater invasive species* (pp. 343–355). London: Earthscan.

Arner, D. H., & Hepp, G. R. (1989). Beaver pond wetland: A southern perspective. In L. M. Smith, R. L. Pederson, & R. M. Kaminski (Eds.), *Habitat management for migrating and winter waterfowl in North America* (pp. 117–128). Lubbock: Texas Technical University Press.

Arner, D. H., Baker, D. W., & Herring, B. (1969). An inventory and study of beaver impounded water in Mississippi. In *Proceedings 23. Annual conference of the Southeastern Association of Game and Fish Commissioners* (pp. 110–128).

Arshad, M. A. (1982). Influence of the termite Macrotermes michaelseni (Sjöst) on soil fertility and vegetation in a semi-arid savannah ecosystem. *Agro-Ecosystems, 8*, 47–58.

Baker, B. W., & Hill, E. P. (2003). Beaver (Castor canadensis). In G. A. Feldhamer, B. C. Thompson, & J. A. Chapman (Eds.), *Wild mammals of North America: Biology, management, and conservation* (2nd ed., pp. 288–310). Baltimore: The Johns Hopkins University Press.

Barnes, W. J., & Dibble, E. (1988). The effect of beaver in riverbank forest succession. *Canadian Journal of Botany, 66*, 40–44.

Baskin, L. M. (2011). Predators as determinants of beaver alertness and shelter-making behavior. In G. Sjöberg & J. P. Ball (Eds.), *Restoring the European beaver: 50 years of experience* (pp. 271–280). Sofia/Moscow: Pensoft.

Beard, E. (1953). The importance of beaver in waterfowl management at the Seney National Wildlife Refuge. *Journal of Wildlife Management, 17*, 392–436.

Becker, T. (2001). Die Feenkresie des Kaokolandes. *Naturwissenschaftliche Rundschau, 54*(8), 430–431.

Becker, T., & Getzin, S. (2000). The fairy circles of Kaokoland (North-West Namibia) – Origin, distribution, and characteristics. *Basic and Applied Ecology, 1*, 149–159.

Beier, P., & Barret, R. H. (1987). Beaver habitat use and impact in Truckee River Basin, California. *Journal of Wildlife Management, 51*, 794–799.

Bennett, A. F. (1999). *Linkages in the landscape. The role of corridors and connectivity in wildlife conservation*. Gland/Switzerland/Cambridge, UK: IUCN.

Bergstrom, D. (1985, October). *Beaver: Biologists 'rediscover' a natural resource*. USDA Forest Service, Forestry Research West (pp. 1–5). Fort Collins: Colorado.

Bishoa Menea, K., & Boloy, N. (1995). Termitary soil and died peanut straw as market-garden fertilisers in Yangambi (Zaire). *Cahiers Agricultures, 4*(2), 125–128.

Black, H. I. J., & Okwakol, M. J. N. (1997). Agricultural intensification, soil biodiversity and agrosystem function in the tropics: The role of termites. *Applied Soil Ecology, 6*, 37–53.

Boyer, P. (1975). Etude particulère des trois termitières de Bellicosithermes et leur action sur les sols tropicaux. *Annales des Sciences Naturelles, Zoologie, 17*, 273–496.

Brown, L. (1972). *The life of African plains*. New York: McGraw-Hill.

Brown, J. H., & Heske, E. J. (1990). Control of a desert-grassland transition by a keystone rodent guild. *Science, 250*, 1705–1707.

Bulluck, J. F., & Rowe, M. P. (2013). The use of southern Appalachian wetlands by breeding birds, with a focus on neotropical migratory birds. *The Wilson Journal of Ornithology, 118*(3), 399–410.

Butler, D. R. (1995). *Zoogeomorphology: Animals as geomorphic agents* (2nd ed., 2008). Cambridge: Cambridge University Press.

Butler, D. R. (2012). Characteristics of beaver ponds on deltas in a mountain environment. *Earth Surface Processes and Landforms*, Special Issue on Geomorphology and Ecosystems, accepted for publication. doi:10.1002/esp.3218.

Butler, J. H. A., & Buckerfield, J. C. (1979). Digestion of ligin by termites. *Soil Biology and Biochemistry, 11*, 507–513.

Butler, D. R., & Malanson, G. P. (1994). Canadian landform examples. Beaver landforms. *Canadian Geographer, 38*, 76–79.

Buxton, R. D. (1981). Termites and the turnover of dead wood in an arid tropical environment. *Oecologia, 9*, 379–384.

Coaton, W. G. H. (1958). *The homotermid harvester termites of South Africa* (Science Bulletin 375). Union of South Africa, Department of Agriculture, Division of Entomology, Pretoria.

Coaton, W. G. H. (1962). The origin and development of massive vegetated termite hills in Northern Rhodesia. *African Wildlife, 16*, 159–166.

Coaton, W. G. H., & Sheasby, J. L. (1972). Preliminary report on a survey of the termites (Isoptera) of South West Africa. *Cimbebasia Memoir, 2*, 1–129.

Collen, P., & Gibson, R. J. (2001). The general ecology of beavers (Castor spp.), as related to their influence on stream ecosystems and riparian habitats, and the subsequent effects on fish – A review. *Reviews in Fish Biology and Fisheries, 10*, 439–461.

Collins, N. M. (1977). *The population ecology and energetics of Macrotermes bellicosus (Smeathman) Isoptera*. Dissertation, London.

Collins, N. M. (1981a). The role of termites in the decomposition of wood and leaf litter in the southern Guinea savanna of Nigeria. *Oecologia, 51*, 389–399.

Collins, N. M. (1981b). The utilization of nitrogen resources by termites (Isoptera). In J. A. Lee, S. McNeill, & I. H. Rorison (Eds.), *Nitrogen as an ecological factor* (pp. 381–412). Oxford: Blackwell.

Cooke, H. A., & Zack, S. (2008). Influence of beaver dam density on riparian areas and riparian birds in shrubsteppe of Wyoming. *Western North American Naturalist, 68*(3), 365–373.

Cosar, H. G. (1934). *Die Termiten in der afrikanischen Landschaft*. Beiheft, Mitteilungen der Geographischen Gesellschaft Rostock 2. Rostock: Verlag Leopold.

Cox, W. T. (1940). The beaver – Friend of the forest. *American Forests 46*, 448–540 and 476–477.

Crome, F. J. H., & Bentruppenbaumer, J. (1993). Special people, a special animal and a special vision: The first step to restoring a fragmented tropical landscape. In D. A. Saunders, R. J. Hobbs, & P. R. Ehrlich (Eds.), *Nature conservation 3. The reconstruction of ecosystems* (pp. 267–279). Chipping Norton: Surrey Beatty & Sons Pty Ltd.

Darlington, J. P. E. (1982). The underground passages and storage pits in foraging a nest of the termite Macrotermes michaelseni in Kajiado, Kenya. *Journal of Zoology (London), 198*, 237–247.

Darlington, J. P. E. C. (1984). Two types of mounds built by the termite Macrotermes subhyalinus in Kenya. *Insect Science Applications, 5*, 481–491.
Darlington, J. P. E. C. (1985). Lenticular soil mounds in the Kenya highlands. *Oecologia, 66*, 116–121.
Denney, R. N. (1952). *A summary of North American beaver management* (Current Report 28). Denver: Colorado Fish and Game Department.
Dietrich, U. (1985). Beobachtungen am Kanadischen Biber Castor canadensis in einem Einbürgerungsgebiet auf der Insel Feuerland, Südamerika. *Säugetierkundliche Mitteilungen, 32*, 241–244.
Djoshkin, W. W., & Safonov, W. G. (1972). *Die Biber der Alten und Neuen Welt* (Neue Brehm-Bücherei 437). Wittenberg Lutherstadt: Ziemsen.
Donkor, N. T., & Fryxell, J. M. (1999). Impact of beaver foraging on structure of lowland forests of Algonkin Provincial Park, Ontario. *Forest Ecology and Management, 118*, 83–92.
Donovan, S. E., Eggleton, P., Dubbin, W. E., Batchelder, M., & Dibog, L. (2001). The effect of soil-feeding termite, Cubitermes fungifaber (Isoptera; Termitidae) on soil properties: Termites may be an important source of soil microhabitat heterogeneity in tropical forests. *Pedobiologia, 45*, 1–11.
Durka, W., Babik, W., DuCroz, J. F., Heidecke, D., Rosell, F., Samjaa, R., Saveljev, P., Stubbe, A., Ulevičius, A., & Stubbe, M. (2005). Mitochondrial phylogeography of the Eurasian beaver (Castor fiber L.). *Molecular Ecology, 14*, 3843–3956.
Eggleton, P., Bignell, D. E., Sands, W. A., Mawdsley, N. A., Lawton, J. H., Wood, T. G., & Bignell, N. C. (1996). The diversity, abundance, and biomass of termites under different levels of disturbance in the Mbalmayo Forest Reserve, southern Cameroon. *Philosophical Transactions: Biological Sciences, 351*, 51–68.
Elkins, N. Z., Sabol, G. V., Timothy, J. W., & Whitford, W. G. (1986). The influence of subterranean termites on the hydrological characteristics of a Chihuahuan desert ecosystem. *Oecologia (Berlin), 68*, 521–528.
Fields, M. J., Coffin, D. P., & Gosz, J. R. (1999). Burrowing activities of kangaroo rats and patterns in plant species dominance at a shortgrass steppe-desert grassland ecotone. *Journal of Vegetation Science, 10*, 123–130.
Flemming, P. A., & Loveridge, J. P. (2003). Miombo woodlands termite mounds: Resource islands for small vertebrates? *Journal of Zoology (London), 259*(2), 1–8.
Freeland, W. J., & Kanzen, D. H. (1974). Strategies in herbivory by mammals. The role of plant secondary compounds. *The American Naturalist, 108*, 269–289.
Gard, R. (1961). Effects of beaver on trout in Sagehen Creek. *Journal of Wildlife Management, 25*, 221–242.
Gill, D. (1972). The evolution of a discrete beaver habitat in the Mackenzie River Delta, Northwest Territories. *The Canadian Field Naturalist, 86*, 233–239.
Glover, P. E., Trump, E. C., & Wateridge, L. E. D. (1964). Termitaria and vegetation patterns on the Loita plains of Kenya. *Ecology, 52*, 367–377.
Godoy, J. C. (1963). *Fauna Silvestre* (Evaluación de los recursos naturales de la Argentina 8). Buenos Aires: Consejo Federal de Investiones.
Goudie, A. S. (1988). The geomorphological role of termites and earthworms in the tropics. In H. A. Viles (Ed.), *Biogeomorphology* (pp. 166–192). Oxford: Blackwell.
Grasse, J. E. (1949). Beaver and trout. *Wyoming Wildlife 13*(11), 4–13, 34.
Grasse, J. E. (1951). Beaver ecology and management in the Rockies. *Journal of Forestry, 49*(1), 3–6.
Grassé, P. P. (1984). *Termitologia: Foundation et construction. II*. Paris: Masson.
Grassé, P. P., & Noirot, C. (1958). Le meule des termites championnistes et sa significance symbiotique. *Annales des sciences naturelles. Zoologie et biologie animale, 20*(11), 113–128.
Guernell, A. M. (1998). The hydrogeomorphological effects of beaver dam-building activity. *Progress in Physical Geography, 22*, 167–189.
Gumnior, M., & Thiemeier, H. (2003). Der Einfluss zweier Termitenarten auf Genese und Stoffhaushalt von Sandböden in Nordost-Nigeria. *Geoöko, 24*, 321–336.

Hall, E. R. (1971). *Ecology of beaver and selection of prey by wolves in central Ontario*. M.S. Thesis, University of Toronto.

Hall, E. R., & Kelson, K. R. (1959). *Mammals of North America*. New York: Ronald Press.

Halley, D. J. (2011). Sourcing Eurasian beaver Castor fiber stock for reintroductions in Great Britain and Western Europe. *Mammal Review, 41*(1), 40–53.

Harris, W. V. (1961). *Termites: Their recognition and control*. London: Longmans.

Harthun, M. (1999). Zur Bedeutung der Biberwiesen in der mitteleuropäischen Urlandschaft. In B. Gerken & M. Görner (Eds.), *Europäische Landschaftsentwicklung mit großen Weidetieren* (Natur und Kulturlandschaft 3, pp. 146t–155t). Höxter/Jena: Huxaria Druck und Verlag.

Harthun, M. (2000). Einflüsse der Stauaktivität des Bibers (Castor fiber albicaulis) auf physikalische und chemische Parameter von Mittelgebirgs-Bächen (Hessen, Deutschland). *Limnologica, 30*, 21–35.

Hartman, G. (1996). Habitat selection by European beaver (Castor fiber) colonizing a boreal landscape. *Journal of Zoology (London), 240*, 317–325.

Hawkins, L., & Nicoletto, P. F. (1992). Kangaroo rat burrows structure the spatial organization of ground dwelling animals in a semiarid grassland. *Journal of Arid Environments, 23*, 199–208.

Hay, K. G. (2009). Succession of beaver ponds in Coloradao 50 years after beaver removal. *Jounal of Wildlife Management, 74*(8), 1732–1736.

Heidecke, D. (1985). Ergebnisse und Probleme beim Schutze des Elbebibers. *Naturschutzarbeit in Berlin und Brandenburg, 21*, 6–12.

Heim, R. (1963). Les termitomyces de la Répiblique Centrafricain I. *Cahiers Maboke, 1*, 20–26.

Hesse, P. R. (1955). A chemical and physical study on the soils of termite mounds in East Africa. *Journal of Ecology, 43*, 449–461.

Heth, G. (1991). The environmental impact of subterranean mole rats (Spalax ehrenbergii) and their burrows. *Symposium of the Zoological Society London, 63*, 265–280.

Hogdon, H. E., & Lancia, R. A. (1983). Behavior of the North American beaver, Castor Canadensis. *Acta Zoologica Fennica, 174*, 99–103.

Holtmeier, F.-K. (1966). Die ökologische Funktion des Tannenhähers im Zirben-Lärchenwald und an der Waldgrenze im Oberengadin. *Journal für Ornithologie, 4*, 337–345.

Holtmeier, F.-K. (1986). Ökologische Studien in Lappland und Südfinnland. *Gesellschaft zur Förderung der Westfälischen Wilhelms-Universität, 1984/85*, 22–26.

Holtmeier, F.-K. (1987). Biber und Steinwild. Ökologische Aspekte der Wiederansiedlung von Tieren in ihren ehemaligen Lebensräumen. *Münstersche Geographische Arbeiten, 26*, 99–117.

Holtmeier, F.-K. (1999). Tiere als landschaftsökologische Faktoren. Arbeiten aus dem Institut für Landschaftsökologie der Westfälischen Wilhlems-Universitat 6.

Holtmeier, F.-K. (2002). *Tiere in der Landschaft. Einfluss und ökologische Bedeutung*. Stuttgart: UTB 8230.

Holtmeier, F.-K. (2009). *Mountain timberlines. Ecology, patchiness, and dynamics* (Advances in Global Change Research 36). Dordrecht: Springer Science+Business Media B. V.

Holtmeier, F.-K. (2012). Impact of wild herbivorous mammals and birds on the altitudinal and northern treeline ecotones. *LandscapeOnline, 30*, 1–28. doi:10.3097/LO.201230.

Howard, R. J., & Larson, J. S. (1985). A stream habitat classification system for beaver. *Journal of Wildlife Management, 49*(1), 19–25.

Huntly, N. (1995). How important are consumer species to ecosystem functioning? In C. G. Jones & J. H. Lawton (Eds.), *Linking species and ecosystem* (pp. 72–83). New York: Chapman & Hall.

Huntly, N., & Inouye, R. (1988). Pocket gophers in ecosystems: Patterns and mechanisms. *BioScience, 38*(11), 786–793.

Ives, R. I. (1942). The beaver-meadow complex. *Journal of Geomorphology, 5*, 191–203.

Jenkins, S. H. (1975). Food selection by beavers. A multidimensional contingency table analysis. *Oecologia, 21*, 157–173.

Jenkins, S. H. (1979). Seasonal and year-to-year differences in food selection by beavers. *Oecologia, 44*, 112–116.

Jenkins, S. H. (1980). A size-distance relation in food selection by beavers. *Ecology, 61*(4), 740–746.
Joachim, A. W. R., & Kandiah, S. (1940). Studies on Ceylon soils XVI. A comparison of soils from termite mounds and adjacent land. *Tropical Agriculturist Magazine of the Ceylon Agricultural Society, 194*, 119–129.
Johnson, C. E. (1927). The beaver in the Adirondacks: Its economics and natural history. *Roosevelt Wildlife Bulletin, 4*(4), 501–541, Syracuse.
Johnston, C. A., & Naiman, R. J. (1987). Boundary dynamics at the aquatic-terrestrial interface: The influence of beaver and geomorphology. *Landscape Ecology, 4*(1), 5–19.
Johnston, C. A., & Naiman, R. J. (1990a). Browse selection by beaver: Effects in riparian forest composition. *Canadian Journal of Forest Research, 20*, 1036–1043.
Johnston, C. A., & Naiman, R. J. (1990b). Aquatic patch creation in relation to beaver population trend. *Ecology, 71*(4), 1617–1621.
Jones, J. A. (1990). Termites, soil fertility, and carbon cycling in dry tropical Africa: A hypothesis. *Journal of Tropical Ecology, 6*, 291–305.
Josens, G. (1992). The soil fauna of tropical savannas. In F. Bourlière (Ed.), *Ecosystems of the world* (Vol. 13, pp. 505–524). Amsterdam: Elsevier.
Jürgens, N. (2013). The biological underpinnings of Namib Desert fairy circles. *Science, 339*, 1618–1621.
Kämpfer, M. (1967). Biber (Castor) – Biologie, Ökologie, Vorkommen, Wiederreinbürgerung. Bundesanstalt für Vegetationskunde, Naturschutz und Landschaftspflege, Bibliographie Nr. 5.
Keya, S. O., Mureria, N. K., & Arshad, M. A. (1982). Population dynamics of soil microorganisms in relation to proximity of termite mounds in Kenya. *Journal of Arid Environments, 5*, 353–359.
Konate, S., Merdaci, K., Lepage, M., Le Roux, X., & Tessier, D. (1998, June 20–26). *Effect of termitaria on spatio-temporal variations of soil water in a tropical savanna.* Proceedings 16. World Congress of Soil Science, Montpellier, France, Symposium No. 18, Contribution 1129.
Lal, R. (1987). *Tropical ecology and physical edaphology.* Chichester/New York/Brisbane/Toronto/Singapore: Wiley.
Leary, R. J. (2012). *Landscape and habitat attributes influencing beaver distribution* (MNR Capstone Project). Logan: Utah State University.
Lee, K. E., & Wood, T. G. (1971). *Termites and soil.* London/New York: Academic.
Lepage, M., Abbadie, I., Konate, S., Merdaci, K., & Quedraogo, P. (1998, August 20–26). Structures related to termite activity and organic matter dynamics at different spatio-temporal scales. In *Proceedings 16. World Congress of Soil Science*, Montpellier, Symposium No. 9, Contribution 1143.
Leuthold, W. (1977a). *African ungulates.* Berlin/Heidelberg/New York: Springer.
Leuthold, W. (1977b). Spatial organization and strategy of habitat utilization of elephants in Tsavo National Park, Kenya. *Zeitschrift fur Saugetierkunde, 42*(6), 358–379.
Lobry de Bryn, L. A., & Conacher, A. J. (1990). The role of termites and ants in soil modification: A review. *Australian Journal of Soil Research, 28*, 55–93.
Löffler, E. (1996). *Bioturbation als Faktor der Reliefgenese in den wechselfeuchten Tropen.* Leipzig: Klett-Perthes.
Logan, F. G. (1992). Termites (Isoptera) – A pest or resource for small farmers in Africa. *Tropical Science, 32*, 71–79.
López-Hernández, D. (2001). Nutrient dynamics (C, N. and P) in termite mounds of the Orinoco Llanos (Venezuela). *Soil Biology and Biochemistry, 23*, 747–753.
Lovegrove, B. (1993). *The living deserts of Southern Africa.* Vlaeberg: Fernwood Press.
MacDonald, D. (1956). *Beaver carrying capacity of certain mountain streams in North Park, Colorado.* M.S. Thesis, Colorado State University, Fort Collins.
MacMahon, J. A., & Wagner, F. H. (1985). The Mojave and Chihuahuan deserts of North America. In M. Evenari and I. Noy-Meir (Eds.), *Warm deserts ecosystems* (Ecosystems of the World 12, pp. 105–202). Amsterdam: Elsevier.

Matsumoto, T. (1976). The role of termites in an equatorial rain forest ecosystem of West Malaysia. *Oecologia, 22*, 153–178.
Medin, D. E., & Torquemada, K. E. (1988). *Beaver in western North America: An annotated bibliography 1966 to 1986* (General Technical Report INT 242). Ogden: USDA Forest Service, Intermountain Research Station.
Mertens, R. (1961). Tier und Landschaft. Zoologische Unterlagen zur Landschaftskunde. *Frankfurter Geographische Hefte, 37*, 31–38.
Milne, G. V. (1947). A soil reconnaissance journey through parts of Tanganyka Territory, December 1935 to February 1936. *Journal of Ecology, 35*, 192–264.
Moll, E. (1994). Fairy rings in Kaokoland. In *Proceedings of the 13th Plenary Meeting ETFAT, Zoma, Malawi*, pp. 1203–1210.
Müller, P. (1977). *Tiergeographie*. Stuttgart: Teubner.
Naiman, R. J. (1988). Animal influences on ecosystem dynamics. *BioScience, 38*, 750–752.
Naiman, R. J., Melillo, J. M., & Hobbie, J. E. (1986). Ecostem alteration of a boreal forest stream by beaver (Castor canadensis). *Ecology, 67*, 1254–1269.
Naiman, R. J., Johnston, C. A., & Kelley, J. C. (1988). Alteration of North American streams by beaver. *BioScience, 38*(11), 753–762.
Naiman, R. J., Pinnay, G., Johnston, C. A., & Pastor, J. (1994). Beaver influences on the long-term biogeochemical characteristics of boreal forests drainage networks. *Ecology, 75*, 905–921.
Nash, M. H., & Whitford, W. G. (1995). Subterranean termites: Regulators of soil organic matter in the Chihuahuan Desert. *Biology and Fertility of Soils, 19*, 15–18.
Neff, D. J. (1957). Ecological effects of beaver habitat abandonment in the Colorado Rockies. *Journal of Wildlife Management, 21*, 80–84.
Nyamapfene, K. W. (1986). The use of termite mounds in Zimbabwe peasant agriculture. *Tropical Agriculture (Trinidad), 63*(2), 191–192.
Nye, P. H. (1955). Some soil-forming processes in the humid tropics. IV. The action of the soil fauna. *Journal of Soil Science, 6*, 73–83.
O'Brien, D. F. (1938). *A qualitative and quantitative food habitat study of beavers in Maine*. M.S. thesis, University of Maine.
Odum, E. (1991). *Prinzipien der Ökologie. Lebensräume, Stoffkreisläufe, Wachstumsgrenzen*. Heidelberg: Spektrum der Wissenschaft.
Okullo, P., Greve, P. M. K., & Moe, S. R. (2013). Termites, large herbivores, and herbaceous plant dominance structure small mammal communities in savannahs. *Ecosystems, 16*, 1002–1012.
Oliveira, L. A., & Paiva, W. O. (1985). Use of termite nests and chicken manure as fertilizers for lettuce in red yellow podzolic soils of the Manaus region. *Acta Amazonica, 15*(1–2), 13–18.
Paine, R. T. (1966). Food web complexity and species diversity. *American Naturalist, 100*, 65–75.
Paine, R. T. (1969). A note on trophic complexity and community stability. *American Naturalist, 103*, 91–93.
Pendleton, R. I. (1941). Some results of termite activity in Thailand soils. *Thai Science Bulletin, 3*(2), 29–35.
Pendleton, R. I. (1942). Importance of termites in modifying certain Thailand soils. *Journal of the American Society of Agronomy, 34*, 340–344.
Perry, D. A. (1994). *Forest ecosystems*. London/Baltimore: Johns Hopkins University Press.
Persico, L., & Meyer, G. (2009). Holocene beaver damming, fluvial geomorphology, and climate in Yellowstone National Park, Wyoming. *Quaternary Research, 71*, 340–353.
Pilleri, G. (1960). Biber. *Umschau, 14*, 420–424.
Pollock, M. M., Naiman, R. J., Erickson, H. E., Johnston, C. A., Pastor, J., & Pinay, G. (1995). Beaver as engineers: Influences on biotic and abiotic characteristics of drainage basins. In C. G. Jones & J. H. Lawton (Eds.), *Linking species and ecosystem* (pp. 117–126). New York: Chapman & Hall.
Pomeroy, D. E. (1976). Some effects of mound-building termites on soils in Uganda. *Journal of Soil Science, 27*, 377–394.
Pullan, R. A. (1979). Termite hills in Africa: Their characteristics and evolution. *Catena, 6*, 267–291.

Reid, K. A. (1951). Planning for wildlife on a managed forest. *Journal of Forestry, 49*(6), 436–439.

Remillard, M. M., Gruendling, G. K., & Bogucki, D. J. (1987). Disturbance in beaver (Castor canadensis Kuhl) and landscape heterogeneity. In M. G. Turner (Ed.), *Landscape heterogeneity and disturbance* (pp. 103–123). New York: Springer.

Remmert, H. (1980). *Ökologie*. Berlin/Heidelberg/New York: Springer.

Richard, P. B. (1983). Mechanisms and adaptation in the constructive behavior of the beaver (C. fiber L.). *Acta Zoologica Fennica, 174*, 105–108.

Roller, S. (2001). Gewässergütedaten – Auswertungsmöglichkeiten am Beispiel des Bibers (Castor fiber). *Gewässer-Info, 20*(Ferbruar 2002), 106–108.

Romašov, B. V. (1992). Krankheiten der Biber. In *Semiaquatische Säugetiere* (Wissenschaftliche Beiträge der Universität Halle, pp. 199–203).

Rosell, F., Bozsér, O., Collen, P., & Parker, H. (2005). Ecological impact of beavers Castor fiber and Castor canadensis and their ability to modify ecosystems. *Mammal Review, 35*(3/4), 248–276.

Ruedemann, R., & Schoonmaker, W. J. (1938). Beaver as geological agents. *Science, 88*(2292), 523–525.

Rutherford, W. H. (1955). Wildlife and environmental relationships of beavers in Colorado forest. *Journal of Forestry, 53*, 803–806.

Rutherford, W. H. (1964). *The beaver in Colorado, its biology, ecology, management and economics*, (Colorado Game, Fish and Parks Department, Technical Publication 17, pp. 1–49), Denver.

Salyer, J. C. (1935). Preliminary report on beaver-trout investigations. *American Game, 24*(1), 6 and 13–15.

Sands, W. A. (1969). The association of termites and fungi. In K. Krishna & F. M. Weesner (Eds.), *Biology of termites* (Vol. 1, pp. 495–524). New York/London: Academic.

Scheffer, P. M. (1938). The beaver as an upstream engineer. *Soil Conservation, 3*(7), 178–181.

Schneider, E. (1985). Erfahrungen zum Management lokaler Vorkommen des Bibers Castor fiber L. in der Bundesrepublik Deutschland. *Angewandte Zoology, 1–2*, 191–203.

Schneider, E., & Schulte, R. (1985). Befunde zu den Habitatansprüchen des Europäischen Bibers (Castor fiber L.) aus einem Wiederansiedlungsversuch an einem Mittelgebirgsbach der nördlichen Eifel. *Zeitschrift für Angewandte Zoologie, 72*(1/2), 167–179.

Schott, C. (1934). Kanadische Biberwiesen - ein Beitrag zur Frage der Wiesenbildung. *Zeitschrift der Gesellschaft für Erdkunde*, 9/10, 370–374, Berlin.

Seastedt, T. R., Reichmann, O. J., & Todd, T. C. (1986). Microarthropods and nematodes in kangaroo rat burrows. *Southwestern Naturalist, 31*, 114–116.

Sharp, L. A., & Barr, W. F. (1960). Preliminary investigation of harvester ants on Southern Idaho rangeland. *Journal of Rangeland Management, 13*, 131–134.

Sinclair, A. R. E. (1979). The Serengeti ecosystem. In A. R. E. Sinclair & M. Norton-Griffiths (Eds.), *Serengeti. Dynamics of an ecosystem* (pp. 31–45). Chicago/London: University of Chicago Press.

Slough, B. G., & Sadleir, R. M. F. S. (1977). A land capability classification system for beaver (Castor canadensis Kuhl). *Canadian Journal of Zoology, 55*, 1324–1355.

Smith, B. H. (1980). *Riparian willow management: Its problems, potentials, within the scope of multiple use on public lands*. Lander: Shrub Ecology Workshop.

Smith, D. W., Murphy, S. C., Phillips, M. K., & Crabtree, R. (1996). *Beaver survey: Yellowstone National Park* (USDI: Report on File). Mammoth: National Park Service.

Snodgrass, J. W. (1997). Temporal and spatial dynamics of beaver-created patches as influenced by management practices in a south-eastern North American landscape. *Journal of Applied Ecology, 34*, 1043–1056.

Stegeman, L. C. (1954). The production of aspen and its utilisation by beaver on the Huntington Forest. *Journal of Wildlife Management, 18*, 348–358.

Stocker, G. (1985). *Biber (Castor fiber L.) in der Schweiz*. Eigenössische Anstalt für das forstliche Versuchswesen, Berichte 274. Teufen: Kommissionsverlag Flück-Wirth.

Stoops, G. (1964). Application of some pedological methods to the analysis of termite mounds. In A. Boullion (Ed.), *Études sur les termites Africains* (pp. 379–398). Léopoldville: Lépoldville University.

Tansley, A. G. (1939). *The British Isles and their vegetation*. Cambridge: Cambridge University Press.

Trapnell, C. G., Friend, M. T., Chamberlain, G. T., & Birch, H. F. (1976). The effects of fire and termites on a Zambia woodland soil. *Journal of Ecology, 64*, 577–588.

Troll, C. (1936). Termitensavannen. In H. Louis & W. Panzer (Eds.), *Landeskundliche Forschung, Festschrift Norbert Krebs* (pp. 275–312). Stuttgart: Engelhorns Nachf.

Troll, C. (1963). Geographische Luftbildinterpreatation. *Archiv für Internationale Photogrammetrie, 14*, 266–274.

Troll, C. (1970). Landschaftsökologie (Geoökologie) und Biogeocoenologie, eine terminologische Studie. *Revue Roumaine de Géologie, Géophysique et Géographie, Series Géographie, 14*(1), 9–18.

Tschinkel, W. (2012). The life cycle and life span of the Maibian fairy circles. *PLoS One, 7*(6), e38056. doi:10.371/journal.pone.0038056.

Vásquez, D. P. (2002). Multiple effects of introduced mammalian herbivores in a temperate forest. *Biological Invasions, 4*, 175–191.

Warren, E. R. (1926). Notes on the beaver colonies in the Longs Peak Region of Estes Park, Colorado. *Roosevelt Wildlife Annals, 1*(1–2), 192–234.

Watson, R. M. (1967). *The population ecology of wildebeest (Connochaetes taurinus albojubatus Thomas) in the Serengeti*. Ph.D. thesis, Cambridge University.

Watson, J. P. (1977). The use of mounds of the termite Macrotermes falciger (Gerstäcker) as a soil amendment. *Journal of soil Science, 28*(Suppl.), 664–672.

Watt, A. D., Stork, N. E., McBeath, C., & Lawson, G. L. (1997). Impact of forest management on insect abundance and damage in a lowland tropical forest in southern Cameroon. *Journal of Applied Ecology, 34*, 985–998.

Whitford, W. G. (1991). Subterranean termites and productivity of desert rangelands. *Sociobiology, 19*, 235–242.

Whitford, W. G. (1999). Comparison of ecosystem processes in the Nama-Karoo and other deserts. In W. R. J. Dean & S. J. Milton (Eds.), *The Karoo ecological patterns and processes* (pp. 292–302). Cambridge: Cambridge University Press.

Whitford, W. G., Steinberger, Y., & Ettershank, G. (1982). Contribution of subterranean termites to the 'economy' of Chihuahuan desert ecosystem. *Oecologia, 55*, 298–302.

Whitford, W. G., Stinnet, K., & Anderson, J. (1988). Decomposition of roots in a Chihuahuan desert ecosystem. *Oecologia, 75*, 8–11.

Wilsson, L. (1971). Observations and experiments on the ethology of the European beaver (Castor fiber L.). *Viltrevy, 8*(3), 113–266.

Woo, M.-K., & Waddington, J. M. (1990). Effects of beaver dams on subarctic wetland hydrology. *Arctic, 43*(3), 223–230.

Wood, T. G., & Sands, W. A. (1978). Their role of termites in ecosystems. In M. Vaughan (Ed.), *Production ecology of ants and termites* (pp. 245–292). Cambridge: Cambridge University Press.

Wood, T. G. (1988). Termites and the soil environment. *Biology and Fertility of Soil, 6*, 228–236. Springer.

Wright, J. P., Jones, C. G., & Flecker, A. S. (2002). An ecosystem engineer, the beaver, increases species richness at the landscape scale. *Oecologia, 132*, 96–101.

Yeager, L. E., & Rutherford, W. H. (1957). *An ecological basis for beaver management in the Rocky Mountain region*. Transactions of the 22nd North American Wildlife Conference, March 5 and 6, 1957. Washington, DC: Wildlife Management Institute.

Zahner, V. (1997). *Einfluß des Bibers auf gewässernahe Wälder*. München: Herbert Utz Verlag.

Chapter 5
Introductions and Re-introductions of Animals

Faunal change due to environmental change, emergence or disappearance of land bridges, climate change and the resulting shift of vegetation zone, origin or extinction of species and invasion by alien species, is a typical evolutionary process. The immigration of the brown rat and house mouse from the Far and Middle East into Europe as well as the invasion of North America by animal species from Asia across the area of the present Bering Sea or the expansion of placental animals from North America across the Middle American land bridge to South America, where they replaced the native marsupials almost completely, are well-known examples (Huston 1994, further literature therein).

We are also witnessing such natural processes at present. The collared turtle dove (*Streptopelia decaoto*), for example, which is originally native to the area from the Mediterranean to the Indian Subcontinent, has been spreading all over western Europe throughout the last 70 years, and has meanwhile arrived in Lapland and Iceland (e.g., Bezzel 1985; Kasparek 1996). During the twentieth century, the cattle egret (*Bubulcus ibis*) expanded within a short time from southern Europe and Africa to South and North America, Australia and to some remote islands (e.g., Austin 1963). A strong storm from the southeast in 1937 carried field fares (*Turdus pilarius*) to southern Greenland, where they have built up a viable population (Sedlag 1995). The serin (*Serinus serinus*), a species that could only be found in the northern Mediterranean (a small Swiss area included) around 1800 now occurs all over Central Europe, and has also arrived in Denmark and southwestern England (Nowak 1977). It is one of the most common songbirds to sit on telephone lines and antennas when singing (Feldmann 1988).

By contrast, natural expansion of moose, for example, from Poland and the Czech Republic into eastern Germany and Bavaria over the last few decades, is considered to be a recolonization of the original habitats where moose lived centuries ago (Nowak 1977; Steiner and Kraus 1993). Moose is a classical example of a rapidly spreading large mammal.

During historical times, translocation of animal species by man has been of major importance. There are presumably only a few regions on earth in which the

F.-K. Holtmeier, *Animals' Influence on the Landscape and Ecological Importance*,
DOI 10.1007/978-94-017-9294-3_5

fauna consists only of species that have occupied their habitats without human assistance (e.g., Roots 1976; Nowak 1981; Nowak and Zsivanovits 1982; Di Castri 1989; Atkinson 1989; Goudie 1994). Since the middle of the nineteenth century, so-called 'acclimation societies' systematically organized the introduction of exotic animals to many European and overseas countries (Podglayen-Neuwall 1982; Scanlon 1989). Some of these societies were active until the 1970s. About 30 exotic ungulate species (Barbary sheep, *Ammotragus lervia*; wild goat, *Capra aegragrus*; Siberian ibex, *Capra ibex sibirica*; oryx, *Oryx gazella*; greater Kudu, *Tragelaphus strepsiceros*; Mongolian gazelle, *Procapra subgutturosa*) have been introduced to New Mexico between 1950 and 1970 to fill up 'open niches' and to make areas that were originally lacking large game animals more attractive for hunters, particularly to solvent Americans (Podglayen-Neuwall 1982). For the same reason, mountain goats (*Oremanos americanus*) were introduced to the Olympic Mountains (Washington) during the 1920s (Sect. 6.1).

Introduction of animals has often resulted in considerable disturbances of ecosystems and their native biocoenoses. The new environments were not 'prepared' for the newcomers for several possible reasons. In many cases, native animals were not able to adjust their behavior rapidly enough to the new competitors for food and to predators, or they suffered from introduced diseases. Balanced predator-prey relationships that had developed over millennia became disturbed. Occasionally, the physical conditions (climate, food supply) did not limit population growth of the introduced animals to the same extent as was the case in their original area of distribution. In other cases, the introduced species had a greater ecological valency, a greater reproduction rate or their competitive capability was higher than in the native animals. Therefore, the native fauna often lost essential habitat structures.

In many cases, the newcomers ran out of control (e.g., Milne and Milne 1960; Tisdell 1982; Eldridge 1983; King 1990a; Bonino 1995; Eisfeld and Fischer 1996; Jaksic et al. 2002; Holtmeier 2002; Hawai'i Conservation Alliance 2007). The after-effects of the newcomers' impact have often lasted for many decades or even longer (e.g., Flook 1962; Holtmeier 1999; Gill 2000; Tanentzap et al. 2009). In a global view, about 50 % of introduced mammals and birds are considered to be pests (Bomford 2003). However, species that were re-introduced to their native habitats have also caused unexpected problems, in particular as their original habitats have changed since these species declined or became extinct. In particular, intense human utilization of landscape often impedes or even prevents integration of the 'homecomers'.

Great efforts have been and are being made to get the problems resulting from introduction and reintroduction of animal species under control. Unforeseeable responses of cause-effect relationships often require new control measures, or 'to patch up old patches' (Milne and Milne 1965). The great helplessness often felt when confronted with these problems is hardly better reflected than in the following phrase taken from a website: '*It is ironic that while we have been so successful at unintentional termination of native species, we have been remarkably unsuccessful at pest eradication*'. The complex problems will be addressed by way of examples in the following sections.

5.1 Newcomers

Animals were introduced intentionally to new environments for enrichment of the local fauna, for hunting purposes, meat and fur production, or as biological pest control (e.g., Di Castri 1989; Bonino 1995; Holtmeier 2002; Bell and Dieterich 2010), as, for example, the marine toad (*Bufo marinus*; Sect. 5.1.2.2), which was translocated from South and Central America into many countries cultivating sugarcane. In Australia, it has become a serious pest. The North American bullfrog (*Rana catesbeiana*), which was released into the backwaters of the Upper Rhine River (Germany), reproduces rapidly and feeds on almost everything: insects, other frogs, young fishes, eggs of birds, young birds, and also small mammals. The bullfrog particularly threatens the fish fauna and native amphibians (Landesanstalt für Umweltschutz Badenwürttemberg, Press information 14 August 2001). Many 'aliens' escaped or were released from private game parks.

Red deer as well as other deer species (e.g., Sika deer, *Cervus nippon*, fallow deer, *Dama dama*; white-tailed deer, *Odocoileus virgianus*; Columbian black-tailed deer, *Odocoileus hemionus columbianus*; Axis deer, *Axis axis*) have been translocated to many regions of the northern and southern hemisphere, mainly for trophy hunting, meat and skins. In many introduction areas, the aliens have caused or are still causing severe damage to the ecosystems (e.g., Eisfeld and Fischer 1996; Ratcliffe 1989). In Point Reyes National Seashore (California), for example, 28 fallow deer were introduced (between 1942 and 1954) for sport-hunting. In addition, a local landowner released eight axis deer obtained from the San Francisco zoo (1947/1948). Both fallow and axis deer have rapidly multiplied. About 275 axis deer and 875 fallow deer were estimated in 2005 (Fellers and Osbourn 2007). They damage streamside vegetation, cause erosion and thereby affect water quality (nutrient input). Moreover, they compete with native Columbian black-tailed deer (*Odocoileus hemionus comlumbianus*), re-introduced native tule elk (*Cervus elaphus nannodes*) and livestock for food. Not least, non-native deer are vectors of diseases (paraturberculosis = Johne's disease; pediculosis = louse attack or hair-loss syndrome; see Bildfell et al. 2004), that may be lethal to the native deer species (National Parks Conservation Association 2009). In 2007, the park management began exterminating non-native deer by lethal means and contraception – not without furious opposition of animals' advocates. The plan is to completely remove non-native deer from the park by 2021.

In England, introduced Sika deer (*Cervus Nippon*) expanded their range sixfold over the last three decades of the previous century, while Muntjac (*Muntiacus reevesi*) have increased ten-fold (Dolman et al. 2010). Damage to crops and trees (bark stripping) is correspondingly also increasing. In general, damage is less than with the other deer species. At high muntjac density, however, impact on preferred forage plants can be intense.

Many animals have colonized their new environment more accidentally, as, for example, ship rats (European house rats, *Rattus rattus*) that came from anchoring or wrecked ships. In the past, sailors released useful animals (mainly pigs, goats, sheep and rabbits) as 'living stock' on islands located on the travel route or where ships had landed by chance and left the animals to fend for themselves. The descendants of European wild boar, for example, are now living on all continents, with the exception of the Antarctic, and on many ocean islands. In more than 90 % of all cases, the introduction of foreign species negatively affected the new environment and the native biocoenoses (Roots 1976), and often ended in an 'ecological catastrophe'. Native species such as flightless birds, for example, became extinct by the introduction of rats, feral pigs and wild boars, dogs, cats, foxes, marten and other predators. In an extensive freshwater basin marsh (Savannas Preserve State Park, Florida), for example, where shallow waters and exposed terrain are dominated by herbs and shrubs that make the area very attractive to feral pigs, these have become a serious threat to aquatic plants and animal communities (wildland and agricultural land; Engeman et al. 2004).

The many numerous feral ungulates, particularly feral goats and pigs, as well as feral cats and dogs, have had long-lasting effects just as important as those caused by the introduced wild animals (Wodzicki 1963; Schweinfurth 1966; Challies 1975). All regions 'explored' during the age of discovery and colonization became affected in one or the other way. In particular, on the many small oceanic (Atlantic and Pacific) islands, biocoenoses and habitats experienced irreversible changes, as almost no refugia were available to the native fauna and the vegetation was not adapted to the newcomers.

Besides introduced cervids, wild boars, feral pigs and feral goats have led to the most profound ecosystem degradation and biodiversity loss. Goats were introduced by Europeans during the eighteenth and nineteenth century to many oceanic islands. Low metabolism, an efficient digestion system, and low water requirement enable goats to survive in harsh environments that are unsuitable for many other mammalian herbivores (Silanikove 2000). Goats eradicated their preferred forage plants almost completely, thus promoting expansion of other species. Trampling caused or enhanced soil erosion. Removal of feral goats has turned out to be the only efficient instrument for ecosystem restoration. Great efforts have been and are being made to free islands from feral goats. On James Island, Pinta and Santiago islands (Galápagos), for example, where the goat population was reduced or completely removed, plant species suppressed by goats for a long time could recover (Hamann 1979, 1993; Calvopina 1985). Eradication of feral goats was most successful during the last three to four decades, mainly through improved hunting techniques (e.g., helicopter hunting, use of Judas goats and specially trained hunting dogs). Up through the early twenty-first century, goats were removed from 120 islands in different regions of the world. Eradication campaigns are still underway (Campbell and Donlan 2005 and literature therein). On many large islands (e.g., Hawaiian and Australian islands), where many people still subsist on feral goats, using them economically (subsistence, recreational hunting), complete eradication has been recommended to prevent continued loss of biodiversity and further landscape degradation. Even though such eradication campaigns would be expensive, they would result in considerably conservation gains (Campbell and Donlan 2005).

Photo 5.1 Beaver habitat at Moat, Tierra del Fuego. *Nothofagus pumilio* died off due to impoundment and raised ground water table. T. Kleinebecker, February 2013

On Tierra del Fuego, beavers were introduced at the end of the 1940s (Godoy 1963; Dietrich 1985; Vásquez 2002; Anderson et al. 2012). Since then, the beaver population has increased to about 200,000 at present. Beavers have expanded across the archipelago, crossed the Straits of Magellan and begun to colonize Brunswick Peninsula (Chile). The newcomers have rapidly adapted to their new habitats and become a serious problem. Beavers use southern beech ('Lenga', *Nothofagus pumilio*) for food and dam construction. They clear-cut riparian forests by felling trees of different diameters. After having exploited the tree stands within their reach, beavers move to still undisturbed beech forests via 'beaver-made' branch channels (length often >150 m). In contrast to willows and aspen in the northern hemisphere beaver habitats, southern beech does not tolerate being flooded (Photo 5.1). Thus, the beavers have been destroying the unique original riparian southern beech forests (Jaksic et al. 2002; Choi 2008; Novillo and Ojeda 2008; Lizzaralde et al. 2004). The natural habitat carrying capacity has already been considerably exceeded long ago. If beavers are reduced, new beavers will invade across the Argentine border. However, in view of the tremendous impact on the riparian forests, agricultural land, native fauna and biodiversity, complete eradication of beavers is being planned (Malmierca et al. 2011; Anderson et al. 2012). The muskrat (*Ondatra zibethicus*) was translocated to Tierra dal Fuego simultaneously with the beaver and has also rapidly built up large populations. It is causing severe damage to shoreline embankments and dikes by burrowing (Bonino 1995; Butler 1995). Since 1954, it has been considered a 'pest' (Fabbro 1989).

The Hawaiian islands are among the regions that have experienced the introduction of a multitude of alien animal species, mainly from temperate, subtropical and tropical regions. Introduced wild sheep (e.g., mouflon, *Ovis musimon*), spotted deer (Axis deer, *Axis axis*) and black-tailed deer (*Odocoileus hemionus columbianus*), for example, damage native plants and agricultural crops, affect forests and other habitats of native birds, cause soil erosion and disperse zoonotic diseases. High reproduction rates and the absence of predators let the wild ungulate populations grow rapidly. Axis deer show annual population growth rate of 20–30 %. Mouflon numbers can double in 3–4 years (Scowcroft and Giffin 1983; Hess et al. 2006). Mouflon and feral mouflon hybrids impede regeneration of Mamane forests (*Sophora chrysophylla*), particularly at the treeline level (Scocroft and Sakai 1983). Control of wild ungulates is very difficult, and eradication over wide areas seems to be impossible at present (Hawai'i Conservation Alliance 2007).

In New Zealand, for example, and on many subantarctic islands, where herbivorous mammals were originally absent, vegegation could not adjust to the sudden impact of domestic and wild ungulates arriving in vast numbers with the European settlers. In Australia, the impact of introduced European rabbits (*Oryctolagus cuniculus*) has reached continental dimensions, and many other introduced species have impacted the native fauna and ecosystems lastingly. Overgrazing and trampling by introduced ungulates destroyed the plant cover and caused severe soil erosion.

Introduced predators, such as cats, for example, often had disastrous effects. Cats were intentionally introduced to many islands to eradicate house mice that had escaped from the ships of whalers and sealers. However, cats neglected their 'job' and turned to the easier available seabirds nesting on the ground or in burrows. On Marion Island (Prince Edward Islands, Sect. 2.6.1.2.2), more than 200 cats had killed about 450,000 seabirds per year until 1970, when the feline panleukopenia virus (feline infectious enteritis) was introduced, which radically reduced the number of cats (Odening 1984).

On Macquarie Island (Sub-Antarctic, 54° 30′ S, 158° 57′ E), rabbits were introduced in 1878. In 1978, the rabbit population was estimated at 130,000 (Copson and Whinam 2001). The rabbits grazed down natural vegetation and changed complex plant communities into short lawns and often exposed mineral soil (Costin and Moore 1960). Moreover, rabbits destroyed petrel burrows. Consequently, petrels became increasingly preyed upon by brown skuas (*Catharacta lonnbergi*) (Bergstrom et al. 2009). After the myxoma virus was introduced in 1978, rabbit numbers declined to about 20,000 (Copson and Whinam 2001). Feral cats, which had been released 60 years prior to introduction of rabbits and previously subsisted mainly on rabbits, switched to seabirds as an alternative prey and heavily impacted seabird population (Copson and Whinam 2001). Thus, a cat eradication program was implemented, and in 2000, the last cat was shot. Afterwards, rabbit numbers rapidly increased again. However, it would be too simplistic to attribute recovery of the rabbit population to eradication of cats alone, as the supply of myxoma virus, which had been released almost annually (probably because of unfavorable environmental conditions) on the island since 1978, became irregular from 1999 onwards (Copson 2002) and ran out in 2006 (Dowding et al. 2009). Moreover, the virus might have become less effective. Thus, eradication of cats and reduced myxoma impact overlapped. Anyway, relieved from predator pressure and myxoma virus, rabbits are increasingly affecting the island's vegetation and soils again. Moreover, rats and mice which have profited from eradication of cats may become a risk to other small animals (e.g., birds, invertebrates) and plants.

On the Galapagos Islands, Santa Cruz, Floreana, Santiago, San Cristobal and Isabela, rats, cats and dogs drastically reduced the population of the dark-rumped petrel (*Pterodroma phaeopygia*) breeding on these islands (Coulter et al. 1985). In addition, feral goats, cattle and donkeys have compacted soil to such an extent that burrowing is impossible, and the petrels have to move to rocky sites for breeding, where they fall prey to rats. Rats also influence petrels in an indirect way, namely when the rats themselves are killed by larger predators. As a result, the predator population and predation pressure on the petrels increases (Atkinson 1985). For quite a long time, attempts have been made to save the petrels through specific management.

The small Indian mongoose (*Herpestes auropunctatus*) was introduced to most cane-growing islands in the Caribbean, Pacific and Indian oceans at the end of the nineteenth century and in the early twentieth century in order to control rats in sugar-cane plantations. The newcomer did its 'job' in the beginning to everyone's satisfaction, but after a while, it increasingly affected native animals and became a pest itself (Milne and Milne 1960; King 1984; Long 2003). Mongoose were also introduced to several subtropical Japanese islands (Amami-Oshima, Okinawa; Yamada and Sugimura 2004), Martinique, St. Lucia and several Adriatic islands (Croatia) to keep venomous snakes under control (Barun et al. 2010a). Mongoose, however, preferred native small animals and also caused damage to crop production. On the Adriatic islands, mongoose may have become the main predator of native amphibians, reptiles and poultry probably because rats became more nocturnal to escape predation pressure (Barun et al. 2010b).

Interestingly, there is some evidence from studies on northern New Zealand that invasive ship rats (*Rattus rattus*), which have contributed considerably to local extinction of native vertebrate species (including the endemic pollinators) and a recent colonist bird (*Zosterops lateralis*, Passerinae), are now partially filling the role of pollinators for three forest plant species (*Metrosideros excelsa*, *Knightia excelsa* and *Veronica macrocarpa*). Without these vertebrates, pollination would probably be insufficient for these plants (Pattymore and Wilcove 2011).

Predators that escaped from pelt farms or were liberated by animal conservationists have built up viable populations (e.g., mink, racoon) and become a threat to native animals, in particular to ground-nesting birds. In Scotland, for example, mink probably killed more seabirds since the 1980s than fell victim to all oil tanker accidents during this period of time (Craik 1995). On Iceland and some islands just off the coast, mink, which is a semiaquatic predator, had already caused great losses to nesting birds by the 1930s. Together with the loss of habitats and drainage measures, mink is very likely one of the main causes of the decline of water rail (*Rallus aquaticus*). Ducks and birds breeding at the coast, such as black guillemot (*Cephus grylle*), have changed selection of nesting places to avoid mink predation. Black guillemots now build their nests at greater distance from lake shores or in colonies of black-headed gull (*Larrus ridibundus*) and Arctic tern (*Sterna paradisea*) (Skirnisson 1992).

On the small islands off the Finnish coast, introduced American mink considerably reduced the populations of small duck species and rails. They recovered after the minks were removed. Mink did not affect the bigger duck species, geese and swans (Nordström et al. 2002). Breeding colonies of common gull (*Larus canus*), black-headed gull (*Larus ridibundus*), black tern (*Chlidonias niger*) and common tern (*Sterna hirundo*) are particularly vulnerable to mink predation (Bräsecke 1989; Hartmann 2002; Zschille et al. 2004).

On Uist island, west of the Scottish coast, hedgehogs (*Ericaceus europaeus*) were released in 1974. They rapidly spread from there over the neighboring islands. In the rest of Scotland, hedgehogs are native to the mammalian fauna and are considered not to cause much harm. On the islands, however, these insectivores seriously impair or even prevent breeding success of many ground-nesting bird species (lapwing, *Vanellus vanellus*; snipe, *Gallinago gallinago*; dunlin, *Calidris alpina*; redshank, *Tringa totanus*) by destroying the clutches of eggs. Consequently, the wader population has been declining. Apparently, these birds have not been able to adjust to the egg predators (Jackson 2001).

In the early 1890s, the brown tree snake (*Boiga irregularis*) arrived, probably via cargo at Guam, the southernmost island of the Marianas. The snake killed more than a half of the native forest avifauna, as well as lizards and fruitbats, within a few decades (Wiles 1987; Savidge 1987; Fritz 1988; Rodda and Fritts 1992). Very likely, eradication of the snake will be impossible. From Guam, the snake has expanded via ships and airplanes, and has even reached far distant islands such as Hawaii and Diego Garcia (Indian Ocean; Fritts 1987).

On the Spanish island of Gran Canaria, introduced non-venomous Californian kingsnakes (*Lampropeltis getula*) escaped from terraria or were released to the wild. In 1998, kingsnakes were first seen in the Gran Canarian landscape. The snakes, benefitting from the mild maritime climate and the absence of predators, have rapidly multiplied and become a serious threat to native lizards, geckos, skinks, birds, and to the island's biodiversity at all (Asenjo 2014). Fighting the snakes has not been successful so far.

Fish, in particular, have been translocated worldwide to new areas to improve food supply or only for sport fishing. These newcomers have often outcompeted the native species and caused profound changes in terrestrial aquatic ecosystems (e.g., Moyle 1980; Willwock 1993; Moyle and Li 1994; Lelek 1996; Löffler 1996b).

Also, invertebrates have been intentionally or accidentally introduced to new environments. In the United States, the European gipsy moth (*Lymantria dispar*) escaped from an experiment and has become a serious threat to hardwood forests in the northeast. The Asian gipsy moth (*Lymantria dispar asiatica*) first arrived in the Northwest in the early 1900s from Russia's far east. It is considered to cause even more damage than the European gipsy moth. The larch casebearer (*Coleophora laricella*) was incidentally introduced to the American east during the mid-1800s, and has expanded to Oregon and Washington, causing severe damage to the forests.

Balsam woolly adelgid (*Adelges picea*) came from Europe around 1900, and has since spread over much of the United States and Canada (Hayes and Ragenovich 2001). The Asian longhorned beetle (*Anoplophora glabripennis*), for example, which is originally native to East Asia (China, Japan, Korea), was also incidentally introduced. It very likely arrived by wooden pack material, and is now affecting deciduous forests in its new habitats (e.g., Hu et al. 2009). Moreover, exotic wood-borers have been detected (Mudge et al. 2001). The growing shortage of qualified taxonomists to identify the insects sampled in surveys has been considered to be a major problem for successful control (Hayes and Ragenovich 2001).

The red palm weevil (*Rhynchophorus ferrugineus*), originally coming from southern Asia and Melanesia, where it is a serious threat to coconut trees (*Cocos nucifera*), has spread westwards with palm plant material (offshoots) and reached Egypt. The beetle was introduced accidentally with the date palm (*Phoenix dactylifera*) from Morocco or Egypt to the Spain mainland and has spread to other Mediterranean areas (e.g., Majorca, Italy); there, it has been killing thousands of palm trees (date palm, and Canary Island date palm, *Phoenix canariensis*) during the last decades (Ferry and Gómez 2003). A palm tree once attacked by the beetle will die. Eradication of the red palm weevil has failed up to the present.

In the early 1890s, the large earth bumblebee (*Bombus terrestris audax*) was observed near Hobart on Tasmania for the first time. It was presumably brought in unintentionally or drifted with strong winds from New Zealand, where it had been introduced about 100 years before (Semmens et al. 1993). The bumblebee will probably occupy most of Tasmania, with the exception of the temperate rain forest in the west of the island that appears to be a barrier. Conditions are most favorable for the bee in the settled areas, where gardens, unploughed sides of fields, strips along roads, and meadows provide ample food. Many of the introduced plant species, such as birdsfoot trefoil (*Lotus corniculatus*), profited from pollination by bumblebees. In areas where bumblebees occur, birdsfoot trefoil, as well as some other plants, produce more seeds and expand more rapidly, at the cost of the native flora (Goulson and Stout 1999). The large earth bumblebees use the same flowering plants as the native bees and, most importantly, the large bumblebees are more numerous. Moreover, they begin earlier in the morning to feed on nectar and stop later in the evening. They also appear to collect nectar faster than their native food competitors. Long-term consequences for the biocoenoses are likely. However, as knowledge on Tasmanian bees is still insufficient – new species are discovered almost every day – we rely on more or less plausible assumptions (Goulson and Stout 1999). It may be safely assumed that the large earth bumblebees will expand to Australia.

The invasive garden ant (*Lasius neglectus*) and the red fire ant (*Solenopsis invicta*) have colonized many parts of the world. The invasive garden ant, which originates from around the Black Sea, is rapidly spreading through Europe and Asia and may become a global problem. In houses, it may spoil stored human food and cause short-circuits in the electric systems. 'Milking' aphids for great quantities of sugar-rich 'honey dew' the ants support sooty moulds to grow on leaves. Trade and exchange in hothouse plants are considered to be an important driving factor (Cremer et al. 2008). The red fire ant was probably introduced with soil used in ships' ballasts from South America to the southern United States during the 1930s/1940s, from where it has spread over other continents and islands (Ascune et al. 2011). Red fire ants damage agricultural crops. Just like invasive garden ants, they may also destroy electric cables and components. Both invasive ants compete with native ant species and other insects. Moreover, red fire ants affect ground-nesting rodents and birds, livestock, pets and may also be a threat to human health.

The crazy yellow ant (*Anoplolepis gracilepis*) was accidentally introduced to northern Australia, probably from India, about 70–80 years ago. As crazy yellow ants displace native animals, they are considered a serious threat to the ecosystems when they occur at extremely high densities. On Christmas Island, for example, ants have eliminated the island's famous red land crab (CSIRO 2003a, b). On the Tiwi Islands off northern Australia, the African big-headed ant (*Pheidole megacephala*), the Ginger ant (*Solenopis germinata*), and the Singapore ant (*Monomorium destructor*) have become real pests that should be eradicated as soon as possible (CSIRO 2003a, b).

During the last century, the highly competitive and polyphagous harlekin ladybird (*Harmonia axyridis*), originally native to central and eastern Asia, was introduced to America (1916) and Europe (1995) for biological control of aphids in green houses. Very soon, however, it rapidly invaded the landscape and became a serious threat to native ladybirds and many useful insects (Brown et al. 2008).

Near Banaue on the northern Philippines, vast numbers of up to 50 cm long earthworms (*Pheritima darleiensis benguetensis*) apparently accelerate the decay of the 2,000-year-old rice-cultivation terraces, which are a famous tourist attraction (inscribed to the UNESCO World Heritage List in 1995). The terraces are located on very steep and rugged mountain slopes and are highly vulnerable to erosion if not carefully maintained. Heavy rains (>20 mm h^{-1}) are typical of the area. The earthworms were probably introduced with more productive rice varieties. Another explanation is that the worms moved from the forests to the rice terraces due to warming temperatures (Ngidlo 2013). The earthworms have really riddled the terraces with holes. No wonder, as a shuffle with terrace soils contains about 20 of these worms. Rain water drains away in the many earthworm tunnels, thus undermining the terraces. These are threatened anyway due to insufficient tending and irrigation (Calderon et al. 2009). In the 1990s, however. the situation has worsened, as a result of decreasing water supply on the poorly maintained terraces. With a little drier conditions, the earthworms reproduce better than before (Calderon et al. 2009). Saving the rice terraces is a complex problem difficult to be solved (Ngidlo 2013). Economic change, population pressure, sociocultural development, rural exodus, restoration of the irrigation systems, geology (slope stability, landslides, earthquakes) and climate (heavy rains, typhoons) as well are the most relevant factors to be considered. The giant-earthworms' impact must not be ignored, however.

The changes of ecological conditions induced by introduced species have often reached landscape dimension, while simultaneous human impact on landscape (e.g., land use, clear-cuts) had exacerbating effects. There is a wealth of literature on this issue (reviews, for example in Riney 1964; Milne and Milne 1960; Holdgate 1967; Roots 1976; Hutchins et al. 1982; Mooney and Drake 1986; Di Castri 1989; Goudie 1994; Sedlag 1995). A few selected examples will be considered in detail in the following sections.

5.1.1 Introduction of Norwegian Reindeer to South Georgia and Other Subantarctic Islands

The possible resilience of ecosystems to non-native mammalian herbivores differs, as is evident, for example, from translocation of Norwegian reindeer to some subarctic (St. Paul, St. Matthew; Scheffer 1951; Klein 1968) and subantarctic (South Georgia, Kerguelen) islands (Lindsay 1973; Leader-Williams 1980, 1988: Leader-Williams et al. 1987; Chapuis et al. 1994; Bell and Dieterich 2010). Typically, the number of reindeer grew rapidly right after introduction. On the arctic islands, reindeer population collapsed after depletion of lichens. On St. Paul, for example, the reindeer population, which had considerably increased during the early 1930s, exceeded the carrying capacity of the area threefold (Scheffer 1951). By contrast, reindeer still exist on South Georgia at high densities, although vegetation could not adapt during evolution to large mammalian herbivores, as these had been absent.

Between 1911 and 1925, Norwegian reindeer were released on two peninsulas (Barff- and Busen Peninsula) at the east coast of South Georgia (Fig. 5.1). A third herd formed in the Royal Bay area, located south of Barff Peninsula, after reindeer had left the original area of introduction. As is typical of primary 'colonizers'

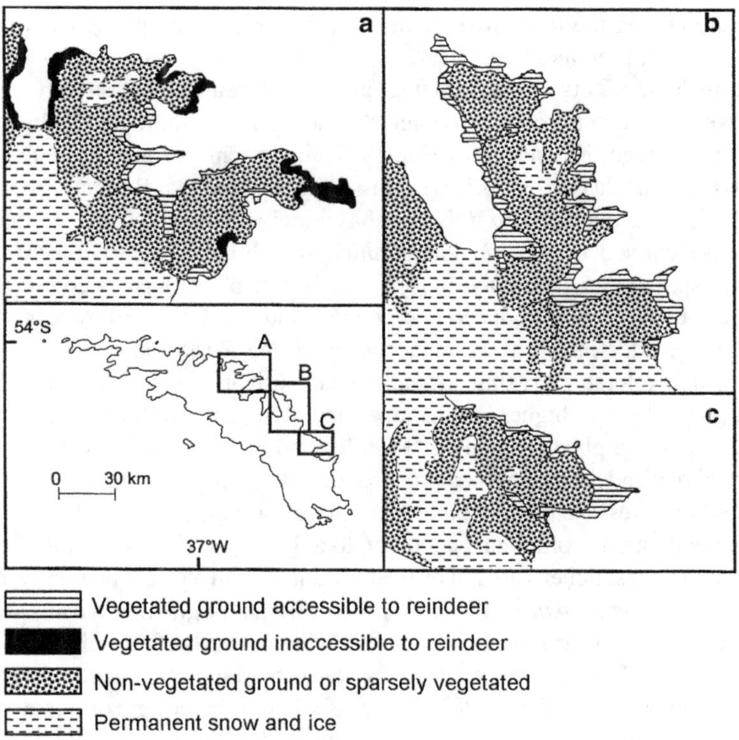

Fig. 5.1 Reindeer distribution areas on South Georgia. *A* – Busen Peninsula, *B* – Barff Peninsula, *C* – Royal Bay (Modified from Leader-Williams et al. (1987))

(Riney 1964; Caughley 1970; Challies 1975), reindeer rapidly reached a population peak and then decreased. Population levelled out at about 40–80 reindeer per square kilometre of vegetated terrain (Leader-Williams et al. 1987). At present, the herds number about 2,600 individuals in total. Thus, reindeer density is 10–20 times as much as in the totally over-grazed reindeer habitats in northernmost Finland, for example, where reindeer density hardly exceeds 3 reindeer km^2 (Kumpula and Nieminen 1992), or 5–10 reindeer km^2 in more intensively grazed areas (Oksanen et al. 1995; see also Sect. 3.3.2).

Topographic barriers restrict reindeer herds to a few areas (cf. Fig. 5.1) at the east coast (Leader-Williams et al. 1987). Contrary to arctic regions, South Georgian reindeer herds do not migrate seasonally. Consequently, grazing pressure is very high all year round. Reindeer have adjusted to an almost lichen-free diet (Leader-Williams 1988). In the coastal area, they feed mainly on tussock grasses, particularly on the about 2-m-high evergreen *Poa flabellata*, which is rich in energy and very productive (Gunn and Walton 1985). *Poa flabellata* is the main winter forage plant. This grass tolerates moderate grazing very well, as it is capable of reproducing vegetatively. However, it may be destroyed by female Antarctic fur seals (*Arctocephalus gazella*), which are used to resting on top of the tussocks (Bonner 1985). During summer, reindeer also feed on other plants particularly rich in nutrients. Grasses, such as *Deschampsia antarctica* and the introduced *Poa annua*, and the dwarf shrub *Acaena magellanica*, are intensively grazed. *Deschampsia* and *Acaena* are among the few native plants rich in nitrogen and phosphorus (Walton and Smith 1979; Pratt and Smith 1982).

Nevertheless, effects of over-grazing due to high reindeer density are apparent almost everywhere in the coastal lowlands (Lindsay 1973; Smith and Walton 1975). Driven by shortage of winter forage in 1957, many reindeer unexpectedly moved from Barff Peninsula (cf. Fig. 5.1, B) across a glacier to Royal Bay (cf. Fig. 5.1, C). Even in this area, tolerant *Poa flabellata* has widely declined, as well as the originally dense carpets of *Acaena magellanica* (height 10–25 cm), macro-lichens (*Cladonia* sp., *Cetratria* sp.) and bryophytes. Native plants have become replaced by mosses or swards of introduced *Poa annua* (Lindsay 1973), and by plant species rejected by reindeer, such as *Festuca contracta* and *Rostkovia magellanica. Poa annua*, which tolerates grazing, trampling and urinary excretion, could rapidly expand, particularly in higher coastal areas and along river valleys, when competition with the native plants decreased (Kightley and Smith 1976; Vogel et al. 1984).

Survival of reindeer in South Georgia for decades at high population density and severe grazing pressure can be explained only by a high grazing tolerance of the vegetation, although evolutional adaptation to ungulate grazing is lacking. Contrary to the arctic islands, lichens are not critical for reindeer survival. Apparently, of both *Poa annua* and *Poa flabellata*, the latter is less tolerant of grazing, having provided sufficient forage all year round until present. Thus, a crash of the reindeer population, which would be comparable to the failures on the arctic islands, is not very likely (Leader-Williams et al. 1987). In areas where reindeer grazing is excluded, the tussock grasses recover more rapidly compared to the northern lichen carpets that need many decades before providing sufficient forage again (Fig. 5.2). If rein-

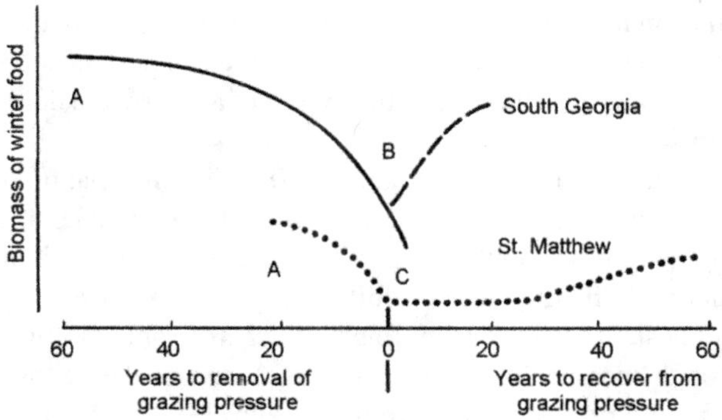

Fig. 5.2 Years to recover of lichen vegetation on St. Matthew Island and of tussock grassland (*Poa flabellata*) on South Georgia after exclusion from grazing or complete decline of reindeer on St. Matthew Island. *A* – Time of introduction, *B* – Exclusion from grazing, *C* – Decline of reindeer (Modified from Leader-Williams et al. (1987))

deer numbers were partly reduced, suppressed plants that were characteristic of the original plant communities would probably expand again. Change would likely be most conspicuous in areas that were grazed over the longest period of time (Leader-Williams et al. 1987).

Despite the considerable impact on plant cover and soils, reindeer are currently protected by law (Bonino 1995; Bell and Dieterich 2010). Nevertheless, the government has planned eradication of one of the herds (Pasteur and Walton 2006). In 2001, reindeer were also successfully translocated from South Georgia to the Falkland Islands for establishment of a commercial reindeer herd and preservation of the genetic resources of the South Georgia reindeer (Bell and Dieterich 2010), whereas translocation to Tierra del Fuego failed mainly due to hunting and acclimatization problems.

5.1.2 *Animals Introduced to Australia*

Introduction of exotic animals has caused dramatic changes in Australian landscapes, biocoenoses and biodiversity over the last 200 years (e.g., Woinarski 2001). Great efforts have been made to get the newcomers and their effects under control, with varying degrees of success. However, attempts to completely eradicate well established exotic vertebrates have failed (Bomford 2003).

5.1.2.1 European Rabbit

The impact of introduced European rabbits (*Oryctolagus cuniculus*) has reached continental dimensions. Although rabbits were considered a 'pest' in England long before the nineteenth century, they were released wherever possible, not only in

Australia but also on many islands in the Atlantic, Pacific and Indian Oceans. In the remote outposts of the British Empire, rabbits were an important source of protein. However, 'introduction of rabbits' has often been equal to 'severe degradation of the new environments'.

The first five rabbits arrived in Australia in 1788 via ships that brought prisoners from England to this continent most distant from their homeland. Rabbits were introduced to improve food supply and to establish a basis for rabbit hunting, which was very popular with the English. In addition, introduction possibly helped a little in alleviating homesickness for distant England, where rabbits occurred all over the parkland. Although more rabbits were introduced during the following years, no unusually strong multiplication occurred (Creagh 1992). The true invasion of Australia by rabbit began in 1859, when the farmer Thomas Austin released 24 English rabbits near the present Geelong located in the vicinity of Melbourne. Obviously, all females were already pregnant on arrival. Very soon, rabbit numbers really 'exploded', driven by the great reproductive capacity and favorable environmental conditions (dry and warm climate, only a few predators). Such an immense population growth had never been experienced before. Up to 7 l per year of five or more young each are possible. Under favorable conditions, one pair of rabbits could theoretically produce 13 million descendants within a 3-year period. However, such approximations do not meet reality, because 80 % of the young rabbits normally die from parasites and diseases or are killed by predators (Price 1972). Many rabbits do not get older than 1 year (Learmonth and Learmonth 1968). Anyway, population growth is enormous, despite such losses. Rabbits expanded at a speed which had never been observed in other introduced animal species.

In 1870, rabbits were released in greater numbers near Kapunda, north of Adelaide. Only 10 years later, the rabbit population of Geelong and Kapunda already overlapped and mixed. It is impossible to find out how often and in how many places rabbits were released. However, as rabbit hunting and trapping turned out to be very profitable (meat, fur, felt hats), rabbits were brought over hundreds of kilometers to even remote unsettled areas to improve the economic base of existence. About 750 million rabbits were living in Australia before myxomatosis was introduced (Roots 1976). Myxomatosis reduced rabbit population to 'only' 200 or 300 million (Creagh 1992). Rabbits occupy more than half of the continent (7.7 million km^2) and are gradually expanding in spite of rigid control.

The 'explosion' of the rabbit populations had disastrous effects on agriculture and native ecosystems. Cattle range became devastated. Soil erosion by wind and surface run-off increased, and agricultural productivity drastically decreased. Rabbits also compete directly with cattle for food. Sixteen rabbits consume as much as one sheep (Short 1985). The area, which carried 15 million sheep prior to the introduction of rabbits, supported only half the number afterwards. Damage caused by rabbits to agriculture increased to almost 95 million Australian dollars (Creagh 1992).

Nevertheless, the impact of rabbits on the ecological conditions is probably more serious. Native animals were not able to keep up with the enormous multiplication of rabbits. Thus, they could not compensate for the enormous losses due to food competition. Bilbys (*Thalocomys lagotis*), bettongs (*Betongia* sp.) and wombats

(*Lasiorhinus latifrons*) decreased because rabbits occupied their burrows. In view of the imminent catastrophe, rabbits had already been declared a 'pest' by 1888, and a royal commission for the development of rabbit control measures was established. Everything was tried to get the rabbits under control: extensive poisoning campaigns (usually by the so-called 1080 poison, sodium fluoroacetate), enforced hunting pressure, trapping, breaking up or blasting the burrows, gassing, release of foxes, ferrets and cats, and the setting up of a several thousand kilometre-long fence, supposed to be 'rabbit-proof'.

In the end, many of these control measures affected the native animals more than the rabbits. As an additional side effect of poisoning, many insectivorous birds and small mammals were killed. Certainly, rabbits experienced great losses, but they could rapidly compensate through high reproduction. After poisoning campaigns, the decomposing rabbit carcasses provided optimal food to blowfly (Calliphoridae) maggots, which, again, severely bothered sheep. Poisoning of insectivorous birds and small mammals has been blamed for having increased locust plagues. However, it may be questioned as to whether all insect predators together would be able significantly to influence or even prevent mass outbreaks of locusts. Although many additional examples of unexpected consequences of rabbit control campaigns could be given, it will be left at that.

It was the myxoma virus which almost wiped out pest rabbits. This virus had been found first in the South American rabbit (*Sylvilagus brasiliensis*) and was introduced to Australia by CSIRO (Commonwealth Scientific and Industrial Research Organisation) in 1950. Whereas the South American rabbit is immune to the virus, European rabbit is extremely vulnerable to myxomatosis. Anyway, the first attempts to introduce the myxoma virus to Australia at the beginning of 1950 failed, as the mosquitoes which are the vector of the myxoma virus were not yet active. Not much later, however, the rabbit population crashed from 600 million to less than 1 million. Vegetation recovered, and pastures that could support just one sheep had enough grass for two cows. Meanwhile, the efficiency of the virus has weakened. It mutated, and some of the new forms are not lethal anymore. Dependent on the virus's virulence, only 30–80 % of the rabbit population fell victim to myxomatosis.

Frantic efforts have been made to develop new biological control techniques. The rabbit flea (*Spilopsyllus cuniculi*), for example, introduced in 1968, has enhanced the efficiency of myxomatosis in some areas. Contrary to mosquitoes, the rabbit flea does not depend on open waters for regeneration. However, it requires a certain amount of moisture and is therefore restricted to regions with annual precipitation above 250 mm. As rabbit control is most difficult in the drier regions, where *Spilopsyllus cuniculi* and mosquitoes will not survive, experiments with rabbit fleas (*Xenopsylla* sp.) from Spain were carried out (Cooke 1990).

The calicivirus, which causes rabbit haemorrhagic disease (RHD), had significantly reduced rabbit populations in Europe, particularly in the drier regions. Since the mid-1980s, RHD was recognized almost all over Europe. Its distribution, however, was uneven. The virus is highly contagious and presumably specific to rabbits. Typically, it may kill 90 % of infected rabbits within 2–3 days of infection.

The virus has been spread by worldwide trade in domestic rabbits. It is not quite clear whether the virus originated in China or Europe. Chinese scientists (cited in Cooke 2002) think that a stock of Angora rabbit introduced into China (1984) from the former German Democratic Republic has spread RHD in China. In North-Rhine Westfalia, which is richest in rabbits of all Länder of the Federal Republic of Germany, wild rabbits have drastically declined due to myxomatosis and rabbit haemorrhagic disease. While the rabbit has often been considered a nuisance animal (e.g., by gardeners, farmers and foresters), hunters and nature conservationists are seriously concerned about the rabbit's future.

The tremendous efficiency of the virus was followed with great interest in Australia, and in 1991, RHD was introduced. Field experiments with the rabbit calicivirus were carried out by CSIRO on Wardang Island off the coast of South Australia. Although the virus was strictly quarantined, it escaped accidentally in 1995 and reached the Australian mainland. Very likely bush flies (*Musca vetustissima*) and some of the larger blowflies, such as *Caliphora dubia*, were the vectors. More than 95 % of the rabbits died within a short time – similar to the crash of rabbit populations after the introduction of myxomatosis. However, RHD failed in the wetter coastal regions. Moreover, young rabbits (less than 10 weeks old) produce antibodies, which make them immune. However, very young rabbits not yet weaned die due to starvation or are killed by predators. Thus, it cannot be taken for sure that the viruses will lead to a decline of rabbits in the long-term. However, a very low rabbit population density due to myxomatosis and RHD might drastically reduce infection by viral diseases.

In addition, genetically modified myxoma viruses are used to sterilise rabbits. Studies in southern West Australia showed, however, that even sterility of 60–80 % in rabbit population is not enough for efficient rabbit control (Twigg et al. 2000). In all likelihood, 90 % of rabbits have to be killed to successfully manage rabbit numbers (Williams et al. 1995: Hone 1999).

Altogether, numerous and often still unknown risks are linked with biological control measures. The knowledge about possible effects on non-target animals (e.g., pigs, cats, marsupials) is still insufficient. Consequently, the public attitude to biological control is very critical. Although it must be accepted that rabbits cannot be completely removed from Australia, there are some good reasons to continue control, particularly in the arid areas, to relieve natural biocoenoses and agriculture a little from the rabbit impact. However, a too drastic decline of rabbits may result in an increased impact of their predators on the remaining native fauna. Both introduced and native predators, such as foxes, cats, ferrets (*Mustela potorius furo*), dingoes (*Canis lupus dingo*), and wedge-tailed eagles (*Aquila audax*) have meanwhile adjusted to the abundant food source, and will have a hard time with low rabbit numbers. This was also the case after introduction of myxomatosis (Milne and Milne 1960). Dingoes (immigrated about 10,000 years ago), for example, which had increased parallel to the expansion of rabbits, increasingly switched over to lambs as alternative prey. Again, several thousand kilometres long and 2-m-high fences were set up to protect sheep herds from dingo attacks, and poisoned baits were dropped from airplanes to kill them. Dingo control, on the other hand, has favored the spread of foxes. Their diet consists of 80 % young rabbits. Foxes avoid areas used by dingoes, as these not only prey on foxes but also compete with them for food (O'Neill 2002; Mitchell and Banks 2005; Johnson and Van Der Wal 2009). Thus, the numbers of foxes and other predators must be adjusted to rabbit control to prevent negative effects on the biocoenoses (Jarman 1986; Creagh 1992). Based on a model, Pech and Hood (1998) speculated that reduction of rabbit population by the calicivirus might be followed by a fox decline, which would allow recovery of the native prey mammals of foxes. However, in view of the great complexity of the foodweb and the multitude of predators (see above) such an assumption appears not to meet reality.

It has become evident from studies on the relative effects of sheep, rabbits, feral goats and kangaroos on vegetation in eastern South Australia that the impact by intense sheep grazing exceeds the influences of all other herbivores combined. Consequently, long-term conservation of threatened plants and biodiversity requires not only rabbit control but also creating a mosaic of grazed land interspersed with ungrazed areas (Tiver and Andrews 1997). New Zealand is facing similar problems.

A new psychological campaign against rabbits has the objective of preserving the 'rights' of the native fauna. In West Australia, nature conservationists have transferred the function of the Easter bunny, which had been represented until then by the physiognomically quite similar rabbit, to the long-eared bilby (*Macrotis lagotis*, a kind of bandicoot), which is near extinction. Slogans like '*Remember bilbies not bunnies when purchasing chocolates for Easter*' are quite popular. Children can be easily convinced that the bilby, having a marsupial pouch, is much better equipped than bunnies for carrying sufficient loads of chocolates and Easter eggs. Meanwhile, the number of chocolate bilbies probably exceeds by far the number of wild rabbits.

5.1.2.2 Marine Toad

The rabbit pest certainly is the best known problem of the Australian continent, but not the only one resulting from introduction of alien animals (Kitching 1986). Almost all exotic animal species in Australia – and there is a long list – have had unexpected adverse effects on native biocoenoses. Compared to the impact of rabbits, their influence is more locally restricted, however. This also holds true for the marine toad (*Bufo marinus*), for example.

In view of the positive experience in Caribbean sugarcane plantations, the marine toad, which was originally native of a region from Northwest Mexico to Central Brazil, has been introduced to almost all sugarcane-cultivating countries for biological control of insects and rats causing damage to sugar cane. In 1935, the marine toad (length up to 25 cm, weight about 1 kg; Photo 5.2) was translocated from Hawaii to Queensland for cane beetle (*Dermolepida albohirtum*) control. Its larvae feed on sugarcane. Toads were released in the northern area of Queensland near Gordonvale. Until the late 1930s, toads were introduced with the permission of the government to 11 sugarcane cultivation areas. However, as the dry sugarcane

Photo 5.2 Marine toad (*Bufo marinus*). M. Tyler

plantations were obviously not the right habitat and the cane beetles not the right food for marine toads, these left the area of introduction and migrated to more humid regions.

There, toads apparently enjoy 'paradisiacal' conditions. They have built up populations up to tenfold larger than those in their South American region of origin, and they colonize quite different habitats, provided that sufficient water is available and temperatures are not too low (Sutherst et al. 1995). Obviously, marine toads fare particularly well in the cultivated landscape (Easteal and Floyd 1986). Marine toads are characterized by a high reproductive capacity. Moreover, there are no parasites and natural predators that would impede their expansion. Neither large snakes nor monitors (Varanidae), herons and predatory birds are a real threat to toads, as these are able successfully to defend themselves. Consequently, toads have very rapidly spread following the water courses. At present, more than 75 years after introduction, toads are omnipresent in eastern Queensland from Brisbane to Cape York. Toad population on York peninsula has been estimated to be about 100 million individuals. In the 1980s, toads had already arrived in the Northern Territory, and they have occurred near Darwin at the Australian north coast. It will not take much time until toads will have colonized the coastal areas further west. In March 2001, they occupied the wetlands of Kakadu National Park (UNESCO world heritage) about 170 km east of Darwin (CSIRO 2001).

Marine toads prey on nearly every animal they can get: small mammals as big as a mouse, other amphibians, young birds and insects, among them the scarab beetle (*Ontophagus gazella*), which was introduced from Africa to decompose cattle dung (Sect. 2.1.1). One toad may consume up to 80 beetles within a single night (Waterhouse 1974). Thus, diurnal dung beetles were looked for, which would escape the crepuscular and nocturnal toads. Introduction of African giant dung beetles (*Heliocopris* spp.; same size as a golf ball) has also been considered. These beetles, when swallowed by a toad, are strong enough to break through the toads' skin to the outside (Waterhouse 1974).

Right after introduction, toads were found in the feeding bowls of dogs and cats. When attacking the very poisonous toads, these domestic animals were bitten and usually died. The toad's skin is also poisonous and may be fatal to the predators (e.g., predatory birds, dasyures, dogs, cats, etc.). To humans, the toad poison (*bufotoxin*) causes irritation of the skin and cauterization of the eyes. In addition, the toads poison ponds and swimming pools. Over time, they have become a real pest. They edge out native amphibians and reptiles, and thus cause profound change in many characteristic biocoenoses. Similar impacts can be observed in New Guinea (Roots 1976).

Although having occupied Kakadu National Park just recently, marine toads are already likely to become a threat to tourism. Under the pressure of the toad problem, the research organization CSIRO has received extensive financial support from the Australian government to develop effective biological control measures. A main objective is to identify a gene which is essential for toad development and to modify it in a way that will prevent toad metamorphosis. Native frogs and fish infected with weakened and non-lethal native amphibian viruses are supposed to spread these manipulated genes among the marine toads (CSIRO 2001). Viruses from Venezuela turned out to be inappropriate, as they killed not only marine toads but also native frogs (information CSIRO 22 July 1998). A solution to the problem in the short term seems unlikely. It will rather take 10 or more years until this control method can be applied in practice without affecting other members of the biocoenoses. Plans to prevent expansion of toads by a 'toad fence' were not realized because of too expensive maintenance and little effectiveness (see also rabbit and dingo control). Thus, up to the present, a practicable and highly efficient control method is lacking. According to a press report (Westfälische Nachrichten, 5 June 2010), the mayor of Darwin, obviously in sheer desperation, has suggested toad-killing as a fun activity for tourists.

Despite the incalculable adverse effects of the marine toad, public opinions considerably differ. People often keep toads as 'pets', which remove cockroaches and other annoying vermin. In Florida, for example, release of toads on private ground, in particular, is supposed to have promoted their rapid spread and establishment of new populations (Krakauer 1970).

In addition, toads have a certain economic value. Everybody is allowed to use toads commercially without an extra permit. In some states and territories (e.g., Northern Territories), introduction is prohibited, however. Toad poison is freeze-dried, pulverized and sold at high prices (1 g for about 100 dollar) to Asian countries, where it is applied in traditional medicine. The skin is used for making high-quality leather for fashionable articles, such as handbags, purses and wallets. Production of toad leather is also becoming increasingly important in other countries harboring large toad populations, such as the Philippines. Only big individuals provide enough leather and butoxin that can be used economically. As such individuals are relatively rare, toads are bred in farms for profitable production of leather and bufutoxin. Taxidermists have discovered a niche and sell great numbers of padded toads as souvenirs. The discussion of the toad problem is also controversial among scientists. Some scientists consider the development of the toad population in Kakadu National Park to be a 'gigantic and fascinating field experiment' rather than a threat to the ecosystem and biocoenoses. They suggest carefully following the processes and not to start panicking and jumping to conclusions (Gearin 2000). Others consider the 'wait and see' attitude to be 'environmental vandalism' that cannot be tolerated, in particular with regard to the enormous loss of quality of human life and having the complete occupation of northern Australia in prospect (Tyler 1996; information M. J. Tyer 23 June 2001).

5.1.2.3 Feral Animals

Feral animals such as donkeys, horses, Asiatic water buffalos, goats, pigs, camels and others, also cause severe problems. Only a few of them can be considered here.

Between 1925 and 1843, water buffalos (*Bubalus bubalis*) were introduced from South Asia to the so-called 'Top End' of Australia (Arnhem Land, Melville Island) to supply meat and milk to the remote northern settlements. After these were abandoned in the mid-1900s, released or escaped buffalo spread across the north, despite being hunted for meat, hides and trophies. Losses due to drought periods, as well as to dingoes and crocodiles, which prey on young calves, could not prevent exponential buffalo population increase. During the 1980s, Australia's water buffalo population peaked at around 350,000 animals. Temporarily, it made up to about 90 % of the global population of free-living water buffalos (McKnight 1971; Graham et al. 1982).

Feral buffalos considerably altered the northern floodplain landscape, monsoon rainforests, savanna woodlands and open forests by trampling, dung, wallowing and fostering soil erosion (e.g., Jesser and Markula 2008). Possibly, feral pigs were also involved. Relocated material silted up water holes upon which many other animals rely during the dry season. By altering vegetation structure and composition (e.g., Petty et al. 2007), buffalo affect habitat conditions of other animals. As buffalo regularly move between wetlands, they create 'swim channels'. Tidal saltwater can intrude upon freshwater habitats via these channels and destroy salt-intolerable plants and animals (Jesser and Markula 2008). Thus, crocodiles and freshwater turtles declined, for example. Buffalo also compete with native wildlife for food. Not least, feral buffalo carry diseases such as bovine tuberculosis (*Mycobacterium bovis*) and brucellosis (*Brucella abortus*) that can affect native animals and livestock. Spreading foot-and-mouth-disease (*Aphtae epizooticae*) is also a potential risk.

As a preventive measure, the 'Brucellosis and Tuberculosis Eradication Campaign' (BTEC) was started (1978), which drastically reduced buffalo numbers between 1985 and 1995. Apparently, helicopter hunting was most successful. In Kakadu National Park, for example, buffalo numbers decreased from about 20,000 to less than 250 animals in 1996. In the park, aborigines are permitted to maintain a small herd of domesticated buffalo as a food source. On Cobourg peninsula, a small population is maintained as a tourist attraction (information Q. Hart 20 April 1998).

However, buffalo population recovered from occasional losses to natural factors (e.g., drought) and culling. In response to a series of wetter years, for example, buffalo population may increase up to 34 animals km^{-2}. The current buffalo density across northern Australia has been estimated at 150,000 animals (80,000, Jesser and Markula 2008). Due to successful control, the buffalos' impact on landscape ecosystems seems no longer to be playing an important role (communication J. McIllroy, 16 April 1998). Queensland has been less affected by buffalo than the Northern Territory due to comparatively low buffalo density (Petty et al. 2007). About 15,000–20,000 buffalos are farmed for meat (low cholesterol), hides and also live export.

Feral horses (*Equus caballus*), donkeys (*Equus asinus*) and camels (*Camelus dromedarius*) are found in many areas of Australia, where they have built up large populations. Horses arrived with the first fleet in 1788, and the first donkeys in 1866. The latter came mainly from Spain, India and Mexico (Bough 2006). In 1840, the first camel was brought to Australia. Thousands of camels were introduced between 1890 and 1907, mainly for transport through dry inland Australia. When motorized vehicles and railroads replaced donkeys and camels, many of them were released into the wild, where they built up large populations that fluctuate considerably due to the great variability of the climate and food and water supply. The distribution of feral horses is comparatively patchy. Donkeys can be found mainly in the Northern Territory. Feral camels occur in the semiarid regions of West Australia, South Australia and in the Northern Territory.

As it is generally difficult to obtain exact data on feral animals, we rely on a few rough estimations. In the late 1980s, about 300,000–600,000 feral horses ('brumbies') were living in Australia. Since then, periods of drought and rigid control have drastically reduced the population (information Q. Hart, 20 April 1998). However, about 265,000 feral horses and 165,000 donkeys have been estimated more recently. The total Australian camel population has been supposed at 740,000–800,000. In the Northern Territory, camel numbers have more than doubled during the 1990s. At the end of the previous century, about 200,000 camels were living there (Wurst and Saalfeld 1994). Feral horses, donkeys and camels are considered to damage vegetation and to foster soil erosion. In general, however, camels that have padded hoofs (like llama) cause less trampling damage than hard-hoofed horses and donkeys. Moreover, like water buffalo, they damage and foul water holes and carry seeds of weeds in their dung. Impact on vegetation close to waterholes is most important during periods of drought when many other animals also depend on the remaining drinkable water. However, the effects on the environment are not well-documented. Not least, feral horses and donkeys may carry diseases that can affect domestic horses and cattle. The most effective control measure is helicopter shooting, particularly in difficult terrain. However, culling often meets stiff public opposition, as, for example, in 2013, when the State of Queensland planned to shoot over 10,000 feral horses, claiming that they damage fragile national park ecosystems ('The Independent', 4 February 2013). However, the assumption that the impacted areas will fully recover after removal of the feral horses might be a little optimistic (Csurhes et al. 2009).

Camels have become a pest in many regions. In 2009, the Australian Feral Camel Management Project became established. Its objective was to reduce feral camel densities to about one camel per 10 km^2 by culling. During the last 4 years about 160,000 camels were culled (ABC Rural 2013) using ground based and aerial techniques. In addition, many were mustered for commercial use. Aerial shooting was most effective. The total population has been brought down to about 300,000. About 100,000 feral camels died in the Simpson Desert due to starvation and drought (SBS, 21 November 2013). On the other hand, camels are considered a valuable economic resource, which may yield hundreds of millions of dollars per year. Stock farmers in the Northern Territory are developing a sustainable camel industry and built a big slaughter house in the land interior (RIRDC 2000). Camel meat may be served as a delicacy in upscale restaurants in Australia, Europe, the United States and Japan or as animal food. Camel meat has a low fat and cholesterol content, which makes it particularly attractive for nutrition-conscious consumers. In addition, camel milk (rich in calcium), the wool and hides, as well as beauty cream made of the camels' hump fat, may be put on the global market (http://www.nachrichten.t-online.de/c/13/85/05/14/13850514.html, 2 February 2008).

Feral pigs are widespread all over the continent, with the exception of the dry interior and South Australia. They are most numerous in New South Wales and Queensland, whereas their area of distribution along Australia's west coast and in Victoria is sporadic. In North- and West-Australia, feral pigs concentrate in riparian habitats (river valleys) and around billabongs. During the wet season, however, they roam about more widely. Lack of water and shady habitats appear to have been a limiting factor (Tisdell 1982).

The feral pigs originated from domestic pigs, which were first introduced in 1788 by Europeans. Due to back-crossing, the feral pigs look very similar to the European wild boar (*Sus scrofa*). Until the mid-1800s, pigs were commonly free-ranging, as they were in Central Europe during the Middle Ages. Later, fences kept them on private ground. Straying pigs became wild. Due to their great adaptability, high reproductive capacity and lack of affect by natural predators, they could spread rapidly. It is almost impossible to estimate their numbers reliably, in particular because of strong annual fluctuations in response to drought and other environmental influences. The available data range between 1 and approximately 14 million (Hone 1990, communication J. Hone, 16 April 1998) or even more (23.5 million, Choquenot et al. 1996).

Over time, feral pigs have become a 'pest' that severely affects agriculture and stock farming. Tisdell (1982) estimated feral pig damage to be about 50 million dollars per year. Feral pigs turn agricultural fields and pastures upside down, break through fences (rabbit and dingo fences included), destroy irrigation systems (pipes, dams, etc.) and stock waterers. They also considerably contribute to erosion. Moreover, they prey not only on small mammals, birds, and rabbits, but also kill newly born lambs and goats (Choquenot et al. 1996, 1997). In addition, they carry a couple of pathogens and promote expansion of leptospirosis, anthrax, brucellosis, tuberculosis and foot-and-mouth disease. In short, feral pigs are often a serious threat to stock farming. Not least, feral pigs destroy the nests of ground-breeding birds and eradicate them, as well as native reptiles and small mammals. Vegetation has also considerably changed under the pigs' impact.

Although the influences of feral pigs on the environment are apparent and often incalculable, little precise knowledge exists about their impacts on the native fauna, in particular. Generally speaking, competition for food, predation and alteration of habitat structures are most important. Resultant loss of refuges and safe breeding places make native animals more vulnerable to other predators and impede successful regeneration. For example, feral pigs reduce breeding possibilities of the ground-breeding Australian brush turkey (*Alectura lathami*) by destroying tall-growing riparian vegegation along water courses and around billabongs. However, the clutches of ground-nesting birds and the birds themselves are not destroyed by feral pigs, but only by foxes, feral dogs and cats. During the dry season, numerous waterbirds and pigs compete for food at the sporadic water holes. Feral pigs feed on lilies, tuberous and other green plants, and also on frogs, which are the only food available to waterbirds at this time of the year. In addition, pigs prey on small mammals living in the riparian zone and destroy their burrow systems. In this way the populations of false water rat (*Xeromys myoides*), which nests in piles of leaves and mud at the rim of swamps, and of the Australian water rat (*Hydromus chrysogaster*), which builds their burrows in the banks of rivers and lagoons, are seriously threatened. In the rain forests, habitat destruction by and food competition with feral pigs threaten the existence of the rufus kangaroo-rat (*Aeprymus rufescens*) and probably of the musk-rat kangaroo (*Hypsiprymnodon moschatus*), too.

In any case, the impact of feral pigs on the native biocoenoses (plant and animal communities) has reached an extent that systematic control is imperative. Control measures include trapping, poisoning, infection with swine fever, and helicopter shooting (Choquenot et al. 1996). Dingoes, often considered efficient predators, have turned out to be of minor importance in decimating feral pigs. This might be at least partly explained by the fact that the dingoes themselves are under rigid control, which, however, is not the only reason. Dingoes depend on pig numbers and not the other way around. When in a field experiment, water buffalos, competing with feral pigs for food, were removed, the number of pigs strongly increased, and consequently, pigs made up a greater portion of the dingoes' diet (Corbett 1995). In any case, the feral pig problem is not yet solved.

The particular situation of Australia is not only the result of the isolated evolution of fauna and flora, but also depends on the climatically caused high variability of life conditions (long-lasting periods of drought, flooding, bush fires, etc.). Thus, the original animal populations fluctuated very strongly (so-called 'boom and bust patterns') and probably have never been constant (Ratcliffe 1959). The collapse of kangaroo populations (*Macropus rufus*, *Macropus fuliginosus*, *Macropus giganteus*) in the eastern part of Central Australia between April 1982 and March 1983 can be considered a typical example (Caughley et al. 1985). It was caused by one of the irregular severe drought periods. In an area of more than 1 million km^2, about 40 % of kangaroos died, most of them during the summer. However, such a population crash, which looks like a 'catastrophe', only reflects adaptation of kangaroo population dynamics and vegetation to the variable environmental conditions

(Ratcliffe 1959; Bomford 1990). A population crash 'adjusts' kangaroo numbers to the reduced habitat carrying capacity during drought periods. Between droughts, the kangaroo numbers increase again. Remarkably, kangaroos, profiting from stock waterers and rigid regulation of dingo, are more numerous in the originally waterless regions than prior to the arrival of the Europeans (Adam and Fox 1982). Kangaroos are competing with sheep and cattle for pasture and water resources, and may destroy crops and fences (Hercock and Tonts 2004).

By contrast, the herds of the many introduced herbivorous mammals have often have considerably reduced the carrying capacity over extended rangeland areas by continuous overgrazing (density, composition and structure of the vegegation, soil erosion). The situation might possibly stabilize at a lower level of productivity, with a vegegation less vulnerable to overgrazing and variability of life conditions. However, only continuous population control will prevent the extinction of native animal species and a decline of agricultural productivity (Hutchins et al. 1982). Traditional lethal methods of animal control have become increasingly disreputable in public opinion. Chemosterilisation is considered to be more appropriate (Bomford 1990). It is exactly the Australian example that shows that biological control measures (e.g., introduction of predators of rabbit or marine toad), generally believed to be 'natural', have often brought about unexpected and incalculable consequences that might have been avoided if better knowledge had been available.

5.1.3 Animals Introduced to New Zealand

Even more exotic animal species have been introduced to New Zealand than to Australia. The first European settlers arrived at the end of the eighteenth century. The real colonization, however, began about 50 years later. The settlers brought about 40 species of wild mammals to New Zealand (Wodzicki 1950, 1963). More than 90 % of all terrestrial mammals living in New Zealand are aliens (Kitching 1986). Almost 40 of about 130 bird species have built up viable populations. The fish fauna is also dominated by exotic species (Sedlag 1995). Prior to European colonization of New Zealand, the native mammalian fauna consisted of only two bat species and the Polynesian rat ('kiore', *Rattus exulans*), which, however, had arrived with Polynesian settlers, probably from the Marquesas, Cook or Society Islands (Davidson 1984) about 1,000–1,200 years ago.

There are many wild game animals among the introduced mammals (Table 5.1), such as red deer (*Cervus elaphus scoticus*), American elk (Wapiti, *Cervus elaphus nelsoni*), moose (*Alces alces andersoni*), fallow deer (*Dama dama*), Virginia deer (*Odocoileus virgianus*), muledeer (*Odocoileus hemionus*), Sika deer (*Cervus nippon*), rusa deer (*Rusa timorensis*), sambar deer (*Rusa unicolor*), chamois (*Rupicapra rupicapra*), and Himalayan tahr (*Hemitragus jemlahicus*), as well as several wallaby species (*Macropus* sp., *Petrogale penicillata*), brush-tailed possum (Australian opossum, *Trychosurus vulpecla*), rabbit (*Oryctolagus cuniculus*) and brown hare (*Lepus europaeus*) (King 1990a). A complete list of introduced animals is given by King (1990b). Most species were translocated from Great Britain more than 100

Table 5.1 Mammals introduced to New Zealand

	Species	Area of origin
Ungulates	Red deer (*Cervus elaphus scoticus*)	Europe
	Wapiti (*Cervus canadensis*)	North America
	Sambar deer (*Cervus unicolor*)	India, Southeast Asia
	Rusa deer (*Cervus timorensis*)	Indonesia
	Sika deer (*Cervus nippon*)	Eastern Asia, Japan
	White-tailed deer (*Odocoileus virgianus*)	North America
	Moose (*Alces alces andersoni*)	North America
	Fallow deer (*Dama dama*)	Britain, Tasmania
	Chamois (*Rupicapra rupicapra*)	Austria
	Tahr (*Hemitragus jemlahicus*)	Asia, Himalayas
Lagomorphes	Brown hare (*Lepus europaeus*)	Britain
	Wild rabbit (*Oryctolagus cuniculus*)	Britain
Rodents	Ship rat (*Rattus rattus*)	Accidentally (ships)
	Norway rat (*Rattus norvegicus*)	Accidentally (ships)
	Polynesian rat, Kiore (*Rattus exulans*)	Polynesia, accidentally or deliberately (?)
	House mouse (*Mus musculus*)	Accidentally (ships)
	Gray chipmunk (*Tamias striatus*)	North America
	Brown California squirrel (*Sciurus niger*)	North America
Carnivores	Ferret (*Mustela putorius furo*)	Britain
	Common weasel (*Mustela nivalis*)	Britain
	Stoat (*Mustela ermine*a)	Britain
	Cats (*Felis catus*)	Europe
	Dogs (*Canis lupus familiaris*)	Europe, Polynesia
Insectivores	Hedgehog (*Erinaceus europaeus*)	Britain
Marsupials	Brushtail (Australian) possum (*Trichosurus vulpecula*)	Australian main land, Tasmania, some offshore islands
	Wallaby (several species)	Australia
Feral animals	Sheep (*Ovis aries*)	Europe, Australia
	Cattle (*Bos primigenius*)	Europe
	Goats (*Capra hircus*)	Europe
	Pigs (*Sus scrofa*)	Europe
	Horses (*Equus caballus*)	Europe

After Wodzicki (1963) and King (1990b)

years ago (56.1 %). The 'rest of Europe' contributed 3.2 %, Asia 13.9 %, America 11.3 % and Australia 6.1 % (Wodzicki 1950). While most species were introduced between 1850 and 1910, no new species were added until the 1980s. Afterwards, some were introduced for diversification of farming stock. None of them were released into the wild, however (King 1990a). Many of the introduced wild animals, as well as the feral animals (e.g., wild pigs, goats, cats, etc.), have considerably influenced their new environment. A few examples are given in the following sections.

5.1.3.1 Red Deer and Other Wild Ungulates

Red deer have been introduced in great numbers since 1861. Most deer came from English game parks and the Scottish Highlands. About a third were translocated from Australia, where a population had already become established (Logan and Harris 1967; Challies 1985). Fallow deer were also introduced from English game parks, although in relatively small numbers. Fallow deer and other exotic cervids have reached relatively high densities. The distribution of the relatively small herds, however, is very localized compared to red deer (Holloway 1973). Fallow deer and white-tailed deer have locally altered forest structures in the Blue Mountains (South Island) and on Stewart Island south of the New Zealand South Island (Wodzicki 1963). The temperate, relatively humid climate with mild winters, a luxuriant evergreen and species-rich vegetation, abundant open space and lack of native predators provide favorable living conditions to the wild mammalian grazers and browsers.

Red deer (*Cervus elaphus scoticus*; Challies 1990), however, have been most successful in occupying the new habitats and built up the largest population of all of the introduced wild ungulates (cf. Table 5.1). Contrary to the introduced tropical cervids (Rusa deer, *Rusa timorensis*; Sambar deer, *Rusa unicolor*), red deer rapidly adapted rutting and calving time to the southern hemisphere seasons, thus reducing losses during the southern winter. Subsequently, the red deer population irrupted despite strong hunting pressure. At present, red deer are found everywhere from the mountains to the coastal lowlands of the South and North Island, with the exception of a few regions in its western and northern parts. Red deer prefer the upper and lower boundaries of the mountain forests.

Until the second half of the twentieth century, red deer severely impacted the environment. The effects on landscape are probably irreversible in some areas. The red deer population rapidly increased, supported by abundant forage. After two to three decades, however, deer numbers exceeded the habitat carrying capacity and declined due to depletion of food, increased competition and 'self-regulation' (Sect. 2.2.1) (see also Fig. 2.9). Severe weather conditions, for example, during the winters from 1928 to 1931 and 1939 to 1942, had an aggravating effect and accelerated the decline of deer (Clarke 1976).

Herbaceous plants suffered most from heavy grazing pressure. In addition, many shrub and tree species in the forest understorey were severely damaged and often destroyed. Red deer impeded or completely prevented natural regeneration of the forests in many areas (Holloway 1950). However, as to the intensity of the red deer impact, composition and successional stage of the forests, and thus, local and regional peculiarities, were the decisive factors (e.g., Wardle et al. 1971; Wardle 1974; Mark and Baylis 1975).

By the 1920s, the oversized red deer population had already become a problem of great economic significance (e.g., Cockayne 1926). Besides damage caused by browsing, barkstripping, and fraying their antlers or beating them against the tree stems, red deer also increased soil erosion (Photo 5.3). Everybody familiar with the New Zealand mountains has observed intense erosion in many places. However, erosion is omnipresent and a normal feature of the mountain landscape in this very

Photo 5.3 Windscarps above climatic treeline on Arthur's Pass (Craigieburn Range, New Zealand, South Island), which have developed after the plant cover was fragmented by red deer (*Cervus elaphus*). F.-K. Holtmeier, 29 November 1979

tectonically active region (Griffith 1979; Adams 1980). It is almost impossible, with a few exceptions, to reliably assess the contribution of red deer to soil erosion, particularly in view of overgrazing by sheep and regular burning (Schweinfurth 1966; communication P. Wardle, 23 April 1998).

In the early 1930s, big campaigns started to reduce red deer through intensive hunting and – to a lesser extent – poisoning. Nobody, however, had expected the great difficulties of ground hunting on rugged mountains. Thus, it is not very astonishing that deer hunting became less popular when deer numbers were declining. Even commercial helicopter hunting (since 1965; export of hides and meat to Europe, mainly to Germany) also turned out to be very difficult and risky on steep mountain terrain, not least because of the rapidly changing weather conditions.

The Department of Conservation tried to eradicate red deer from several areas on both North- and South-Island and to prevent further dispersal (Saunders 2000; http://www.doc.govt.nz/conservation/threats-andimpacts/). Commercial hunting is still the most important method for regulating numbers of wild-living red deer, particularly on the South Island, while recreational hunting is more common on the North Island (Nugent et al. 2001). Commercial deer hunting, however, has become less attractive due to an expanding deer farming industry that produces more venison at lower costs. Nevertheless, since the mid-1960s, red deer numbers were reduced by 75–95 %, and alternative control measures were not needed anymore (Challies 1985). Secondary vegetation, which emerged after the red deer decline, now provides a very productive food source to the remaining comparatively small herds.

In the alpine zone of the Fiordland (South Island of New Zealand), however, the long-term effects on the native fauna of overgrazing of the tussock grassland (*Chionochloa pallescens*) by red deer are still omnipresent. The alpine tussock has only partly recovered, although red deer numbers were considerably reduced in the 1960s (Rose and Platt 1987). Full recovery of the snow tussock may take about 30 years or so. The tussock grassland is an important summer food source for the takahé (*Noctornis hochstetteri* (*mantelli*)). This flightless rail feeds selectively on the protein-rich meristems of tussock grass (Williams et al. 1976). Consequently, survival of the takahé will only be possible at low red deer densities (Mills and Mark 1977; Lee et al. 2000).

In addition to red deer, wapiti (*Cervus canadensis*), moose (*Alces alces andersoni*), Himalayan tahr (*Hemitragus jemlahicus*) and chamois (*Rupicapra rupicapra*) were also introduced to the Fiordland (e.g., Caughley 1970; Douglas 1977; Clarke 1978). American elk, which is closely related to European elk, has not been as successful, probably because it was released onto the extremely difficult terrain of the Fiordland, to which it is also restricted at present. European red deer, however, have also gradually expanded into this area (Challies 1985). At present, red deer-wapiti hybrids are more common in Fjordland than pure-bred wapiti (Nugent et al. 2001). Wapiti and hybrids have modified their habitat in a similar way to red deer (Challies 1985, 1990), while no comparable effects for moose have been reported.

As to habitat conditions, both chamois and Himalayan tahr are highly flexible and are particularly able to move safely on extremely steep and rocky alpine terrain. Chamois were translocated from the European Alps to Mt. Cook (South Island) at the beginning of the twentieth century (1907 and 1914). They rapidly multiplied and spread in small groups over the South Island. They now occupy very different habitats, such as the high-alpine zone and the low mountains in the coastal areas. Chamois are still expanding, although hunting pressure is very strong. Chamois appear to be the most widely spread wild ungulate species in New Zealand next to red deer (Clarke 1990).

The tahr is a wild goat native to the Himalayas, where it occurs over a broad altitudinal range from the broadleaved forests (2,500 m) into the alpine meadow zone (4,400 m) (Schaller 1973). The first five tahrs were released near Mt. Cook in 1904. Fifteen animals were added in 1919 to support the herd's development. This, however, was probably unnecessary, as the population rapidly increased, benefitting from the abundant food. Herds of 50–70 animals could already be observed after a short while. At present, tahrs are found at elevations from 1,400 to 1,700 m, i.e., the zone of alpine shrub (*Dracophyllum* sp.) and tussock grassland above the upper tree- and bush limit (Douglas 1977). Tahrs apparently prefer steep slopes (>40°) and ranges with crests above 1,800 m elevation (Caughley 1970).

Chamois are less gregarious than tahrs. They usually graze in comparatively small groups, roaming about wide areas (Holloway 1973), which makes hunting more difficult. In winter, chamois occasionally concentrate on wind-exposed topography without snow (see also Photo 2.17). In such sites, plant cover is usually very sparse and particularly vulnerable to grazing and trampling. Moreover, chamois destroy the grass vegetation on narrow structural benches and steep rocky terrain, as well as on debris slopes, where the plant cover is very scattered anyway. Chamois may contribute considerably to soil erosion in such places (Clarke 1978). Tahrs, on the other hand, usually live in large herds, causing severe damage to the alpine tussock grassland (*Chionochloa* sp., *Poa colensoi*) (Holloway 1973), where they thin out the original plant cover within a short time. Grazing is most critical in winter, when low-growing plants are buried under snow and the tussock grasses (height 60–90 cm) project beyond the snowpack (Caughley 1970). Snow tussock (*Chionochloa* sp.) is particularly sensitive to grazing (Mark 1965). After tussock grasses are grazed down, the shade-demanding smaller plants between the tussocks become rapidly eliminated and finally replaced by low-growing creeping plant species (Evans 1973). Snow tussocks are followed by lower growing grasses (*Poa* spp., *Festuca* spp.), which, however, will also decline under continued severe grazing pressure. The shrub zone (*Dracophyllum* sp.) has become completely destroyed. Soil erosion increases parallel to fragmentation of the plant cover. Once the humus-rich top soil has eroded, the most important nutrient source is lost. Needle ice formation and frost-heaving accelerate soil erosion. In many places, the bedrock has become exposed.

Shortly after chamois and tahr had been 'successfully' released, their rapidly increasing populations and resultant impact on vegegation in the Alpine had already caused general concern. As complaints increased, hunting of both species was permitted. In view of the 'indescribable extent' (Douglas 1977) of damage, the government decided in 1936 to reduce the numbers of both mountain animals drastically. During the first campaign, about 3,000 individuals were culled, and some 10,000 in the following years. Tahr population was considerably reduced by commercial hunting booming in the early 1970s. However, commercial hunting declined a few years later because of low profit (Tustin 1990). In the end, however, the problem had not been not solved.

Both tahr and chamois have been classified as 'pests'. On the other hand, chamois and tahr (as well as red deer and the other cervid species) have an economic value (meat, trophies, recreational hunting, etc.). No wonder that the dispute between those seeing tahr and chamois as an economic resource and those that consider them a pest has increased (Hughey and Wason 2005). The discussion about what is the 'right' management will probably never end. The interests of different parties (highland farmers, hunters, Department of Conservation, conservationists, etc.) have to be considered. Highest priority has been given to conservation of what is left of New Zealand's original biodiversity (e.g., Ministerial Panel 2008). The management objective is to prevent animal population growth beyond the natural carrying capacity, or, in other words, to prevent further destruction of the native alpine grass- and shrub land (Department of Conservation 1993).

Successful control of tahr and chamois should go beyond just simply harvesting 'surplus' individuals. In addition, differing sex-specific and age-specific mobility and behavior of the animals in utilization of their habitats must be considered. Reliable data on sex ratio and age structures are an indispensable precondition for a clear management plan. Herewith, the difficulties begin (Forsyth 1999, 2000).

Moreover, hunting of both tahr and chamois on rugged mountain terrain is extremely difficult, time-consuming and strenuous.

In locations of high conservational value, the animal numbers are controlled by the Department of Conservation, assisted by recreational, professional and government-funded hunters (Forsysth and Hickling 1997). Shooting from helicopter gunships was most effective, but expensive. Anyway, ungulate populations could be reduced and expansion to other areas still free of these herbivores has been prevented so far. Their impact, however, often has lasting after-effects (e.g., Schweinfurth 1966; Rose and Platt 1987; Holtmeier 1999, 2002; Gill 2000). Not least, the impacts of overabundant wild-living ungulates on plant cover and soil erosion in the New Zealand Alps have to be put in perspective, as overgrazing, in particular, by sheep, and careless burning (e.g., U. Schweinfurth 1966; communication P. Wardle, 23 April 1998) have probably been and still are at least as important as the effects of red deer, tahr and chamois. Repeatedly, a reduction of grazing by domestic cattle has been postulated (e.g., Douglas 1977). The situation has become particularly critical in the greywacke areas, which are highly susceptible to erosion. As the soils on this substrate are very poor in nutrients, recolonization of eroded terrain by vegetation is very slow. Conditions are different on the more humid west side of the New Zealand Alps (Sect. 5.1.3.3). In areas from which tahr were removed, the plant cover has recovered, and snow tussock (*Chionochloa pallescens*) has become established again in the dense carpets of blue tussock (*Poa colensoi*) that had developed under the impact of tahr. As ever, the local conditions have to be carefully considered if assessing the carrying capacity of an area. A generally applicable rule cannot be set up.

5.1.3.2 European Rabbit

European rabbits have also impacted New Zealand. Rabbits were released in several places between 1830 and 1860 to lay the foundation for rabbit hunting, which was a very popular 'sport' among the English settlers, and to establish a fur industry. Between 1920 and 1929, about 150 million rabbit pelts were exported (Gibb and Williams 1990). In the late 1940s, exported pelts still amounted to 17 million.

On the North Island, higher human population density and different forms of agricultural cultivation have impeded expansion of rabbit. On the South Island, however, the use of fire to remove tussock grass and shrub vegetation, seeding of non-native grass species and intensive sheep grazing supported rabbits, which rapidly multiplied and spread like the plague, with similar consequences as in Australia. Sheep could not find food anymore. Many farmers went bankrupt and had to abandon their land. Erosion increased. However, some authors hold the view (e.g., Bell and Williams 1981) that erosion, usually ascribed to the rabbit's impact, is less important than natural erosion, and that the combined effect of overgrazing by sheep and great rabbit numbers is the real cause of omnipresent erosion damage (Photo 5.4). In 1952 and 1953, efforts were made to eradicate rabbits with myxomatosis (Williams 1983), which, however, were of no avail. The reason

Photo 5.4 Soil erosion initiated by introduced European wild rabbits (*Oryctolagus cuniculus*) at about 750 m on the Pisa Range, New Zealand (South Island). The trees of which the roots have become exposed are Snow Totara (*Podocarpus nivalis*). F.-K. Holtmeier, 29 November 1979

for the failure was unclear in those days. Nowadays, one knows that it was the lack of an appropriate vector.

Rabbits are most numerous in the dry mountains and hill countries of the South Island (Central Otago: Gibb and Williams 1990). They are the second largest pest after the brush-tailed opossum (*Trichosurus vulpecula*). In these areas, which are characterized by a relatively continental climate (warm summers, cold winters, mean annual precipitation 330 mm) and brown-greyish steppe soils covered by an often degraded vegetation (overgrazing by sheep, frequent fires), rabbits meet optimal conditions (Kerr 1989). Rabbits are combatted by all possible means: poisoning, predators, destroying their burrows, fences, biological control, and change of agricultural cultivation with temporarily fallow-lying land in rotation with grazing (e.g., Ross and Arthur-Worsop 1987; Kerr 1989; France 1991). On pastures from which rabbits were removed, productivity promptly increased threefold to tenfold, depending on the season. It also became apparent that successful long-term rabbit control (mainly poisoning) can only be achieved when rabbit numbers have been declining anyway, due to adverse weather conditions, for example. Complete eradication, however, is practically impossible, regardless of the unacceptably immense costs. Cultivation methods that brought about a serious deterioration of habitat conditions for rabbits were most efficient in rabbit control (Bell and Williams 1981; Williams 1983). In some areas, regular direct control (shooting, poisoning) was no longer even needed.

Nevertheless, in New Zealand, the era of the European rabbit has not yet ended, as was supposed in view of the initial success (e.g., Milne and Milne 1960). Meanwhile, successful rabbit control by the 'Integrated Land Management Strategy' has become increasingly disputed. In the opinion of farmers living outside the areas covered by the 'Rabbit and Land Management Programmes', the rabbit problem has even been exacerbated.

Under the pressure of the immense and continuously increasing costs of rabbit control, a renewed introduction of myxomatosis and their specific vectors (rabbit fleas) has been postulated since the beginning of the 1980s (France 1991). Still, in 1982, this was rejected by the government, probably because they were afraid of the opposition from the general public. However, there is no doubt that all possible means must be applied to get the rabbits under control in the particularly rabbit-prone semi-arid regions of the South Island (see also Kerr 1989). Besides 'traditional' methods, such as poisoning, fencings, habitat management, destruction of burrows and release of predators, myxomatosis is considered, as ever, to be highly efficient and economically acceptable (Ross and Arthur-Worsop 1987). Meanwhile, the calicivirus has also arrived in New Zealand. Since the 1990s, research has been conducted to find out how rabbit haemorrhagic disease can be efficiently used without having adverse effects on other organisms (MAF 1997).

5.1.3.3 Brush-Tailed Opossum

No other animal has been introduced to New Zealand that has caused more problems to nature and the economy than the brush-tailed opossum (*Trichosurus vulpecula*) or possum, as this animal is called by the New Zealanders. The possum belongs taxonomically to the phalangers (Phalangeridae) and should not be confused with the true opossum (Didelphidae) of South and North America.

Possum were introduced during the mid-1850s, at the latest. Very likely, a few animals had already been released before 1840 (Pracy 1962; Cowan 1991) for 'sport' (hunting) and as a fur-bearer. Enthusiastic about a future flourishing fur trade and unlimited possibilities for 'sport', possum liberations were handled very casually, although caution would have been appropriate in view of their impact on the Australian and New Zealand environments. Moreover, at that time, possum liberations were not prepared and supported by scientific research.

Most possums were introduced between 1890 and 1930. However, until 1959, possums were released legally or illegally into new areas (Fig. 5.3). Keeping a soft and cuddly possum as a pet was and still is very popular. There are about eight subspecies or varieties. The so-called 'Tasmanian blacks' and 'red-blacks' were introduced first because of their excellent pelt quality. Since about 1910, possums with grey fur have been released.

Nowadays, possum completely occupy New Zealand (>90 % of the total area): the warm-tropical north, as well as the subantarctic south, the extremely humid west coast, the dry regions of Otago and Canterbury in the east, and some offshore islands. A New Zealander being asked what a possum looks like may perhaps

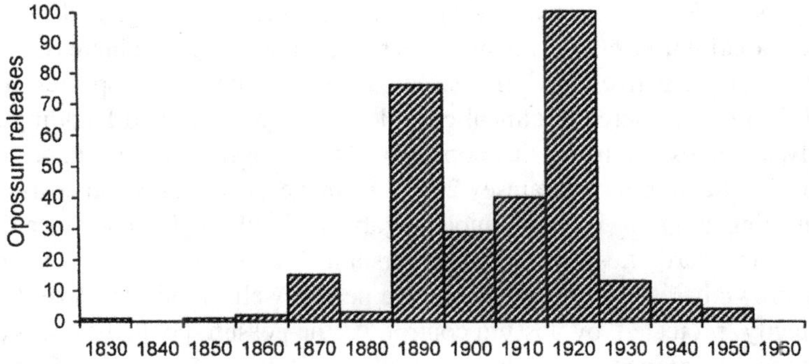

Fig. 5.3 Releases of Australian opossum (*Trichosurus vulpecula*) in New Zealand (Modified from Aspinall (1977))

answer, '*A possum is that flat brown animal you can always find on the road*'. In the 1980s, the total number of possums in New Zealand was estimated to be about 50–70 million (Cowan 1991). As a result of strict control, possums dropped to about 30 million in 2009, which is an overall reduction below natural carrying capacity (Warburton et al. 2009). Habitat carrying capacity in the absence of possum control is estimated to be 48 million. As estimates of the carrying capacity of different possum habitats are not available, estimates of tolerable possum numbers are rather speculative (Warburton et al. 2009).

Possums affect trees and other plants, as well as the indigenous fauna. They feeds on leaves, shoots and fruits of more than 100 native and numerous introduced plants. The diet, however, differs by region. Within the different areas, possums feed on only a few species (Green 1984). In addition, possums also prey on invertebrates, birds, and their eggs, and also scavenge (Brown et al. 1993). Damage ascribed to possums is immense and hardly calculable. In short, the possum has become a 'pest'. The impact of possums exceeds by far damage caused by any other introduced animal. Possums do not only affect the native forests, but also severely damage economic forests consisting of exotic species (Farmer 1973). Moreover, possums devastate orchards, flower gardens and pastures. Not least, like the introduced cervids, feral goats and pigs, possums transmit bovine tuberculosis. At present, on about 25 % of the land area possums are infected with tuberculosis. Tuberculosis is apparently most frequent in overgrazed areas with high possum density, i.e., where cattle and possums compete directly for food (Aspinall 1977). Possum is a serious economic threat to the New Zealand economy, which largely relies on the export of meat and dairy products. Export is worth more than 12 billion dollars per year. Annual economic losses due to bovine tuberculosis are said to amount to 500 million dollars (MAF 1996).

Despite permanent strong control, the possum population is flourishing. Though commercial trapping has been encouraged by the government, it is not very profitable any more, in particular, as the value of possum fur has drastically decreased due to anti-pelt campaigns in Europe and the United States, resulting in the boycott of all furs. Poisoning by aerially sown baits (e.g., carrots) impregnated with 1080 (sodium- monofluoroacetate) is more 'effective', though there are increasing public concerns about negative side effects. For example, non-target species may be affected. Moreover, there are ethical considerations (Morgan and Hickling 2000). Currently, methods are being developed to reduce possum fertility as a possible alternative in the long-term (Ramsey 2005). Damage by possums and costs of control (poisoning, trapping, culling) amount to about 35 million dollars (Cowan 1991). Possum control partly ties up money and personnel needed for the control of rabbits and also brown hares, which have become a pest as well. In other words, the latter are indirectly 'protected' by possum control. As the possum population was able to rapidly recover and increase to high levels about 10 years after control (Rose et al. 1992), possum control is a never-ending task.

With regard to the message of this book – to highlight effects of animals on their habitats – the situation in the western New Zealand Alps is of particular interest. Already by the end of the nineteenth century, increasing canopy dieback was observed in the 'rata-kamahi' forests ('rata' = *Metrosideros umbellata*; 'kamahi' = *Weinmannia racemosa*) of Westland (South Island) (Photo 5.5). Canopy dieback and an increasing number of possums very soon turned out to be closely linked to each other. At that time, however, the chance of founding a lucrative fur industry directed by the government obviously played a major role compared to the growing concerns about possum damage. Kirk (1920) and Cockayne (1926), for example, believed that the returns from fur production would exceed the economic losses caused by possums

Photo 5.5 Canopy dieback in rata trees (*Metrosideros umbellata*) caused by possum (*Trichosurus vulpecula*) near Rough Creek (Westland National Park, New Zealand, South Island) at about 1,800 m. L. E. Burrows, January 1982

to the forest. Thus, these authors indirectly 'recommended' support of the possum population (Pracy 1962). Perhaps the interactions between possum and red deer and their adverse effects on the New Zealand forest were not yet obvious or were simply overlooked, which is likely, as Cockayne (1926) and other researchers had already emphasized the impact of an oversized red deer population on the forests.

Canopy dieback had considerably increased since the 1940s (Chavasse 1955: Wardle 1971). Tall mature trees obviously suffer more from possum than younger trees. While some trees become completely defoliated, others remain untouched. This more patchy distribution of possum damage in forests has also been found on the North Island (Elder 1965; Meads 1976). Veblen and Stuart (1982; see also Stewart and Veblen 1982a, b), by contrast, point to young vital stands of 'rata' and 'kamahi' in the damaged area, and ascribe widespread tree mortality first of all to natural decline of overaged trees rather than to defoliation by possum. In the view of Mosley (1978), who studied erosion in the southern Ruahine Range (North Island), forest decline had become predetermined by overaging and possibly by insect calamities and unfavorable weather conditions during the 1930s and 1940s. The possums, together with red deer and feral goats, had probably only triggered forest decline and also retarded recovery of the tree stands.

Other observations (Pekelharing 1979; Pekelharing and Reynolds 1983; Payton 1987), however, provide evidence that, in areas with continuous high possum density, forest stands, being vital at a young age, succumb in the end to defoliation by possums. The same views are held by Batcheler (1983), based on his thorough review of this issue. If metal collars put around the stems of 'rata' trees prevent possums from climbing them, these will not become affected, and declining defoliated trees may recover again, even in areas where possum density is high (Meads 1976; Coleman et al. 1985). Stewart (1992), on the other hand, referring to his studies in southern West (South Island), takes the view that reliable statements on the role of possums in dieback 'rata-kamahi' forest stands are impossible, as the knowledge of the forests' past is usually insufficient.

About 30 % of all forests appear to be susceptible to canopy dieback. As of yet, the hypothesis that the possums destroy complete forests (Schweinfurth 1966) has not been proven by long-term studies so far (communication L. Burrows, Landcare Research NZ, 11 May 1998). However, possums probably do accelerate tree mortality in over-aged stands (e.g., Veblen and Stewart 1982) and in forests already affected by other factors (e.g., insects, goats, unfavorable climatic conditions; Mosley 1978). Anyway, it has become evident that high possum density and tree mortality are often positively correlated, and that no forest dieback has occurred in possum-free regions (Batcheler 1983).

In 1955, the New Zealand Forest Service (Chavasse 1955, cited in Payton 1987) was already emphasizing that possum was not the only cause of forest damage and that oversized numbers of red deer were no less important. Schweinfurth (1966) described in detail the combined effects of both animal species on New Zealand mountain forests. Red deer were already a problem animal, particularly on rugged mountain terrain (Sect. 5.1.3.1). During the first years after introduction, red deer enjoyed relatively favorable conditions at lower elevations. However, due to increasing hunting pressure in the coastal lowlands, red deer moved to the upper zone of the mountain forests and to the alpine tussock grassland. As the dark, cool and wet virgin mountain forests are unsuitable for red deer as habitat, it would probably not have become seriously affected. It was the concurrence with the possum impact that led to increased tree mortality and forest decline (Chavasse 1955, cited in Payton 1987; Schweinfurth 1966).

In central Westland (South Island), possums feeding mainly on the leaves of 'mahoe' (*Melyctus ramiflorus*), 'rata' (*Metrosideros umbellata*) and 'kamahi' (*Weinmannia racemosa*) affected about 90 % of the total forest canopy (e.g., Rose et al. 1992, 1993), and even in southern Westland, canopy dieback was conspicuous (16 %), even though less palatable *Nothofagus* species prevail. The leaves of 'rata' and 'kamahi' make up to 70 % of the possums' diet.

Possums apparently prefer sunny and warm habitats. There, they first cause gaps in the otherwise compact forest canopy. After canopy dieback, more sunlit patches and higher temperatures on the forest floor stimulate growth of forbs, herbs and shrubs in the forest understorey, thus providing an attractive food source for red deer. Red deer leave the tussock grassland at higher elevation and enter the gaps in the mountain forests to forage. They destroy the understorey vegetation and impede tree regeneration, thus accelerating senescence and subsequent decay of forest stands. Possums benefit from the now warmer conditions and better food supply in the opened forest canopy. Possums alone would kill many trees, but natural regeneration could perhaps compensate the possums' impact. In this context, Schweinfurth mentioned observations in the mountain forests on Mt. Egmont (North Island) that also harbor large numbers of possum. As red deer, however, are absent and feral goats were drastically reduced, natural regeneration is possible.

Trampling by red deer initiates or enhances soil erosion in the openings, no longer protected from the often heavy rains (>5,000 mm). As mountain forests decline due to the combined impact of possums and red deer, subalpine scrub and tussock grassland expand to lower elevation (Rogers and Leathwick 1997). Prospering red deer also cause severe damage to the alpine tussock grassland when grazing there under good weather conditions in summer.

Meanwhile, the possum-red deer problem has turned out to be even more complicated than described by Schweinfurth (1966). Regional differences in susceptibility of the forests to possum damage and subsequent erosion occur. These may be ascribed to the history of possum occupation, forest composition and age structure and other local factors, such as substrate-dependent (slate, grewacke, granite) susceptibility to erosion, frequency of landslides, and, not least, the impact of secondary infestation, all of them influencing forest structure, composition and succession. Probably, geological stability and frequency of landslides play a major role in this context (e.g., Payton 1987), particularly because of the high tectonic activity in this region.

Thus, careful regional and local analyses are required to assess and differentiate the problem. The studies by Payton (1987) on canopy dieback in Westland are very instructive in this respect. Payton found that not all 'rata-kami' forests are equally affected by possums. Damage was most conspicuous and serious in the particularly geologically and geomorphologically unstable areas east of the Alpine Fault, which are built up by schist. The Alpine Fault, which separates the Pacific and Indian plates, is located west of the main crest of the New Zealand Alps and oriented parallel to the mountain range. Comparatively little possum damage occurs in granite areas. Dead trees can only be found in about 10–11 % of the mature forests, whereas on shale, dead trees typically occur in 24 % of the mature forest stands at lower elevations and in 35 % of the forests at higher locations. This difference between granite and shale areas is probably due to the fact that on easily eroding shale, new soil may develop within a comparatively short time after slope erosion and landslides. Both rapid soil formation and a comparatively mild climate support the establishment of very dense even-aged secondary forest stands, which are characterized by a high portion of seral hardwoods. These provide optimal habitats for possums. These dense forest stands are relatively insensitive to possums during the first decades until competition between the trees and/or defoliation by possums cause first gaps in the compact forest canopy. Consequently, red deer will promptly get in, and the permanent strong winds may cause stem breakage and windthrow, followed by insect infestation, fungal attacks, and increasing erosion.

The situation is completely different in southeastern Otago (Catlin District). Low mountains formed by sandstone, phyllite and conglomerates merge upwards into almost level or gently sloping highlands. Compared to the steep mountain terrain of the New Zealand Alps in the west, gravitational energy and susceptibility to erosion are relatively low. The 'rata-kamahi' forests have not substantially changed, although the first possums had already been released in 1890. The forests, exhibiting a mixed age structure and relative low portions of seral hardwoods, are very similar to those on granite in Westland.

As to the extent of possum impact, the more or less great geological stability of landscape and their influence on forest composition and structure must obviously be considered to be crucial factors. Possum impact can only be assessed if considered as one agent within the locally varying interactions of many landscape-ecological factors. Thus, local findings should be transferred to other areas with great reservation, if ever.

5.1.3.4 Wallabies

Wallabies, the smallest members of the kangaroo group (Macropodidae family), are the last example in this book of animals intentionally translocated to New Zealand. Altogether, eight species were introduced, most of them from Australia. Five of them still occur locally. These are Dama wallaby (*Macropus eugenii*), Bennett's wallaby (*Macropus rufogriseus*), parma wallaby (*Macropus parma*), brushtailed rock wallaby (*Petrogale penicillata*), and swamp wallaby (*Wallabia bicolor*). In contrast to red deer and other cervids, for example, which were translocated from the northern hemisphere, wallabies did not need to adjust their seasonal rhythm of life. With the exception of Bennett's wallaby, the small kangaroos live on small islands (Kawau, Rangitoto and Motutapo) in the Hauraki gulf of the North Island

and (Dama wallaby) around Roturua (North Island.). The Bennett's wallaby occurs on the South Island only and is the most widely distributed of all wallaby species in New Zealand.

The first Bennett's specimen came from Tasmania and some islands located in the Bass Strait. They were released near Waimate (southeastern South Island) in 1874 (Catt 1975). Until 1976 (!), additional wallabies were released. They adapted without any problem to the new habitats, which are totally different from their Tasmanian homeland, and rapidly multiplied, thanks to their high reproductive capacity. Already in the 1920s, there were increasing complaints about wallabies, which competed with sheep for forage, polluted pastures, destroyed snow tussocks and impeded regeneration of native bush and tree vegetation. At the same time, there also were efforts to protect them. Since the 1940s, however, wallabies have been under rigid control (shooting, poisoning). Between 1947 and 1956, about 70,000 individuals were killed by the Forest Service. Private hunters shot an additional 30,000 specimens (Warburton 1986). About 95 % of the total wallaby population were eradicated during a poisoning campaign (1080 poison) in 1960. However, wallabies rapidly recovered and are considered as big a pest as ever. Commercial hunting (skins, meat) is not very lucrative. Hunting wallabies for sport, however, enjoys great popularity. Escaping the strong hunting pressure, wallabies have increasingly spread over wider areas. There is no doubt that overabundant wallaby numbers are a threat to livestock and to 'landscape stability' in catchment areas (Graham 1977).

The other wallaby species have also sustainably affected the environment in their comparatively small areas of distribution (see above), in particular vegetation. On Kawau, for example, parma wallabies, together with other species and with the omnipresent possums, have impaired forest regeneration and eliminated many native plant species. For a time, the kangaroos were intensively hunted and also poisoned to an extent, such that the parmas even became protected by law in 1969. In 1984, protection was removed (Warburton 1986). After farming was abandoned in 1973, however, control has not continued.

The brushtailed rock wallabies on Motutapu are considered to be competing with cattle for food. Moreover, they are reported to exacerbate erosion on steep terrain through burrowing. On Rangitoto, wallabies adversely influence vegetation that became established after a volcanic eruption around 1400 AD and that is of international interest for botanists. As the island is open to the public, there is some concern that tourists will take the 'sweet' wallaby babies home to other islands and to the main island, thus creating new centres of wallaby expansion (Warburton and Sadleir 1990).

In New Zealand and Australia, the general public and, in particular, ranchers and farmers are directly affected by often unexpected consequences of exotic animal introduction, as, in most cases, 'repairing' the mistakes that were made in the past is extremely expensive – and the costs must by paid by the community. In view of the continuous discussion about the 'right' animal control measures and the demands for reduction of the exotic animals to a low level, it is often overlooked that the combined impact of inconsiderate overgrazing (particularly by sheep), continued careless use of fire and introduced ungulates allowed the problems (soil erosion, in particular) to run out of control. In any case, the idea of careful and sustainable use of the natural resources – and these also include pastures – has not played a role for a long time in New Zealand, which still bears features of a pioneer country in many areas (see also Schweinfurth 1966).

5.1.4 Animals Introduced to Europe

In contrast to Australia and New Zealand, the European landscape had been intensively used for centuries or even longer before most exotic animals were introduced. Besides many mammals, bird and fish species have also been brought from other countries to Europe (Niethammer 1963). Many have been released intentionally, for example, some cervid species and the mouflon (*Ovis musimon*). As the native populations had just become 'successfully' eradicated, one refrained from introducing predators. The foreign carnivores (racoon, *Procyon lutor*; mink, *Mustela vison*) escaped from fur farms, and the muskrat (*Ondatra zibethicus*) and nutria (*Myocastor coypus*) got away from private preserves. The raccoon dog (*Nyctereutes procynoides*), originally native to East Asia, had immigrated to Central Europe from the former Soviet Union, where it had been introduced for fur farming.

There are endless discussions on the impacts of these introductions and to what extent they are tolerable. Anyway, a few remarks on these discussions may be allowed. Many of the introduced species have successfully occupied an ecological niche and built up viable populations. This, however, has often happened at the cost of native species. In England, for example, the grey squirrel (*Sciurus carolinensis*), which has been introduced to many places since the late nineteenth century up to 1929, has replaced the native red squirrel (*Sciurus leucurus*) (Niethammer 1963) in the deciduous forests as well as in the conifer forests (southern England) that are the red squirrels' preferred habitat (Corbet 1974). More recently, grey squirrels have expanded into Switzerland from Italy (Piedmont), where two pairs were released in 1948 in a private park near Torino. Further spread over Europe and even Eurasia is likely in the long-term, and the grey squirrel might outcompete the red squirrel over wide areas.

Nevertheless, as suggested by Reichholf (1977), from an ecological view, the 14 introduced mammalians currently living in England may be considered an enrichment of the fauna, in particular, as none of the newcomers adversely affects the ecosystem function as much as the oversized populations of the native ungulate species. The occupation of an ecological niche – disregarded whether by a native or an exotic animal species – is usually disadvantageous to the less competitive species. It may be questioned, however, whether this is a reason to eliminate introduced species that have successfully occupied ecological niches, with the objective of preventing faunal falsification and restoring original biocoenoses. This would be very much the same as tilting at windmills.

Even in a country like New Zealand, 'blessed' with alien species, one is ready – possibly because of a certain helplessness felt when facing the many problems caused by the alien animals – to accept the newcomers as part of the country's present nature, in tolerable numbers at least. In Europe, the damage caused by exotic animals has not assumed such alarming proportions as in Australia, New Zealand and Hawaii. Often, only experts are able to recognize changes of the biocoenoses and their real causes. For example, modelling the expansion of the parapox-virus disease (Poxviridae; Liebermann 1992) in the red squirrel provided some evidence that this virus may be the cause of red squirrel decline rather than competition between the red squirrel and the introduced grey squirrel (Wauters and Gurnell 1999; Rushton et al. 2000). While grey squirrels, which also transmit the virus, are not seriously affected, the disease decimates red squirrels. This hypothesis is substantiated by the fact that in Ireland, for example, where the parapox-virus has not been discovered up to now, red squirrels and grey squirrels have shared the same habitats for more than half a century (Duff et al. 1996).

In general, the introduced ungulates adversely affect vegetation first of all (browsing, grazing, trampling, barkstripping, etc.). Altogether, however, the impact by often overabundant numbers of native red deer, roe deer and fallow deer (already introduced by the Romans) exceeds the damage done by the more recently introduced ungulates. Nevertheless, sika deer (Ratcliffe 1989; Ueckermann 1992; Hespeler 1999) or even mouflons (Türcke and Schmincke 1965; Hofmann 1977; Türcke 1981; Eisfeld and Fischer 1996) in locally high densities may cause considerable damage (browsing and barkstripping). Moreover, chamois introduced (reintroduced ?) to the Vosges Mountains (1955/1956) and to the Black Forest (1935) may put a heavy strain on local vegetation, particularly on Norway spruce (Knaus and Schröder 1975; Wotschikowsky 1981a).

All introduced predators, such as the mink, raccoon dog and raccoon, which have a particularly bad reputation, as they prey on eggs and young birds, are viewed very suspiciously. Obviously, emotions play an important role in this respect. Apprehensions that these predators would become a serious threat to the native fauna and biocoenoses and to the 'ecological balance' have turned out to be unfounded (Dunstone and Ireland 1989; Harrison and Symes 1989; Hespeler 1995; Lutz 1996; Michler and Michler 2012).

5.1.4.1 Muskrat

Muskrats (*Ondatra zibethicus*) regularly conflict with human interests. Although called a 'rat', the muskrat is a vole (Microtinae). The muskrat is originally native to North America, where its area of distribution extends from high northern latitude to the northern border of New Mexico. This wide distribution reflects the great adaptability of muskrat to different environments. In this respect, muskrat is by far superior to the nutria (*Myocastor coypus*), which was already held in cages in 1927 (Germany). The first muskrats were very likely released to the eastern Bohemian Forest. In 1905, additional animals were liberated at Dobris, west of Prague. Later, muskrats were also introduced to other European countries for fur farming. Over and over, many animals escaped (Nowak 1977). Muskrat have rapidly spread over

Europe (with the exception of southern France and the Iberian Peninsula) and northern Eurasia. Great adaptability to the 'new' environmental conditions, high fecundity, lack of competitors, and the dense stream network, including canals and drainage channels, have promoted muskrat expansion (Frank and Härle 1967; Gersdorf and Frank 1969). Natural predators (e.g., otter, fox, polecat, stoat, mink, several predatory birds, great owl) could not prevent muskrat expansion. However, reduction of muskrat population by introduced North American mink has been reported from eastern Germany (Bräsecke 1989) and also from some Karelian lakes (Danilov 1992). Mink also prey on ground vole (*Arvicola terrestris*) and affect populations of small duck species and rails (Nordström et al. 2002). Soon after introduction, the muskrat had already been considered an unwelcome immigrant and declared a 'pest'.

The muskrat is an semiaquatic animal which goes on land only when foraging or when streams, ponds, and drainage channels dry up. Although the muskrat does not roam about more than the ground vole (*Arvicola terrestris*), its closest relative, it needs more food and space, as it is not very social during the reproductive season. Its unsocial behavior is probably one reason for rapid spreading, besides dispersal by floods and 'travelling' on ice floes (KWK-DVWW 1977).

The muskrat subsists mainly on vegetarian food, such as aquatic plants and soft parts of terrestrial plants. Thanks to its long blind gut, the muskrat is able to digest plants poor in protein. Animal food, such as fresh water mussels, crayfish, fish and fry (Akkermann 1975; further reading therein) makes up a comparatively small portion of its diet. Occasionally, however, larger losses in fresh water mussels and swan mussels have been reported (Schmidt 2001). Muskrats in large numbers may influence the composition and succession of the vegetation in their aquatic habitats. Some species, such as white water lily (*Nymphaea alba*), yellow water lily (*Nuphar lutea*), water chestnut (*Trapa natans*) and marsh trefoil (*Menyanthes trifoliata*), decline under muskrat impact, while others, such as bulrush (*Typha* spp.) and common reed (*Phragmites communis*) expand (Brandner 1951; Akkermann 1975; Danell 1977a, b; therein further references). Muskrats use different plant species for different purposes. In reed communities, for example, muskrats use sprouts and leaves of broadleaf cattail (*Typha latifolia*) and narrow leaf cattail (*Typha angustifolia*) for construction of its lodges, whereas they forage on common club-bulrush (*Schoenoplectus lacustris*) (Bernhart and Schröpfer 1992). Club-bulrush, however, could gradually expand, whereas cattails decline. Germination and establishment of some species benefit from muskrats' burrowing (Brandner 1951; Akkermann 1975).

The muskrat may strongly affect its habitats by selective feeding on river bank vegetation and aquatic plants, for example, and, in a point of major importance, by burrowing holes and multiple branching tunnels into the banks of ponds, lakes, and rivers, as well as in dikes and dams. Parallel to growing muskrat population and the development of superfamilies, the holes and tunnels increase, lying side by side and/or on top of each other (Fig. 5.4). In addition, there are numerous escape tunnels, nest holes and also 'training holes' dug by young muskrats. Occasionally, muskrats may undermine roads and railroads (cf. for beaver). Damage caused by burrowing to fish ponds and to water supplies and distribution has been and still is

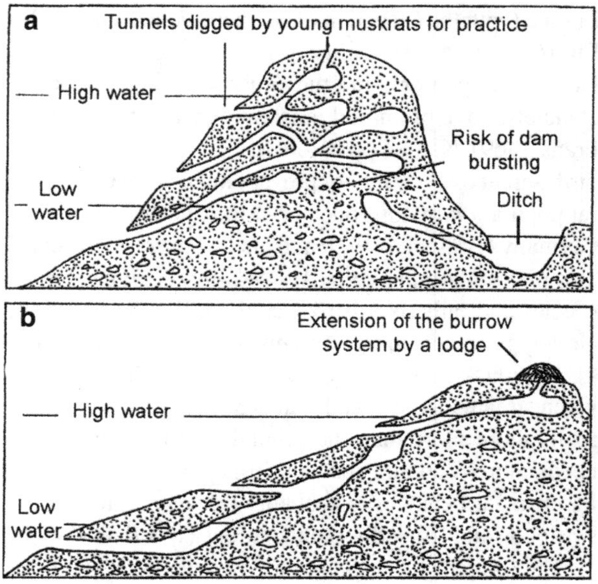

Fig. 5.4 Muskrat (*Ondatra zibethicus*) burrow systems. (**a**) In a steep-sided dam. As the nesting chambers and tunnels are lying on top of each other, the risk of dam bursting is relatively high. (**b**) in a gently sloping dam. The risk of dam bursting is relatively low, as the nesting holes and tunnels are not lying on top of each other (Drawings by F.-K. Holtmeier after a figure in KWK-DVWW (1977))

of major concern in some areas. Not least, the muskrat is a carrier of different kinds of tapeworms (dog, fox, cat), which may also affect humans (e.g., Diemer 1996).

Along rivers, the risk of muskrat damage is particularly high where flow velocity is low and loamy soils provide sufficient stability to the burrow systems. Moreover, reeds covering the zone which is influenced by the fluctuating water level are attractive to muskrats. Although muskrats prefers ponds, lakes, rivers and drainage channels with little fluctuation of the water level (20 cm or less. at best), it is able, similar to the beaver, to adjust to greater differences in water level at steep banks by establishing its burrows at different floors.

In sections undermined by an extended burrow system, the embankments may break off or even collapse, particularly during floods. Water moves within the muskrat burrows and multiple-branched tunnels, according to the principle of communicating tubes, and makes the ground soggy. Occasionally, agricultural equipment maneuvering close the river bank or grazing cattle may break through.

In view of the often really ‘dramatic’ reports, one could get the impression that muskrat damage is permanently a ‘risk to life and limb’. However, those who profited from muskrats in one way or another have much exaggerated the problem. Muskrat hunters and trappers (private and official) interested in higher premiums, and also official institutions (e.g., chambers of commerce, administration and other offices, associations for pest management, soil conservation, water management, etc.) involved in muskrat control have seen a good chance to get more money for stocking up on personnel. Daily press and TV often present real ‘horror stories’ on muskrat and nutria damage. Although from an ecological view, muskrat control may look like a job-generating measure (Reichhof 1993), muskrat damage may cause local concern.

In the Netherlands, which are partly located below sea level, burrowing by muskrat is considered a permanent threat to agricultural land and to the people living in these areas. About 8 % of the total land area is covered by polders, lakes, ponds and rivers. Omnipresent drainage channels measure several thousand kilometres in length. Thus, muskrats enjoy optimal habitat conditions. They may easily propagate and find banks, dams and dikes suitable for burrowing almost everywhere. Moreover, there is plenty of food. Consequently, without rigid control, muskrat populations may rapidly grow to densities that are intolerable with regard to the safety of people and cultivated land. The same holds true for nutria. The costs for repairing damage caused by muskrats to river embankments and dikes, in particular, exceed the costs for muskrat control by far. At least a half of the spring population must be removed to keep muskrat numbers at a tolerable level (Doude van Troostwijk 1976).

In Germany, the situation was exacerbated during the 1960 and the 1970s when muskrats rapidly colonized the northern lowlands and were considered to be a serious threat to the dikes along the coast and rivers. Later, however, this suspicion turned out to be unfounded (Pelz 1996). During this period of time, trapping bags drastically increased. In Germany, 365,000 muskrats were caught. However, it has not been proved that even such mass exterminations can reduce muskrat damage to embankments. Despite extensive control (Doude van Troostwijk 1976; Pelz 1996) through trapping, poisoning, shooting, and domestic ferrets (*Putorius putorius*), the muskrat could not be eradicated. It was only on the British Isles that muskrat could be removed about 70 years ago (Gosling and Baker 1989).

Being realistic, muskrat cannot be tolerated in 'unstable' riparian and lacustrine habitats. However, for economic and also ecological reasons, control measures must focus on the most endangered objects (dikes, dams, roads, etc.). No constructions can make waterways completely 'muskrat-proof'. Thus, muskrat will continue to cause local problems in the intensely used agricultural landscapes of Central and Western Europe. Very likely, 'self-regulation' will limit population growth exceeding the natural carrying capacity of the habitats (cf. Fig. 6.1). Although one cannot wait in each case for natural regulation, trapping hundreds or thousands of muskrats per year, just to be on the safe side, does not make much sense, in particular, as non-target species often also get caught in the traps (Pelz 1996).

In less densely populated regions of Europe, such as Finland, Sweden and Russia, muskrats enjoy a much better reputation. By feeding on cat tail (*Typha*) and reeds (*Phragmites*), they facilitate expansion of less competitive species, thus creating mosaics of different successional stages that influence biodiversity (Brandner 1951; Danell 1978; Kowarik 2003). The resulting new structures provide niches to a species-rich submersed vegetation, which again attracts waterfowl, mammals and large invertebrates (Beecher 1942; Weller and Spatcher 1965; Weller and Fredrickson 1973). Apparently, in the northern landscapes, the positive effects of muskrat on its environment outweigh its negative impacts.

However, in Central and Western Europe, a more differentiated view of the muskrat's influence has also developed (e.g., Pelz 1996). It has become apparent that embankments of rectified stream sections deprived of their original vegetation are particularly prone to muskrat burrowing, whereas undisturbed sections are less affected (e.g., Looschen 1986). Thus, restoration of natural embankment structures and vegetation might reduce or even prevent muskrat damage. In the past, light-protective alder (*Alnus glutinosa*) were often planted in great numbers alongside water courses and drainage channels in order to prevent overgrowth (Lohmeyer and Krause 1974). Alder typically occurs in natural bank vegegation. Monoculture-like stands, however, adversely affect biodiversity. On the other hand, the strong root system of alder stabilizes embankments, reduces erosion and muskrat burrowing. However, one cannot be sure that muskrats will not move to more attractive places. Muskrat will continue causing conflicts in the intensively utilized cultural landscapes of Central and Western Europe. In Germany, the government has largely retired from muskrat control because of limited financial resources and has left it partly to associations for soil conservation and water management, as well as to licensed people. For a long time, top priority was given to the development of control methods, while basic research on muskrat was of minor importance. This has probably been the main obstacle to effective muskrat control (Pelz 1996).

As to the ecological aspect, the muskrat should be accepted as an animal species that has successfully established itself in its new habitats. In the view of Ellenberg (1980), muskrat, filling the niche between the much smaller ground vole (*Arvicola terrestris*) and the much larger beaver, has enriched the native fauna of Middle and Western Europe and should be taken for an integral member of the biocoenoses of river banks, ponds and lake shores. However, those who are affected by muskrat damage (e.g., farmers, garden owners) might hardly agree.

5.1.4.2 Nutria (Coypu)

The nutria (*Myocastor coypus*) originally comes from South America. About 80 years ago, the nutria was introduced to Louisiana and kept in cages. A few animals escaped. As the reproductive capacity of the nutria is very high, a population of several million built up. This was the basis of a lucrative fur industry during the 1960s and 1970s. About one million pelts were sold per year. After the fur industry had declined, nutrias could rapidly multiply and expand with almost no control. They cause severe damage in the coastal regions (Linscombe et al. 1981).

In Central Europe, permanent populations could develop only very locally because of the relatively unfavorable climate. Very strong winters may cause high mortality in nutria (Niethammer 1963; Klemann 1997; Pelz et al. 1997). During the late 1990s, however, nutrias considerably multiplied and propagated. The nutria usually produces 2 l with six young each per year. Very likely, several subsequent mild winters, abundant food supply on cultivated fields and in gardens (vegetables, fruits, windfall) in close proximity to the aquatic habitats, and completely insufficient control were the main driving factors. In urbanized areas, winter feeding by people has also supported nutrias. In contrast to muskrat, nutria enjoy great sympathy in the public (Heidecke et al. 2001).

Nutria damage tree plantations (browsing, girdling), agricultural crops and gardens next to water courses, as well as embankments, where they may enlarge abandoned muskrat burrows, for example (Müller-Using 1965). Moreover, nutria destroy submersed aquatic plants around their burrows, whereas perennial herbs are usually not lastingly affected (Schmidt 2001). During the 1960s, nutria impact was still of minor importance. However, it has increased parallel to growing nutria numbers and expansion. The magnitude of damage beats everything observed before. As to water supply and distribution, nutria impact appears to be even more important than damage by muskrats (Heidecke et al. 2001). Locally, nutria-carried parasites might also be a risk to other animals, such as beavers, wild ungulates and cattle. However, one might get the impression that the situation has been a little overblown (cf. muskrat).

In the face of recent developments, the fastest possible eradication of nutria has been recommended (Heidecke et al. 2001). Elimination, however, appears to be impossible. Even severe winters have not completely eradicated nutria population. The survivors reproduce all year round, rapidly compensating for the losses. Thus, an effective reduction could be achieved, if ever, only through trapping, shooting and poisoning.

In England, however, nutria was eradicated by the end of the 1980s, i.e., 50 years after introduction. Thanks to the mild maritime climate, nutria had rapidly spread over the wetlands in the east and caused enormous damage to agricultural crops. Undermining of dams and dikes, however, was even more problematic, particularly in the lowland of eastern England. In a big campaign (1962), the nutria population was reduced by about 90 %. It must be added, however, that most nutria fell victim to the extreme winter of 1962/1963 (depletion of fat reserves, shortage of food, miscarriages, and reduced fertility; Gosling 1981). Anyway, during the following years, the nutria population increased again considerably. Consequently, a new campaign was started in 1980, the concept and implementation of which was closely oriented to results of research on population ecology. Thus, the 'mistake' of introduction was successfully corrected (Gosling 1981, 1989).

5.1.4.3 Racoon

Racoon (*Procyon lutor*), which is originally native to North America, has been introduced as a fur-bearing wild game species to many countries and is now distributed all over Europe (Britain included) and several Asian countries. During the 1920s, the first racoons were introduced in great numbers to Germany. In 1927, some were released at Eder Lake (Hesse) to enrich game animals. Racoon caps and coats were very popular and sold at a good price those days. The present main distribution area of racoon in Germany is still Hesse (Lagoni-Hansen 1981). From there, raccoon have spread almost concentrically in all directions (Röben 1976). East of Berlin, raccoons which had escaped in 1945 from a bombed fur farm near the city of Wolfshagen have built up a stable population (Grummt 1973; Stubbe 1975; Hohmann 2001) numbering at least 5,000–6,000 individuals. In 1966, a few raccoons that had been kept as mascots by a US Air Force Unit near the town of

Laon (Dépertement Aisne, France) were released into surrounding forests (Hohmann and Bartussek 2001), where they successfully multiplied. Raccoons are occasionally kept as pets (e.g., Hohmann and Bartussek 2001). As raccoons become very aggressive and difficult to manage when they reach sexual maturity, their owners often just release them into the landscape. Raccoon population has rapidly grown. No other comparable wild animal reaches such densities. The largest urban raccoon population exists in the city of Kassel (50–150 racoons km^{-2}). The total population in Germany is estimated to be 600,00–800,000 individuals. Since 1980, the raccoon is considered a native species (Lagoni-Hansen 1981).

Raccoons subsist mainly on vegetarian food (>30 %, Lutz 1981) and small animals. It may affect fruit plantations, gardens and cornfields. The raccoon's animal diet consists mainly of invertebrates, small (young) animals and birds' eggs. It may also occasionally prey on poultry and on species that mean a lot to hunters and nature conservationists, in particular ground breeding waterfowl, capercaillie (*Tetrao urogallus*), black grouse (*Lyrurus tetrix*), hazel-grouse (*Tetrastes bonasia*), and partridge (*Perdix perdrix*). However, birds nesting on trees (e.g., common heron) or hole breeders may also be affected (e.g., Helbig 2011). Not least, the raccoon is considered adversely to affect efforts to improve small game populations. No wonder that in hunting literature, a one-sided view of the raccoon prevails (e.g., Kampmann 1972, 1973; see also Lagoni-Hansen 1981 for further references). However, until the 1960s and 1970s, popular predictions that predation and/or competition (e.g., with wild cats) by raccoons would seriously impact the native game animals has not been substantiated (Michler and Michler 2012), possibly because the lack of suitable resting and sleeping places, as well as places to rear their young, limits raccoon density (Lutz 1996; Hohmann 2001). However, at high raccoon density, damage is not exceptional. On the other hand, raccoons killing muskrats and rabbits might cause them to be considered 'useful' animals.

As to the assessment of the raccoon's impact on its habitats in Central Europe, the results of studies in America cannot be directly transferred to Central Europe. In Europe, basic research on this issue is still scarce, and it's possible that not everybody would like to hear that supporting wild game such as pheasants or wild ducks (Hespeler 1990) by winter feeding practically 'sets the table' for the newcomer. In Germany, and also in other European countries, reports in daily newspapers and on TV about 'raccoon pests' somewhere in North America or in the Far East (e.g., Japan) are promptly used by certain people (mainly by hunters) to claim the need for a more rigid control of raccoons.

Expansion of raccoons is continuing at great intensity, and neither hunting nor natural predators (great horned owl, *Bubo bubo*; fox, Vulpes *vulpes*) nor losses to vehicles will stop it. Hunting bags have increased from a few hundred in the 1960s to above 70,000 in 2011 (Fig. 5.5). Until present, little has been known about raccoon diseases and parasites in Europe (Michler and Köhnemann 2009). In the United States, the raccoon is considered to be an important vector of rabies. During the period from 1975 to 1985, three quarters of the raccoon population were affected (Jenkins and Winkler 1987). Moreover, the raccoon carries dog distemper and round worms (*Ascaris*), which may also affect humans (Ludwig et al. 2000). As the

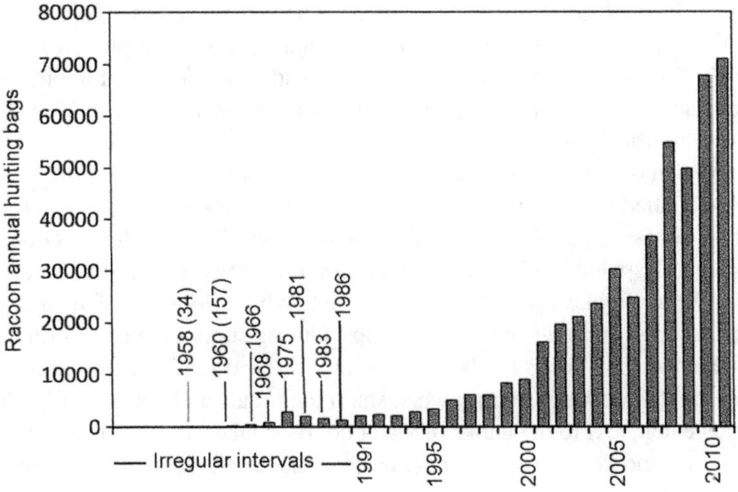

Fig. 5.5 Racoon annual hunting bags in Germany (Data gratefully received from F.-U. Michler, Dresden, University of Technology)

raccoon has become 'urbanized' (Sect. 6.4) and its population is rapidly increasing, a problematic development might be on the way.

As in many other countries, the racoon has also become a pest as in Japan. It was the animated TV-series 'Araiguma Rasukaru' (= 'Rascal, the racoon') which made the racoon very popular during the 1970s and motivated many Japanese to keep racoons as pets. About 1,500 racoons were annually imported from the United States as pets at that time. After the boom had died down, many racoons were discarded and released into the wild where they rapidly multiplied enjoying favorable habitat conditions. In 2004, racoons already occupied 42 of 47 Japanese prefectures (Ikeda et al. 2004). Farmers, in particular, complain of racoon-damage. Racoons also compete successfully with foxes and the native racoon-dog. Great efforts are now made in several areas to eradicate or, at least, to reduce racoons to tolerable numbers. Racoon control, however, is facing similar problems as elsewhere, and eradication is unlikely.

5.1.4.4 Racoon Dog

During the first half of the twentieth century, the raccoon dog (*Nyctereutes procyonoides*) was introduced from the Amur and Ussuri regions (Siberia, China) to many places of the former Soviet Union as a fur bearer. Via eastern Europe, it has rapidly invaded Central and Northern Europe up to the Nordic Circle (Nowak 1977; Hespeler 1995; Sedlag 1995; Lutz 1996; Kauhala and Kowalczyk 2011), profiting from its high ecological adaptability, great migratory ability and high regenerative capacity. Moreover, in Central Europe, predators are almost absent (e.g., Drygala et al. 2000). In some areas, there was a time-lag between first observations and rapid population growth, as in Sweden and Germany, for example (Kauhala and Kowalczyk 2011). In Germany, raccoon dogs suddenly increased about 30 years after they had been observed for the first time (Ansorge and Stiebling 2001), despite continuous hunting pressure.

The raccoon dog prefers riparian landscapes characterized by a mosaic of water bodies surrounded by reed belts, wet meadows and swamp forest stands difficult for humans to access. It does not colonize areas above 700 m. The raccoon dog is omnivorous (Hespeler 1995; Drygala et al. 2000; for review, see Sutor et al. 2010). Plants, however, make up to 80 % of its diet, which consists of fruits, berries, chestnuts, acorns, nuts and agricultural crops. During summer, the raccoon dog uses maize fields, where it feeds on the not yet fully ripened seeds. In addition, raccoon dogs prey on invertebrates, amphibians, small mammals and occasionally also fish, which they catch next to the banks of water courses, lakes and ponds. Occasionally, waterfowl, meadowbirds (e.g., ducks, snipes, black-tailed godwits, curlews) and their clutches fall victim to it (Möckel 2000). Not least, the raccoon dog scavenges on carcasses and also checks kitchen waste for food.

Soon after this 'terrible robber' had arrived, the extinction of small game, particularly of grouse, was predicted. Although this prognosis has not been realized (Hespeler 1995; Drygala et al. 2000; Kauhala et al. 1993), efforts to eradicate the raccoon dog have continued with particular regard to preservation of the 'last' groundbreeding waders (Möckel 2000). Daily newspapers have often contributed to the bad reputation of the raccoon dog, presenting it as a general threat to the native fauna. In particular, it is blamed for eliminating singing birds. This appears to be a more popular and psychologically conclusive justification for killing raccoon dogs than the hunters' lobby's undisguised claim of saving small game populations would be. The raccoon dog is also blamed for expelling foxes and badgers, which are now, surprisingly enough, presented as predators essential for ecosystem function. This might be genuinely accepted by the nowadays well-informed general public. However, this 'concern' looks a little hypocritical in view of rigid elimination of these very predators in the past.

On the other hand, one should not ignore that the impact of the raccoon dog, and also of the raccoon, on the native biocoenoses and ecosystems is likely to increase. During recent times, the number of raccoon dogs has continuously increased. It is possible that the raccoon dog has profited from the rabies immunisation of wild animals by inoculated baits (Tschirch 2001). The raccoon dog itself (like the raccoon) is a carrier of rabies and some other diseases and parasites (e.g., tapeworms, parasitic nematodes). This is very likely the most serious effect of this newcomer. In any case, caution is appropriate. If raccoon dog densities continued to increase, one could hardly close one's mind to the demands of nature conservationists, hunters and also epidemiologists for radical reduction. However, in the end, it will be impossible to eradicate the raccoon dog, or the raccoon or muskrat (Hespeler 1995; Möckel 2000; Kauhala and Kowalczyk 2011).

Apparently, campaigns to eradicate well established populations of introduced vertebrates have failed on all continents (e.g., Dahlsten 1986; MacDonald et al. 1989;

Bomford 2003). Thus, any attempt to eradicate viable populations of alien animals might be questioned (Dahlsten 1986). On the other hand, removal of pest animals was occasionally successful on some islands, as, or example, eradication of nutria in England (see Sect. 5.1.4.2), feral cat on Macquarie Island and feral goats on Galápagos Islands (see Sect. 5.1). Anyway, as the numbers of intentionally or accidentally introduced exotic animals increase due to ever-expanding global trade, new problems can be expected.

5.2 'Homecomers'

Numerous efforts have been and are being made to bring animal species back to the original habitats from which they were removed by direct persecution (hunting, trapping) and/or landscape change under the influence of man. The reasons for reintroduction are different: for example, conservation of species, biological variety and high genetic variety (e.g., Arzdorf 1990, therein further references), restoration of impoverished biocoenoses, or simply compensation for losses caused by human impact in the past. Moreover, reintroduction is supposed to improve ecosystem function and restore the 'ecological balance', whatever this means and whatever it would look like. In most cases, it has turned out to be difficult or impossible to predict whether the re-introduced animals will really 'do their job' as expected. Thus, in many cases, justification for reintroduction of animals is nothing but 'verbal-ecology'. Protection and reintroduction of capercaillie (*Tetrao urogallus*), for example, are not a precondition for so-called 'landscape functioning'. In Central Europe, capercaillie was also too rare in the past to influence vegetation and landscape ecosystem function sustainably (Scherzinger 1981). Compared to the present time, black grouse (*Lyrurus tetrix*) were more frequent only on cultivated moorland and unwooded mountain areas. In the Alps, black grouse have profited from clear-cutting and pastoral use of the upper high-elevation forests (Eiberle 1974). The black grouse population has continuously declined over the last 100 years due to changes caused by human land use. With regard to conservation of the native grouse populations and their genetic variety and species-rich environment, i.e., mainly for ethical reasons, effective protective measures and eventual support of the remaining population by release of specimens have been pleaded for (Scherzinger 1981).

Some authorities have even considered reduction of the natural predators (goshawk, *Accipiter gentilis*; carrion crow, *Corvus corone*; fox, *Vulpes vulpes*) to be a precondition for the success of grouse conservation (Kalchreuter 1981; Kalchreuter and Wagner 1981; Ammermann 1998), which appears to be hardly tenable from the ecological view. In its main distribution areas (Scandinavia, Siberia), the existence of capercaillie is not threatened.

One might suppose that reintroduction of animals into their original habitats is less problematic than introduction of alien species to areas they never would have reached through natural expansion. Many reintroductions have been successful in the biological view, insofar as the homecomers have built up viable populations.

This applies, for example, to the reintroduction of the Hawaiian goose (*Branta sanvicensis*) to Hawaii and of the European bison (*Bison bonasus*) to Bialowieza National Park in Poland. Just as successful has been the reintroduction of the ibex (*Capra ibex*) to several areas of the European Alps (Holtmeier 1987a, b). In addition, beaver and lynx, re-introduced to many European countries, have successfully re-established themselves (Festetics 1981; Arzdorf 1990).

From the landscape-ecological view, however, some reintroductions have to be assessed more critically. Certainly, reintroduced species have built up viable populations, but not all have adjusted themselves well to the present 'tolerable carrying capacity' (Sect. 6.2) of their original habitats, which had often considerably changed under historical human impact (Holtmeier 1987a, c). All the time, new conflicts between the animals' habitat requirements and landscape utilization by humans (agriculture, forestry, water supply, etc.) become apparent. Therefore, in addition to habitat suitability, potential compatibility of the 'homecomers' with the present more or less modified environment must be carefully analysed before reintroduction.

5.2.1 Reintroduction of the Beaver

Beaver had been trapped close to extinction because of its high-quality fur (12,000 coarse hairs cm^{-2}, 23,000 hairs cm^{-2} in the underfur), its meat and castoreum. Only the fur of otters is thicker than that of beavers. In North America, beaver fur was a regular 'currency unit' in countertrade with the Indians (Table 5.2). In addition, trade in castoreum was very lucrative. Castoreum, which is a glandular secretion by which beavers mark their territory, was considered a remedy for almost all kinds of aches and pains, and was also thought to be an aphrodisiac. Moreover, felt hats made of beaver hairs were believed to protect from diseases and to improve memory. The yellow-reddish incisors were supposed to ease teething when hung around babies' necks. As a 'living pharmacy', the beaver shared a common fate with some other animals, such as ibex (*Capra ibex*) and Saiga (*Saiga tartarica*). In Europe, tasty beaver meat was highly appreciated, particularly by clericals. Beavers, because they live in the water and have a scaly tail, were declared to be a fish, and thus, could

Table 5.2 Beaver pelt as a currency unit in countertrade with the Indians

Item	Beaver pelts
1 Gun	20
1 Copper kettle	16
12 lb of glass beads	4
1 Axe	1
100 g Tobacco	1
10 Flints	1
1 Horn comb	1

From Horn (1968)

continue to be eaten during Lent (Bobak 1961). The same holds true for otter meat, by the way (Weinzierl 1973; Kurt 1982). Thus, the monks' communities were anxious to preserve this highly appreciated 'fish' source.

Relic populations had survived in some regions of North America, Russia, southern Norway, eastern Germany and southern France. Soon after beaver had almost been eradicated, it was reintroduced to many of its original habitats for restoration of the riparian ecosystems. In northern Germany, for example, first efforts had already been made around 1600 AD (Hinze 1950). Later, beavers were released to many other areas. Most of the reintroductions failed, however (Bobak 1961). Nowadays, the Eurasian beaver can be found in isolated but expanding populations from eastern France to Lake Baikal and Mongolia, and also in far-eastern Russia (Khabarovsk; MacDonald 2001). In the north, its area of distribution extends along wooded riparian habitats into the forest tundra, in the south into the forest steppe, and even touches on the semi-desert zone. Canadian beavers (*Castor canadensis*) were also introduced to many European countries, in addition to the reintroduced European beavers (e.g., Schneider and Rieder 1981; Härkönen 1999; Halley and Rosell 2002; Parker et al. 2012). Thus, in contrast to the European beaver, which is a homecomer, the American beaver is new to the Old World.

In North America, the surviving beaver populations rapidly recovered when beaver fur and beaver hats fell out of fashion and hunting and trapping became unprofitable. Great efforts were made to bring the beaver back to its original habitats. Hunting was strictly controlled, and the new populations could develop almost unaffected by humans. The number of beavers living in the USA has been estimated to be 2–6 million (Naiman et al. 1986). Anyway, this number makes up only a fraction of the original population, which has been estimated at 40–60 million (Seton 1929). Population development of reintroduced beavers has been similar to that of other animals introduced to new environments. A rapid increase in the beginning was followed by a marked decrease (Hartmann 1994; see also Riney 1964; Caughley 1970, 1976).

In case of the beaver, which may radically alter its habitat, the possible consequences of reintroduction have to be carefully considered, particularly potential conflicts with landscape use by humans. The varying local conditions (nature, landscape use) play an important role in this respect.

5.2.1.1 North America

In the mountains of the western United States, for example, where flooding and erosion had considerably increased as a result of extensive clear cutting, overgrazing, wild fires and inappropriate agricultural cultivation practices by the European settlers, restocking with beaver has had many positive effects (e.g., Apple et al. 1984; Parker et al. 1985; Naiman et al. 1986; Holtmeier 1987a; Butler and Malanson 1994; Meentemeyer et al. 1998; Meentemeyer and Butler 1995). Beaver dams and ponds reduce flow velocity (Johnston and Naiman 1987; see also Fig. 4.5), and thus erosion downstream of the dams, particularly during snowmelt and after heavy rainfall. Overbank deposition of fluvial gravel and finer material on cultivated land has decreased. While the influence of a single dam on stream flow may be neglected, a series of dams may have considerable effects (e.g., Grasse 1951). Moreover, beaver ponds accumulate organic matter and mineral sediments and reduce up-filling downstream lakes and reservoirs with suspended material. Beaver ponds can retain >50 % of the sediment load per water volume of a stream (Gorshkov 2011).

During droughts, grasses and other forage plants profit from a rising ground water table, thus improving the food source for livestock and wild game in the riparian landscape. Occasionally, little streams that had usually dried in the summer showed permanent run-off after reintroduction of beavers to the upper stream course (Denney 1952). Farmers use the water of beaver ponds for irrigation and to fight forest fires. Beaver ponds also give shelter to wild living animals from wild fires. Moreover, beavers may increase species richness and diversity of riparian habitats (e.g., Wright et al. 2002). Beaver ponds may harbor many fish species (Sect. 4.2.2; Collen and Gibson 2001), and thus provide excellent fishing. Valleys settled by beavers are also very popular among hunters and trappers as they attract great numbers of wild game and water fowl (cf. Fig. 4.5). This aspect is often of major importance to beaver management.

In general, the positive effects of beavers on landscape usually outweigh their negative effects. Locally, however, serious problems may arise. Along high mountain streams of higher order, slow plant growth and comparatively small size of soft wood stands (willows, aspen) within the beavers' reach and steep topography set a limit to beaver population growth. As soon as beavers exceed the local habitat carrying capacity, they usually move upstream to 'exploit' new food sources. If this is impossible, the beaver population will probably collapse. Increased felling of coniferous trees is a first sign of food shortage (e.g., Packard 1947; MacDonald 1956; Retzer et al. 1956; Yeager and Rutherford 1957). Preconditions for reintroduction of beaver and build-up of a persistent population are more suitable in the lower stream sections, where flow velocity is lower and more food is available. Conditions may be critical, however, on unstable ground such as argillaceous schist, for example (1955; Yeager and Rutherford 1957). Raised groundwater table and heavy load of impounded water may cause the collapse of the whole beaver pond system, resulting in disastrous flooding and erosion of the valley fillings. Local problems may also arise, such as in a case of an occasional outburst flood when a beaver dam fails (Butler 1991a, b), due to excessive rain, run-off at rapid snow melt (flash floods) or after beavers left their habitat (c.f. Photo 4.11).

No area, however, even if providing optimal habitat conditions for beavers, would allow unlimited population growth without the food source becoming depleted in the long-term. When the habitat (e.g., too steep terrain) prevents surplus individuals from moving to other places, the situation may become critical relatively soon. Therefore, beaver population and the numbers of deer (elk, muledeer) competing with beavers for winter forage (e.g., fresh shoots of soft woods) must be managed in a way that allows for regular and sufficient regrowth of the food plants. This is the only way to prevent beavers from occupying river sections where they cannot be tolerated.

In densely populated river sections, people often complain of beaver damage. Beavers cut aspen or fruit trees on private ground, other times they flood roads or block water supply to irrigation systems. In upper mountain valleys, such damage is usually spatially restricted and can be prevented by installation of drainage pipes in the beaver dams, for example, or by trapping surplus animals. The better way, however, is careful beaver management from the beginning. In the mountain valleys, where beaver control, carefully adjusted to the tolerable habitat carrying capacity (Sect. 6.2), has precluded or reduced conflicts with human interests to a tolerable extent, beaver have become successfully integrated into the present riparian ecosystems.

In other areas, however, damming and tree-cutting activities of reintroduced beavers and the resulting effects have run into serious conflict with current land use, forestry, river control, function of drainage or irrigation systems, power lines, etc. In the Rocky Mountain foreland, for example, beavers cause severe damage to local cultivated land. A peaceful coexistence of beaver and man seems to be impossible in these areas. As Rutherford (1964) already wrote, '*… and the plain truth is that under present-day conditions, the beaver does not have a place there*'. In Canada, large beaver populations have caused serious damage to cultivated land and other private ground. Efforts have been made to prevent such damage by trapping and, if necessary, by destroying the beaver dams (Photo 5.6).

Although public attitude towards beavers is generally positive, it usually becomes negative as soon as beaver damage exceeds the local limit of tolerance (e.g., Jonker et al. 2006). In the southeastern United States, for example, in Alabama, Louisiana, Mississippi and Georgia, beaver populations have grown and spread to an extent that beaver is considered a 'pest' (Arner et al. 1969; Hill 1976; King et al. 1998). Economic forests (hardwoods, veneer), as well as agricultural fields and pastures, are adversely affected by raised ground water table and flooding. Economic losses amount to several million dollars. Damage is particularly severe in the coastal plain of the Mexican Gulf (Arner, no year). The main reason is the almost unlimited availability of forage and space. Relatively low beaver dams may cause extensive flooding of cultivated land (agricultural fields, forests). High beaver densities force beavers into suboptimal habitats where they may promote stream bank erosion by burrowing into stream banks (Meentemeyer et al. 1998). Bank dens and burrows also occasionally destabilize dams, dikes, railroads and other traffic lines. All means have been utilized to try to get the 'beaver pest' under control. In Alabama and Mississippi, beaver was removed from the list of protected animal species (Arner et al. 1969). Beavers are shot and trapped all year round, poisoned and chemically sterilized (Hill 1976). Moreover, beaver dams are often blasted (cf. Photo 5.6). In some places, alligators (*Alligator mississippiensis*) have been released to kill the beavers, without much success, however. When, ironically enough, alligator attacks on swimmers increased, this kind of 'biological' beaver control was ceased (Hill 1976). In the end, trapping has turned out to be the best method of beaver control. Attempts are being made to make trapping attractive again and to open a new market for beaver pelt and meat. When the price for one beaver pelt ranged around 43 $, more than 600,000 beavers were trapped in Canada from 1970/1980.

Photo 5.6 This beaver dam in Gander Lake (Newfoundland) was blasted to prevent further rise of the ground water level causing damage to the upper forest . G. Leydag

Never before were so many beavers trapped in the country's history (Larson and Gunson 1983). Moreover, trapping has been recommended for outdoor recreation.

5.2.1.2 Northern Europe

In Sweden and Finland, the last beavers became extinct in the late nineteenth century (Lahti and Helminen 1974; Hartmann 2011). Current beaver populations in Fennoscandia originate mainly from reintroduction, with the exception of Norway, where a relic population of about 100 individuals had survived. Beaver population has increased to about 70,000 individuals at present (Rosell and Parker 2011) and continues to expand to the north and west to habitats, where beavers have not been present for 200–300 years (Rosell and Parker 2011). Most of the present Swedish beavers are descendants of beavers that were reintroduced from Norway (Collett 1897). Since the release of the first beavers in 1922, beaver population has gradually increased (Lavsund 1975, 1983) to about 100,000 individuals (Hartmann 1994, 1999). The total European beaver population is estimated to exceed 600,000 individuals (Halley and Rosell 2002).

During the 1930s, 17 Norwegian and 7 Canadian beavers were introduced to Finland. The Canadian beavers were released in two places in southeastern Finland. They multiplied successfully. Many Canadian beavers trapped in the areas of early release were translocated to northern and easternmost Finland, from where they spread into Russia. The current (2001) total number of beaver in Finland is estimated to be about 13,500. About 12,000 of them are of North American origin

(Nummi 2001). Thus, in the case of the American beavers, it has been an introduction of a closely related substitute species rather than a real reintroduction of the native species. Not all restocking programs have been successful. Thus, introductions of Canadian beavers to northern Finland and European beavers to northern Norway largely failed (Parker 2005, Danilov et al. 2011a).

In Finland, the Canadian beaver is spreading more rapidly than the European beaver (Lahti and Helminen 1974, 1980). As there is a complete niche overlap and niche partitioning, and competitive coexistence appears to be unlikely, the Canadian beaver might outcompete the European beaver (Nummi 1996, 2001; Lahti 1997). In Karelia, on the other hand, European beavers have completely replaced Canadian beavers, which had been released into this area earlier (Danilov et al. 2011b). Obviously, the particular local and regional conditions play an important role in this respect.

Therefore, and in order to prevent the further spread of the Canadian beaver, some authors have suggested eradicating it. Although complete elimination of the Canadian beaver throughout Eurasia seems unlikely, local eradication might materialize (Parker et al. 2012). In Austria and Poland, for example, Canadian beavers will probably become extinct (Halley and Rosell 2002). From the landscape-ecological view, however, the resulting 'multi-species mix' does not seem to be a problem, as the various provenances and races influence their habitat in a similar way. Their diet depends on the availability of preferred food, and the construction activities of both species are very much the same (Danilov et al. 2011a, b).

In northern Europe, beaver damage has increased parallel to increasing beaver population (cf. North America). While flooding is generally of minor importance, it may become a serious problem in areas where valuable commercial forests are affected, particularly by raised ground water (Feldmann 1980; Lahti and Helminen 1974; Holtmeier 1987a). Beaver damage has turned out to be particularly important to private forest owners (Härkönen 1999). It must be mentioned, however, that many of these forests became established on drained moorland. In Sweden, beavers have caused severe local damage to embankments through burrowing and to agricultural fields through flooding. Trees cut by beavers occasionally block roads or interrupt powerlines and telephone lines (Lavsund 1975). In Norway, Sweden and Finland, attempts have been made to prevent such damage through intensified hunting. Hunting, however, is strongly regimented. In Sweden, for example, beaver hunting is permitted from the beginning of October to the middle of May. Landowners affected by beaver are allowed to remove beaver dams (May–October) (Hartmann 2011). Thus, beavers may be removed locally, but the existence of beaver is not threatened in general. Destruction of beaver dams is legal only in case of acute flooding problems. While beaver populations in Sweden and Norway have continuously increased despite strict regulation through hunting, the number of beavers in Finland has slightly declined due to increased hunting pressure (e.g., 1975–1980) (Ermela et al. 1989).

Severe damage has also been reported in Estonia. In the Karula National Park, impoundments by reintroduced beavers resulted in flooding of major areas along the border to Latvia. In the City of Järvselja, roads undermined by beavers have collapsed under the load of heavy vehicles. Hunters and landowners who are affected by flooding are allowed to shoot or trap beavers. These people, however, do not make much use of it. Actually, there is not much interest in hunting beavers, as there is no market for beaver pelts. In Latvia, meanwhile, introduced beavers (1927, 1935, 1952) are considered 'noxious' animals, because they conflict with land owners' interests. By 1985, more than 25,000 animals were already estimated to be living in that country. Additional beavers are invading the country from the east. There is some predator pressure, as beavers have become an alternative prey for wolves in summer, when ungulate populations are low (Andersone 1999). Although the natural habitat carrying capacity is reported to be sufficient for about 150,000 beavers, no more than 50,000 may be tolerated in view of the potential conflicts with land use by humans (Balodis 1992).

Altogether, most reintroductions of beaver have been successful from the biological viewpoint. Persistent populations have developed and the existence of the species is no longer threatened. Not everywhere, however, have beavers become 'harmoniously' integrated into landscapes subjected to increased human utilization. As beaver may completely modify riparian and lacustrine landscapes and seriously conflict with present human use of floodplains and their adjacent terrain, the discussion as to whether beaver is a 'useful' or a 'noxious' (pest) animal has been revived. Nevertheless, if the local conditions allowed for both a persistent beaver population and effective control of them to prevent intolerable damage, reintroduced beavers should be conceded the 'right of residence'. Cox (1940) previously wrote in his article, 'The beaver – friend of the forest': *Beavers do not belong in thickly settled communities, since their flooding operations may become troublesome to meadows and roads. In the wild forest country, they do little harm and an immense amount of good.*

5.2.1.3 Central Europe

Keeping the above in mind, the reintroduction of beaver to Central Europe is considered, with special regard to Germany as a very densely populated country. In Germany, beavers still widely occurred about 200 years ago (Nitsche 1994), as is evidenced by many local names of villages, fields and water courses (e.g., Biberach, Bevern, Bevergern and others). By the end of the nineteenth century, the beaver was almost extinct. The town of Marburg, for example, had employed 50 beaver hunters at the same time (Remmert 1999). While beavers were re-introduced to North America, Russia and also Northern Europe because of their valuable fur and positive influence on hydrology, the animal was brought back to Germany exclusively to restore the original fauna and biocoenoses and to increase biodiversity of riparian habitats (e.g., Weinzierl 1973; Schneider and Rieder 1981). Reintroduction should compensate a little for the negative historical human impact on river systems and lakes (Reichholf-Riehm 1981) and to right the wrong of extinction (appeal for funds, BUND 1992). It even has be argued that reintroduced beavers would not cause problems but rather point out the mistakes that were made by historical land use (Frobel 1994; Schwab et al. 1994) and hydrological works (Allgöver 1995).

Reintroduction has also been justified by the beaver's excellent job as a 'hydraulic engineer'. In this context, the many successful reintroductions to North America and Russia have been put forward as examples worth following (e.g., Weinzierl 1973), without, however, any differentiation of the specific local conditions and taking note of the negative experiences. Such comparisons of densely populated Central European landscapes with still near-natural habitats (e.g., Rocky Mountains, extended river plains in Russia) do not work (see also Stocker 1985), and may lead to wrong rather than to helpful conclusions and decisions. European cultural landscapes have considerably changed under the massive impacts of river control and human use of the riparian habitats after the beaver became extinct (Holtmeier 1987a). They are used down to the last square meter (e.g., Germany, The Netherlands, Switzerland). In general, the original habitats are fragmented and will not carry any but relatively small populations, with a few exceptions. Thus, it would have probably been wiser to justify beaver reintroduction with the ethical obligation, which undoubtedly is the most convincing reason.

Finding the 'right' beaver species for reintroduction projects has been a difficult problem. In Germany, for example, beavers of different provenances (Russia, Poland, Sweden, Finland, France, Germany) have been released into former beaver habitats (e.g., Schneider and Rieder 1981). We will not discuss this in more detail because the origin of beavers seems to be of minor importance with regard to their effects on their riparian environment.

Beavers were first released in Bavaria near the City of Neustadt at the Danube River (1966; Weinzierl 1973; Frobel 1994). Since that time, beavers have been reintroduced into many other areas of Germany and neighboring countries (France, Belgium, Luxembourg, Switzerland, Austria, Poland, The Netherlands). Most restocking programs have been successful inasmuch as beavers built up viable self-reproducing populations in many areas and are spreading from the original places of release. In the northern Eifel mountains, for example, five beavers were released more than 30 years ago. The present population (about 70 individuals) has been expanding west to the Belgian border ever since (e.g., LÖBF 2000). While the beaver population at the river Elbe (about 4,000; Heidecke 1997) goes back to surviving native beavers, the Bavarian beaver colonies originated from reintroduced animals only. At present, about 6,000 beavers are estimated to be living in 1,500 colonies between the town of Bayreuth and Munic. Both the Elbian and Bavarian populations are considered to be stable from the population-ecological view. Most of the other beaver colonies appear to be too young to make far-reaching predictions (Nitsche 1995b).

The beaver's great adaptability is of major importance in expansion dynamics (Sect. 4.2.2). Beavers tolerate even severe water pollution and traffic, and have colonized suburban and urban habitats without any problems. Moreover, they are capable of adjusting their foraging to the particular local vegetation. When soft woods (willows, aspen, poplars) are not available, they may even switch to beech (Schneider and Schulte 1985), oaks and conifers (Djoshkin and Safonov 1972;

Jenkins 1979; Schwab et al. 1994; Zuppke 1995; Dahlbeck et al. 2008) or subsist exclusively on grasses and herbs, as, for example, in the lower Rhône area, where winters are mild (Richard 1981). Such arguments, however, put forward by operators and advocates of further reintroduction, do not really meet the problem, as it is usually not the point as to whether the 'home-coming' beavers will find suitable habitats in the present landscape, but whether humans will tolerate them in the long-term (Nitsche 1995a).

Anyway, it is exactly the great adaptability of beavers to different habitat conditions (e.g., Frobel 1994; Schwab et al. 1994) that may cause problems, because it is hard to predict reliably how the beavers will modify their environment. Surprisingly enough, beavers that had been released at a tributary of the river Inn near Eggenfelden (Lower Bavaria, Germany) moved after only a short while to the lower Inn river, where they preferred to colonize the Inn reservoirs (Reichholf 1976) – which later turned out to be a very positive development. On the other hand, so-called 'agrarian beavers' (Maier 1994), which switched to agricultural crops (e.g., sugar beet, maize, corn) next to their colony as soft woods were lacking, are hardly compatible with the original objective of beaver reintroduction and are one reason for increasing opposition from farmers and other people affected by beavers. Thus, it is no wonder that recent reintroduction of beavers (from Norway) to Scotland and England also is controversially debated, as lairds, farmers and fisheries are anxious that beavers could cause damage to farmland, commercial forests and trout and salmon rivers (e.g., Carrel 2009; Davies 2011).

In Biesbosch (The Netherlands), beavers were reintroduced beginning in 1988 to the riparian area to reduce the prevailing willows and support the development of species-rich mixed forest stands. However, instead of 'fulfilling their job', within 1 year the beavers had felled ash trees, hazels and alders, stripping off their bark and effectively destroying these trees that were relatively rare in this area (Nolet et al. 1994). One should not expect that beavers will build dams exactly in the place where water engineers would like them.

Although the home-comers are generally welcome, reintroduction can be considered successful only when beavers became integrated into the riparian ecosystem without causing major conflicts. This, however, has apparently not been possible in all former habitats. In Switzerland, for example, reintroduced beavers have occupied the optimal habitats. Young beavers hardly find new places to establish self-sustaining persistent colonies (Stocker 1985; Rahm 1997) without eventually causing intolerable damage. Almost 30 years ago, newspapers reported on a 'beaver war', after beavers that had moved from their original place of release at the Lake of Hüttwil (Canton Thurgau) to a channel of the river Thur area started cutting trees on private ground. Some 'nuisance beavers' were trapped and moved to another channel (Sihl river). After a short while, however, new beavers arrived in the Thur area, which demonstrates the beaver's great capability for rapid expansion even in intensely used cultural landscapes. In the view of Stocker (1985), a natural development of a contiguous beaver population in the Swiss cultural landscape appeared not to be very likely. On the other hand, in the opinion of the same author, the management objective should be to establish groups as large as possible with the ability to interchange despite obstacles and inappropriate sections of the riparian environment. At that time, the World Wildlife Fund (WWF) still considered the existence of the beaver in Switzerland to be threatened and preservation of the beaver to have priority over the interests for forest owners. However, with regard to limited appropriate habitats and permanent conflicts evolving in the long-term perspective, it seems hard to follow these arguments (Holtmeier 1987a). The present beaver population has been estimated at about 1,600 individuals (Angst 2010). The colonies should be protected as well as possible, but additional beavers should not be released (Rahm 1997; Schulte 1998).

There are measures that could possibly help solve problems caused by beavers reintroduced to their former and also new habitats in the cultural landscape (e.g., Heidecke and Klenner-Fringes 1992). There is certainly no lack of suggestions in this respect, however, practical implementation has often failed or has only been possible locally. Thus, for example, a 20–30 m wide strip along riverbanks and lake shores should not be used and left to natural succession. Plantations of aspen, poplar, willow and other soft woods preferred by beavers with a capacity to reproduce through root suckers or from stump sprouts (e.g., elm, *Ulmus*; hawthorn, *Crategus*; dogwoold, *Cornus*) could help to improve the food source, and thus relieve commercial forest stands next to the riparian habitat from the beavers' impact (e.g., Zahner 1997). 'Valuable' solitary trees can be protected by beaver-proof fences or wire netting (e.g., Zuppke 1995). In case of stream bank erosion due to burrowing activities, protective measures, such as pavement, sheet piling and lattice works, may prevent damage (e.g., Schwab et al. 1994; Geiersberger 1997). Such measures are expensive, however. The list of suggestions could easily be continued. However, it might be questioned in this context as to why such big efforts are made to eradicate the muskrat. Is it just lack of sympathy for the newcomer, or is it the public's positive attitude to the 'good' and 'eager' beaver (e.g., Reichholf 1997)?

Occasionally, however, suggestions for avoidance of possible conflicts and limitation of beaver damage may pervert nature conservation. This is the case, for example, if installation of artificial beaver burrows consisting of concrete is recommended (Rahm and Baettig 1996) to support beaver colonization of unstable river banks in danger of collapsing anyway (Fig. 5.6). The extent to which people will not only tolerate the reintroduced beavers but also support their further expansion depends very much on ecological education and understanding – peoples' increased sensibility in this respect nowadays should not be underestimated – and on whether the costs of improvement of habitat conditions and their long-term maintenance will be acceptable.

Beavers have been successfully reintroduced into many of their former habitats. In addition, beavers have also been introduced to new places (e.g., reservoirs, channels) supposed to provide suitable habitat conditions. Beavers are still expanding and will probably still occupy beaver-free habitats during the years to come (Schwab 2003). On the other hand, in the Elbe area, where beaver populations had considerably increased, population growth has tapered off in all colonies due to increased mortality and depletion of food sources. Obviously, the habitat carrying capacity has been reached, as can be concluded from emigration of young beavers to

Fig. 5.6 Artificial beaver burrow installed in an unstable river bank (Drawing by F.-K. Holtmeier after Rahm and Baettig (1996))

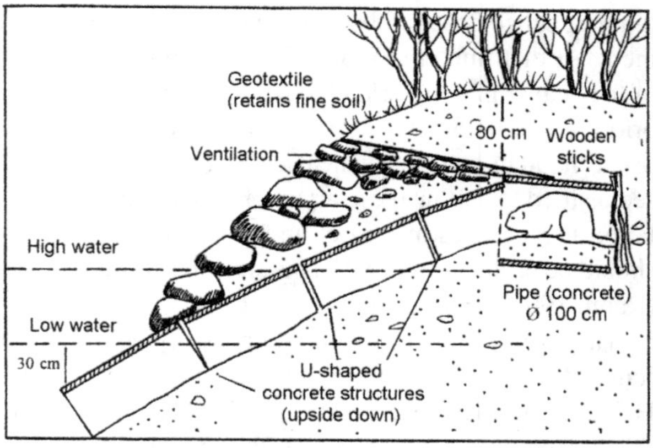

suboptimal habitats, such as small lakes far distant from the main streams, and increasing damage on cultivated land. Beavers are even expanding into suburban areas of big cities, such as Frankfurt on the Main (Germany), where they have been felling trees at the Nidda River (Westfälische Nachrichten, 23 March 2013). Only careful beaver management may reduce conflicts, and thus ensure continued existence of the beavers (Heidecke 1997). In order to prevent growing beaver impacts, several hundred surplus Bavarian beavers have been translocated to Croatia and Hungary, for example. The Hungarian beaver population (700–900 animals) is growing (Bajomi 2011). The continuing expansion of reintroduced beavers in Central Europe (e.g., Heidecke 1997; Zahner 1997; Blanke 1998) is absolutely comparable to the natural or near-natural beaver population dynamics in the Rocky Mountains, for example, or in Northern Europe. Obviously, only man can set a limit. This should also persuade nature conservationists engaged in restoration of riparian environments. Reintroduction just makes sense and is also in the beavers' 'interest', provided that potential habitats are sufficiently connected to each other and allow for the natural 'conquest' of landscape by beavers (Nitsche 1995b; Heidecke 1997). Anyway, control will be necessary to guarantee peaceful coexistence of beavers and humans. Habitat aptitude tests may be a useful instrument in this respect (Hugo 2001).

Beavers in France all originate from the relic population (<100 animals; Rouland 1997) of the Rhône River. They gradually re-colonized the Rhône river system. In addition, beavers were released to other river catchments (Rouland and Migot 1990). More recently, beavers from Germany and Belgium have expanded to the northeastern part of the country and are still on the way. Beavers are also colonizing the Ardennes and will very likely spread south (Ruys et al. 2008). Reintroduction of beaver into France has been very successful, although there are local and regional differences (Dewas et al. 2012). The current beaver population has been estimated at about 14,000. This is more than a 100-fold increase within one century. Some river systems seem to be saturated, and beavers are colonizing marginal and less suitable habitats, where conflict with human activities are more likely (Dewas et al. 2012).

The development in many areas of North America, Northern Europe and also Russia (Saveljev and Safonov 1999) show that it is better to prevent beavers from becoming 'nuisance beavers' by restriction of reintroduction to appropriate habitats right from the beginning, rather than applying drastic and surely less popular control measures later (culling, poisoning, blasting beaver dams, etc.; see also Wilde et al. 1950), in particular as beaver is an important sympathetic animal in nature conservation (Sect. 6.4.). In any case, the shift of public opinion from a positive attitude at the beginning of beaver reintroduction to an increasingly critical attitude should make us think (Zahner 1997; Bloch 2000). Recently, beavers, which had climbed to safe terrain on riverbanks and dikes projecting above the flooded marshland of the Oder river (Oderbruch; eastern Germany/Poland) during the historic 2013-flood disaster, have been blamed for having locally destabilized soaked dikes by digging burrows into the upper parts. Beavers are not well reputed in the affected areas at present (abundant information in the local media; e.g., Brzoska 2014), and, understandably, flood prevention is given priority over the beaver's welfare. Thus, effective beaver control appears to be inevitable, in particular as comparable flood disasters can also be expected for the future. No wonder that it is controversially debated whether or not and how beaver control has to be materialized (e.g., translocation, destruction of the burrows, shooting, trapping). Nature conservationists imply that the farmers are blowing up beaver damage in order to receive compensation from the administration.

5.2.2 Reintroduction of Ibex into the European Alps

The ibex colony at Pontresina (1,820 m a.s.l., Grison, Switzerland) is an additional, though more local, example, showing the complex interrelationships of the hysteresis of grazing history and the impact of a restored ibex population on its former habitat (Fig. 5.7; Holtmeier 1968, 1969a, b, 1987). In the Alps, ibex (*Capra ibex*; Photo 5.7) had already been eradicated from most of its habitats by the middle of the sixteenth century. In the canton Grison, the last ibex were shot in the early 1700s. The last ibex in Switzerland was shot in Einfisch Valley (Canton Valais) in 1809 (Desax 1978).

The main reasons for ibex decline were overhunting for meat and trophies, as well as the belief in alleged therapeutic and aphrodisiacal effects of certain parts of the ibex body (e.g., ground horns, Bezoar balls) and ibex blood. Bezoar balls are small concretions consisting of licked hairs and partly undigested resinous plant remains found in the ibex stomach. The Bezoar balls were gilded and sold at high prices by pharmacies. They were believed to protect from dizziness, fainting, jaundice, dysentery and melancholy. The blood was supposed to prevent bladder stones. The heart and marrow were used as tonics, and the ground horns were believed to protect from convulsions, poisoning and colics. Ossified parts of the aorta (so-called 'heart crosses' in common parlance) were considered lucky charms, and the heels an aphrodisiac (Schnitter 1965). In other words, ibex was a 'walking pharmacy', as

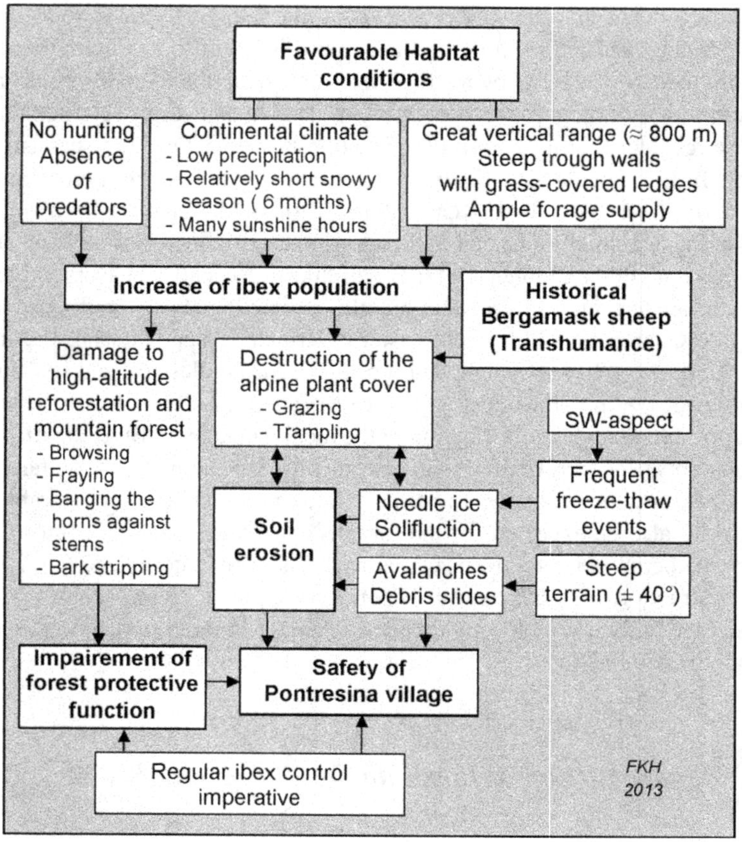

Fig. 5.7 Influence of habitat conditions on the ibex population at Pontresina and the impact of ibex and after-effects of historical transhumant Bergamask sheep on high-altitude reforestation, subalpine forest and soil erosion. Soil erosion is enforced by natural physical factors. This situation may threaten the safety of Pontresina village (Scheme by F.-K. Holtmeier, modified from Holtmeier 2003 (see also Holtmeier 2012))

was the beaver (Sect. 5.2.1). Moreover, enormous numbers of sheep and goats also adversely affected ibex populations, driving them back from the best grazing areas to unfavorable slopes. Ibex also hybridized with domestic goats and consequently suffered from the goats' parasites, diseases and epidemics (Schnitter 1965). As ibex is not very shy and only climbs to steep inaccessible 'safe' terrain if pursued, it was easy to shoot with crossbows and firearms.

Shortly after extinction, ibexes were already reintroduced to many of their original habitats, with the good intention of righting the wrongs of the past and restoring the native fauna. In Switzerland, the first ibexes were released in 1911. These animals were descended from ibexes which had been stolen in 1906 in a nighttime raid of a Swiss action group from the Gran Paradiso game reserve of the king of Italy, where the last pure-bred alpine ibexes were living. A previous official request from

Photo 5.7 Male ibex (*Capra ibex*) on the southeast-facing slope of Munt da la Bês-cha (Sheep mountain) at about 2,750 m. F.-K. Holtmeier, August 1973

Switzerland had been rejected. The current total ibex population in Switzerland is estimated at about 17,000.

The origin of the ibex reserve on the slopes of Munt da la Bês-cha ('Sheep Mountain') and Piz Albris above Pontresina village in the Upper Engadine dates back to the early 1920s, when two female ibexes that had likely escaped poachers in the Swiss National Park (Lower Engadine) appeared in the Piz Albris area. After that, an additional 11 individuals from Swiss game parks at Interlaken and St. Gallen were released to support the development of the ibex population. Due to the optimal local habitat conditions (cf. Fig. 5.7), the ibex population had grown to more than 700 individuals by the end of the 1960s (Ratti 1998). From the point of release, ibex spread northwest following the mountain range (over Piz Languard, Las Suors, Munt da la Bês-cha and Muottas Muragl). In the valleys north of these areas (Val Chamagna, Val Chamuera), small herds and solitary individuals can regularly be observed (Campell 1958; Nievergelt 1966; Bisaz 1968; Holtmeier 1968, 1969a, b, 1987). Nowadays, the Pontresina ibex reserve is the largest in Switzerland.

The habitat (Photo 5.8; cf. Fig. 5.7) is characterized by a continental high-mountain climate with low precipitation (Pontresina just above 800 mm), many sunshine hours and high solar radiation intensity. The snowy season is relatively short (6 months) compared to the same elevation in the northern Alps (Turner 1961). The steep trough walls and grass-covered ledges are often snow-free in winter and provide food and shelter to the ibexes. The great vertical range (about 800 m) of the habitat guarantees ample forage up to the upper alpine zone during the growing season. Hunting was restricted and predators have been absent, with the exception

Photo 5.8 View of the steep southwest exposed slope of Munt da la Bês-cha, where ibex enjoy optimal living conditions (cf. Fig. 5.7). Pontresina village located in the foot zone of the mountain. Dry stone walls (horizontal structures) that were built to stabilize the winter-snow are recognizable above the forest limit. F.-K. Holtmeier, 14 October 1967

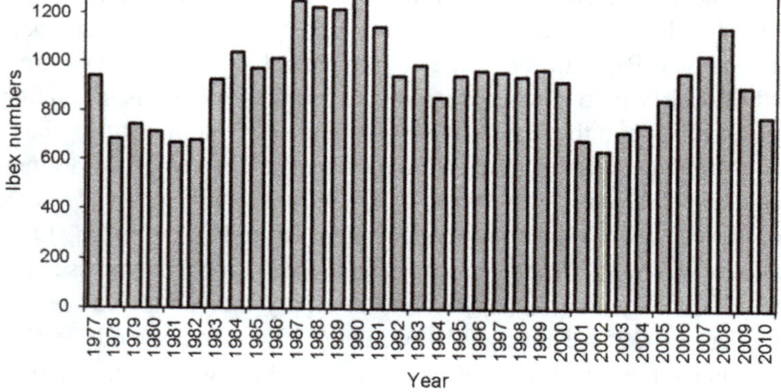

Fig. 5.8 Fluctuations in numbers of the ibex population on Munt da la Bês-cha (including Val Chamuera) (Data gratefully received from 'Amt für Jagd- und Fischerei Graubünden', 2011)

of the golden eagle (*Aquila chrysaetos*), which may occasionally kill ibex kids (Haller 1996). These optimal habitat conditions and prohibition of hunting were the main factors driving rapid growth of the ibex population. Highest density was reached around 1990 (Fig. 5.8).

Photo 5.9 Swiss stone (*Pinus cembra*) damaged by ibex in high-altitude afforestation on Munt da la Bês-cha at 2,300 m. F.-K. Holtmeier, 5 June 1967

Ibex impact on the habitat increased parallel to growing animal numbers. Ibexes damage trees in the treeline ecotone by browsing and bark stripping as well as by rubbing and banging their horns against the tree stems (Photo 5.9). Young trees (height up to 130 cm) are most affected (Holtmeier 1965, 1969a, 1972, 1978, 2009; Feuerstein 1997). Their elastic stems, bouncing back under the ibexes' attacks, possibly stimulate ibex aggressiveness (Feuerstein 1997). High-altitude reforestation carried out up to 2,500 m altitude is most adversely affected. Damage by banging the horns against the tree stems and fraying prevail. These kinds of damage are more serious than those caused by browsing (see also Ten Houte de Lange 1978; Feuerstein 1997). Reforestation is part of extensive and expensive measures that have been carried out since the end of the nineteenth century in order to reduce the danger of avalanches threatening Pontresina village and roads located in the footzone of the steep mountain slopes (Bisaz 1968; Holtmeier 1969b, 1976). In addition, on the avalanche-prone slopes, extensive snow-stabilizing structures (dry stone walls, structures made of steel or concrete) have been installed above the closed forest (cf. Photo 5.8). The dry stone walls are covered by thick grass sods to reduce damage by falling rocks. Ibexes often use these dry stone walls as a resting place and gradually destroy the protective sod cover through trampling.

Ibex frequent high-altitude afforestation mainly during winter when the snowpack prevents access to ground vegetation. Stem injuries render the trees more vulnerable to the harsh high-altitude climate and expose them to pathogens. Moreover, ibexes destroy alpine vegetation by grazing and trampling, thus initiating

Photo 5.10 Soil erosion on the mountain above Pontresina village at about 2,600 m. Several ibex are in the middle of the picture. Vegetation probably was destroyed first by historical Bergamask sheep. Fragmentation continued under the impact of the at least temporarily over-sized ibex population, combined with needle ice formation, solifluction, and wind. F.-K. Holtmeier, 8 October 1967

or enhancing soil erosion (Holtmeier 1968, 1969a, b, 1976, 1987). Grazing, however, is possibly of minor importance compared to trampling effects (Ten Houte de Lange 1978). In many places on Munt da la Bês-cha, Piz Albris and Piz Languard, alpine plant cover has become very fragmented (Photo 5.10).

Mainly on the southwest-facing slope, ibexes have exposed the mineral soil over wide areas. In such places, erosion is enhanced by frequent needle ice formation (Photo 5.11). The ice needles forming at sub-zero temperatures (about – 4 °C) in moist top soil loosen the top soil and lift up the upper soil particles. More than 13 of fine material per square meter and even stones weighing several kilograms may become uplifted (unpublished data from the present author). As soon as temperatures increase after sunrise, the ice needles fall over, and soon the uplifted soil has already been translocated downslope by a few centimeters. When the ice is melting, the fine saturated material starts flowing downhill (so-called 'needle ice solifluction'). The water equivalent of 2–3 cm long needle ice per square meter amounts to about 20 l. Meltwater seeping into the ground becomes a source of new needle ice after air and soil temperatures have dropped by several degrees below freezing. Once dried up, loose fine material on the surface is prone to deflation. Wind-driven mineral particles may act as an erosive agent, undercutting the remaining vegetation patches (e.g., Holtmeier 1996; Holtmeier et al. 2004; Holtmeier and Broll 2010). Recurrent needle ice formation and run-off are also involved in this erosive process. In the course of time, wind scarps develop. Undercut overhanging sods break off and slide down the steep slopes (Photo 5.12).

Photo 5.11 Needle ice on exposed mineral soil on the southwest-facing slope of Languard Valley at 2,580 m. F.-K. Holtmeier, 6 July 1967 (10.30 AM)

Photo 5.12 Fragmented plant cover on the southwest-exposed slope of Languard Valley (Upper Engadine) at 2,800 m. Wind in combination with needle ice undermine the remained vegetation patches. Overhanding sods (*Carex curvula*) break off and slip downslope (cf. Fig. 5.7). F.-K. Holtmeier, 23 September 1967

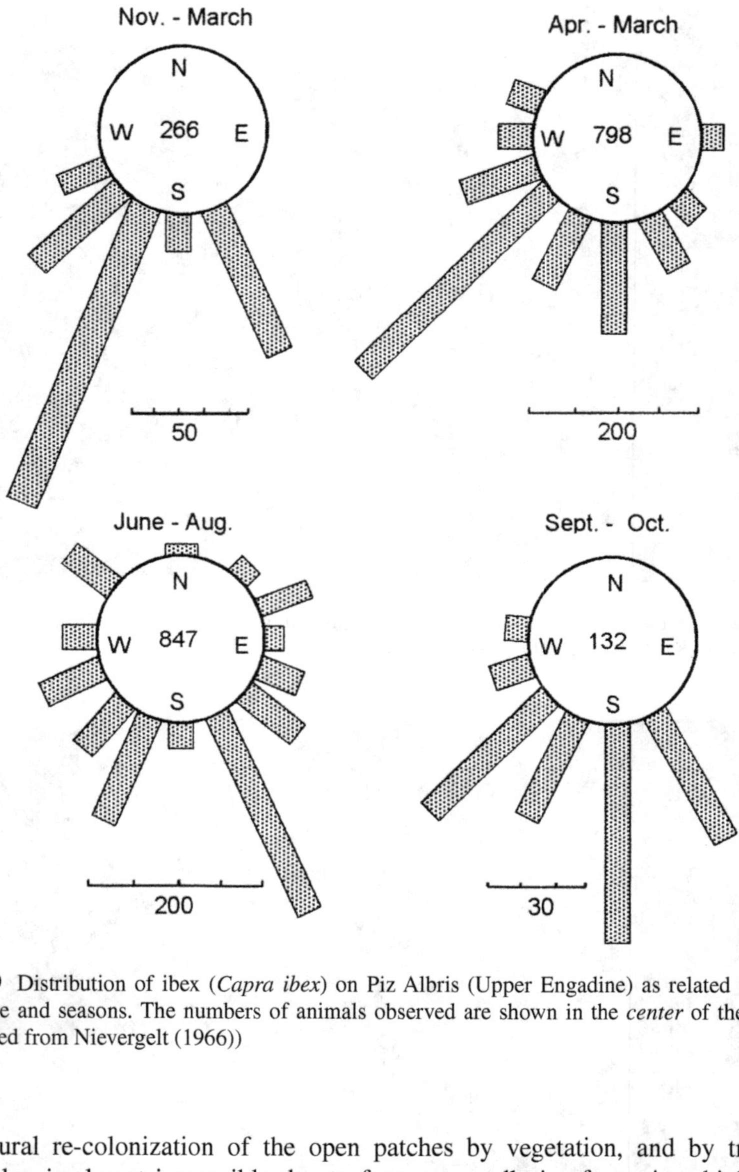

Fig. 5.9 Distribution of ibex (*Capra ibex*) on Piz Albris (Upper Engadine) as related to slope exposure and seasons. The numbers of animals observed are shown in the *center* of the *circles* (Modified from Nievergelt (1966))

Natural re-colonization of the open patches by vegetation, and by trees in particular, is almost impossible due to frequent needle ice formation, high solar radiation loads (great altitude, southwest exposure) and rapid draining due to the lack of an organic layer. Needle ice may push seeds and seedlings out of the soil. Sun-exposed terrain is particularly affected, as freeze-thaw frequently occurs in such places. Moreover, in winter, ibex can usually only be found on such exposures, and they also prefer them during autumn and spring. Thus, their impact is comparatively high on these slopes. In high summer (July, August), when the winter snow has gone, ibex move to the northern exposures (Fig. 5.9) where they find fresh forage and may escape the midday heat. Trampling and grazing effects on vegetation

Photo 5.13 Trampling effects in a sheltered site within the subalpine forest (*Larix decidua*, *Pinus cembra*) on the steep SW-facing slope above Pontresina village. Ibex frequent such sites during adverse weather conditions. In winter, snow often slips off thus making forage plants easily accessible.
F.-K. Holtmeier, 1967

and soil are also common within the forest, where varied geomorphic structures in steep, often snow-free slope sections not only provide forage but also shelter from bad weather in winter and the other seasons (Photo 5.13).

Severe soil erosion above the mountain forest, however, cannot be attributed to the ibexes alone. It was initiated by large transhumant herds of Bergamask sheep that had regularly used the alpine pastures for centuries (therefore called 'Munt da la Bês-cha' in the Rhaetoromanic language). Leasing of alpine pastures was an important income source for the mountain people who lived near the subsistence level before the economy completely changed due to tourism. Leasing parts of the summer pastures was the only way to profit from the extensive alpine pasture area, as the number of local livestock was not optimal due to the limited winter forage supply. At the beginning of the twentieth century, Bergamask shepherds were forbidden to graze their sheep in the Upper Engadine because of spreading foot-and-mouth disease. Reintroduced ibexes, however, continued the destruction of alpine vegetation and enforced soil erosion that became a permanent threat to Pontresina village.

During the 1950s, when ibex damage could no longer be ignored, an attempt was made to keep the ibexes away from the high-altitude afforestations by fence constructions and strongly smelling cloths attached, soaked in carbolineum. Dogs were also used (Campell 1958; Bisaz 1968). The retarded succession of generations, which had already been noticed during the 1960s, possibly indicated the beginning of natural population decrease, which, however, was only temporary. In 1973, the ibex population had more than doubled. About 1,400 ibex were estimated (communication Jagd- und Fischereiinspetorat Graubünden, 25 April 1982). Subsequently, reduction by culling, trapping, decreasing reproduction, mortality, accidents, and excessively strong winters (e.g., 1999, 2001, 2009) temporarily reduced ibex numbers. Nevertheless, despite control operations and extreme winters, the ibex population regularly recovered and peaked at more than 1,200 individuals between the late 1980s and early 1990s (Holtmeier 2012; cf. Fig. 5.8). The severe winter of 2001 caused a drastic decline of ibex numbers to about 600 animals. The last peak was reached in 2008. The following severe winter claimed many victims again. However, even if such extremes prevented the ibex population from growing to the natural carrying capacity of the habitat, even the lower numbers might be too high with regard to the precarious situation on the steep slopes above Pontresina village (Holtmeier 2012).

Too many ibexes leave alpine vegetation no chance of recovery. Soil erosion is irreversible anyway. Thus, with respect to the safety of Pontresina, a regular control of ibex numbers is needed. On the other hand, the existence of the ibex population must not be put at risk. When removing 'surplus' ibexes, age structure, sex ratio and social behavior have to be considered. If too many old males were to be harvested – which may be suspected in view of the much coveted trophies – the troops of ibexes crowded around the remaining strong males might exceed the carrying capacity of the spatially limited sun-exposed winter habitats (steep trough walls) and increase damage by trampling and grazing in these places (cf. Photo 5.13). In order to guarantee a professional ibex control, only hunters specially advised by game wardens were permitted to shoot ibex according to a culling plan of the canton's government (Ratti 1981). The tolerable level of ibex population in the Pontresina reserve is probably around 600 individuals. As the population regularly recovered after reduction by control operations and extreme winters, keeping the ibex numbers at this level would probably not threaten the continued existence of ibex in this area. As has also become apparent in the case of beaver reintroduction, the example of the ibex reserve at Pontresina shows that present tolerable habitat carrying capacity is often reached before natural control mechanisms will limit animal population growth. Therefore, careful ecological management (wildlife and landscape) is needed that minimizes conflicts between animals' habitat requirements and human landscape use (pastures, afforestation, avalanche protection) in the long-term. Ecologically adequate control, in particular, when combined with economic profit (sale of ibex meat, sale of trapped individuals for reintroduction into other areas), will likely support the continued existence of ibex, even at conditions given in the ibex reserve at

Pontresina. With regard to the considerable increase of ibex reintroduced to other areas of Switzerland (e.g., Piz Lagrev near Sils and Val Bever in the upper Engadine; Brienzer Rothorn, the Bernese Oberland; see also Zuber et al. 2001; Lüps 2001), where all suitable habitats are occupied (Brosi 2007), and also to Austria and other Alpine countries, where about a total of 40,000 ibexes have been estimated (Giacometti 2011), this aspect is worth particular consideration.

5.2.3 Reintroduction and Natural Invasion of Large Predators

Although predators do not usually influence landscape directly, with the exception of burrowers (e.g., badgers, foxes), they may have indirect effects on the habitats by influencing numbers, social structures, reproduction, foraging behavior and migration of their prey animals. This is why predators are considered here in the context of reintroduction and natural invasion.

In densely populated regions, for example, in Central Europe, large predators such as bear, wolf and lynx were almost extinct, with the exception of a few small and isolated remainder populations. The same holds true for many smaller mammalian predators and large predatory birds. A hunting instruction of 1931 (Diezels 1931), for example, explicitly encouraged hunters 'to eradicate marten by all means'. Fox, goshawk and even peregrine falcon were also notorious as awful and bloodthirsty 'murderers' and killed whenever possible. After the predators had been eradicated, hunters were 'perfectly justified' as absolutely indispensable regulators of big and small game.

However, even in Yellowstone and other national parks of North America, the large predators (in particular, wolves, coyotes and mountain lions) were systematically eliminated under the pressure of the ranchers outside the parks and also of the public in order to protect park employees and visitors, as well as wild game populations (Calahane 1939; Musselman 1969). Consequently, wolves were removed from Yellowstone, Rocky Mountain National Park and Glacier National Park, and mountain lions had become very rare. After the almost complete extinction of the large predators, authorities turned with 'utmost care' and comparable success to lynxes, bobcats, badgers, minks, martens, otters and wolverines. The same 'attention' was paid to many predatory bird species. On Molly Island, located in Yellowstone Lake, even the white pelican (*Pelecanus erythrorhynchos*), which had been suspected to be a serious threat to local trout fishing (*Onchorhyncus clarkii*), was drastically reduced through destruction of the clutches and killing of the nestlings and young birds. This clearly reflects the prevailing attitude of the park management towards 'predators' in those days (Wright 1992). Such actions, however, are part of history, and as a result of better ecological education and training, the attitude towards predators has become a little less rigid, although cattle and sheep farmers are suspicious as ever of the increasing number of mountain lions and wolves within and also outside the park. Understandably, however, people may get hysterical when a mountain lion kills a dachshund in a front garden, shows obvious interest in joggers or breaks into llama corrals in suburban areas. From a 'mountain lion's view', however, this kind of easy prey just saves energy. In the Rocky Mountains, llamas are increasingly used as pack animals (McClaran and Cole 1993).

In Central Europe, efforts have been made for a while to reintroduce larger predators such as lynx, wild cat, wolf and also bear into their former habitats or to support their natural comeback from relic areas. The driving force behind this is nature conservation. Reintroductions often cause violent discussions in the public, in particular, if people feel threatened by freely roaming wolves or bears. Lynx or wolf, for example, may easily adjust to cultivated landscapes, provided that humans tolerate them, which, however, is not always the case. A lot of information and education is still needed to create awareness of the complex problems. On the other hand, the argument, often brought forward by nature conservationists, that wolf were indispensable for ecosystem function (e.g., WWW press information from 23 August 2001) is too simplistic and hardly appropriate to justify its reintroduction as an 'ecological must'.

It is the plain truth that the possible consequences of the many well-mentioned and often laborious and costly efforts to reintroduce predators into the present biocoenosis have not been sufficiently considered. They could have been foreseen, however, if not only the ecological requirements of the predators but also possible conflicts with human interests (e.g., livestock, land use, outdoor recreation, etc.) had been taken into consideration. Often, a great potential of unnecessary conflicts has been generated which cannot simply be 'discussed away'. The problems must be solved through predator control (e.g., Breitenmoser 1998c), for example, prevention of predator damage (watching or fencing in cattle, sheep and other endangered domestic animals) and compensation for damage. However, public understanding, tolerance and support of such measures has its limits, particularly in times of critical economic conditions. Recent reports about the development of bear, lynx and wolf populations in the Austrian Alps (e.g., Komposch and Gutleb 1999), for example, point to this precarious situation.

Decreasing tolerance of wolves and other large predators in those countries where they have survived hunting pressure and eradication campaigns (e.g., Poland, Scandinavia, Transylvania) should also give us pause. To most people, predators mean a public good, having an ethical value per se. However, the predators also prey on sheep and reindeer. Losses have really increased more or less parallel to the increasing number of predators (e.g., Asheim and Mysterud 2004; Swedish Wildlife Damage Centre 2007). Thus, sheep and reindeer owners consider large predators, and wolves in particular, to be a serious threat to their economy. In the northern reindeer-herding areas, wolves are not and will not be tolerated at all. In Norway, for example, smear campaigns against wolves have become quite usual these days, and wolf opponents are losing no time in calling them 'man-eating monsters'. Sheep herders, reindeer owners and hunters postulate rigid control. In Sweden, the numerous recreational hunters crusade against 'the big four' (bear, wolf, lynx and wolverine), which prey on the hunters' most favorite wild game, the moose. The argument that wolf reduction is inevitable for guaranteeing the safety of the many mushroom gatherers (mushroom gathering is very popular in Sweden) and even of the hunters' hounds (communication H. Zimmermann, 'Djurens rätt', Askerby, Sweden) is rather flimsy. In Norway, reduction of wolf population has started. The wolverine, which has been expanding, is also a focus of the sheep farmers' lobby. In the south-western part of the country, more shooting of wolverine has been claimed to reduce unacceptable losses in sheep over the long-term. Intensified shooting, however, may threaten the existence of wolverine. In the following sections, a few examples will illustrate the complex problem of the large predators' reintroduction and expansion with special reference to Europe.

5.2.3.1 Lynx

Under natural conditions, lynx prey on roe deer, reindeer, young or sick red deer, chamois, young wild boars, hares, voles and similar small mammals. In rural landscapes, however, lynx may also kill domestic sheep, particularly lambs, and goats. While lynx had been eradicated in Central Europe, relic populations survived in northern Europe, in the Baltics, Russia, the Carpathians and the Balkan Mountains.

Lynx have been successfully reintroduced to many areas in Central Europe (e.g., Carinthia, Switzerland, the Harz Mountains [Germany], Bohemian Forest and Slovakian Kocesko). Even though strictly protected by law, many lynx have been killed illegally (e.g., Reidt 1995). In a socio-economical case study on the acceptance of lynx by the local people in the Simmen Valley (Switzerland), it was found that they tolerate only a moderate lynx density. Rural people, who might be more affected, consider growing numbers of lynx with increasing reservation, in contrast to urban people who usually live far distant from the places of possible conflict. Urban people often hold a more sentimental view of nature and usually advocate this predator's comeback (Breitenmoser 1998a; Hunziker 2000). About a third of local Bavarian politicians, for example, that were asked for their opinion said that lynx should be supported in reoccupying its original habitats, but none of them liked having them near their homes (Gernhäuser 1991). Possibly, reintroduction of lynx to the Bavarian Forest has failed for this reason (Dahl et al. 2000). As ever, discussion between nature conservationists, huntsmen, forest owners and farmers (sheep herders) is controversial, and although lynx and other large predators are protected by law, their chance of becoming integrated into the biocoenoses do not look very good (ANL 2000). In some areas, for example, in the Bernese Oberland, the conflict between opponents and advocates of lynx reintroduction is even aggravating the situation. Lynx are 'successfully' competing with hunters for prey, as they obviously kill more roe deer than are shot. The advocates hope that lynx will be better accepted by the next human generation, which is already familiar with their presence and less prejudiced (Baur and Hunziker 2001). In a survey on the acceptance of lynx, bear and wolves in Austria, for example, people >60 years turned out to hold a more critical view of these homecomers compared to younger people (Wechselberger and Leizinger 2005).

Although lynx density has reached the habitat carrying capacity in the north-western Alps, lynx have not spread into new areas, and up to the present, only about

50 % of the potentially suitable habitats are occupied. Apparently, lynx population pressure is still insufficient to trigger expansion from the densely populated centers of distribution into new habitats. Endorsement of such a high lynx density can, however, hardly be justified politically (Breitenmoser and Baumgartner 2001).

In any case, there is no ecological reason to reintroduce lynx. Extinction of lynx (and also of bear) had no major impact on the ecosystems, and reintroduction is unlikely to bring about greater change (Wotschikowsky 2001). Although lynx may positively influence the biocoenoses, a significant qualitative and quantitative effect on ungulates cannot be expected (Loos 1972; Wotschikowsky 1981b, 1990, 2001; Hespeler 1995; Jobin 1998). Lynx alone would not be able to reduce oversized wild ungulate populations to a tolerable level. The decline of mouflon colonies in Carinthia and Slovenia and of roe deer in a climatically extreme high-mountain valley in the Valais (Switzerland) that has been attributed to lynx predation is exceptional, and may be ascribed to the absurd introduction of these wild game species into completely unsuitable habitats (Wotschikowsky 1990). On the other hand, it is possible that lynx, which profit from the present high ungulate population, will switch to other prey animals if red deer and roe deer populations are reduced to tolerable numbers. The same effect might occur as a result of rapid adaptation of wild game (red deer, roe deer, chamois) to the 'unknown' predator (see also Gossow and Honsing-Erlenburg 1986; Haller 1992). Assessment of the influence of lynx predation on the biocoenoses, and thus on ecosystems, is difficult, as the lynx is polyphagous and depends on the population density of all available prey animals, which, again, are controlled by forage supply and competition (Sect. 2.2.1). There is some evidence that the less close the prey populations are to natural conditions, the stronger is the lynx's impact.

In each case, however, the local situation needs to be carefully considered, as other factors such as habitat structures, for example, may also play an important role. In some small areas of the French Jura, for example, lynx heavily impacted sheep herds (so-called 'hot spots' sensu Stahl et al. 2001), whereas the herds in other areas (>70 %) were only occasionally attacked or even not at all. That is why increased losses in sheep cannot be ascribed to the growing total lynx population, but must be attributed to a few individuals preying in the 'hot spot' areas. This seems to be corroborated by the stop of sheep-kills after lynx had been removed. After a while, several lynx came back and increasingly killed sheep again, although other prey (roe deer) was abundant. The problem appears to be multi-faceted. Very likely, the particular landscape structures, such as wooded mountain slopes, undisturbed resting and rearing places, and also the type of sheep herding, are the essential reasons for the local concentration of lynx impact and specialization on sheep as prey. In those regions of the French Jura where lynx kill sheep only occasionally, financial compensation of sheep farmers may help to ensure the continued existence of lynx. In areas, however, where single 'naughty' lynxes regularly prey on sheep, there is nothing left than to remove the predator and/or better survey the herds. The best way is to improve shepherding techniques, including the use of guard dogs (Stahl et al. 2002).

5.2.3.2 Wolf

Similar problems have been experienced with the comeback of the wolf. In Central Europe, wolves are increasingly expanding from Poland to the west. Although native wolves had been eradicated in Germany more than 200 years ago, single individuals (so-called 'migrating wolves', Hespeler 1995) were regularly observed and also occurred in Austria and Switzerland. There are good reasons to assume that all of them were killed. When in 1976, nine wolves escaped from reserves in the Bavarian National Park, people stirred up by the media became really hysterical. Police and even German soldiers moved out to kill the 'dangerous beasts'.

Since then, wolves from Poland have invaded the 'Muskauer Heide' (Saxony), a former military training area in eastern Germany. Two wolfpacks have successfully brought up whelps. There is no doubt that these highly adaptive predators would colonize wider areas in Germany and other Central European regions if man tolerated them. These days, hunters and farmers are complaining of growing wolf impact on red deer, fallow deer and mouflon in (fenced!) game reserves and nature parks. This group requires wolf control to prevent damage to their favorite game species and sheep (Berliner Morgenzeitung, 27 May 2013). Recently, wolves killed five sheep in a corral in a nature park in Brandenburg. The corral was protected by an electric fence (Berliner Morgenzeitung, 3 October 2013). Mouflon, introduced from the Mediterranean for hunting purposes, is particularly threatened, because it is not adapted to this predator and not fast enough to escape wolves. Most recently, wolves have also been observed in western Germany (Niedersächsisches Ministerium für Umwelt, Energie und Klimaschutz 2013). As effective protective measures, such as fencing and keeping strong gard-dogs, are very expensive and compensation payment is considered to be insufficent, many sheep farmers might give up. This woul have negative effects on *Calluna*-heathland conservation (Sect. 6.3).

In the Italian Abruzzi Mountains, wolves partly subsist on garbage dumps in close vicinity to human settlements, and in the suburban area of Rome, they search for food on garbage dumps. People do not worry much about the presence of wolves (Meyer 1996). Although this situation is surely not desirable, it shows that these predators may exist as typical opportunists under conditions that are completely different from those in their original habitats. It also becomes obvious that direct persecution has been the cause of extinction rather than landscape change (Mech 1070).

The main cause of intolerance is probably the deep-seated hereditary fear of the large predators (Lopez 1978; Zimen 1997). Wolves, however, can hardly be considered a serious threat to humans. As has been formulated a little sloppily by Hespeler (1995), '*In any case, wolf is less dangerous than drunken car drivers, politicians or environmental pollution*'. Thus, it seems more likely that active opposition to wolves comes from people affected by material losses, such as hunters, sheep and cattle owners, which is certainly understandable. In the Val Ferret and Val d'Entremont north of the Great St Bernhard (Switzerland), two wolves that had immigrated from Italy probably killed about 130 sheep between 1994 and 1996 (Landry 1997; Breitenmoser 1998b). In Tuscany, Italy, nature conservationists were suspected by sheep farmers of having introduced wolves illegally. In the view of sheep owners, wolves have no place in sheep country. In those days, the government, on the other hand, tried hard to make coexistence of humans and predators easier (Schlamp 2001).

About 50 % of the European wolf populations are living in southern Europe (including areas that previously belonged to the Soviet Union). The populations, however, vary in size and their distribution is scattered. Sheep herders and hunters in this region also consider wolves to be noxious animals and they kill them whenever possible, even in areas where they are protected by law. In Italy and on the Iberian Peninsula, wolves preferably prey on wild ungulates, particularly wild boar, rather than on livestock (Meriggi and Lovar 1996; therein further references). However, when the wild ungulate population is low or consists of one species only, wolves subsist mainly on sheep, calves and other prey animals (Meriggi et al. 1996), as has also been observed in other regions (e.g., Pulliainen 1965; Mech 1970). Simultaneous reintroduction of wild ungulate species might alleviate the problem, as, if one species were at a population low, wolves could switch to another species. Losses of livestock could probably be reduced if the wild ungulates were so numerous that collisions with wolves would be much more likely than with sheep and other domestic grazers. However, the relevant studies leave it open to question whether the huntsmen could be integrated into such a strategy and whether such high wild ungulate numbers would not be tolerable in cultivated landscapes.

The comeback of wolves to southern France has also caused violent discussions. While the attitude of most urban people is positive, sheep farmers reject wolf comeback without any concession. Their opposition has been successful, insofar as the government passed the plan 'Loup' ('wolf'), which allows for shooting wolves in critical situations (Neff 2001). The fact that more sheep are killed by thunderstorms (lightning, heavy hail) and other inevitable natural catastrophes than by wolves will hardly ever convince a sheep farmer to tolerate this predator.

Even though wolf attacks on domestic animals are absolutely exceptional, provided that roe deer and red deer are overabundant, as in many regions of Central Europe (Reidt 1995), this is not a very good argument for tolerating wolves, because the high ungulate density is not natural and wolves will switch to other prey if wild ungulate numbers decline for whatever reason (Fig. 5.10). Thus, accompanying measures are needed to reduce losses in sheep. This could possibly be achieved by grazing more agile sheep races, which might escape predators more easily than heavier sheep. In addition, careful herding would reduce or even prevent predator attacks (Landa et al. 1999). Not least, better surveillance of freely roaming herds is needed. In southern France, for example, the almost forgotten strong Pyrenean guard dogs (so-called 'Patous' in French, 'Great Pyrenees' in English) can help to protect sheep better and minimize conflicts (Neff 2001). Anyway, alarming reports of sheep kills by wolves are regularly spread by the media. If wolves killed a person in France, the public opinion would escalate and wolf advocates would have a hard time in the future (a park manager's view, Mercantour National Park, cited by S. Ulrich in Süddeutsche Zeitung, 19 July 2013).

A partly dramatic decline of wild ungulates (with the exception of wild boar) parallel to increased numbers of predators has been reported in Bulgaria (Obretenov

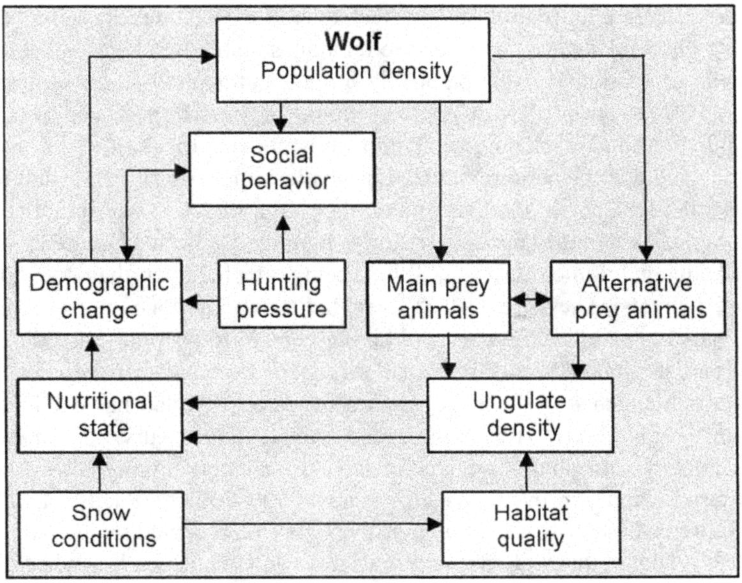

Fig. 5.10 Dynamics of a wolf population in Canada (Model modified from Keith (1983))

et al. 2001). Wolves and common jackals (*Canis aureus*), in particular, are considered to have reached intolerable densities. The wolf population has almost doubled since 1995, and the numbers of jackals, feral dogs and foxes increased tenfold up to the beginning of the present century, while bear numbers have not significantly changed. One should not ignore, however, that during the 1950s–1960s, wolves had been close to extinction due to massive hunting pressure and poisoning campaigns (strychnine), and is still not granted legal protection. The 'explosion' of predators has been ascribed to the protests of allegedly 'unqualified' nature conservation organizations against eradication of predators, the removal of the barbed wire entanglements along the state border, and prohibition of the use of anesthetics in game management. Rigid control has been required in order to maintain a profitable hunting economy (Obretenov et al. 2001; Markov 2012). However, removal of the top predator wolf may result in an accelerated increase in jackals, continuing or even exacerbating predator pressure on ungulates.

Reintegration of wolves into the strongly fragmented and intensely utilized Central European landscape without running into conflicts with human interests is not very likely. In the United States and Canada, wolf control – in other words, killing 'surplus' wolves – is considered to be inevitable. However, recreational hunting and trapping of wolves, as well as commercial trapping, have turned out to be ineffective, although hunting regulations were generously handled. Consequently, wolf control to a 'tolerable level' is done mainly by professional hunters and poisoning. Although pretending to kill wolves only when they are 'too numerous' or have already caused losses in livestock, 'careful' preventive culling is very common in some regions (e.g., Manitoba, British Columbia, Alberta), as it also has been in the past (Gunson 1983). In Central Europe, such operations would meet violent opposition in the public. However, as available habitats suitable for the establishment of viable wolf populations are comparatively limited and depletion of prey animals is not unlikely, regulation of wolf numbers might become imperative. Isolation of local populations and resultant genetic impoverishment in the not-too-distant future cannot be excluded. Genetic revitalization would be necessary from time to time (Theberge 1983). Genetic impoverishment is even a threat to wolf populations in North America, where they occupy territories ranging from some hundred to several thousand square kilometres in size, thus being much larger than the potential European habitats. Theberge (1983) even believes that none of the American national parks or wildlife reserves will be big enough to prevent inbreeding if the territories are not linked to each other and no genetic exchange between the populations is possible. Since the wolf population in western Poland has declined from more than 50 individuals to two to three packs, the long-term survival of the wolf in eastern Germany, for example, apparently depends on natural immigration of wolves from eastern Poland, i.e., over a distance of about 500–700 km (Sächsische Zeitung 2004). Active release of wolves would probably not be accepted by the public.

There is no doubt that wolf and lynx have left a gap in biocoenoses when brought close to extinction. However, one should not expect that these reintroduced and invading homecomers will resume their functional role as predators in the current ecosystems exactly in the way man has intended for them. The same also holds true, in principle, for the bear.

5.2.3.3 Bear

In Central, southern and southwestern Europe, a few small populations of brown bear (*Ursus arctos*) have survived in inaccessible mountains, such as the Cantabrian Mountains (Spain), the Pyrenees (France), the Abruzzi (Italy), the Italian Alps (Trentino) and the mountains of southern Bulgaria and northeastern Greece (see also Jakubiec 1993). Thus, it was a sensation when the first bears from Slovenia pushed ahead to the Austrian Alps. They had left Slovenia because of high bear population pressure. About 250–300 bears were living in the relatively small country (20,251 km^2) (Sorger 1995) and an additional 700 in adjacent Croatia (Wiegand et al. 1998).

A natural recolonization of their original alpine habitats, including the western Alps, by these large predators appears to be not unlikely. The last surviving alpine bear population living in the present Adamello-Brenta National Park (620 km^2, Trentino, Italy) has been considered to have an important connective function within a possible future bear habitat network. It is the geographical location of this area in the center of the Alps that actually suggests this idea. However, the habitat conditions in the Brenta area are anything than favorable for bears. Thus, in Niethammer's opinion (1963), the Brenta-bears just managed to vegetate miserably and would die of starvation if they could not prey on a sheep or goat once in a while. In 1998, probably only three native individuals lived there. The bear population would not have survived if not massively supported by the introduction of bears from other regions (Meyer 1996; Breitenmoser 1998b). About 40–50 individuals should be added to establish a viable population. Between 1999 and 2002, 10 Slovenian bears were released in the area (Kaczensky and Knauer 2001) with unanimous approval.

Moreover, release of Slovenian bears in the western French Alps is planned. The objective is to encourage genetic exchange by natural migration. Therefore, migration corridors must be provided or former ones restored. Wildlife crossings and underpasses have to be built that allow bears and also other wild animals safe crossing of heavily frequented traffic lines, such as roads, highways and railways. As other wildlife will also profit from such crossings, it may be questioned, however, why safe migration routes were not built before the 'migrating bears' appeared, particularly as bear cannot be considered an indispensable keystone species in alpine ecosystems.

In the Alps, extensive, sparsely populated and hardly utilized woodland as potential bear habitat is still abundant. However, it is unlikely that bears will stay forever in the areas of release. Being highly adaptive and learning quickly, bears will very soon 'appreciate' the benefits of alpine rural landscapes, as already happened in Austria, for example. Bears looted beehives, went for winter feeding places of deer and fishponds, and also killed a few sheep. These bears were definitely no 'rowdies', as occasionally assumed (e.g., Meyer 1996). Instead, they behaved quite naturally when using easily available food, thus saving energy. If bears are not hunted or bothered in another way, they will get habituated to people, for example, due to frequent encounters from a safe distance. Moreover, bears are often conditioned by people's food or garbage, and it is the combination of habituation and food-conditioning (Herrero 1985) that may make bears 'problem animals'. Problems with grizzlies and black bears regularly occur in the national parks in North America (e.g., Yellowstone National Park, Glacier National Park, Banff National Park, Yosemite National Park) – although bear habitats are extensive compared to the potential bear habitats in the European Alps (Table 5.3) and also provide abundant natural food other than hikers' supplies and garbage at camping sites.

In view of the experience with bears in North America, it seems a little strange that stands of sweet chestnut (*Castanea satviva*) in the Ticino (Italian Alps) have been considered to be an appropriate potential bear habitat (Meyer 1996). It may be questioned, however, why the chestnut forests in other similarly densely populated and intensively used valleys of the southern Alps were not considered on this occasion. Anyway, suggesting these chestnut forests regularly harvested by the local people as potential bear habitats reflects a little blurred view of reality.

The Austrian mountain landscape, for example, has been considered to provide suitable natural habitats for about 70 bears, whereas no more than 30 individuals can be tolerated with regard to possible conflicts (Sorger 1995). Great efforts have been and are being made to correct the still widely held view of bears as 'aggressive beasts'. In most cases, however, humans themselves are the actual cause of conflicts. In the American national parks, for example, most problems result from hikers' supplies left in tents, backpacks, or cars and from human food thrown away. Moreover, sudden encounters play an important role. Specific information for the

Table 5.3 Size of some national parks

National Park	Area [km^2]
Banff-Jasper National Park (Canada)	17,514
Yellowstone National Park (USA)	~9,000
Glacier National Park (USA)	>4,000
Yosemite National Park (USA)	3,108
Rocky-Mountain National Park (USA)	1,062
Etosha National Park (Africa)	22,270
Tsavo National Park (Africa)	20,182
Serengeti National Park (Africa)	16,500
Mara Wildlife Reserve (Africa)	1,672
Fiordland National Park (New Zealand)	12,500
Egmont National Park (New Zealand)	335
Kakadu National Park (Australia)	19,804
Kosciuszko National Park (Australia)	690
Yugyd Va National Park (Russia)	18,917
Vodlozersky National Park (Russia)	4,280
Cairngorms National Park (Scotland)	4,528
Snowdonia National Park (Wales)	2,140
Adamello-Brenta National Park (Italy)	620
Swiss National Park (Switzerland)	169
Müritz National Park (Germany)	320
Bavarian Forest National Park (Germany)	80
Berchtesgaden National Park (Germany)	70

From Holtmeier (2002), completed

park visitors may help to prevent such risky situations. However, bears' behavior is hardly predictable.

In Glacier National Park (Montana), unexpected encounters of bears habituated with hikers and other tourists have turned out to be less hazardous than running into bears that are unfamiliar with humans (Jope 1985). On the other hand, habituation of bears to people has increased critical encounters (Martinka 1982). Problems have increased parallel to growing numbers of park visitors (Wright 1992, therein further literature). However, despite information, education measures and strategies of bear avoidance, bear management has remained very much the same. As ever, 'problem bears' are translocated inside the park or to other areas. Bears which have become a permanent threat to people are shot.

The risk of bear attacks also varies in dependence of the availability and amount of natural bear food. In Yellowstone, for example, the amount of whitebark (*Pinus albicaulis*) pine seeds plays an important role. Whitebark pine seeds are an important food for grizzly bears, particularly in the subalpine zone. In case of poor or lacking seed production, bears searching for alternative food move to lower elevations, where the likelihood of conflicts with humans 'automatically' increases. Compared to mast years, six times as many 'problem bears' must be trapped, and the mortality of female bears (souws) doubles, while that of adolescent bears may even triple (Mattson et al. 1992, 2001).

Almost every week, the daily press reports conflicts between bears and humans. In the Lake Tahoe area (California), for example, black bears have started to hibernate in hollow spaces beyond wooden houses, in sheds and garages. To many bears, these 'new' activities are often fatal. In order to de-escalate the critical situation, members of a Bear League have excavated artificial hibernation hollows outside the residential area. They 'homely furnish' them with conifer twigs and hope that the bears will appreciate them and give up 'infiltrating' civilization.

From experiences with 'town bears' in Finland, a negative development is more likely. For about 20 years, increasing numbers of hungry bears get into towns after hibernation to search for edible garbage. The same has been reported in Colorado, Wyoming, New Mexico, and the Canadian provinces of Alberta, Ontario and Québec. Reports in newspapers on marauding bears and bear attacks on people in eastern Europe (Slovenia, Croatia, Romania, Poland, Russia) were relatively frequent even before alarming news increased in Central Europe (Austria, Switzerland, Bavaria). The story of the 'problem bear' named Bruno, which was shot in Bavaria on 26 June 2006, caused quite a stir in the public. Bruno (a male) was born in the Adamello-Brenta National Park. It left the park and commuted between the Italian, Austrian and German Alps. Bruno was the first wild bear to occur in Bavaria for 170 years. Dead Bruno even became a political issue when Bavaria and Italy had a strong argument as to the ownership of Bruno, a dispute that occasionally looked a little burlesque.

Anyway, in a case in which reintroduced and invading bears build up persistent populations in the European Alps, they have to be restricted to defined areas and must be kept away from regions where conflicts with humans are inevitable. However, bears are used to roaming over wide areas. The home range of old male grizzly bears, for example, may extend over 2,500 km^2 (Craighead et al. 1995). Migrating bears often leave their reserves, as is regularly the case in Yellowstone (Fig. 5.11) or in the Canadian Banff National Park (Craighead 1980; Craighead et al. 1995; Locke 1997). Outside the parks, many of them are shot by hunters or die in collisions with cars or trains. The existence of isolated bear populations, in particular, is threatened more by such losses beyond the park borders than by natural accidents inside the reserves (Woodroffe and Ginsberg 1998). The extension of the protected zones around the parks or, better, the implementation of a concept that plans to connect the present national parks and smaller reserves as the core of a large protected habitat network, including the northern Rocky Mountains between Yellowstone and the Yukon by migration corridors (Locke 1997), could surely help to prevent losses of roaming bears. Such a comprehensive concept, however, would hardly work in the European Alps, where the human population is comparatively high and open unused space is rare. Moreover, the idea of creating linear corridors that would connect isolated habitats appears to be too simplistic, as the different landscape matrix (i.e., spatial distribution and configuration of suitable habitats and barriers) matters, creating different levels of 'landscape permeability' for different species (Ricketts 2001; Wiens 2001; Singleton et al. 2002).

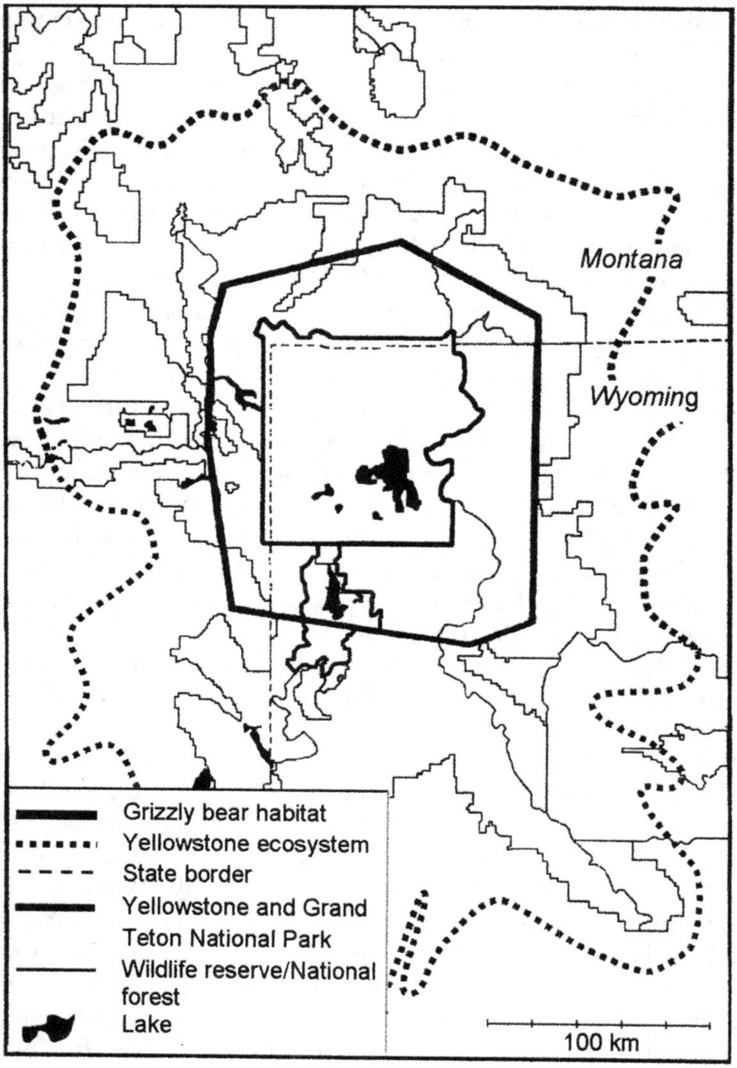

Fig. 5.11 Distribution of grizzly bear (*Ursusarctoshorribilis*) in the Yellowstone area (Modified from Craighead et al. (1995))

A network of relic habitats may look very easy on paper, but its implementation into practice is often very difficult, as the areas between the present isolated populations are subject to manifold use, such as nature conservation areas, economic forestry, intensive agriculture, mining industry, access roads, etc. Therefore, cooperation of all land users (state, countries, communities, private land owners) is needed for successful implementation of necessary measures (Salwasser et al. 1987). This also applies to the brown bear in the Austrian Alps, for example. The present bear population is estimated at 20–30 individuals. However, without release of additional bears, the population is probably not viable in the long-term, in particular, because of too few female bears (Zedrosser et al. 1999). As additional releases have met violent opposition in the public, the future development partly depends on occasional invasion of female bears from Slovenia. For a while, Slovenia was called the 'world champion of bear exports'. However, immigration into Austria decreased, because the Slovenian bear population, which has drastically increased since the mid-1990s, is being reduced as the result of public pressure through legal shooting (Kaczensky and Knauer 2001).

In the Alps, more conflicts with problem bears will be inevitable at increasing bear densities. Practicable solutions are not readily available. Translocation of problem bears is often impossible due to limited availability of suitable habitats. Deterrent measures may be efficient in random cases, at least in the short term. In the end, however, culling might be the only way to solve problems, otherwise, people feeling threatened by problem bears will find other ways to get rid of the 'beasts'. According to models of bear habitat use and of potential conflicts, conflicts are likely to increase to an extent, as is already typical even in the more marginal areas of the brown bear population in Slovenia (Wiegand et al. 1998).

Altogether, in contrast to the large and small herbivores, birds and insects, neither bear nor wolf as top predators play a comparable major role in the function of landscape ecosystems, although they may occasionally influence ungulate populations (physical conditions, foraging behavior, migration routes, etc.). In general, however, the role of predators as a prey-controlling factor has been overestimated. Nevertheless, would this be a reason to desist from more reintroductions (see also Ellenberg 1980)? The same question may be asked in view of species reinvading their original (former) habitats by themselves, such as wolves coming back to Germany (Brandenburg, Saxony and Western Pomerania) from Poland, or brown bears from Slovenia immigrating into the Austrian Alps.

References

ABC Rural. (2013, October 29). *Feral camel cull reaches 160.000 as project winds down*. Brisbane: ABC Rural.

Adam, D. A., & Fox, M. D. (1982). Change in Australian vegetation since European settlement. In M. B. Smith (Ed.), *A history of Australasian vegetation* (pp. 109–160). Sydney: McGraw-Hill.

Adams, J. (1980). High sediment yields from major rivers of western Southern Alps, New Zealand. *Nature, 284*, 88–89.

Akkermann, R. (1975). Untersuchungen zur Ökologie und Populationsdynamik des Bisams (Ondatra zibethicus L.). II. Nahrung und Nahrungsaufnahme. *Zeitschift für Angewandte Zoologie, 62*, 173–218.

Allgöver, R. (1995). Stellungnahmen zu "Biber und Hochwasser". *Natur und Landschaft, 70*(4), 1981.

Ammermann, D. (1998). Zur Problematik von Wiedereinbürgerungen am Beispiel des Birkhuhnprojektes in Oberschwaben. *Natur und Landschaft, 12*, 519–522.

Anderson, C. B., Soto, N., Cabello, J. L., Pastur, G. M., Lencinas, M. V., Wallem, P. K., Atúnez, D., & Davis, E. (2012). Castor canadensis Kuhl (North American Beaver). Building effective alliances between research and management to mitigate the impacts of an invasive ecosystem engineer: Lessons from the study and control of Castor canadensis in the Fuegian Archipelago. In R. A. Francis (Ed.), *A handbook of global freshwater invasive species* (pp. 343–355). London: Earthscan.

Andersone, Z. (1999). Beaver: A new prey of wolves in Latvia? Comparison of winter and summer diet of Canis lupus Linnaeus, 1758. In P. E. Busher & R. M. Dzieciolowski (Eds.), *Beaver protection, management and utilization in Europe and North America* (pp. 103–108). New York: Kluwer Academic/Plenum Publishers.

Angst, C. (2010). Vivre avec le castor. Recensement National de 2008. In Office fédéral de l'environnement et Centre des Cartographie de la Faune, Berne, Switzerland (Ed.), *Perspectives pour la cohabitation avec le castor en Suisse*. Bern: Office fédéral de l'environment et Centre de Cartographie de la Faune.

ANL. (2000). Kommen Elche, Wölfe und Bären wieder nach Bayern? *Naturschutz und Landschaftsplanung, 32*(2–3), 85–86.

Ansorge, H., & Stiebling, U. (2001). Population biology of the raccoon dog Nyctereutes proyonoides in eastern Germany: Immigration strategy of an alien species? *Beiträge zur Jagd- und Wildforschung, 26*, 247–254.

Apple, L. L., Smith, B. H., Dunder, J. D., & Baker, B. W. (1984, February 8–10). The use of beavers for riparian/aquatic habitat restoration of cold desert, gully-cut stream systems in southwestern Wyoming. In *Proceeding of the 18th annual meeting of American Fisheries Societies* (pp. 123–310). Logan.

Arner, D. H. (1963). Production of duck food in beaver ponds. *Journal of Wildlife Management, 27*, 78–81.

Arner, D. H., Baker, D. W., & Herring, B. (1969). An inventory and study of beaver impounded water in Mississippi. In *Proceedings 23. Annual conference of the Southeastern Association of Game and Fish Commissioners* (pp. 110–128).

Arzdorf, W. (1990). Stand, Möglichkeiten und Grenzen von Zucht und Auswilderung bedrohter Tieraten als Beitrag zum Artenschutz. Gefördertes Forschungs- und Entwicklungsvorhaben des Bundemisteriums für Umwelt, Naturschutz und Reaktorsicherheit (Zoologischer Garten Köln, ed.), Köln.

Ascune, M. S., Yang, C.-C., Oakey, J., Calcaterra, L., Wu, W.-J., Shih, C.-J., Goudet, J., Ross, K. G., & Shoemaker, D. (2011). Global invasion history of the fire ant (Solenopsis invita). *Science, 25*(331, no. 6020), 1066–1068. doi:101126/science.1198734

Asenjo, I. (2014, May 6). La serpiente rey albino, una amenaza en Gran Canaria. *ABC.es/Natural, 13*, 40.

Asheim, L. J., & Mysterud, I. (2004). Economic impact of protected large carnivores on sheep farming in Norway. *Sheep & Goat Research Journal, 19*, 88–96.

Aspinall, J. C. (1977). The opossum threat. *Journal of the Tussock Grasslands & Mountain Lands Institute, Review, 34*, 47–50.

Atkinson, I. (1985). The spread of commensal species of Rattus to oceanic islands and their effects on island avifauna. In P. J. Moors (Ed.), *Conservation of island birds: Case studies for the management of threatened island birds* (ICBP Technical Publications 3, pp. 35–81). Cambridge: International Council for Bird Preservation.

Atkinson, I. (1989). Introduced animals and extinctions. In D. Western & M. C. Pearl (Eds.), *Conservation for the 21st century* (pp. 54–75). New York: Oxford University Press.

Austin, O. L. (1963). *Die Vögel der Welt*. München/Zürich: Droemersche Verlagsanstalt, TH. Knaur Nachf.

Bajomi, B. (2011). *Reintroduction of the Eurasian beaver (Castor fiber) in Hungary*. Budapest: Danube-Drǻva National Park Directorate.

Balodis, M. (1992). Die Biber in Lettland. In R. Schröpfer, M. Stubbe, & D. Heidecke (Eds.), *Säugetiere* (pp. 121–129). Wissenschaftliche Beiträge der Universität Halle. Halle: Martin-Luther Universität.

Barun, A., Hanson, C. C., Campbell, K. J., & Simberloff, D. (2010a). A review of small Indian mongoose management and eradication on islands. In C. R. Veitch, M. N. Clout, & D. R. Towns (Eds.), *Island invasives: Eradication or management* (pp. 17–25). Gland: IUCN.

Barun, A., Simberloff, D., & Budinski, I. (2010b). Impact of the small Indian mongoose on native amphibians and reptiles on the Adriatic islands, Croatia. *Animal Conservation, 13*(6), 549–555.

Batcheler, C. L. (1983). The possum and rata-kamahi dieback in New Zealand: A review. *Pacific Science, 37*(4), 415–426.

Baur, H., & Hunziker, M. (2001). Welche Rolle spielen die "Luchsadvokaten" im "Luchskonflikt"? Ergebnisse einer sozialwissenschaftlichen Pilotstudie. *Informationsblatt Forschungsbereich Landschaft, 50*, 4–6.

Beecher, W. J. (1942). *Nesting birds and the vegetation substrate*. Chicago: Chicago Ornithological Society.

Bell, V. M., & Dieterich, R. A. (2010). Translocation of reindeer from South Georgia to the Falkland Islands. *Rangifer, 30*(1), 1–8.

Bell, J. H., & Williams, J. M. (1981). Where are we with rabbit control? In *Proceedings of the 1981 Hill & Country Seminar* (pp. 33–44). Lincoln College, Tussock Grasslands and Mountain Land Institute.

Bergstrom, D. M., Luciier, A., Kiefer, K., Wasley, J., Belbin, L., Pedersen, T. K., & Chown, S. L. (2009). Indirect effects of invasive species removal devastate world heritage island. *Journal of Applied Ecology, 46*, 73–81.

Bernhart, K.-G., & Schröpfer, R. (1992). Einfluß des Bisams auf die Vegetation. Untersuchungen im Ersatzbioptop Geeste im Emsland. *Naturschutz und Landschaftplanung, 24*(1), 20–26.

Bezzel, E. (1985). *Kompendium der Vögel Mitteleuropas Bd. 1, Nonpasseriformes*. Wiesbaden: Aula-Verlag.

Bildfell, R. J., Martins, J. W., Mortenson, J. A., & Cottam, D. F. (2004). Hair-loss syndrome in black-tailed deer of the Pacific Northwest. *Journal of Wildlife Diseases, 40*(4), 670–681.

Bisaz, O. (1968). Das Lawinenverbauungs- und Aufforstungsprojekt "Munt da la Bêcha" (Schafberg) der Gemeinde Pontresina. *Bündnerwald, 21*(5), 201–207.

Blanke, D. (1998). Biber in Niedersachsen. *Informationen zum Naturschutz in Niedersachsen, 18*(2), 29–35.

Bloch, J.-M. (2000). Wiederansiedlung mit Folgen. *Die Pirsch, 12*, 12–14.

Bobak, A. W. (1961). Wiedereinbürgerungsversuche mit Bibern in alter und neuer Zeit. *Natur und Lanschaft, 36*(7), 130–132.

Bomford, M. (1990). *A role for fertility control in wildlife management* (Bureau of Rural Resources Bulletin 7). Canberra: Australian Government Publishing Service.

Bomford, M. (2003). *Risk assessment for the import and keeping of exotic vertebrates in Australia*. Canberra: Bureau of Rural Sciences.

Bonino, N. (1995). *Introduced mammals into Patagonia, southern Argentina: Consequences, problems and management strategies* (Communicacion Technica 73, 7 pp). Insituto Nacional de Tecnologi a Agropecuaria, Centro Regional Patagonia Norte, Estación Experimental Agropecuaria Bariloche.

Bonner, W. N. (1985). Impact of fur seals on the terrestrial environment at South Georgia. In W. R. Siegfried, P. R. Condy, & R. M. Laws (Eds.), *Antarctic nutrient cycles and food webs* (pp. 641–646). Berlin/Heidelberg: Springer.

Bough, J. (2006). From value to vermin: A history of the donkey in Australia. *Australian Zoologist, 33*(3), 388–397.

Brandner, T. (1951). *Bidrag till kännedomen om bisamrattan (Ondatra z. zibethica L.) kägebobyggen* (Memoranda Societas Fauna Flora Fennica 26). Helsinki: Societas pro Fauna et Flora Fennica.

Bräsecke, R. (1989). Zur Verbeitung und Ökologie des Mink (Mustela vison SCHREBER) im Bezirk Schwerin. *Wissenschaftliche Beiträge der Martin-Luther-Universität Halle-Wittenberg, 37*(2), 333–346.

Breitenmoser, U. (1998a). Grundzüge eines alpenweiten Schutzkonzeptes für den Luchs. *Schriftenreihe des Landesjagdverbandes Bayern e, 5*, 7–10.

Breitenmoser, U. (1998b). Recent development of carnivore population in Switzerland and adjacent countries. *KORA Bericht Nr, 3*, 7–8.

Breitenmoser, U. (1998c). Large predators in the Alps: The fall and rise of man's competitors. *Biological Conservation, 83*(3), 279–289.

Breitenmoser, U., & Baumgartner, H. (2001). Die Rückkehr des Luchses in die Schweiz. *Nationalpark, 1*, 28–32.

Brosi, G. (2007). *30 Jahre Steinwildjagd in Graubünden*. Chur: Amt für Jagd und Fischerei Graubünden.

Brown, J. H., Innes, J. G., & Shorten, R. M. (1993). Evidence that possum prey on and scavenge birds' eggs, birds, and mammals. *Notornis, 40*, 169–177.

Brown, P. M. J., Adriaens, T., Bathon, H., Cuppen, J., Golderazena, A., Hägg, T., Kenis, M., Klausnitzer, B. E. M., Kovář, I., Loomans, A. J. M., Majeras, M. E. N., Nedved, O., Pedersen, J., Rabitsch, W., Roy, H. E., Tenois, V., Zakharov, I. A., & Roy, D. B. (2008). Harmonia axyridis in Europe: Spread and distribution of a non-native coccinellid. *Biological Control, 53*, 5–21.

Brzoska, I. (2014, February 1). *Teure Biber*. Berlin: Berliner Zeitung.

BUND (Bund für Umewlt und Naturschutz Deutschland). (1992). *Der BUND bringt Biber wieder auf den Damm*. BUNDruf 1, Februar.

Butler, D. R. (1991a). Beavers as agents of biogeomorphic change: A review and suggestions for teaching exercises. *Journal of Geography, 90*, 210–217.

Butler, D. R. (1991b). The reintroduction of the beaver into the South. *Southeastern Geographer, 31*, 39–43.

Butler, D. R. (1995). *Zoogeomorphology: Animals as geomorphic agents* (2nd ed., 2008). Cambridge: Cambridge University Press.

Butler, D. R., & Malanson, G. P. (1994). Canadian landform examples. Beaver landforms. *Canadian Geographer, 38*, 76–79.

Calahane, V. H. (1939). The evolution of predator control policy in National Parks. *Journal of Wildlife Management, 3*(3), 229–237.

Calderon, M. M., Dizon, J. T., Sajise, A. J. U., Andadra, R. T. II., Bantyan, N. C., & Salavador, M. G. (2009). *Towards the development of a sustainable financing mechanism for the conservation of the Ifuago rice terraces in the Philippines*. Economy and Environment Program for Southeast Asia (EEPSEA), Report No. 2008-RR10.

Calvopina, L. (1985). The impact and eradication of feral goats on the Galapagos Islands. In P. J. Moors (Ed.), *Conservation of island birds: Case studies for the management of threatened island birds* (ICBP Technical Publications 3, pp. 157–158). Cambridge: International Council for Bird Preservation.

Campbell, K., & Donlan, C. J. (2005). Feral goat eradication on islands. *Conservation Biology, 19*(5), 1362–1374.

Campell, E. (1958). Steinwild im Engadin. *Bündner Wald, 11*, 287–294.

Carrel, S. (2009, May 29). Will beavers return to British waters for first time in 200 years? *The Guardian*.

Catt, D. (1975). *Growth, reproduction and mortality in Bennett's Walleby (Macropus rufogriseus fruticus) in South Canterbury, New Zealand*. M.Sc. thesis, Department of Zoology, University of Canterbury (cited in Graham 1977).

Caughley, G. (1970). Eruption of ungulate populations, with emphasis on Himalayan Tahr in New Zealand. *Ecology, 51*(1), 53–72.

Caughley, G. (1976). Management and the dynamics of ungulate populations. In T. H. Coaker (Ed.), *Applied biology* (Vol. 1, pp. 183–246). London/New York: Academic.

Caughley, G., Griggs, G. C., & Smith, L. (1985). The effect of drought on kangaroo populations. *Journal of Wildlife Management, 49*, 679–685.

Challies, C. N. (1975). Feral pigs (Sus scrofa) on Auckland Island: Status and effects on vegetation and nesting sea birds. *New Zealand Journal of Zoology, 2*(4), 479–490.

Challies, C. N. (1985). Establishment, control and commercial exploitation of wild deer in New Zealand. In P. F. Fennessy & K. R. Drew (Eds.), *Biology of deer production* (Royal Society of New Zealand Bulletin 22, pp. 23–36). Wellington: Royal Society of New Zealand.

Challies, C. N. (1990). Red deer. In C. M. King (Ed.), *The handbook of New Zealand mammals (paperback edition 1995)* (pp. 436–458). Auckland/Melbourne/Oxford/New York: Oxford University Press.

Chapuis, J. L., Boussès, P., & Barnaud, G. (1994). Alien mammals, impact and management in the French subantarctic islands. *Biological Conservation, 67*, 97–104.

Chavasse, C. G. R. (1955). *Mortality in rata/kamahi protection forest – Westland*. Unpublished Report New Zealand Forest Service, cited by Payton (1987).

Choi, C. (2008). The beaver must die. *Nature, 453*. doi:10.1038/453968a.

Choquenot, D., McIllroy, J., & Korn, T. (1996). *Managing vertebrate pests: Feral pigs*. Canberra: Australian Government Publication Service.

Choquenot, D., Lukins, B., & Curran, G. (1997). Assessing lamb predation by feral pigs in Australia's semi-arid rangelands. *Journal of Applied Ecology, 34*, 1445–1454.

Clarke, C. M. H. (1976). Eruption, deterioration and decline of the Nelson red deer herd. *New Zealand Journal of Forestry Science, 5*(3), 235–249.

Clarke, C. M. H. (1978). *How far do chamois range?* What's new in Forest Research? 60. Rotorua: Forest Research Institute.

Clarke, C. M. H. (1990). Chamois. In C. M. King (Ed.), *The handbook of New Zealand mammals* (pp. 380–393). Melbourne/Oxford/New York: Oxford University Press.

Cockayne, L. (1926). Monograph on the New Zealand beech forests. *New Zealand State Forest Service Bulletin 4*(1).

Coleman, J. D., Green, W. Q., & Polson, J. G. (1985). Diet of brushtail possums over a pasture-alpine gradient in Westland, New Zealand. *New Zealand Journal of Ecology, 8*, 21–35.

Collen, P., & Gibson, R. J. (2001). The general ecology of beavers (Castor spp.), as related to their influence on stream ecosystems and riparian habitats, and the subsequent effects on fish – A review. *Reviews in Fish Biology and Fisheries, 10*, 439–461.

Collett, R. (1897). Bæveren I Norge, dens utbredelse og leevemaade (1996). *Bergens Musum Aarbog, 1*, 1–139.

Cooke, B. D. (1990). Notes on the comparative reproductive biology and the laboratory breeding of the rabbit flea Xenospylla cunicularis Smit (Siphonatera: Pulicidae). *Australian Journal of Zoology, 38*, 527–534.

Cooke, B. D. (2002). Rabbit haemorrhagic disease: File epidemiology and the management of wild rabbit populations. *Revue scientifique et technique, Office international des epizooties, 21*(2), 347–358.

Copson, G. R. (2002). *Integrated vertebrate pest management on subantarctic Macquarie Island 1997–2002* (Final Report). Hobart: Department of Primary Industries, Water and Environment.

Copson, G. R., & Whinam, J. (2001). Review of ecological restoration programme on subantarctic Macquarie Island: Pest management progress and future directions. *Ecological Management and Restoration, 2*, 129–138.

Corbet, G. B. (1974). The distribution of mammals in historic times. In D. L. Hawksworth (Ed.), *The changing flora and fauna of Britain* (pp. 179–202). London/New York: Academic.

Corbett, L. (1995). Does dingo predation or buffalo competition regulate feral pig population in the Australian Wet-Dry Tropics? An experimental study. *Wildlife Research, 22*, 65–74.

Costin, A. B., & Moore, D. M. (1960). The effects of rabbit grazing on the grasslands of Macquarie Island. *Journal of Ecology, 48*, 729–732.

Coulter, M. C., Cruz, F., & Cruz, J. (1985). A programme to save the dark-rumped petrel, Pterodroma phaeopygia, on Floreana Island, Galapagos, Ecuador. In P. J. Moors (Ed.), *Conservation of island birds: Case studies for the management of threatened island birds*

(ICBP Technical Publications 3, pp. 177–180). Cambridge: International Council for Bird Preservation.

Cowan, P. (1991). *The ecological effects of possums on the New Zealand environment*. International Symposium on Turberculosis, Veterinary Continuing Education, Massey University Publication 132, pp. 73–89.

Cox, W. T. (1940). The beaver – Friend of the forest. *American Forests 46*, 448–540 and 476–477.

Craighead, J. J. (1980). A proposed delineation of critical grizzly bear habitat in the Yellowstone region. *International Conference on Bear Research Management, 4*, 379–399.

Craighead, J. J., Summer, J. S., & Mitchell, J. A. (1995). *The grizzly bear population in the Yellowstone ecosystem, 1959–1992*. Washington, DC: Island Press.

Craik, J. C. A. (1995). Mink predation at seabird colonies. *Mammal News, 100*, 10.

Creagh, C. (1992). New approaches to rabbit and fox control. *Mammal News, 100*, 10.

Cremer, S., Ugelvig, L. V., Drijfhout, F. P., Schlick-Steiner, B. C., Steiner, F. M., Seifert, B., Hughes, D. P., Schulz, A., d'Ettore, P., Aktac, N., Eilenberg, J., Jones, G. R., Nash, D. R., Pedersen, J. S., & Boomsma, J. J. (2008). The evolution of invasiveness in garden ants. *PLoS ONE, 2*(12), e3838.

CSIRO. (2001). *CSIRO cane toad research*. Information sheet, last updated 14 June 2001. Collingwood: CSIRO Publishing.

CSIRO. (2003a, February 28). *Yellow crazy ants invade Northern Australia*. CSIRO Media Release, Ref 2003/41.

CSIRO. (2003b, December 3). *Exotic ants threaten Aboriginal communities*. CSIRO Media Release, Ref 2003/213.

Csurhes, S., Paroz, G., & Markula, A. (2009). *Pest animal risk assessment. Feral horses Equus caballus*. Biosecurity Queensland, Queensland Primary Industries and Fisheries. Department of Employment, Economic Development and Innovation, Queensland Government.

Dahl, H.-J., Altmüller, R., Grave, E., Kaufmanmn, W., Südbeck, P., & Bierhals, E. (2000). Artenschutz. In K. Buchwald & W. Engelhardt (Eds.), *Umweltschutz – Grundlagen und Praxis* (pp. 1–172). Bonn: Economica Verlag.

Dahlbeck, L., Fink, D., & Landvogt, M. (2008). 25 Jahre Biber in der Eifel. *Natur in NRW, 3*, 30–34.

Dahlsten, D. L. (1986). Control of invaders. In H. A. Mooney & J. A. Drake (Eds.), *Ecology of biological invasions of North America and Hawaii* (pp. 275–302). New York: Springer.

Danell, K. (1977a). Dispersal and distribution of the muskrat (Ondatra zibethica L.) in Sweden. *Viltrevy, 19*, 1–26. Stockholm.

Danell, K. (1977b). Short-term plant successions following the colonization of a northern Swedish lake by the muskrat, Ondatra zibethica. *Journal of Applied Ecology, 14*, 933–947.

Danell, K. (1978). *Ecology of the muskrat in Northern Sweden*. Dissertation Department of Ecological Zoology. Umea University, Umea, Sweden.

Danilov, P. I. (1992). *Introduction of North American semiaquatic mammals in Karelia and its consequences for aboriginal species* (pp. 267–276). Wissenschaftliche Beiträge der Martin-Luther- Universität Halle-Wittenberg, Kongress- und Tagungsberichte.

Danilov, P., Kanshiev, V., & Fyodorov, F. (2011a). History of beavers in eastern Fennoscandia from the neolithic to the 21st century. In G. Sjöberg & J. P. Ball (Eds.), *Restoring the European beaver: 50 years of experience* (pp. 27–38). Sofia: Pensoft Publishers.

Danilov, P., Kanshiev, V., & Fyodorov, F. (2011b). Characteristics of North American and Eurasian beaver ecology in Karelia. In G. Sjöberg & J. P. Ball (Eds.), *Restoring the European beaver: 50 years of experience* (pp. 55–72). Sofia: Pensoft Publishers.

Davidson, J. (1984). *The prehistory of New Zealand*. Auckland: Longman Paul.

Davies, E. (2011, June 1). Britain's beaver reintroduction stirs controversy. *BBC Nature Features*.

Denney, R. N. (1952). *A summary of North American beaver management* (Current Report 28). Denver: Colorado Fish and Game Department.

Department of Conservation. (1993). *Himalayan tahr control plan* (Canterbury Conservation Management Series 3, pp. 1–38). Canterbury: Department of Conservation.

Desax, C. (1978). Die Wiedereinbürgerung des Steinwildes in der Schweiz. In C. Desax (Ed.), *Arbeitstagung über Steinwild vom 9. - 11. Februar 1978 in Pontresina der Kommission Großwild Europa-Asien des internationalen Jagdrates zur Erhalung des Wildes* (pp. 24–35). Bern: Bundesamt für Forstwesen.

Dewas, M., Herr, J., Schley, L., Angst, C., Manet, B., Landry, P., & Catusse, M. (2012). Recovery and status of native and introduced beavers Castor fiber and Castor canadensis in France and neighbouring countries. *Mammal Review, 42*(2), 144–164.

Di Castri, F. (1989). History of biological invasions with special emphasis on the Old World. In J. A. Drake, F. Mooney, F. Di Castri, R. H. Growver, F. J. Kruger, M. Remancek, & M. Williamson (Eds.), *Biological invasions: A global perspective* (pp. 1–30). Chichester: Wiley.

Diemer, B. (1996). Der Bisam (Ondatra zibethicus) in Baden Württemberg. In Verein der Freunde und Förderer der Akademie für Natur-und Umweltschutz (Umweltakademie) beim Ministerium für Umwelt und Verkehr Baden-Württemberg (Ed.), *Neophyten, Neozoen – Gefahr für die heimische Natur?* (Beiträge der Akademie für Natur-und Umweltschutz Baden-Württemberg 22, pp. 182–186). Stuttgart.

Dietrich, U. (1985). Beobachtungen am Kanadischen Biber Castor canadensis in einem Einbürgerungsgebiet auf der Insel Feuerland, Südamerika. *Säugetierkundliche Mitteilungen, 32*, 241–244.

Diezel, C. E. (1931). *Diezels Niederjagd* (14 ed., E. Kluge ed.). Berlin: Parey.

Djoshkin, W. W., & Safonov, W. G. (1972). *Die Biber der Alten und Neuen Welt* (Neue Brehm-Bücherei 437). Wittenberg Lutherstadt: Ziemsen.

Dolman, P., Fuller, R., Gill, R., Hooton, D., & Tabor, R. (2010). Escalating ecological impacts of deer in lowland woodland. *British Wildlife, 10*, 242–254.

Doude van Troostwijk, W. J. (1976). *The musk-rat (Ondatra zibethicus L.) in the Netherlands, its ecological spects and the consequences for man*. Dissertation Rijksuniversitet Leiden, RIN Verhandeling 7.

Douglas, M. H. (1977). The warning whistle of tahr. *Journal of the Tussock Grassland & Mountain Lands Institute, 35*, 57–66.

Dowding, J. E., Murphy, E. C., Springer, K., Peacock, A. J., & Krebs, J. (2009). Cats, rabbits, Myxoma virus, and vegetation on Macquarie Island: A comment on Bergstrom et al. (2009). *Journal of Applied Ecology, 46*, 1129–1132.

Drygala, F., Mix, H. M., Stier, N., & Roth, M. (2000). Preliminary findings from ecological studies of the raccoon dog (Nyctereutes procyonoides) in eastern Germany. *Zeitschrift für Ökologie und Naturschutz, 9*, 147–152.

Duff, J. P., Scott, A., & Keymer, I. F. (1996). Parapox virus infection of the grey squirrel. *The Veterinary Report, 138*(21), 527.

Dunstone, N., & Ireland, M. (1989). The mink menace? A reappraisal. In R. J. Putnam (Ed.), *Mammals as pest* (pp. 224–241). London/New York: Chapman and Hall.

Easteal, S., & Floyd, R. B. (1986). The cane toad – An amphibian weed. In R. L. Kitching (Ed.), *The ecology of exotic animals and plants. Some Australian case histories*. Brisbane/New York/ Chichester/Toronto/Singapore: Wiley.

Eiberle, K. (1974). Waldkundliche Aspekte der Forschung an Rauhfusshühnern. *Schweizerische Zeitschrift für Forswesen, 125*(3), 147–170.

Eisfeld, D., & Fischer, U. (1996). Das Management eingebürgerter Huftierarten – widerstreitende Interessen, fehlende Konzepte. In H. Gebhart, R. Kinzelbach, & S. Schmidt-Fischer (Eds.), *Gebietsfremde Tierarten. Auswirkungen auf ein heimische Arten, Lebensgemeinschaften und Biotope, Situationsanalyse* (pp. 287–295). Landsberg: Ecomed Verlagsgesellschaft AG & Co KG.

Elder, N. L. (1965). Vegetation of the Ruahine range: An introduction. *Transactions of the Royal Society of New Zealand (Botany), 3*, 13–66.

Eldridge, W. (1983). Impacto ambiental, alimentación y conducta social de el ciervo rojo a y dama en el sur de Chile. *Boletín Técnico (Corporación Nacional Forestal, Chile), 9*, 1–53.

Ellenberg, H. (1980). Für und Wider der Einbürgerung von Großtieren in Mitteleuropa. *Jahrbuch des Vereins zum Schutz der Bergwelt, 45*, 43–63.

Engeman, R. M., Smith, H. T., Severson, R., Severson, M. A., Shwiff, S. A., Constantin, B., & Griffin, D. (2004). The amount and economic cost of feral swine damage to the last remnant of a basin marsh system in Florida. *Journal for Nature Conservation, 12*, 143–147.

Ermela, A., Helminen, M., & Lahti, S. (1989). Majaremme levinneisyyden ja runsauden raihteluista sekä tulevaisuuden näkymistä (Summary: Some aspects of the occurrence, abundance and future of the Finnish beaver population. *Suomen Riista, 35*, 108–118.

Evans, G. R. (1973). *Animals in the alpine grasslands* (New Zealand Forest Service Reprint No. 856). Wellington: New Zealand Forest Service.

Fabbro, E. (1989). *Fauna autóctona e introducida de la Tierra del Fuego, Boletin 2*. Ushuaia: Dirección de Recursos Naturales, Gobierno del Territorio de la Tierra del Fuego Press.

Farmer, A. P. (1973). *Assessment of animal damage in exotic forests of the central North Island* (New Zealand Forest Service Reprint No. 859). Wellington: New Zealand Forest Service.

Feldmann, K. B. (1980). Zu: Neuer Lebensraum für den Baumeister unter den Tieren. *Niedersächsischer Jäger, 25*(15), 799.

Feldmann, R. (1988). Neubürger in der Wirnbeltierfauna Westfalens. *Natur und Landschaftskunde, 24*, 79–86.

Fellers, G. M., & Osbourn, N. S. (2007). Abundance and impacts of fallow deer leks at Point Reyes National Seashore. *California Fish and Game, 93*(3), 149–160.

Ferry, M., & Gómez, S. (2003). The red palm weevil in the Mediterranean area. *Palms, 46*(4), 172–178.

Festetics, A. (1981). Die Wiederansiedlung des Luchses am Beispiel der Ostalpen. *Natur und Landschaft, 56*(4), 120–122.

Feuerstein, G. C. (1997). *Analyse von Stammverletzungen durch den Alpensteinbock (Capra i. ibex) in einem subalpinen Lärchen-Arvenwald*. Diplomarbeit, Abteilung für Forstwissenschaft (VI), ETH Zürich, Samedan.

Flook, D. R. (1962). Range relationships of some ungulates native to Banff and Jasper National Parks, Alberta. In J. D. Criso (Ed.), *Grazing in terrestrial and marine environments* (pp. 119–128). Oxford: Blackwell.

Forsysth, D. M., & Hickling, G. J. (1997). An improved technique for indexing abundance of Himalayan tahr. *New Zealand Journal of Ecology, 21*(1), 97–101.

Forsyth, D. M. (1999). Long-term harvesting and male migration in a New Zealand population of the Himalayan tahr Hemitragus jemlahicus. *Journal of Applied Ecology, 36*, 351–362.

Forsyth, D. M. (2000). Habitat selection and coexistence of the Alpine chamois (Rupicapra rupicapra) and Himalayan tahr (Hemitragus jemlahicus) in the eastern Southern Alps, New Zealand. *Journal of Zoology (London), 242*, 215–225.

France, A. J. (1991). The rabbit and land management programme – Will it achieve its objectives? A runholder's perspective. *Journal of the New Zealand Mountain Lands Institute Review, 48*, 1–7.

Frank, F., & Härle, A. (1967). Derzeitiger Stand und voraussichtliche Entwicklung des Bisambefalles (Ondatra zibethicus) in der Bunderepublik Deutschland von 1957 bis 1963. *Nachrichtenblatt des Deutschen Pflanzenschutzdienstes (Braunschweig), 16*, 145–147.

Fritts, T. H. (1987). Movement of snakes via cargo in the Pacific region. *Elepaio, 47*(2), 17–18.

Fritz, T. H. (1988). *The brown tree snake, Boiga irregularis, a threat to Pacific islands* (U.S. Fish and Wildlife Service Biological Report 88). Washington, DC: U.S. Fish and Wildlife Service Biological Report 88.

Frobel, K. (1994). Die Wiedereinbürgerung des Bibers in Bayern durch den Bund für Naturschutz. *Schriftenreihe des Bayerischen Landsamtes für Umweltschutz, 128*, 61–65.

Gearin, M. (2000). Can toads march on Kakadu. Landline 10/12/00, *Australia Rural Affairs Weekly*. www.abc.net.au/landline/stories/s220858.htm.

Geiersberger, I. (1997). Biber und Mensch – Konflikte in der Landnutzung. In Ministerium für Umwelt, Energie und Verkehr, Saarbrücken (Ed.), *Der Biber in der Kulturlandschaft – eine Illusion? Internationales Fachsymposium zur Wiederansiedlung des Bibers im Saarland, August 1994, Saarbrücken* (pp. 49–53).

Gernhäuser, S. (1991). *Ein Meinungsbild zum Luchs in Bayern* (Mitteilungen aus der Wildforschung 110, pp. 1–4). München: Wildbiologische Gesellschaft.

Gersdorf, J. A., & Frank, F. (1969). *Der Bisam (die Bisamratte) Ondatra zibethicus* (L.). Stuttgart: Ulmer.

Giacometti, M. (2011). *Vor 100 Jahren: Beginn der Wiederansiedlung des beinahe ausgerotteten Alpensteinbocks in der Schweiz*. Wildvet Projects, Stampa.

Gibb, J. A., & Williams, J. M. (1990). European rabbit. In R. L. Kitching (Ed.), *The ecology of exotic animals and plants. Some Australian case histories* (pp. 138–160). Brisbane/New York/Chichester/Toronto/Singapore: Wiley.

Gill, R. (2000, August). *The impact of deer on woodland biodiversity* (Information Notes 36, pp. 1–6). Edinburgh: Forestry Commission.

Godoy, J. C. (1963). *Fauna Silvestre* (Evaluación de los recursos naturales de la Argentina 8). Buenos Aires: Consejo Federal de Investiones.

Gorshkov, D. (2011). Utilizing beaver dam-building activity to reduce lake sedimentation. In G. Sjöberg & P. Ball (Eds.), *Restoring the European beaver, 50 years of experience* (pp. 205–217). Sofia/Moscow: Pensoft.

Gosling, M. (1981). Climatic determinants of spring littering by feral coypus, Myocastor coypus. *Journal of the Zoological Society London, 195*, 281–288.

Gosling, L. M. (1989). Extinction to order. *New Scientist, 4*, 44–49.

Gosling, L. M., & Baker, S. J. (1989). The eradication of muskrat and coypus from Britain. *Biological Journal of the Linnean Society, 38*, 39–51.

Gossow, H., & Honsing-Erlenburg, P. (1986). Management problems with re-introduced lynx in Austria. In S. D. Miller & D. D. Everett (Eds.), *Cats of the world: Biology, conservation and management* (pp. 77–83). Washington, DC: National Wildlife Federation.

Goudie, A. (1994). *Mensch und Umwelt*. Heidelberg: Spektrum Akademischer Verlag.

Goulson, D., & Stout, J. C. (1999). Impacts of introduced bumblebees in Tasmania. *Bulletin of the British Ecological Society, 30*(4), 39–40.

Graham, P. (1977). The war against the wallaby. *Tussock Grassland & Mountain Lands Institute, Review, 36*, 5–15.

Graham, A., Begg, R., Graham, P., & Raskin, P. (1982). *An aerial survey of buffalo in the Northern Territory: A technical report*. Darwin: Parks and Wildlife Unit, Conservation Commission of the Northern Territory.

Grasse, J. E. (1951). Beaver ecology and management in the Rockies. *Journal of Forestry, 49*(1), 3–6.

Green, W. Q. (1984). A review of ecological studies relevant to management of the common brushtail possum. In A. P. Smith & A. P. Hume (Eds.), *Possums and gliders* (pp. 483–499). Chipping Norton: Surrey Beatty & Sons.

Griffith, G. A. (1979). High sediment yields from major rivers in the western Southern Alps, New Zealand. *Nature, 282*, 61–63.

Grummt, W. (1973). Der Waschbär. In H. Stubbe (Ed.), *Das Buch der Hege* (Vol. 1, pp. 220–226). Berlin: Deutscher Landwirtschaftsverlag.

Gunn, T. C., & Walton, D. W. H. (1985). Storage carbohydrate production and overwintering strategy in a winter-green tussock grass on South Georgia (Sub Antarctic). *Polar Biology, 4*, 237–242.

Gunson, J. R. (1983). Wolf predation of livestock in western Canada. In L. N. Carbyn (Ed.), *Wolves in Canada and Alaska, Proceeding of the Wolf Symposium*, 12–14 May 1981, Edmonton (Canadian Wildlife Service, Report Series 45, pp. 102–105).

Haller, H. (1992). *Zur Ökologie des Luchses Lynx lynx im Verlauf seiner Wiederansiedlung in den Alpen*. Hamburg/Berlin: P. Parey.

Haller, H. (1996). Prädation und Unfälle beim Steinbock Capra ibex im Engadin. *Zeitschrift für Jagdwissenschaft, 42*, 26–35.

Halley, D. J., & Rosell, F. (2002). The beaver's reconquest of Europe: Status, population development and management of a conservation success. *Mammal Review, 32*, 153–178.

Hamann, O. (1979). Regeneration of vegetation on Santa Fe and Pinta Islands, Galapagos, after the eradication of goats. *Biological Conservation, 15*, 215–236.

Hamann, O. (1993). On vegetation recovery, goats and giant tortoises on Pinta Island, Galápagos, Ecuador. *Biodiversity and Conservation, 2*, 138–151.

Härkönen, S. (1999). Forest damage caused by the Canadian beaver (Castor canadensis) in South Savo, Finland. *Silva Fennica, 33*(4), 247–259.

Harrison, M. D. K., & Symes, R. G. (1989). Economic damage by feral American mink (Mustela vison) in England and Wales. In R. J. Putnam (Ed.), *Mammals as pest* (pp. 242–250). London/New York: Chapman and Hall.

Hartmann, G. (1994). *Ecological studies of a reintroduced beaver population*. Dissertation Uppsala, Sveriges Landbruksuniversitet, Institutionen för Viltekologi Rapport 25.

Hartmann, G. (1999). Beaver management and utilization in Scandinavaia. In P. Busher & R. Dzieciolowski (Eds.), *Beaver protection, management, and utilization in Europe and North America* (pp. 1–6). New York: Kluwer Academic/Plenum Publishers.

Hartmann, F. (2002). *Zur Problematik der Raubsäuger in ausgewählten Küstenvogelschutzgebieten Mecklenburg-Vorpommern*. Diplomarbeit Fachhochschule, Eberswalde.

Hartmann, G. (2011). The beaver (Castor fiber) in Sweden. In G. Sjöberg & J. P. Ball (Eds.), *Restoring the European beaver: 50 years of experience* (pp. 13–17). Sofia: Pensoft Publishers.

Hawai'i Conservation Alliance. (2007). *Controlling wild sheep and deer on conservation lands in Hawai'i* (Position Paper 2007-01).

Hayes, J. L., & Ragenovich, I. (2001). Non-native invasive forest insects of eastern Oregon and Washington. *Northwest Science 75*(Special issue), 77–84.

Heidecke, D. (1997). Erfahrungen bei der Ansiedlung des Elbebibers. In Ministerium für Umwelt, Energie und Verkehr, Saarbrücken (Ed.). *Biber in der Kulturlandschaft – eine Illusion? Internationales Fachsymposium zur Wiederansiedlung des Bibers im Saarland, August 1994, Saarbrücken* (pp. 69–78).

Heidecke, D., & Klenner-Fringes, B. (1992). Studien über die Habitatausnutzung des Bibers in der Kulturlandschaft. In R. Schröpfer, M. Stubbe, and D. Heidecke (Eds.), *Semiaquatische Säugetiere* (pp. 215–265). Wissenschaftliche Beiträge der Universität Halle. Halle/Saale: Martin-Luther-Universität Halle-Wittenberg.

Heidecke, D., Stubbe, M., & Königsfeld, T. (2001). Status der Nutria Myocastor coypus (Molina 1982) in Deutschland. *Beiträge zur Jagd- und Wildforschung, 26*, 321–338.

Helbig. D. (2011). Untersuchungen zum Waschbären (Procyon lutor LINNÈ, 1958). *Naturschutz im Land Sachsen-Anhalt 48*(1,2), 3–19.

Hercock, M., & Tonts, M. (2004). From the rangelands to the Ritz. Geographies of kangaroo management and trade. *Geography, 89*(3), 214–225.

Herrero, S. (1985). *Bear attacks. Their causes and avoidance*. New York: Nick Lyons Books/Winchester Press.

Hespeler, B. (1990). *Jäger wohin? Eine kritische Betrachtung deutschen Waidwerks*. München/Wien/Zürich: BLV.

Hespeler, B. (1995). *Raubwild heute. Biologie, Lebensweise, Jagd*. München/Wien/Zürich: BLV.

Hespeler, B. (1999). *Wildschäden heute. Vorbeugung, Feststellung, Abwehr*. München/Wien/Zürich: BLV.

Hess, S. C., Kawakami, B., Okita, D., & Medeiros, K. (2006). *A preliminary assessment of mouflon abundance at the Kahuku Unit of Hawaii Volcanoes National Park* (US Geological Service Open File Report Of, 2006–1143). Reston: USGS.

Hill, E. P. (1976). Control methods for nuisance beaver in the southwestern United States. In C. C. Siebe (Ed.), *Proceedings of the 7th Vertebrate Pest Control Conference, March 9–11, Monterey*, pp. 85–98.

Hinze, G. (1950). *Der Biber. Körperbau und Lebensweise, Verbreitung und Geschichte*. Berlin: Akademie Verlag.

Hofmann, R. (1977). Wildbiologische Erkenntnisse, ein Hilfsmittel zur Minderung der Waldschäden. *Allgemeine Forstzeitung, 32*(5), 111–115.

Hohmann, U. (2001). Stand und Perspektiven der Erforschung des Waschbären in Deutschland. *Beträge zur Jagd- und Wildforschung, 26*, 181–186.

Hohmann, U., & Bartussek, I. (2001). *Der Waschbär*. Reutlingen: Oertel+Sprörer.

Holdgate, M. W. (1967). The influence of introduced species on the ecosystems of temperate oceanic islands. In *Towards a new relationship of man and nature in temperate islands, Part III: Changes due to introduced species*. Morges: IUCN.

Holloway, J. T. (1950). Deer and the forest of Western Southland. *New Zealand Journal of Forestry, 6*, 123–137.

Holloway, J. T. (1973). *The status quo in animal control and management: A research assessment* (New Zealand Forest Service Reprint No. 863, New Zealand Forest Service, Forest Research Institute, Symposium No. 14).

Holtmeier, F.-K. (1965). *Die Waldgrenze im Oberengadin in ihrer physiognomischen und ökologischen Differenzierung*. Dissertation, Mathematisch-Naturwissenschaftliche Fakultät der Rheinischen Friedrich-Wilhelms-Universität, Bonn.

Holtmeier, F.-K. (1968). Ergänzende Beobachtungen in der Steinwildkolonie am Schafberg und Piz Albris bei Pontresina. *Bündnerwlad, 21*, 244–249.

Holtmeier, F.-K. (1969a). Das Steinwild in der Landschaft von Pontresina. *Natur und Museum, 99*(1), 15–24.

Holtmeier, F.-K. (1969b). Die Landschaft von Pontresina im Luftbild. *Erdkunde, 23*(2), 133–142.

Holtmeier, F.-K. (1972). The influence of animal and man on the alpine timberline. *Erdwissenschaftliche Forschung, IV*, 93–97.

Holtmeier, F.-K. (1976). Pontresina - landschaftliche Bindung und aktuelle Probleme einer Oberengadiner Gemeinde. *Westfälische Geographische Studien, 33*, 329–349.

Holtmeier, F.-K. (1978). Die bodennahen Winde in den Hochlagen der Indian Peaks Section, Colorado Front Range, USA. *Münstersche Geographische Arbeiten, 3*, 3–47.

Holtmeier, F.-K. (1987). Biber und Steinwild. Ökologische Aspekte der Wiederansiedlung von Tieren in ihren ehemaligen Lebensräumen. *Münstersche Geographische Arbeiten, 26*, 99–117.

Holtmeier, F.-K. (1996). Die Wirkungen des Windes in der subalpinen und alpinen Stufe der Front Range, Colorado, U.S.A. *Arbeiten aus dem Institut für Landschaftsökologie, Westfälische Wilhelms-Universität, Münster, 1*, 19–45.

Holtmeier, F.-K. (1999). Tiere als landschaftsökologische Faktoren. Arbeiten aus dem Institut für Landschaftsökologie der Westfälischen Wilhlems-Universitat 6.

Holtmeier, F.-K. (2002). *Tiere in der Landschaft. Einfluss und ökologische Bedeutung*. Stuttgart: UTB 8230.

Holtmeier, F.-K. (2003). Der Einfluss der Tiere in der Landschaft. *Berichte der Reinhold-Tüxen-Gesellschaft, 15*, 31–44.

Holtmeier, F.-K. (2009). *Mountain timberlines. Ecology, patchiness, and dynamics* (Advances in Global Change Research 36). Dordrecht: Springer Science+Business Media B. V.

Holtmeier, F.-K. (2012). Impact of wild herbivorous mammals and birds on the altitudinal and northern treeline ecotones. *LandscapeOnline, 30*, 1–28. doi:10.3097/LO.201230.

Holtmeier, F.-K., & Broll, G. (2010). Wind as an ecological agent at treelines in North America, the Alps, and the European Subarctic. *Physical Geography, 31*(3), 203–233.

Holtmeier, F.-K., Broll, G., & Anschlag, K. (2004). Winderosion und ihre Folgen im Waldgrenzbereich und in der alpinen Stude einiger nordfinnischer Fjelle. *GeoÖko, 25*, 204–224.

Hone, J. (1990). How many feral pigs in Australia? *Australian Wildlife Research, 17*, 571–572.

Hone, J. (1999). On rate of increase: Pattern of variation in Australian mammals and the implication for wildlife management. *Journal of Applied Ecology, 36*, 709–718.

Horn, E. (1968) *Das Pelzbuch*. München.

Hu, J., Angeli, S., Schuetz, S., Luo, Y., & Hajek, A. E. (2009). Ecology and management of exotic and endemic Asian longhorned beetle Anoplophora glabripennis. *Agricultural and Forest Entomology, 11*(4), 359–375.

Hughey, K. F. D., & Wason, K. M. (2005). *Management of the Himalayan tahr in New Zealand. High country farmer perspective and implications* (Research Report No. 276). Canterbury: Agribusiness and Economics Research Unit, Lincoln University.

Hugo, R. (2001). Integrierte Gewässerinformationssysteme - der Biber. *KA-Wasserwirtschaft, Abwaser, Abfall, 48*(5), 694–699.

Hunziker, M. (2000). *Warum sind Raubtiere willkommen - warum nicht?* (Das Jahr 1999, Jahresbericht der Eidgenössischen Forschungsanstalt WSL, pp. 22–23). Birmensdorf: EAVW.

Huston, M. A. (1994). *Biological diversity. The coexistence of species in changing landscapes* (2nd ed.). Cambridge: Cambridge University Press.

Hutchins, M., Stevens, V., & Atkins, N. (1982). Introduced species and the issue of animal welfare. *International Journal of Studies on Animal Problems, 3*, 318–336.

Ikeda, T., Asano, M., Matoba, Y., & Abe, G. (2004). The present status of invasive Alien raccoon and its impact in Japan. *Global Environmental Research, 8*(2), 125–131.

Jackson, D. B. (2001). Experimental removal of introduced hedgehogs improves wader nest success on the Western Isles, Scotland. *Journal of Applied Ecology, 38*, 802–812.

Jaksic, F. M., Iriate, J. A., Jiménez, J. E., & Martínez, D. R. (2002). Invaders without frontiers: Cross-border invasions of exotic animals. *Biological Invasions, 4*, 157–173.

Jakubiec, Z. (1993). Ursus arctus Linneus, 1758 – Braunbär. In M. Stubbe & F. Knapp (Eds.), *Handbuch der Säugetiere Europas, Raubsäuger* (pp. 254–300). Wiesbaden: Akademische Verlagsgesellschaft.

Jarman, P. (1986). The red fox – An exotic predator. In R. L. Kitching (Ed.), *The ecology of exotic animal and plants. Some Australian case histories* (pp. 45–61). Brisbane/Chichester/Toronto/Singapore: Wiley.

Jenkins, S. H. (1979). Seasonal and year-to-year differences in food selection by beavers. *Oecologia, 44*, 112–116.

Jenkins, S. R., & Winkler, W. G. (1987). Descriptive epidemiology from an epizootic of raccoon rabies in the Atlantic States, 1982–1983. *American Journal of Epidemiology, 126*(3), 429–437.

Jesser, P., & Markula, A. (2008). *Pest animals risk assessment: Water buffalo Bubalus bubalis* (pp. 1–14). Brisbane: Queensland Government, Department of Primary Industries and Fisheries.

Jobin, A. (1998). Der Einfluß des Luchses auf die Beutepopulation. *Schriftenreihe des Landesjagdverbandes Bayern e.V., 5*, 11–12.

Johnson, C. N., & Van Der Wal, J. (2009). Evidence that dingoes limit abundance of a mesopredator in eastern Australian forests. *Journal of Applied Ecology, 46*, 641–646.

Johnston, C. A., & Naiman, R. J. (1987). Boundary dynamics at the aquatic-terrestrial interface: The influence of beaver and geomorphology. *Landscape Ecology, 4*(1), 5–19.

Jonker, S. A., Muth, R. M., Organ, J. F., Zwick, R. R., & Siemer, W. F. (2006). Experiences with beaver damage and attitudes of Massachusetts residents toward beaver. *Wildlife Society Bulletin, 34*(4), 1009–1021.

Jope, K. L. (1985). Implications of grizzly bear habituation to hikers. *Wildlife Biology, 13*, 32–37.

Kaczensky, P., & Knauer, F. (2001). Wiederkehr des Braunbären in den Alpen – Erfahrungen mit einem anspruchsvollen Großräuber. *Beiträge zur Jagd- und Wildforschung, 26*, 67–75.

Kalchreuter, H. (1981). Erste Erfolge der Birk- und Auerhuhn-Projekte in Baden-Württemberg. *Natur und Landschaft, 56*(4), 129–130.

Kalchreuter, H., & Wagner, E. (1981). Ausbürgerung von Auerwild – Projektbericht. *Der Jäger in Baden-Württemberg, Mai, 81*, 3–4.

Kampmann, H. (1972). *Der Waschbär in Deutschland.* Dissertation, Forstliche Fakultät der Georg-August-Universität, Göttingen.

Kampmann, H. (1973). Der Waschbär, ein in Deutschland unerwünschtes Raubtier. *Wild und Hund, 75*(21), 123–124.

Kasparek, M. (1996). Dismigration und Brutarealexpansion der Türkentaube (Streptopelia decaoto). *Journal für Ornithologie, 137*(1), 1–33.

Kauhala, K., & Kowalczyk, R. (2011). Invasion of the raccoon dog Nyctereutes procyonoides in Europe: History of colonization, features behind its success, and threats to native fauna. *Current Zoology, 57*(5), 584–598.

Kauhala, K., Kaunisto, M., & Helle, E. (1993). Diet of the raccoon dog, Nyctereutes procyonoides, in Finland. *Zeitschrift fur Saugetierkunde, 58*, 129–136.

Keith, L. B. (1983). Role of food in hare population cycles. *Oikos, 40*, 385–395.

Kerr, I. G. C. (1989). What are they doing about rabbits? *Journal of the New Zealand Mountain Lands Institute, Review, 46*, 11–24.

Kightley, S. P. J., & Smith, R. I. L. (1976). The influence of reindeer on the vegetation of South Georgia. I. Long-term effects of unrestricted grazing and the establishment of exclosure experiments in various plant communities. *British Antarctic Survey Bulletins, 44*, 57–76.

King, C. (1984). *Immigrant killers. Introduced predators and the conservation of birds in New Zealand*. Auckland/Melbourne/Oxford: Oxford University Press.

King, C. M. (Ed.). (1990a). *The handbook of New Zealand mammals* (paperback edition 1995). Auckland/Melbourne/Oxford/New York: Oxford University Press.

King, C. M. (1990b). Introduction. In C. M. King (Ed.), *The handbook of New Zealand mammals* (paperback edition 1995) (pp. 3–21). Auckland/Melbourne/Oxford/New York: Oxford University Press.

King, S. L., Keeland, B. D., & Moore, J. L. (1998). Beaver lodge distributions and assessments in a forested wetland ecosystem in the southern United States. *Forest Ecology and Management, 108*(1), 1–7.

Kirk, B. H. (1920). *Opossums in New Zealand*. New Zealand Parliament: 1st session, 20th Parliament, Appendix H, 28 to Journals of the House of Representatives (cited in Pracy 1962).

Kitching, R. L. (1986). Exotics in Australia and elsewhere. In R. L. Kitching (Ed.), *The ecology of exotic animals and plants. Some Australian case histories* (pp. 1–5). Brisbane/New York/Chichester/Toronto/Singapore: Wiley.

Klein, D. R. (1968). The introduction, increase and crash of reindeer on St. Matthew Island. *Journal of Wildlife Management, 32*, 350–367.

Klemann, N. (1997). *Das "Europareservat Rieselfelder" als Habitat der allochthonen Nutria (Myocastor coypus)*. Diplomarbeit, Institut für Landschaftsökologie, Westfälische Wilhelms-Universität, Münster.

Knaus, W., & Schröder, E. (1975). *Das Gamswild. Naturgeschichte, Verhalten, Ökologie, Hege und Jagd, Krankheiten*. Hamburg/Berlin: Parey.

Komposch, B., & Gutleb, B. (1999). Bestandessituation von Wolf, Bär und Luchs in den südöstlichen Kalkalpen (Österreich). *Artenschutzreport, 9*, 42–44.

Kowarik, I. (2003). *Biologische Invasionen: Neophyten und Neozoen in Mitteleuropa*. Stuttgart: Verlag Eugen Ulmer.

Krakauer, T. (1970). The invasion of toads. *The Florida Naturalist, 1970*, 12–14.

Kumpula, J., & Nieminen, M. (1992, October 6–8). The inventory of reindeer pastures in Finland by satellite imageries. In E. Tikkanen, M. Varmola, & T. Katermaa (Eds.), *Extended abstracts: Symposium on the state of the environment and environmental monitoring in Northern Fennoscandia and Kola Peninsula*, Rovaniemi, pp. 294–296.

Kurt, F. (1982). Das comeback. *Natur, 11*, 83–91.

KWK-DVWW. (1977). *Empfehlungen für bisamsicheren Ausbau von Gewässern, Deichen und Dämmen* (KWK-DVVW Regeln zur Wasserwirtschaft 107). Hamburg/Berlin: Parey.

Lagoni-Hansen, A. (1981). *Der Waschbär*. Mainz: Verlag Dieter Hoffmann.

Lahti, S. (1997, September 15–19). *Development of population, distribution problems and prospects of Finnish beaver populations (Castor fiber L. and Castor Canadensis Kuhl). !*. European beaver symposium, Bratislava, unpublished.

Lahti, S., & Helminen, M. (1974). The beaver Castor fiber (L.) and Castor canadensis (Kuhl) in Finland. *Acta Theriologica, 19*, 177–189.

Lahti, S., & Helminen, M. (1980). Suomen majavien levinneisyyden muutokset vuosina 1965–1975. Summary: The status of European and Canadian beavers in Finland in 1965–1976. *Suomen Riista, 27*, 70–77.

Landa, A., Gudvangen, K., Swenson, J. E., & Røskaft, E. (1999). Factors associated with wolverine Gulo gulo predation on domestic sheep. *Journal of Applied Ecology, 36*, 963–973.

Landry, J.-M. (1997). *La bête du Val Ferret*. KORA Bericht Nr. 1, Kora. Muri, Switzerland.

Larson, J. S., & Gunson, J. R. (1983). Status of the beaver in North America. *Acta Zoologica Fennica, 174*, 91–93.
Lavsund, S. (1975). *Bävern I Sverige*. Stockholm: Statens naturvardsverk.
Lavsund, S. (1983). Beaver management and economics. Europe except the USSR. *Acta Theriologica, 174*, 133–135.
Leader-Williams, N. (1980). Population dynamics and mortality of reindeer introduced into South Georgia. *Journal of Wildlife Management, 44*, 640–657.
Leader-Williams, N. (1988). *Reindeer on South Georgia. The ecology of an introduced population*. Cambridge: Cambridge University Press.
Leader-Williams, N., Smith, R. I. L., & Rothery, P. (1987). Influence of introduced reindeer on vegetation of South Georgia: Results from a long-term exclusion experiment. *Applied Ecology, 24*, 801–822.
Learmonth, A. T. A., & Learmonth, A. I. (1968). Rabbits. In *Encyclopaedia of Australia* (p. 442). London: Frederick Warne & Co.
Lee, W. G., Fenner, M., Loughnan, A., & Lloyd, K. H. (2000). Long-term effects of defoliation: Incomplete recovery of New Zealand alpine tussock grass, Chionochloa pallescens, after 20 years. *Journal of Applied Ecology, 37*, 348–355.
Lelek, A. (1996). Die allochthonen und die beheimateten Fischarten unserer großen Flüsse – Neozoen der Fischfauna. In H. Gebhardt, R. Kinzelbach, & S. Schmidt-Fischer (Eds.), *Gebietsfremde Tierarten* (pp. 197–215). Landsberg: Ecomed.
Liebermann, H. (1992). *Lehrbuch der veterinärmedizinischen Virologie*. Jena/Stuttgart: G. Fischer.
Lindsay, K. (1973). Effects of reindeer on plant communities in the Royal Bay area of South Georgia. *British Antarctic Survey Bulletin, 35*, 101–109.
Linscombe, G., Kinler, N., & Wright, V. (1981). Nutria population density and vegetative changes in brackish marsh in coastal Louisiana. In J. A. Chapman & A. Purseley (Eds.), *Worldwide Furbearer Conference proceedings* (Vol. 1, pp. 129–141). Frostburg: Worldwide Furbearer Conference.
Lizzaralde, M., Escobar, J., & Deferrari, G. (2004). Invader species in Argentina: A review about the beaver (castor canadensis) population situation on tierra del fuego ecosystem. *Interciencia, 29*(7), 352–356.
LÖBF (Landesanstalt für Ökologie, Bodenbewertung und Forsten NRW). (2000). Ansiedlung von Bibern in der Eifel erfolgreich. *LÖBF-Mitteilungen, 4*, 9.
Locke, H. (1997). The role of Banff National Park as protected area in the Yellowstone to Yukon mountain corridor of western North America. In J. G. Nelson & R. Serafin (Eds.), *NATO ASI series G* (Ecological Sciences 40, pp. 117–124). Berlin: Springer.
Löffler, E. (1977). *Geomorphology of Papua New Guinea*. Canberra: CSIRO/ANU.
Löffler, E. (1996a). *Bioturbation als Faktor der Reliefgenese in den wechselfeuchten Tropen*. Leipzig: Klett-Perthes.
Löffler, H. (1996b). Neozoen in der Fischfauna Baden-Württembergs. ein Überblick. In H. Gebhardt, R. KInzelbach, & S. Schmidt-Fischer (Eds.), *Gebietsfremde Tierarten* (pp. 217–226). Landsberg: Ecomed.
Logan, P. C., & Harris, L. H. (1967). *Introduction and establishment of red deer in New Zealand* (New Zealand Forest Service, Information Series 55). Wellington: New Zealand Forest Service.
Lohmeyer, W., & Krause, A. (1974). Über den Gehölzbewuchs an kleinen Fließgewässern Nordwestdeutschlands und seine Bedeutung für den Uferschutz. *Natur und Landschaft, 49*, 323–330.
Long, J. (2003). *Introduced mammals of the world: Their history, distribution and influence*. Collingwood: CSIRO Publishing.
Loos, R. (1972). Raubwild in Gefahr. *Die Pirsch, 24*, 601–605.
Looschen, R. (1986). *Verbreitung und Auzswirkungen des Bisams (Ondatra zibethica L.) in Westfalen*. Master thesis, Institut für Geographie, Westfälische Wilhelms-Universität, Münster.
Lopez, B. H. (1978). *Of wolves and men*. New York: Charles Scribner's Sons.
Ludwig, N., Gebhardt, H., Ludwig, H. W., & Schmidt-Fischer, S. (2000). *Neue Tiere und Pflanzen in der heimischen Natur*. München: BLV.

Lüps, P. (2001). Steinbock Capra ibex, Luchs Lynx lynx und viele andere: Wunschdenken und Wirklichkeite bei Wiedereinbürgerungen in der Schweiz. *Beiträge zur Jagd- und Wildforschung, 26*, 15–22.

Lutz, W. (1981). *Untersuchungen zur Nahrungsbiologie des Waschbären (Procyon lutor Linné, 1758) und zum Einfuß auf andere Wildarten in seinem Lebensraum*. Dissertation, Universität Heidelberg, Heidelberg.

Lutz, W. (1996). Goldschakal, Marderhund: Waschbär: Nachtaktive Neubürger. *Rheinisch-Westfälischer Jäger, 8/96*, 36–38.

MacDonald, D. (1956). *Beaver carrying capacity of certain mountain streams in North Park, Colorado*. M.S. Thesis, Colorado State University, Fort Collins.

MacDonald, D. (Ed.). (2001). *The new encyclopedia of mammals*. Oxford: Oxford University Press.

MacDonald, I. A. W., Loope, L. L., Usher, M. B., & Hamann, O. (1989). Wildlife conservation and the invasion of natural reserves by introduced species. A global perspective. In J. A. Drake, H. A. Mooney, F. di Castri, R. H. Groves, F. J. Kruger, M. Rejmanek, & M. W. Williamson (Eds.), *Biological invasions: A global perspective* (pp. 215–255). Chichester: Wiley.

MAF. (1996). Ministry of Agriculture and Fisheries. Technical Paper 96 (4).

MAF. (1997). *MAF Policy information paper 17*. Christchurch: Ministry of Agriculture and Fisheries.

Maier, P. (1994). Der Biber aus wasserwirtschaftlicher Sicht. Probleme und Lösungsmöglichkeiten. *Schriftenreihe des Bayerischen Landesamtes für Umweltschutz, 128*, 51–56.

Malmierca, L., Menvielle, M. F., Ramadori, D., Saavedra, B., Saunders, A., Volkhart, N. S., & Schiavini, A. (2011). Eradication of beaver (Castor Canadensis), an ecosystem engineer and threat to southern Patagonia. In C. R. Veitch, M. N. Clout, & D. R. Towns (Eds.), *Island invasives: Eradication and management* (pp. 87–90). Gland: IUCN.

Mark, A. F. (1965). Flowering, seeding and seedling establishment of narrow-leaved snow tussock (Chionocloa rigida). *New Zealand Journal of Botany, 3*, 180–193.

Markov, G. (2012). Golden jackal (Canis aureus L.) in Bulgaria. What is going on? *Acta Zoologica Bulgarica, 64*(Suppl. 4), 67–71.

Mark, A. F., & Baylis, G. T. (1975). Impact of deer on Secretary Island, Fiordland, New Zealand. *Proceedings New Zealand Ecological Society, 22*, 19–24.

Martinka, C. J. (1982). Rationale and options for management in grizzly bear sanctuaries. *Transactions of the North American Wildlife and Natural Resources Conference, 47*, 470–475.

Mattson, D. J., Blanchard, B. M., & Knight, R. R. (1992). Yellowstone grizzly bear mortality, human habituation, and whitebark pine seed crops. *Journal of Wildlife Management, 56*, 432–442.

Mattson, D. J., Kendall, K. C., & Reinhart, D. P. (2001). Whitebark pine, grizzly bears, and red squirrels. In D. F. Tomback, S. F. Arno, & R. E. Keane (Eds.), *Whitebark pine communities. Ecology and restoration* (pp. 121–136). Washington, DC/Covelo/London: Island Press.

McClaran, M. P., & Cole, D. N. (1993). *Packstock in wilderness: Use, impacts, monitoring, and management* (Technical Report INT-301). Ogden: USDA Forest Service, Intermountain Research Station.

McKnight, T. L. (1971). Australia's buffalo dilemma. *Annals of the Association of American Geographers, 61*, 754–773.

Meads, M. J. (1976). Effects of possum browsing on northern rata trees in New Zealand. *New Zealand Journal of Zoology, 3*, 127–139.

Mech, L. D. (1970). *The wolf: The ecology and behavior of an endangered species*. New York: Natural History Press (Doubleday Publishing Company).

Meentemeyer, R. K., & Butler, D. R. (1995). Hydrogeomorphic effects of beaver dams in Glacier National Park, Montana. *Physical Geography, 20*(5), 436–446.

Meentemeyer, R. K., Vogler, J. B., Hill, C., & Butler, D. R. (1998). The geomorphic influences of burrowing beavers on streambanks, Bolin Creek, North Carolina. *Zeitschrift für Geomorphologie, Neue Folge, 42*(4), 453–468.

Meriggi, A., & Lovar, S. (1996). A review of wolf predation in southern Europe: Does wolf prefer wild prey to livestock? *Journal of Applied Ecology, 33*, 1561–1571.
Meriggi, A., Brangi, A., Mateucci, C., & Sacchi, O. (1996). The feeding habit of wolves in relation to large prey availability in northern Italy. *Ecography, 19*, 287–295.
Meyer, T. (1996). Die Rückkehr der Raubtiere. *Natur, 7/96*, 44–45.
Michler, F.-U., & Köhnemann, B. A. (2009). Maskierte Langfinger auf dem Vormarsch – Waschbären in Mecklenburg-Vorpommern. Aktueller Wissensstand über potentielle Auswirkungen der Waschbärenbesiedlung und Hinweise zur Bejagung. In M. Stubbe & V. Böhning (Eds.), *Neubürger und Heimkehrer in der Wildtiierfauna* (pp. 61–61). Halle/S: Damm/S.
Michler, F.-U. F., & Michler, B. A. (2012). Ökologische, ökonomische und epidemiologische Bedeutung des Waschbären (Procyon lutor) in Deutschland. *Beiträge zur Jagd- und Wildforschung, 37*, 387–395.
Mills, J. A., & Mark, A. F. (1977). Food preference of Takahé in Fiordland National Park, New Zealand, and the effect of competition from introduced red deer. *Journal of Animal Ecology, 467*, 939–958.
Milne, L. J., & Milne, M. (1960). *The balance of nature*. New York: Alfred A. Knopf.
Ministerial Panel. (2008, March). *Managing numbers of deer, chamois, tahr and wild pigs*. (Report of the Ministerial Panel). Wellington.
Mitchell, B. D., & Banks, P. B. (2005). Do wild dogs exclude foxes? Evidence for competition from dietary and spatial overlaps. *Austral Ecology, 30*, 581–591.
Möckel, R. (2000). Der Marderhund in der Niederlausitz. *Naturschutz und Landschaftspflege in Brandenburg, 9*(1), 19–32.
Mooney, H. A., & Drake, J. A. (Eds.). (1986). *Ecology of biological invasions of North America and Hawaii* (Ecological Studies 58). New York: Springer.
Morgan, D., & Hickling, G. J. (2000). Techniques used for poisoning possum. In T. L. Montague (Ed.), *The brushtailed possum: Biology, impact and management of an introduced marsupial* (pp. 143–153). Lincoln: Manaaki Whenua Press.
Mosley, M. P. (1978). Erosion in the south-eastern Ruahine Range: Its implication for downstream river control. *New Zealand Journal of Forestry, 23*, 21–48.
Moyle, P. B. (1980). Fish introduction into North America: Patterns and ecological impact. In H. A. Mooeney & J. A. Drake (Eds.), *Ecology of biological invasions of north America* (pp. 27–43). New York: Springer.
Moyle, P. B., & Li, H. W. (1994). Good report but should go much further. *Fisheries, 19*, 22–23.
Mudge, A. D., LaBonte, J. R., Johnson, K. J. R., & LaGasa, E. H. (2001). Exotic woodboring Coleoptera (Micromalthidae, Scolytidae) and Hymenoptera (Xiphydriidae) new to Oregon and Washington. *Proceedings of the Entomological Society of Washington, 103*(4), 1011–1019.
Müller-Using, D. (1965). Das Vorkommen der Nutria in Deutschland. *Zeitschrift für Jagdwissenschaft, 11*(3), 161–165.
Musselman, L. K. (1969). *Rocky Mountain National Park, 1915–1965: An administrative history*. Dissertation, University of Denver.
Naiman, R. J., Melillo, J. M., & Hobbie, J. E. (1986). Ecostem alteration of a boreal forest stream by beaver (Castor canadensis). *Ecology, 67*, 1254–1269.
National Parks Conservation Association. (2009). Point Reyes National Seashore. A resource assessment. Washington, DC: National Parks Conservation Association.
Neff, C. (2001). Der rezente Landschaftswandel im westlichen Mediterranraum – Herausforderungen für Natur- und Landschaftsschutz. *Petermanns Geographische Mitteilungen, 145*, 72–83.
Ngidlo, R. T. (2013). Assessment of climate impacts, vulnerabilities and adaptation strategies in the traditional rice terraces of the Cordillera Region, northern Philippines. *Asian Journal of Science and Technology, 4*(11), 180–186.
Niethammer, G. (1963). *Die Einbürgerung von Säugetieren und Vögeln in Europa*. Hamburg/Berlin: Parey.
Nievergelt, B. (1966). *Der Alpensteinbock (Capara ibex L) in seinem Lebensraum. Mammalia depicta*. Hamburg/Berlin: Verlag Paul Parey.

Nitsche, K.-A. (1994). Biber-Ausrottung, Schutz und Wiederansiedlung in Deutschland. *Säugetierkundliche Mitteilungen, 34*(2), 83–178.

Nitsche, K. A. (1995a). Biber zwischen den Fronten. *Naturschutz heute, 2*, 338–339.

Nitsche, K.-A. (1995b). Warum keine neuen Wiederansiedlungen von Bibern? *Säugetierkundliche Mitteilungen, 36*(1), 43–48.

Nolet, B. A., Hoekstra, A., & Ottenheim, M. M. (1994). Selective foraging on woody species by the beaver Castor fiber and its impact on a riparian willow forest. *Biological Conservation, 70*, 117–128.

Nordström, M., Högmander, J., Nummelin, J., Laine, J., Laanettu, N., & Koppimäki, E. (2002). Variable response of waterfowl breeding populations to long-term removal of introduced American Mink. *Ecography, 25*, 385–394.

Novillo, A., & Ojeda, R. A. (2008). The exotic mammals of Argentina. *Biological Invasions*. doi:10.1007/s10530-007-9208-8.

Nowak, E. (1977). *Die Ausbreitung der Tiere. Die Neue Brehm-Bücherei*. Wittenberg Lutherstadt: Ziemsen.

Nowak, E. (1981). Geschichtliches und Rezentes über die aktive Rolle des Menschen bei der Bereicherung bzw. Erhaltung der Fauna durch Tieraussetzungen. *Akademie für Naturschutz und Landschaftspflege, Tagungsbericht, 12/81*, 19–38.

Nowak, E., & Zsivanovits, A. (1982). *Wiereinbürgerung gefährdeter Tierarten: Wissenschaftliche Grundlagen, Erfahrungen und Bewertungen* (Schriftenreihe für Landschaftspflege und Naturschutz 23). Bonn/Bad Godesberg: Bundesforschungsanstalt für Naturschutz und Landschafsökologe.

Nugent, G., Fraser, K. W., Asher, G. W., & Tustin, K. G. (2001). Advances in New Zealand mammalogy 1990–2000: Deer. *Journal of the Royal Society of New Zealand, 31*(1), 263–298.

Nummi, P. (1996). Wildlife introduction to mammal-deficient areas: The Nordic Countries. *Wildlife Biology, 2*, 221–226.

Nummi, P. (2001). Alien species in Finland. *The Finnish Environment, 466*, 36–37, Ministry of the Environment, Helsinki, Finland.

O'Neill, A. (2002). *Living with the dingo*. Annandale: Envirobook.

Obretenov, A., Russev, J., & Markoff, I. (2001). Ermittlung eines tragbaren Raubwildbestands in Bulgarien. *Beiträge zur Jagd- und Wildforschung, 26*, 77–83.

Odening, K. (1984). *Antarktische Tierwelt. Einführung in die Biologie der Antarktis*. Leipzig/Jena/Berlin: Urania Verlag.

Oksanen, L., Moen, J., & Helle, T. (1995). Timberline patterns in northernmost Fennoscandia. Relative importance of climate and grazing. *Acta Botanica Fennica, 153*, 93–107.

Packard, F. M. (1947). A survey of the beaver population of the Rocky Mountain National Park, Colorado. *Journal of Mammalogy, 28*(3), 219–227.

Parker, H. (2005). *The state of the Eurasian beaver Castor fiber population in northern Norway in 2005* (6 pp). Department of Environmental and Health Studies, Telemark University Colleg, Boe i Telemark, Norway.

Parker, M., Wood, F. J., Smith, B. H., & Elder, R. G. (1985, April 16–18). *Erosional downcutting in lower order riparian ecosystems: Have historical changes been caused by the removal of beavers?* Paper presented at the First North American Riparian Conference, The University of Arizona, Tucson. Duplicated Manuscript.

Parker, H., Nummi, P., Hartman, G., & Rosell, F. (2012). Invasive North American beavers Castor canadensis in Eurasia: A review of potential consequences and a strategy for eradication. *Wildlife Biology, 18*(4), 354–365.

Pasteur, E. C., & Walton, D. W. H. (2006). *South Georgia: Plan for progress. Managing the environment 2006–2010*. Government of South Georgia and the South Sandwich Islands. Cambridge: British Antarctic Survey.

Pattymore, D. E., & Wilcove, D. S. (2011). Invasive rats and recent colonist birds partially compensate for the loss of endemic New Zealand pollinators. *Proceedings of the Royal Society B*. doi:10.1098/rspb.2011.2036.

Payton, I. J. (1987, September 5–13). Canopy dieback in the rata (Metrosideros)-kamahi (Weinmannia) forests of Westland, New Zealand. In T. Fujimori & M. Kimura (Eds.), *Human impacts and management of mountain forests, Proceedings of the International Workshop, Ibaraki* (pp. 123–136). Ibaraki: Forestry and Forest Products Research Institute.

Pech, R. P., & Hood, G. M. (1998). Foxes, rabbits, alternative prey and rabbit calicivirus disease: Consequences of a new biological control agent for an outbreaking species in Australia. *Journal of Applied Ecology, 35*, 434–453.

Pekelharing, C. J. (1979). Fluctuation in opossum populations along the north bank of the Taramaku catchment and its effect on the forest canopy. *New Zealand Journal of Forestry Science, 9*(2), 212–224.

Pekelharing, C. J., & Reynolds, R. N. (1983). Distribution and abundance of browsing mammals in Westland National Park in 1978, and some observation on their impact on the vegetation. *New Zealand Journal of Forestry Science, 13*(3), 247–265.

Pelz, H.-J. (1996). Zur Geschichte der Bisambekäpfung in Deutschland. *Mitteilungen der Biologischen Bundesanstalt für Land- und Forstwirtschaft, Berlin Dahlem, 317*, 219–234.

Pelz, H.-J., Klemann, N., & Giesemann, R. (1997). Zur Entwicklung der Nutriabestände in Westfalen. *Abhandlungen aus dem Westfälischen Museum für Naturkunde, 59*(3), 97–105.

Petty, A. M., Werner, P. A., Lehmann, C. E. R., Riley, J. E., Banfal, D. S., & Elliot, L. P. (2007). Savannah responses to feral buffalo in Kakadu National Park, Australia. *Ecological Monographs, 77*(3), 441–463.

Podglayen-Neuwall, I. (1982). Einbürgerung exotischer Huftiere in Neu Mexico. 1950–1979. *Zoologischer Garten, N. F., 52*, 195–232.

Pracy, L. T. (1962). *Introduction and liberation of the Opossum (Trichosurus vulpecula Kerr) into New Zealand* (New Zealand Forest Service Information Series 45, pp. 3–15). Wellington: New Zealand Forest Service.

Pratt, R. M., & Smith, R. I. L. (1982). Seasonal trends in the chemical composition of reindeer forage plants on South Georgia. *Polar Biology, 1*, 13–32.

Price, A. G. (1972). *Island continent, aspects of the historical geography of Australia and its territories*. Sydney/London/Melbourne/Brisbane/Singapore: Angus and Robertson.

Pulliainen, E. (1965). Studies on the wolf (Canis lupus) in Finland. *Annales Zoologici Fennici, 2*(4), 214–259.

Rahm, U. (1997). Der aktuellle Stand der Biber in der Schweiz. In Ministerium für Umwelt, Energie und Verkehr, Saarbrücken (Ed.), *Internationales Fachsymposium zur Wiederansiedlung des Bibers im Saarland*, August 1994, Saarbrücken (pp. 29–33).

Rahm, U., & Baettig, M. (1996). Der Biber in der Schweiz. Bestand, Gefährdung, Schutz. BUWAL, BWV, Schriftenreihe Umwelt 249. Bern: Bundesamt für Umwelt, Wald und Landschaft (BUWAL).

Ramsey, D. (2005). Population dynamics of brushtail possum subject to fertility control. *Journal of Applied Ecology, 42*, 348–360.

Ratcliffe, F. N. (1959). The rabbit in Australia. In A. Keast, R. L. Crocker, & C. S. Christian (Eds.), *Biogeography and ecology in Australia* (pp. 545–564). The Hague: Dr. W. Junk Publishers.

Ratcliffe, P. R. (1989). The control of red and sika populations in commercial forests. In R. J. Putnam (Ed.), *Mammals as pest* (pp. 98–155). London/New York: Chapman and Hall.

Ratti, P. (1981). Zur Hege des Steinwildes in Graubünden. *Zeitschrift für Jadgwissenschaft, 27*(1), 41–57.

Reichholf, J. (1976). Zur Wiederansiedlung des Bibers (Castor fiber L.), biologische Grundlagen und Erfahrungen von den Innstauseen. *Natur und Landschaft, 51*(2), 41–44.

Reichholf, J. (1977). Zur Ein- und Wiedereinbürgerung von pflanzenfressenden Säugetieren. *Zeitschrift fur Saugetierkunde, 42*, 189–196.

Reichholf, J. (1993). *Das comeback der Biber*. München: Beck.

Reichholf, J. (1997). Der Biber - eine Schlüsselart der Fließgewässer. In Ministerium für Umwelt, Energie und Verkehr, Saarbrücken (Ed.), *Biber in der Kulturlandschaft - eine Illusion? Internationales Fachsymposium zur Wiederansiedlung des Bibers im Saarland*, August 1994, Saarbrücken (pp. 9–14).

Reichholf-Riehm, H. (1981). Noch eine Chance für den Biber. *Nationalpark, 31*(2), 6–9.

Reidt, L. (1995). Auswilderung, Heimkehr in ein Feindesland. *Natur, 5*, 38–42.

Remmert, H. (1999). Das Leitbild für Schutz, Pflege und Gestaltung der Landschaft Mitteleuropas - der Blickwinkel eines Ökologen (Überarbeitetes Manskript eines Vortrages auf dem Internationalen Symposium 1993 in Neuenhaus, Solling). In B. Gerken & M. Görner (Eds.), *Europäische Landschaftsentwicklung mit großen Weidetieren. Geschichte, Modelle und Perspektiven* (Natur- und Kulturlandschaft 3, pp. 115–117). Höxter/Jena: Huxaria Druck und Verlag.

Retzer, J. L., Swope, H. M., Remington, J. D., & Rutherford, W. (1956). *Suitability of physical factors for beaver management in the Rocky Mountains of Colorado* (Technical Bulletin 2). Denver: State of Colorado, Department of Game and Fish.

Richard, B. (1981). *Les castors. Collection Faune et Flore*. Paris: Editions Balland.

Ricketts, T. H. (2001). The matrix matters: Effective isolation in fragmented landscapes. *The American Naturalist, 158*(1), 87–99.

Riney, T. (1964). *The impact of introductions of large herbivores on tropical environment* (New Series 4, pp. 261–273). Gland: IUCN Publications.

RIRDC (Rural Industries Research & Development Corporation). (2000). Development of a sustainable camel industry. RIRDC Publication 99/118. Barton/Canberra.

Röben, P. (1976). Veränderungen des Säugetierbestandes der Bundesrepublik Deutschland und deren Ursachen. *Schriftenreihe für Vegetationskunde, 10*, 239–254.

Rodda, G. H., & Fritts, T. H. (1992). The impact of the introduction of the colubrid snake Boiga irregularis on Guam's lizards. *Journal of Herpetology, 26*, 166–174.

Rogers, G. M., & Leathwick, J. R. (1997). Factors predisposing forests to canopy collapse in the southern Ruahine Range, New Zealand. *Biological Conservation, 80*, 325–338.

Roots, C. (1976). *Animal invaders*. Newton Abbot/London/Vancouver: Universe Books.

Rose, A. B., & Platt, K. H. (1987). Recovery of northern Fjordland alpine grassland after reduction in the deer population. *New Zealand Journal of Ecology, 10*, 23–34.

Rose, A. B., Pekelharing, C. J., & Platt, K. H. (1992). Magnitude of canopy dieback and implications for conservation of southern rata-kamahi (Metrosideros umbellata-Weinmannia racemosa) forests, central Westland, New Zealand. *New Zealand Journal of Ecology, 16*, 23–32

Rose, A. B., Pekelharing, G. C. J., Platt, K. H., & Woolmore, C. B. (1993). Impact of invading brushtail possum population on mixed beech-hardwood forests, South Westland, New Zealand. *New Zealand Journal of Ecology, 17*, 19–28.

Rosell, F., & Parker, H. (2011). A population history of the beaver Castor fiber in Norway. In G. Sjöberg & J. P. Ball (Eds.), *Restoring the European Beaver: 50 years of experience* (pp. 19–25). Sofia/Bulgaria: Pensoft Publishers.

Ross, K. C., & Arthur-Worsop, M. J. (1987). *A preliminary economic evaluation of rabbit control* (Resource Use Paper 3/87, MAF corp). Christchurch: Ministry of Agriculture and Fisheries.

Rouland, P. (1997). *Le castor dans le sud-est de la France*. Paris: Office National de la Chasse et de la Faune Sauvage.

Rouland, P., & Migot, P. (1990). La reintroduction du castor (Castor fiber) en France. Essai de synthèse et réflexion. *Revue d'Écologie: la Terre et la Vie, 5*(Suppl.), 145–158.

Rushton, S. P., Lurz, P. W. W., Gurnell, J., & Fuller, R. (2000). Modelling the spatial dynamics of parapox-virus disease in red and grey squirrels: A possible cause of the decline in the red squirrels in the UK? *Journal of Applied Ecology, 37*, 997–1012.

Rutherford, W. H. (1964). *The beaver in Colorado, its biology, ecology, management and economics*. (Colorado Game, Fish and Parks Department, Technical Publication 17, pp. 1–49). Denver.

Ruys, T., Kneveler, M., & Bernez, I. (2008, September 8–12). *Importance of riparian survey for river management and restoration in relation with the presence of three aquatic rodents species: Muskrat, Coypu and European Beaver*. 6th European Conference on Ecological Restoration, Ghent, pp. 1–4.

Sächsische Zeitung. (2004, July 28). *Lausitzer Wölfe brauchen frisches Blut*. www.smul.de

Salwasser, H., Schonewald-Cox, C., & Baker, R. (1987). The role of interagency cooperation in managing for viable populations. In M. E. Soulé (Ed.), *Viable populations for conservation* (pp. 159–173). Cambridge: Cambridge University Press.

Saunders, A. (2000). *A review of Department of Conservation mainland restoration in projects and recommendations for further action*. Wellington: Department of Conservation.

Saveljev, A. P., & Safonov, V. G. (1999). The beaver in Russia and adjoining countries. In P. E. Busher & R. Dzieciolowski (Eds.), *Beaver protection, management, and utilization* (pp. 17–24). New York: Kluwer Academic/Plenum Publishers.

Savidge, J. A. (1987). Extinction of an island avifauna by an introduced snake. *Ecology, 63*(3), 660–668.

Scanlon, P. F. (1989). Introduced species in the United States as damage causing species. *Transactions Congress International Union of Game Biology, 19*, 543–548.

Schaller, G. B. (1973). Observations on Himalayan Tahr (Hemitragus jemlaicus). *Journal of the Bombay Natural History Society, 70*(1), 1–24.

Scheffer, V. B. (1951). The rise and fall of a reindeer herd. *Scientific Monthly, 73*, 356–362.

Scherzinger, W. (1981). Chancen der Wiedereinbürgerung von Waldhühnern in Deutschland. *Natur und Lanschaft, 56*(4), 131–132.

Schlamp, H.-J. (2001). Wölfe vom Himmel. *Der Spiegel, 34*, 131.

Schmidt, E. (2001). Bisam und Nutria. Neubürger an urbanen Gewässern. *Unterricht Biologie, 25*(269), 53–56.

Schneider, E., & Rieder, N. (1981). Wiederansiedlung des Bibers in der Bundesrepublik Deutschland. *Natur und Landschaft, 56*(4), 118–120.

Schneider, E., & Schulte, R. (1985). Befunde zu den Habitatansprüchen des Europäischen Bibers (Castor fiber L.) aus einem Wiederansiedlungsversuch an einem Mittelgebirgsbach der nördlichen Eifel. *Zeitschrift für Angewandte Zoologie, 72*(1/2), 167–179.

Schnitter, M. (1965). *Der Steinbock* (Schriften aus dem Zoologischen Museums der Universität Zürich 3). Zoologisches Museum Zürich.

Schulte, R. (1998). Perspektiven und Zukunftsaufgaben des Biberschutzes in Deutschland. *Naturschutz und Landschaftspflege, 30*(10), 331–332.

Schwab, G. (2003). *Biberverbreitung und Bestand in Deutschland*. Manuskript zur Bibertagung der ANN, Schneverdingen, März 2003.

Schwab, G., Dietzen, W., & Von Lossow, G. (1994). Biber in Bayern. *Schriftenreihe Bayerisches Landesamt für Umweltschutz, 128*, 9–44.

Schweinfurth, U. (1966). *Neuseeland, Beobachtungen und Studien zur Pflanzengeographie und Ökologie der antipodischen Inselgruppe* (Bonner Geographische Abhandlungen 36). Bonn: Ferdinand Dümmlers Verlag.

Scocroft, P. G., & Sakai, J. G. (1983). Feral herbivores suppress the regeneration of mamane and other browse species on Mauna Kea, Hawaii. *Journal of Range Management, 36*, 638–645.

Scowcroft, P., & Giffin, J. G. (1983). Feral herbivores suppress Mamane and other browse species on Mauna Kea, Hawaii. *Journal of Range Management, 36*(5), 638–645.

Sedlag, U. (1995). *Tiergeographie*. Leipzig/Jena/Berlin: Urania Verlag.

Semmens, T. D., Turner, E., & Buttermore, R. (1993). Bombus terrestris (L.) (Hymenoptera, Apidae) now established in Tasmania. *Journal of Australian Entomological Society, 32*, 346.

Seton, E. (1929). *Lives of game animals*. Garden City: Doubleday.

Short, J. (1985). The functional response of kangaroos, sheep, and rabbits in an arid grazing system. *Journal of Applied Ecology, 22*, 435–447.

Silanikove, N. (2000). The physiological adaptation in goats to harsh environments. *Small Ruminant Research, 35*, 181–193.

Singleton, P. H., Gaines, W. L., & Lehmkuhl, J. F. (2002). *Landscape permeability for large carnivores: A geographic information system weighted-distance and least-cost corridor assessment* (Research Paper PNW-RP 549). USDA Forest Service, Pacific Northwest Research Station, Portland, Oregon.

Skirnisson, K. (1992). Zur Biologie der isländischen Minkpopulation. In R. Schröper, M. Stubbe, & D. Heidecke (Eds.), *Semiaquatische Säugetiere*. Wissenschaftsliche Beiträge Universität Halle 1991, pp. 277–295. Halle-Wittenberg: Martin-Luther-Universität.

Smith, R. I. L., & Walton, D. W. H. (1975). South Georgia, Subantarctic. In T. Rosswall & O. Heal (Eds.), *Structure and function of tundra ecosystems* (Ecological bulletin 20, pp. 399–423). Stockholm: Swedish Natural Science Research Council.

Sorger, H.-P. (1995). *Der Bär ist wieder da! Konflikte und Chancen*. Graz/Stuttgart: Stocker.

Stahl, P., Vandel, J. M., Herrenschmidt, V., & Migot, P. (2001). Predation on livestock by an expanding reintroduced lynx population: Long-term and spatial variability. *Journal of Applied Ecology, 38*, 647–687.

Stahl, P., Vandel, J. M., Ruette, S., Coat, L., Coat, Y., & Balestra, L. (2002). Factors affecting lynx predation on sheep in the French Jura. *Journal of Applied Ecology, 39*, 204–216.

Steiner, E., & Kraus, E. (1993). Des einen Freund, des anderen Leid: Elche in Österreich. *Nationalpark, 3/93*, 19–22.

Stewart, G. H. (1992). *Forest structure and regeneration in conifer-broadleaved hardwood forests, Westland. Monitoring rata-kamahi forest canopy condition and stand dynamics, Copland Valley* (Contract Report FEW 92/31). Christchurch: Forest Research Institute.

Stewart, G. H., & Veblen, T. T. (1982a). Regeneration patterns in southern rata (Metrosideros umbellata)-kamahi (Weinmannia racemosa) forest in central Westland, New Zealand. *New Zealand Journal of Botany, 20*, 55–72.

Stewart, G. H., & Veblen, T. T. (1982b). A commentary on canopy tree mortality in Westland rata-kamahi protection forests. *New Zealand Journal of Forestry, 27*, 168–188.

Stocker, G. (1985). *Biber (Castor fiber L.) in der Schweiz*. Eigenössische Anstalt für das forstliche Versuchswesen, Berichte 274. Teufen: Kommissionsverlag F. Flück-Wirth.

Stubbe, M. (1975). Der Waschbär Procyon lutor (L., 1758) in der DDR. *Hercynia, Neue Folge, 12*, 80–91.

Sutherst, R. W., Floyd, R. B., & Maywald, G. F. (1995). The potential geographical distribution of the cane toad, Bufo marinus L., in Australia. *Conservation Biology, 9*(6), 294–299.

Sutor, A., Kauhala, K., & Ansorge, H. (2010). Diet of the raccoon dog Nyctereutes procyonoides: A canid with an opportunistic foraging strategy. *Acta Theriologica, 55*, 165–176.

Swedish Wildlife Damage Centre (SWD). (2007). *Viltskadestatistik 2006*, Swedish University of Agricultural Sciences, Sweden. http://www.viltskadecentre.com

Tanentzap, A. J., Burrows, L. E., Lee, W. G., Nugent, G., Maxwell, J. M., & Coomes, D. A. (2009). Landscape-level vegetation recovery from herbivory: progress after four decades of invasive red deer control. *Journal of Applied Ecology, 46*, 1064–1072.

Ten Houte de Lange, S. M. (1978). Zur Futterwahl des Alpensteinbocks (Capra ibex L.). Eine Untersuchung an der Steinbockkolonie am Piz Albris bei Pontresina. *Zeitschrift für Jagdwissenschaft, 24*, 113–138.

Theberge, J. B. (1983). Considerations in wolf management related to genetic variability and adaptive range. In L. N. Carbyn (Ed.), *Wolves in Canada and Alaska. Proceedings of the Wolf Symposium*, Edmonton, 12–14 May 1981 (Canadian Wildlife Service, Report Series 45, pp. 86–99).

Tisdell, C. A. (1982). *Wild pigs: Environmental pest or economic resource?* Sydney/Oxford/New York/Toronto/Paris/Frankfurt: Pergamon Press.

Tiver, F., & Andrew, M. H. (1997). Relative effects of herbivory by sheep, rabbits, goats and kangaroos on recruitment of shrubs and trees in eastern South Australia. *Journal of Applied Ecology, 34*, 903–914.

Tschirch, W. (2001). Die Bedeutung von Luchs, Wildkatze, Waschbär und Marderhund in der Tollwutepidemiologie. *Beträge zur Jagd- und Wildforschung, 26*, 281–298.

Türcke, F. (1981). Einbürgerung des Mufflon, ein Beispiel für Zielgemeinschaft der Jäger und Naturschützer. *Natur und Landschaft, 56*(4), 137.

Türcke, F., & Schmincke, S. (1965). *Das Muffelwild. Naturgeschichte, Hege, Jagd*. Hamburg/Berlin: Parey.

Turner, H. (1961). Standortsuntersuchungen in der subalpinen Stufe: Die Niederschlags- und Schneeverhältnisse. *Mitteilungen der forstlichen Bundesversuchsanstalt Mariabrunn, 59*, 265–315.

Tustin, K. G. I. (1990). Himalayan tahr. In C. M. King (Ed.), *The handbook of New Zealand mammals (paperback edition 1995)* (pp. 392–406). Auckland/Melbourne/Oxford/New York: Oxford University Press.

Twigg, L. E., Lowe, T. J., Martin, G. R., Wheeler, A. G., Gray, G. S., Griffin, S. L., O'Reilly, C. M., Robinson, D. J., & Hubach, P. H. (2000). Effects of surgically imposed sterility on free-ranging rabbit population. *Journal of Applied Ecology, 37*, 16–39.

Tyler, M. (1996). Cane toad control. *Australian Biologist, 9*(4), 120–121.
Ueckermann, E. (1992). *Das Sikawild. Vorkommen, Naturgeschichte und Bejagung*. Hamburg/ Berlin: Parey.
Vásquez, D. P. (2002). Multiple effects of introduced mammalian herbivores in a temperate forest. *Biological Invasions, 4*, 175–191.
Veblen, T. T., & Stewart, G. H. (1982). The effects of introduced wild animals on New Zealand forests. *Annals of the Association of American Geographers, 72*, 372–397.
Vogel, L. M., Remmert, H., & Smith, R. I. L. (1984). Introduced reindeer and their effects on the vegetation and the epigeic invertebrate fauna of South Georgia (subantarctic). *Oecologia, 62*, 102–109.
Walton, D. W. H., & Smith, R. I. L. (1979). The chemical composition of South Georgian vegetation. *British Antarctic Survey Bulletins, 49*, 117–135.
Warburton, B. (1986). *Wallabies in New Zealand: History, current status, research and management needs* (Forest Research Bulletin No. 114). Christchurch: Protection Forestry Division, Forest Research Institute.
Warburton, B., & Sadleir, R. M. F. (1990). Parma walleby. In C. M. King (Ed.), *The handbook of New Zealand mammals* (pp. 51–57). Melbourne/Oxford/New York: Oxford University Press.
Warburton, B., Cowan, P., & Shepherd, J. (2009). *How many possums are now in New Zealand following control and how many would there be without it?* (Landcare Research Contract Report: LC0910/060). Lincoln: Landcare Research, Manaaki Whenua.
Wardle, P. (1971). Biological flora of New Zealand. 6 Metrosideros umbellata Cav. [syn. M. lucida (Forst.f.) A. Rich.] (Myrtaceae) southern rata. *New Zealand Journal of Botany, 9*, 645–671.
Wardle, J. (1974). Influence of introduced mammals on the forests and shrublands of the Grey River headwaters. *New Zealand Journal of Forestry Science, 4*(3), 459–486.
Wardle, J., Hayward, J., & Herbert, J. (1971). Influence of ungulates on the forests and scrublands of South Westland. *New Zealand Journal of Forestry Science, 3*(1), 3–36.
Waterhouse, D. F. (1974). The biological control of dung. *Scientific American, 230*, 100–109.
Wauters, L. A., & Gurnell, J. (1999). The mechanism of replacement of red squirrels by grey squirrels: A test of the interference competition hypothesis. *Ethology, 105*, 1053–1071.
Wechselberger, M., & Leizinger, D. (2005). *Endbericht. Die Akzeptanz von Bär, Wolf und Luchs in Österreich* (pp. 1–42). WWF Austria and Institut für Wildbiologie und Jagdwirtschaft (IWJ). Wien.
Weinzierl, H. (1973). *Projekt Biber. Wiedereinbürgerung von Tieren* (Kosmos Bibliothek 279). Stuttgart: Franckh.
Weller, M. W., & Fredrickson, L. H. (1973). Avian ecology of a managed glacial marsh. *Living Bird, 12*, 269–291.
Weller, M. W., & Spatcher, C. E. (1965). *Role of habitat in the distribution and abundance of marsh birds* (Special Report Iowa Agricultural and Home Economics Experiment Station 43). Ames: State University Agricultural and Home Economics Experiment Station.
Wiegand, T., Knauer, F., Kaczensky, P., Naves, J., Strümpeler, M., & Wissel, C. (1998). *Einwanderungen von Braunbären in die Ostalpen* (Jahresbericht 1996–1997, pp. 68–77). Leipzig/ Halle/ Magdeburg: Umweltforschungszentrum Leipzig-Halle.
Wiens, J. A. (2001). The landscape context of dispersal. In J. Clobert, E. Danchin, A. A. Dhondt, & J. D. Nichols (Eds.), *Dispersal* (pp. 98–122). New York: Oxford University Press.
Wilde, S. A., Youngberg, C. T., & Hovind, J. H. (1950). Changes in composition of ground water, soil fertility and forest growth produced by the construction and removal of beaver dams. *Journal of Wildlife Management, 14*(2), 123–128.
Wiles, G. J. (1987). The status of fruit bats on Guam. *Pacific Science, 41*(1–4), 148–157.
Williams, J. M. (1983). The impact of biological research on rabbit control politics in New Zealand. *Acta Zoologica Fennica, 174*, 79–83.
Williams, P. A., Cooper, P., Nes, P., & O'Connor, K. F. (1976). Chemical composition of tall tussocks in relation to the diet of the takahe (Notornis mantelli Owen), on the Murchison mountains, Fiordland, New Zealand. *New Zealand Journal of Botany, 14*, 55–61.

Williams, K., Parer, I., Coman, B., Burley, J., & Braysher, M. (1995). *Managing vertebrate pests: Rabbits*. Canberra: Australian Government Publishing Service.

Willwock, W. (1993). Die Titicaca-See Region auf dem Altiplano von Peru und Bolivien und die Folgen eingeführter Fische für Wildarten und ihren Lebensraum. *Naturwissenschaften, 80*, 1–8.

Wodzicki, K. (1950). *Introduced mammals in New Zealand* (New Zealand DSIR Bulletin 98). Wellington: Department of Scientific and industrial Research.

Wodzicki, K. (1963). Introduced mammals in New Zealand forests. *Acta Biologica Cracoviensia Series Zoologica, 6*, 11–134.

Woinarski, J. (2001). Background paper 1: A review of changes in status and threatening processes. In P. Whitehead, J. Woinarski, A. Fischer, R. Fensham, & K. Beggs (Eds.), *Developing an analytical framework for monitoring biodiversity in Australia's rangelands. A report by the Tropical Savannas CRC for the National Land and Water Resources Audit* (pp. 71–310). Darwin: Tropical Savannas CRC.

Wood, T. G. (1988). Termites and the soil environment. *Biology and Fertility of Soil, 6*, 228–236. Springer.

Woodroffe, R., & Ginsberg, J. R. (1998). Edge effects and the extinction of populations inside protected areas. *Science, 280*, 2126–2128.

Wotschikowsky, U. (1981a). Jagdwesen und Naturschutz. Konflikte und Gemeinsamkeiten bei Tieraussetzungen. *Akademie für Naturschutz und Landschaftspflege, Tagungsbericht, 12*, 29–31.

Wotschikowsky, U. (1981b). Der Luchs - aus der Theorie und Praxis der Wiedereinbürgerung. *Natur und Landschaft, 65*(4), 122–124.

Wotschikowsky, U. (1990). Wiedereinbrügerung des Luchses. *Allgemeine Forstzeitung, 12/13*, 313–314.

Wotschikowsky, U. (2001). Luchs zurück in Deutschland. *Beiträge zur Jagd- und Wildforschung, 26*, 173–178.

Wright, R. G. (1992). *Wildlife research and management in the national parks*. Urbana/Chicago: University of Illinois.

Wright, J. P., Jones, C. G., & Flecker, A. S. (2002). An ecosystem engineer, the beaver, increases species richness at the landscape scale. *Oecologia, 132*, 96–101.

Wurst, D., & Saalfeld, W. K. (1994). *Aerial survey of feral camels and other large vertebrate species in the Alice Springs region of the Northern Territory. A report to the Australian Nature Conservation Agency.* Alice Springs: Conservation Commission of the Northern Territories.

Yamada, F., & Sugimura, K. (2004). Negative impact of an invasive small Indian mongoose Herpestes javanicus on native wildlife species and evaluation of a control project in Amami-Oshima and Okinanwa Islands, Japan. *Global Environmental Research, 8*(2), 117–124.

Yeager, L. E., & Rutherford, W. H. (1957). *An ecological basis for beaver management in the Rocky Mountain region*. Transactions of the 22nd North American Wildlife Conference. Washington, DC: Wildlife Management Institute.

Zahner, V. (1997). *Einfluß des Bibers auf gewässernahe Wälder*. München: Utz Verlag

Zedrosser, A., Gersti, N., & Rauer, G. (1999). *Brown bears in Austria. 10 years of conservation and actions for the future* (Monographien M-117). Wien: Umweltbundesamt Österreich, WWF Österreich.

Zimen, E. (1997). *Der Wolf, Verhalten, Ökologie und Mythos*. München: Knesebeck.

Zschille, J., Heidecke, D., & Stumm, M. (2004). Verbreitung und Ökologie des Minks – Mustela vison SCHREBER, 1777 (Carnivora, Mustelidae) – In Sachsen-Anhalt. *Hercynia, Neue Folge, 37*, 103–126.

Zuber, M., Blöchlinger, B., & Lüps, P. (2001). Bewirtschaftung des Steinwildes Capra i. ibex im Berner Oberland (Schweiz): Erfahrungen aus den ersten 20 Jahren (1980–1999). *Beiträge zur Jagd- und Wildforschung, 26*, 33–42.

Zuppke, H. (1995). Der Einfluß des Elbebibers auf Waldbestände und forstwirtschaftliche Konsequenzen. *Hercynia, Neue Folge, 29*, 349–380.

Chapter 6
Animals in Landscape Management and Nature Conservation

6.1 Objectives and Problems

The impacts of oversized native game populations (e.g., cervids, wild boars) as well as the influence of introduced alien and reintroduced animals on habitats and biocoenoses require continual control somewhere. Successful control, however, needs clear objectives – and this seems to be the main problem. What shall be managed, for what purpose and when? Black grouse (*Lyrurus tetrix*), for example, is known for destroying buds and terminal shoots in high-altitude afforestation (Streule 1973). Should black grouse therefore be reduced to 'tolerable numbers' (and what constitutes 'tolerable'?), or should afforestation be left to natural selection? Any attempt to reduce black grouse impact would appear strange in view of parallel efforts to re-establish viable populations in the moorlands of Upper Swabia (Sect. 5 2).

As should have become obvious from the previous sections, objectives are often rather diffuse. Usually, different people, groups, associations or institutions are involved in developing management concepts. They all may have different ideas and opinions that are hardly or not compatible, partly because they are focused on different ecosystem levels, animal species, a particular biocoenosis or a landscape. The neverending discussions among nature conservationists, as well as between them, forest owners, farmers and huntsmen, are typical examples of conflicting group interests. Ethical aspects, traditions, occasionally romanticised perceptions of man in harmony with nature (did this ever exist?; see Sanders and Webster 1994; Isenberg 2000) and economic interests may be involved. Moreover, unpredictable events and processes (natural disturbances, political and economic decisions, economic change) may require modification of management objectives.

On the other hand, there is no doubt that the long-term effects of landscape management need to be carefully considered. Present landscape spatial and functional structures (spatial pattern, connectivity) have increasingly become influenced by man during history, and landscape will continue to change. The question is, how landscape development can be managed in a way that will guarantee a sound basis for the existence of biocoenoses and for humans.

F.-K. Holtmeier, *Animals' Influence on the Landscape and Ecological Importance*,
DOI 10.1007/978-94-017-9294-3_6

Which criteria may serve as a reliable basis for landscape management? Actually, most current environmental concepts are oriented towards ancient landscapes supposed to have been 'better' from an ecological view than present landscapes. In Central Europe, the historical agricultural landscapes that were characterized by great patchiness, species richness and a great biodiversity (e.g., Zoller and Bischoff 1980; Jedicke 1994; Sturm 2000) have been and still are favorite guidelines (e.g., 'Natur 2000' in MURL 1990). Such landscapes have been supposed to guarantee 'ecological stability' (whatever this means) and a 'healthy environment' (e.g., UNCED-Conference in Rio de Janeiro 1992), including for the future, although this causal relationship is controversial as ever (May 1973; Dahl 1982; Pflug 1987; Duelli 1995; Huston 1997, further references therein). It does not make much sense to justify all intentions and measures in nature conservation by the alleged necessity of maintaining ecosystems' function and natural balance, as ecosystems will function as long as life on earth exists. The often-heard demand for coming as close as possible to 'natural conditions' is also unfit for landscape management and development as long as it is not clear which near 'natural state' we are headed for. In the end, more than 'furnishing' the continuously and often rapidly changing present landscapes with the elements of the historical cultural landscape is needed to create conditions for maintenance and development dynamics of self-sustaining biocoenoses, and also to ensure essential resources to human livelihood.

Within the scope of the manifold problems considered in the context of this book, the question is how to deal with the animals as part of the biocoenoses. Obviously, they are not simply 'standing about' (Remmert 1973). The influence of some species on their habitats is hardly perceivable, while the effects of others may even be spectacular. Many of these animal species, although not listed in the Red Data Book, have been able to occupy their niche and build up self-sustaining viable populations often characterized by high individual density. Among them are not only native species, but also introduced species, such as muskrat and raccoon in Europe or most of the alien species in Australia and New Zealand. It must be accepted as a matter of fact that biocoenoses and habitats in most regions of the world are more or less disturbed by human impacts. In many cases, the effects of natural regulation processes are not sufficiently effective or have occasionally even been negative. Often, they are simply out of order. The reintroduced species (e.g., ibex, beaver, predators) must also be seen from this angle.

Contrary to wide-spread opinion (e.g., Van Wieren 1995), all efforts, well-intentioned as they may be, to shape present ecosystems, taking the more near-natural and gradually grown agricultural landscape prior to industrialization as an example, will fail, because landscape development – like evolution – is irreversible and unpredictable, at least in the long-term. Models, however, occasionally give this impression. Models are very useful tools, as they allow us to check alternative possibilities of landscape development and the potential effects of various management practices on the animals' habitats. However, models are necessarily based on the present state of knowledge of ecological interactions, the quality of which depends on the given local conditions and their history. Models may help us make decisions. Anyway, decision-making requires clear objectives. The objectives again depend first of all on political, economic and, not least, social interests and decisions. These point the way for further manipulation of the natural human environment.

The longer the period of time and the larger the areas under consideration, the more predictions that might apply to the local situation and within the short-term turn out to be speculative, because nobody knows the future political and economic circumstances that will control animal and landscape management in 40 or even a 100 years. It is only the natural causality of processes that is a constant factor, whereas the quality of ecosystems, their structure and the relative importance of their interacting physical and biological components, as well as the species spectrum, diversity and fluxes of matter and energy (quantity, turnover rates, etc.), are subject to minor or major change. Moreover, the functional role of the ecosystems within the landscape context will change due to changing landscape structures and utilization (e.g., due to agriculture, forestry, hydraulic works, etc.). This is why the tolerable game density, no matter how determined, applies only to a certain landscape condition and a limited period of time (Holtmeier 1999; see also Senn 2000). This aspect must also be considered in wildlife management and nature conservation as well.

Wildlife management is applied population ecology, which may, however, have different objectives. When an oversized animal population has to be reduced, for example, in order to prevent damage to forests or agricultural crops, the removal of surplus animals must not threaten population survival. Browsing damage, for example, can be mitigated by extra fields, providing food for wild game and distracting it from the forest. This might help to reduce shooting and to keep the number of animals above the minimum level necessary for population survival. Another objective of wildlife management could be to optimize management in order to guarantee profitable harvesting, which is worth the effort in the long-term (Caughley 1976; Bergström and Hjeljord 1987). In this respect, nutrient-rich winter forage supply must be ensured, which does not mean extra winter feeding is needed. The quality of winter forage (*Agropyron spicatum*, *Festuca idahoensis*) of American elk, for example, can be improved by grazing sheep on elk winter range in spring (Clark et al. 2000). In general, population response to control measures is the focus, whereas the potential effects of animals on their habitats play a minor role, even though the animals' influence would be of primary interest for landscape ecology and also modern nature conservation (biotope conservation). Nevertheless, research in wildlife management provides valuable fundamentals for the planning and realization of nature conservation measures.

The main objective must be to control demographic structure and dynamics of animal populations in a way that serious damage to the environment, in particular to vegetation and soil (erosion), will be prevented. This, however, is often easier said than done. Mountain goats (*Oreamnos americanus*), which were translocated from Alaska and Canada to the Olympic mountains (Washington) during the 1920s, may be considered a typical example. When the Olympic Mountains National Park was founded (1938), goat hunting was prohibited. Subsequently, mountain goat numbers increased from about a dozen to more than 1,000 individuals in the 1980s (Houston et al. 1986). Possibly, the number was overestimated (report Ramsey in

Noss et al. 2000). Nevertheless, destruction of the alpine vegetation, which covers about 30 % of the total park area (Moorhead 1989), and increasing wind erosion had already become apparent during the 1960s (Pfitsch 1980; Pike 1981). Mountain goats also affected subalpine fir (*Abies lasiocarpa*) at the treeline by feeding on its foliage. During the 1970s, a reduction in goat numbers through trapping was attempted (Wright 1992b). When only about a third of the population could be removed this way by 1982, and trapping had to be drastically reduced because it was too expensive, the park management considered radical reduction of the mountain goats by shooting to be the only efficient measure for solving the problem. The park management justified this action by emphasizing the prevailing opinion that mountain goats were not native to the Olympics.

This drastic measure, however, met with violent opposition (e.g., report of Ramsey in Noss et al. 2000). In the view of the opponents, the vegetation was not so seriously affected as to justify eradication of mountain goat in the park area. Rather, they ascribed local concentrations and resultant extent of erosion to numerous salt licks, which park personnel had installed for better watching and easier catching of the goats (Anunsen and Anunsen 1993; Wagner et al. 1995). The same effect was attributed to the many urine patches left by the tourists (Moorhead 1977a, b, 1981). Moreover, erosion due to snow slides, frost action (e.g. needle ice), other ungulates (e.g. mule deer) and wind could not be distinguished from erosion that could be unambiguously ascribed to trampling and wallowing by the mountain goats (Burger 1987; reports of Williams and Whitford in Noss et al. 2000). The controversial discussion on whether mountain goats are exotic animals or whether they had been native to the Olympics before they became extirpated by man during the Holocene is still going on (e.g., Wright 1992a; 1998; Anunsen and Anunsen 1993; Hutchins 1995; Lyman 1998). Be that as it may, after having built up a self-maintaining vital population for about 90 years, the mountain goats should be considered essentially to be native animals, as in other areas into which mountain goat has been introduced (e.g., Mt. Evans and Mt Shavano, Colorado). In the Olympic National Park, the present carrying capacity of the mountain goat habitat ranges between 300 and 500 individuals (report McCullough in Noss et al. 2000). In 2004, mountain goat numbers were reduced by trapping and shooting to about 300 (Mason County Daily News from 5 August 2011).

The situation is similar to the situation in the upper Danube valley (Southern Germany) where grazing, trampling and eutrophication by introduced chamois are threatening the existence of the local xerothermic vegetation characterized by alpigenous relic species. Although it cannot be excluded that chamois once emigrated from the Alps, the present population goes back mainly to massive introduction during the 1960s. Protection of the xerothermic vegetation seemingly requires drastic reduction of chamois. Even nature conservationists postulate eradication. This, however, meets the opposition of hunters and the public, not least because chamois are very attractive to the tourists. It is open to question whether nature conservationists and advocates of chamois will come to an agreement that would guarantee the existence of both the xerothermic vegetation and the chamois in the long-term. At present, reliable data on the number of chamois are still not available.

After all, wildlife management presents many problems. Management of the people involved, however, might be an even more difficult task (Leopold 1933 [1996]).

6.2 Natural and Tolerable Habitat Carrying Capacity

As has repeatedly been shown in the previous chapters, the 'tolerable carrying capacity' (Fig. 6.1) usually ranges below the natural habitat carrying capacity. In the lowlands of western and northwestern Germany, for example, and in some other regions of western Europe, the agricultural landscape provides abundant food to migratory and overwintering wild geese. In North Rhine-Westfalia, for example, goose damage to agricultural crops was first reported in the mid-1970s (Mildenberger 1971). Since the 1980s, it has increased to an extent (Mooij 1999a) considered intolerable by the farmers. Considerably improved forage quality and quantity due to intensification of agriculture have been the main causes (e.g., Bergmann 1999; Mooij 1999b; Südbeck and Königstedt 1999). There was an attempt to solve the problem through the shooting of geese, financial compensation for damage, and extra fields providing food to distract the geese from the valuable agricultural fields. As the quality of the overwintering and resting habitats, the number and distribution of geese, and the disturbances affecting them are closely related to the local conditions, control measures must be specifically adjusted in a way that will bring the farmers' demands for reduced goose impacts and the requirements of nature conservation measures in line as closely as possible. However, as local and regional

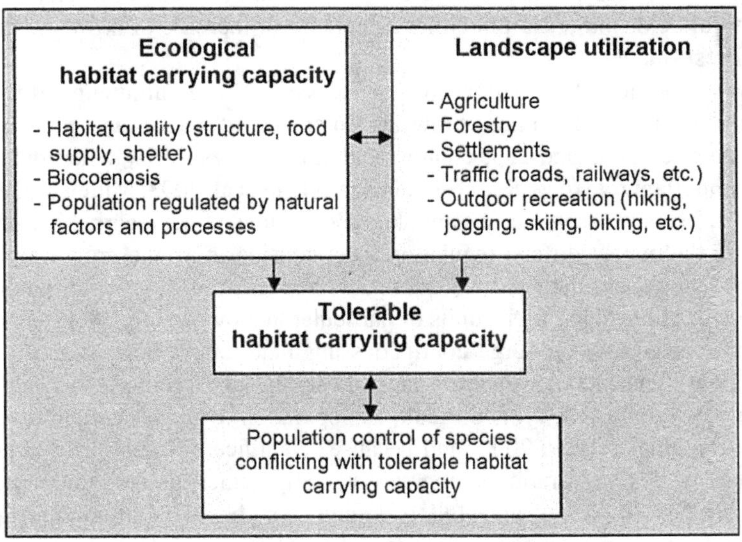

Fig. 6.1 Ecological and tolerable habitat carrying capacity (Scheme by F.-K. Holtmeier)

measures must be integrated in a broad supraregional concept that considers the manifold multiple interrelationships between the various goose populations in Europe (Mooij 1999b), in one or more places, the farmers' profit might be of secondary importance.

The impact of 'oversized' wild ungulate populations on forests and their ecological, economical and social functions appears to be a more serious problem. Although ungulate population growth decreases approaching the ecological habitat carrying capacity, the ungulates' impact might already be intolerable before the possible natural carrying capacity is reached. As to the possible management measures – ranging from intensified shooting to total modification of present silviculture that would allow for wild game populations not harmful to the forests or the environment – opinions differ widely (see discussion in Scherzinger 1996, for example). From a landscape ecological view, harvesting wild game should be considered a contribution to sustainable land use (Leopold 1933 [1996]), rather than a recreational sport (trophy hunting) of a privileged group. Hunters sufficiently trained in ecology have become aware that they have to play an active part in forest and other environmental management and to adjust hunting methods accordingly (e.g., Ökologischer Jagdverein 1994; Seidenschnur 1998; Emmer 2001).

In cultural landscapes characterized by a mosaic of various partial habitats (e.g., summer and winter grazing areas, structures providing shelter and resting places, etc.), not only is effective wildlife management inside the forested areas needed, but the distribution of the forest stands in the open agricultural landscape must also be considered. Outside forest stands, along forest edges, on agricultural fields, unploughed strips and grassland, abundant forage is available. It was due to optimal forage supply (and also to over-protection of game) that the roe deer population could grow to numbers too high for the partial habitat 'forest' (Fig. 6.2; Leibundgut 1975; Kurt 1977; David 1998). Even minor changes in cultivation of agricultural fields may result in either a decrease or increase of browsing damage in the forest stands.

The precondition of the habitats (e.g., successional stage, minimum size, spatial structures, plant communities, browsing tolerance of the forage plants), their functional role in the landscape context, and attractiveness to large herbivores also play an important role (e.g., Reimoser 2003; Suchant et al. 2003; Millett et al. 2006). Moreover, in many cases, various ungulate species may affect their habitats concurrently. As the impact differs, regulation must consider both the proportion of the different species and their particular impact. Not least, as large wild mammalian herbivores do not adhere to the units of particular landownership, wildlife management must be adjusted to large contiguous ungulate ranges (e.g., Ratcliffe 1987; Petrak 1999; Senn 2000). However, any adjustment of large herbivore density to 'tolerable carrying capacity' is difficult, as population dynamics are almost unpredictable (Holtmeier 1999, 2000) and because scientifically sound data on population structures (age, sex ratio, demography) and dynamics, in particular, long-term studies, are rare. Many decisions follow dogma or are based on intuition and practical experience rather than on substantiated data. Often, wildlife professionals confuse scientific and ethical judgements when justifying management decisions

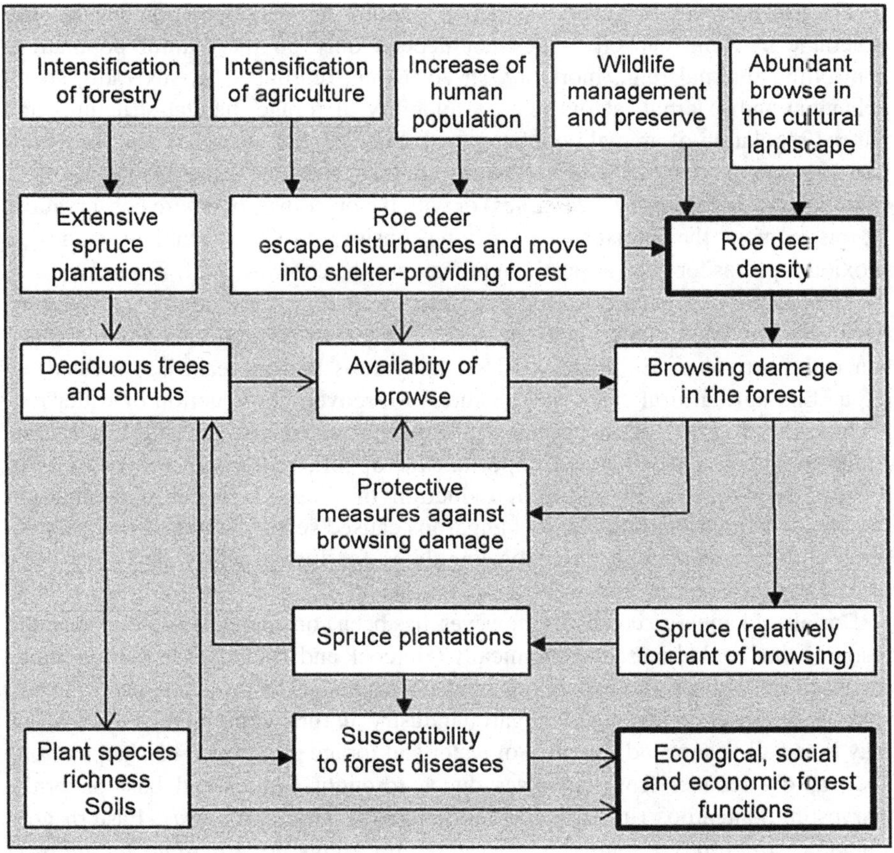

Fig. 6.2 Causes and effects of high roe deer (*Capreolus capreolus*) density (Modified from Leibundgut (1975))

(Decker et al. 1991). Missing clear concepts and conflicting interests often make a well-balanced management considering both ecological and economic aspects difficult or impossible (see also Eisfeld and Fischer 1996).

Anyway, populations of large mammalian herbivores have to be adjusted to the present 'tolerable carrying capacity' of the landscape under multiple-use in order to minimize conflicts between land use and herbivores ('de-facto carrying capacity'; Holtmeier 2002) (cf. Fig. 6.1). This is possibly the only chance to guarantee the continued existence of animal populations, as has been demonstrated in the Section on reintroduction of ibex, for example (Sect. 5.2.2). While ibex is not a problem animal in most of its present area of distribution, it is the particular situation on the steep slopes above Pontresina village (cf. Fig. 5.7) that requires regular control of the ibex population in order to reduce soil erosion and prevent damage to the subalpine forest and high-altitude afforestation.

The first problem, however, is to define tolerable animal population density. The 'tolerable carrying capacity' does not depend only on the natural population-controlling internal (e.g., mortality, survival rate, reproduction, sex ratio, social behavior) and external factors (e.g., availability of forage, habitat structures and natural regulation of animal populations; cf. Fig. 2.9) but also on when, for example, large populations of wild herbivores conflict with landscape use (agriculture, livestock grazing, forestry, hydraulics) or with the objectives of nature conservation. At this moment, the animal species may be considered to be a 'problem species', a 'noxious species' or even a 'pest', and thus, must be controlled (cf. Fig. 6.1).

The status of kangaroos, for example, in the pastoral region of Western Australia has been changing from pest to resource (Griggs and Pople 2001). Since the arrival of European settlers, kangaroos had increased as a result of expanding agricultural crops and pastures. Moreover, construction of water reservoirs and troughs for sheep and cattle that provided a water supply all-year-round has been and still is an important factor (cf., red-billed weaver birds in Africa; Sect. 2.4.4). In addition, reduction of dingoes probably resulted in increasing kangaroo numbers. As kangaroos caused severe damage to crops, pastures and competed with domestic ungulates for forage and water, they were drastically reduced through culling operations.

Burials of kangaroo carcasses, however, has been considered waste of a valuable resource that could be used economically (Hercock and Tonts 2004). Falling prices for mutton and beef and expanding markets for kangaroo products, especially in Europe, has resulted in a vital 'kangaroo industry' in rural communities. Kangaroos may breed all year round, supplies of water and forage provided. Their populations also rapidly recover from collapses due to drought ('boom-and bust pattern'). Harvesting kangaroos (in particular red kangaroo, *Macropus rufus*; Eastern grey kangaroo, *Macropus giganteus* and Western grey kangaroo, *Macropus fuliginosus*) (Hercock and Tonts 2004) for export of meat and other products (skins, leather) is not only a sustainable natural economic source, but also helps to control kangaroo populations, which have doubled since the 1990s (Kelly 2003). Harvesting is regulated by a strict annual quote system. However, there are regional differences. In Victoria, for example, the commercial harvesting of kangaroos is prohibited, whereas in South and Western Australia, kangaroo meat for human consumption has been available in supermarkets for more than 10 years (Hercock and Tonts 2004). Nutrition-conscious people appreciate kangaroo meat, as it is rich in iron, protein and unsaturated fatty acids (Omega 3). Beef and lamb producers, however, are distrustful of this development and fear competition. The deep-rooted traditional attitude towards kangaroos as pest animals will probably prevent general acceptance of 'industrial' kangaroo meat production (Boom and Ben-Ami 2010).

Wild boar, which has rapidly increased in the entirety of Europe over the last few decades (Sáez-Royuela and Tellería 1986; Geisser and Reyer 2004, 2005; Bieber and Ruf 2005; Herrero et al. 2008; Apollonio et al. 2010; Magnusson 2010), is also an impressive example. While overabundant cervids mainly impact forests, wild boars may cause damage to agricultural fields, vineyards, and grassland by digging for roots, tubers, bulbs, invertebrates and small vertebrates (Schley and Roper

2003), as well as through consumption of agricultural crops and fruits. In many areas, wild boars have invaded urban areas, where they are devastating public parks, gardens, golf courses and cemeteries. Having lost their natural shyness, they also rummage for discarded food items through trashcans and occasionally get into houses. Moreover, vehicle collisions with wild boars have considerably increased. Not least, wild boars are a reservoir and vector of swine fever, bovine tuberculosis, foot-and-mouth disease and some other diseases affecting domestic animals (Ruiz-Fons et al. 1997; Wilson 2005; Meng et al. 2012).

The main driving factor of wild boar eruption is the large-scale cultivation of maize and other so-called 'energy plants' (bio-fuel), such as rapeseed (*Brassica napus*; Canola) and giant miscanthus (*Miscanthus × giganteus*). Moreover, fructification in oak and beech has become more frequent due to climatic warming and increased emissions, in particular, of nitrogen. Acorns and beech nuts provide ample energy-rich food in autumn. Under these circumstances, reproduction of wild boars is very high. They may produce 2 l per years with up to 10 piglets each. Mild winters have increased their chance of survival. Wild boar numbers may triple within 1 year. Very likely, wild boar populations have profited more from the modern agricultural landscape than any other wild ungulates, and have exceeded tolerable habitat carrying capacity by far. On the other hand, wild boars at moderate numbers may be beneficial to woodland (Sect. 2.6.2.2). By rooting, they expose the soil, thus facilitating seedling establishment. Disturbed patches also provide favorable insect microclimates.

In most of wild boar distribution areas, their numbers must be drastically reduced. Reduction, however, is difficult, particularly because high reproduction rates may rapidly compensate for losses through hunting. Moreover, hunting wild boar is usually harder than hunting roe deer or hare. Although wild boar are active all-day in general, they become nocturnal if frequently disturbed. In addition, extensive and still increasing fields of maize and giant miscanthus provide optimal shelter to the animals. Within such fields, wild boar are almost out of the hunters' reach. Not least, hunting wild boar seems to be less profitable compared to hunting roe deer or red deer, for example, as venison gets better prices. Moreover, compensation for damage caused by wild boars to cultivated land has become increasingly expensive to tenant hunters, and as a result, leasing of a hunting ground has become impossible for many of them.

All over Europe, reproduction of wild boars surpasses hunting kills. Effectiveness of hunting practices could be improved by more drive hunts over extensive contiguous areas unrestricted by borders of individual hunting grounds. About 80 % of the annual piglets and additional young highly reproductive females should be harvested to regulate the wild boar population (Keuling 2011). The wild boar problem cannot be solved by more shooting alone, however. Trapping and fertility control (immune-contraception) may contribute to wild boar regulation. Exclosure by fencing, chemical repellents, as well as supplementary feeding (e.g. maize, oats, potatoes) to distract wild boars from cultivated fields, can help to prevent or reduce damage to cultivated land. On the other hand, supplementary feeding, for example, as bait for effective hunting, improves the conditions for wild boars, particularly as too great quantities are often exposed. Thus, feeding may exacerbate rather than mitigate the problem (Bieber and Ruf 2005; Geisser and Reyer 2004, 2005). Among the possible control measures, large-scale restriction of maize expansion would probably be most effective in the long-run. As maize, however, is an important and very profitable ‘energy plant’ (bio-gas) and high-energy food in animal meat production, there is no real chance at present to reduce maize cultivation.

The second problem is the usually controversial attitudes of different groups (hunters, animal rights activists, wildlife managers, nature conservationists, farmers, ranchers, forest owners, etc.) to adjusting animal population to the ‘tolerable carrying capacity’. Foresters, for example, appreciate lower red deer and roe deer densities that would allow for sustainable natural forest regeneration, whereas hunters usually prefer comparatively great numbers. In some regions, wild ungulates may be an important value in a rural economy.

In northern Europe, for example, where overabundant moose affect young afforestation, harvesting more moose might alleviate the problem. However, moose control would only be effective if hunting practice changed, and more females and calves would be killed rather than adult males preferred for their enormous antlers (trophies). In addition, the combined effect of large native predators (wolf, lynx, wolverine, bears) that have recovered during the last 50 years from near extinction (e.g., Sandström et al. 2009; Wotschikowski 2010) could possibly help in adjusting moose numbers to a tolerable size. The mechanisms regulating moose population, however, are not as simple as they may appear at first sight. A complex of interacting factors is involved, such as availability of alternative prey to the predators, competition between different predator species, plant productivity and browsing tolerance, forage quality (e.g., nutrients, defensive substances), forest composition and age structures, and, not least, climate (e.g., Van Ballenberghe and Ballard 1994; Messier 1994; Bowyer et al. 1997; Hörnberg 2001). Moreover, control by predators is double-edged. While most people consider predators to be a public good, having an ethical value per se, sheep and reindeer owners do not tolerate large predators at all, and moose hunters argue for rigid reduction of the remaining great large carnivores because these take the hunters’ favorite prey. The harvest yield from moose is often important to local economics. Hunting success will decrease at lower moose density, and the outfitters would suffer income loss. Therefore, Finland plans to reduce lynx population by about 22 % (Saavalainen 2010).

Moorlands in England are popular for sport shooting of red grouse (*Lagopus lagopus scoticus*). Grouse population is artificially maintained at high densities by rigid control of predators (fox, *Vulpes vulpes*; carrion crow, *Corvus corone*). Grouse managers, however, do not make much difference, and also kill protected predatory birds, such as hen harrier (*Circus cyaneus*) and peregrine falcon (*Falco perigrinus*) (Thirgood et al. 2002). The number of predators, however, would very likely decline at lower grouse densities. Thus, competition for prey between ‘sportsmen’ and predatory birds appears to be the real problem. The conflict between grouse hunters, stakeholders and advocates of birds of prey has not be resolved so far (Redpath and Thirgoods 2009; Thompson et al. 2009; Sotherton et al. 2009), although the parties have been striving to reach a compromise.

As landscape use, and thus, landscape spatial pattern and functional role, depends on economic conditions and political decisions that cannot be predicted in a long-term perspective, tolerable animal density will change over the course of time, and can thus be determined in whatever way for a limited period and area only (Holtmeier 1999, 2002; Senn 2000). In a case in which political boundaries must also be considered along with the kind of land use, wildlife management may be particularly difficult, as efficient international cooperation is needed to solve the problems. A project in the area along the state boundary between Germany and Belgium (Nature Park Nordeifel-Hohes Venn), for example, has been successful in this respect. It considers both cross-border preservation and hunting of red deer, as well as red deer behavioral response to the influences of forestry and tourism in the nature park and its surroundings (Petrak 1999). Red deer density has been reduced to a tolerable level and forest damage has decreased accordingly. Deciduous trees such as mountain ash, birch and aspen are prospering, and habitat conditions for red deer have improved. As a result of landscape and red deer management, the total red deer habitat has increased more than tenfold since the early 1980s. In the view of silviculture, damage by browsing and bark stripping are kept within tolerable limits at current red deer densities of three to five individuals per 100 ha. It was the bundle of measures that created a balance of red deer population numbers and its habitat under manifold use. This would never have been possible if the development had been left to self-regulation.

Perhaps in national parks, animal population and habitat development might be left to self-regulation, as the parks became established to preserve a piece of undisturbed nature or at least near-natural conditions. However, the national parks are also facing serious problems, as has already been previously shown in this book. Even in the large American and African national parks, development is 'natural' only insofar as it follows natural causality. The park ecosystems, however, were not really 'natural' in the sense of 'original' or 'undisturbed by humans' from the time they became established. External factors such as immigration of ungulates escaping severe hunting pressure and competition with livestock and agriculture outside the parks have increasingly influenced the parks. In Zimbabwe, Tsavo National Park and Mara Wildlife Reserve (Kenya), for example, the ecological habitat carrying capacity has already been exceeded long ago due to high reproduction in the elephants and, in particular, as a result of mammalian herbivores immigrating from outside the parks and wildlife reserves. The elephants continue to destroy their own base of existence and that of many other members of the park biocoenosis. When under the pressure of public opinion, radical culling had to be stopped, the ecosystem functioned 'naturally' in a very effective way, which, however, brought some nature conservationists down to earth again, as during the severe drought 1970/1971 (Sect. 3.1.2), mainly cow elephants and young elephants died. Thus, this natural elephant decline was much more effective than any management by culling could have been (Leuthold 1978). It is not without its irony that the vehement and emotional opposition to a drastic reduction contributed to an unexpected ecological solution to the elephant problem, at least for a while.

In Krueger National Park, 7 % of the elephant population has been killed per year to prevent damage to the park ecosystem. Such culls, however, are controversial. Contraception has been suggested as an alternative. About three quarters of the adult female elephants had to be treated to stop population growth, while numerous physiological and behavioral side effects were disregarded (Whyle et al. 1998). The discussion concerning population control through culling continues with undiminished intensity. Studies on four elephant sub-populations in Krueger National Park (South Africa) suggest reducing culling in this area to those cases in which the elephant density exceeds 0.37 elephant km^{-2} for 2 subsequent years. Earlier culling is expected to cause population increase; and this will be due to the immigration of elephants from uncontrolled herds rather than to higher reproduction. As a result, elephant damage is likely to increase in the areas where culling has been practiced, while the neighboring areas are relieved from elephant impact (Van Aarde et al. 1999).

The efforts to protect and maintain the elephant habitats are numerous. However, they must be extended by incorporating the seasonally used partial habitats outside the national parks' borders. In addition, poaching has to be drastically reduced. Although poaching for ivory is generally focused upon (Sect. 3.1.3), poaching for meat is no less important. In the Serengeti, for example, about 200,000 animals are killed per year to supply more than one million native people living west of the park with meat. As long as it is not possible to meet the demand of this rapidly growing human population with wild game meat from other sources, to control poaching effectively and to open alternative chances for more than 30,000 hunters to earn some money, the future of the parks and game reserves will be quite uncertain (Arcese et al. 1995; Campbell and Borner 1995; Campbell and Hofer 1995). Successful implementation of such measures, however, is more unlikely due to increasing human population, increasing famines and expanding landscape use (Vorlaufer 1998; Job 1999). The attempts to motivate native people through explanation and education about conservation of fauna and flora does also not appear to be very promising. The best way would probably be to stimulate people's material interest in animals and nature in general. Thus, people might realize their natural environment to be 'valuable' and worthy of protection.

The development of these large national parks and wildlife reserves cannot be turned back. However, just maintaining them in the long-term without further reduction in size would already be a great success, although they would have to be considered an 'outdoor zoo' rather than undisturbed wilderness. 'Umbrella species', such as wildebeest and elephants, are of major importance in this respect. In contrast to the usually intensively used European landscapes, disappearance of the large herbivores in the African national park landscapes could probably bring about far-reaching changes even now, without it even being possible to predict how they would change in the future. Natural reforestation of extended areas is only one possible alternative.

6.3 Large Herbivores as Instruments in Landscape Management

In the intensely cultivated landscapes of Europe, it is often open to question as to how to manage increasing fallow land sustainably (e.g., Ganzert 1994). Should fallow land be kept open or left to natural succession? In the latter case, shrubs and forest will become established. There are good reasons for both alternatives. Keeping fallow land open would benefit meadow birds and preserve a high landscape diversity. Conservation of biodiversity and landscape diversity are often given high priority. On the other hand, spontaneous reforestation would mean letting nature take its course towards a 'near-natural' environment (potential natural vegetation). Woodland is often considered to be more like the conventional idea of the original natural landscape than grassland. To the meadow birds, on the one hand, it surely does not make any difference whether the grassland is of natural origin or secondary, provided that it meets their habitat demands. In any case, fallow land left to natural succession ('nature total', Wagner 1993) would considerably reduce their habitats. Besides artificial removal of invading shrubs and trees, both large wild and domestic ungulates may be appropriate 'tools' for keeping fallow land and grassland open (e.g., Wallis de Vries 1995; Hofmann et al. 1998).

In Central Europe, domestic ungulates such as sheep, including German grey heath and the white polled heath (moorland sheep), cattle and horses have been used for a long time to keep landscape open (e.g., Oosterveld 1975, 1979; Scherrer and Surber 1978; Thalen 1981, 1984; Zimmermann and Woike 1982; Wibbels 1990, 1994; Van Wieren 1991, 1995; Lammers 1994; Nitsche and Nitsche 1994; Fischer et al. 1995, 1996; Beinlich et al. 1997; Bunzel-Drüke 1997; Martin 1997; Völkl 1997; Oppermann and Luick 1999). Heck cattle (Photo 6.1) and Heck horses (Koniks, Photo 6.2) are used to prevent shrub and tree establishment on flood plains and other wetlands, which are important habitats of meadow birds. Heck cattle resulted from attempts to breed back the extinct aurochs (*Bos primigenius*). Koniks are backcrosses of the European wild horse or tarpan (*Equus ferus ferus*).

German grey heaths help to preserve the extended *Calluna* heath areas in the Lüneburg Heath (Germany), for example (Photo 6.3). Since the 1980s, white polled heaths have been used to maintain heath peatland in northern Germany (Nitsche and Nitsche 1994) and on both sides of the state border between Germany and the Netherlands. White polled heaths remove purple moor grass (*Molinia caerula*), for example, and prevent establishment of birch (Lambeck and Schwöppe 1987). Purple moor grass suppresses other vegetation almost completely. In a 10-year study on the aptitude of alternative management practices (mowing once in autumn, lie fallow, sheep grazing) for maintaining or establishing species-rich calcareous grassland in the Netherlands (Limburg Province), grazing by sheep turned out to be the most effective measure (Willems 1983). Sheep grazing is also used on the Swabian Jura to maintain calcareous grassland, which originated, by the way, under the influence

Photo 6.1 Heck cattle are used to maintain open meadows in the sewage irrigation field area of Münster (Germany), which are an important resting place for migratory birds in Europe. W. Holtmeier, September 2012

Photo 6.2 Heck horses (Koniks) in the floodplain of the Haase River (NW Germany). A few Scottish highland cattle are in the background. Both grazers have been introduced to this area to impede encroachment of shrub and trees on the floodplain. W. Holtmeier, March 2012

Photo 6.3 Moorland sheep (German Grey Heaths) suppress tree and bush seedlings and keep the moorland open. Amtsvenn, northwestern Germany. F.-K., Holtmeier, 23 September 1999

of traditional grazing by sheep (Beinlich et al. 1997). On the other hand, Schreiber (2005) found that extensive grazing by sheep, cattle or horses could not prevent an 'explosion-like' invasion of pastures by bird-dispersed tree and shrub species in different parts of Southern Germany. Species-rich grasslands in the riparian zone of Scottish upland conifer forests are likely to be colonized by scrub and trees after cessation of cattle grazing. Cattle grazing has turned out to be an appropriate tool to keep the former pastures open and to prevent a decline in species richness (Humphrey and Patterson 2000).

Moreover, itinerant flocks of sheep turned out to be effective dispersers of diaspores and also of small animals (e.g., grasshoppers, snails, spiders and beetles), thus contributing to the exchange between isolated areas and preservation of species diversity (Fischer et al. 1996). Since 1993, efforts have been made in northern Spain to revive the former seasonal migration of sheep (transhumance) between the Estremadura and summer pastures in the Cantabrian Mountains. The broad old migration routes, the so-called 'cañadas', are characterized by a species-rich fauna and flora that has to be attributed mainly to the dispersal of seeds and insects by transhumant sheep (Bauschmann 2000).

On Hiddensee, an island off the German coast of the Baltic Sea, mouflons (*Ovis musimon*) were released in 1988. These have efficiently contributed to conservation of the dune heath. In view of this positive effect in landscape management, it has been suggested to increase the mouflon population by about 30–60 individuals at least (Jeschke 1997). However, it may be questioned as to why exactly this species (originally native to the Corsican Mountains) of all wild sheep has been introduced to Hiddensee. Sheep from Gotland (Sweden) which had been brought to the neighboring Fähr Island were just as effective in conservation of the coastal heaths. However, in contrast to Gotland sheep, mouflon are a favorite wild game animal, which guarantees pleasant hunting.

Red deer may be used as an alternative tool for grassland management only in comparatively small areas and in the short-term. In the nature conservation areas of the 'Perlebach Valley' (western Eifel mountains, Germany), for example, red deer have considerably delayed establishment of shrubs and trees on the former oligotrophic cattle paths (Petrak 1992) and contributed to the conservation of these valuable grassland communities. However, additional supporting measures are necessary. The cattle paths rich in spignel (*Meum athamanticum*), in particular, must be mown every 2–3 years, and the mown crop has to be removed. Roe deer will not hold up or even prevent succession on fallow land (Völkl 1997, 1999). Anyway, both ungulate species are no alternative for near-natural grassland cultivation.

By contrast, in the Swiss National Park (Lower Engadine, Switzerland), for example, conservation of botanical diversity will only be possible if browsing pressure is sufficient (cf. Fig. 3.14). Current great botanical diversity is the result of forest removal and pastoralism over hundreds of years, and thus, is not natural. Increasing browsing pressure may, however, impede natural regeneration and reduce tree species richness, as well as diversity. If conservation of plant diversity is the main objective, which might be put into question, the red deer population will have to be adjusted accordingly by harvesting.

Above the present forest limit, abandonment of pastoral use and increasing encroachment of shrubs and trees in the former pastures (e.g., Holtmeier 2009, further literature therein) have led to an increase of avian diversity during early succession. In the following, however, at increasing tree and shrub density, habitat conditions will become unfavorable for grassland bird species (e.g., Laiolo et al. 2004). Thus, moderate grazing by livestock or red deer may be appropriate for maintaining the mosaic of open grassland, scattered tree groves and shrubs. This would be beneficial to both grassland and woodland species as well as to species using different habitat types.

In the high mountains of Córdoba (Argentina), domestic grazers were excluded (1998) from many areas to reduce omnipresent soil erosion and to restore *Polylepis* woodland. Prior to the introduction of domestic livestock about 400 years ago, the mountains were grazed by native large herbivores such as guanacos (*Lama guanicoe*), rheas (*Rhea americana*, *Pterocnemia pennata*) and deer (species unknown). These became locally extinct before the early twentieth century. Field surveys (Garcia et al. 2008, further references therein) have shown, however, that complete exclusion of domestic herbivores not only reduced plant diversity but also bird abundance and diversity. While woodland restoration would be beneficial to typical woodland species, it would adversely affect the bird communities that have developed in the mountain landscape grazed for thousands of years. In areas where pastoral use is declining or has ceased, re-introduction of guanacos might have positive effects. Soil erosion would probably be less than under the impact of livestock, and, in addition, guanacos would not threaten the persistence of plants and birds. If re-introduction, however, is not feasible moderate livestock grazing could be an alternative to preserve small-scale habitat heterogeneity, biocoenoses and biological diversity (Garcia et al. 2008). On the other hand, livestock production should not become unprofitable. Therefore, well-balanced management measures are needed.

In Provence (France), llamas are being tried to remove shrub undergrowth in forests in order to reduce the risk of forest fires. Although the llama is an alien species, it appears to be more efficient than native sheep in removing shrubs from the forest understory. In addition, the padded llama hooves do not affect forest soils, in contrast to the hard claw feet of sheep. In Germany, use of New World camelides in landscape management has also been considered. As keeping llamas, alpacas and guanacos is not difficult, these animals are becoming increasingly popular as domestic animals ('pets'). Depending on the landscape management objective, mixed grazing with llamas, sheep, goats, cattle or horses appears to be promising (Gerken 1999). Nevertheless, even more than in the case of the mouflons introduced to Hiddensee, it may be questioned as to whether the native ungulates really need support by the aliens to do their job in landscape management or whether llama are more of a hobby of a few ecologists.

It seems hard to imagine wild boars as 'landscape managers'. Their manifold effects often conflict with management objectives and make it almost impossible to integrate them into management concepts. Wildboar may considerably influence vegegation dynamics and contribute to the enrichment of diversity. They may activate seedbanks located at great depth (Sect. 2.6.2.2) and create suitable seed beds for certain plant species. For the same reasons, however, they may support expansion of woody vegetation into fallow land (Völkl 1999). As their populations have been rapidly increasing to excessive numbers over recent decades, they have become a 'pest animal' in many areas and a risk factor for domestic pigs because of the spread of swine fever.

Integration of domestic and particularly wild ungulates into landscape management – in a certain sense comparable to the original multi-species system (different foraging types; Fig. 6.3) – appears to be reasonable from a functional view, might reduce costs and is less unfavorable for the biocoenoses (e.g., fauna living in the litter, soil fauna) than sod removal, burning, or mechanical and chemical removal of birch, for example. The herbivores could be considered as 'natural' disturbance factors which influence composition and structure of the ecosystems and increase biodiversity (e.g., White and Pickett 1985; Hobbs and Huennecke 1992). Anyway, not much is known at the present as to how such a landscape management with large herbivores as integrated agents will influence succession of vegetation, nutrient balance and species diversity in the long-term (Bokdam and Gleichman 2000). Additional management measures will probably also be needed in the future. It seems unlikely that the *Calluna* heathland (*Calluna vulgaris*), for example, a characteristic historical landscape in northern and northwestern Germany, can be maintained only through sheep grazing, deer and rabbits, if not invading copses (shrubs and trees) will be regularly removed by humans. In the long-term, half-open pastoral woodland would develop (Pott and Hüppe 1991; Bokdam and Gleichman 2000).

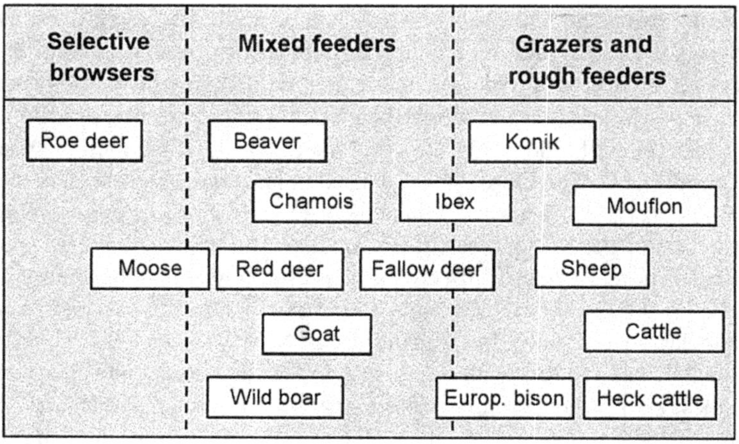

Fig. 6.3 Foraging types of large herbivores in Central Europe. (1) New breed of the European forest horse, (2) New breed of the aurox (Modified from Hofmann et al. (1998))

Multi-species grazing, largely mimicking the role of the original grazers and browsers, meets the demand for 'protection and reintroduction of dynamic processes' (Riecken et al. 1998) and contrasts with 'static' conservation of nature. It may be questioned, however, why dynamic processes have to be 'reintroduced', as they were never absent. The main objective has not changed, however, as, in the end, great efforts will be made to preserve the conditions achieved by 'modern' management, which presumably will be characterized by great biodiversity. The only new idea is the integration of the large ungulates' functions into landscape structural and dynamic development. As to the successful implementation of such concepts, relatively large and continuous areas are required.

Large herbivores may be considered to be so-called 'umbrella species' which modify their environment in a way that also benefits many other species (e.g., Wallis de Vries 1995, 1999; Popp 1999). The landscape development of the Oostvaardersplassen in Flevoland (The Netherlands) is of particular interest in this respect. The Oostvaarderplassen are the result of land reclamation (polder). The original plan was to establish an industrial complex in the present Oostvaardersplassen area. In 1982, however, they were designated a large nature reserve with a total area of 5,600 ha. The Oostvaardersplassen are the largest field experimental area in Central Europe and belong to the game-richest places in the Netherlands (Kampf 2000; Vera 2009). In 1989, the reserve was declared a Ramsar wetland site. The main objective is preservation of threatened wetland and grassland species. Koniks and Heck cattle, as well as red deer and roe deer, were introduced to keep the outer margins of the area free of shrubs and trees. Red deer were translocated from the Veluwe, the largest continuous forested area in the Netherlands, or introduced from Scotland and the Czech Republic to the Oostvaardersplassen (Kampf 2000).

The introduced large herbivores rapidly multiplied. In 2011 (autumn), the Oostvaardersplassen harbored 1,145 Koniks, 356 Heck cattle and 3,304 red deer (communication J. Kuipers. Staatsbosbeheer, 26 March 2013), whereas the roe deer population has been comparatively small (<100). Most probably, the fence that was set up around the nature reserve to prevent red deer from moving to the surrounding agricultural area and a railway nearby had very negative effects on the roe deer. The different foraging behavior of the large herbivores and grey-lag geese living there in great numbers has resulted in a varied vegetation mosaic that provides suitable habitat conditions for many other species. The press has already been talking of the '*Serengeti behind the dikes*' (Der Spiegel 32, 2001). However, as the carrying capacity of the reserve area, fenced off from the modern Dutch countryside, is limited, control of both wild and domestic herbivores has turned out to be necessary (Krüger 1999). Grey-lag geese are feeding on the surrounding agricultural countryside and return to the Oostvaardersplassen for resting during night. There, they contribute to excessive eutrophication of the aquatic ecosystem.

Regulation of Heck cattle and Koniks may be a problem, because when their meat is marketed as human food, regular veterinarian control is required. Therefore, it has been suggested to use only wild large herbivores for wetland management. Game can be marketed without any difficulty, and continuous removal of the cattle and Konik cadavers could be avoided (Lumeij and Oosterbaan 2000). Implementation of the ideas behind the Ooostervaardersplassen-concept might be difficult, because of fragmentation and multiple use of cultural landscape. Nevertheless, Heck cattle and Koniks have increasingly become popular in many areas as tools of landscape management and nature conservation.

Occasionally, the development of savanna grassland in the large African national parks and wildlife reserves (c.f. Table 5.3) under the influence of ungulate herds and elephants, in particular, is referred to as an example of the functional role of large herbivores acting as 'umbrella species' under natural conditions. This example, however, is not only 'far-fetched' in the literal sense. At high large herbivore density, savanna tends towards open grassland and considerable increase of typical grassland species in the animal communities. At low herbivore density, the grassland will change into tree savanna. Both situations are certainly results of 'natural' ecosystem response (natural causality) to grazing and browsing impact. However, it should not be ignored that African savannas also have considerably changed due to human disturbances (Introduction and eradication of the rinderpest, fighting sleeping sickness, burning, poaching, livestock grazing and cutting fire wood), even though the human impact might have been less severe than in the European cultural landscapes.

In Australia, African grama grass (*Andropogon guyanus*) was introduced in the 1930s to increase forage for livestock. Grama grass grows to 4 m in height with tussocks up to nearly 1 m in diameter. Growing mature, it gradually becomes woody and is no longer used by cattle, buffalo and kangaroos. The grass has been spreading all over northern and western Australia. Being highly competitive, it has become a serious threat to native grasses, and thus, to biodiversity. Moreover, grama grass produces large amounts of fuel, and thus, increases the risk of bush fire hazards. The fires also destroy trees and cause long-term landscape change. As conventional methods of preventing grama grass advance (e.g., aerial spraying of herbicides, mechanical destruction) have turned out to be almost completely inefficient, introduction of elephants and rhinos has recently been suggested in all seriousness to control grama grass (Bowman 2012). They are the only large herbivores that would be able to feed on and open dense grama grass vegetation by trampling it down, as they do in African swamp savannas (Sect. 3.1.2). This idea, however, is controversial among ecologists, nature conservations and in the Australian and international public. It has even been called stupid, which is easily understandable in view of Australia's already existing problems with alien species (Sect. 5.1.2). Presumably, farmers and aborigines in the outback would not be very enthusiastic about such big newcomers. Who would control their population (helicopter shooting?), and would elephants really do their job as intended, or would they prefer savanna trees and shrubs for food? The same author also suggests introducing predators to control feral animals, as all other efforts have been extremely expensive and ineffective. Introduction of both large herbivores and predators would be an experiment the outcome of which cannot be predicted. Anyway, at present, introduction of alien species into the wild is prohibited by law, and illegal introduction of such big animals might hardly be possible. Fortunately, introduction of elephants and rhinos into Australia as instruments of landscape management is unlikely to become implemented.

In many European countries as well as in America, attempts have been and are being made to restore the wilderness (so-called 're-wilding'), with core habitats and corridors for exchange between the habitats. Re-wilded areas may also allow profitable eco-tourism and wildlife watching. Re-wilding, however, should not be confused with restoration of original habitats. Instead, the present landscapes and wild animals have to be managed in a way that animal populations will not exceed the current tolerable habitat carrying capacity as well as to ensure that human disturbance does not jeopardize the existence of viable animal populations. Although re-wilding is often driven by the actions of ambitious nature conservationists, nevertheless it is sentiment and emotions rather than soundly based ecological experience that too often drives many of these well-funded programs.

6.4 The Role of Public Attitude to Landscape and Nature Conservation Measures

Whether wild living herbivorous mammals do 'good' or 'bad' depends on whether or not they run into conflict with land use (see also Decker et al. 1991; Holtmeier 2002; Reimoser 2003). In general, public attitude to wild animals is positive, but it may turn into rejection as soon as people are negatively affected by animals' activities.

Manipulation of our environment regularly requires decisions as to which species shall be supported, conserved, reduced or even eradicated (pest animals). The local conditions are of major importance in this respect. In many cases, such decisions are often influenced by emotion and/or political calculation rather than being based on substantial ecological facts. When politicians, for example, speak out in favor of enriched biodiversity, it is seldom clear what they really mean and what this biodiversity should look like.

In general, the environmental (landscape) situation also depends on conditions besides the ecological (e.g., politics, economy, kind and intensity of land use), which need to be considered before implementing management objectives. In a case in which the preservation of the curlew (*Numenius arquata*), for example, or the great bustard (*Otistrada*), in our intensely used landscape is declared to be the objective of nature conservation, this may be accepted or rejected for various reasons, but it cannot be explained to be mandatory for ecological reasons.

The same holds true for the recurrent demand for rigid control of natural predators in order to support reintroduction of black grouse (*Lyrurus tetrix*) into the moorlands of Upper Swabia (southwestern Germany). As to the success of reintroduction, predators may appear to be too numerous. In a broader view, however, their numbers seem in balance with the present landscape, being a product of historical and current human impact. The same goes for allegedly too high populations of natural predators of the brown hare (*Lepus europaeus*). Landscape change due to intensive agriculture, environmental pollution, increased outdoor activities, road traffic, and weather conditions (e.g., wet springtime) rather than predators has been the main reason for considerable hare decline (e.g., Stubbe 1999). Moreover, brown hare numbers have always been subject to natural fluctuations. The impact of predators, huntsmen included, increases parallel to declining hare populations (Sect. 2.2.2). In the hunters' view, predators simply are 'too numerous'. Foxes (*Vulpes vulpes*), in particular, are said to have reproduced 'explosion-like'. Beech martens (*Martes foina*), hooded crow (*Corvus corone cornix*) and common raven (*Corvus corax*) have increased to an 'intolerable extent'. Altogether, the 'proven', putative or 'likely' impact of the many predators (Goretzky 1999) is supposed to seal the brown hare's fate in our landscapes. Thus, the demand for reduction of the hare's natural enemies may be understandable. Hespeler (1990) got to the point when he wrote, '*Everybody, me included, wants to eat it*'. On the other hand, the successful reintroduction of the bearded vulture (*Gypaetus barbatus*) into the European Alps has considerably profited from great acceptance, after traditional prejudices were cleared up. In the past, bearded vulture had been blamed for killing lambs, goat kids and even young children. In the Alps, it was eradicated. However, as the vulture is a scavenger and does not prey on living wild or domestic ungulates, it has been relatively easy to characterize it as the natural 'health police' of the mountains, having a positive effect on the ecosystems.

The efforts towards species conservation will also be confronted with many problems and different attitudes in the future. Making people aware that preservation of species and their habitats is an ethical obligation of human society is of greater need than endless discussions, which often refer to scientifically garnished 'pseudoecological' arguments rather than to scientifically well-substantiated results. Successful implementation of management measures is hardly possible without public acceptance and support. Acceptance, however, is founded on understanding, insight and, not least, sympathy. Therefore, nature conservationists use so-called 'sympathetic animals' to promote acceptance on the emotional track. Certainly, not every animal which is sympathetic is also essential for ecosystem function and is not necessarily an 'umbrella species', such as the beaver, wildebeest or elephant. It is easily understandable, however, that measures which may protect the habitats of the common tree frog (*Hyla arborea*), which is considered a 'sympathetic animal' and a 'character species', will also be beneficial to other members of the biocoenosis, competitors and predators included. 'Sympathetic animals' may convey objectives of species and habitat conservation even to people not familiar with ecology.

Although measures to improve the habitat network may conflict with land use, acceptance must not necessarily dwindle as long as possible economic losses, for example, will be compensated. However, when animal populations must be regulated through culling or poisoning so as not to exceed the tolerable habitat carrying capacity, the situation may rapidly heat up. Debates often are polemic and then it depends more on emotions than on rational consideration whether or not necessary measures will be accepted. When a fire destroys a forest, people will usually regret the dead animals more than the burnt trees and feel urged to help the singed animals rather than to save the aboreous vegetations. Indispensable culling of red deer in the Rocky Mountain National Park met vehement opposition from the tourists, which come to the park to enjoy red deer herds easily visible from the park roads. 'Peacefully' grazing deer obviously suggest something like 'undisturbed nature'. On the other hand, hardly a park visitor becomes aware of the severe browsing damage in willows and aspen stands. To avoid increasing protests against culling, salt licks were installed outside the park boundaries and away from the tourist routes to attract reed deer and to kill them there hidden from the tourists' eyes. Translocation is not only too expensive, but may also be a risk, as red deer and bison may transmit brucellosis (*Brucella abortus*). Brucellosis was probably already introduced by the Spanish in the sixteenth century (e.g., Peterson 1991; Aguirre and Starkey 1994; Simonetti 1995; Cunningham 1996; Marcot et al. 1998).

In general, people are more touched by measures affecting mammals that are phylogenetically relatively close to humans – in particular, if they come up to 'cuddly animal babies' or 'Bambi' – than by eradication of noxious insects (Fig. 6.4). Launching stirring pictures of mother elephants with their cute elephant babies (Photo 6.4) into the media is probably the easiest way to shock people and to provoke public debates on the spirit and purpose of elephant culling. Reduction of koala bears (*Phascolarctos cinereus*), which are exceeding the carrying capacity of the remaining *Eucalyptus* forests, has met with stiff public opposition – particularly from the tourism industry in Europe and North America (Der Spiegel 32, 1998). Reduction very likely is the only way to prevent starvation of many of the cuddly koalas after they have depleted their shrinking food source. In contrast, people are not usually worried about anti-toad and anti-rabbit campaigns.

The furious protests of animal conservationists against hunting for bison, which regularly move beyond Yellowstone National Park boundaries despite repeated population control (Meagher 1989a, b; Yellowstone National Park 1997), have been motivated by the deep moral feeling of having to protect this impressive and highly symbolic animal, which became almost extinct in real massacres more than

Fig. 6.4 People's relative concern (decreasing from the center to the periphery) about control (removal, culling, poisoning) of animals decreasing parallel to increasing phylogenetic distance from man. *1* – Man, *2* – Mammals, *3* – Other vertebrates, *4* – Invertebrates, *5* – Plants. Compared to the concern about mammals concern about plants appears to be almost negligible (Scheme by F.-K. Holtmeier)

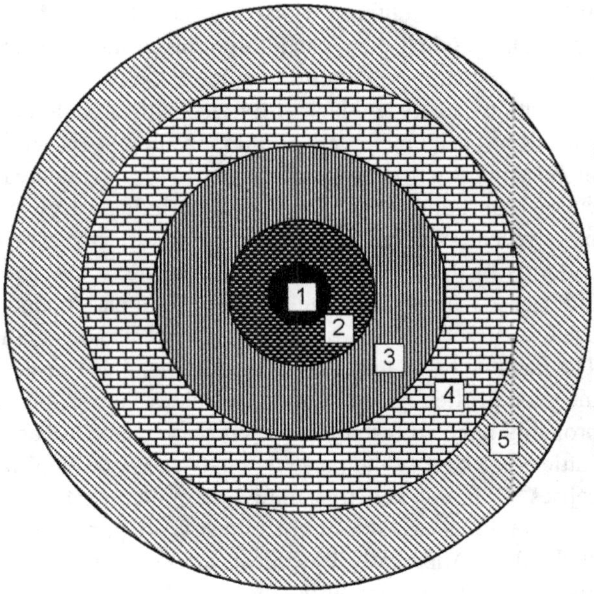

Photo 6.4 Mother elephant and its baby, Namibia. A. DeLella-Benedict 2012

100 years ago. Hunting for bison outside the park has become very popular these days. The good physical condition of bison shot outside the park's boundary seems to indicate that the animals do not leave the park during winter because of too high population density and lack of forage – which could justify regular reduction in the park. It has been hypothesized that traveling bison profit from the numerous snowmobile routes where walking to the winter grazing areas at lower elevations is less energy-consuming than clearing paths through deep snow (Meagher 1989a, b). A more recent study (Bruggeman et al. 2009), however, has shown that roads may certainly facilitate bison travel in certain areas, but that road grooming in winter is unlikely to support bison migration beyond park boundaries. Instead, bison mainly used 'self-groomed' off-road trails, as they also do in other areas (Gates et al. 2001). The fact that bison may spread brucellosis, thus threatening domestic cattle outside the park, is often ignored. However, it may be considered an irony of fate that cattle probably infected bison with this disease (Meagher and Meyer 1994). Since 1910, cattle were kept in the park to supply the park employees with meat. Brucellosis causes stillbirths. It is transmitted by oral contact with dead foetuses and infected placenta. In the absence of brucellosis bison population could grow by 29 % (Fuller et al. 2007). Apparently, brucellosis does not threaten the existence of bison, as cow bison become resistant during the second or third gestation period. Afterwards, stillbirths are rare.

As to the acceptance of wildlife control measures, it is of minor importance whether a species is native or has been introduced, as long as there is a 'good feeling' that they fit 'harmoniously' into the environment. Nature purists may be resentful to the newcomers and insist on preserving pure pristine nature, as, for example, in the case of the mountain goats in the Olympic Mountains or of the rheas (*Rhea americana*) which escaped in 2002 from private ground in northern Germany (Groß Schönau, close to the Schleswig-Holstein border). The latter have moved eastwards and multiplied successfully, profiting from favorable habitat conditions (abundant food, no effective control by predators). At present, about 120 rheas are living east of the Wakenitz River (Mecklenburg), and there is a good chance of continued population growth and expansion (news, nordkurier, 5 April 2013). Whereas some farmers are concerned about damage to agricultural corps, most local people consider these ratites to be a welcome tourist attraction in the economically weak Mecklenburg region (Communication R. Kinzelbach, 27 July 2013).

Public acceptance may gradually increase due to habituation of alien animal species. Traditional prejudices and fixed opinions have to be overcome, and it may take one or even more human generations to get there. Waiting for habituation may be a useful strategy for the advocates of reintroduction of large predators, for example (see also Baur and Hunziker 2001).

References

Aguirre, A. A., & Starkey, E. E. (1994). Wildlife disease in U. S. national parks: Historical and coevolutionary perspectives. *Conservation Biology, 8*(3), 654–661.

Anunsen, C. S., & Anunsen, R. (1993). Response to Scheffer. *Conservation Biology, 7*, 954–957.

Apollonio, M., Andersen, R., & Putman, R. (2010). Present status and future challenges for European ungulate management. In M. Apollonio, R. Andersen, & R. Putman (Eds.), *European ungulates and their management in the 21st century* (Vol. 27, pp. 578–604). Cambridge: Cambridge University Press.

Arcese, P., Hando, J., & Campbell, K. (1995). Historical and present-day-anti-poaching efforts in Serengeti. In A. R. E. Sinclair & P. Arcese (Eds.), *Serengeti II* (pp. 506–533). Chicago/London: University of Chicago Press.

Baur, H., & Hunziker, M. (2001). Welche Rolle spielen die "Luchsadvokaten" im "Luchskonflikt"? Ergebnisse einer sozialwissenschaftlichen Pilotstudie. *Informationsblatt Forschungsbereich Landschaft, 50*, 4–6.

Bauschmann, G. (2000). Alte Wege, neue Chancen – die Transhumanz. *Naturschutz und Landschaftspflege, 32*(10), 315–316.

Beinlich, B., Hampicke, U., Plachter, H., & Tampe, K. (1997). Erhaltung großflächiger Kalkmagerrasen und magerer Wirtschaftswiesen auf der Schwäbischen Alb. *Schriftenreihe für Landschaftspflege und Naturschutz, 54*, 53–76.

Bergmann, H. H. (1999). Winterökologie arktischer Gänse in Deutschland. *NNA Berichte, 12*(3), 105–112.

Bergström, R., & Hjeljord, O. (1987). Moose and vegetation interactions in northwestern Europe and Poland. *Swedish Wildlife Research Supplement, 1*, 213–228.

Bieber, C., & Ruf, T. (2005). Population dynamics in wild boar Sus scrofa: ecology, elasticity of growth rate and implications for the management of pulsed resource consumers. *The Journal of Applied Ecology, 42*(6), 1203–1213.

Bokdam, J., & Gleichman, M. (2000). Effects of grazing by free-ranging cattle on vegetation dynamics in a continental north-west European heathland. *Journal of Applied Ecology, 37*, 415–431.

Boom, K., & Ben-Ami, D. (2010). *Shooting our wildlife: An analysis of the law and policy governing the killing of kangaroo*. Sydney: THINK, the Kangaroo think tank, University of Technology.

Bowman, D. D. (2012). Conservation: Bring elephants to Australia? *Nature, 482*(7383), 30. PMID 22297953.

Bowyer, R. T., Van Ballenberghe, V., & Kie, J. G. (1997). The role of moose in landscape processes: Effects of biogeography, population dynamics and predation. In J. A. Bisonette (Ed.), *Wildlife and landscape ecology. Effects of pattern and scale* (pp. 265–287). New York: Springer.

Bruggeman, J. E., Garrott, R. A., Whitze, P. J., Bjornlie, D. B., Watson, F. G. R., & Borkowski, J. J. (2009). Bison winter road travel: Facilitated by road grooming or a manifestation of natural trends? In R. A. Garrott, P. J. White, & F. G. R. Watson (Eds.), *The ecology of large mammals in central Yellowstone* (pp. 603–621). Amsterdam/Boston/Heidelberg/London/New York/Oxford/Paris/San Diego/San Francisco/Singapore/Sydney/Toronto: Academic.

Bunzel-Drüke, M. (1997). Großherbivore und Naturlandschaft. *Schriftenreihe für Landschaftspflege und Naturschutz, 54*, 109–128.

Burger, J. (1987). A mountainous appetite: Olympic searches for the best way to rid the park of exotic mountain goats. *National Parks, 61*(1), 28–31.

Campbell, K., & Borner, M. (1995). Population trends and distribution of Serengeti herbivores: Implications for management. In A. R. E. Sinclair & P. Arcese (Eds.), *Serengeti II*. Chicago/London: University of Chicago Press.

Campbell, K., & Hofer, H. (1995). People and wildlife: Spatial dynamics and zones of interaction. In A. R. E. Sinclair & P. Arcese (Eds.), *Serengeti II* (pp. 534–570). Chicago/London: University of Chicago Press.

Caughley, G. (1976). Management and the dynamics of ungulate populations. In T. H. Coaker (Ed.), *Applied biology* (Vol. 1, pp. 183–246). London/New York: Academic.

Clark, P. E., Krueger, W. C., Bryant, L. D., & Thomas, D. R. (2000). Livestock grazing effects on forage quality of elk winter range. *Journal of Range Management, 53*, 97–105.

Cunningham, A. A. (1996). Disease risk of wildlife translocation. *Conservation Biology, 10*(2), 349–353.

Dahl, H.-J. (1982). Ökologie pur. *Natur, 12*, 74–79.

David, A. (1998). *Das Reh – "Freund und Leidwild" deutscher Jagd* (Jagdbuch-Report 98/99, pp. 5–10). Wiesbaden: Quelle & Meyer.

Decker, D. J., Shanks, R. E., Nielsen, L. A., & Parsons, G. R. (1991). Ethical and scientific judgements: Beware of blurred distinctions. *Wildlife Society Bulletin, 19*, 523–527.

Duelli, P. (1995). Biodiversität erhalten und fördern: Was sagt die ökologische Forschung dazu? In Eidgenössische Forschungsanstalt für Wald, Schnee und Landschaft (Eds.), *Forum für Wissen* (pp. 13–21). Birmensdorf.

Eisfeld, D., & Fischer, U. (1996). Das Management eingebürgerter Huftierarten – widerstreitende Interessen, fehlende Konzepte. In H. Gebhart, R. Kinzelbach, & S. Schmidt-Fischer (Eds.), *Gebietsfremde Tierarten. Auswirkungen auf einheimische Arten, Lebensgemeinschaften und Biotope, Situationsanalyse* (pp. 287–295). Landsberg: Ecomed Verlagsgesellschaft AG & Co KG.

Emmer, E. (2001). Ökologischer Jagdverband e. V. (ÖJV). Wald und Wild. *Natur und Landschaft, 76*(5), 232–234.

Fischer, S. F., Poschlod, P., & Beinlich, B. (1995). Die Bedeutung der Wanderschäferei für den Artenschutz zwischen isolierten Schaftriften. *Beiheft zu den Veröffentlichungen zur Naturschutz und Landschaftspflege, Baden-Württemberg, 83*, 229–256.

Fischer, S. F., Poschlod, P., & Beinlich, B. (1996). Experimental studies on the dispersal of plants and animals on sheep in calcareous grasslands. *Journal of Applied Ecology, 33*, 1206–1222.

Fuller, J. A., Garrott, R. A., White, O. J., Aune, K. E., Roffe, T. J., & Rhyan, J. C. (2007). Reproduction and survival of Yellowstone bison. *Journal of Wildlife Management, 10*, 2365–2372.

Ganzert, C. (1994). *Umweltgerechte Landwirtschaft, nachaltige Wege für Europa*. Bonn: Economica.

Garcia, C., Renison, D., Cingolani, A. M., & Fernández-Juricic, E. (2008). Avifaunal changes as a consequence of large-scale livestock exclusion in the mountains of Central Argentina. *Journal of Applied Ecology, 45*, 351–360.

Gates, C. C., Mitchell, J., Wierzchowski, J., & Giles, I. (2001). *A landscape evaluation of bison movements and distribution in Northern Canada*. Calgary: AXYS Environmental Consulting Ltd.

Geisser, H., & Reyer, H. U. (2004). Efficacy of hunting, feeding and fencing to reduce crop damages by wild boars. *Journal of Wildlife Management, 68*(4), 939–946.

Geisser, H., & Reyer, H. U. (2005). The influence of food and temperature on population density of wild boar Sus scrofa in the Thurgau (Switzerland). *Journal of Zoology, 267*(1), 89–96.

Gerken, B. (1999). Neuweltkameliden in der Landschaftspflege. In B. Gerken & B. Görner (Eds.), *Landschaftsentwicklung mit großen Weidetieren. Geschichte, Modelle und Perspektiven* (Natur- und Kulturlandschaft 3, pp. 236–240). Höxter/Jena: Huxaria Druck und Verlag.

Goretzky, J. (1999, June 16). Einfluß von Beutegreifern auf die Feldhasenpopulation. In Bundesminister für Ernährung, Landwirtschaft und Forsten (Ed.), *Feldhasensymposium, Informationsveranstaltung am*, Bonn.

Griggs, G. C., & Pople, A. R. (2001). Sustainable use and pest control in conservation: Kangaroos as a case study. In J. D. Reynolds, G. M. Mace, K. H. Redford, & J. G. Robinson (Eds.), *Conservation of exploited species* (pp. 403–423). Cambridge: Cambridge University Press.

Hercock, M., & Tonts, M. (2004). From the rangelands to the Ritz. Geographies of kangaroo management and trade. *Geography, 89*(3), 214–225.

Herrero, J., García-Serrano, A., & García-Gonzales, R. (2008). Reproductive and demographic parameters in the Iberian wild boar Sus scrofa populations. *Acta Theriologica, 53*(4), 355–364.

Hespeler, B. (1990). *Jäger wohin? Eine kritische Betrachtung deutschen Waidwerks*. München/Wien/Zürich: BLV.

Hobbs, R. J., & Huennecke, I. F. (1992). Disturbance, diversity amd invasion: Implications for conservation. *Conservation Biology, 6*, 324–337.

Hofmann, R., Cornelius, R., Linder, R., & Scheibe, K. (1998). Wiedererstehen des postglazialen Großsäugerspektrums als landschaftsgestaltendes Multi-Spezies-System: ein Pilotversuch zur "Redynamisierung" halboffener Landschaften. *Schriftenreihe für Landschaftspflege und Naturschutz, 56*, 301–315.

Holtmeier, F.-K. (1999). Tiere als landschaftsökologische Faktoren. Arbeiten aus dem Institut für Landschaftsökologie der Westfälischen Wilhlems-Universitat 6.

Holtmeier, F.-K. (2000). *Die Höhengrenze der Gebirgswälder* (Arbeiten aus dem Institut für Landschaftsökologie 8). Münster: Westfälische Wilhelms-Universität.

Holtmeier, F.-K. (2002). *Tiere in der Landschaft. Einfluss und ökologische Bedeutung*. Stuttgart: UTB 8230.

Holtmeier, F.-K. (2009). *Mountain timberlines. Ecology, patchiness, and dynamics* (Advances in Global Change Research 36). Dordrecht: Springer Science+Business Media B. V.

Hörnberg, S. (2001). Changes in population density of moose (Alces alces) and damage to forest in Sweden. *Forest Ecology and Management, 149*, 141–151.

Houston, D. B., Moorhead, B. B., & Olsen, R. W. (1986). An aerial census of mountain goats in the Olympic Mountain Range. *Northwest Science, 60*, 131–136.

Humphrey, J. W., & Patterson, G. S. (2000). Effects of late summer grazing on the diversity of riparian pasture vegetation in an upland conifer forest. *Journal of Applied Ecology, 37*, 986–996.

Huston, N. A. (1997). Hidden treatments in ecological experiments: Re-evaluating the ecosystem function of biodiversity. *Oecologica, 110*, 449–460.

Hutchins, M. (1995). Olympic mountain goat controversy continues. *Conservation Biology, 9*(5), 1324–1326.

Isenberg, A. C. (2000). *The destruction of the bison. An environmental history, 1750–1920*. Cambridge: Cambridge University Press.

Jedicke, E. (1994). *Biotopverbund. Grundlagen und Maßnahmen einer neuen Naturschutzstrategie*. Stuttgart: Ulmer.

Jeschke, L. (1997). Pflege einer Küstenheide auf Hiddensee durch Wildschafe. *Schriftenreihe für Landschaftspflege und Naturschutz, 54*, 177–188.

Job, H. (1999). Probleme afrikanischer Großschutzgebiete - die Situation Kenias und das Fallbeispiel Samabura National Reserve. *Petermanns Geographische Mitteilungen, 143*(1), 3–15.

Kampf, H. (2000). The role of large grazing animals in nature conservation – A Dutch perspective. *British Wildlife, 12*, 37–46.

Kelly, J. (2003). *The kangaroo industry, its image and market: improving one by improving the other*. Canberra: Rural Industries Research and Development Corporation.

Keuling, O. (2011). Managing wild boar – Considerations for wild boar management based on some biological data. In *8th European vertebrate pest management conference* (Julius-Kühn-Archiv 432, pp. 215–216). Quedlingburg: JKI.

Krüger, U. (1999). Das niederländische Beispiel: Die Oostvaardersplassen - ein Vogelschutzgebiet mit Großherbivoren als Landschaftsgestalter. *Natur und Landschaft, 74*(10), 428–435.

Kurt, F. (1977). *Wildtiere in der Kulturlandschaft*. Erlenbach/Zürich/Stuttgart: Rentsch Verlag.

Laiolo, P., Dondero, F., Ciliento, E., & Rolando, A. (2004). Consequences of pastoral abandonment for the structure and diversity of the alpine avifauna. *Journal of Applied Ecology, 41*, 295–304.

Lambeck, S., & Schwöppe, W. (1987). Schafbeweidung zur Pflege von Heiden und Mooren. *Unsere Heimat, Jahrbuch des Kreises Borken*, 73–80.

Lammers, I. (1994). De grote grazers van de nieuwe wildernis. *Panda, 30*(2), 4–11.

Leibundgut, H. (1975). *Wirkungen des Waldes auf die Umwelt des Menschen*. Erlenbach-Zürich/Stuttgart: Eugen Rentsch.

Leopold, A. (1933). *Game management* (reprinted in 1996). Dehra Dun: Natraj Publishers.

Leuthold, W. (1978). Kann Afrikas Großwild überleben? *Vierteljahresschrift der Naturforschenden Gesellschaft Zürich, 123*, 333–340.

Lumeij, J. T., & Oosterbaan, J. (2000). Gote grazers in het Nederlands wetland "Oostervardersplassen". *Tidschrift voor Diergeneeskunde, 127*(7), 230–234.

Lyman, R. L. (1998). *White goats white lies: The abuse of science in the Olympic National Park.* Salt Lake City: The University of Utah Press.

Magnusson, M. (2010). *Population and management models for the Swedish wild boar (Sus scrofa).* Swedish University of Agricultural Sciences, The Faculty of Natural Resources and Agricultural Sciences, Department of Ecology, Master's thesis 2010, 18, Grimsö Wildlife Research Station.

Marcot, B. G., Croft, L. K., Lehmkuhl, J. F., Naney, R. H., Niwa, C. G., Owen, W. R., & Sandquist, R. E. (1998). *Macroecology, paleoecology, and ecological integrity of terrestrial species and communities of the interior Columbia Basin and northern portions of the Klammath and Great Basin* (USDA Forest Service, Pacific Northwest Research Station: Portland OR. General Technical Report PNW-GTR 410). Portland: USDA Forest Service, Pacific Northwest Research Station.

Martin, D. (1997). Erfahrungen mit der Extensiv-Haltung von Fjellrindern im Müritz Nationapark. *Schriftenreihe für Landschaftspflege und Naturschutz, 54*, 161–175.

May, R. M. (1973). *Stability and complexity in model ecosystems.* New York: Princeton University Press.

Meagher, M. M. (1989a). Range of expansion by bison of Yellowstone National Park. *Journal of Mammalogy, 70*(3), 670–675.

Meagher, M. M. (1989b). Evaluation of boundary control for bison of Yellowstone National Park. *Wildlife Society Bulletin, 17*, 15–19.

Meagher, M. M., & Meyer, M. E. (1994). On the origin of Brucellosis in bison of Yellowstone National Park: A review. *Conservation Biology, 8*, 645–653.

Meng, X. J., Lindsay, D. S., & Sriranganathan, N. (2012). Wild boars as sources for infectious diseases in livestock and humans. *Philosophical Transactions of the Royal Society B, 364*, 2697–2707.

Messier, F. (1994). Ungulate population models with predation: A case study with the North American moose. *Ecology, 75*(2), 478–488.

Mildenberger, H. (1971). Wildschäden durch Gänse. *Charadrius, 7*, 13–15.

Millett, J., Hester, A. J., Millard, P., & McDonald, A. J. S. (2006). How do different competing species influence the response of Betula pubescens Erh. to browsing? *Basic and Applied Ecology, 7*, 123–132.

Mooij, J. H. (1999a). Übersicht über die Bestandessituation und Bestandesentwicklung der Gänse in Deutschland und der westlichen Paläarktis. *NNA Berichte, 12*(3), 113–126.

Mooij, J. H. (1999b). Kann die Jagd zur Verringerung der Gänseschäden beitragen? *NNA Berichte, 12*(3), 164–172.

Moorhead, B. B. (1977a). The enigmatic mountain goat. *Pacific Search, 11*(8), 23.

Moorhead, B. B. (1977b). *Status and management of the mountain goat in the Olympic Mountains, Washington.* Prepared for status and management of the mountain goat in North America by Rolf Johnson. First International Mountain Goat Symposium, Northwest Section Meeting, The Wildlife Society of Kalispell, Montana.

Moorhead, B. B. (1981). Olympic N P stages well-run removal of exotic goats. *Park Science, 2*(1), 5.

Moorhead, B. B. (1989). Non-native mountain goat management undertaken at Olympic National Park. *Park Science, 9*(3), 10–11.

Nitsche, S., & Nitsche, L. (1994). *Extensive Grünlandnutzung*. Radebeul: Neumann.

Noss, R. F., Graham, R., McCullough, D. R., Ramsey, F. L., Seavey, J., Whitlock, C., & Williams, M. P. (2000). *Review for scientific material relevant to the occurrence, ecosystem role, and tested management options for mountain goats in Olympic National Park.* Corvallis: Conservation Biology Institute.

Ökologischer Jagdverein. (1994). Waldgrechte Jagdmethoden. *Natur und Landschaft, 64*(9), 425.

Oosterveld, P. (1975). Beheer en ontwickeling van natuurreservaten door begrazing. *Natuur en Landschap, 29*, 161–171.

Oosterveld, P. (1979). Maaien, grazen of stuiven. Via naturrbeheer naar meer natuur. *Duin, 2*, 3–8.
Oppermann, R., & Luick, R. (1999). Extensive Beweidung und Naturschutz. Charakterisierung einer dynamischen und umweltverträglichen Landnutzung. *Natur und Landschaft, 74*(10), 411–419.
Peterson, M. J. (1991). Wildlife parasitism, science and management policy. *Journal of Wildlife Management, 55*, 782–789.
Petrak, M. (1992). Rotwild (Cervus elaphus Linné, 1758) als Pflegefaktoir für bärwurzreiche Magertriften (Arnicetum montanae Schw. 1974 = Meso-Festucetum) in der Nordeifel. *Zeitschrift für Jadgwissenschaft, 38*, 221–234.
Petrak, M. (1999). Grenzüberschreitende Rotwildhege im deutsch-belgischen Naturpark Nordeifel-Hohes Venn. *LÖBF-Mitteilungen, 4*, 28–31.
Pfitsch, W. A. (1980). *The effect of mountain goats on the subalpine plant communities of Klahane Ridge, Olympic National Park, Washington*. M.S. thesis, University of Washington.
Pflug, W. (1987). Der Naturschutz und die Natur. *Naturschutzzentrum NRW, Seminarbericht, 1*(1), 5–10.
Pike, D. K. (1981). *Effects of mountain goats on three plant species unique to the Olympic Mountains, Washington*. M.S. thesis, University of Washington.
Popp, A. (1999). Zur Habitatnutzung von Wisenten (Bison bonasus bonasus) im Gehege und im Freiland. *Artenschutzreport, 9*, 16–21.
Pott, R., & Hüppe, R. (1991). *Die Hudelandschaften Nordwestdeutschlands* (Abhandlungen aus dem Westfälischen Museum für Naturkunde 53 (1/2)). Münster: Landschaftsverband Westfalen-Lippe.
Ratcliffe, P. R. (1987). *The management of red deer in upland forests* (Forestry commission bulletin, 71). London: HMSO.
Redpath, S., & Thirgood, S. (2009). Hen harriers and red grouse: Moving towards consensus. *Journal of Applied Ecology, 46*, 961–963.
Reimoser, F. (2003). Steering the impacts of ungulates on temperate forests. *Journal of Nature Conservation, 10*, 243–252.
Remmert, H. (1973). Über die Bedeutung warmblütiger Pflanzenfresser für den Energiefluß in terrestrischen Ökosystemen. *Journal für Ornithologie, 114*, 227–249.
Riecken, U., Finck, P., Klein, M., & Schröder, E. (1998). Schutz und Wiedereinführung dynamischer Prozesse als Konzept des Naturschutzes. *Schriftenreihe für Landschaftsplanung und Naturschutz, 56*, 7–19.
Ruiz-Fons, F., Segalés, J., & Gortázar, C. (1997). A review of viral diseases of the European wild boar: Effects of population dynamics and reservoir role. *Veterinary Journal, 176*(2), 158–169.
Saavalainen, H. (2010, November 10). Finland plans to reduce the number of European lynx. *Helsingin Sanomat* – International Edition, Helsinki.
Sáez-Royuela, C., & Tellería, J. L. (1986). The increases population of wild boar (Sus scrofa) in Europe. *Mammal Review, 16*(2), 97–101.
Sanders, W., & Webster, D. (1994). Preindustrial man and environmental degradation. In K. C. Kim & D. Weaver (Eds.), *Biodiversity and landscapes* (pp. 77–104). Cambridge: Cambridge University Press.
Sandström, C., Pellikka, J., Ratamäki, O., & Sande, A. (2009). Management of large carnivores in Fennoscandia: New patterns of regional participation. *Human Dimensions of Wildlife, 14*, 37–50.
Scherrer, H. U., & Surber. E. (1978). *Behandlung von Brachland in der Schweiz* (Eidgenössische Anstalt für das forstliche Versuchswesen, Berichte 189). Bimensdorf: Eidgenössische Anstalt für das forstliche Versuchswesen.
Scherzinger, W. (1996). *Naturschutz im Wald, Qualitätsziele einer dynamischen Waldentwicklung*. Stuttgart: Ulmer.
Schley, L., & Roper, T. J. (2003). Diet of the wild boar in western Europe, with particular reference to consumption of agricultural crops. *Mammal Review, 33*, 43–66.
Schreiber, K.-F. (2005). *Offenhaltunf von Landschaften – Wissenschaftliche Begleituntersuchungen auf Brachflächen mit unterschiedlichem Management in Baden-Württemberg* (Laufener Seminarbeiträge 1/04) (pp. 33–53). Ausgburg: Bayerische Akademie für Naturschutz und Landschaftspflege.

Seidenschnur, W. (1998). *Rehwild im naturnahen Waldbau* (Jagdbuch-Report 98/99, pp. 20–24). Wiesbaden: Quelle & Meyer.

Senn, J. (2000). Huftiere und Verjüngung im Gebirgswald: eine Geschichte mit vielen Variablen und noch mehr Interaktionen. *Schweizerische Zeitschrift für Forstwesen, 151*, 99–106.

Simonetti, J. A. (1995). Wildlife conservation outside parks is a disease-mediated task. *Conservation Biology, 9*(2), 454–456.

Sotherton, N., Tapper, S., & Smith, A. (2009). Hen harriers and red grouse: Economic aspects of red grouse shooting and the implication for moorland conservation. *Journal of Applied Ecology, 46*, 955–960.

Streule, A. (1973). Schäden in Gebirgsaufforstungen durch das Birkhuhn (Lyrurus tetrix). *Bündnerwald, 26*(8), 249–254.

Stubbe, C. (1999, June 16). Zur Situation des Feldhasens in Deutschland. In Bundeminsiterium für Ernährung, Landwirsctahft und Forsten (Ed.), Feldhasensyposium, Informationsveran-staltung am, Bonn, pp. 5–10.

Sturm, P. (2000). *Vom Aussterben bedroht: Situation und Bestandesentwicklung hochgradig gefährdeter Arten in Bayern* (Laufener Seminararbeiten 3/00, pp. 37–43). Laufen/Salzach: Bayerische Akademie für Naturschutz und Landschaftspflege.

Suchant, R., Baritz, R., & Braunisch, V. (2003). Wildlife habitat analysis – A multidimensional habitat management model. *Journal for Nature Conservation, 10*, 253–268.

Südbeck, P., & Königstedt, B. (1999). Gänseschadensmanagement in Niedersachsen. *NNA Berichte, 12*(3), 145–151.

Thalen, D. C. P. (1981). Grote grazers en snelle snoeiers - het beheer an vegetaties. In *Verslag van het Veluwe-Syposium gehouden op 9 mei 1980 te Amsterdam* (pp. 21–29), Arnheim.

Thalen, D. C. P. (1984). Large mammals as tools in the conservation of divers habitats. *Acta Zoologica Fennica, 172*, 159–163.

Thirgood, S. J., Redpath, S. M., Campbell, S., & Smith, A. (2002). Do habitat characteristics influence predation on red grouse? *Journal of Applied Ecology, 39*, 217–225.

Thompson, P. S., Amar, A., Hoccom, D. G., Knott, J., & Wilson, J. D. (2009). Resolving the conflict between driven-grouse shooting and conservation of hen harriers. *Journal of Applied Ecology, 46*, 950–954.

Van Aarde, R., Whyte, L., & Pim, S. (1999). Culling and the dynamics of the Kruger National Park African elephant population. *Animal Conservation, 2*, 287–294.

Van Ballenberghe, V., & Ballard, W. B. (1994). Limitation and regulation of moose population: The role of predation. *Canadian Journal of Zoology, 72*, 2071–2077.

Van Wieren, S. E. (1991). The management of populations of large mammals. In I. F. Spellenberg, G. M. Goldsmith, & M. G. Moser (Eds.), *The scientific management of temperate communities for conservation* (pp. 103–127). Oxford: Blackwell Scientific.

Van Wieren, S. E. (1995). The potential role of large herbivores in nature conservation and extensive land use in Europe. *Biological Journal of the Linnean Society, 56*(Suppl), 11–23.

Vera, F. W. M. (2009). Large-scale nature development – The Osstvaardersplassen. *British Wildlife, 20*(5), 28–36.

Vökl, W. (1999). Schalenwild und Sukzession: Welche Rolle können einheimische Wildtiere beim Erhalt der Kulturlandschaft spielen? In B. Gerken & M. Görner (Eds.), *Europäische Landschaftsentwicklung mit großen Weidetieren. Geschichte, Modelle und Perspektiven* (Natur und Kulturlandschaft 3, pp. 330–317). Höxter/Jena: Huxaria Druck und Verlag.

Völkl, W. (1997). Die Offenhaltung von Grünland in Mittelgebirgen – Problematik und Möglichkeiten anhand eines Beipiels aus dem Fichtelgebirge. *Schriftenreihe für Lanschaftspflege und Naturschutz, 54*, 85–91.

Vorlaufer, K. (1998). Die Selous Game Reserve/Tanzania – Naturschutz, Jagdtourismus und nachhaltige Entwicklung in Afrikas größtem Wildreservat. *Erdkunde, 52*, 315–329.

Wagner, H. (1993). "Natur total" – Biotoppflege oder Sukzession? In Landsamt für Umweltschutz Baden-Württemberg (Ed.), *Landschaftspflege – Quo vadis?* (pp. 38–52). Karlsruhe.

Wagner, F. H., Foresta, R., Gill, R. B., McCullough, D. R., Pelton, M. R., Porter, W. F., & Salwasser, H. (1995). *Wildlife policies in the U. S. National Parks*. Washington, DC: Island Press.

Wallis de Vries, M. (1995). Large herbivores and the design of large-scale nature reserves in western Europe. *Conservation Biology, 9*, 25–33.

Wallis de Vries, M. (1999). The dilemma facing natural conservation and the role of large herbivores. In B. Gerken & M. Görner (Eds.), *Europäische Landschaftsentwicklung mit großen Weidetieren. Geschichte, Modelle, Perspektiven* (Natur und Landschaft 3, pp. 24–31). Höxter: Universität-Gesamthochschule Paderborn, Abt. Höxter, Lehrgebiet Tierökologie.

White, P. S., & Pickett, S. T. A. (1985). Natural disturbance and patch dynamics: An introduction. In S. T. A. Pickett & P. S. White (Eds.), *The ecology of natural disturbance and patchy dynamics* (pp. 3–13). Orlando: Academic Press.

Whyle, I., van Aarde, R., & Pimm, S. L. (1998). Managing the elephants of Krueger National Park. *Animal Conservation, 1*, 77–83.

Wibbels, V. L. (1990). *Wetland management in the Oostvaardersplassen*. Ministerie van Verkeer en Waterstaad, Directoraat - General Rijkswaterstaat, Directoraat Flevoland, Center for Integrated New-Land Development.

Wibbels, V. L. (1994). De edelherten van Oostvaardersplassen. *Zoogdier, 5*(3), 4–10.

Willems, J. H. (1983). Species composition and above ground phytomass in chalk grassland with different management. *Vegetatio, 52*, 171–180.

Wilson, C. J. (2005). *Feral wild boar in England. Status, impact and management*. Defra European Wildlife Division. www.defra.gov.uk

Wotschikowski, U. (2010). Schweden eröffnet die Jagd auf Wölfe. *Naturschutz und Landschaftsplanung, 42*(7), 218–224.

Wright, R. G. (1992a). *Wildlife research and management in the national parks*. Urbana/Chicago: University of Illinois.

Wright, R. G. (1992b). *Case study: Removal of exotic mountain goats from Olympic National Park*. Problem statement presented at Strategic Ecological Research Workshop, National Park Service, 24–26 February 1991, University of New Mexico, Albuquerque.

Wright, R. G. (1998). Review on White goats white lies: The abuse of science in Olympic National Park (Lyman, R. L. 1998). *Northwest Science, 73*(3), 236–238.

Yellowstone National Park. (1997). *Yellowstone's northern range complexity and change in a wild land ecosystem*. Mammoth Hot Springs: National Park Service.

Zimmermann, P., & Woike, M. (1982). Das Schaf in der Landschaftspflege. *LÖBF Mitteilungen, VII*(2), 1–3.

Zoller, H., & Bischoff, N. (1980). Stufen der kulturintensität und ihr Einfluß auf Artenzahl und Artengefüge der Vegetation. *Phytocoenologia, 7*, 35–51.

Chapter 7
Potential Effects of Climate Change on Animal-Habitat Interactions

Not much fantasy is needed to predict that substantial climate change will have far-reaching and lasting effects on ecosystems (structures, function, dynamics) and their biocoenoses (e.g., Parmesan and Yohe 2003; Root et al. 2003; Hulme 2005). Although in the winter-cold regions warming winters will probably have a stronger impact on plants than on the northern ungulates (e.g., red deer, reindeer, muskox; Post and Stenseth 1999), continued warming and stabilization of the thermal conditions at a higher level than at present will also bring about change in animal phenology (e.g., breeding, hatching, whelping, mating), forage quality, foraging behavior, social behaviour and competition (Fig. 7.1). In addition, climate change may influence the timing of seasonal migration, migration routes, and distances between wintering grounds and summer habitats. Not least, changing climate may influence survival/mortality rates, population cycles, and thereby, the animals' effects on vegetation composition, structure, and succession through grazing or browsing (consumption of green phytomass, selectivity, nutrient flow, etc.), trampling and burrowing (soil erosion, bioturbation). Feedback on changing habitat conditions for animals would also change (e.g., food supply, seasonal changes in forage quality, shelter-giving structures, availability of water, etc.). The interrelationships between the many factors will not linearly change.

Information on current interrelationships between climate, ecosystem and biocoenoses is abundant (e.g., seasonal migrations, range expansion, shifts of distribution limits, competitiveness of species, etc.) (e.g., review by Walther et al. 2002). However, information is often anecdotal or too general. Comparable systematic long-term studies and sound data are rare or even missing. Predictions usually focus on the influence of climate change on animals, whereas potential effects of animal populations on their changing habitats are marginally considered; with the exception of 'pest animals' or animals that are likely to become a pest (e.g., outbreak insects, voles) in the warmer future.

Predictions of climate change are based on computation models. Although these have become perfected considerably during recent years, there still are – and will probably always be – great uncertainties in the prediction of future climate,

F.-K. Holtmeier, *Animals' Influence on the Landscape and Ecological Importance*,
DOI 10.1007/978-94-017-9294-3_7

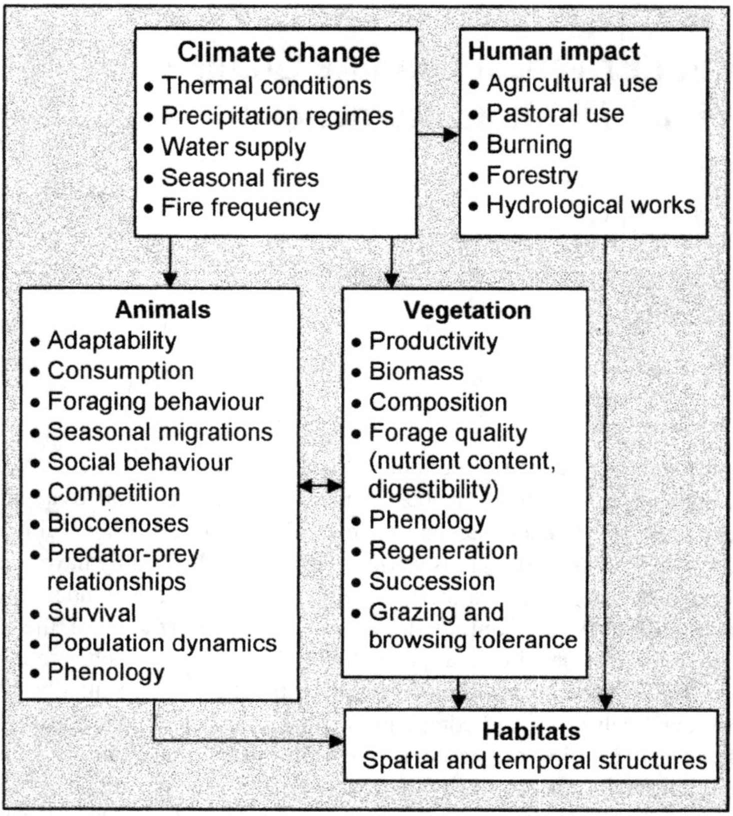

Fig. 7.1 Climate change and its possible effects (simplified) (Scheme by F.-K. Holtmeier)

particularly at the regional or even smaller scales, as in mountain systems with varied topography, for example (e.g., Giorgi and Hewitson 2001; Nogués-Bravo et al. 2007). Assessment of ecosystems' response to climate change and of animals' influence on their habitats is even more difficult, not only because of the uncertainties in the predictions of the future climate but also because of the greater complexity of the often inscrutable and hardly quantifiable interrelationships between organisms and their environment, even if compared to the many variables in the climate system.

Not least, we do not know whether the present interrelationships of the many biological and physical factors and their relative intensity will be the same in a warmer climate (Holtmeier 2000; Giorgi and Hewitson 2001; Holtmeier and Broll 2005). Adaptability of animals (e.g., population dynamics, food selection, competition, predator avoidance) and plants (e.g., productivity, sensitivity of herbivore impact, regeneration) to the future conditions will play an important role in this respect. In intensively cultivated landscapes, human impact often overlaps, and thus disguises possible influences of climatic change. Even in remote areas (e.g., Antarctic, subantarctic islands; see below), this aspect has to be considered. Thus, hypotheses on possible consequences of climate change (e.g., ACIA 2004) for the interrelationships between animal populations and their habitats are very general and vague.

Mechanisms influencing population dynamics will possibly change in a warming climate. The effects of warming climate on animals and their effects on landscape might be strongest in relatively cold regions, and in regions where drought periods are likely to increase. In the Arctic, climate warming has been about two times higher than on the global average. Winter temperatures have more increased than temperatures in summer. As a result, many species might benefit from higher overwintering survival, increased population growth and a longer growing season. On the other hand, the relatively strong influence of warmblooded herbivores (reindeer/caribou, muskox, moose, geese, small mammals) on phytomass and nutrient turnover (Sect. 2.1.1) might decrease, whereas decomposition rates and availability of nutrients (e.g., nitrogen, phosphorus), and thus, plant productivity (forage supply) would probably increase due to higher temperatures.

Increased insect harassment during warmer and longer summers may cause higher mortality rates in reindeer and caribou. Moreover, a positive relationship between infection of reindeer by parasites such as nematodes (e.g., brain worm, *Elaphostrongylus rangiferi*; causing cerebrospinal elaphostrongyloses, CSE) and summer temperature is apparent. The brain worm was introduced to Newfoundland in the early 1900s when infected reindeer were translocated from Norway. Since then, it has spread over the entire caribou range. An increase in disease and death of reindeer and caribou due to parasites has been predicted (Halvorsen 2012).

In addition, muskoxen become increasingly infected with the protostrongylid muskox lung worm (*Umingmakstrongylus pallikuukensis*; Kutz et al. 2005). As even a moderate increase of temperature (1 °C) may shorten the 2-year cycle for the development of the lungworm larvae to a 1-year cycle, infection with this parasite is likely to increase in the future (Gallana et al. 2013). Vulnerability, however, of the muskox populations will be different dependent on whether they have little or no previous exposure (immunity; Kutz et al. 2005). Anyway, in case of locally increasing muskox mortality, vegetation patterns as described from Ellesmere Island (Sect. 3.3.1; Photo 3.8), for example, will probably undergo considerable change.

Reindeer and caribou may also change their grazing behaviour. On wind-exposed terrain that would be increasingly frequented by the ungulates to escape molestation by insects, increase of trampling damage and subsequent erosion is likely, and may prevent climatically-driven expansion of dwarf shrubs, such as downy birch (*Betula nana*) and greyleaf willow (*Salix glauca*) into such locations (den Herder et al. 2008; Ost and Pedersen 2008; Zöckler et al. 2008; Olofsson et al. 2009). On the other hand, increasing warm-spells and freeze-thaw in the winter will result in an often icy snow surface which prevents reindeer, caribou and muskoxen as well from feeding on the ground vegetation. Mortality in the ungulates may increase due to malnutrition. Consequently, their impact on plant cover and soil will decrease. Nevertheless, according to resent studies near Abisko (Sweden), Kilpisjärvi (Finland), Dovre fjell and Finnmarksvidda (Norway) moderate reindeer grazing might impede climatically driven expansion of lowland plants into the mountain tundra, as lowland plants, typically taller and also richer in nutrients than tundra plants, become more easily detected and eaten by reindeer. Thus, rapid changes in plant cover due to climate change appear to be more likely in ungrazed areas (Umeå universitet 2014; see also Sect. 3.4.2).

Climatic warming will probably bring about changes in snow depth, although local contrasts in distribution and depth of the snowpack would not change much, as they depend on microtopography (relocation of snow by wind; Holtmeier 2005; Holtmeier and Broll 2012). Varying snow depth would influence food availability and foraging behaviour of the willow grouse (*Lagopus lagopus*). At increasing snow depth, willow grouse and ptarmigan would switch from dwarf shrubs to willow shrubs (*Salix* spp.), clipping buds, staminate catkins and fresh shoots projecting beyond the snow surface (Pulliainen and Iivanainen 1981). If willows were completely covered with snow, the birds would feed on mountain birch not covered with snow (Hakkarainen et al. 2007).

A warming climate, influencing forage quantity and quality, is also likely to influence population cycles of small rodents (lemmings, voles, hares) and their predators in the Subarctic/Arctic. Changes in the subnivean conditions (temperature, humidity, moisture) due to increasing freeze-thaw cycles play a major role. Reduced thermal insulation by the winter snowpack adversely affects lemmings and other small rodents. Moreover, freezing on the ground deprives them of food and, in addition, they become more vulnerable to predators. Maybe that cyclicity will become less pronounced (Ims and Fuglei 2005; Ims et al. 2008; Kausrud et al. 2008). In the Norwegian Alpine, for example, regular mass-outbreaks have not occurred during the last two decades. In a changing climate, which may bring about more precipitation and higher temperatures, lemming cycles are likely to cease. As voles, lemmings snowshoe hares and insects are important agents in the terrestrial food web, declining rodent cycles will probably have a far-reaching impact on the ecosystems (Henttonen and Wallgren 2001; Ims et al. 2008). Thus, the decline of arctic foxes and snowy owls in Scandinavia is probably related to less regularly occurring rodent peaks and increasing moss cover, which can spread because of reduced rodent impact (Ims and Fuglei 2005). In Taymyr peninsula (Russia), higher summer temperatures and stronger droughts are expected to cause dramatic decline in lemming populations. Their predators (foxes and predatory birds) would switch to alternative prey such as red-breasted goose (*Branta ruficollis*) and dark-bellied Brent goose (*Branta bernica bernicla*) and affect the geese populations. Changes in plant cover are likely in case mammalian and avian herbivores decline.

Although rodent cycles are fading in most parts of Europe, rising temperature may encourage expansion of rodents such as the common vole (*Microtus arvalis*) and field vole (*Microtus agrestis*) in agricultural landscapes. In many cases, however, other factors than climate appear to be the main drivers. In the northwest of Spain, for example, common vole, typically preferring moist grassland, has expanded from humid mountains to adjacent agricultural plains. Since the beginning of the 1980s, unprecedented outbreaks have occurred at intervals of about 5 years in spite of increasing climatic aridity. Extension of irrigation systems probably compensates for dry climatic aridity and facilitates expansion of voles (Luque-Larena et al. 2013). In any case, damage to agricultural crops has increased and local outbreaks of tularemia have been reported (see also Jacob and Tkadlec 2010).

Insects in particular are likely to benefit from a warmer climate (e.g., Solbreck 1991; Williams and Liebhold 1995). At higher latitudes and elevations, forest damage by bark beetles (Sect. 2.4.3.2), for example, and some other noxious forest insects is likely to increase, as has recently occurred across western North America (Bentz et al. 2012). Trees stressed before by drought or defoliation are usually highly susceptible to bark beetle attacks. Increased tree mortality can influence forest ecosystems and biocoenoses in the long-term. However, other factors, such as a possible higher fire frequency (as a result of drought periods) and/or predators also profiting from warmer conditions, may have adverse effects. In many areas, however, wildfires are increasingly suppressed by systematic fire-fighting. Consequently, trees may grow older and bigger in diameter, thus providing favourable conditions (thick phloem) for bark beetle reproduction (see also Figs. 2.37 and 2.38).

Mountain pine beetle populations are expanding to higher latitude and elevation (Aukema et al. 2008; Mitton and Ferrenberg 2012). The pine beetles respond to warming climate through faster development, with the flight season beginning more than a month earlier and lasting twice as long than previously. As the eggs are deposited earlier, they may develop into adult pine beetles during the warmest month of the year. In the Colorado Rocky Mountains, for example, the upper limit of bark beetle damage has advanced from about 2,700 m to above 3,350 m elevation. In other words, the mountain pine beetle now occupies the entire mountain forests up to the treeline (e.g., Holtmeier 1978, 1996). As trees at high-elevations are stressed by harsh climatic conditions anyway (Holtmeier 2009, further literature therein), they are particularly vulnerable to bark beetle attacks. Anomalous large outbreaks of the spruce beetle (*Dendroctonus rufipennis*) in both the Kluane National Park and Reserve (Yukon) and the Kenai Peninsula (Alaska) from 1994 to 2004, as well as the current epidemic of the mountain pine beetle (*Dendroctonus ponderosae*) extending from the southwestern United States to the Yukon Territory (Sect. 2.4.3.2.1; Logan and Powell 2001; Taylor et al. 2006; Mitton and Ferrenberg 2012), may give an idea of what is likely to happen in the warmer future (Berg et al. 2006). In central British Columbia (Canada), for example, the current mass-outbreak has killed about 10 million ha of lodgepole pine, resulting in immense economic loss and serious ecological problems (Natural Resources Canada 2005). There is some evidence that severe mountain pine beetle infestation might cause an increase in surface temperature of 1 °C in this region. Consequences for the climate (latent and sensitive heat fluxes, cloud cover, precipitation) are likely (Maness et al. 2013).

Modelled egg mortality in dependence on the local climates in the Kevo area (Virtanen et al. 1998) provides an interesting forward look at possible changes in the areas that will be prone to *Epirrita* attacks during warmer winters. In this case, with winter temperatures increased as predicted by different scenarios (Räisänen 1994: further literature in Virtanen et al. 1998), the area characterized by extremely low winter temperatures and high egg mortality would shrink by a third by the middle of the present century (Neuvonen et al. 2001). At the end of the century, only a tenth of the area would be left. Thus, more birch forests would succumb to defoliation and a rise of the birch treeline due to the warming climate would be less likely. The influence of higher summer temperatures is still controversially debated and not well understood (Neuvonen et al. 2001). Opposite factors might buffer or even prevent increase of mass-outbreaks (Bylund 1997, 1999; Virtanen and Neuvonen 1999). Even if birch forests recovered more rapidly after defoliation under warmer climatic conditions, it would take a long time, as ever, until forest structures would have developed that meet the requirements for mass-outbreaks, and forest species composition would not completely change (Bylund 1999). Moreover, in a warming climate, the influence of the predators and parasites of the autumnal moth would increase and reduce larvae density (e.g., Laine and Niemelä 1980; Karhu and Neuvonen 1998; Virtanen and Neuvonen 1999). It is uncertain how early leafing will influence hatching date and further development of the caterpillars (Bylund 1999). Altogether, there is considerable scope for speculation. As the varying local topography will also modify the thermal conditions in a warmer future, any prediction has to be considered with great reserve.

The winter moth (*Operophtera brumata*) will probably continue expanding its area of distribution in northern Norway and northern Finland (Sect. 2.4.1). Interspecific competition will not stop it (Ammunet et al. 2010). The long-term consequences of overlapping winter moth and autumnal moth have been modelled using empirical data from the recent outbreak (Ammunet et al. 2014). Accordingly, radical changes in host plant quality appear to be unlikely, while stressed birches may increasingly become affected by pathogens (Ammunet et al. 2011). Due to the great complexity of the interrelationships between the defoliators, climate, predators, parasites, pathogens and host trees, the long-term response of subarctic birch forests to the moths' impact under the influence of climate change is difficult to predict. In any case, the outbreaks of defoliating insects have been and will continue to be major factors in birch forest and treeline dynamics (e.g., Holtmeier and Broll 2006; Holtmeier 2009).

In Central Europe, warming climate and mild winters may promote expansion of nutria, for example, which does not tolerate extreme winter cold. However, abundant food supply (agricultural crops, gardens), insufficient control and winter feeding by people will very likely be of greater importance. Mild winters reducing young wild boar mortality, together with increased fructification in oak and beech due to warmer conditions, will also support wild boar population growth. The positive effect of maize cultivation on wild boar populations, however, may overlap the influence of climate change as a driving factor. Milder and shorter winters may also positively influence survival rate in red deer, roe deer and exotic deer species. Increasing damage to forests is likely. Therefore, careful game and landscape management will be necessary to adjust cervid numbers to the future tolerable habitat carrying capacity.

In tropical savannas, increasing CO_2-fertilization, changing precipitation and fire regimes will change vegetation, and thereby, habitat conditions. As it happened in Tsavo, where the elephant population collapsed after the extreme drought in 1970/1971 (Corfield 1973; Leuthold 1978; Cobb 1980; Botkin et al. 1981), and in Gonarezhow National Park (Zimbabwe), where 1,500 elephants died after a severe drought in 1991/1992 (Gandiva and Zisadza 2010), increasing extreme droughts would probably become a serious threat to the elephant populations and also to many other wildlife species. However, even a slight increase in droughts' frequency and/or intensity is expected to considerably reduce forage and water availability and thus ungulate populations (e.g., Ogutu and Owen-Smith 2003, 2008; Duncan et al. 2012) Response to climate change will vary among savanna types (e.g., Lehmann et al. 2014). Open woodland may replace grassland, for example (e.g., Wigley et al. 2010; Buitenwerf et al. 2012). This shift would probably vary locally and also be asynchronous (Higgins and Scheiter 2012). Browser feeding on the leaves of shrubs and trees may get an advantage compared to grazers subsisting on the ground vegetation. Consequently, present animal communities and their influences on the environment will change. Moreover, seasonal migration routes of the large ungulate herds (e.g., Serengeti-Mara ecosystem) and their far-reaching effects on vegegation, nutrient transport and turnover may also change. Information on these issues is rare, however (e.g., Kaeslin et al. 2012 for review), and general predictions cannot be given. Nevertheless, human impact (rapidly growing human population and increasing land use, poaching) may become a more serious threat than climate change to elephants and other large herbivores, as it will prevent them from using the migration routes to their traditional seasonal foraging areas.

In Australia's savannas rapidly expanding pastoralism and livestock production (e.g., Williams et al. 2005) also overlap with climate change. An increase of 2 °C in temperature has been predicted kangaroo populations in northern Australia to decline by about 50 %. Kangaroos' response to rising temperatures would, however, differ by species. Antilopine wallaby (wallaro; *Macropus antilopinus*), which is adapted to humid tropical climate, would be most affected and decrease by about 90 %. If temperature increased by 6 °C (what seems very unlikely) all macropods would be reduced almost completely, and antilopine wallaby would become extinct (Ritchie and Bolitho 2008). Kangaroos would probably be more affected by reduced availability of water (as a result of higher evaporation) than by direct influence of higher temperatures.

In temperate grasslands (e.g., North American prairie, Eurasian steppes, Patogonia etc.) large grazers will probably be negatively affected by sustained hotter and drier conditions that may reduce forage quality (protein-rich forage; e.g., Craine 2013). In Central Asia, freely roaming ungulates (cf. Sect. 3.2.1), their habitats and migration routes are threatened more by unprecedented exploitation of mineral resources than by climate change. Compared to these human impacts the animals' influence on steppe vegetation and soils will be of minor importance.

In the tropical oceans, sea-level rise may flood low atolls, and thus deprive seabirds of nesting sites (Brooke 2004). In the Californian Current, a rise in sea temperature of 0.5 °C and a consequent reduction in plankton have possibly contributed to an almost complete (90 %) loss of millions of sooty shearwaters between 1987 and 1994 (Veit et al. 1996, 1997). Climate change has been hypothesized to be the driving factor. The situation may be aggravated in the future. In New Zealand, where temperatures have been rising for the last 20 years, heat stress and blowfly attacks have increased nestling mortality of Northern Royal Albatross (*Diomedia sanfordi*) (Robertson 1998).

Along the arid coast of Peru and northern Chile, El Niño events adversely affect guano birds (Sect. 2.1.2) as the shoals of fish (anchovy, *Engraulis encrasicholous*) in the upper ocean layers decline. The birds cannot compensate for energy expended on catching fish at greater depth. The young, in particular, suffer from malnutrition and die due to starvation. The impact of the severe El Niño of 1957, for example, had a drastic impact on the guano bird population, which dropped from estimated 27 million to 6 million. Afterwards, the population gradually recovered. Since the mid-1950s, however, a general decrease in guano birds is obvious (Clarke 2008).

In the Antarctic and on subantarctic islands, considerable increase in temperature and extensive melting of the ice shelves have occurred over the past 50 years (e.g., Croxall et al. 2002; Gille 2002; Weimerskirch et al. 2003). Climate change will alter the marine food web (e.g., Everson 1977; Laws 1985). The birds' food at sea might diminish, or it might shift towards the poles, increasing the distance between bird colonies and food (Brooke 2004). Changes in predator-prey interactions are already affecting seabird populations. However, the effects differ regionally and between species. In different (ice-tolerant and ice-intolerant) penguin species, for example, poleward shifts and shrinking, as well as expanding ranges, have been observed during the recent decades (e.g., Liu et al. 2005; Forcada and Trathan 2009, further literature therein).

A 1,300-year record of penguin population (Gentoo, chinstrap and Adélie penguins) on Ardley Island (South Shetland Islands) has documented a relative population stability prior to the last two centuries. Afterwards the population declined, probably due to a general cooling. During the twentieth century, however, the population decline has speeded up. The abnormal warming in the Antarctic Peninsula might have been the driving factor (Liu et al. 2005). In some areas, ice-intolerant gentoo penguins (*Pygoscelis papua*) and chinstrap penguins (*Pygoscelis antarctica*) have expanded their range southward, where the ice is retreating rapidly. However, the new available breeding places do not always meet all habitat demands of the penguins (Forcada and Trathan 2009). The worldwide largest temperate colony of Magellanci penguins (*Spheniscus magellanicus*) in Punta Tombo Reserve (Patagonia) has declined from about 500,000 to 350,000 due to increased reproductive failure. Climate change, in particular more frequent violent storms and increased precipitation, as well as ocean variability and overfishing have been suggested to be the main causes (Boersma and Rebstock 2014).

Due to the changes in penguin populations nutrient fluxes between the sea and coastal areas (cf. Figs. 2.3 and 2.4) may change locally. The decrease of burrowing seabirds (e.g., shearwaters, Magellan penguins) may have lasting effects on soils and micro-geomorphology (Sect. 2.6.1.2.2). Anyway, commercial fishing (pelagic drift-nets, long-line fisheries), ballast water dumping, and most likely climate change are possibly the biggest threats to these birds (e.g., Scott et al. 2008). Fishing, the whaling industry and harvesting of seabirds and fur seals have already caused major disturbances to the marine and coastal ecosystems long before climate warming had become a matter of concern. Therefore, the after-effects of historical human impact (e.g., habitat manipulation, introduced predators; Jones 2000; Brooke 2004) must also be considered. However, as to the understanding of the complex interactions between climate impact (increased ocean temperatures, seasonally changing sea-ice), seabirds and their environment in the Arctic/Subantarctic, uncertainties prevail and hypotheses often are equivocal (Croxall et al. 2002; Siegel 2006), in particular as the effects of climate change and human impact overlap.

The response of animal populations and biocoenoses to climate change and their effects on the environment will also be influenced by predator density. Both a loss or an increase in predatatory activity can have lasting effects. While in most regions, the populations of large predators have declined or are continuously shrinking, mainly due to conflicts with stock farmers and cattle drovers, they have increased in a few areas as a result of natural invasion or conservation measures. Such a case is seen in Yellowstone National Park, for example, where vegetation has profited from reduced browsing and grazing pressure after wolves were re-introduced (cf. Sect. 3.4.3 and 5.2.3). Loss of predators and a subsequent increase in ungulates, on the other hand, will affect ecosystem structures and function (Estes et al. 2011; Ripple et al. 2014, further literature therein). In such cases a reduction in plant growth, productivity, regeneration and carbon storage may be expected. Reduced carbon storage and changing vegetation surface (coverage, albedo, roughness, turbulent heat exchange, etc.), in particular, are all likely to contribute to climate change. Restoring the populations of large predators may help to regulate wild ungulates. On the other hand, the order of magnitude of such indirect effects of predators' appears to be relatively small compared to the influences of the many other variables involved in climate change. The huge number of domestic grazers certainly is a much bigger problem, especially as ruminants produce significant amounts of greenhouse gases (methane, dinitrogen, carbon dioxide). Thus, a global reduction in domestic ruminants is more likely to contribute significantly to mitigation of climate change (Ripple et al. 2014). In view of the rapidly growing human population, however, a decrease of livestock seems unlikely.

Altogether, prediction of the influence of climate change on future interactions between animals and their habitats usually are and will always be rather speculative, although scenarios may allow for assessing possible alternatives. Due to the many uncertainties, the future will probably produce many surprises. This applies in particular to the long-term effects of current wildlife and landscape management, as well as to introduction of alien species and re-introduction of native species eradicated in the past, which may take a different course than was originally intended.

References

ACIA. (2004). *Impacts of a warming Arctic. Arctic climate impact assessment* (pp. 1–144). Cambridge: Cambridge University Press.

Ammunet, T., Heisswolf, A., Klemola, N., & Klemola, T. (2010). Expansion of the winter moth outbreak range: No restrictive effects of competition with the resident autumnal moth. *Ecological Entomology, 35*, 45–52.

Ammunet, T., Klemola, T., & Saikkonen, K. (2011). Impact of host plant quality on geometrid moth expansion on environment and local population scales. *Ecography, 34*, 848–855.

Ammunet, T., Klemola, T., & Parvinen, K. (2014). Consequences of asymmetric competition between resident and invasive defoliators: A novel empirically bases modelling approach. *Theoretical Population Biology, 92*, 107–117.

Aukema, B. H., Carroll, A. L., Zheng, Y., Zhu, J., Raffa, K. F., Moore, R. D., Stahl, K., & Taylor, S. W. (2008). Movement of outbreak populations of mountain pine beetle: Influence of spatiotemporal patterns and climate. *Ecography, 31*, 348–358.

Bentz, B. J., Régnière, J., Fettig, C. J., Hansen, E. M., Hayes, J. L., Hickie, J. A., Kelsey, R. G., Negrón, J. F., & Seybold, S. J. (2012). Climate change and bark beetles of the Western United States and Canada: Direct and indirect effects. *BioScience, 60*(8), 602–613.

Berg, E. E., Henry, J. D., Fastie, C. L., De Volder, A. D., & Matsuoka, S. M. (2006). Spruce beetle outbreaks on the Kenai Peninsula, Alaska, and Kluane National Park and Reserve, Yukon Territory: Relationship to summer temperatures and regional differences in disturbance regimes. *Forest Ecology and Management, 227*, 219–232.

Boersma, P. D., & Rebstock, G. A. (2014). Climate change increases reproductive failure in Magellanic penguins. *PLoS One, 9*(1), e85602. doi:10.1371/Journal.pone0085602.

Botkin, D. B., Mellilo, J. M., & Wu, L. S.-Y. (1981). How ecosystem processes are linked to mammal population dynamics. In C.-W. Fowler & T. D. Smith (Eds.), *Dynamics of large mammal populations* (pp. 373–387). New York: Wiley.

Brooke, M. D. L. (2004). *Albatrosses and petrels across the world*. Oxford: Oxford University Press.

Buitenwerf, R., Bond, W. J., Stevens, N., & Trollope, W. S. W. (2012). Increased tree densities in South African savannas: >50 years of data suggests CO_2 as a driver. *Global Change Biology, 18*, 675–684.

Bylund, H. (1997). Stand age-structure influence in a low population peak of Epirrita autumnata. *Ecography, 20*, 319–329.

Bylund, H. (1999). Climate and the population dynamics of two insect outbreak species in the north. *Ecological Bulletins, 47*, 54–62.

Clarke, A. (2008). *Introduction to the dynamics of El Nino and the southern oscillation* (p. 324). Waltham: Academic Press.

Cobb, S. (1980). Tsavo, the first thirty years. *Swara, 3*, 12–15.

Corfield, T. F. (1973). Elephant mortality in Tsavon National Park, Kenya. *East African Wildlife Journal, 11*, 339–368.

Craine, J. M. (2013). Long-term climate sensitivity of grazers performance: A croo-site study. *PLoS ONE*. doi:10.1371/journal.pone.0067065.

Croxall, J. P., Trathan, P. N., & Murphy, E. J. (2002). Environmental change and Antarctic seabird populations. *Science, 297*, 1510–1514.

den Herder, M., Virtanen, R., & Roininen, H. (2008). Reindeer herbivory reduces willow growth and grouse forage in a forest-tundra ecotone. *Basic and Applied Ecology, 9*, 324–331.

Duncan, C., Chauvenet, L. M., McRae, L. M., & Pettorelli, N. (2012). Predicting the future impact for droughts on ungulate population in arid and semi-arid environments. *PLoS One*, Research Article. doi:10.1371/journal.pone.0051490.

Estes, J. A., Terborgh, J., Brashares, J. S., Power, M. E., Berger, J., Bond, W. J., Carpenter, S. R., Essington, T. E., Holt, R. D., Jackson, J. B., Marquis, R. J., Oksanen, L., Oksanen, T., Paine, R. T., Pikitch, E. K., Ripple, W. J., Sandin, S. A., Scheffer, M., Schoener, T. W., Shurin, J. B., Sinclair, A. R. E., Soulé, M. E., Virtanen, R., & Wardle, D. A. (2011). Trophic downgrading of Planet Earth. *Science, 333*, 301–306.

Everson, I. (1977). *The living resources of the Southern Ocean*. Rome: UN Development Program, Food and Agricultural Organization.

Forcada, J., & Trathan, P. N. (2009). Penguin responses to climate change in the Southern Ocean. *Global Change Biology, 15*, 1618–1630.

Gallana, M., Ryser-Degiorgis, M.-P., Wahli, T., & Segner, H. (2013). Climate change and infectious diseases of wildlife. Altered interactions between pathogens, vectors and hosts. *Current Zoology, 59*(3), 427–437.

Gandiva, E., & Zisadza, P. (2010). Wildlife management in Gonarezhow National Park, Southeast Zimbabwe: Climate change and implications for management. *Nature et Faune, 25*(1), 95–104.
Gille, S. T. (2002). Warming of the southern ocean since the 1950s. *Science, 295*, 1275–1277.
Giorgi, F., & Hewitson, B. (2001). Regional climate information – Evaluation and projections. In J. T. Houghtons, Y. Ding, D. J. Griggs, P. Noguer, P. Van der Linden, X. Dai, K. Maskell, & C. A. Johnsons (Eds.), *Climate change 2001. The scientific basis. Contributions of working group 1 to the third assessment report of the Intergovernmental Panel on Climate Change* (pp. 583–638). Cambridge: Cambridge University Press.
Hakkarainen, H., Virtanen, R., Honkanen, J. O., & Roininen, H. (2007). Willow bud and shoot foraging by ptarmigan in relation to snow level in NW Finnish Lapland. *Polar Biology, 30*, 619–624.
Halvorsen, O. (2012). Reindeer parasites, weather and warming of the Arctic. *Polar Biology, 35*, 1749–1752.
Henttonen, H., & Wallgren, H. (2001). Rodent dynamics and communities in the birch forest zone of northern Fennoscandia. In F. E. Wielgolaksi (Ed.), *Nordic mountain birch ecosystems* (Man and the Biosphere Series 27, pp. 261–278). Paris/New York/London: Parthenon Publishing.
Higgins, S. I., & Scheiter, S. (2012). Atmospheric CO2 forces abrupt vegetation shifts locally, but not globally. *Nature, 488*, 209–212. doi:10.1038/nature11238.
Holtmeier, F.-K. (1978). Die bodennahen Winde in den Hochlagen der Indian Peaks Section, Colorado Front Range, USA. *Münstersche Geographische Arbeiten, 3*, 3–47.
Holtmeier, F.-K. (1996). Die Wirkungen des Windes in der subalpinen und alpinen Stufe der Front Range, Colorado, U.S.A. *Arbeiten aus dem Institut für Landschaftsökologie, Westfälische Wilhelms-Universität, Münster, 1*, 19–45.
Holtmeier, F.-K. (2000). *Die Höhengrenze der Gebirgswälder* (Arbeiten aus dem Institut für Landschaftsökologie 8). Münster: Westfälische Wilhelms-Universität.
Holtmeier, F.-K. (2005). Relocation of snow and its effects in the treeline ecotone – With special regard to the Rocky Mountains, the Alps and Northern Europe. *Die Erde, 136*(4), 343–373.
Holtmeier, F.-K. (2009). *Mountain timberlines. Ecology, patchiness, and dynamics* (Advances in Global Change Research 36). Dordrecht: Springer Science+Business Media B. V.
Holtmeier, F.-K., & Broll, G. (2005). Sensitivity and response of northern hemisphere altitudinal and polar treelines to environmental change at landscape and local scales. *Global Ecology and Biogeography, 14*, 395–410.
Holtmeier, F.-K., & Broll, G. (2006). Radiocarbon-dated peat and wood remains from the Finnish Subarctic – Evidence of treeline and landscape history. *The Holocene, 16*(5), 743–751.
Holtmeier, F.-K., & Broll, G. (2012). Landform influences on treeline patchiness and dynamics in a changing climate. *Physical Geography, 33*(5), 403–437.
Hulme, P. (2005). Adapting to climate change: Is there scope for ecological management in face of a global threat? *Journal of Applied Ecology, 42*, 784–794.
Ims, R. A., & Fuglei, E. (2005). Trophic interaction cycles in tundra ecosystems and the impact of climate change. *BioScience, 55*, 311–322.
Ims, R. A., Henden, J.-A., & Killengren, S. T. (2008). Collapsing population cycles. *Trends in Ecology and Evolution, 23*(2), 79–86.
Jacob, J., & Tkadlec, E. (2010). Rodent outbreaks in other regions: A search for generalities. In G. R. Singleton, S. R. Belmain, P. R. Brown, & B. Hardy (Eds.), *Rodent outbreaks: Ecology and impacts* (pp. 207–223). Los Baños: International Rice Research Institute (IRRI).
Jones, C. (2000). Sooty shearwater (Puffinus griseus) breeding colonies on mainland South Island, New Zealand: Evidence of decline and predictors of persistence. *New Zealand Journal of Zoology, 27*, 327–334.
Kaeslin, E., Redmond, I., & Dudley, N. (Eds.). (2012). *Wildlife in a changing climate* (FAO Forestry Paper 167). Rome: FAO.
Karhu, K. J., & Neuvonen, S. (1998). Wood ants an geometrid defoliator of birch: Predation outweighs beneficial effects through the host plant. *Oecologia, 113*, 509–516.

Kausrud, K. L., Mysterud, A., Steen, H., Vik, J. O., Østbye, E., Cazelles, B., Framstad, E., Eikeset, A. M., Mysterud, I., Solhøy, T., & Stenseth, N. C. (2008). Linking climate change to lemming cycles. *Nature, 456*, 93–97.

Kutz, S., Hoberg, E. P., Polley, L., & Jenkins, R. E. (2005). Global warming is changing the dynamics of Arctic host-parasite systems. *Proceedings of the Royal Society B, 272*, 2571–2576.

Laine, K., & Niemelä, P. (1980). The influence of ants on the survival of mountain birch during an Oporinia autumnata (Lep. Geometridae) outbreak. *Oecologia, 47*, 39–42.

Laws, R. M. (1985). The ecology of the Southern Ocean. *American Scientist, 73*, 26–40.

Lehmann, C. E. R., Anderson, T. M., Sankaran, M., Higgins, S. I., Archibald, S., Hoffmann, W. A., Hanan, N. P., Williams, R. J., Fensham, R. J., Felfili, J., Hutley, L. B., Ratnam, J., Jose, J. S., Montes, R., Franklin, D., Haidar, R., Bowman, D. M. J., & Bond, W. J. (2014). Savanna vegetation-fire-climate relationships differ among continents. *Science, 343*, 548–552.

Leuthold, W. (1978). Kann Afrikas Großwild überleben? *Vierteljahresschrift der Naturforschenden Gesellschaft Zürich, 123*, 333–340.

Liu, X., Sun, L., Xie, U., Yin, X., & Wang, Y. (2005). A 1300-year record of penguin populations at Ardley Island in the Antarctic, as deduced from geochemical data in the ornithogenic lake sediments. *Arctic, Antarctic, and Alpine Research, 37*(4), 490–498.

Logan, J. A., & Powell, J. A. (2001). Ghost forests, global warming, and the mountain pine beetle (Coleoptera: Scolytidae). *American Entomologist, 47*, 160–173.

Luque-Larena, J. J., Mougeot, F., Viñuela, J., Jareño, D., Arroyo, L., Lambin, X., & Arroyo, B. (2013). Recent large-scale expansion and outbreaks of the common vole (Microtus arvalis) in NW Spain. *Basic and Applied Ecology*. http://dx.doi.org/1016/j.baae.2013.04.006

Maness, H., Kushner, P. J., & Fung, I. (2013). Summertime climate response to mountain beetle disturbance in British Columbia. *Nature Geoscience, 6*, 65–70.

Mitton, J. B., & Ferrenberg, S. M. (2012). Mountain pine beetle develops an unprecedented summer generation in response to climatic warming. *The American Naturalist 1979*(5), 9 pp. doi:10.1086/665007.

Natural Resources Canada. (2005). *Mountain pine beetle initiative interim report*. Victoria: Canadian Forest Service, Pacific Forestry Centre.

Neuvonen, S., Ruohomäki, K., Bylund, H., & Kaitaniemi, P. (2001). Insect herbivores and herbivory effects on mountain birch dynamics. In F. E. Wielgolaski (Ed.), *Nordic mountain birch ecosystems* (Man and the Biosphere Series 72, pp. 207–239). Paris/New York/London: UNESCO/Parthenon Publishing Group.

Nogués-Bravo, D., Araújo, M. B., Errea, M. P., & Martínez-Rica, J. P. (2007). Exposure of global mountain systems to climate warming during the 21st century. *Global Environmental Change, 17*, 420–428.

Ogutu, J. O., & Owen-Smith, N. (2003). ENSO, rainfall and temperature influences on extreme population declines among African savanna ungulates. *Ecological Letters, 6*(5), 412–419.

Ogutu, J. O., Piepho, H. P., Dublin, H. T., Bhola, N., & Reid, R. S. (2008). Rainfall influences on ungulate population abundance in the Mara-Serengit ecosystem. *Journal of Animal Ecology, 77*, 814–829.

Olofsson, J., Oksanen, L., Callaghan, T., Hulme, P. E., Oksanen, T., & Suominen, O. (2009). Herbivores inhibit climate-driven shrub expansion on the tundra. *Global Change Biology, 15*, 2681–2693.

Ost, E., & Pedersen, C. (2008). Opposing plant community responses to warming with and without herbivores. *Proceedings of the National Academy of the Science USA, 105*(34), 12353–12358.

Parmesan, C., & Yohe, G. (2003). A globally coherent fingerprint of climate change impacts across natural systems. *Nature, 241*, 37–42.

Post, E., & Stenseth, N. C. (1999). Climate variability, plant phenology and northern ungulates. *Ecology, 80*, 1322–1339.

Pulliainen, E., & Iivanainen, J. (1981). Winter nutrition of the willow grouse (Lagopus lagopus L.) in the extreme north of Finland. *Annales Zoologici Fennici, 18*, 263–269.

Räisänen, J. (1994). A comparison of the results of seven CGM experiments in northern Europe. *Geophysica, 3*, 3–30.

Ripple, W. J., Estes, J. A., Beschat, R. L., Wilmers, C. C., Ritchie, E. G., Hebblewhite, M., Berger, J., Elmhagen, B., Letnic, M., Nelson, M. P., Schmitz, O. J., Smith, D. W., Wallach, A. D., & Wirsing, A. J. (2014). Status and ecological effects of the world's largest carnivores. *Science, 343*, 1241484. doi:10.1126/science,1241484.

Ritchie, E. G., & Bolitho, E. E. (2008). Australia's savanna herbivores: Bioclimatic distribution and an assessment of the potential impact of regional climate change. *Physiological and Biochemical Zoology, 81*(6), 880–890.

Robertson, C. J. R. (1998). Factors influencing the threatened performance of the Northern Royal Albatross. In G. Robertson & R. Gales (Eds.), *Albatross biology and conservation* (pp. 99–104). Chipping North: Surrey Beatty & Sons.

Root, T. L., Price, J. T., Hall, K. R., Schneider, S. H., Rosenzweig, C., & Pounds, J. A. (2003). Fingerprints of global warming in animal and plant populations. *Nature, 421*(2), 57–60.

Scott, D., Scofield, P., Hunter, C., & Fletcher, D. (2008). Decline of sooty shearwater, Puffinus griseus, on the Snares, New Zealand. *Papers and Proceedings of the Royal Society of Tasmania, 142*(1), 185–196.

Siegel, V. (2006). Der Einfluss von Fischerei und Klima auf die Bestände des antarktichen Krill. In J. L. Lozán, H. Graßl, H. W. Huberten, P. Hupfer, L. Karbe, & G. Piepenburg (Eds., in Kooperation mit GEO). *Warnsignale auf den Polarregionen. Wissenschaftliche Fakten. Wissenschaftliche Auswertungen* (pp. 121–125). Hamburg: Springer.

Solbreck, C. (1991). Predicting insect faunal dynamics in a changing climate – A north European perspective. In J. I. Holten, G. Paulsen, & W. C. Oechel (Eds.), *Impacts of climate change on natural ecosystems with emphasis on boreal and arctic/alpine ecosystems*, Trondheim, 27–29 November 1990, pp. 176–185.

Taylor, S. W., Carroll, A. L., Alfaro, R. I., & Safranyik, L. (2006). Forest, climate and mountain pine beetle outbreak dynamics in western Canada. In L. Safranyik & B. Wilson (Eds.), *The mountain pine beetle: A synthesis of biology, management and impacts in lodgepole pine* (pp. 67–94). Victoria: Natural Resources Canada, Canadian Forest Service.

Umeå universitet. (2014, February 13). Reindeer counteract certain effects of climate change. *ScienceDaily*. www.sciencedaily.com/releases/2014/02/140213084052.htm

Veit, R. R., Pyle, P., & McGowan, J. A. (1996). Ocean warming and long-term change to pelagic bird abundance within the California Current system. *Marine Ecology Progress Series, 139*, 11–18.

Veit, R. R., McGowan, J. A., Ainley, D. G., Wahls, T. R., & Pyle, P. (1997). Apex marine predator declines ninety percent in association with changing oceanic climate. *Global Change Biology, 3*, 23–28.

Virtanen, T., & Neuvonen, S. (1999). Performance of moth larvae on birch in relation to altitude, climate, host quality and parasitoides. *Oecologia, 120*, 92–101.

Virtanen, T., Neuvonen, S., & Nikula, A. (1998). Modelling topoclimatic patterns of egg mortality of Epirrita autumnata (Lepidoptera: Geometridae) with Geographical Information Systems: Predictions of current and warmer climate scenarios. *Journal of Applied Ecology, 32*, 1–12.

Walther, G.-R., Post, E., Convey, P., Menzel, A., Parmesan, C., Beebee, T. J., Fromentin, J. M., Guldberg, H.-G., & Bairlein, F. (2002). Ecological responses to recent climate change. *Nature, 416*, 389–395.

Weimerskirch, H., Inchausti, P., Guinet, C., & Barbraud, C. (2003). Trends in bird and seal populations as indicators of a system shift in the Southern Ocean. *Antarctic Science, 15*(2), 249–256.

Wigley, B. J., Bond, W. J., & Hoffman, M. T. (2010). Thicket expansion in a South African savanna under divergent land use: Local vs. global drivers? *Global Change Biology, 16*, 964–976.

Williams, D. W., & Liebhold, A. M. (1995). Forest defoliators and climate change: potential changes in spatial distribution of outbreaks of Western spruce budworm (Lepidoptera: Tortricidae) and Gypsy moth (Lepidoptera: Lymantriidae). *Environmental Entomology, 24*, 1–9.

Williams, R. J., Carter, J., Duff, G. A., Woinarski, J. C. Z., Cook, G. D., & Farrer, S. L. (2005). Carbon accounting, land management, science and policy uncertainties in Australian savanna landscapes: Introduction and overview. *Australian Journal of Botany, 53*, 583–588.

Zöckler, C., Liles, L., Fish, L., Wolf, A., Rees, G., & Danks, F. (2008). Potential impact of climate change and reindeer density on tundra indicator species in the Barents Sea region. *Climate Change, 87*, 119–130.

Chapter 8
Epilogue

I come back to the objectives of this book. I have tried to assess the multi-facetted interactions of animal populations and their habitats combining autecological, synecological, population-ecological and landscape ecological aspects. In the last chapters, a few practical implications for landscape management and nature conservation have been considered as well as possible responses to climate change.

The effects of animals on landscape and the feedback are closely related to the specific local physical conditions (natural environment) and landscape history (e.g., human impact). Careful local studies are needed to approach the often inscrutable interactions of animals' impact and habitat response. Measures and concepts that have locally been successful can be applied to other areas and habitats only with great reservation, if ever. The effects will usually be different in different environments. Conclusions by analogy – i.e. applying the results of local studies, and also of models, to other areas (habitats, landscapes) seemingly characterized by similar conditions at first sight – may disguise possible problems and thus prevent in-depth analyses. Unexpected problem are likely. This hold particularly true for nature conservation and landscape planning. Not least, local studies and their results have to be integrated into regional and supra-regional concepts (e.g., summer and winter grazing areas of ungulates, resting and moulting places of migratory birds, traditional migration routes, etc.). Developing such concepts is difficult, and implementation may even be harder because of state borders, for example, and/or due to controversial policy of the people involved.

As the environmental conditions are changing almost continuously, long-term planning may even be unrealistic. It does not make sense (see also Götamark 1992) to develop scenarios of potential biocoenoses and ecosystems oriented in a period of time before the pristine landscape became altered by human influence. Since the second half of the Holocene at the latest, human impact on landscape pattern, vegetation, biocoenoses, soils and hydrology has been almost continuously increasing. Alteration of the habitats will also occur in the future (e.g., Schmidt 1993a, b), and animal species that are able to adjust to the changing environment will be at an advantage.

F.-K. Holtmeier, *Animals' Influence on the Landscape and Ecological Importance*,
DOI 10.1007/978-94-017-9294-3_8

'Urbanization' of many species is a typical example (e.g., Klausnitzer 1987, 1989). Birds originally breeding in rock cliffs, such as swift (*Apus apus*), jackdaw (*Corvus monedula*) and kestrel (*Falco tunnincolus*), have occupied urban areas. The same holds true for the so-called 'town fox' (*Vulpes vulpes*), beech marten (*Martes foina*) and racoon (*Procyon lutor*). In some areas of Germany, the racoon density is as high as in some American cities (one racoon per hectare) and many times higher than in the wild (Hohmann 2001; Hohmann and Bartussek 2001). Abundant food (gardens, human waste), favorable conditions for rearing young and, not least, feeding of the funny small bears by people are the main driving factors.

Wild boars have also 'discovered' the urban environment, where they regularly devastate private and public gardens, urban greens, sport grounds, graveyards and cemeteries when searching for food. Many wild boars have already been born in towns. The Grunewald, a woodland-like parkland (about 3,200 ha) in the city of Berlin, which is intensively used for outdoor recreation and leisure time activities, harbors about 1,000 of these highly adaptive animals. Some of these 'town boars' have even rapidly learned to exploit a new food source by begging sandwiches from the schoolchildren during the breaks. However, although wild boars has become a 'pest animal' reduction, in particular by shooting females and piglets, can be sure to face the opposition of animals' advocates.

Common heron (*Ardea cinera*) – usually a very shy bird – enter increasingly residential areas for fishing in garden ponds. Magpies (*Pica pica*) and jays (*Garrulus glandarius*) have expanded into urban areas, where parks and gardens provide favorable habitat conditions (e.g., Knief and Borkenhagen 1999; Bellebaum and Nottemeyer-Linden 1998; Kooiker 1998; Mäck and Jürgens 1999). The loss of natural habitats in the impoverished landscape and the attractiveness of the urban area are the main factors driving occupation of urban habitats. In addition, abundant food is usually available all year round, and the urban habitats provide numerous favorable breeding places. Not least, predation pressure is of minor importance compared to the countryside, and hunting is not allowed in residential areas.

In the meantime, however, predators have also adjusted to the new conditions. In Cologne, for example, 25–40 % of the 'town goshwaks' (*Accipiter gentilis*) prey are magpies. Vast numbers of feral homing pigeons (*Columba domestica*), collared doves (*Stretopelia decaoto*), ring doves (*Columba palumbus*), blackbirds (*Turdus merula*), song thrushes (*Turdus philomelos*), starlings (*Sturnus vulgaris*), exotic rose-ringed parakeets (*Psittacula krameri*), and wild rabbits (*Oryctolagus cuniculus*) complete the 'luxuriant' source of food. In the green belt of Cologne, goshawk is breeding more successfully than in the rural countryside, and the number of fledged young is above the average of the state of North Rhine-Westphalia (Martin 1998). It is impossible to predict reliably which and how many species will be able to occupy their niches successfully in this continuously changing environment and what the consequences will be for urban and suburban biocoenoses.

In the end, the main objective of all nature conservation measures and landscape management will be to ensure human livelihood (unpolluted water and atmosphere, uncontaminated and fertile soils, uncontaminated food) in the long-term. In densely populated and intensively used areas such as Central Europe, for example, high priority should be given to maintenance and improvement of the welfare functions of the forests. The indispensable adjustment of the wild ungulate populations to the tolerable carrying capacity has been repeatedly addressed in the foregoing chapters. However, the problem is, as ever, to define the tolerable carrying capacity – with special consideration of the local characteristics and their influence on 'natural fluctuations'.

Conservation of soil fertility and soil water filtering capacity is not less important. In this respect, efficient protection of soil biota plays an important role, as they strongly influence soil's physical and chemical conditions (decomposition, nutrient turnover rates, translocation, availability of nutrients). Although their effects are fundamental for ecosystem function, they are not sufficiently considered in nature conservation. Occasionally, one may get the impression that nature conservation is an 'above ground activity' and is focused mainly on species that are enjoying people's sympathy or are threatened by extinction (Red Data Book). Preservation of species is only possible through habitat conservation, and this holds particularly true for the soil habitat. In contrast to the polluted atmosphere and water, 'cleaning' and restoration of the soil is almost impossible (Ehrnsberger 1993).

The insufficient consideration of soil biocoenoses on the one hand, and the strong efforts to prevent the development of a new business park and even of a lignite power plant near Cologne (an investition of about two billion €; Chilla 2006) in order to save an 'assumed' occurrence of the common hamster (*Cricetus cricetus*) on the other hand, show that people's and decision makers' understanding of ecological interactions is usually still insufficient to solve the ecological problems. The same goes for the 'hamster-adequate' utilization of the building terrain of the centre for molecular biosciences (University of Göttingen, Germany; Hunger 1999). It was hoped that the solution of the 'hamster problem' lay in translocation of the hamsters to a naturally structured 'reserve area' (5.6 ha) immediately adjacent to the new buildings, where the hamsters were expected to build up a viable population. Although this compensation area meets the hamsters' ecological requirements, the project is now likely to fail.

In reality, however, even an improved education in ecological issues would probably not prevent those profiting from landscape use (agricultural cultivation, construction ground, business parks, etc.) from implementing their ideas by means of policy. If it appears opportune to them, they will usually not loose time also using ecological arguments as a 'vehicle' for reaching their goal. In most cases, however, these arguments exist on a 'pseudo-ecological' level and are untenable from the scientific view. Thus, as in case of the 'hamster problems', substitute areas compensating for habitat loss are often readily presented, even though their substitute function cannot be guaranteed.

On the other hand, nature conservation and ecological management focused on the animals only, are unrealistic without the demands of actual and future landscape utilization also being considered. Highly engaged environmentalists often ignore this. Finally, the long-term consequences of nature conservation measures cannot reliably predicted – in particular as the objectives will change in time. As has become obvious from the serious problems resulting from introduction and re-introduction of animal species, permanent control is inevitable to prevent avoidable feedbacks – or in other words, new patches have to be put on the older ones once again and again. This goes, in particular, for assessing possible alterations of the relationships between animals and their environment under the influence of climate change. The only safe but quite general assumption is that ecosystems will respond to climate change in one way or another. Changes of migration pattern, competition, adaptation, ecosystems' character and the effects on the interactions of animal populations and the environment may be expected. However, regional and local differences and their possible effects are less clear. Thus, in some areas, the natural range of animal populations, for example, has already expanded while it has shrunk in other regions. We have to be aware that our ideas on animal's impact on the landscape ecosystems and ecosystem's response under the influence of the warming climate mainly refer to plausible hypotheses rather than to well-founded data. Not least, human impact (land use, exploitation of natural resources, pollution, political conflicts) on the present environment, even in remote areas, may have broader implications than climate and overrule its possible effects, at least in the short- and medium-term. Altogether, the possible consequences for animals' impact on and ecosystem's response are not predictable in the long-term. The more a careful landscape use and dealing with the animals' impact on and in response to the changing environment are imperative, as it is as the philosopher Immanuel Kant (1787) already said:

Our decision reaches farther than our knowledge

References

Bellebaum, J., & Nottemeyer-Linden, K. (1998). Gibt es "Überpopulationen" von Elster, Rabenkrähe und Eichelhäher in Nordrhein-Westfalen? *LÖBF-Mitteilungen, 1*, 29–34.

Chilla, T. (2006). Der Feldhamster und die Milliardeninvestitionen. Naturschutz und die Bedeutung von Diskursen. *Natur und Landschaft, 81*(6), 327–331.

Ehrnsberger, R. (1993). Bodenzoologie und Agrarökosysteme. In R. Ehrnsberger (Ed.), *Bodenmesofauna und Naturschutz* (Informationen zu Naturschutz und Landschaftsschutz in Nordwestdeutschland 6, pp. 11–41). Cloppenburg: G. Runge.

Götamark, F. (1992). Naturalness as an evaluation criterion in nature conservation: A response to Anderson. *Conservation Bulletin, 6*(3), 455–458.

Hohmann, U. (2001). Stand und Perspektiven der Erforschung des Waschbären in Deutschland. *Beträge zur Jagd- und Wildforschung, 26*, 181–186.

Hohmann, U., & Bartussek, I. (2001). *Der Waschbär*. Reutlingen: Oertel &Spörer.

Hunger, U. (1999, October 7). Hamsterdrama zu Göttinmgen. Wie das Schicksal einer Feldhamsterkolonie eine Stadt in Atem hält. *DIE ZEIT 41*, 39.

Kant, I. (1787). *Krik der reinen Vernunft* (2nd ed.). Riga: Johann Friedrich Hartknoch.

Klausnitzer, B. (1987). *Ökologie der Großstadtfauna*. Stuttgart/New York: Fischer.

Klausnitzer, B. (1989). *Verstädterung von Tieren. Die neue Brehm-Bücherei 579*. Ziemsen: Wittenberg-Lutherstadt.

Knief, W., & Borkenhagen, P. (1999). Ist eine Bestandesregulierung von Rabenkrähen und Elstern erforderlich? Ein Untersuchungsbeispiel aus Schleswig-Holstein. *Natur und Landschaft, 68*(3), 102–107.

Kooiker, G. (1998). Warum werden Elstern immer noch getötet. Rabenvogelstreit. *Der Falke, 45*, 122–126.

Mäck, U., & Jürgens, M. E. (1999). *Aaskrähe, Elster und Eichelhäher in Deutschland*. Bonn: Bundesamt für Naturschutz.

Martin, W. (1998). Stadt-Habichte im Aufwind. *Wild und Hund, 16*, 22–23.

Schmidt, K. (1993a). Winter ecology of nonmigratory Alpine red deer. *Oecologia, 95*, 226–233.

Schmidt, M. (1993b). Naturschutzmanagement – Nowendigkeit oder Übereifer? In Landesamt für Umweltschutz Baden Württemberg (Ed.), *Landschaftspflege - quo vadis?* (pp. 138–142). Karlsruhe.

References

Arner, D. H. (no year). *Research and a practical approach need in management of beaver and beaver habitat in the southern United States*. Duplicate of a manuscript provided by D. H. Arner to the present author

Batzli, G. O. (1983). Responses of arctic rodent populations to nutritional factors. *Oikos, 40*, 396–406.

Bibikov, D. I. (1988). *Der Wolf: Camus lupus* (Die Neue Brehmbücherei 587). Wittenberg Lutherstadt.

Blyth, C., Campbell, C., Campbell, I., & McAndrew, J. (1994). Bison extirpation may have caused aspen expansion in western Canada. *Ecography, 17*, 360–362.

Boom, K., Ben-Armi, D., Croft, D. B., Chushing, N., Ramp, D., & Boronyak, L. (2012). 'Pest' and resource: A legal history of Australia's kangaroos. *Animal Studies Journal, 1*(1), 17–40.

Brown, K. P., Innes, J. G., & Shorten, R. M. (1993). Evidence that possums prey on and scavange birds' eggs, birds, and mammals. *Notornis, 40*, 169–177.

Brühne, M., Mooij, J. H., Schwöppe, M., & Wille, V. (1999). Projekt zur Minderung von Gänsefraßschäden am Unteren Niederrhein in Nordrhein-Westfalen. *NNA Berichte, 12*(3), 156–162.

Chavasse, C. G. R. (1955). *Mortality in rata/kamahi protection forest – Westland*. New Zealand Forest Service, Unpublished Report (cited in Payton 1987).

Cox, G. W., Gakahu, C. G., & Waithaka, J. M. (1989). The form and small stone content of large earth mounds constructed by mole rats and termites in Kenya. *Pedobiologia, 33*, 307–314.

Crouch, G. L. (1969). *Animal damage to conifers in national forests in the Pacific Northwest region* (USDA Forest Service, Research Bulletin PNW-28). Portland: Pacific Northwest Forest and Range Experimental Station.

Dean, W. R. J., & Milton, S. J. (1991). Disturbances in semi-arid shrubland and arid grassland in the Karoo, South Africa: Mammal diggings as germination sites. *African Journal of Ecology, 29*, 11–16.

Detling, J. K. (1988). Grassland and savannas; Regulation of energy flow and nutrient cycling by herbivores. In L. R. Pomeroy & J. J. Alberts (Eds.), *Concepts of ecosystem ecology* (pp. 131–148). New York: Springer.

Fritz, T. H. (1987). Movements of snakes via cargo in the Pacific regions. *Elepaio, 47*(2), 17–18.

Geiger, A., Steven, M., Glandt, D., Kronshage, A., & Schwartze, M. (2000). Laubfroschschutz im Münsterland. Das Kooperationsprojekt "Eine König sucht sein Reich" im Artenschutzprogramm NRW. *LÖBF-Mitteilungen, 4*, 16–34.

Glutz von Blotzheim, U. N., & Bauer, K. M. (1993). *Handbuch der Vögel Mitteleuropas 13 (III)*. Wiesbaden: Aula Verlag.

Halley, D., & Rosell, F. (2003). Population and distribution of European beavers (Castor fiber). *Lutra, 46*, 91–101.

Hartnett, D. C., Hickman, K. R., & Fischer, W. L. E. (1996). Effects of bison grazing, fire and topography on floristic diversity in tallgrass prairie. *Journal of Range Management, 49*, 413–420.

F.-K. Holtmeier, *Animals' Influence on the Landscape and Ecological Importance*,
DOI 10.1007/978-94-017-9294-3

Hofmann, R., & Stewart, D. R. M. (1972). Grazer or browser: A classification based on the stomach structure and feeding habits of East African ruminats. *Mammalia, 36*(2), 226–240.

Holdo, R. M., Holtz, R. D., & Fryxell, J. M. (2009). Grazers, browsers, and fire influence the extent and spatial pattern of tree cover in the Serengeti. *Ecological Application, 19*, 95–109.

Holtmeier, F.-K. (1980). Influence of wind on tree physiognomy at the upper timberline in the Colorado Front Range. *New Zealand Forest Service Technical Paper, 70*, 247–261.

Hoogesteeger, J., & Karlsson, P. S. (1992). Effects of defoliation on radial growth and photosynthesis in the mountain birch (Betula pubescens ssp. tortuosa). *Functional Ecology, 6*, 317–323.

Jacibiec, Z. (1993). *Ursus arctos Linnaeus, 1758 – Braunbär*. In M. Stubbe & F. Krapp (Eds.), *Handbuch der Säugetiere Europas. Raubsäuger (Teil I)* (pp. 254–300). Wiesbaden: Aula Verlag.

Jope, K. L. (1985). Implications of grizzly bear habituation to hikers. *Wildlife Biology, 13*, 32–37.

Kareiva, P. (1996). Diversity and sustainability on the prairie. *Nature, 379*, 673–674.

Lieberman, D., Lieberman, M., & Martin, C. (1987). Notes on seeds in elephant dung from Bio National Park, Ghana. *Biotropica, 19*(4), 356–369.

Linhart, Y. B., & Tomback, D. F. (1995). Seed dispersal by nutcrackers causes multi-trunk growth forms in pines. *Oecologia, 67*, 107–111.

Löffler, H. (1996c). Neozoen in der Fischfauna Baden-Württembergs - ein Überblick. In H. Gebhardt, R. Kinzelbach, & S. Schmidt-Fischer (Eds.), *Gebietsfremde Tierarten* (pp. 217–226). Landsberg: Ecomed.

MacLean, D. A. (1988). Effects of spruce budworm on vegetation structure and succession of balsam fir forests on Cape Breton Island, Canada. In M. J. A. Werger, P. J. M. Van der Aart, H. J. During, & J. T. A. Verhoven (Eds.), *Plant, form and vegetation structure* (pp. 253–262). The Hague: SPB Academic Publishing bv.

MacLulick, D. A. (1937). Fluctuations in numbers of varying hare (Lepus americanus). *University of Toronto Studies, Biological Series, 43*, 1–136.

McNaughton, S. J., & Georgiadis, N. J. (1986). Ecology of African grazing and browsing animals. *Annual Review of Ecology and Systematics, 17*, 39–65.

Michigan Tech News/Media. (2004). Isle Royale wolf-moose study report. http://www.admin.mtu.edu/urel-cgi-bin/print.Pl?story=211

Munroe, E. (1979). Lepidoptera. In H. V. Danks (Ed.), *Canada and its insect fauna* (Entomological Society of Canada, Memoirs 108, pp. 427–481). Ottawa: Entomological Society of Canada.

MURL (Ministerium für Umwelt, Raumordnung und Landwirtschaft des Landes Nordrhein-Westfalen). 1990. *Natur 2000 in Nordrhein-Westfalen, Leitlinien und Leitbilder für Natur und Landschaft im Jahr 2000*. Düsseldorf.

Niedersächsisches Ministerium für Umwelt, Energie und Klimaschutz. (2013). *Wolfsmonitoring bringt ungewöhnliche Familienbande ans Licht und neue Nachweise im Emsland und Wendland*. Pressemitteilung 28. Hannover: Niedersächsisches Ministerium für Umwelt, Energie und Klimatschutz.

Oguto, J. O., & Owen-Smith, N. (2003). ENSO, rainfall and temperature influences on extreme population declines among African savanna ungulates. *Ecology Letters, 5*, 412–419.

Oosterveld, P. (1975). Beheer en ontwikkeling van natuurreservaten door begrazing. *Natuur Landschap, 29*, 161–171.

Persico, L., & Meyer, G. (2012). Natural and historical variability in fluvial processes, beaver activity, and climate in the greater Yellowstone ecosystem. *Earth Surface Processes and Landforms*. doi:10.1002/esp.3349.

Peterson, R. O., & Page, R. E. (1988). The rise and fall of Isle Royale wolves, 1975–1986. *Journal of Mammalogy, 69*, 88–89.

Puth, L. M., & Wilson, K. A. (2001). Boundaries and corridors as a continuum of ecological flow control: Lessons from rivers and streams. *Conservation Biology, 15*, 21–30.

Ratti, P. (1968). Das Steinwild am Piz Albris. *Bündnerwald, 21*(5), 176–183.

Redpath, S., & Thirgood, S. (2009). Hen harriers and red grouse: Moving towards consensus. *Journal of Applied Ecology, 46*, 961–963.

Remmert, H. (1985). Der vorindustrielle Mensch in den Ökosystemen der Erde. *Naturwissenschaften, 72*, 627–632.

Robbins, C. T. (1983). *Wildlife feeding and nutrition*. New York: Academic.

SBS (Special Broadcasting Service). (2013, November 21). *Feral camels brought under control*. SBS.

Schaller, G. B. (2003, September). Sterben für die Mode. Warum die Tschirus gejagt werden. *National Geographic, 9*, 148–149.

Schneider, H.-J. (1993). *Der Biber in Mitteleuropa. Ein Beitrag zur Ökologie der Art und Chancen einer flächenhaften Wiederansiedlung*. Diplomarbeit Gesamthochschule Paderborn, Abteilung Höxter, Studiengang Landschaftspflege, Fachbereich 7, Lehrgebiet Tierökologie.

Tenow, O., & Holmgren, B. (1987). Low winter temperatures and an outbreak of Epirrita autumnata along a valley of Finnmarksvidda, the "cold pole" of Fennoscandia. In H. Alexandersson & B. Holgren (Eds.), *Climatological extremes in the mountains* (UNGI Rapport 65, pp. 203–216). Uppsala: Department of Physical Geography.

The Independent. (2013, February 4). Uproar in Australia over wild horse cull. *The Independent*.

Ulrich, S. (2013, July 19). Räuber und Gendarm. Süddeutsche Zeitung Nr. 165, p. 3.

Van Soest, P. J. (1985). *Nutritional ecology of the ruminant*. Corvallis: O & O Books.

Wallis de Vries, M. (1999). The dilemma facing natural conservation and the role of large herbivores. In B. Gerken & M. Görner (Eds.), *Europäische Landschaftsentwicklung mit großen Weidetieren. Geschichte, Modelle und Perspektiven* Natur- und Kulturlanschaft 3, pp. 24–31. Höxter/Jena: Huxaria Druck und Verlag.

Weimer, J. (1975). Model of the energy budget of an adult roe deer. *Polish Ecological Studies, 1*, 103–119.

Westrich, P. (1989). *Die Wildbienen Baden-Württembergs*. Stuttgart: Verlag Eugen Ulmer.

White, T. C. R. (1993). *The inadequate environment. Nitrogen and the abundance of animals*. Berlin/Heidelberg: Springer.

White, P. S., and Pickett, S. T. A. (1972). Natural disturbance and patch dynamics: An introduction. In S. T. A. Pickett & P. S. White (Eds.), *The ecology of natural disturbances and patch dynamics* (pp. 3–13). Orlando: Academic.

Woods, R. W. (1970). The avian ecology of a tussock island in the Falkland Islands. *Ibis, 112*(1), 15–25.

Animal Index

F.-K. Holtmeier, *Animals' Influence on the Landscape and Ecological Importance*,
DOI 10.1007/978-94-017-9294-3

Subject Index

F.-K. Holtmeier, *Animals' Influence on the Landscape and Ecological Importance*,
DOI 10.1007/978-94-017-9294-3

D

E

F

N

O

P

T

www.ingramcontent.com/pod-product-compliance
Lightning Source LLC
LaVergne TN
LVHW021716270926
848991LV00059B/98

9789401792936